Global Change – The IGBP Series

Josep G. Canadell · Diane E. Pataki · Louis F. Pitelka (Eds.)

Terrestrial Ecosystems in a Changing World

With 104 Figures

Springer

Editors

Josep G. Canadell

Global Carbon Project
CSIRO Marine and Atmospheric Research
GPO Box 3023
Canberra, ACT 2601, Australia

Diane E. Pataki

Department of Earth System Science
and Department of Ecology & Evolutionary Biology
University of California
Irvine, CA 92697, USA

Louis F. Pitelka

Appalachian Laboratory
University of Maryland
Center for Environmental Science
Frostburg, MD 21532, USA

Library of Congress Control Number: 2006922797

ISSN 1619-2435
ISBN-10 3-540-32729-0 **Springer Berlin Heidelberg New York**
ISBN-13 978-3-540-32729-5 **Springer Berlin Heidelberg New York**

Springer is a part of Springer Science+Business Media
springer.com

Cover design: Erich Kirchner, Heidelberg
Cover photos, credits: (1) Michael Ryan, (2) CSIRO, (3) Richard Norby, (4) David Tilman
Typesetting: Stasch · Bayreuth (stasch@stasch.com)
Production: Christine Adolph, Heidelberg

Printed on acid-free paper 30/2133/CA – 5 4 3 2 1 0

Preface

The GCTE project was born on the day the Berlin wall fell (November 10, 1989). It was the final day of the Planning Committee meeting for the IGBP, at Berlin's Wissenschaftskolleg. Co-ordinating Panel 4 had presented its recommendations to the IGBP Planning Committee during the preceding week, and on that day they were accepted as the basis for GCTE, ratified later at the general IGBP inaugural meeting in Paris, in 1990.

The first full meeting of GCTE was its Open Science meeting in Brighton, England, in February 1991. Much good science was defined and put into effect at that meeting. But one point remained unresolved – an appropriate name and acronym. Everyone was agreed that GCTE was dreadful and could never work. The topic was debated at (the end of) each subsequent meeting of the Steering Committee, and no agreement on a better (publishable) name and acronym was ever reached. Its clumsiness eventually made it distinctive and so, 15 years later, it is finally put to rest, with the project.

From its inception GCTE was marked by, and most fortunate in having, a group of outstanding scientists to lead its defined Activities. They constituted the GCTE Steering Committee and their performance and stature attracted the best researchers joining what has been an exemplary (yet essentially voluntary) scientific effort.

The evolution of GCTE is an interesting reflection of scientific progress and increasing awareness of what is needed to understand the functioning of the Earth System. It began with three "Foci", ecophysiology (at various scales), vegetation dynamics (again at scales from patches to the globe) and agro-ecosystems. A fourth Focus, on biodiversity, but also involving linkages across the other three, was added as results from initial studies and models began to emerge. The evolution to a more integrated approach continued and the results presented in this volume show the level of awareness that has now been achieved.

Perhaps the most important achievement of GCTE has been to demonstrate the critical role that terrestrial ecosystems play in the functioning of the Earth System. When GCTE began, it was widely assumed that Earth System dynamics were dominated by the ocean-atmosphere system, and that terrestrial systems were just the recipients of changes in the dynamics of these two great fluids. Now the picture is much different, as the following examples demonstrate.

Terrestrial processes in the carbon cycle. Until very recently, projections of future atmospheric CO_2 concentration were determined only by estimated emissions from fossil fuel combustion and land-use change. Research within GCTE and elsewhere has elucidated the important role that feedback processes in terrestrial ecosystems – heterotrophic respiration, wildfires, permafrost melting – will play in determining the trajectory of atmospheric CO_2 concentration over the next few decades and centuries. This work has contributed to the issue of 'sink saturation' and the possibility that the terrestrial will switch later this century from being a net sink to a net source of carbon.

Nonlinearities in the Earth System. Within the IGBP framework, GCTE took the lead in analyzing the nature of nonlinear change in Earth System functioning. This work played a central role in the emergence of abrupt change, surprises and extreme events as unifying themes in the second phase of IGBP research.

Dynamic Global Vegetation Models (DGVMs). When GCTE began its implementation in 1991, the terrestrial surface was treated as a 'big leaf' or a 'green slime' in global climate models. One of the project's highest priorities was to change this situation. Several research groups associated with GCTE produced prototype DGVMs by the mid-1990s, a model intercomparison was implemented later in the decade, and now DGVMs are recognized as an essential component – as important as the oceans and the atmosphere – in Earth System models.

Complementing this Earth System perspective, the last phase of GCTE also placed emphasis on the consequences of global change for the things that matter to people – captured in this book in the section on "Ecosystem Services". These consequences are mostly reflected at regional scales and the regions under most stress are discussed in the final section of the book.

We are delighted and honored to have been asked to write this Preface. Brian was GCTE's first Chair and Will the first full time Scientific Officer, before he took over as IGBP Director. We both greatly enjoyed our involvement, benefiting from it enormously, and this was in large measure thanks to all the fine people who were involved. We cannot mention them all but we want to acknowledge one person in particular, Rowena Foster, for the prodigious effort she has put in, throughout the 15 years of its existence, to making GCTE work. We know that every scientist who was involved in one of the many workshops organized by Rowena will join us in thanking her.

GCTE's research over the past 15 years provides a sound base for the new Global Land Project, and the community that GCTE has built will make many further contributions to the GLP. This book highlights the exciting work that was carried out during the second half of GCTE and points towards the new challenges to be undertaken under the GLP banner. We congratulate the authors and editors on a fine effort. We thank the GCTE community for its many achievements and wish the GLP all the best for the future.

Brian Walker and *Will Steffen*
November 2006
Canberra

Acknowledgments

The implementation and success of GCTE was possible thanks to the commitment and contributions of many scientists from around the world who have volunteered for leading roles to drive activities, established networks, and run workshops and synthesis efforts for almost 15 years. Over 100 scientists played roles as members of the scientific steering committee, and as activity and task leaders. To all of them we want to show our appreciation and gratitude for their valuable time and intellectual contributions. Their willingness to contribute well beyond their own interest provided an invaluable service to the development of a globally coordinated understanding of science. Particular thanks go to the chairs of GCTE: Brian Walker, Ian Noble, and Louis Pitelka; and Harold Mooney for being such a motivating and inspiring leader.

We also want to thank to a smaller group of individuals who invested their careers in the roles of executive and project officers to support the implementation of the GCTE science plan. Without those individuals who were able to be full time facilitators, coordinators, and leaders, the GCTE would have not been able to operate successfully. Those individuals include: William Batista, Pep Canadell, Sara Duke, Pablo Inchausti, John Ingram, Elisabeth Huber-Sannwald, George Koch, Diane Pataki and Will Steffen. We also want to thank Rowena Foster who supported with great proficiency the International Project Office (IPO) in Canberra for the entire life of the project.

No GCTE activity or office would have been possible without the engagement and long term commitment of the many funding agencies which supported the offices and the development of networks, workshops, and synthesis efforts.

These long term funding relationships were key to the success of GCTE, enabling the establishment of an IPO and several focus offices that continuously supported the operations of GCTE. The IPO was based in Canberra, Australia and funded by the Australian Greenhouse Office (AGO) of the Department of the Environment and Heritage, and the Australian Commonwealth Scientific and Research Organization (CSIRO); initially in the Division of Sustainable Ecosystems and later on in the Division of Marine and Atmospheric Research. Both AGO and CSIRO Marine and Atmospheric Research are continuing their invaluable support to international research coordination through their support to the Global Carbon Project, a joint project of the Earth System Science Partnership (IGBP, IHDP, WCRP, and Diversitas).

The US-National Science Foundation (NSF), the National Aeronautics and Space Administration (NASA), Stanford University, and the University of Utah in Salt Lake City supported the focus 1 office on ecosystem physiology and biogeochemistry; the US-Department of Energy (DOE) funded many activities related to ecosystem physiology. The Natural Environment Research Council (NERC) through the Centre for Ecology and Hydrology in Wallingford, UK supported the focus 3 office on agroecology and production systems. The Inter-American Institute for Global Change based in Brazil, the University of Buenos Aires, and the CNRS-Ecole Normale Superieure in Paris supported the focus 4 office on functional biodiversity.

On behalf of the GCTE and its sponsor program, the International Geosphere-Biosphere Program, we want to express our appreciation and thanks for the long term commitment and significant contributions to the coordination of international science. This support has enabled the GCTE to leave behind a legacy of improved understanding of the effects of global change on terrestrial ecosystems, and a large community with the scientific capacity to continue this work into a new phase.

Finally we want to thank all the authors of chapters in this book for their time and valuable contributions towards this final GCTE effort.

Josep Canadell, Diane Pataki, Louis Pitelka
The editors

Contents

1 **Global Ecology, Networks, and Research Synthesis** 1
1.1 Introduction 1
1.2 Carbon and Water Cycles in the 21st Century 2
1.3 Changing Biodiversity and Ecosystem Functioning 3
1.4 Landscapes under Changing Disturbance Regimes 3
1.5 Managing Ecosystem Services 4
1.6 Regions under Stress 4
1.7 The Way Forward 4
References 5

Part A
Carbon and Water Cycles in the 21st Century 7

2 **CO_2 Fertilization: When, Where, How Much?** 9
2.1 Carbon a Limiting Plant Resource? 9
2.2 Long-Term Biomass Responses and Carbon Pools 10
2.2.1 Time Matters 10
2.2.2 Nutrients and Water Determine Biomass Responses at Elevated CO_2 11
2.2.3 Scaling from Growth to Carbon Pools 13
2.3 Carbon to Nutrient Ratios and Consumer Responses 13
2.3.1 The C to N Ratio Widens 13
2.3.2 Consequences for Herbivory, Decomposition and Plant Nutrition 14
2.4 Plant Water Relations and Hydrological Implications 14
2.5 Stress Resistance under Elevated CO_2 16
2.6 Biodiversity Effects May Outweigh Physiology Effects 16
2.6.1 Hydrology Implications of Elevated CO_2 Depend on Species Abundance 16
2.6.2 Biodiversity Effects on Forest Carbon Stocking and Grassland Responses 16
2.7 Summary and Conclusions 17
References 18

3 **Ecosystem Responses to Warming and Interacting Global Change Factors** 23
3.1 The Multiple Factor Imperative in Global Change Research 23
3.2 Ecosystem Responses to Experimental Warming 23
3.2.1 The GCTE-NEWS Synthesis 24
3.2.2 The ITEX Synthesis 25
3.2.3 The Harvard Forest Soil Warming Experiment 26

3.3 Temperature and CO_2 Interactions in Trees: the TACIT Experiment 26
3.3.1 Experimental Design 26
3.3.2 Growth Responses 27
3.3.3 Higher-Order Responses 28
3.3.4 TACIT Summary 28
3.4 More Than Two Factors: the Jasper Ridge Global Change Experiment 28
3.4.1 Experimental Design 28
3.4.2 Net Primary Productivity 29
3.4.3 Community Composition 29
3.4.4 JRGCE Summary 30
3.5 Modeling Temperature, CO_2 and N Interactions in Trees and Grass 30
3.5.1 Global Change Simulations for a California Annual Grassland 30
3.5.2 Comparing Forest and Grassland with G'DAY 32
3.6 Summary and Conclusions 33
Acknowledgments 34
References 34

4 Insights from Stable Isotopes on the Role of Terrestrial Ecosystems in the Global Carbon Cycle 37
4.1 Introduction 37
4.2 Ecosystem Carbon Cycles 37
4.3 The Global Carbon Cycle 40
4.4 Future Directions 42
Acknowledgments 42
In Memoriam 42
References 43

5 Effects of Urban Land-Use Change on Biogeochemical Cycles 45
5.1 Introduction 45
5.2 Urban Land-Use Change 46
5.3 Urban Environmental Factors 47
5.3.1 Climate and Atmospheric Composition 47
5.3.2 Atmospheric and Soil Pollution 49
5.3.3 Introductions of Exotic Species 49
5.4 Disturbance and Management Effects 50
5.4.1 Lawn and Horticultural Management 50
5.4.2 Management Effort 51
5.5 Effects of Built Environment 52
5.6 Assessing Biogeochemical Effects – the Importance of Scale 54
5.7 Summary and Conclusions 55
Acknowledgments 56
References 56

6 Saturation of the Terrestrial Carbon Sink 59
6.1 Introduction 59
6.2 Location of the Current Terrestrial Carbon Sinks 59
6.3 Dynamics of Processes that Contribute to Carbon Sink Saturation 60
6.4 Processes Contributing to Terrestrial Carbon Sink Saturation 60
6.4.1 Processes Driven by Atmospheric Composition Change 60
6.4.2 Processes Driven by Climate Change 64
6.4.3 Processes Driven by Land-Use Change and Land Management 66
6.5 Integration and Model Predictions 71
6.6 Summary and Conclusions 73
Acknowledgments 74
References 74

Part B
Changing Biodiversity and Ecosystem Functioning ... 79

7 Functional Diversity – at the Crossroads between Ecosystem Functioning and Environmental Filters ... 81
7.1 Introduction ... 81
7.2 Environmental Filters Affect FD ... 82
7.3 FD effects on Global Change Drivers ... 82
7.3.1 The Traits of the Dominants ... 82
7.3.2 The Role of Interactions ... 87
7.4 Summary and Conclusions ... 89
Acknowledgments ... 89
References ... 90

8 Linking Plant Invasions to Global Environmental Change ... 93
8.1 Introduction ... 93
8.2 Plant Invasions and Elevated CO_2 ... 93
8.3 Plant Invasions and Climatic Change ... 95
8.4 Plant Invasions and Land Eutrophication ... 96
8.5 Plant Invasions and Changes in Land Use/Cover ... 97
8.6 Multiple Interactions ... 98
8.7 Summary and Conclusions ... 99
Acknowledgments ... 99
References ... 99

9 Plant Biodiversity and Responses to Elevated Carbon Dioxide ... 103
9.1 Ten Years of GCTE Research: Apprehending Complexity ... 103
9.1.1 Effects of CO_2 on Plant Diversity Through Alterations of the Physical Environment ... 103
9.2 Temporal Variation and Response to Elevated CO_2 ... 105
9.2.1 Reproductive and Evolutionary Aspects of the Response to Elevated CO_2 105
9.2.2 Communities at Equilibrium Versus Dynamic Systems ... 105
9.3 Biodiversity Loss and Response to Elevated CO_2 ... 107
9.3.1 Species Diversity and Response to Elevated CO_2 ... 107
9.3.2 Ecosystem C Fluxes in a Species-Poor World ... 108
9.4 Summary and Conclusions ... 110
References ... 111

10 Predicting the Ecosystem Consequences of Biodiversity Loss: the Biomerge Framework ... 113
10.1 Biodiversity and Ecosystem Functioning: a Synthesis ... 113
10.1.1 Why Biodiversity Matters to Global Change Ecology ... 113
10.1.2 Linking Change in Biodiversity with Change in Ecosystem Functioning 114
10.1.3 Lessons Learned from Early Debates ... 114
10.1.4 What We Have Learned about the Relationship between Biodiversity and Ecosystem Function ... 115
10.1.5 The Scientific Framework for Linking Biodiversity and Ecosystem Functioning ... 115
10.2 The BioMERGE Framework ... 117
10.2.1 The BioMERGE Structural Sub-Framework ... 117
10.2.2 The BioMERGE BEF Sub-Framework: an Expansion of the Vitousek-Hooper Framework ... 117
10.2.3 The BioMERGE Research Implementation Sub-Framework ... 119
10.3 Discussion: Towards a Large Scale BEF ... 122
Acknowledgments ... 123
References ... 123

Part C
Landscapes under Changing Disturbance Regimes 127

11 Plant Species Migration as a Key Uncertainty in Predicting Future Impacts of Climate Change on Ecosystems: Progress and Challenges 129
11.1 Introduction 129
11.2 Will Migration Be Necessary for Species Persistence? 130
11.2.1 Vegetation-Type Models 131
11.2.2 Species-Based Models 132
11.3 Measurements and Models of Migration Rates 133
11.4 Linking Migration and Niche Based Models 134
11.5 Summary and Conclusions 135
Acknowledgments 135
References 135

12 Understanding Global Fire Dynamics by Classifying and Comparing Spatial Models of Vegetation and Fire 139
12.1 Introduction 139
12.2 Background 140
12.3 Model Classification 140
12.4 Model Comparison 141
12.4.1 The Models 141
12.4.2 The Comparison Design 143
12.5 Results and Discussion 144
12.5.1 Model Classification 144
12.5.2 Model Comparison 145
12.6 Summary and Conclusions 145
Acknowledgments 146
References 146

13 Plant Functional Types: Are We Getting Any Closer to the Holy Grail? 149
13.1 In Search of the Holy Grail 149
13.2 Individual Plant Structure and Function 149
13.3 Traits and Environmental Gradients 152
13.3.1 Plant Functional Response to Mineral Resource Availability 152
13.3.2 Plant Functional Response to Disturbance 152
13.3.3 Projecting Changes in Plant Functional Traits in Response to Global Change 154
13.4 Scaling from Individual Plants to Communities: from Response Traits to Community Assembly 155
13.5 Scaling from Communities to Ecosystems: from Response Traits to Effect Traits 156
13.6 So, Are We Getting Closer to the Holy Grail? Scaling beyond Ecosystems 157
13.6.1 Plant Functional Traits and Landscape Dynamics 157
13.6.2 Regional to Global Models – Revisiting the Early Functional Classifications 157
13.6.3 Validation: the Contribution of Paleo-Data 158
13.7 Summary and Conclusions 159
Acknowledgments 159
References 159

14 Spatial Nonlinearities: Cascading Effects in the Earth System 165
14.1 Introduction 165
14.2 Conceptual Framework 166
14.3 Insights to Global Change Issues 166
14.3.1 Historical Example: the Dust Bowl of the 1930s 166
14.3.2 Wildfire 168
14.3.3 Invasive Species and Desertification 171
14.4 Forecasting Spatial Nonlinearities and Catastrophic Events 172
14.5 Summary and Conclusions 173
Acknowledgments 173
References 173

15 Dynamic Global Vegetation Modeling: Quantifying Terrestrial Ecosystem Responses to Large-Scale Environmental Change 175
15.1 Introduction 175
15.2 Historical Antecedents and Development of DGVMs 175
15.2.1 Plant Geography 176
15.2.2 Plant Physiology and Biogeochemistry 176
15.2.3 Vegetation Dynamics 177
15.2.4 Biophysics 177
15.2.5 Human Intervention 178
15.3 Principles and Construction of DGVMs 178
15.3.1 Model Architecture 178
15.3.2 Net Primary Production 179
15.3.3 Plant Growth and Vegetation Dynamics 179
15.3.4 Hydrology 180
15.3.5 Soil Organic Matter Transformations 180
15.3.6 Nitrogen (N) Cycling 180
15.3.7 Disturbance 180
15.4 Evaluating DGVMS 181
15.4.1 Net Primary Production 181
15.4.2 Remotely Sensed "Greenness" and Vegetation Composition 181
15.4.3 Atmospheric CO_2 Concentration 181
15.4.4 Runoff 182
15.4.5 CO_2 and Water Flux Measurements 182
15.5 Examples of Applications of DGVMS 182
15.5.1 Holocene Changes in Atmospheric CO_2 182
15.5.2 Boreal "Greening" and the Contemporary Carbon Balance 182
15.5.3 The Pinatubo Effect 183
15.5.4 Future Carbon Balance Projections 183
15.5.5 Carbon-Cycle Feedbacks to Future Climate Change 183
15.5.6 Effects of Land-Use Change on the Carbon Cycle 185
15.6 Some Perspectives and Research Needs 185
15.6.1 Comparison with Field Experiments 185
15.6.2 Plant Functional Types 185
15.6.3 The Nitrogen Cycle 185
15.6.4 Plant Dispersal and Migration 186
15.6.5 Wetlands 186
15.6.6 Multiple Nutrient Limitations 186
15.6.7 Agriculture and Forestry 186
15.6.8 Grazers and Pests 186
15.6.9 Biogenic Emissions of Trace Gases and Aerosol Precursors 187
15.7 Summary and Conclusions 187
References 187

Part D
Managing Ecosystem Services 193

16 Wheat Production Systems and Global Climate Change 195
16.1 Introduction 195
16.2 Global Atmospheric Change, Climate and Yields 197
16.3 Impacts on Wheat Productivity 199
16.4 Addressing the Yield Gap 200
16.5 The Protein Gap 200
16.6 The Rice-Wheat System 201
16.7 The Effect of Climate Change on the Rice-Wheat System 202
16.8 The Contribution of the Rice-Wheat System to Climate Change 202
16.9 Carbon Dioxide 203
16.10 Methane 203
16.11 Nitrous Oxide 204
16.12 Comparative Greenhouse Gas Budgets for Rice-Wheat Farming Systems 204
16.13 Summary and Conclusions 207
References 208

17 Pests Under Global Change – Meeting Your Future Landlords? 211
17.1 Introduction 211
17.2 Methods 211
17.2.1 IPCC Processes 211
17.2.2 Monitoring, Benchmarks and Indicators for Measuring Impacts 212
17.2.3 Estimating Impacts 213
17.3 Impacts 216
17.3.1 Atmospheric CO_2 and Climate 216
17.3.2 Land Use, Land Cover and Biodiversity 219
17.3.3 Trade and Travel 219
17.4 Adaptation 220
17.4.1 Natural Adaptations 220
17.4.2 Adaptive Management Options 220
17.4.3 Adaptation of Control Measures in Response to Global Change 221
17.4.4 Threats to Sustainability of Adaptation Options 221
17.5 Vulnerability 222
17.6 Summary and Conclusions 222
References 223

18 Greenhouse Gas Mitigation Potential in Agricultural Soils 227
18.1 Introduction 227
18.1.1 Soil Carbon and Carbon Dioxide 227
18.1.2 Trade-Offs between GHGs in Agriculture 227
18.2 What Is Meant by GHG Mitigation Potential? 229
18.3 Regional Case Studies 230
18.3.1 Sustainable Soil Management in the Moscow Region to Enhance Soil Carbon 230
18.3.2 Soil Carbon Sequestration Potential in the US 231
18.4 Carbon Sequestration in the Future 232
18.5 Win-Win Strategies for Greenhouse Gas Mitigation by Agricultural Soils 232
18.6 Future Challenges 233
18.6.1 Improving Carbon Sequestration 233
18.6.2 Monitoring Soil Carbon Sequestration 233
18.7 Summary and Conclusions 234
References 235

19 Carbon and Water Tradeoffs in Conversions to Forests and Shrublands 237
19.1 Introduction 237
19.2 Afforestation 237
19.2.1 Afforestation: Carbon Storage Potential 237
19.2.2 Afforestation: Evapotranspiration and Water Yield 238
19.2.3 Afforestation: Potential Atmospheric Feedbacks 239
19.3 Woody Encroachment and Agriculture 240
19.3.1 Grassland Conversions with Woody Plant Encroachment and Agriculture 240
19.3.2 Processes Controlling Soil C Storage: Grassland Vs. Woodland 241
19.3.3 Uncertainties in Water and Carbon Balances with Woody Plant Encroachment 242
19.4 Summary and Conclusions 243
References 244

20 Natural and Human Dimensions of Land Degradation in Drylands: Causes and Consequences 247
20.1 Introduction 247
20.2 Drylands, Desertification, Drivers, and Scales 247
20.2.1 Distribution of People and Land-Cover Types 247
20.2.2 Defining Land Degradation and Desertification 248
20.2.3 What Drives Land Degradation and Desertification? 249
20.2.4 Estimating the Extent of Desertification 249
20.2.5 Consequences of Desertification 250
20.2.6 Scale and Hierarchy 251
20.3 Joint GCTE-LUCC Desertification Initiative 252
20.3.1 Dahlem Desertification Paradigm 252
20.3.2 Initiatives to Test the Dahlem Desertification Paradigm 253
20.4 Management of Desertified Drylands 254
20.4.1 Avoidance 254
20.4.2 Monitoring 254
20.4.3 Restoration 254
20.5 Summary and Conclusions 255
Acknowledgments 255
References 255

Part E Regions under Stress 259

21 Southeast Asian Fire Regimes and Land Development Policy 261
21.1 Introduction 261
21.2 Underlying Causes of Land Fires 262
21.2.1 Explaining Fire Occurrence 262
21.2.2 Land Development Policies 262
21.2.3 Land Management Practices 263
21.2.4 Property Rights and Conflicts 265
21.3 Landscape, Regional and Global Interactions 265
21.3.1 Ecosystem Dynamics 265
21.3.2 Regional Haze Episodes 266
21.3.3 Greenhouse Gas Emissions 266
21.3.4 Interactions with Climate Variability and Change 267
21.4 Human Well-Being 267
21.4.1 Economic and Health Impacts 267
21.4.2 Livelihoods 267

21.5 Informed Decision-Making and Better Governance 268
21.5.1 Role of Expertise 268
21.5.2 Regional Cooperation 269
21.6 Summary and Conclusions 269
Acknowledgments 270
References 270

22 Global Change Impacts on Agroecosystems of Eastern China 273
22.1 Introduction 273
22.2 Chinese Terrestrial Transects 274
22.3 Physiological and Plant Responses to Multiple Global Change Forcing .. 275
22.4 Productivity and Its Responses to Global Change 276
22.5 Carbon Budget and Its Responses to Global Change 278
22.6 Summary and Conclusions 282
Acknowledgments 282
References 282

23 Terrestrial Ecosystems in Monsoon Asia: Scaling up from Shoot Module to Watershed 285
23.1 Introduction 285
23.2 Responses of Plant Communities to the Global Change: Scaling from Leaf to Landscape Through Individual Plant 285
23.2.1 Competition among Individual Plants in Even-Aged Monospecific Stands at Elevated CO_2 286
23.2.2 Shoot-Module-Based Simulator As a Tool of Individual Tree Response 288
23.2.3 Modeling the Shift of Forest Zonation 289
23.3 Carbon Budget at the Forest Watershed Scale 289
23.3.1 Carbon Exchange between Atmosphere-Forest-Stream Boundaries . 290
23.3.2 Transport of Dissolved Organic Carbon Associated with Dissolved Nitrogen from Terrestrial to Aquatic Ecosystems 290
23.3.3 Dynamics of Dissolved Organic Carbon at the Interface of Stream and Lake Ecosystems 291
23.4 Carbon Budget and Functions of the Lake Biwa Ecosystem 292
23.4.1 Carbon Budget in the Lake 292
23.4.2 Metabolism in the Lake Sediments 293
23.4.3 Terrestrial Environment and Function of Lake Ecosystems 293
23.5 Summary and Conclusions 294
References 294

24 Responses of High Latitude Ecosystems to Global Change: Potential Consequences for the Climate System 297
24.1 Introduction 297
24.2 Recent Changes in Climate, Disturbance Regimes, and Land Cover 297
24.3 Responses of Radiatively Active Gases 300
24.3.1 General Issues 300
24.3.2 Responses of CO_2 Exchange to Climatic Change 300
24.3.3 Responses of CH_4 Exchange to Climatic Change 302
24.3.4 Responses to Changes in Disturbance and Land Cover 303
24.4 Responses of Water and Energy Exchange 304
24.4.1 General Issues 304
24.4.2 Responses to Changes in Climate, Disturbance, and Land Cover 304
24.5 Delivery of Freshwater to the Arctic Ocean 305
24.5.1 General Issues 305
24.6 Summary and Conclusions 305
References 306

Part F
Future Directions: the Global Land Project ... 311

25 The Future Research Challenge: the Global Land Project ... 313
25.1 Introduction ... 313
25.2 Research Objectives ... 314
25.3 Emergent Concepts ... 315
25.3.1 Land-Use Decision Making and Adaptive Management ... 315
25.3.2 Ecosystem Services ... 316
25.3.3 Vulnerability and Sustainability Science ... 316
25.4 Research Framework ... 317
25.4.1 Theme 1: Dynamics of Land System ... 317
25.4.2 Theme 2: Consequences of Land-System Change ... 318
25.4.3 Theme 3: Integrating Analysis and Modeling for Land Sustainability ... 318
25.5 Implementation Strategy ... 319
25.6 Summary and Conclusions ... 320
References ... 321

Index ... 323

List of Contributors

Craig D. Allen

U.S. Geological Survey,
Fort Collins Science Center,
Jemez Mountains Field Station,
Los Alamos, NM 87544, USA

Mike Apps

Pacific Forestry Centre, Canadian Forest Service,
Victoria, BC, Canada

Richard H. A. Baker

Central Science Laboratory,
Sand Hutton, York YO41 1LZ, UK

Lawrence E. Band

University of North Carolina,
Chapel Hill, NC 27599, USA

Kenneth T. Belt

USDA Forest Service,
Northern Research Station,
c/o Center for Urban Environmental Research and Education,
Baltimore, MD 21227, USA

Brandon T. Bestelmeyer

USDA ARS, Jornada Experimental Range,
Las Cruces, NM 88003, USA

Jagtar Bhatti

Northern Forestry, Centre,
Canadian Forest Service,
Edmonton, AB, Canada

Alberte Bondeau

Potsdam Institute for Climate Impact Research (PIK),
Telegrafenberg,
14412 Potsdam, Germany

Terry Callaghan

Abisko Scientific Research Station, Sweden,
Department of Animal and Plant Sciences,
University of Sheffield, UK

Josep G. Canadell

Global Carbon Project,
CSIRO Marine and Atmospheric Research,
GPO Box 3023,
Canberra, ACT 2601, Australia

Geoffrey J. Cary

Research School of Biological Sciences,
and The Bushfire Cooperative Research Center,
Australian National University,
Canberra, ACT 0200, Australia

F. Stuart Chapin III

Institute of Arctic Biology, University of Alaska,
Fairbanks, AK 99775, USA

Torben R. Christensen

Geobiosphere Science Centre, Lund University,
Lund, Sweden

Joy S. Clein

Institute of Arctic Biology, University of Alaska,
Fairbanks, AK 99775, USA

Stella Coakley

Department of Botany and Plant Pathology,
Oregon State University,
Corvallis, OR 97331, USA

Robert Colwell

Department of Ecology and Evolutionary Biology,
University of Connecticut,
Storrs, CT 06269-3043, USA

Jeffrey D. Corbin

Department of Integrative Biology,
University of California,
Berkeley, CA 94720-3140, USA

J. Hans C. Cornelissen

Department of System Ecology,
Institute of Ecological Science, Vrije Universiteit,
Amsterdam, The Netherlands

Wolfgang Cramer

Potsdam Institute for Climate Impact Research (PIK),
Telegrafenberg,
14412 Potsdam, Germany

Ian D. Davies

Ecosystem Dynamics,
Research School of Biological Sciences,
Australian National University,
Canberra, ACT 0200, Australia

Stephen J. Del Grosso

Natural Resources Ecology Laboratory,
Colorado State University,
Fort Collins, CO 80523-1499, USA

Sandra Díaz

Instituto Multidisciplinario de Biología Vegetal
(CONICET - UNC) and FCECFyN,
Universidad Nacional de Córdoba
Córdoba, Argentina

Jeffrey S. Dukes

Department of Biology, University of Massachusetts,
Boston, MA 02125, USA

James R. Ehleringer

Department of Biology, University of Utah,
Salt Lake City, UT 84112, USA

Pete Falloon

Hadley Centre, UK Meteorological Office,
Exeter, UK

Kathleen A. Farley

Department of Geography,
San Diego State University,
San Diego, CA 92182-4493, USA

Michael D. Flannigan

Canadian Forest Service,
Sault Ste Marie, ON P6A 2E5, Canada

Uwe Franko

Centre for Environmental Research,
UFZ-Leipzig-Halle, Germany

Masami Fukuda

Institute of Low Temperature, Hokkaido University,
Sapporo, Japan

Robert H. Gardner

Appalachian Laboratory,
University of Maryland Center for Environmental Science
Frostburg, MD 21532, USA

Eric Garnier

Centre d'Ecologie Fonctionnelle et Evolutive,
34293 Montpellier Cedex 5, France

Roger Gifford

CSIRO Plant Industry, GPO Box 1600,
Canberra, ACT 2601, Australia

Andrew Gonzalez

Department of Biology, McGill University,
Montreal, PQ H3A 1B1, Canada

Peter R. Grace

Queensland University of Technology,
Brisbane, QLD, Australia

Karl Grigulis

Laboratoire d'Ecologie Alpine, Université Joseph Fourier, BP 53X,
38041 Grenoble Cedex 9, France

Peter M. Groffman

Institute of Ecosystem Studies,
Millbrook, NY 12545, USA

Diego E. Gurvich

Instituto Multidisciplinario de Biología Vegetal
(CONICET - UNC) and FCECFyN,
Universidad Nacional de Córdoba
Córdoba, Argentina

Richard Harrington

Rothamsted Research,
Harpenden, Hertfordshire, AL5 2JQ, UK

Sandy P. Harrison

School of Geographical Sciences, University of Bristol,
Bristol, BS8 1SS, UK

Kris M. Havstad

USDA ARS, Jornada Experimental Range,
Las Cruces, NM 88003, USA

Thomas Hickler

Department of Physical Geography & Ecosystems Analysis,
Lund University, Sölvegatan 12,
223 62 Lund, Sweden

Steven I. Higgins

Lehrstuhl für Vegetationsökologie,
Technische Universität München,
D-85350 Freising-Weihenstephan, Germany

Kouki Hikosaka

Graduate School of Life Sciences, Tohoku University,
Sendai, 980-8578, Japan

William A. Hoffmann

Department of Botany, North Carolina State University,
Raleigh, NC 27695, USA

John Hom

USDA Forest Service, Northern Research Station,
Newtown Square, PA 10973, USA

Richard A. Houghton

Woods Hole Research Center,
Falmouth, MA 02540, USA

Jennifer B. H. Martiny

School of Biological Sciences, University of California,
Irvine, CA 92697, USA

Robert B. Jackson

Nicholas School of the Environment and Earth Sciences,
Duke University,
Durham, NC 27708-1000, USA

Pete D. Jamieson

New Zealand Institute for Crop and Food Research Ltd.,
Christchurch, New Zealand

Esteban G. Jobbágy

IMASL, Universidad Nacional de San Luis & CONICET,
San Luis 5700, Argentina

Claire Jouseau

Department of Ecology,
Evolution and Environmental Biology,
Columbia University,
New York, NY 10025, USA

Robert E. Keane

USDA Forest Service, Rocky Mountain Research Station,
Missoula Fire Sciences Laboratory,
Missoula, MT 59808, USA

Charles D. Keeling

Scripps Institutions of Oceanography, University of California,
La Jolla, CA 92093, USA

Paul R. Kemp

Department of Biology, University of San Diego,
San Diego, CA 92110, USA

Takashi Kohyama

Graduate School of Environmental Earth Science,
Hokkaido University,
Sapporo, 060-0810, Japan

Eiichi Konohira

Graduate School of Environmental Studies,
Nagoya University,
Nagoya, 464-8601, Japan

Christian Körner

Institute of Botany, University of Basel, Schönbeinstrasse 6,
4056 Basel, Switzerland

Martin Körschens

Centre for Environmental Research,
UFZ-Leipzig-Halle, Germany

Darren J. Kriticos

Forest Biosecurity and Protection Unit, Ensis,
Private Bag 3020, Rotorua, New Zealand

Chun-Ta Lai

Department of Biology, San Diego State University,
San Diego, CA 92182, USA

Rattan Lal

School of Natural Resources, Ohio State University,
Columbus, OH, USA

Eric Lambin

Universite Catholique de Louvain,
Louvain-la-Neuve, Belgium

Sandra Lavorel

Laboratoire d'Ecologie Alpine, CNRS UMR 5553,
Université Joseph Fourier, BP 53X,
38041 Grenoble Cedex 9, France

Paul Leadley

Ecologie des Populations et Communaute,
Universite Paris-Sud XI,
URA CNRS 2154, Orsay Cedex, France

Louis Lebel

Unit for Social and Environmental Research,
Faculty of Social Sciences, Chiang Mai University,
Chiang Mai, 50200, Thailand

James M. Lenihan

USDA Forest Service, Pacific Northwest Research Station,
Corvallis Oregon, 97330, USA

Chao Li

Canadian Forest Service,
Edmonton, AB T6H 3S5, Canada

Wolfgang Lucht

Potsdam Institute for Climate Impact Research,
Telegrafenberg,
14412 Potsdam, Germany

Yiqi Luo

Department of Botany and Microbiology, University of Oklahoma,
Norman, OK 73019, USA

Fernando T. Maestre

Unidad de Biodiversidad y Conservación,
E.S.C.E.T., Universidad Rey Juan Carlos,
Móstoles, Spain

Trofim Maximov

Institute for Biological Problems of Cryolithozone,
Siberian Branch, Russian Academy of Sciences,
Yakutsk, Russia

William J. McConnell

Department of Fisheries and Wildlife,
Michigan State University,
East Lansing, MI 48824-1222, USA

Rebecca L. McCulley

Nicholas School of the Environment and Earth Sciences,
Duke University,
Durham, NC, USA

A. David McGuire

U.S. Geological Survey,
Alaska Cooperative Fish and Wildlife Research Unit,
University of Alaska,
Fairbanks, AK 99775, USA

Susan McIntyre

CSIRO Sustainable Ecosystems,
Canberra, ACT 2601, Australia

Ross E. McMurtrie

School of Biological, Earth and Environmental Sciences,
University of New South Wales,
Sydney, NSW 2052, Australia

Guy F. Midgley

Global Change Research Group,
South African National Biodiversity Institute,
Claremont, 7735 Cape Town, South Africa

Emilio Moran

Anthropological Center for Training and Research on Global Environmental Change (ACT),
Indiana University,
Bloomington, IN 47405-6101, USA

Jack Morgan

Rangeland Resources Research Unit, USDA-ARS,
Fort Collins, CO 80526-2083, USA

Peter Morin

Department of Ecology, Evolution & Natural Resources,
Rutgers University,
New Brunswick, NJ 08901, USA

Stuart Munson-McGee

Department of Chemical Engineering,
New Mexico State University,
Las Cruces, NM 88003, USA

Jun Murase

Graduate School of Bioagricultural Sciences, Nagoya University,
Nagoya, Japan

Daniel Murdiyarso

Centre for International Forestry Research,
Jl. CIFOR, Sindangbarang,
Bogor Barat, 16680, Indonesia

Shahid Naeem

Department of Ecology, Evolution and Environmental Biology,
Columbia University,
New York, NY 10025, USA

Richard Norby

Environmental Sciences Division,
Oak Ridge National Laboratory,
Oak Ridge, TN 37831-6422, USA

Dennis S. Ojima

Natural Resources Ecology Laboratory, Colorado State University,
Fort Collins, CO 80523-1499, USA

Alexander Onuchin

V.N. Sukachev Institute of Forestry,
Siberian Branch of Russian Academy of Sciences,
Krasnoyarsk, Russia

William J. Parton

Natural Resources Ecology Laboratory, Colorado State University,
Fort Collins, CO 80523-1499, USA

Diane E. Pataki

Department of Earth System Science
and Department of Ecology & Evolutionary Biology,
University of California,
Irvine, CA 92697, USA

Juli G. Pausas

Centro de Estudios Amientales del Mediterráneo,
E-46980 Paterna, València, Spain

Keith Paustian

Department of Soil and Crop Sciences, and Natural Resource Ecology, Laboratory,
Colorado State University, USA

Shaolin Peng

South China Institute of Botany,
Chinese Academy of Sciences
and College of Life Sciences,
Sun Yat Sen (Zhongshan) University,
Guangzhou, PR China

David A. Pepper

School of Biological, Earth and Environmental Sciences,
University of New South Wales,
Sydney, NSW 2052, Australia

Natalia Pérez-Harguindeguy

Instituto Multidisciplinario de Biología Vegetal
(CONICET – UNC) and FCECFyN,
Universidad Nacional de Córdoba, Argentina

Owen Petchey

Department of Animal and Plant Sciences,
University of Sheffield, Western Bank,
Sheffield, S10 2TN, UK

Debra P. C. Peters

USDA ARS, Jornada Experimental Range,
Las Cruces, NM 88003, USA

Roger A. Pielke, Sr.

Department of Atmospheric Sciences, Colorado State University,
Fort Collins, CO 80523, USA

Joan Pino

Center for Ecological Research and Forestry Applications
08193 Bellaterra, Catalonia, Spain

Louis F. Pitelka

Appalachian Laboratory, University of Maryland,
Center for Environmental Science,
Frostburg, MD 21532, USA

John R. Porter

Royal Veterinary and Agricultural University,
2630 Taastrup, Denmark

Catherine Potvin

Department of Biology, McGill University,
Montreal, PQ H3A 1B1, Canada

Richard V. Pouyat

USDA Forest Service, Northern Research Station,
c/o Center for Urban Environmental Research and Education,
Baltimore, MD 21227, USA

David Powlson

Agriculture and Environment Division,
Rothamsted Research, UK

I. Colin Prentice

QUEST, Department of Earth Sciences, University of Bristol,
Wills Memorial Building,
Bristol, BS8 1RJ, UK

Michael R. Raupach

Global Carbon Project,
CSIRO Marine and Atmospheric Research,
GPO Box 3023,
Canberra, ACT 2601, Australia

Peter Reich

Department of Forest Resources,
University of Minnesota,
St. Paul, MN 55108-6112, USA

Jim Reynolds

Nicholas School of the Environment and Earth Sciences,
and Department of Biology
Duke University,
Durham, NC 27708-0340, USA

Vladimir Romanenkov

Russian Institute of Agrochemistry,
Moscow, Russia

Catherine Roumet

Centre d'Ecologie Fonctionnelle et Evolutive,
CNRS UMR 5175, 1919 route de Mende,
34293 Montpellier Cedex 5, France

Jacques Roy

Center of Functional Ecology and Evolution,
Department of Ecosystem Function,
34293 Montpellier Cedex 5, France

Scott Rupp

University of Alaska Fairbanks,
Department of Forest Sciences,
Fairbanks, AK 99775, USA

Lindsey E. Rustad

USDA Forest Service,
Durham, NH 03824, USA

Harald Scherm

Department of Plant Pathology, University of Georgia,
Athens, GA 30602, USA

Ludmila Shevtsova

Russian Institute of Agrochemistry,
Moscow, Russia

Hideaki Shibata

Field Science Center for Northern Biosphere,
Hokkaido University,
Tokuda, Nayoro 096-0071, Japan

Anatoly Shvidenko

International Institute for Applied Systems Analysis,
Laxenburg, Austria

Stephen Sitch

Met Office (JCHMR), Maclean Building,
Crowmarsh-Gifford, Wallingford, OX10 8BB, UK

Stanley D. Smith

Department of Biological Sciences, University of Nevada,
Las Vegas, NV 89154-4004, USA

Ben Smith

Department of Physical Geography & Ecosystems Analysis,
Lund University, Sölvegatan 12,
223 62 Lund, Sweden

Pete Smith

School of Biological Sciences,
University of Aberdeen,
Aberdeen, AB24 3UU, UK

Jo Smith

School of Biological Sciences,
University of Aberdeen, UK

Mark Stafford-Smith

CSIRO Sustainable Ecosystems,
PO Box 284, Canberra, Australia

Will Steffen

Centre for Resource and Environmental Studies,
Australian National University,
Canberra, ACT 0200, Australia

Robert W. Sutherst

CSIRO Entomology, Indooroopilly, QLD 4068,
and School of Integrative Biology, University of Queensland,
St Lucia, Qld 4067, Australia

Martin T. Sykes

Department of Physical Geography & Ecosystems Analysis,
Lund University, Sölvegatan 12,
223 62 Lund, Sweden

Paula A. Tecco

Instituto Multidisciplinario de Biología Vegetal
(CONICET - UNC) and FCECFyN,
Universidad Nacional de Córdoba,
5000 Córdoba, Argentina

Wilfried Thuiller

Laboratoire d'Ecologie Alpine,
UMR-CNRS 5553,
Université J. Fourier, BP 53,
38041 Grenoble, France

Billie L. Turner III

George Perkins Marsh Institute/Geography,
Clarke University, 950 Main Street,
Worcester, MA 01610-1477, USA

Jotaro Urabe

Graduate School of Life Sciences, Tohoku University,
Sendai, 980-8578, Japan

Carlos Urcelay

Instituto Multidisciplinario de Biología Vegetal
(CONICET – UNC) and FCECFyN,
Universidad Nacional de Córdoba, Argentina

Eugene Vaganov

V.N. Sukachev Institute of Forestry,
Siberian Branch of Russian Academy of Sciences,
Krasnoyarsk, Russia

Montserrat Vilà

Center for Ecological Research and Forestry Applications,
Barcelona, Catalonia, Spain

Eitaro Wada

Japan Agency for Marine-Earth Science and Technology,
Yokohama, 236-0001, Japan

Yuhui Wang

Laboratory of Quantitative Vegetation Ecology,
Institute of Botany, the Chinese Academy of Sciences,
Beijing, PR China

Christian Wirth

Max-Plank Institute for Biogeochemistry,
D-07745 Jena, Germany

Justin Wright

Department of Biology, Duke University,
Durham, NC 27708, USA

Zhenzhu Xu

Laboratory of Quantitative Vegetation Ecology,
Institute of Botany,
Chinese Academy of Sciences,
Beijing, PR China

Takahito Yoshioka

Research Institute of Humanity and Nature,
Kamigyo-ku,
Kyoto 602-0878, Japan

Guangsheng Zhou

Laboratory of Quantitative Vegetation Ecology,
Institute of Botany, the Chinese Academy of Sciences,
Beijing, PR China
and
Institute of Atmospheric Environment,
China Meteorological Administration,
Shenyang, PR China

Li Zhou

Laboratory of Quantitative Vegetation Ecology,
Institute of Botany, the Chinese Academy of Sciences,
Beijing, PR China

Chapter 1

Global Ecology, Networks, and Research Synthesis

Louis F. Pitelka · Josep G. Canadell · Diane E. Pataki

1.1 Introduction

The Earth System, including its terrestrial and marine ecosystems, is being altered as a result of human activities. These global changes have been well-documented (e.g., Vitousek et al. 1997; Mooney and Canadell 2001; Steffen et al. 2004). They include changes in the composition of the atmosphere, including increasing concentrations of greenhouse gases; changes in global and regional climate; habitat destruction and land cover change; increases in the amounts of reactive nitrogen compounds in the biosphere; increases in species extinction rates; and increases in the number and impacts of exotic invasive species. The extent and magnitude of human-caused changes has led Nobel laureate Paul Crutzen (2002) to propose that the Earth has entered a new geologic era, the Anthropocene – an era of human domination of the Earth System, in contrast to the previous billions of years of Earth's history when natural forces dominated the Earth.

Nowhere are the impacts of human activities more apparent than in terrestrial ecosystems, with even the most remote and pristine terrestrial systems experiencing the effects of global change. The terrestrial portion of our planet obviously supplies many of the important ecosystem services upon which our society depends, including fresh water and much of our food and fiber. Because global change has the potential to significantly alter the structure and functioning of terrestrial ecosystems, considerable resources have been invested over the past decade on research to understand the effects of global change on ecosystems. In addition, since terrestrial systems are active components of a dynamic Earth System, research has also focused on whether those changes in terrestrial ecosystems are resulting in positive or negative feedbacks to the atmosphere and physical climate system.

Since 1991, the Global Change and Terrestrial Ecosystems (GCTE) core project of the International Geosphere-Biosphere Program (IGBP) has played a leadership role in developing and coordinating an international agenda on the two research areas mentioned above (Steffen et al. 1992). Over 100 synthesis papers and 25 special issues and books were published as a result of GCTE activity, not counting the primary research-data papers that stemmed from the various efforts. A synthesis of the first five years of GCTE research was published in 1999 (Walker et al. 1999) which followed an initial project-wide effort to further define the scientific scope of GCTE (Walker and Steffen 1997). This new volume, unlike the previous ones, describes examples of research conducted under the umbrella of GCTE during the second half of the project and does not intend to provide a comprehensive overview of all research. This chapter places these contributions in a historical context and explains how they fit into the broader GCTE endeavor.

This volume also marks the end of GCTE. After more than a decade of focusing on linkages between geophysical and biological research, the IGBP has entered a new, second phase aimed at describing and understanding the interactive physical, chemical, biological, and socio-economic processes that regulate the total Earth System and the changes that are occurring in this system. Because of the tremendous advances in disciplinary research on specific components of the Earth System made during phase I of IGBP, the global change research community is well poised to undertake the types of more interdisciplinary research that will be required to advance to the next level of understanding of the Earth System. As part of the transition to IGBP II, major restructuring of the core projects has taken place. A number of projects have ended, while new projects that bring together elements of the original projects have been initiated. GCTE is now closed, but IGBP research on terrestrial ecosystems will continue under the new Global Land Project (GLP), a joint project of IGBP and IHDP (the International Human Dimensions Project), built upon the foundations of GCTE and the Land-Use and Land-Cover Change (LUCC) project. The science objectives of the GLP are described in Chap. 25 (Ojima et al. 2007, Chap. 25 of this volume).

The overarching goals of GCTE were

- to predict the effects of changes in climate, atmospheric composition, and land use on terrestrial ecosystems;
- to determine how these effects lead to feedbacks to the atmosphere and physical climate system.

GCTE addressed these goals primarily by playing a coordinating and networking role among new and existing research projects of individual investigators. This role took a variety of forms. Through workshops and conferences involving broad representation from the international global change science community, GCTE developed and implemented a research agenda to address the goals of the project. Another successful strategy used by GCTE to advance global science was to facilitate the development of international research networks to enhance communication and integration. Workshops organized by these networks have produced a large number of synthesis books and journal articles. Still other GCTE activities have included model inter-comparisons and the development of critical databases.

GCTE was originally comprised of four foci: Ecosystem Physiology; Ecosystem Structure; Agriculture, Forestry, Soils; and Biodiversity. It is worth reflecting on the fact that, in the early 1990s, it was necessary and made sense for these four foci to operate fairly independently. At that point, there were models and experiments that focused only on responses of ecosystem physiology (e.g., NPP) to global change (e.g., elevated CO_2). Researchers were interested in how ecosystem structure (e.g., vegetation type) would respond to a changing climate without simultaneously considering effects on ecosystem physiology. Research on agricultural and forestry systems was distinguished from research on more natural ecosystems. And research on the interactions between biodiversity and ecosystem functioning (physiology) was just beginning. A measure of the progress made over the past decade is the extent to which research activities now integrate across two or more of these previously separate subdisciplines, and we will highlight examples in this chapter.

1.2 Carbon and Water Cycles in the 21st Century

Research on the effects of rising atmospheric CO_2 concentrations on terrestrial ecosystems has been a component of GCTE since its inception. At the time GCTE was founded, much of the literature on elevated CO_2 effects focused on individual plants grown under ideal conditions of adequate water and nutrient availability (Mooney et al. 1999). To promote and synthesize ecosystems-based research on the effects of elevated CO_2 on whole communities in the field, the Elevated CO_2 Network began to hold a series of workshops on the role of community responses, belowground processes, interactions with environmental stress, and other key uncertainties in understanding the responses of the terrestrial biosphere to elevated CO_2. An increasing number of experiments have addressed these uncertainties in a variety of biomes from grasslands and forests to drylands, agroecosystems, and more. As a result, our understanding of the variability and complexity of ecosystem responses to elevated CO_2 has improved dramatically over the past decade, particularly with regard to water and nutrient limitations to CO_2 responses, belowground carbon pools, and community scale dynamics (see Chap. 2 by Körner et al.).

Changes in atmospheric CO_2 concentrations in the 21st century will occur in concert with changes in temperature, precipitation, nitrogen deposition, and other aspects of global change. Studies of ecosystem responses to perturbations such as elevated CO_2 or temperature can improve future projections of critical attributes of the terrestrial biosphere such as the carbon balance (see Chap. 6 by Canadell et al.). In addition to the Elevated CO_2 Network, the Network of Ecosystem Warming Studies greatly added to our understanding of global change ecology by synthesizing the results of experimental manipulations of temperature in a variety of ecosystems (Rustad et al. 2001; Shaver et al. 2000; Chap. 3 by Norby et al.). However, interactive effects among multiple disturbances may not be additive and merit further investigation in the next generation of manipulative and modeling experiments. In Chap. 3, Norby et al. review ecosystem responses to warming, interactive effects of CO_2 and temperature, and effects of multiple of aspects of global change as assessed by experimental and modeling results. Relative to responses of individual perturbations, our understanding of interactive effects of multiple global change drivers is at a fairly early stage, and will continue to remain a critical aspect of understanding terrestrial ecosystems responses in the coming decades.

Another important contribution of GCTE was the initiation of the Biosphere-Atmosphere Stable Isotope Network (BASIN), which is still ongoing. As described by Pataki et al. in Chap. 4, the application of measurements of the isotopic composition of atmospheric trace gases in ecosystem-scale studies has provided new information about ecosystem physiology and its role in ecosystem, regional, and the global carbon cycle. Isotopes integrate physical and biological processes over space and time, and are increasingly measured in a variety of ecosystems ranging from virtually unmanaged to highly managed and human dominated. In general, human dominated ecosystems are increasingly of interest to the ecological and global change communities due to their large influence on the atmosphere and climate system, resource availability, and ecosystem services and human welfare. The effects of urbanization on biogeochemistry are reviewed by Pouyat et al. in Chap. 5 as an emerging area of critical importance in quantifying impacts of land use and land cover change, as well as the dynamics of coupled human and natural systems, which will be a focus of the new Global Land Project (see Chap. 25 by Ojima et al.).

1.3 Changing Biodiversity and Ecosystem Functioning

The interest in establishing relationships between biodiversity and ecosystem function arose from the realization that rates of species loss are accelerating due to human activities, particularly due to habitat lost and fragmentation among others. Given that species have different functional traits, major changes in biodiviersity could result in significant changes in biogeochemistry and other processes that regulate the Earth System (Loreau et al. 2001; see Chap. 7 by Diaz et al., and Chap. 10 by Naeem et al.). This GCTE activity quickly moved to include the effects of atmospheric composition and climate change on biodiversity (see Chap. 9 by Potvin et al.), and the effects of rapid expansions of exotic species which can profoundly change the structure and function of many terrestrial and freshwater ecosystems (see Chap. 8 by Vilà et al.).

The GCTE established a focus on this topic in partnership with the emerging program of Diversitas, leading the development of what was a completely new field of research early in the 1990s. It soon became clear that functional changes due to species loss, addition or replacement could have significant impacts on ecosystem properties such as productivity, decomposition rates, nutrient cycling, and resistance and resilience to perturbations. Numerous reviews were instrumental in the development of this new field (Sala et al. 1999; Loreau et al. 2001, 2002; Hooper et al. 2005). These syntheses described some fundamental patterns (Hooper et al. 2005): *(i)* Certain combinations of species are complementary in their patterns of resource use and can increase average rates of productivity and nutrient retention, *(ii)* Susceptibility to invasion by exotic species is strongly influenced by species composition and, under similar environmental conditions, generally decreases with increasing species richness, *(iii)* Having a range of species that respond differently to different environmental perturbations can stabilize ecosystem process rates in response to disturbances and variation in abiotic conditions.

Finally, GCTE also led the development of new biodiversity scenarios with a first attempt at mapping changes in the global distribution of species due to climate and land use change (Sala et al. 2000).

1.4 Landscapes under Changing Disturbance Regimes

Disturbances such as fire, pests, windthrow, and harvesting are important in determining the structure and functioning of terrestrial ecosystems. Global warming and other forms of global change will further perturb ecosystems over the next decades, and will interact with the other forms of disturbance in complex ways. The chapters in this section are examples of research conducted over the past decade to better understand how changing disturbance regimes will interact with and affect ecosystems.

One of the key effects of climate change on ecosystems will be adjustments in the ranges of plant and animal species, and accordingly, of communities and biomes. It is well known from the paleoecological literature that species have migrated in the past in response to changes in climate (e.g., Davis and Shaw 2001), and relatively recent evidence demonstrates that species ranges are already beginning to change in response to contemporary climate change (Parmesan and Yohe 2003; Walther et al. 2002). A critical question given the expected rate of climate change over the next decades is whether and which species will be able to migrate fast enough to keep pace with the changing climate. The answer to this question is relevant to issues as diverse as the conservation of species and the carbon balance of ecosystems. It has been recognized for some time that models of ecosystem response to climate change must include migration, but the challenges of incorporating these effects into models are immense. A GCTE activity was devoted to this issue (Pitelka et al. 1997; Neilson et al. 2005), and in Chap. 11, Midgley et al. review our current understanding and the hurdles that still face ecologists in understanding and predicting future migrations of plants.

Migration is actually a two-part process, with one being dispersal and the second successful establishment and population growth. Disturbance by fire could be one of the most important mechanisms that would facilitate and speed up the successful establishment of migrant species. Because fire occurs across much of the Earth's land surface, it also has great significance for the carbon budget of terrestrial ecosystems. Thus, it is critically important for ecologists to understand how fire regimes might change with global change and incorporate this understanding in models of ecosystem response, especially to climate change. Keane et al. (Chap. 12) review models of vegetation dynamics that incorporate fire and describe our current understanding of global fire dynamics.

A major concern throughout global change science is the potential for non-linearities in responses of ecosystems or other components of the Earth System (Steffen et al. 2004). Fire is an example of a process that can spread non-linearly across space and generate cascading effects (see Chap. 14 by Peters et al.). Unless scientists can identify where and when non-linear responses are likely, human society will like be unprepared for their occurrence, and thus the consequences are likely to be far greater than if the response had been identified and anticipated. In Chap. 14, Peters et al. describe a general framework for understanding and predicting spatially non-linear responses to global change.

A GCTE activity that built on the understanding developed in other areas of GCTE research, including the efforts referred to above, is the development of Dynamic Global Vegetation Models or DGVMs. Early in the history of GCTE entirely different classes of models were employed to simulate how climate change affects ecosystem structure and composition vs. ecosystem processes. These models were equilibrium models that could not predict how long it would take to reach a future state or the vegetation dynamics of the process. The next development was a simple linking of vegetation and ecosystem models such that the ecosystem models used the future vegetation distribution predicted by vegetation models as inputs (VEMAP Members 1995). Meanwhile, ecosystem modelers recognized the need to fully couple vegetation and ecosystem models and make them dynamic, i.e., to develop DGVMs. Prentice et al. (Chap. 15) provide a review of DGVMs, including their successes in simulating various phenomena, as well as their limitations.

DGVMs and most other types of models described in this section do not simulate individual plant species but rather group species into a relatively small number of *plant functional types*, or PFTs, with similar morphological and physiological traits. Lavorel et al. (Chap. 13) describe the efforts to develop useful and logical classifications of plants into PFTs.

1.5 Managing Ecosystem Services

Many of the GCTE activities directly contributed to a better understanding of the provision of goods and services for the well-being of human societies, and how those goods and services were being altered by changes in atmospheric composition, climate and land use. However, the single biggest contribution came from the focus on "agroecology and production systems" which dealt with the effects of global change on the production of food as a key component in global food security. A number of networks and consortia were established to study individual crops including wheat (see Chap. 16 by Porter et al.), rice, potato, pastures, and forest plantations. Key outcomes from this research were a better understanding of the benefits of increasing atmospheric CO_2 on crop production and how warming, nutrients, and water modulate that response, either by dampening it or by enhancing it depending on the particular combination. Temperature thresholds for crop failure were also identified for some of the rice varieties growing under elevated CO_2 and warming. This research naturally included the role of agricultural pests in reducing food production and how climate change is likely to alter pest-plant dynamics (see Chap. 17 by Sutherst et al.).

The work in agroecosystems also extended to the integration of food/fiber production and biogeochemistry to fully account for plant-soil feedbacks, and with that the study of soil carbon sequestration which yielded one of the important legacies of GCTE, a global soil organic matter database and a number of carbon sequestration assessments under different climate and management scenarios (SOMNET; see Chap. 18 by Smith et al.).

A key development which required higher level integrative research was the one on complex agroecosystems and farm-level integration of the management of single units of land with multiple crops, as separate subsystems, or as a single interconnected complex systems. The latter are still the systems that much of the world's population relies on for food and fiber despite an overall push for intensification and simplification.

Two additional chapters complement what is perhaps one of the most critical trade-offs in managing ecosystem goods and services, that between carbon and water. Chap. 19 by Jackson et al. focuses on this trade off in conversions to forests and shrublands, and Chap. 20 by Reynolds et al. focuses on a new paradigm to explain the natural and human dimensions of land degradation in arid and semi-arid regions.

1.6 Regions under Stress

The GCTE made a substantial effort to initiate research on the effects of global change on terrestrial ecosystems in parts of the world where little research or capacity existed despite playing key roles in the functioning of the Earth System.

Early in the 1990s the GCTE developed the transect approach for global change research, establishing 15 transects in critical regions of the world to cover most environmental conditions and biomes/ecotones with special attention to highly sensitive regions (e.g., high latitudes and tropical regions) (Canadell et al. 2002).

The transect approach has evolved over the years to comprehensive regional studies building upon the initial research undertaken during the 1990s. In this book we present research focusing on several key sensitive regions in the world including tropical Asia (see Chap. 21, Murdiyarso et al.), regions affected by the Monsoon climate (see Chap. 22 by Zhou, and Chap. 23 by Kohyama et al.), and high latitude ecosystems (see Chap. 24, McGuire et al.).

1.7 The Way Forward

As noted earlier, the GCTE has now ended, but clearly there is still much to do to understand the effects of global change on terrestrial ecosystems. Research on terrestrial ecosystems will continue under the auspices of the new Global Land Project (GLP), a joint project of the IGBP and the International Human Dimensions Program (IHDP). The GLP will integrate research elements from both GCTE and the Land-Use and Land-Cover Change (LUCC) project, the latter already jointly sponsored by the IGBP and IHDP. A major goal is to treat human activities and decision-making as an integral part of the terrestrial system, the so called human-environment system. The final chapter of this book, Chap. 25 by Ojima et al., describes the science objectives of the GLP and the innovative research agenda on land for the next decade (GLP 2005).

In addition, GCTE has also contributed a great deal to the development of the scientific agendas of the Global Carbon Project (GCP) and the Global Environmental Change and Food Systems (GECAFS), both joint projects of the new Earth System Science Partnership (ESSP). The ESSP has brought together IGBP, IHDP, the World Climate Research Program, and Diversitas, and has developed a highly interdisciplinary and integrative research agenda to support Earth System sustainability.

References

Canadell J, Steffen W, White PS (eds) (2002) IGBP/GCTE Terrestrial transects: dynamics of terrestrial ecosystems under environmental change. Journal Vegetation Science 13:297–448

Crutzen PJ, Stoermer EF (2002) The Anthropocene. IGBP Newsletter 41:17–18

Davis MB, Shaw RG (2001) Range shifts and adaptive responses to Quaternary climate change. Science 292:673–679

GLP (2005) Science Plan and Implementation Strategy. IGBP Report No. 53/IHDP report no. 19. IGBP Secretariat, Stockholm, 64 pp

Hooper DU, Chapin FS, Ewel JJ, Hector A, Inchausti P, Lavorel S, Lawton JH, Lodge DM, Loreau M, Naeem S, Schmid B, Setala H, Symstad AJ, Vandermeer J, Wardle DA (2005) Effects of biodiversity on ecosystem functioning: a consensus of current knowledge. Ecological Monographs 75:3–35

Loreau M, Naeem S, Inchausti P, Bengtsson J, Grime JP, Hector A, Hooper DU, Huston MA, Raffaelli D, Schmid B, Tilman D, Wardle DA (2001) Biodiversity and ecosystem functioning: current knowledge and future challenges. Science 294:804–807

Loreau M, Inchausti P, Naeem S (2002) Biodiversity and ecosystem functioning: synthesis and perspectives. Oxford University Press, Oxford

Mooney HA, Canadell JG (2001) The Earth System: biological and ecological dimensions of global environmental change. Encyclopedia of Global Environmental Change Vol. 2, Wiley, London

Mooney H, Canadell JG, Chapin FS, Ehleringer J, Körner Ch, McMurtrie R, Parton W, Pitelka L, Schulze D-E (1999) Ecosystem physiology responses to global change. In: Walker BH, WL Steffen, J Canadell, JSI Ingram (eds) The terrestrial biosphere and global change. Implications for natural and managed ecosystems. Cambridge University Press, Cambridge, pp141–189

Neilson RP, Pitelka LF, Solomon AM, Nathan R, Midgley GF, Fragoso JMV, Lischke H, Thompson K (2005) Forecasting regional to global plant migration in response to climate change. BioScience 55:749–759

Parmesan C, Yohe G (2003) A globally coherent fingerprint of climate change impacts across natural systems. Nature 421: 37–42

Pitelka LF, Gardner RH, Ash J, Berry S, Gitay H, Noble IR, Saunders A, Bradshaw RHW, Brubaker L, Clark JS, Davis MB, Sugita S, Dyer JM, Hengeveld R, Hope G, Huntley B, King GA, Lavorel S, Mack RN, Malanson GP, McGlone M, Prentice IC, Rejmanek M (1997) Plant migration and climate change. American Scientist 85:464–473

Rustad LE, Campbell JL, Marion GM, Norby RJ, Mitchell MJ, Hartley AE, Cornelissen JHC, Gurevitch J, GCTE-NEWS (2001) A meta-analysis of the response of soil respiration, net nitrogen mineralization, and aboveground plant growth to experimental ecosystem warming. Oecologia 126:543–562

Sala OE, Chaping III FS, Gardner RH, Laurenroth WK, Mooney HA, Ramakrishnan PS (1999) Global change, biodiversity and ecological complexity. In: Walker B, Steffen W, Canadell J, Ingram J (eds) The terrestrial biosphere and global change. Cambridge University Press, Cambridge, pp 304–328

Sala OE, Chapin III FS, Armesto JJ, Berlow E, Bloomfield J, Dirzo R, Huber-Sanwald E, Huenneke LF, Jackson RB, Kinzig A, Leemans R, Lodge DM, Mooney HA, Oesterheld M, Poff NL, Sykes MT, Walker BH, Walker M, Wall DH (2000) Biodiversity – global biodiversity scenarios for the year 2100. Science 287:1770–1774

Shaver GR, Canadell J, Chapin III FS, Gurevitch J, Harte, Henry G, Ineson P, Jonasson S, Melillo J, Pitelka L, Rustad L (2000) Global warming and terrestrial ecosystems: a conceptual framework for analysis. BioScience 50:871–882

Steffen WL, Walker BH, Ingram JSI, Koch GW (1992) Global change and terrestrial ecosystems: the operational plan. IGBP Report No. 21, Stockholm, 97 pp

Steffen W, Sanderson A, Tyson PD, Jager J, Matson PA, Moore III B, Oldfield F, Richardson K, Schellnhuber HJ, Turner II BL, Wasson RJ (2004) Global change and the Earth System. A planet under pressure. The IGBP Series, Springer, Berlin

VEMAP members (1995) Vegetation/ecosystem modeling and analysis project: comparing biogeography and biogeochemistry models in a continental-scale study of terrestrial ecosystem responses to climate change and CO_2 doubling. Global Biogeochem Cycles 9:407–438

Vitousek PM, Mooney HA, Lubchenco J, Melillo JM (1997) Human domination of Earth's ecosystems. Science 277:494–499

Walker BH, Steffen WL (1997) Global change and terrestrial ecosystems. IGBP Book Series. Cambridge University Press, Cambridge

Walker BH, WL Steffen, J Canadell, JSI Ingram (1999) The terrestrial biosphere and global change. Implications for Natural and Managed Ecosystems. Cambridge University Press, Cambridge

Walther GR, Post E, Convey P, Menzel A, Parmesan C, Beebee TJC, Fromentin JM, Hoegh-Guldberg O, Bairlein F (2002) Ecological responses to recent climate change. Nature 416:389–395

Part A

Carbon and Water Cycles in the 21st Century

Chapter 2

CO_2 Fertilization: When, Where, How Much?

Christian Körner · Jack Morgan · Richard Norby

2.1 Carbon a Limiting Plant Resource?

Among the certain facets of global change, the steady CO_2 enrichment of the atmosphere affects plants directly. CO_2 is a substrate for photosynthesis, and for most (C_3) plant species CO_2 concentrations will remain below saturation levels of photosynthesis even at twice-current concentrations. Thus, the photosynthetic CO_2 response curve (a saturation function with a near-linear initial part and a plateau reached at close to 1 000 ppm) became the corner stone of modeling the biological consequences of rising of CO_2, because it was assumed to represent the degree of CO_2-fertilization in plants. In fact, models built around this response function forecast an approximately 290 Gt carbon sequestration into the terrestrial biosphere (biomass and soils) within the 21st century (e.g., Cao and Woodward 1998; IPCC 2001), raising hopes for a significant biotic mitigation of atmospheric CO_2 enrichment, which currently measures 3–4 Gt C yr^{-1}.

Thirty years of research, the most active part in the GCTE decade (1993–2003), have produced an estimated more than 4 000 publications (Körner 2000) in essence testing this or related hypotheses. It turned out that the rate of CO_2 uptake per unit leaf area is a rather unreliable predictor of plant growth and the cascade of follow up ecosystem processes. This limits predictions of future plant growth and crop production and it also restricts the credibility of estimates of the amounts of carbon that could be sequestered to the biosphere as a result of 'CO_2-fertilization'.

Still not fully resolved, the discrepancy between photosynthetic stimulation by elevated CO_2, which is almost always observed, and actual plant growth responses, which vary considerably, became a central problem in predicting both the biosphere's future in a CO_2-rich world and the potential of biological mitigation of atmospheric CO_2 enrichment. A related quandary has been the observed stimulation of growth in the less common but important C_4 class of plants through elevating CO_2 above present ambient levels (Wand et al. 1999; Ghannoum et al. 2000), since C_4 photosynthesis was thought to be saturated at present atmospheric CO_2 concentrations (Bazzaz 1990; Bowes 1993). A weak quantitative linkage between photosynthesis and growth (or yield) has long been known among crop breeders (Evans and Dunstone 1970; Saugier 1983; Wardlaw 1990), and there are many reasons for the mismatch between the CO_2-stimulation of photosynthesis and the actual biomass responses, but the most important ones are:

(1) The photosynthetic uptake of CO_2 depends on the total photosynthetically active surface of a plant and its integrated CO_2 flux activity over time (not easily quantified in discontinuous leaf-level gas exchange measurements) relative to its respiratory losses by heterotrophic tissue, (2) CO_2 release by respiration, which likewise depends not only on tissue specific rates but on the contribution of various, differentially active tissues to total plant biomass, and (3) the unknown export of assimilated carbon through roots, either directly to the rhizosphere microbe community or to mycorrhizal fungi.

With these and some other minor components of the internal carbon housekeeping, plants co-control their carbon investment in addition to their control over photosynthesis. The relationship between carbon uptake (source activity) and carbon investment (sink activity) drives the net outcome in terms of growth and biomass production.

It is obvious that these carbon investments also depend on resources other than CO_2, in particular mineral nutrients. A common effect of short-term plant exposure to elevated CO_2 is a reduced consumption of nutrients, but also water, per unit of biomass produced (Drake et al. 1997) or a constant consumption at greater biomass per unit land area (Niklaus and Körner 2004). In cases where total nutrient uptake is increased under elevated CO_2 (Finzi et al. 2002) this will deplete soil resources in the long run. In cases where tissue nutrient concentrations are depleted, this will induce cascades of negative ecosystem level feedbacks, which eventually may also cause initial rates of carbon gain to diminish. In many cases, it became questionable whether carbon is a limiting resource at the whole plant or ecosystem level (Körner 2003a). It is worth recalling that all taxa of today's biosphere grew and reproduced successfully with only 180–190 ppm, half the current CO_2

concentration, 18 000 years before present (peak of last glaciation). Based on this reference period, current biota operate already in a double CO_2 atmosphere. In addition, the observed reduction of water consumption per unit land area is likely to induce climatic feedbacks (through a drier atmosphere), not yet accounted for in experiments. Furthermore, any CO_2 enrichment effect on plants will depend on their developmental stage, with younger plants more responsive than older ones (Loehle 1995). Most of the CO_2-enrichment responses for woody species available to date are – for very practical reasons – for young, rapidly expanding life stages, during which carbon is more likely a limiting resource.

All together these interdependencies are making predictions of the biosphere's overall response to elevated CO_2 far more difficult than was thought initially in view of the straightforward photosynthetic CO_2 response of leaves. The experimental evidence exemplified here, points at the need to reconsider the largely source activity dominated model parameterisation and to account for the constraints to carbon sink activity. Using current atmospheric carbon relations, Schimel et al. (2001) find the biosphere roughly in balance, i.e., deforestation being balanced by new C-fixation, part of which is due to forest expansion, hence leaving a much smaller leeway for CO_2-driven C-sequestration compared to what was anticipated during the 1980s.

In contrast to theoretical models, experiments are limited in space and time, and manipulations may disrupt ecosystem processes and thus, can create artifacts, though directions of responses are commonly still identified correctly in such situations (Norby et al. 1999). A major limitation, however, is that more than 80% of the Earth's life biomass is stored in wood, but trees are too big to be studied in conventional test systems, and it is nearly impossible to put a mature forest into an enclosure. Free air CO_2 enrichment (FACE) at such scales entails exorbitant costs, so that tests remained restricted to a few places and mostly fairly young trees. FACE refers to a CO_2-release system that does not need any enclosures, but freely releases (controlled by a computer in connection with an infrared gas analyser) CO_2 over test plots. The situation is far better for grassland and agricultural crops, where the low stature vegetation permits smaller CO_2-enrichment plots and where enclosures create fewer artifacts.

In this summary we will highlight what we think were major achievements during the last decade and we will restrict our considerations to the most realistic field tests, largely but not exclusively, using free air CO_2 enrichment *in situ*. Given the urgent need for forest data, these will receive particular attention. We will not re-review this vast field, but will try to illustrate the major lines of evidence and future needs in this arena of global change research.

2.2 Long-Term Biomass Responses and Carbon Pools

Whether a measured biomass gain under a step increase of CO_2 and otherwise unchanged life conditions reflects a change which might be seen in the real world, depends on three factors: time, nutrients and water. In addition, global change factors in the real, future world like temperature change (apart from effects on water, see Norby et al. 2007, Chap. 3 of this volume), light (change in cloudiness), and biotic interactions, will co-determine CO_2 effects. There have been few attempts at deciphering such interactions under realistic test conditions (see e.g., Rawson 1992; Shaw et al. 2002; Vonder et al. 2004; Wan et al. 2004) and the answers so far are inconclusive, with no clear interactive effect (Pendall et al. 2004). The general impact of a warming climate on the carbon balance of ecosystems will be dealt with elsewhere in this volume (Norby et al. 2007, Chap. 3 of this volume). We will, however, refer to the significance of interactions between CO_2 and light for tree seedlings and lianas in shaded habitats.

2.2.1 Time Matters

Plants with determinate growth (annuals, many crops) may grow faster initially at elevated CO_2, but end up with similar biomass accumulation per individual a couple of days or weeks earlier than controls, except under very fertile conditions. This was so for most of the 25 species of semi-desert winter annuals from the Negev desert of Israel, grown for their full life cycle in elevated CO_2, with only one species, namely one of the five legumes (the most 'mesic' one), taking great advantage of CO_2 enrichment, causing the whole sward to end up with more biomass (Grünzweig and Körner 2001, Fig. 1). In seedlings or saplings of woody species, such initial responses (when space and nutrients are ample and plants are very leafy) will cause future growth to capitalize on size differences induced by CO_2 very early in their life (analogous to compound interest in economics), which has nothing to do with ongoing CO_2 stimulation. Such initial responses simply shift the growth curve in time (Norby et al. 1995; Loehle 1995; Körner 2000; Spinnler et al. 2002) and cause the date of harvest to determine the difference in biomass gain.

Among the many other time dependent drivers of plant and ecosystem CO_2 responses, changes in community composition are important, and soil responses are particularly critical, given the size and slow turnover of soil C pools. Most of the humus carbon stored in soil is of very old age (often >1 000 years) and only a small fraction (commonly <1% in steady state systems) participates in year to year C-turnover. Mass based assessments of soil carbon are thus near to impossible. On the other hand, carbon tracer experiments with true control (also the ambient CO_2 system being labeled at the same rate) are restricted to small scale, highly disturbed test systems (e.g., Leavitt et al. 2001; Niklaus et al. 2001). So, experiments provide rather limited access to the long-term fate of C in soils, but may point at some likely trends.

2.2.2 Nutrients and Water Determine Biomass Responses at Elevated CO_2

Grassland biomass responses to elevated CO_2 reported in the literature (Körner 2002) are clearly related to the fertility of the test conditions as well as to soil water availability. In mesic grasslands, CO_2-induced biomass increases are greater under more fertile conditions (e.g., Hebeisen et al. 1997). Alpine grassland seems to be an exception, since there was no positive growth response to elevated CO_2 even when mineral nutrients were added (Schäppi and Körner 1996). In native and semi-natural grasslands with no fertilizer amendments (Fig. 2.1), the seasonal biomass response to elevated CO_2 appears to be more pronounced in dry years, as a result of the CO_2-induced stomatal reduction of transpiration. This indirect water relations response to CO_2 enrichment increases leaf and canopy-level water use efficiency, extends the effective growing season in drier grasslands, and enhances nutrient availability (Morgan et al. 2004a). Most grasslands are characterized by periods of water shortage (Stephenson 1990; Campbell et al. 1997) and thus experience this indirect water relations benefit. A serious constraint on the usefulness of grassland biomass data produced under elevated CO_2 is the extrapolation of these indirect CO_2 effects to future environments because such water relation responses to CO_2 may not be seen in reality. At the landscape scale, reductions in leaf conductance to water vapor will lead to a drier and warmer atmosphere, which in turn would increase evaporative losses. Warmer future ambient temperatures will enhance such evaporative losses, while predicted higher humidity in some locations might reduce them (IPCC 2001). The final outcome will depend on complicated interactions between the atmosphere and the biosphere that are difficult to predict, and will vary geographically. Thus, much of the presumed direct CO_2 effects on grassland biomass (Fig. 2.1) are in fact indirect water relations responses (Morgan et al. 2004a) that will have to be corrected for future atmospheric humidity conditions and humidity driven biosphere/atmosphere feedback – a major forthcoming challenge for modeling.

With these caveats, the seasonal gain in harvestable biomass of seven different grassland systems is between 0% and 30%, with the Californian grassland showing the largest and the alpine the smallest response. Removing the CO_2-induced moisture effect on biomass, the remaining effect will most likely be smaller than +10% for a doubling of CO_2 in most cases. Here and in the following, it is important to distinguish between net biomass accretion (what is left for harvest at peak season or after a season) and the total biomass which had been produced during a year, irrespective of whether it was preserved or had been recycled i.e., lost before harvest. The latter is called net primary production (NPP), and is much larger than, and not to be confused with, the standing stock of biomass.

Mature forest trees have been exposed to elevated CO_2 only in a few cases, with the majority of data coming from young trees in enclosures. A particularly valuable source of data has come from trees growing around natural sources of CO_2 (Miglietta et al. 1993; Amthor 1995). Oak trees (*Quercus ilex*) grown around natural CO_2 springs in Italy, showed a significant stimulation of tree ring growth during their years of re-growth after coppicing (Hättenschwiler et al. 1997), very similar to the responses seen in scrub-oak after fire (Dijkstra et al. 2002). Such re-growth situations after complete losses of above

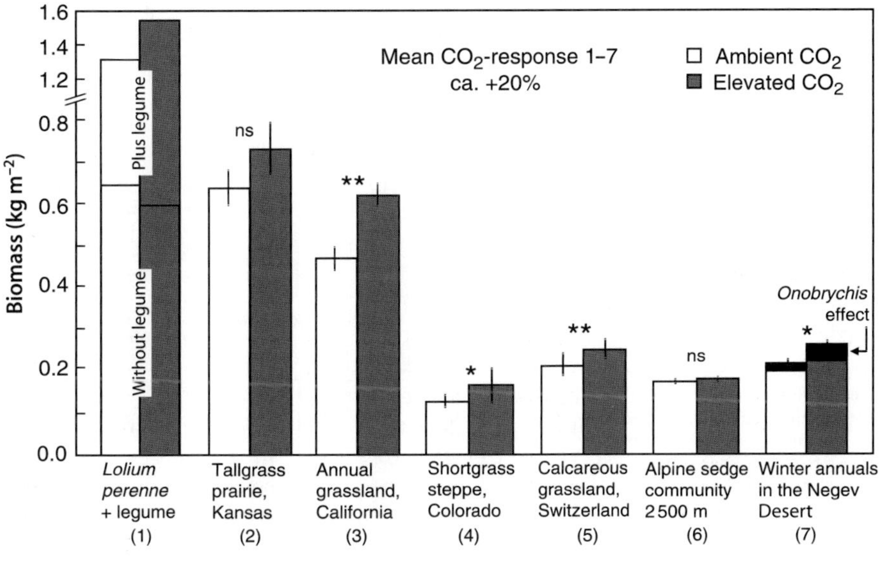

Fig. 2.1.
A comparison of biomass responses to elevated CO_2 in 6 natural grassland ecosystems (all grown on unfertilized ground) plus one artificial *Lolium perenne/Trifolium repens* stand (*1*). The prairie (*2*) and calcareous grassland (*5*) data are multi-year averages, with the positive effect largely introduced by dry years. Bars indicate plus/minus one standard error. References: (*1*) Hebeisen et al. (1997a), 5 cuts per year, 12 g N m^{-2} a^{-1}, (*2*) Owensby et al. (1997), (*3*) Shaw et al. (2002), (*4*) Morgan et al. (2001), (*5*) Niklaus et al. (2001), Körner (2000), (*6*) Körner et al. (1997), (*7*) Grünzweig and Körner (2001)

ground structures represent a life phase for these trees with unlimited nutrient supply and ample space and light, conditions where carbon-assimilation may be the prime limitation of growth. Later, as stands close, the biomass relationships are re-established and excess nutrients are used up, the gain would be expected to vanish. Indeed, the gain in growth of Mediterranean oaks around CO_2 springs was greatly diminished compared to controls after about 25 years of growth under elevated CO_2 (Hättenschwiler et al. 1997).

Trees growing in older stands, which had reached a closed canopy before the onset of CO_2 enrichment may thus show much smaller CO_2 stimulation or more rapid re-adjustment to previous growth rates than was found in seedlings and saplings. Total foliage area (LAI) of fully developed canopies has not increased in response to CO_2 enrichment in any closed canopy system studied so far (Körner and Arnone 1992; Hättenschwiler and Körner 1996; DeLucia et al. 2002; Norby et al. 2003); hence, any growth response must derive from an enhancement of growth per unit leaf area.

Three FACE experiments have been operating in tree stands starting with fully developed canopies: a pine (*Pinus taeda*) stand in the Duke forest in North Carolina, a sweetgum (*Liquidambar styraciflua*) stand at the Oak Ridge National Laboratory in Tennessee, and a mature multi-species deciduous forest near Basel, Switzerland. The young monospecific stands have shown a sustained increase in productivity in response to elevated CO_2, but the trajectory of growth response has differed. Increased productivity in the pine stand has resulted in more stem growth and more leaf litter production (Hamilton et al. 2002; DeLucia and Moore 2005, Fig. 2.2a). Analysis of an additional plot at the Duke FACE site, however, suggested that an N limitation would cause this growth stimulation to disappear (Oren et al. 2001). Productivity in the deciduous sweetgum stand has been 22% higher in elevated CO_2, but the initial stimulation of aboveground growth disappeared after the first year (Norby et al. 2002; Fig. 2.2b). Instead, the additional carbon assimilated in elevated CO_2 has supported a large increase in fine-root production (Norby et al. 2004). The inherent difference in fine-root turnover rates in the pine and sweetgum stands (Matamala et al. 2003) may be an important determinant of whether a CO_2 stimulation of productivity results in increased growth or increased turnover, and this difference affects the long-term fate of C in the ecosystem.

The Swiss forest FACE experiment as been run for 4 years only, and the 30 m tall trees belong to several species, requiring a careful analysis of long tree ring chronologies to overcome obvious replication problems in the case of such large 'experimental units' and their restricted number. However, accounting for pre-treatment differences in vigor of individual trees (via tree ring analysis), the stem growth trajectory is similar to the one seen in the Oak Ridge FACE, showing a stimulation of radial growth in 100 year old beech in the first year which disappeared thereafter (Fig. 2.2c; Körner et al. 2005). Surprisingly, leaf photosynthetic capacity showed no down

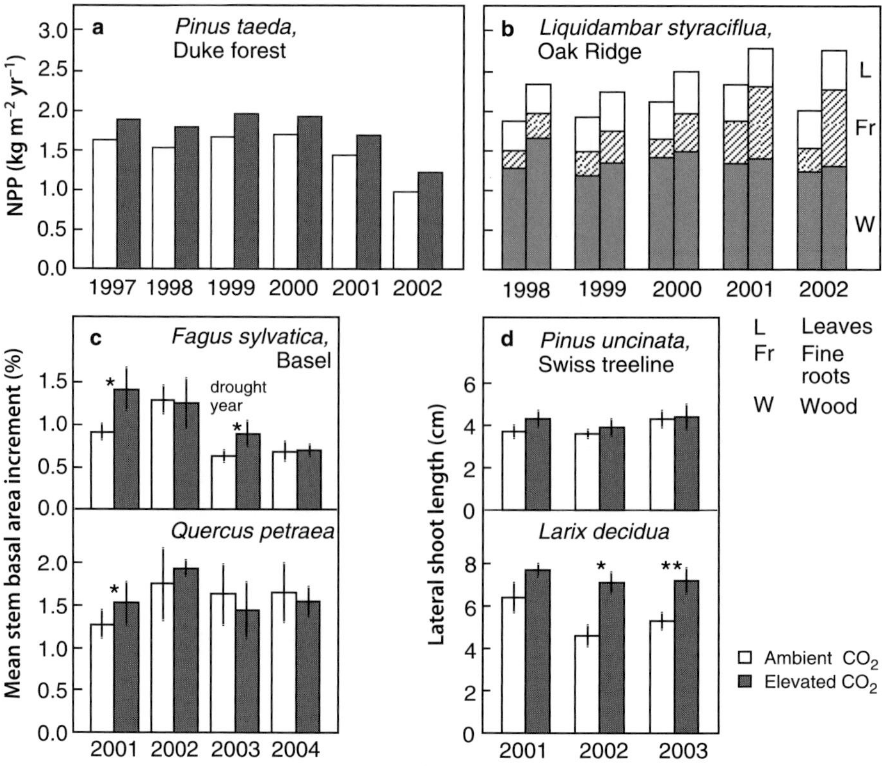

Fig. 2.2. A synthesis of in situ biomass responses of mature trees to Free Air CO_2 Enrichment (FACE). Net primary production (NPP) estimated from basal area increment, leaf litter collection, and minirhizotron analysis for **a** a *Pinus taeda* (loblolly pine) plantation in North Carolina (DeLucia et al. 2005), and **b** for a uniform stand of *Liquidambar styraciflua* (sweet gum) in Tennessee, in this case separated by biomass compartments. Note the exclusive effect on fine roots (Norby et al. 2004). **c** Mean stem basal area increase (+ s.e.) in 2001–2004 of oak and beech at the Swiss Canopy Crane site (Asshoff et al. 2006). **d** as in **c**, but for shoot growth response to three consecutive years of free air CO_2 enrichment in 30 year old larch and pine trees growing at treeline (Swiss Alps, 2 200 m). Mean shoot length was based on five lateral branches per tree ($n = 5$ trees per species/CO_2 level; Handa et al. 2005)

ward adjustment (Zotz et al. 2005), but soil CO_2-emission suggests enhanced carbon throughput in this system as well (Steinmann et al. 2004), which is in line with earlier findings for complex model ecosystems (Körner and Arnone 1992; Hättenschwiler and Körner 1996, 1998). The only case where woody plants showed a near homoeostatic downward adjustment of photosynthesis under elevated CO_2 is the desert FACE experiment in Nevada (Hamerlynck et al. 2002). One may speculate that this has to do with a lack of significant below ground biological activity (e.g., mycorrhiza). A fourth FACE test established at the Swiss high elevation climatic treeline in 30 year old 1.5 to 2 m tall pines (*Pinus uncinata*) and larches (*Larix decidua*) also revealed an initial pine response, which disappeared by year three, but so far, a positive response in larch is persisting (Handa et al. 2005, Fig. 2.2d).

In summary, these data suggest a rather limited growth stimulation in mature trees, after the initial effect of a step-increase in CO_2 concentration is surpassed. However net primary production and the associated carbon turnover become enhanced in a high CO_2 environment, explaining in part the discrepancy between photosynthetic stimulation and biomass responses. The challenge for the future is to identify the fingerprints of enhanced carbon flows through undisturbed soils.

The available data for tall, closed canopy forests contrast with the significant stimulation often reported for seedlings or saplings grown under favorable experimental conditions. However, seedlings growing in deep forest shade, will take great advantage from the reduced light requirements for photosynthesis under elevated CO_2 under natural conditions, as was documented in situ for both a tropical (Würth et al. 1998) and a temperate forest (Hättenschwiler and Körner 2000). Under such conditions, elevated CO_2 is likely to affect tree recruitment in favor of particularly responsive species, and likely could affect future forest composition (see also the effect on lianas below).

2.2.3 Scaling from Growth to Carbon Pools

Great care is advised when translating such plot-based data into landscape carbon gains (Amthor 1995). At these larger scales the mean residence time of carbon per unit land area influences carbon stocks. If trees would grow faster and mature earlier, tree turnover would be greater, but not necessarily the mean C-storage in biomass. In fact, a faster forest rotation may reduce the forest C-pool. There is evidence for accelerated maturation in pine trees grown in elevated CO_2: cones were produced earlier and in greater masses than in controls (LaDeau and Clark 2001). The above described transient nature of tree growth responses to elevated CO_2 and the contribution of CO_2-induced water savings, which may in part be artifacts, because of a lack of atmospheric feedback, set rather stringent limits to expected growth stimulations in a CO_2-rich world. Accelerated development may reduce mean forest carbon stocks (Körner 2003b, 2004), but enhanced C-turnover may also feed more carbon into the humus compartment, a rather uncertain projection, given the high demand for nitrogen and other minerals to form the complex humus aggregates.

Thus, an important message to users of experimental findings is that growth or productivity, i.e., a rate of carbon incorporation, as they are documented in some cases, should not be confused with the amounts of carbon stocked per unit land area. It is the latter which matters for sequestering carbon away from the atmosphere. In terms of carbon storage policy, a high stocking rate (e.g., old growth forests) is desirable rather than a faster forest rotation, as it may become facilitated by elevated CO_2 in plantations on fertile ground. Such growth stimulations, however, can add to the substitution of fossil by renewable fuels and products.

2.3 Carbon to Nutrient Ratios and Consumer Responses

2.3.1 The C to N Ratio Widens

It has been documented many times that exposure of plants to elevated CO_2 causes an increase in non structural carbohydrate concentrations (NSC; e.g., Körner and Miglietta 1994 for natural CO_2 springs, underlining the long term nature of this response) and a reduction in leaf nitrogen concentrations (Cotrufo et al. 1998; King et al. 2004). However, N-depletion is not a universal effect and often is a dilution effect caused by NSC accumulation, disappearing on an NSC-free dry matter basis. In tissues other than leaves, such changes have been seen less frequently. But seeds have also been found N-(protein)depleted in plants grown under elevated CO_2 under natural growth conditions (Fig. 2.3a), and in wheat when not heavily fertilized, with far ranging nutritional consequences (see below). Secondary compounds do respond as well, but in unpredictable direction (Penuelas and Estiarte 1998). An increase in desired mobile carbon compounds could be an asset. For instance the rubber tree might be expected to produce more latex and thus channel excess carbon assimilates in this non-structural biomass compartment. A growth chamber test with a relative of the rubber tree, *Euphorbia lathyris*, one of the so-called Diesel-plants, from which one hopes for fossil fuel substitutes, ended with a big surprise (Häring and Körner 2004). Elevated CO_2 significantly reduced latex production (Fig. 2.3b), unless plants received a lot of N-fertilizer. So far, stem wood produced under elevated CO_2 has not yet been found to contain less N or more NSC. Hence the idea that a widening of the C/N ratio in wood could allow more wood carbon to accumulate in the landscape with a given N supply has so far not been supported by data.

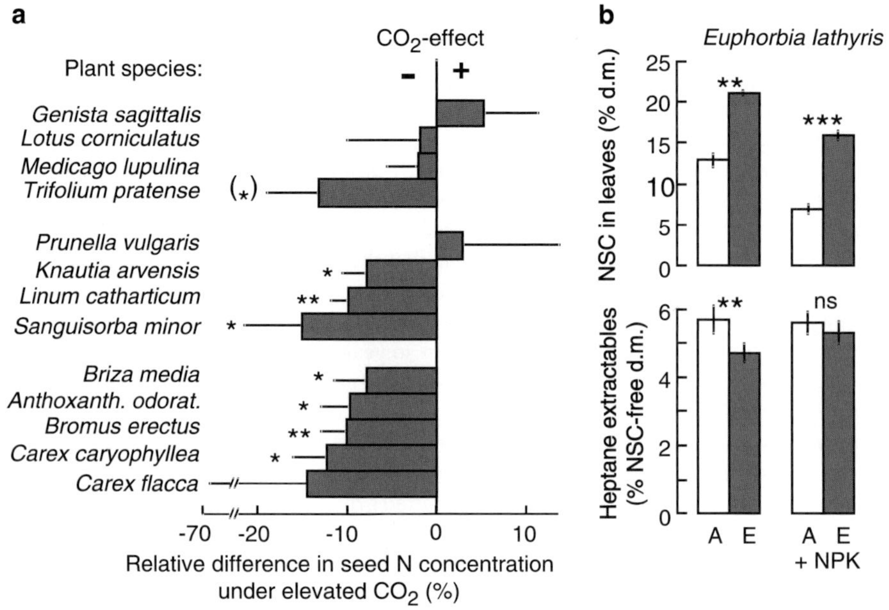

Fig. 2.3. Elevated CO_2 influences plant chemical composition. **a** Relative difference in seed nitrogen-concentration under elevated compared to ambient CO_2 concentration in a calcareous grassland community (Thürig et al. 2003). **b** Effect of elevated CO_2 on non-structural carbohydrate concentration and concentration of heptane extractable hydrocarbon (latex) in *Euphorbia lathyris* leaves (Häring and Körner 2004)

2.3.2 Consequences for Herbivory, Decomposition and Plant Nutrition

A shift from protein to carbohydrates or more generally a wider C/N ratio is commonly considered negative for consumers (herbivores) and litter break down (decomposition). Some animals feed at equal rates and thus have reduced protein uptake rates, some feed less when faced with the N-depleted food, others compensate by feeding more, and others shift to other host plants. One important conclusion is that animals behave differently when given a choice of food plants (for reviews see Lincoln et al. 1993; Kinney et al. 1997). Reduction in forage quality due to CO_2 enrichment is a common theme in grassland CO_2 enrichment work (Owensby et al. 1996; Milchunas et al. 2005), and will likely have important consequences for many grazers. Altered grazing of CO_2-enriched forage may also feed-back on forage quality since re-growth tissue often differs in quality from non-defoliated tissue (Milchunas et al. 1995), and ultimately interacts with the environment to determine plant community composition. The change in seed quality (Fig. 2.3a) propagates effects to the next generation, but seedling success did not reveal strong influences when tested. Cereal grains, besides the nutritional effect as such, had been found to be of reduced baking quality in cases where elevated CO_2 was applied without extra fertilizer (e.g., Kimball et al. 2001).

Initial thoughts that leaf quality changes may translate into litter quality changes and a slowing of decomposition have not been supported by the majority of tests (O'Neill and Norby 1996; Norby et al. 2001; but there are a few exceptions). Effects may depend on the maturity of the test system and timing of the trial. Another key aspect is the export of excess carbohydrates to rhizosphere microbes, which then become competitors for nutrients (e.g., Zak et al. 1993) and may even draw down productivity (Díaz et al. 1993; Hättenschwiler and Körner 1996; Shaw et al. 2002).

2.4 Plant Water Relations and Hydrological Implications

As briefly addressed in the biomass section, elevated CO_2 influences plant water relations through its effect on stomata. In turn plant water consumption influences soil moisture and atmospheric humidity, both feeding back on plants. Soil moisture feedback can be controlled and documented, or even simulated experimentally, but atmospheric feedback cannot. Hence all assumptions on influences of atmospheric CO_2 on hydrology depend on the nature of this feedback, which needs large-scale climate models to estimate. We recall, that between 65 and 70% of all vapor loss from vegetated land masses passes through leaf stomates (Maniak 1988; Gerten et al. 2005), thus causing any CO_2-driven effects to have substantial impact on the water balance.

Currently available data for natural ecosystems suggest a remarkable balance between stomatal responses and aerodynamic resistance which operates in a way so that water savings for a doubling of preindustrial CO_2 (ca. 550 ppm) are surprisingly similar for grassland, deciduous forests and conifer forests, despite large differences in stomatal responsiveness (Fig. 2.4). Although, signals differ a lot with species, with some responding strongly and others not at all, average stomatal conductances in grassland and crops may drop by 30 and 50% of controls (Knapp et al. 1996; Lauber and Körner 1997; Ainsworth et al. 2003; Kimball et al. 2002; LeCain et al.

Fig. 2.4.
A synthesis of water relations responses to elevated CO_2 for three types of vegetation. *Round arrows* symbolize the degree of canopy ventilation (a distillate from various sources, references in the text)

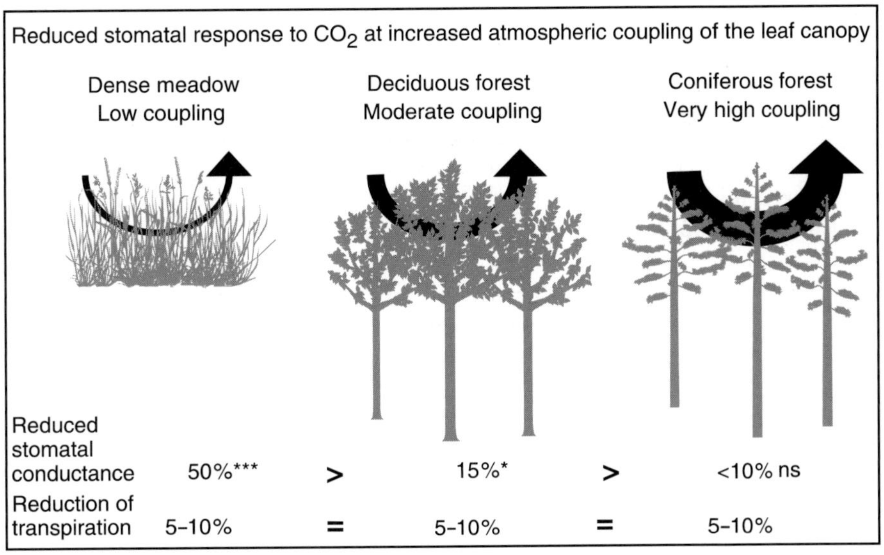

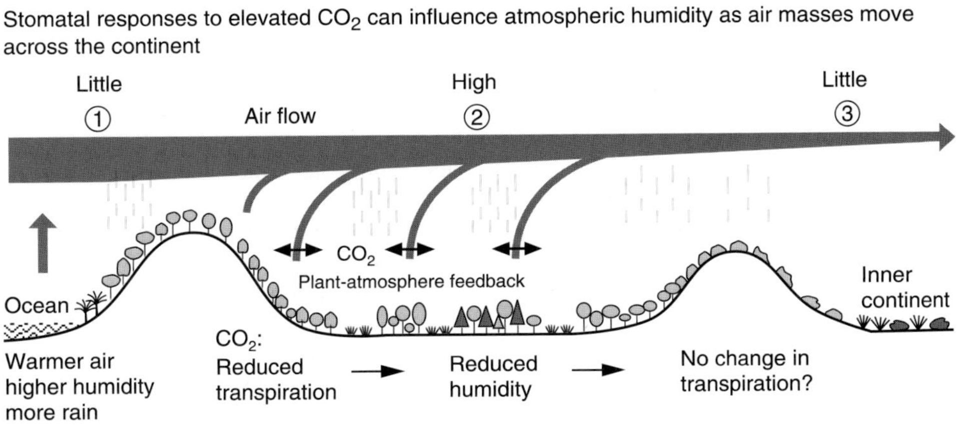

Fig. 2.5. Schematic landscape scale vegetation-atmosphere interaction resulting from stomatal responses to elevated CO_2 (Fig. 2.4). Coastal regions *(1)* are exposed to ocean driven atmospheric humidity (no or little vegetation-climate feedback). As air masses move across the continent, they take up moisture from transpiring vegetation, but based on available experimental data 5–10% less (Fig. 2.4) when elevated CO_2 reduces stomatal opening, with cascading downwind consequences for humidity *(2)* which feed back on transpiration to a point, where transpiration becomes very low in the semiarid continental interior *(3)* with minor vegetation influence on the atmosphere. Horizontal arrows symbolize regions of intense CO_2-driven interactions between vegetation and climate

1993; Bunce 2004). In deciduous forest trees, signals average at around –20% (Saxe et al. 1998; Medlyn et al. 2001). It was a big surprise that none of the adult conifers ever tested produced the theoretically expected stomatal signal, which means, if there is a response, it would be within measurement error, i.e., smaller than 10% (e.g., Barton et al. 1993; Pataki et al. 1998; Ellsworth 1999).

This sequence of responsiveness at the stomata level is the reverse of atmospheric coupling, with dense grassland most strongly decoupled and conifers very well coupled to air circulation. Actual water flux studies in non-chambered grassland (e.g., Stocker et al. 1997) and cereals (e.g., Pinter et al. 1996), in deciduous forest trees (Wullschleger et al. 2002; Hungate et al. 2002; Cech et al. 2003) and in conifers (Schäfer et al. 2002), almost all arrive at seasonal savings between 5 and 10%, more close to 5%. Such a uniform effect is good news for model parameterization and is in line with an earlier proposition of such a balance of stomatal and aerodynamic controls of transpiration (Meinzer and Gantz 1991; Meinzer et al. 1993). Effects have been larger in some closed or semiclosed systems in which atmospheric coupling may have been enhanced, (e.g., Owensby et al. 1997). At the canopy scale, a best current guess is a reduction of evapotranspiration in this range, yet, without accounting for atmospheric feedback, which will reduce the effect. So, a realistic signal for hydrological considerations may be a 5% reduction in transpiration for twice pre-industrial CO_2 concentrations, irrespective of plant functional type. It is noteworthy that for plants it matters *when* these savings happen. Otherwise small differences in water consumption can improve drought survival. In wet conditions they may also enhance runoff. An important consensus from experimental results is that effects are strongest under high humidity and decline rapidly as the vapor pressure deficit exceeds 10–15 hPa (Wullschleger and Norby 2001; Cech et al. 2003), an observation confirmed by earlier measurements in trees around natural CO_2 vents (Tognetti et al. 1996). Transferred to a landscape or continental scale, any reduction of vapor loss will change atmospheric moisture loading, which in turn, would affect (increase) vapor loss from vegetation (Fig. 2.5).

2.5 Stress Resistance under Elevated CO_2

Exposing plants to elevated CO_2 can have beneficial effects under conditions of stress such as water shortage and air pollution, but it can also diminish frost resistance.

Since stomates lose less water for a given amount of CO_2 absorbed when CO_2 concentrations are elevated, plants may do better under drought. More than 100 controlled experiments have explored this field and nearly all found drought mitigating effects of elevated CO_2 (e.g., Tolley and Strain 1984; Hibbs et al. 1995). However, as several authors have shown (e.g., Samarkoon and Gifford 1995; Volk et al. 2000) the secondary effects of high CO_2 on soil water (as discussed above) exert a strong confounding influence, particularly in pot studies with limited rooting volume. Wullschleger et al. (2002) consider the actual benefits in the field as minor. In the desert FACE experiment in Nevada, it was the moist period which permitted the greatest CO_2 effect on shrubs and none during a drought period, somewhat dampening perspectives for a greening of dry lands in a high CO_2 world (Smith et al. 2000). No conclusive benefits for Mediterranean shrubs around CO_2 springs could be detected during drought (Tognetti et al. 2000), but crops grown under FACE profited more from elevated CO_2 (in relative terms) when passing through a drought (Conley et al. 2001). Future research will have to clearly separate effects of soil moisture savings from actual drought × CO_2 effects.

In the light of the greater carbohydrate supply of leaves grown in elevated CO_2 (including osmotically active sugar), it came as a surprise that frost resistance declines in plants exposed to a CO_2-rich atmosphere. This was found independently in plants as different as a temperate herbfield community (Obrist et al. 2001), Australian eucalypts (Lutze et al. 1998) and Douglas fir (Guak et al. 1998). The changes may be within the natural safety margin (Repo et al. 1996) and roughly match predicted global warming, but nevertheless affect one of the basic determinants of global plant distribution.

Elevated CO_2 concentrations have also been shown to mitigate the impact of ozone and other airborne pollutants, in part by reducing diffusive uptake (stomata) and in part by compensating damage (reviews by Sullivan 1997; Volin et al. 1998; Poorter and Perez-Soba 2001). A combined $O_3 \times CO_2$ FACE experiment with aspen confirmed these earlier findings and showed that certain realistic ozone levels may offset positive CO_2 effects (Isebrands et al. 2001). In summary, CO_2 enrichment has the potential to mitigate certain environmental stresses.

2.6 Biodiversity Effects May Outweigh Physiology Effects

We want to close this brief assessment by highlighting the key significance of species identity for the long-term outcome of atmospheric CO_2 enrichment (see also Potvin et al. 2007, Chap. 9 of this volume). Beyond the well known differential influence of elevated CO_2 on C_3 and C_4 species, recent studies have revealed a far more subtle differentiation of species that does not relate to any conventional groupings of plants into functional types and for which we often have no plausible explanation. We will illustrate this problem with a few examples.

2.6.1 Hydrology Implications of Elevated CO_2 Depend on Species Abundance

Imagine a mixed forest in a catchment, with some species saving water under elevated CO_2 and others not. This had been documented in a pine-sweetgum forest in N-Carolina (*Liquidambar* responding, *Pinus* not; Schäfer et al. 2002) and in a deciduous forest in Switzerland (*Carpinus* responding, *Fagus* not; Cech et al. 2003). The relative contribution of these species to the total forest canopy will determine whether the forest is saving water and how much. Furthermore, if one species saves, their non-saving neighbors may profit, which will alter the competitive balance and is most likely to induce a long-term change in community structure.

2.6.2 Biodiversity Effects on Forest Carbon Stocking and Grassland Responses

The longest term CO_2-enrichment experiments to data, the one in a wetland ecosystem, a sub-estuary of the Chesapeak Bay (Maryland; Hymus et al. 2003), all other grassland CO_2 studies (some examples in Fig. 2.1), the two forest FACE experiments in N-Carolina and Tennessee, but also two newer tree FACE experiments in Switzerland (Fig. 2.2), clearly demonstrate that CO_2 exerts species specific effects. The two US forest FACE studies, for instance, arrived at similar responses of net primary production, but in the case of pine, the additional biomass largely ended up in stems, whereas in the case of sweet-gum, it ended up in fine root turnover.

As mentioned in Sect. 2.2, accelerated growth may enhance tree turnover and lead to a lower rather than greater steady state C-pool in the landscape. One functional type of plants, lianas, the life cycle of which often

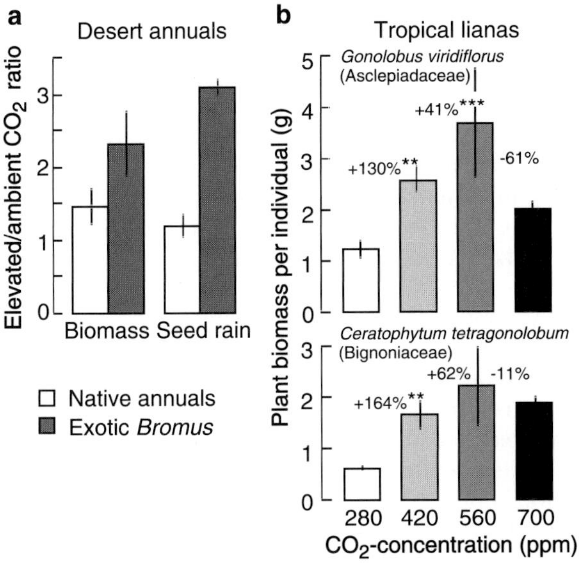

Fig. 2.6. Biodiversity–CO_2 interactions. **a** Mojave Desert native winter annuals and the exotic *Bromus madritensis* ssp. *rubens* (Nevada desert FACE, Smith et al. 2000). **b** the plant functional group of lianas. *Bars* indicate plus/minus one standard error (data by Granados and Körner 2002)

commences in deep shade, turned out to be particular responsive to elevated CO_2 (Granados and Körner 2002; Fig. 2.6). This has to do with a generally greater relative effect of elevated CO_2 on photosynthesis in deep shade. In the case of lianas, however, this stimulation greatly enhances their chance to reach the top of the forest. Well known to be the major driver of forest dynamics in the humid tropics (Schnitzer and Bongers 2002), such a greater vigor of lianas may cause the whole system to become more dynamic and store less C because trees do not live as long (Körner 2004). Demographic evidence suggests that such a process may be under way in Amazonia (Phillips et al. 2002) and had recently been reported for Panama as well (Wright et al. 2004). This example illustrates that there is no straightforward way to predict forest responses to elevated CO_2 from photosynthetic response curves without accounting for such biodiversity effects.

Similar examples have been shown in non-forest systems. For instance in the Mojave desert, an exotic annual *Bromus* grass species responds more strongly to CO_2 than native annuals (Fig. 2.6) enhancing the risk of fire (Smith et al. 2000). In the Negev winter-annuals' productivity (Fig. 2.1), it was a single species' response, which drove the whole community response. In the Colorado shortgrass steppe, recruitment of a single species under elevated CO_2 was largely responsible for a 41% production response (Morgan et al. 2004b). What if these species had been absent or were studied alone? Quite clearly species identity and species abundance have a major influence on the impact of elevated CO_2 on ecosystems.

2.7 Summary and Conclusions

Experimental CO_2 research with plants and whole ecosystems has made it clear that there is no straightforward relationship between the generally observed stimulation of leaf photosynthetic rates by elevated CO_2 concentrations and growth or productivity. The large number of tests with a broad spectrum of species and growth conditions have made it obvious that the extent to which greater availability of carbon to plants will translate into more structural growth depends on nutrient availability, either directly or via soil moisture conditions. The realism of projections derived from experimental works thus depends on the realism of nutrient and water regimes provided during tests (Körner 2003c). For the vast majority of non-agricultural ecosystems it seems that resources other then CO_2 control growth and productivity to such an extent that CO_2 concentrations above current levels exert little or no long-term stimulation. The major influence of atmospheric CO_2 on biota comes in (1) via differential responses of plant taxa or plant functional groups, (2) effects on water relations, and (3) soil feed back induced by greater C-input, the latter least understood.

Biodiversity effects have been found to be significant, with single species or plant functional types potentially driving whole system responses, and with no good explanations to date why this is so. The example of enhanced vigor of tropical lianas illustrates that functional group responses may in fact reverse the stimulatory effect of CO_2 enrichment into a reduction of carbon stocking in the single greatest biological C-reservoir on the globe. Water feedback may mimic growth stimulations in experiments, which may not be seen in reality when landscape wide feedback by climate and climatic changes are accounted for – a major future challenge for modelers. Nevertheless, the hydrological implications of elevated CO_2 are still a key global change aspect. Among the greatest surprises emerging from experimental works is that the differential responses of leaf stomates seen across taxa seem to level out to a common ca. 5–10% reduction of water loss at the ecosystem scale (evapotranspiration), a signal largely produced during humid periods.

Soil feedbacks other than through moisture include consequences of greater root turnover and microbial stimulation, but whether there will be a long term net input into the recalcitrant humus C pool is still very uncertain. Humus formation also strongly depends on nutrients, N in particular, and thus, competes directly with plant production and new C-input.

Overall, experimental CO_2 research has led to a wider appreciation of feedback constraints at the ecosystem level and the significance of plant taxa for ecosystem behavior (Fig. 2.7). At the ecosystem level, shifts in the presence and abundance of species may outweigh CO_2 effects seen at leaf or single species level. Taken together,

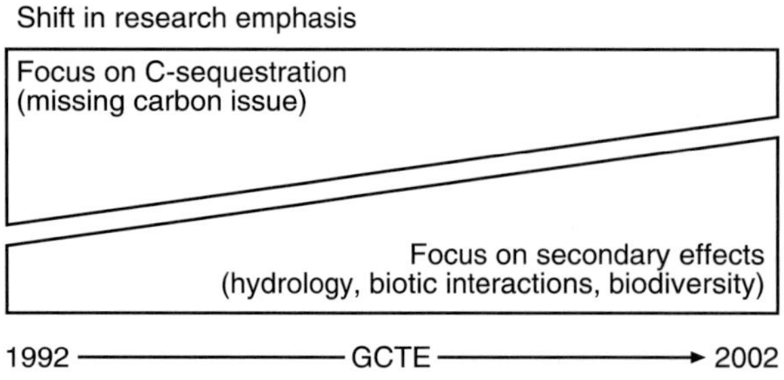

Fig. 2.7. The shift in research emphasis during the GCTE decade

the experimental findings suggest very cautious projections of CO_2 fertilization effects seen in short term tests with nutrient rich systems. Given that the few tests which used lower than current CO_2 (pre-industrial) concentrations as a reference to current or future concentrations revealed the strongest CO_2 effects (Fig. 2.6), the world may have seen or is midway into the most effective phase of CO_2 effects on biota, with a further rise perhaps exerting less additional effects. However, it will take many decades for the past and ongoing CO_2 effects to materialize, for instance, in the form of new plant assemblages.

Major questions to be explored in future biological CO_2 research are:

1. The translation of greater ecosystem carbon throughflow (growth rate, productivity, soil carbon fluxes) into steady state carbon pool sizes (biomass carbon, soil humus). CO_2 experiments revealed a lot of evidenced for enhanced carbon flux under elevated CO_2, but they are intrinsically constrained in estimating related changes in C-stocks. A worthwhile arena to be explored empirically is ecosystems of varying fertility, which exhibit contrasting carbon turnover for reasons unrelated to elevated CO_2. What are the mechanistic links between rates of carbon turnover (both at annual and centennial time scales) and the related pool sizes of biologically tied up carbon?
2. What are the long term consequences of plant species or functional type specific responses to elevated CO_2? It was a major outcome of GCTE related experimental research that biodiversity effects can outweigh physiology based 'average' plant responses at community level, and these consequences need to become focal areas of both further experiments and modeling. Based on this insight, it seems imperative, that future test systems include aspects of biodiversity, also in the light of the enormous costs involved.
3. There is an urgent need to implement realistic feedback between plant and ecosystem water relations under elevated CO_2 and atmospheric feedback in models. This is an arena were regional climate modeling and experimental research have to merge. Given the extent to which secondary effects of elevated CO_2 on water relations affect biomass responses and the potential hydrological implications, we need to arrive at a clear separation of direct, photosynthesis driven CO_2 effects and indirect, water driven CO_2 effects, which reflect certain climatic conditions.
4. Among the many secondary effects of exposing plants and ecosystems to enhanced availability of carbon, those on herbivory, mycorrhiza and rhizosphere food webs and all sorts of other interactions between plants and other organisms will have to remain focal areas.

A grand vision from an experimental point of view, would be an international research consortium which taps resources representing a small fraction of outer space research funding in order to establish three major forest research facilities, one in the boreal zone, one in the temperate zone and one in the tropics. These facilities would test complex forest responses to our future atmosphere, capitalizing on the experience accumulated over the last decades on simpler test systems. These large scale tests would employ the latest CO_2 enrichment technology, applicable for tall and diverse tree stands; would gradually elevate CO_2 concentrations, provide the needed canopy access technology; take a broad interdisciplinary approach; and guarantee long test periods under common protocols. They would represent international focal points of experimental global change research and tap resources from national partner agencies, which otherwise may continue to be diluted into smaller scale enterprises.

References

Ainsworth EA, Davey PA, Hymus GJ, Osborne CE, Rogers A, Blum H, Nösberger J, Long SE (2003) Is stimulation of leaf photosynthesis by elevated carbon dioxide concentration maintained in the long term? A test with *Lolium perenne* grown for 10 years at two nitrogen fertilization levels under Free Air CO_2 Enrichment (FACE). Plant Cell Environ 26:705–714

Amthor JS (1995) Terrestrial higher-plant response to increasing atmospheric CO_2-concentration in relation to the global carbon cycle. Global Change Biology 1:243–274

Asshoff R, Zotz G, Körner C (2006) Growth and phenology of mature temperate forest trees in elevated CO_2. Global Change Biology 12:848–861

Barton CVM, Lee HSJ, Jarvis PG (1993) A branch bag and CO_2 control system for long-term CO_2 enrichment of mature Sitka spruce (*Picea sitchensis* (Bong.) Carr.). Plant Cell Environ 16:1139–1148

Bazzaz FA (1990) The response of natural ecosystems to rising global CO_2 levels. Annu Rev Ecol Syst 21:167–196

Bowes G (1993) Facing the inevitable: plans and increasing atmospheric CO_2. Annu Rev Plant Phsiol Mol Biol 44:309–332

Bunce JA (2004) Carbon dioxide effects on stomatal responses to the environment and water use by crops under field conditions. Oecologia 140:1–10

Campbell BD, Stafford-Smith DM, McKeon GM (1997) Elevated CO_2 and water supply interactions in grasslands: a pastures and rangelands management perspective. Glob Change Biol 3:177–187

Cao MK, Woodward FI (1998) Net primary and ecosystem production and carbon stocks of terrestrial ecosystems and their responses to climate change. Glob Change Biol 4:185–198

Cech PG, Pepin S, Körner Ch (2003) Elevated CO_2 reduces sap flux in mature deciduous forest trees. Oecologia 137:258–268

Conley MM, Kimball BA, Brooks TJ, Pinter PJ, Hunsaker DJ, Wall GW, Adam NR, LaMorte RL, Matthias AD, Thompson TL, Leavitt SW, Ottman MJ, Cousins AB, Triggs JM (2001) CO_2 enrichment increases water-use efficiency in sorghum. New Phytol 151:407–412

Cotrufo MF, Ineson P, Scott A (1998) Elevated CO_2 reduces the nitrogen concentration of plant tissues. Glob Change Biol 4:43–54

DeLucia EH, Moore DJ (2005) Contrasting responses of forest ecosystems to rising atmospheric CO_2: implications for the global C cycle. Global Biogeochemical Cycles DOI 10:1029/2004GB002346

DeLucia EH, George K, Hamilton JG (2002) Radiation use efficiency of a forest exposed to elevated concentrations of atmospheric carbon dioxide. Tree Physiol 22:1003–1010

Díaz S, Grime JP, Harris J, McPherson E (1993) Evidence of a feedback mechanism limiting plant response to elevated carbon dioxide. Nature 364:616–617

Dijkstra P, Hymus G, Colavito D, Vieglais DA, Cundari CM, Johnson DP, Hungate BA, Hinkle CR, Drake BG (2002) Elevated atmospheric CO_2 stimulates aboveground biomass in a fire-regenerated scrub-oak ecosystem. Glob Change Biol 8:90–103

Drake BG, Gonzalez-Meler MA, Long SP (1997) More efficient plants: a consequence of rising atmospheric CO_2? Annu Rev Plant Physiol 48:609–639

Ellsworth DS (1999) CO_2 enrichment in a maturing pine forest: are CO_2 exchange and water status in the canopy affected? Plant Cell Environ 22:461–472

Evans LT, Dunstone RL (1970) Some physiological aspects of evolution in wheat. Aust J Biol Sci 23:725–741

Finzi AC, DeLucia EH, Hamilton JG, Richter DD, Schlesinger WH (2002) The nitrogen budget of a pine forest under free air CO_2 enrichment. Oecologia 132:567–578

Gerten D, Hoff H, Bondeau A, Luch W, Smith P, Zaehle S (2005) Contemporary "green" water flows: simulations with a dynamic global vegetation and water balance model. Phys Chem Earth (in press)

Ghannoum O, Von Caemmerer S, Ziska LH, Conroy JP (2000) The growth response of C_4 plants to rising atmospheric CO_2 partial pressure: a reassessment. Plant Cell Environ 23:931–942

Granados J, Körner Ch (2002) In deep shade, elevated CO_2 increases the vigor of tropical climbing plants. Glob Change Biol 8:1109–1117

Grünzweig JM, Körner Ch (2001) Biodiversity effects of elevated CO_2 in species-rich model communities from the semi-arid Negev of Israel. Oikos 95:112–124

Guak S, Olsyzk DM, Fuchigami LH, Tingey DT (1998) Effects of elevated CO_2 and temperature on cold hardiness and spring bud burst and growth in Douglas-fir (*Pseudotsuga menziesii*). Tree Physiol 18:671–679

Hamerlynck EP, Huxman TE, Charlet TN, Smith SD (2002) Effects of elevated CO_2 (FACE) on the functional ecology of the drought-deciduous Mojave Desert shrub, *Lycium andersonii*. Environ Exp Bot 48:93–106

Hamilton JG, DeLucia EH, George K, Naidu SL, Finzi AC, Schlesinger WH (2002) Forest carbon balance under elevated CO_2. Oecologia 131:250–260

Handa IT, Körner Ch, Hättenschwiler S (2005) A test of the tree-line carbon limitation hypothesis by in situ CO_2 enrichment and defoliation. Ecology 86:1288–1300

Häring DA, Körner Ch (2004) CO_2 enrichment reduces the relative contribution of latex and latex-related hydrocarbons to biomass in *Euphorbia lathyris*. Plant Cell Environ 27:209–217

Hättenschwiler S, Körner Ch (1996) System-level adjustments to elevated CO_2 in model spruce ecosystems. Global Change Biol 2:377–387

Hättenschwiler S, Körner Ch (1998) Biomass allocation and canopy development in spruce model ecosystems under elevated CO_2 and increased N deposition. Oecologia 113:104–114

Hättenschwiler S, Körner Ch (2000) Tree seedling responses to in situ CO_2-enrichment differ among species and depend on understorey light availability. Global Change Biol 6:213–226

Hättenschwiler S, Miglietta F, Raschi A, Körner Ch (1997) Thirty years of *in situ* tree growth under elevated CO_2: a model for future forest responses? Global Change Biology 3:436–471

Hebeisen T, Lüscher A, Zanetti S, Fischer BU, Hartwig UA, Frehner M, Hendrey GR, Blum H, Nösberger J (1997) Growth response of *Trifolium repens* L. and *Lolium perenne* L. as monocultures and bi-species mixture to free air CO_2 enrichment and management. Glob Change Biol 3:149–160

Hibbs DE, Chan SS, Castellano M, Niu CH (1995) Response of red alder seedlings to CO_2 enrichment and water stress. New Phytol 129:569–577

Hungate BA, Reichstein M, Dijkstra P, Johnson D, Hymus G, Tenhunen JD, Hinkle CR, Drake BG (2002) Evapotranspiration and soil water content in a scrub-oak woodland under carbon dioxide enrichment. Glob Change Biol 8:289–298

Hymus GJ, Johnson DP, Dore S, Dijkstra P, Anderson HP, Hinkle CR, Drake BG (2003) Effects of elevated atmospheric CO_2 on net ecosystem CO_2 exchange of a scrub-oak ecosystem. Global Change Biology 9:1802–1812

IPCC (2001) Climate change 2001: the scientific basis. Third Assessment Report. Cambridge University Press, Cambridge, UK

Isebrands JG, McDonald EP, Kruger E, Hendrey G, Percy K, Pregitzer K, Sober J, Karnosky DF (2001) Growth responses of *Populus tremuloides* clones to interacting elevated carbon dioxide and tropospheric ozone. Environ Pollut 115:359–371

Kimball BA, Morris CF, Pinter PJ, Wall GW, Hunsaker DJ, Adamsen FJ, LaMorte RL, Leavitt SW, Thompson TL, Matthias AD, Brooks TJ (2001) Elevated CO_2, drought and soil nitrogen effects on wheat grain quality. New Phytol 150:295–303

Kimball BA, Kobayashi K, Bindi M (2002) Responses of agricultural crops to free-air CO_2 enrichment. Adv Agron 77:293–368

King JY, Mosier AR, Morgan JA, LeCain DR, Milchunas DG, Parton WJ (2004) Plant nitrogen dynamics in shortgrass steppe under elevated atmospheric carbon dioxide. Ecosystems 7: 147–160

Kinney KK, Lindroth RL, Jung SM, Nordheim EV (1997) Effects of CO_2 and NO_3-availability on deciduous trees: phytochemistry and insect performance. Ecology 78:215–230

Knapp AK, Hamerlynck EP, Ham JM, Owensby CE (1996) Responses in stomatal conductance to elevated CO_2 in 12 grassland species that differ in growth form. Vegetatio 125:31–41

Körner Ch (2000) Biosphere responses to CO_2 enrichment. Ecol Appl 10:1590–1619

Körner Ch (2002) Grassland in a CO_2-enriched world. In: Durand JL, Emile JC, Huyghe C, Lemaire G (eds) Multi-function grasslands, quality forages, animal products and landscapes. Imprimerie P. Oudin, Piotiers, pp 611–624

Körner Ch (2003a) Carbon limitation in trees. J Ecol 91:4–17

Körner Ch (2003b) Ecological impacts of atmospheric CO_2 enrichment on terrestrial ecosystems. Phil Trans R Soc Lond A 361:2023–2041

Körner Ch (2003c) Nutrients and sink activity drive plant CO_2 responses – caution with literature-based analysis. New Phytol 159:537–538

Körner Ch (2004) Through enhanced tree dynamics carbon dioxide enrichment may cause tropical forests to lose carbon. Philos Trans R Soc Lond Ser B-Biol Sci 359:493–498

Körner Ch, Arnone JA III (1992) Responses to elevated carbon dioxide in artificial tropical ecosystems. Science 257:1672–1675

Körner Ch, Miglietta F (1994) Long term effects of naturally elevated CO_2 on mediterranean grassland and forest trees. Oecologia 99:343–351

Körner Ch, Asshoff R, Bignucolo O, Hättenschwiler S, Keel SG, Pelaez-Riedl S, Pepin S, Siegwolf RTW, Zotz G (2005) Carbon flux and growth in mature deciduous forest trees exposed to elevated CO_2. Science 309:1360–1362

LaDeau SL, Clark JS (2001) Rising CO_2 levels and the fecundity of forest trees. Science 292:95–98

Lauber W, Körner Ch (1997) In situ stomatal response to long-term CO_2 enrichment in calcareous grassland plants. Acta Oecologica 18:221–229

Leavitt SW, Pendall E, Paul EA, Brooks T, Kimball BA, Pinter Jr PJ, Johnson HB, Matthias A, Wall GW, LaMotte RL (2001) Stable-carbon isotopes and soil organic carbon in wheat under CO_2 enrichment. New Phytol 150:305–314

LeCain DR, Morgan JA, Mosier AR, Nelson JA (1993) Soil and plant water relations determine photosynthetic responses of C_3 and C_4 grasses in a semi-arid ecosystem under elevated CO_2. Ann Bot 92:41–52

Lincoln DE, Fajer ED, Johnson RH (1993) Plant-insect herbivore interactions in elevated CO_2 environments. Trends Ecol Evol 8:64–68

Loehle C (1995) Anomalous responses of plants to CO_2 enrichment. Oikos 73:181–187

Lutze JL, Roden JS, Holly CJ, Wolfe J, Egerton JJG, Ball MC (1998) Elevated atmospheric [CO_2] promotes frost damage in evergreen tree seedlings. Plant Cell Environ 21:631–635

Maniak U (1988) Hydrologie und Wasserwirtschaft – eine Einführung für Ingeniere. Springer-Verlag, Berlin

Matamala R, Gonzalez-Meler MA, Jastrow JD, Norby RJ, Schlesinger WH (2003) Impacts of fine root turnover on forest NPP and soil C sequestration potential. Science 302:1385–1387

Medlyn BE, Barton CVM, Broadmeadow MSJ, Ceulemans R, De Angelis P, Forstreuter M, Freeman M, Jackson SB, Kellomäki S, Laitat E, Rey A, Roberntz P, Sigurdsson BD, Strassemeyer J, Wang K, Curtis PS, Jarvis PG (2001) Stomatal conductance of forest species after long-term exposure to elevated CO_2 concentration: a synthesis. New Phytol 149:247–264

Meinzer FC, Grantz DA (1991) Coordination of stomata, hydraulic, and canopy boundary layer properties: do stomata balance conductances by measuring transpiration? Physiol Plant 83: 324–329

Meinzer FC, Goldstein G, Holbrook NM, Jackson P, Cavelier J (1993) Stomatal and environmental control of transpiration in a lowland tropical forest tree. Plant Cell Environ 16:429–436

Miglietta F, Raschi A, Bettarini I, Resti R, Selvi F (1993) Natural CO_2 springs in Italy: a resource for examining long-term response of vegetation to rising atmospheric CO_2 concentrations. Plant Cell Environ 16:873–878

Milchunas DG, Varnamkhasti AS, Lauenroth WK, Goetz H (1995) Forage quality in relations to long-term grazing history, current-year defoliation, and water resource. Oecologia 101:366–374

Milchunas DG, Mosier AR, Morgan JA, LeCain DR, King JY, Nelson JA (2005) Elevated CO_2 and defoliation effects on a shortgrass steppe: forage quality versus quantity for ruminants. Agriculture, Ecosystems and Environment (in press)

Morgan JA, LeCain DR, Mosier AR, Milchunas DG (2001) Elevated CO_2 enhances water relations and productivity and affects gas exchange in C-3 and C-4 grasses of the Colorado shortgrass steppe. Glob Change Biol 7:451–466

Morgan JA, Pataki DE, Körner Ch, Clark H, Del Grosso SJ, Grünzweig JM, Knapp AK, Mosier AR, Newton PCD, Niklaus PA, Nippert JB, Nowak RS, Parton WJ, Polley HW, Shaw MR (2004a) Water relations in grassland and desert ecosystems exposed to elevated atmospheric CO_2. Oecologia 140:11–25

Morgan JA, Mosier AR, Milchunas DG, LeCain DR, Nelson JA, Parton WJ (2004b) CO_2 enhances productivity, alters species composition, and reduced digestibility of shortgrass steppe vegetation. Ecological Applications 14:208–219

Niklaus PA, Körner Ch (2004) Synthesis of a six-year study of calcareous grassland responses to in situ CO_2 enrichment. Ecol Monogr 74:491–511

Niklaus PA, Glöckner E, Siegwolf R, Körner Ch (2001a) Carbon allocation in calcareous grassland under elevated CO_2: a combined ^{13}C pulse-labelling/soil physical fractionation study. Funct Ecol 15:43–50

Niklaus PA, Leadley PW, Schmid B, Körner Ch (2001b) A long-term field study on biodiversity x elevated CO_2 interactions in grassland. Ecol Monogr 71:341–356

Norby RJ, Wullschleger SD, Gunderson CA, Nietch CT (1995) Increased growth efficiency of *Quercus alba* trees in a CO_2-enriched atmosphere. New Phytologist 131:91–97

Norby RJ, Wullschleger SD, Gunderson CA, Johnson DW, Ceulemans R (1999) Tree responses to rising CO_2 in field experiments: implications for the future forest. Plant Cell Environ 22:683–714

Norby RJ, Cotrufo MF, Ineson P, O'Neill EG, Canadell JG (2001) Elevated CO_2, litter chemistry, and decomposition. A synthesis. Oecologia 127:153–165

Norby RJ, Hanson PJ, O'Neill EG, Tschaplinski TJ, Weltzin JF, Hansen RT, Cheng W, Wullschleger SD, Gunderson CA, Edwards NT, Johnson DW (2002) Net primary productivity of a CO_2-enriched deciduous forest and the implications for carbon storage. Ecological Applications 12:1261–1266

Norby RJ, Sholtis JD, Gunderson CA, Jawdy SS (2003) Leaf dynamics of a deciduous forest canopy: no response to elevated CO_2. Oecologia 136:574–584

Norby RJ, Ledford J, Reilly CD, Miller NE, O'Neill EG (2004) Fine-root production dominates response of a deciduous forest to atmospheric CO_2 enrichment. Proc Nat Acad Sci 101:9689–9693

Obrist D, Arnone III JA, Körner Ch (2001) In situ effects of elevated atmospheric CO_2 on leaf freezing resistance and carbohydrates in a native temperate grassland. Ann Bot 87:839–844

O'Neill EG, Norby RJ (1996) Litter quality and decomposition rates of foliar litter produced under CO_2 enrichment. In: Koch GW, Mooney HA (eds) Carbon dioxide and terrestrial ecosystems. Academic Press, San Diego, pp 87–103

Oren R, Ellsworth DE, Johnsen KH, Phillips N, Ewers BE, Maier C, Schäfer KVR, McCarthy H, Hendrey G, McNulty SG, Katul GG (2001) Soil fertility limits carbon sequestration by forest ecosystems in a CO_2-enriched atmosphere. Nature 411:469–472

Owensby CE, Cochran RC, Auen LM (1996) Effects of elevated carbon dioxide on forage quality for ruminants. In: Körner Ch, Bazzaz FA (eds) Carbon dioxide, populations, and communities. Academic Press, San Diego, pp 363–371

Owensby CE, Ham JM, Knapp AK, Bremer D, Auen LM (1997) Water vapour fluxes and their impact under elevated CO_2 in a C_4-tallgrass prairie. Glob Change Biol 3:189–195

Pataki DE, Oren R, Tissue DT (1998) Elevated carbon dioxide does not affect average canopy stomatal conductance of *Pinus taeda* L. Oecologia 117:47–52

Pendall E, Bridgham S, Hanson PJ, Hungate B, Kicklighter DW, Johnson DW, Law BE, Luo YQ, Megonigal JP, Olsrud M, Ryan MG, Wan SQ (2004) Below-ground process responses to elevated CO_2 and temperature: a discussion of observations, measurement methods, and models. New Phytol 162:311–322

Penuelas J, Estiarte M (1998) Can elevated CO_2 affect secondary metabolism and ecosystem function? Trends Ecol Evol 13:20–24

Phillips OL, Vasquez Martinez R, Arroyo L, Baker TR, Killeen T, Lewis SL, Malhi Y, Monteagudo Mendoza A, Neill D, Nunez Vargas P, Alexiades M, Ceron C, Di Fiore A, Erwin T, Jardim A, Palacios W, Saldias M, Vincenti B (2002) Increasing dominance of large lianas in Amazonian forests. Nature 418:770–774

Pinter PJ jr., Kimball BA, Garcia RL, Wall GW, Hunsaker DJ, LaMorte RL (1996) Free-air CO_2 enrichment: responses of cotton and wheat crops. In: Koch GW, Mooney HA (eds) Carbon dioxide and terrestrial ecosystems. Academic Press, San Diego, pp 215–249

Poorter H, Perez-Soba M (2001) The growth response of plants to elevated CO_2 under non-optimal environmental conditions. Oecologia 129:1–20

Rawson HM (1992) Plant responses to temperature under conditions of elevated CO_2. Aust J Bot 40:473–490

Repo T, Hänninen H, Kellomäki S (1996) The effects of long-term elevation of air temperature and CO_2 on the frost hardiness of Scots pine. Plant Cell Environ 19:209–216

Samarakoon AB, Gifford RM (1995) Soil water content under plants at high CO_2 concentration and interactions with the direct CO_2 effects: a species comparison. Journal of Biogeography 22:193–202

Saugier B (1983) Plant growth and its limitations in crops and natural communities. In: Mooney HA, Godron M (eds) Disturbance and ecosystems: components of response. Ecol Studies 44:159–174, Springer-Verlag, Berlin

Saxe H, Ellsworth DS, Heath J (1998) Tansley Review No. 98 Tree and forest functioning in an enriched CO_2 atmosphere. New Phytol 139:395–436

Schäfer KVR, Oren R, Lai CT, Katul GG (2002) Hydrologic balance in an intact temperate forest ecosystem under ambient and elevated atmospheric CO_2 concentration. Glob Change Biol 8:895–911

Schäppi B, Körner Ch (1996) Growth responses of an alpine grassland to elevated CO_2. Oecologia 105:43–52

Schimel DS, House JI, Hibbard KA, Bousquet P, Ciais P, Peylin P, Braswell BH, Apps MJ, Baker D, Bondeau A, Canadell J, Churkina G, Cramer W, Denning AS, Field CB, Friedlingstein P, Goodale C, Heimann M, Houghton RA, Melillo JM, Moore III B, Murdiyarso D, Noble I, Pacala SW, Prentice IC, Raupach MR, Rayner PJ, Scholes RJ, Steffen WL, Wirth C (2001) Recent patterns and mechanisms of carbon exchange by terrestrial ecosystems. Nature 414:169–172

Schnitzer SA, Bongers F (2002) The ecology of lianas and their role in forests. Trends Ecol Evol 17:223–230

Shaw MR, Zavaleta ES, Chiariello NR, Cleland EE, Mooney HA, Field CB (2002) Grassland responses to global environmental changes suppressed by elevated CO_2. Science 298:1987–1990

Smith SD, Huxman TE, Zitzer SF, Charlet TN, Housman DC, Coleman JS, Fenstermaker LK, Seemann JR, Nowak RS (2000) Elevated CO_2 increases productivity and invasive species success in an arid ecosystem. Nature 408:79–82

Spinnler D, Egli P, Körner Ch (2002) Four-year growth dynamics of beech-spruch model ecosystems under CO_2 enrichment on two different forest soils. Trees 16:423–436

Steinmann K, Siegwolf R, Saurer M, Körner Ch (2004) Carbon fluxes to the soil in a mature temperate forest assessed by C-13 isotope tracing. Oecologia 141:489–501

Stephenson NL (1990) Climatic control of vegetation distribution: the role of water balance. Am Nat 135:649–679

Stocker R, Leadley PW, Körner Ch (1997) Carbon and water fluxes in a calcareous grassland under elevated CO_2. Functional Ecology 11:222–230

Sullivan JH (1997) Effects of increasing UV-B radiation and atmospheric CO_2 on photosynthesis and growth: implications for terrestrial ecosystems. Plant Ecol 128:194–206

Thürig B, Körner Ch, Stöcklin J (2003) Seed production and seed quality in a calcareous grassland in elevated CO_2. Glob Change Biol 9:873–884

Tognetti R, Giovannelli A, Longobucco A, Miglietta F, Raschi A (1996) Water relations of oak species growing in the natural CO_2 spring of Rapolano (central Italy). Ann Sci Forest 53:475–485

Tognetti R, Raschi A, Jones MB (2000) Seasonal patterns of tissue water relations in three Mediterranean shrubs co-occurring at a natural CO_2 spring. Plant Cell Environ 23:1341–1351

Tolley LC, Strain BR (1984) Effects of CO_2 enrichment and water stress on growth of *Liquidambar styraciflua* and *Pinus taeda* seedlings. Can J Bot 62:2135–2139

Volder A, Edwards EJ, Evans JR, Robertson BC, Schortemeyer M, Gifford RM (2004) Does greater night-time, rather than constant, warming alter growth of managed pasture under ambient and elevated atmospheric CO_2? New Phytol 162:397–411

Volin JC, Reich PB, Givnish TJ (1998) Elevated carbon dioxide ameliorates the effects of ozone on photosynthesis and growth: species respond similarly regardless of photosynthetic pathway or plant functional group. New Phytol 138: 315–325

Volk M, Niklaus PA, Körner Ch (2000) Soil moisture effects determine CO_2 responses of grassland species. Oecologia 125: 380–388

Wan SQ, Norby RJ, Pregitzer KS, Ledford J, O'Neill EG (2004) CO_2 enrichment and warming of the atmosphere enhance both productivity and mortality of maple tree fine roots. New Phytol 162:437–446

Wand SJE, Midgley GF, Jones MH, Curtis PS (1999) Responses of wild C_4 and C_3 grass (Poaceae) species to elevated atmospheric CO_2 concentration: a test of current theories and perceptions. Global Change Biol 5:723–741

Wardlaw IF (1990) Tansley Review No. 27. The control of carbon partitioning in plants. New Phytol 116:341–381

Wright SJ, Calderon O, Hernandez A, Paton S (2004) Are lianas increasing in importance in tropical forests? A 17-year record from Panama. Ecology 85:484–489

Wullschleger SD, Norby RJ (2001) Sap velocity and canopy transpiration in a sweetgum stand exposed to free-air CO_2 enrichment (FACE). New Phytol 150:489–498

Wullschleger SD, Gunderson CA, Hanson PJ, Wilson KB, Norby RJ (2002) Sensitivity of stomatal and canopy conductance to elevated CO_2 concentration – interacting variables and perspectives of scale. New Phytol 153:485–496

Würth MKR, Winter K, Körner Ch (1998) In situ responses to elevated CO_2 in tropical forest understorey plants. Funct Ecol 12:886–895

Zak DR, Pregitzer KS, Curtis PS, Teeri JA, Fogel R, Randlett DL (1993) Elevated atmospheric CO_2 and feedback between carbon and nitrogen cycles. Plant & Soil 151:105–117

Zotz G, Pepin S, Körner Ch (2005) No down-regulation of leaf photosynthesis in mature forest trees after three years of exposure to elevated CO_2. Plant Biology (in press)

Chapter 3

Ecosystem Responses to Warming and Interacting Global Change Factors

Richard J. Norby · Lindsey E. Rustad · Jeffrey S. Dukes · Dennis S. Ojima · William J. Parton · Stephen J. Del Grosso Ross E. McMurtrie · David A. Pepper

3.1 The Multiple Factor Imperative in Global Change Research

Increases in atmospheric CO_2 concentration in the coming decades will be accompanied by other global changes. Higher air temperatures, altered precipitation patterns, increased tropospheric ozone concentrations, and N deposition are among the most prominent of the predicted changes that, along with elevated CO_2, have a high potential to affect ecosystem structure and function. Although the effect of elevated atmospheric CO_2 on ecosystem function was the primary focus of much of the GCTE effort in ecosystem physiology, each of these additional factors presents the possibility of altering the response of ecosystems to elevated CO_2 – perhaps negating the CO_2 response, enhancing it, or completely changing the nature of the response. Predictions of future ecosystem metabolism based solely on changes in a single factor are likely to be misleading. Hence, in addition to the elevated CO_2 network, GCTE fostered the development of a network to stimulate and coordinate research on ecosystem responses to climatic warming. Through all of its activities, GCTE promoted an agenda that embraced the mandate for understanding multi-factor interactions.

In the past, many model simulations of ecosystem response to global change were based on changes in climate alone, in part because the effects of elevated CO_2 were considered insignificant or too uncertain (Solomon 1986). Now, ecosystem and global models include multiple factors, particularly climate and CO_2, and the predicted responses can differ significantly from predictions based on changes in a single factor (Melillo et al. 2001; Cramer et al. 2001). It is important to the international global change research agenda that progress in experimental approaches keeps pace with model development. While many of the fundamental relations between ecosystem processes and temperature are well known, it is more difficult to have confidence in predictions of the combined responses to temperature and CO_2. Some interactions have a strong theoretical and empirical foundation: the optimum temperature for photosynthesis increases with increasing CO_2 (Long 1991). Temperature affects all biological processes, however, and the net response of an ecosystem to the combined effects of warming and elevated CO_2 are not so simply described (Norby and Luo 2004). Furthermore, ecosystem responses to CO_2 and temperature are likely to be modified by other environmental factors, especially the availabilities of water and nitrogen, which in turn are modified by CO_2 and temperature (Medlyn et al. 2000; McGuire et al. 2001).

Here, we explore recent experimental approaches to understanding ecosystem responses to global change, focusing on approaches that span an increasing range of complexity. First we will consider warming as a single factor, followed by an example of an experiment investigating interactions between warming and elevated CO_2 (Sect. 3.3). Temperature is more difficult to manipulate in field experiments than is CO_2, and the difficulties – both conceptual and operational – increase when temperature and CO_2 treatments are combined (Norby and Luo 2004). Hence, experiments on CO_2-temperature interaction have been at a smaller scale and in less complex biological systems than larger-scale FACE experiments, particularly those in forest ecosystems. Smaller stature systems, however, do permit more ambitious, whole-ecosystem, multifactor experiments, such as the Jasper Ridge Global Change Experiment, which is discussed in the Sect. 3.4. Results from such studies can be complex and perplexing, and there is a strong need for experimental results to be linked to ecosystem models so that the responses can be better understood and projected over longer time frames, as will be described in the fourth section.

3.2 Ecosystem Responses to Experimental Warming

It is now widely accepted that human-induced increases in greenhouse gas concentrations have already raised mean global temperature by ~0.6 °C during the last century, and unless these emissions are significantly curtailed, will likely result in a continued increase in mean global temperature of ~1.4 to 5.8 °C during the next century (IPCC 2001). Because temperature is one of the fundamental regulators of all chemical and biological processes, climatic warming in combination with other global change drivers – elevated atmospheric CO_2, changes in the quantity and timing of precipitation, and alterations in the global N cycle – is likely to have profound effects on the structure and function of terrestrial ecosystems (Rustad and Norby 2002). For example, in undisturbed systems, temperature is a key factor that regulates many terrestrial biogeochemical processes, such as soil respiration (Raich and Nadelhoffer 1989; Raich and Schlesinger 1992; Raich and Potter 1995; Kirschbaum 1995), litter decomposition (Meentemeyer 1978; Jansson and Berg 1985; Hobbie 1996), N mineralization and nitrification (MacDonald et al. 1995), denitrification (Malhi et al. 1990), CH_4 emission (Crill et al. 1988, 1991; Johnson et al. 1996), fine root dynamics (Boone et al. 1998; Pregitzer et al. 2000; Gill and Jackson 2000), plant productivity (Warren-Wilson 1957; Morison and Lawlor 1999), and plant nutrient uptake (BassiriRad et al. 2000). However, despite this extensive literature, the longer-term response of whole ecosystems to warming remains elusive. This is due, in part, to the complex array of direct and indirect responses of ecosystem processes to changes in temperature, the variable time-scale of response of different ecosystem processes and components to changes in temperature, and the importance of initial conditions (Shaver et al. 2000).

To help disentangle these complex issues and to establish cause-and-effect relationships between warming and ecosystem effects, a growing number of ecosystem warming experiments have been initiated around the world over the past few decades. The accumulating evidence from these experiments has greatly increased our understanding of short-term (i.e., 1–10 yr) responses of terrestrial ecosystems and their components to experimental warming. In this section we highlight results from two syntheses of experimental results (the GCTE-NEWS synthesis, Rustad et al. 2001; and the ITEX synthesis, Arft et al. 2000) and one longer-term study (the Harvard Forest soil warming experiment, Melillo et al. 2002).

3.2.1 The GCTE-NEWS Synthesis

Rustad et al. (2001) used meta-analysis to synthesize results on the response of soil respiration, net organic horizon N mineralization, and aboveground plant productivity to experimental warming from 32 ecosystem warming experiments, representing four broadly defined biomes – high arctic, low arctic (including a Minnesota bog and fen and a Colorado alpine dry tundra), forest, and grassland – associated with the GCTE Network of Ecosystem Warming Studies (GCTE-NEWS). GCTE-NEWS was established in 1999 in response to a perceived need for better synthesis and integration of results from ecosystem warming experiments. The goals of this network are to integrate and foster research on ecosystem-level effects of rising temperature, both alone and in combination with other vectors of global change. Warming methods used in the studies in the meta-analysis included electrical heat-resistance ground cables (Peterjohn et al. 1994; Rustad and Fernandez 1998), infrared heaters (Harte et al. 1995; Bridgham et al. 1999), greenhouses (Shaver et al. 1998; Jonasson et al. 1999), open-top chambers (Marion et al. 1997; Norby et al. 1997), climate-controlled chambers (Tingey et al. 1996), and passive nighttime warming (Alward et al. 1999). The relative merits of these different approaches have been discussed (Shaver et al. 2000). The duration of the experimental studies in the meta-analysis ranged from 2 to 9 years, and the experimental increases in soil or air temperature ranged from 0.3 to 6.0 °C, with a mean increase of 2.4 °C across all 32 studies.

Results for individual sites showed considerable variation in response to warming, as illustrated by the range in effect sizes [*d*] (Fig. 3.1). However, when all sites were considered together, results from the meta-analysis showed that experimental warming significantly increased rates of soil respiration by a weighted mean average of 20% (with a 95% confidence interval of 18 to 22%; $n = 17$), O horizon net N mineralization by a weighted mean average of 46% (with a 95% confidence interval of 30 to 64%; $n = 12$), and plant productivity by a weighted mean average of 19% (with a 95% confidence interval of 15 to 23%; $n = 20$) (Fig. 3.1). Although the number of studies was limited, the response of soil respiration to warming was generally greater in forested ecosystems compared to low arctic and grassland ecosystems ($P < 0.10$), and the response of plant productivity was generally greater in low arctic ecosystems than in forest and grassland ecosystems ($P < 0.01$). With the exception of aboveground plant productivity, which showed a greater positive response to warming in colder ecosystems, meta-analysis did not reveal any other significant relationships between the magnitude of the response of these three ecosystem processes to experimental warming and the other geographic (latitude, longitude), climatic (mean annual and growing season temperature and precipitation, mean frost-free days), or environmental (soil and foliar chemical properties, dominant vegetation type, soil classification, successional status) variables evaluated in this analysis.

These results appear to be consistent with the hypothesis that warming, at least in the short term, directly increases rates of microbial processes including litter decomposition and N mineralization, thereby increasing the availability of nutrients, and, particularly in nutrient-limited ecosystems, increasing plant productivity. Unfortunately, the observations of increased N mineralization and increased plant productivity largely came from different experiments; few experiments took a whole-ecosystem approach with integrated analysis of both above- and belowground responses. Hence, available experimental data cannot be used to evaluate a hypothesis linking warming, nutrient availability, and productivity.

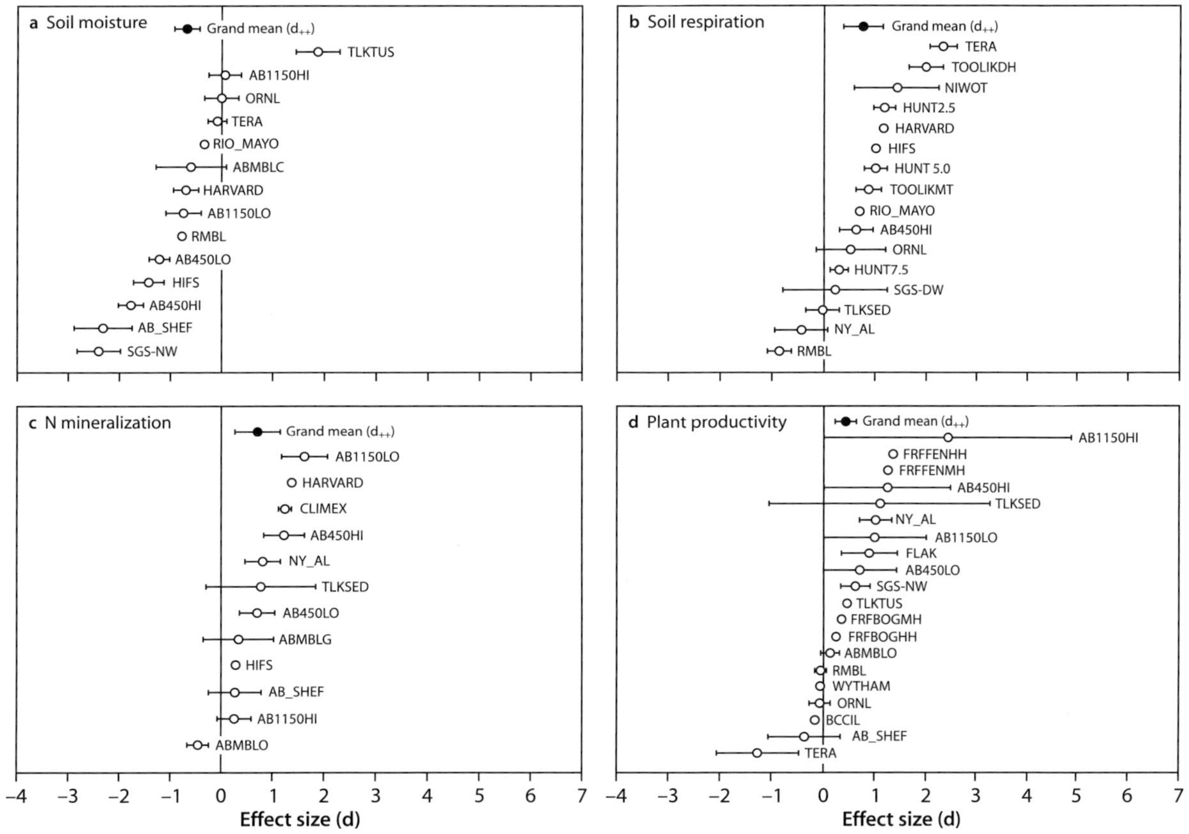

Fig. 3.1. Mean effect size, *d*, and 95% confidence intervals for different experimental sites used in a meta-analysis of responses to ecosystem warming. Measures of response are soil moisture, N mineralization, soil respiration, and plant productivity. If the 95% confidence interval of the grand mean does not include $d = 0$, the effect is considered statistically significant. The key to the different sites and a full analysis of the data are presented in Rustad et al. (2001) (reprinted with permission from Fig. 2 in Rustad et al. 2001)

The stimulation of plant productivity may also be a direct effect of warming on rates of photosynthesis (Lewis et al. 2001) or a warming-induced extension of the growing season (Dunne et al. 2002). However, the absence of strong relationships between the magnitude of the effect size of warming on the soil respiration, O horizon net mineralization, and aboveground plant productivity and the various geographic, climatic, and environmental variables considered in this analysis underscores the need to better understand the relative importance of specific factors (such as temperature, moisture, site quality, vegetation type, successional status, land-use history, etc.) at different spatial and temporal scales.

3.2.2 The ITEX Synthesis

Arft et al. (2000) also used meta-analysis to examine the response of plant phenology, growth, and reproduction to 1–4 years of experimental warming at 13 circumpolar sites associated with the International Tundra Experiment (ITEX). ITEX is a collaborative, multi-site experiment using a common open-top chamber experimental design to evaluate variability in species response to warming in tundra ecosystems. Results showed that (1) key phenological events including leaf bud burst, flowering, and seed dispersal generally occurred earlier in warmed plots compared to control plots throughout the four years of the study; (2) plant productivity was greater in the warmed plots than the control plots only during the first 2–3 years of the experiments (probably due to a depletion of belowground resources); (3) reproductive effort and success increased in later years, particularly in year 4 (probably due to the fact that flower buds are typically formed from one to several seasons before flowering); (4) the vegetative response to warming differed among plant life forms, with the response being generally greater in herbaceous than in woody species; (5) warmer, low arctic sites showed the strongest growth response to warming, and (6) colder, high arctic sites showed the strongest reproductive response to warming. Overall, the responses to warming observed at the ITEX study sites were consistent with those observed in the GCTE-NEWS study sites, and the two syntheses reinforce the observation that although terrestrial ecosystem processes are very sensitive to warming, the magnitude and even direction of response can be highly variable in time and space.

3.2.3 The Harvard Forest Soil Warming Experiment

Results from the aforementioned syntheses and individual warming experiments have greatly increased our understanding of the short-term responses of terrestrial ecosystems and their components to experimental warming. Nevertheless, concern exists that the majority of the existing or completed ecosystem warming experiments are of relatively short duration (i.e., <10 years) and that the initial reported responses may differ in magnitude and even direction compared to longer-term responses, thereby invalidating our ability to extrapolate short-term results to infer longer-term responses. An example of such a shift in response patterns over time is provided by the Harvard Forest soil warming experiment. Peterjohn et al. (1994) initially reported an approximately 40% increase in soil respiration during the first 6 months of the experiment. However, the magnitude of this initial increase diminished over time such that after 10 years of warming soils at 5 °C above ambient, soil respiration rates in the heated plots were not significantly different from rates in the control plots, probably due to depletion over time of temperature-sensitive, labile soil C (Fig. 3.2; Melillo et al. 2002). Net nitrogen mineralization rates, however, remained elevated for the duration of the experiment, resulting in a cumulative release of 41 g N m^{-2}. Assuming that 12.7% of this added N would be taken up by woody tissue (based on results from an associated long-term nitrogen fertilization experiment) and that C/N ratios of woody tissue are ~300/1, then warming-induced nitrogen mineralization could have resulted in a sequestration of ~1560 g C m^{-2} during the course of the experiment, thereby off-setting the estimated 944 g m^{-2} of labile soil C lost from the upper 60 cm of the soil profile. Results from this study underscore the needs for (1) long-term (i.e., >10 year) studies to evaluate changes in response patterns over time, and (2) understanding both direct and indirect effects of warming on ecosystem processes.

An additional consideration should be how the results would have differed in the presence of an elevated atmospheric CO_2 concentration, given that the warming scenario being simulated in this experiment will in reality be associated with a significantly CO_2-enriched atmosphere. CO_2 enrichment of deciduous forests can significantly enhance the flux of C below ground (Norby et al. 2004) and soil respiration (King et al. 2004). Pendall et al. (2004) speculated that the effects of elevated CO_2 and warming on C substrate availability would moderate each other. In the Harvard Forest soil warming experiment the depletion of labile soil C and diminution of the soil respiration response (Melillo et al. 2002) might well not have occurred under a more realistic scenario combining warming and elevated CO_2.

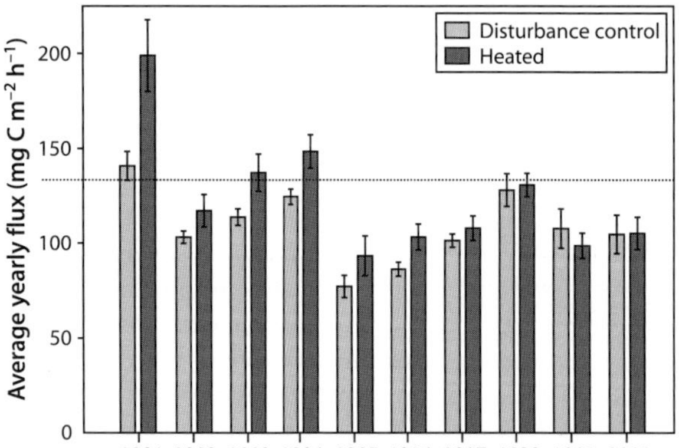

Fig. 3.2. Average yearly fluxes of CO_2 from the heated and disturbance control plots of the Harvard Forest soil warming experiment. Electrical resistance heating cables were buried 20 cm apart at 10 cm depth in the 6 × 6 m plots, and average soil temperature in the heated plots was maintained at 5 °C above ambient. Measurements were made from April through November for the period from 1991 through 2000. *Error bars* represent the standard error of the mean ($n = 6$ plots) between plots of the same treatment (reprinted with permission from Melillo et al. 2002)

3.3 Temperature and CO_2 Interactions in Trees: the TACIT Experiment

As discussed above, the climatic warming that is predicted for the future will be associated with increased atmospheric CO_2 concentrations, and there are strong reasons to suggest that CO_2 will alter the response of plants to temperature. The response of trees to a warmer, CO_2-enriched atmosphere is an important component of integrated analyses of forest response to global change and, therefore, of analyses of the feedbacks between the terrestrial biosphere, the global carbon cycle, and the climate system. Hence, it is important to understand not just the separate effects of elevated CO_2 and warming, but their combined effects and interactions as well.

3.3.1 Experimental Design

A 4-year experiment on temperature and CO_2 interactions in trees (TACIT) was designed to describe the effects of elevated CO_2 and temperature on ecosystem-level processes that control or influence C sequestration in ecosystems and to test specific hypotheses about the differential response of related species in relationship to their geographic ranges and ecological characteristics. TACIT explored the responses to a 4 °C increase in air temperature and a 300 ppm increase in CO_2 concentration in young *Acer rubrum* and *A. saccharum* trees growing in open-top chambers.

The experimental trees were grown in small stands for four growing seasons in open-top field chambers on the Oak Ridge National Environmental Research Park in Roane County, Tennessee (35°54' N; 84°20' W). Twelve

3-meter diameter chambers were modified to control air temperature to be the same as ambient air or to maintain a constant 4 °C elevation in air temperature (Norby et al. 1997). The temperature treatments were combined with ambient or elevated (+300 ppm) CO_2 concentrations. The temperature treatments were maintained throughout the year, but the CO_2 treatments were suspended during the winter (November–March) when the trees were leafless. The four treatments, comprising a 2 × 2 factorial arrangement of the two levels of temperature and CO_2 concentration, were replicated three times in a randomized complete block design. One-year-old seedlings were planted into the soil within the chambers in spring, 1994, and additional seedlings were planted in spring, 1995, for a total of 10 plants per species per chamber. The trees were harvested in September, 1997, ending the experiment. More details about the research site, chamber operation, and environmental conditions were provided by Norby et al. (1997, 2000).

Fig. 3.3. *Acer saccharum* and *A. rubrum* trees in an open-top chamber at the conclusion of the TACIT experiment. The plastic panels of the chambers and lower portions of the shade cloth, which were present throughout the four growing seasons of the experiment, have been removed to facilitate harvesting of the trees

3.3.2 Growth Responses

When the trees were harvested after four growing seasons (3.5 years) in the different combinations of CO_2 and temperature, they were in dense stands 4 m tall and with a leaf area index between 5 and 7 (Fig. 3.3). Elevated CO_2 enhanced growth: stem dry mass was increased by CO_2 enrichment by 34% in ambient temperature and 88% in elevated temperature. Warming, however, retarded growth by 35% in ambient CO_2 but only 9% in elevated CO_2 (Fig. 3.4). The main effects of CO_2 and temperature were statistically significant ($P = 0.001$ and $P = 0.025$, respectively), but the interaction between CO_2 and temperature was not ($P = 0.198$) (Norby and Luo 2004). Fine root biomass at the end of the experiment showed a similar pattern: a 60% reduction caused by the warming treatment in ambient CO_2 and a 27% reduction in elevated CO_2, but the CO_2–temperature interaction was not statistically significant (Wan et al. 2004).

These biomass values at the end of the experiment mask a great deal of complexity. The final harvest data are the net result of both positive and negative effects that changed through time. Although elevated temperature suppressed instantaneous rates of photosynthesis (Gunderson et al. 2000), it also lengthened the growing season by 2–3 weeks (Norby et al. 2003). Relative growth rates were depressed in elevated temperature only during the second year of treatment when a severe hot period caused heat stress, but since the trees were undergoing exponential growth during the course of the experiment, this negative effect on dry matter accumulation during the stress period had a continuing effect on absolute growth rate for the remainder of the experiment despite a recovery in relative growth rate. Similarly, observations of fine root production and mortality during

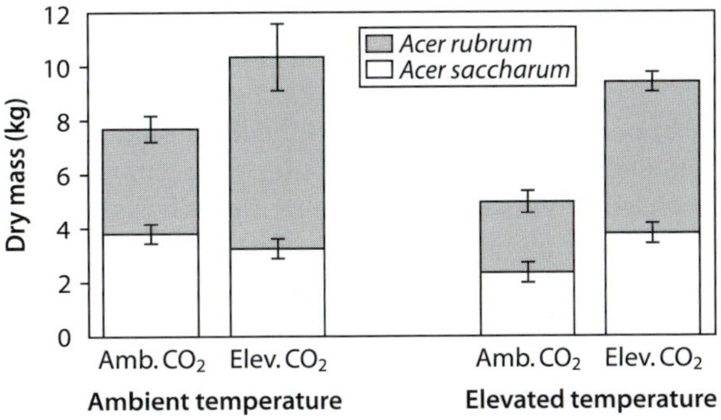

Fig. 3.4. Aboveground woody dry mass of *Acer saccharum* and *A. rubrum* trees after exposure for four growing seasons in open-top chambers with ambient or elevated CO_2 (+300 ppm) in combination with ambient or elevated (+4 °C) air temperature. The data are presented as biomass per chamber and are the means of three chambers per treatment ±SE (reprinted from Norby and Luo 2004 with the permission of New Phytologist Trust)

the third year of treatment did not predict fine root biomass at the final harvest. Both production and mortality were significantly increased by CO_2 enrichment and warming. The discrepancy between these observations of root dynamics and the harvest data are probably associated with drought-induced differential responses of productivity and mortality during the latter half of the growing season (Wan et al. 2004).

3.3.3 Higher-Order Responses

Higher-order responses also were addressed in this experiment. A premise of the experiment was that the two species would respond differently based on their presumed adaptations to temperature. The experimental site was close to the southern limit of the range of *A. saccharum*, whereas the range of *A. rubrum* extends much farther to the south. Although temperature effects were similar in the two species, *A. rubrum* was stimulated by CO_2 enrichment much more than *A. saccharum*. This observation is consistent with the premise that when temperatures are close to optimal, the relative biomass increase caused by increases in CO_2 enrichment is greater than when temperatures are sub- or supra-optimal (Poorter and Pérez-Soba 2001). The species also differed with regard to biotic interactions. Elevated CO_2 reduced growth of gypsy moth larvae feeding on *A. rubrum*, but not *A. saccharum* leaves (Williams et al. 2000). Warming shortened insect development but had no effect on pupal weight (Williams et al. 2003).

3.3.4 TACIT Summary

This relatively simple experiment illustrated many of the issues that must be considered when addressing multi-factor interactions, as well as many of the difficulties in taking up those issues in experimental studies (Norby and Luo 2004). Temperature affects many processes in plants and the net effect of a warmer atmosphere on tree growth will be difficult to predict when both negative and positive responses can occur together. Stochastic events, such as a late-spring frost that altered normal phenology (Norby et al. 2003) and the period of especially hot and dry weather that altered the growth trajectory, can play an important role in the integrated response. The duration of the experiment is an important consideration in this regard: the influence of a single unusual event should decrease with time, but the likelihood of an event occurring at all should increase. The specific combination of factors that influenced final dry mass in this experiment are not likely to be reproduced exactly in another experiment or in the real world, making it more difficult to extrapolate these results directly to an ecosystem scale. Different species – even closely related ones – may respond differently to global change factors, although some of those differences may be predictable based on the species' characteristics. In this experiment the effects of CO_2 and warming were additive; hence, their combined effects were predictable from knowledge of their single-factor effects. However, as will be apparent in the next section, unexpected interactions can also occur.

3.4 More Than Two Factors: the Jasper Ridge Global Change Experiment

Few ecosystem experiments have examined responses to more than two global change factors. Results from a four-factor study, the Jasper Ridge Global Change Experiment (JRGCE), demonstrate the need for this approach if we are to develop a comprehensive understanding of community and ecosystem responses to global environmental changes.

3.4.1 Experimental Design

Starting in 1998, the JRGCE exposed grassland plots in the San Francisco Bay Area (California, USA) to a factorial combination of warming, elevated CO_2, increased precipitation, and increased nitrogen deposition (Fig. 3.5). Circular grassland plots (2 m diameter) experienced either ambient conditions, ~1 °C of warming from infrared heat lamps, CO_2 enhancement of ~300 ppm via a FACE system, or warming and CO_2 enrichment together. Each plot was divided into four quadrants, which received am-

Fig. 3.5. An experimental plot in the Jasper Ridge Global Change Experiment. Four infrared heat lamps suspended over the 2-m diameter plot increase the canopy temperature by ~1 °C. Pure CO_2 is released from a circle of tubing surrounding the plot, enriching the CO_2 concentration in the air by about 300 ppm. Subplots comprise different combinations of precipitation and N deposition

bient or increased precipitation (+50%, including a 20-day extension of the growing season) combined with ambient or increased nitrogen deposition (+7 g N $m^{-2} yr^{-1}$ as $CaNO_3$). Fiberglass barriers below the soil surface prevented roots and resources from escaping the quadrants. Because annual grasses and forbs dominate the grassland, species composition and standing biomass can respond quickly to the step changes in environmental variables. In addition, evolutionary responses to the treatments can occur within the course of a relatively short-term experiment.

3.4.2 Net Primary Productivity

Shaw et al. (2002) described responses of grassland net primary productivity (NPP) to the global change treatments after three years of exposure. At this time, warming, precipitation, and nitrogen all tended to stimulate NPP. The authors did not present a direct test of whether treatment responses were additive, but qualitatively described the interactions. NPP responses to some treatment combinations appeared approximately additive (warming and N deposition), while others were intermediate between treatments (warming and precipitation). The response to elevated CO_2 was even more complex. CO_2 enrichment increased aboveground biomass when all other factors were at ambient levels (paired *t*-test, $P = 0.0003$), but significantly suppressed NPP responses to other global change factors (paired *t*-test, $P = 0.048$). Much of this suppression occurred in the root biomass, which declined by 22% in response to CO_2 across all treatments.

Although this strong interaction appeared in the 2001 season, interactions involving CO_2 or warming were less common in other years. Dukes et al. (2005) found that, over the first five years of the experiment, the only consistent interaction occurred between increased precipitation and N deposition, with each treatment making the response to the other more positive in annual-dominated grassland (repeated measures mixed-model ANOVA, precipitation–N interaction, $P < 0.01$). NPP and aboveground biomass increased significantly in response to N deposition in four of the five years (by 21–42%, mixed-model ANOVA, $P < 0.05$). Precipitation tended to increase shoot growth, but decreased root growth, causing a shift in allocation but no net effect on NPP across the five years. Effects of CO_2 and warming were not significant, with warming tending to slightly increase NPP on average and CO_2 tending to increase NPP in two years and decrease it in three.

In this Mediterranean-climate system, plants germinate in October or November and senesce during May and June, after which they no longer use soil moisture. Zavaleta et al. (2003a) found that warming accelerates this senescence, leaving wetter soils in late spring and early summer. Elevated CO_2 also increased soil moisture during late spring, and the individual effects of warming and CO_2 were approximately additive over the first two years of the experiment (CO_2 effect on volumetric spring soil moisture: 1.6 ±1.0%; warming effect: 1.1 ±0.95%; combined warming and CO_2 effect: 2.7 ±1.0%).

3.4.3 Community Composition

Zavaleta et al. (2003b) found that some of these global changes affected the species richness of the grassland over the first three years of the JRGCE. Elevated CO_2 and nitrogen deposition both decreased species richness by reducing the number of forb species present in the plots, while more forb species were present in the plots receiving increased precipitation. In this case, treatment effects on species richness were almost perfectly additive, as indicated by both a lack of significant interactions in the ANCOVA model and by a "striking" correspondence between observed responses to treatment combinations and predicted responses based on the effects of single factors.

Global change treatments also affected the production and relative abundance of grasses and forbs (Zavaleta et al. 2003c). Although individual species responses were weak, responses of these groups could be more clearly characterized. Few responses were consistently strong across the first three years of the experiment, but a split-plot ANCOVA model identified some general patterns. Nitrogen deposition favored grasses at the expense of forbs. Forbs benefited from increased precipitation and from warming, but grass production was unresponsive to these treatments. Finally, elevated CO_2 had little effect on production of either functional group. Interactions among the treatment effects were even less consistent than the main effects. The treatment combination that featured warming, elevated CO_2, and increased precipitation caused the greatest change in relative abundance of the functional groups, a 50% increase in forbs.

Because these functional groups differ in their chemical makeup, shifts in community composition can affect litter quality. Henry et al. (2005) found that global change treatments affected lignin and nitrogen concentrations in litter, but that some of these direct effects were counteracted by the shifts in community composition. Direct effects included CO_2-induced increases in lignin (which were attenuated by warming in grasses and increased precipitation in forbs) and increases in litter nitrogen concentration in response to nitrogen deposition. This latter effect was dampened by increased precipitation. Although litter quality responded predictably to the global change treatments, the effects on litter quality did not markedly affect decomposition rates over the short term.

Litter decomposition and many other ecosystem functions could be altered by shifts in microbial community structure. Horz et al. (2004) observed shifts in the abundance and community composition of ammonia-oxidizing bacteria in response to some global change treatments. In this case, all responses depended on the level of other factors. Nitrogen deposition affected community structure, but only at ambient temperature and precipitation, and elevated CO_2 affected bacterial abundance, but only under increased precipitation.

3.4.4 JRGCE Summary

Thus far, results from the JRGCE suggest that treatment responses of production, community composition, and other measured variables are not tightly linked. Supplemental nitrate deposition has affected more of these variables than other factors, consistently increasing NPP, decreasing species richness by favoring grasses over forbs, and, under certain background conditions, influencing litter quality and bacterial community structure. The modest warming treatment has had few effects, most notably an acceleration of phenology that led to a small and unexpected increase in spring soil moisture. The precipitation treatment caused shifts in biomass allocation (from belowground to aboveground) and plant species composition (increasing forb diversity and abundance), and had interactive effects on litter quality and bacterial communities. Elevated CO_2 has had small, mostly nonsignificant effects on biomass production (with exceptions in 2001), and positive effects on spring soil moisture. Responses of plant and microbial community composition to elevated CO_2 have mostly been subtle, and CO_2-driven increases in litter lignin concentration were sometimes ameliorated by other treatments. Overall, the complexity of responses in the JRGCE confirms that multifactor, multi-year experiments are critical to forecasting global change responses, and highlights the challenge of determining the mechanisms behind such responses.

3.5 Modeling Temperature, CO_2 and N Interactions in Trees and Grass

The complexity of the ecosystem- and community-level responses to multiple changes in environmental conditions as observed in the JRGCE increases the need for a modeling framework to assist in drawing useful generalizations that are applicable to different ecosystems and over longer time frames. Models are needed to explore questions that cannot be addressed in experiments. Whereas climate-change experiments impose step changes in CO_2 concentration and temperature over periods of several years, models can be used to simulate either a step change or gradual change on the longer (decadal) timescales that are of primary relevance to policy makers.

In one modeling study, six dynamic global vegetation models were used to simulate future global terrestrial net primary production (NPP) and net ecosystem production (NEP) (Cramer et al. 2001). Under gradual CO_2 change alone NPP was predicted to increase by 60% on average by 2100, and NEP was predicted to increase to an average value of approximately 6 Pg C yr^{-1} by the end of the century, with little evidence of terrestrial carbon-sink saturation. Under rising CO_2 plus climate change, the models still predicted a sustained carbon sink, but with smaller increases in NPP (+50%) and NEP. There were, however, wide discrepancies among the six models. Explaining model discrepancies was difficult because of the models' complexity and the global scale of simulations with dynamic vegetation distributions and complicated climate-change scenarios. One factor contributing to discrepancies was that only two of the six models included soil nutrient feedbacks.

It is more straightforward to interpret results from models of individual stands that use simplified climate-change scenarios and simulate ecosystem processes, but not vegetation dynamics. Two ecosystem models that have been used in stand-scale simulations of experimental sites are DAYCENT (Del Grosso et al. 2001; Kelly et al. 2000; Parton et al. 1998) and G'DAY (Comins and McMurtrie 1993; Medlyn et al. 2000). The responses of NPP, NEP, and C storage to gradually increasing CO_2, temperature, and nitrogen input have been simulated for grasslands and forests.

3.5.1 Global Change Simulations for a California Annual Grassland

The Jasper Ridge experiment provided a unique opportunity to compare model results with field data for a complex situation with multiple global change drivers. To have confidence in model projections that are unconstrained by data, it is important that the models can adequately recreate observed responses. Here, we describe how the DAYCENT model was used to simulate responses to increased atmospheric CO_2 concentration, N deposition temperature and precipitation for this site and compare the simulations with data collected during 1999–2002.

3.5.1.1 *DAYCENT Model Description*

DAYCENT is the daily time-step version of the CENTURY model (Parton et al. 1994). DAYCENT simulates exchanges of carbon, nutrients, and trace gases among the atmosphere, soil, and vegetation (Fig. 3.6). The model is of intermediate complexity and requires relatively simple inputs. Site-specific model input data include climate (daily maximum and minimum temperature and precipitation), soil texture and physical properties, vegetation

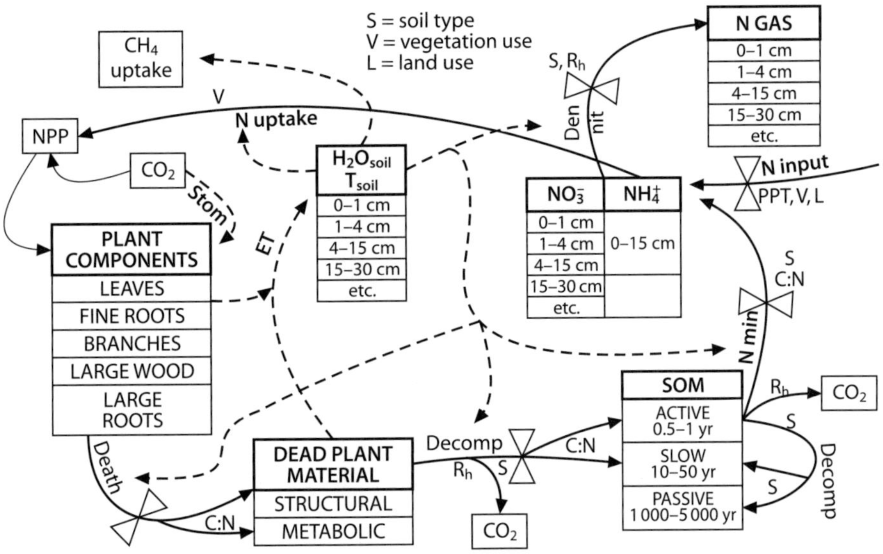

Fig. 3.6.
Pools and fluxes simulated in the DAYCENT model (Parton et al. 1998; Kelly et al. 2000; Del Grosso et al. 2001)

cover, and land management. Maximum plant growth rate is a function of vegetation type and solar radiation, which is calculated from latitude and day of year. Maximum NPP is down-regulated to account for nutrient, water, and temperature limitation. C allocation patterns, C/N ratios of biomass components, and senescence of plant components are functions of plant type, phenology and nutrient and water stress. Decomposition of dead plant material and SOM are driven by the amount of material and C/N ratios of different pools, as well as water and temperature limitation. Decomposition results in N mineralization, which makes N available for plant uptake, NO_3 leaching, and the processes (nitrification and denitrification) that result in N gas losses. DAYCENT includes a fairly complex soil water and temperature submodel (Parton et al. 1998) and also simulates CH_4 uptake in non-saturated soils. The effects of increased CO_2 concentration are implemented in three ways in DAYCENT. Maximum potential NPP rates can increase, C/N of above ground biomass can increase, and transpiration demand can decrease.

3.5.1.2 *DAYCENT Simulations and Results*

DAYCENT was used to investigate the effects of changes in precipitation, temperature, CO_2 concentration, and N deposition on plant growth and soil water content at the Jasper Ridge annual grassland. Model drivers (CO_2 concentration, precipitation, air temperature, and N inputs) were varied to mimic the experiments conducted at this site. Effects of elevated CO_2 were operationally defined for this site as a 10% increase in maximum NPP and C/N of leaf biomass, and a 50% decrease in transpiration demand. To better match observed NPP data, parameters had to be adjusted to decrease late-season belowground C allocation.

These effects capture the major biogeochemical interactions between CO_2, rainfall, heat and N fertilization. The experimental analysis and the model simulations indicate the role that water use and nitrogen availability have in affecting the net response to global change factors. Comparisons between mean simulated and observed changes in ANPP showed that the model did well for the single factor treatments but not for the interactions (Fig. 3.7a). The data showed a larger positive response for CO_2 alone than when CO_2 interacted with the other factors, whereas the model showed an additive response between CO_2 and the other factors.

Both simulated and observed aboveground productivity at the Jasper Ridge study was more dynamic relative to belowground productivity. The observed and simulated interannual and treatment variability tended to be much higher in the aboveground biomass. However, model results for mean BNPP did not agree with the data (Fig. 3.7b). In contrast to the data that showed a negative BNPP response for some treatments, DAYCENT showed a positive BNPP effect for all of the treatments because the model assumes that increased ANPP leads to higher nutrient and water demand so BNPP is enhanced. This indicates that ecosystem responses to global change are sensitive to environmental conditions modifying the belowground allocation, and the model needs improvement in how it represents allocation. Variability in observed data also contributes to the difficulty in modeling this system. Figure 3.7c shows the 4 years of data points that were averaged to obtain the bars in Fig. 3.7a along with the standard errors of the observed data. Both the spatial and temporal variability in the observed data are large and the model shows substantial error for some treatments during some years.

Simulation results indicate that environmental conditions and allocation patterns interact and have indirect effects on N mineralization and N-use efficiency,

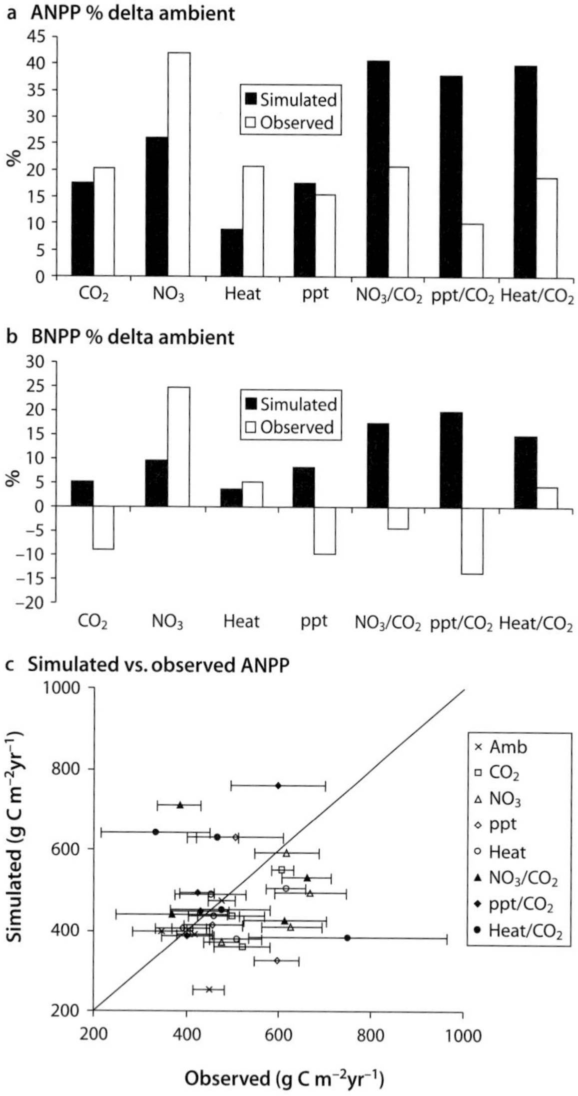

Fig. 3.7. Simulations and measurements of the relative effect of elevated CO_2, rainfall, heat and N fertilization on mean **a** aboveground and **b** belowground net primary productivity (ANPP and BNPP) during 4 years of the Jasper Ridge Global Change Experiment. The relative effect was calculated as (treatment – ambient)/ambient. **c** Simulated vs. observed annual ANPP, standard errors of observations, and the 1–1 line

which alter root-to-shoot allocation of new production. This effect has both short-term and long-term impacts on carbon biogeochemistry of these systems. The long-term storage of carbon will depend on the amount structural carbon formed, and changes in N-use efficiency may favor formation of less labile carbon materials and promote sequestration of carbon.

3.5.2 Comparing Forest and Grassland with G'DAY

DAYCENT and G'DAY were used to simulate responses of NPP, NEP, and C storage to gradually increasing CO_2, temperature, and nitrogen input for a short-grass steppe and a boreal forest site (Pepper et al. 2005). Here we will discuss NPP and NEP results from the G'DAY model. G'DAY (Comins and McMurtrie 1993; Medlyn et al. 2000) is a mechanistic model of C and N cycling and water balance in plant ecosystems. The model simulates plant physiological and soil processes: canopy photosynthesis, respiration, C and N allocation within the plant, and soil N feedbacks associated with decomposition of litter and soil organic matter. Processes included in the water-balance model are canopy interception and soil-surface evaporative water loss, drainage, runoff and transpiration modified by the CO_2 effect on water-use efficiency.

The sites considered are a short-grass steppe in Colorado, USA (40°8' N, 104°45' W; 1625 m a.s.l.) where the perennial C_4-dominant grassland is strongly water- and N-limited (Uresk et al. 1996), receiving 320 mm mean annual precipitation with a 4-month growing season. The forest plantation site is located at Flakaliden, Sweden (64°7' N, 19°27' E; 310 m a.s.l.) where stands of Norway spruce, planted in 1963, are N-limited but not water-limited (Bergh et al. 1999), receiving approx. 600 mm annual precipitation with a 4-month growing season. Both ecosystems are the subject of major long-term (>8 yr) climate-change experiments with CO_2-enrichment treatments at both sites and soil-warming at Flakaliden.

G'DAY was parameterized for each site and then simulated to equilibrium under current climate. Treatments simulated at each site were: gradually (linearly) increasing [CO_2] from 350 to 700 ppm over 100 years (Treatment C); gradually (linearly) increasing daily maximum and minimum air temperatures and soil temperature by 1, 3, and 2 °C, respectively, over 100 years (T), and nitrogen addition that increases linearly from 0 to 1 g m^{-2} yr^{-1} over the century. The four treatments shown in Fig. 3.8 are denoted C, T, CT (rising CO_2 and temperature), and CTN (rising CO_2, temperature and N input). The CT treatment approximated the IS92a IPCC scenario (Houghton et al. 1995). Key results from multi-factor simulations are summarized below.

3.5.2.1 *Simulations with Rising Temperature and CO_2*

The NPP-response to treatment CT is similar in percentage terms at the grassland and forest, but comprises a large CO_2 effect and small negative T-effect at the water-limited grassland, compared with a small CO_2 effect and large temperature effect at the N-limited forest (Fig. 3.8). The explanation for these differences between grassland and forest is that the CO_2-effect is amplified under water limitation but diminished under N limitation, whereas warming enhances water stress at the grassland but stimulates N availability at the forest. NPP-responses to

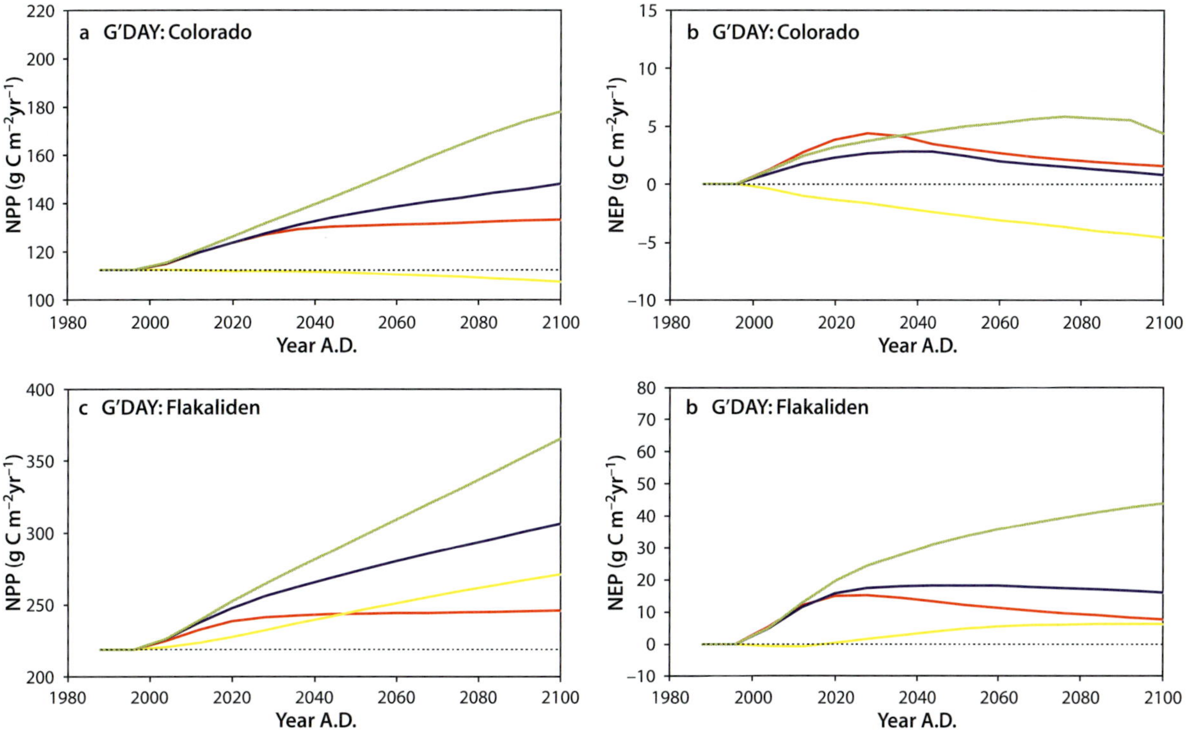

Fig. 3.8. Simulated **a** NPP and **b** NEP for Colorado short-grass steppe and **c** NPP and **d** NEP for Flakaliden boreal forest under single- and multi-factor treatments, where the *dotted line* denotes current equilibrium, *C* (*red*) denotes gradually increasing CO_2 concentration, *T* (*yellow*) denotes warming, *CT* (*blue*) the combination of *C* and *T*, and *CTN* (*green*) denotes the combination of *C* and *T* and increasing N inputs

CO_2 and T are approximately additive for forest, but not for grassland. The modeled CO_2–T interaction is strong at the grassland because of a synergistic effect whereby elevated CO_2 alleviates water-stress that is the major limitation on the grassland's temperature response.

Simulations of NEP under rising CO_2 show evidence of C-sink saturation (NEP declining towards zero) at both sites, probably because of developing N limitation. Under rising CO_2 and temperature, however, the forest shows little evidence of C-sink saturation because warming causes a gradual N transfer from soil with low C/N ratio to wood with high C/N ratio, which promotes C storage.

3.5.2.2 *Simulations with Increased N Input*

CTN simulations for both grassland and forest (Fig. 3.8) show that a relatively modest cumulative N fertilization (50 g m^{-2} over 100 years) is sufficient to sustain large NPP and NEP responses that are comparable in percentage terms to results of Cramer et al. (2001). This modest N addition overcomes the negative soil-N feedback that otherwise occurs at high CO_2 when increased litter input to the soil leads to increased N immobilization and reduced plant N availability. The CTN simulations show no evidence of C-sink saturation for at least 80 years (Fig. 3.8).

3.6 Summary and Conclusions

It is undeniable that ecosystems of the future will be subjected to multiple atmospheric and climatic changes. Research planning documents recognize this certainty and strongly promote the understanding of multi-factor interactions as a research imperative for the future (Committee on Global Change Research 1999). The research that we have highlighted here illustrates many of the challenges of addressing this imperative, but it also shows the importance of attempting to meet those challenges.

The net response to increased CO_2 concentration, warming, altered precipitation, N deposition, and other changes may be a simple additive effect of the separate influences, but experimental results also demonstrate the possibility of complex interactions. Single-factor experiments, such as those in the GCTE Elevated CO_2 network, are highly valuable for advancing understanding of the primary responses and how they are modified by other environmental factors. However, such experiments cannot be interpreted as providing predictions for ecosystem responses in a future climate. Single-factor ecosystem warming experiments are especially problematic, given the temporal and spatial variability of air temperature and the greater uncertainty (compared to CO_2) in future trajectories. Realistic warming treatments also are more difficult to carry out in experiments, especially in forests, and it has been harder to attain an integrated, ecosystem-scale understanding of all of the factors that will influence how ecosystems will respond to a warmer climate. Coupling the uncertain responses to warming to the simultaneous changes in other global change factors would seem to be a daunting task.

The experiments and modeling efforts highlighted in this chapter do not lead to a clear prediction of how ecosystems will be affected by various combinations of global change factors. They do, however, support several general principles about responses to climatic warming and multi-factor interactions:

1. Warming alone affects multiple pools and processes with different rate constants. Hence, responses are likely to change over time, as occurred in the response of soil respiration to soil warming at the Harvard Forest.
2. Whole-ecosystem warming experiments are necessary to address the complex interactions between belowground and aboveground responses. Without a whole-ecosystem perspective, critical hypotheses such as the linkage between warming effects on N availability and aboveground production cannot be evaluated.
3. Stochastic events can strongly alter the trajectory of response to warming. In the TACIT experiment a severe summer heat spell led to a loss in productivity after 4 years despite generally positive effects of warming. Short-term experiments may over emphasize the importance of such events while under representing the possibility of their occurrence.
4. Elevated CO_2 can sometimes ameliorate deleterious effects of warming, and since higher CO_2 concentrations are certain to be associated with future warming scenarios, it is imperative that their combined effects be considered when interpreting data. However, we cannot assume that the responses to elevated CO_2 always are positive, as evidenced by the Jasper Ridge experiment.
5. The responses to combinations of factors often equal the simple additive effects of the individual factors. That being the case, single-factor experiments continue to be very informative and can be the basis for model simulations. However, complex interactions do occur, and they may or may not be predictable. Multifactor experiments are important for reminding us of this complexity.
6. Ecosystem models that incorporate our best understanding of the modes of actions of the individual factors will also capture many of the major biogeochemical interactions. Discrepancies between experimental data and model projections, such as for belowground productivity in the Jasper Ridge experiment, indicate areas where model improvement is needed. The influence of stochastic events and unexplained year-to-year variation in the nature of interactions suggest that simulations should be expected only to provide an envelope of possible future responses.
7. The long-term net effect of elevated CO_2 and temperature may be similar in different ecosystems, but the relative importance of the two global-change factors varies with site factors (e.g., water and N availability). Relatively modest N additions can overcome the soil-N feedback that can otherwise lead to C-sink saturation and loss of continued stimulation of NEP. These model results demonstrate why experiments must be conducted in a range of ecosystems under different conditions.

Acknowledgments

The experiments presented here are but a few of the many that contributed to GCTE Focus 1 activities, and we gratefully acknowledge the participation of all of the researchers who contributed their data, analyses, and ideas to GCTE workshops and syntheses. This paper was prepared at the Oak Ridge National Laboratory, which is managed by UT-Battelle, LLC, for the U.S. Dept. of Energy under contract DE-AC05-00OR22725.

References

Alward RD, Detling JK, Milchunas DG (1999) Grassland vegetation changes and global nocturnal warming. Science 283:229–231
Arft AM, Walker MD, Gurevitch J, Alatalo JM, Bret-Harte MS, Dale M, Diemer M, Gugerli F, Henry GHR, Jones MH, Hollister RD, Jonsdottir IS, Laine K, Levesque E, Marion GM, Molau U, Molgaard P, Nordenhall U, Raszhivin V, Robinson CH, Starr G, Stenstrom A, Stenstrom M, Totland O, Turner PL, Walker LJ, Webber PJ, Welker JM, Wookey PA (1999) Responses of tundra plants to experimental warming: meta-analysis of the international tundra experiment. Ecolog Monogr 69:491–511
BassiriRad H (2000) Kinetics of nutrient uptake by roots: responses to global change. New Phytol 147:155–169
Bergh J, Linder S, Lundmark T, Elfving B (1999) The effect of water and nutrient availability on the productivity of Norway spruce in northern and southern Sweden. For Ecol Manage 119:51–62
Boone RD, Nadelhoffer KJ, Canary JD, Kaye JP (1998) Roots exert a strong influence on the temperature sensitivity of soil respiration. Nature 396:570–572
Bridgham SD, Pastor J, Updegraff K, Malterer TJ, Johnson K, Harth C, Chen J (1999) Ecosystem control over temperature and energy flux in northern peatlands. Ecol Appl 9:1345–1358
Comins HN, McMurtrie RE (1993) Long-term response of nutrient-limited forests to CO_2-enrichment: equilibrium behaviour of plant-soil models. Ecol Appl 3:666–681
Committee on Global Change Research (1999) Global environmental change: research pathways for the next decade. National Academy Press, Washington
Cramer W, Bondeau A, Woodward FI, Prentice IC, Betts RA, Brovkin V, Cox PM, Fisher V, Foley JA, Friend AD, Kucharik C, Lomas MR, Ramankutty N, Sitch S, Smith B, White A, Young-Molling C (2001) Global response of terrestrial ecosystem structure and function to CO_2 and climate change: results from six dynamic global vegetation models. Global Change Biol 7:357–373
Crill PM (1991) Seasonal patterns of methane uptake and carbon dioxide release by a temperate woodland soil. Global Biogeochem Cycles 5:319–334
Crill PM, Bartlett KB, Hariss RC, Gorham E, Verry ES, Sebacher DI, Madzar L, d Sanner W (1988) Methane flux from Minnesota peatlands. Global Biogeochem Cycles 2:371–384

Del Grosso SJ, Parton WJ, Mosier AR, Hartman MD, Brenner J, Ojima DS, Schimel DS (2001) Simulated interaction of carbon dynamics and nitrogen trace gas fluxes using the DAYCENT model. In: Shaffer MJ, Ma L, Hansen S (eds) Modeling carbon and nitrogen dynamics for soil management. Lewis Publishers. Boca Raton, Florida, pp 303–332

Dukes JS, Chiariello NR, Cleland EE, Moore L, Shaw MR, Thayer S, Tobeck T, Mooney HA, Field CB (2005) Responses of grassland production to single and multiple global environmental changes. PLoS Biology 3:1829–1837

Dunne J, Harte J, Taylor K (2002) Response of subalpine meadow plant reproductive phenology to manipulated climate change and natural climate variability. Ecolog Monogr 73:69–86

Gill RA, Jackson RB (2000) Global patterns of root turnover for terrestrial ecosystems. New Phytol 147:13–31

Gunderson CA, Norby RJ, Wullschleger SD (2000) Acclimation of photosynthesis and respiration to simulated climatic warming in northern and southern populations of *Acer saccharum*: laboratory and field evidence. Tree Physiol 20:87–96

Harte J, Torn M, Chang F, Feifarek B, Kinzig A, Shaw MR, Shen K (1995) Global warming and soil microclimate: results from a meadow warming experiment. Ecol Appl 5:132–15

Henry HAL, Cleland EE, Field CB, Vitousek PM (2005) Interactive effects of elevated CO_2, N deposition, and climate change on plant litter quality in a California annual grassland. Oecologia 142:465–473

Hobbie SE (1996) Temperature and plant species control over litter decomposition in Alaskan tundra. Ecolog Monogr 66:503–522

Horz HP, Barbrook A, Field CB, Bohannan BJM (2004) Ammonia-oxidizing bacteria respond to multifactorial global change. Proc Nat Acad Sci 101:15136–15141

Houghton JT, Meira Filho LG, Bruce J, Lee H, Callander BA, Haites E, Harris N, Maskell K (Eds) (1995) Climate change 1994: radiative forcing of climate change and an evaluation of the IPCC IS92 emission scenarios. IPCC, Cambridge University Press, Cambridge

IPCC (2001) Climate change 2001: the scientific basis. Contribution of Working Group I to the Third Assessment Report of the Intergovernmental Panel on Climate Change (Houghton JT, Ding Y, Griggs DJ, Noguer M, van der Linden PJ, Dai X, Maskell K, Johnson CA, eds.). Cambridge UK: Cambridge University Press

Jansson PE, Berg B (1985) Temporal variation of litter decomposition in relation to simulated soil climate long-term decomposition in a Scots pine forest. 5. Can J Bot 63:1008–1016

Johnson LC, Shaver GR, Giblin AE, Nadelhoffer KJ, Rastetter ER, Laundre JA, Murray GL (1996) Effects of drainage and temperature on carbon balance of tussock tundra microcosms. Oecologia 108:737–748

Jonasson S, Michelsen A, Schmidt IK (1999) Coupling of nutrient cycling and carbon dynamics in the Arctic, integration of soil mircrobial and plant processes. Appl Soil Ecol 11:135–146

Kelly RH, Parton WJ, Hartman MD, Stretch LK, Ojima DS, Schimel DS (2000) Intra-annual and interannual variability of ecosystem processes in shortgrass steppe. J Geophys Res-Atmos 105:20093–20100

King JS, Hanson PJ, Bernhardt E, DeAngelis P, Norby RJ, Pregitzer KS (2004) A multi-year synthesis of soil respiration responses to elevated atmospheric CO_2 from four forest FACE experiments. Global Change Biol 10:1027–1042

Kirschbaum MUF (1995) The temperature-dependence of soil organic-matter decomposition, and the effect of global warming on soil organic-C storage. Soil Biol Biochem 27:753–760

Lewis JD, Lucash M, Olszyk D, Tingey DT (2001) Seasonal patterns of photosynthesis in Douglas fir seedlings during the third and fourth year of exposure to elevated CO_2 and temperature. Plant Cell Environ 24:539–548

Long SP (1991) Modification of the response of photosynthetic productivity to rising temperature by atmospheric CO_2 concentrations – has its importance been underestimated? Plant Cell Environ 14:729–739

MacDonald NW, Zak DR, Pregitzer KS (1995) Temperature effects on kinetics of microbial respiration and net nitrogen and sulfur mineralization. Soil Sci Soc Am J 59:233–240

Malhi SS, McGrill WB, Nyborg N (1990) Nitrate losses in soils: effects of temperature, moisture, and substrate concentration. Soil Biol Biochem 22:917–927

Marion GM, Henry GHR, Freckman DW, Johnstone J, Jones G, Jones MH, Lévesque E, Molau U, Mølgaard P, Parsons AN, Svoboda J, Virginia RA (1997) Open-top designs for manipulating field temperatures in high-latitude ecosystems. Global Change Biol 3:20–32

McGuire AD, Sitch S, Clein JS, Dargaville R, Esser G, Foley J, Heimann M, Joos F, Kaplan J, Kicklighter DW, Meier RA, Melillo JM, Moore B, Prentice IC, Ramankutty N, Reichenau T, Schloss A, Tian H, Williams LJ, Wittenberg U (2001) Carbon balance of the terrestrial biosphere in the twentieth century: Analyses of CO_2, climate and land use effects with four process-based ecosystem models. Global Biogeochem Cycles 15:183–206

Medlyn BE, McMurtrie RE, Dewar RC, Jeffreys MP (2000) Soil processes dominate the long-term response of forest net primary productivity to increased temperature and atmospheric CO_2 concentration. Can J For Res 30:873–888

Meentemeyer V (1978) Macroclimate and lignin control of litter decomposition rates. Ecology 59:465–472

Melillo J, Janetos A, Schimel D, Kittel T (2001) Vegetation and biogeochemical scenarios. In: National Assessment Synthesis Team (ed) Climate change impacts on the United States: the potential consequences of climate variability and change. Report for the US Global Change Research Program. Cambridge UK: Cambridge University Press, pp 73–91

Melillo JM, Steudler PA, Aber JD, Newkirk K, Lux H, Bowles FP, Catricala C, Magill A, Ahrens T, Morrisseau S (2002) Soil warming and carbon-cycle feedbacks to the climate system. Science 298:2173–2176

Morison JIL, Lawlor DW (1999) Interactions between increasing CO_2 concentration and temperature on plant growth. Plant Cell Environ 22:659–682

Norby RJ, Luo Y (2004) Evaluating ecosystem responses to rising atmospheric CO_2 and global warming in a multi-factor world. New Phytol 162:281–293

Norby RJ, Edwards NT, Riggs JS, Abner CH, Wullschleger SD, Gunderson CA (1997) Temperature-controlled open-top chambers for global change research. Global Change Biol 3:259–267

Norby RJ, Long TM, Hartz-Rubin JS, O'Neill EG (2000) Nitrogen resorption in senescing tree leaves in a warmer, CO_2-enriched atmosphere. Plant Soil 224:15–29

Norby RJ, Cotrufo MF, Ineson P, O'Neill EG, Canadell JG (2001) Elevated CO_2, litter chemistry, and decomposition: a synthesis. Oecologia 127:153–165

Norby RJ, Hartz-Rubin JS, Verbrugge MJ (2003) Phenological responses in maple to experimental atmospheric warming and CO_2 enrichment. Global Change Biol 9:1792–1801

Norby RJ, Ledford J, Reilly CD, Miller NE, O'Neill EG (2004) Fine root production dominates response of a deciduous forest to atmospheric CO_2 enrichment. Proc Nat Acad Sci 101:9689–9693

Parton WJ, Ojima DS, Schimel DS (1994) Environmental change in grasslands: Assessment using models. Climatic Change 28: 111–141

Parton WJ, Hartman M, Ojima D, Schimel D (1998) DAYCENT and its land surface submodel: description and testing. Global Planetary Change 19:35–48

Pendall E, Bridgham S, Hanson PJ, Hungate B, Kicklighter DW, Johnson DW, Law BE, Luo YQ, Megonigal JP, Olsrud M, Ryan MG, Wan SQ (2004) Below-ground process responses to elevated CO_2 and temperature: a discussion of observations, measurement methods, and models. New Phytol 162:311–322

Pepper DA, Del Grosso SJ, McMurtrie RE, Parton WJ (2005) Simulated carbon sink response of shortgrass steppe, tallgrass prairie and forest ecosystems to rising [CO_2], temperature and nitrogen input. Global Biogeochem Cycles 19: GB1004

Peterjohn WT, Melillo JM, Steudler PA, Newkirk KM, Bowles FP, Aber JD (1994) Responses of trace gas fluxes and N availability to experimentally elevated soil temperatures. Ecol Appl 4:617–625

Poorter H, Pérez-Soba N (2001) The growth response of plants to elevated CO_2 under non-optimal environmental conditions. Oecologia 129:1–20

Pregitzer KS, King JS, Burton AJ, Brown SE (2000) Responses of tree fine roots to temperature. New Phytol 147:105–115
Raich JW, Nadelhoffer KJ (1989) Belowground carbon allocation in forest ecosystems – global trends. Ecology 70:1346–1354
Raich JW, Potter CS (1995) Global patterns of carbon-dioxide emissions from soils. Global Biogeochem Cycles 9:23–36
Raich JW, Schlesinger WH (1992) The global carbon-dioxide flux in soil respiration and its relationship to vegetation and climate. Tellus Ser B-Chem Phys Meteorol 44:81–99
Rustad LE, Fernandez IJ (1998) Soil warming: consequences for litter decay in a spruce-fir forest ecosystem in Maine. Soil Sci Am J 62:1072–1081
Rustad LE, Norby RJ (2002) Temperature increase: effects on terrestrial ecosystems. In: Mooney HA, Canadell JG (eds), The Earth System: biological and ecological dimensions of global environmental change. Vol. 2 in Encyclopedia of Global Environmental Change. John Wiley & Sons, Chichester, pp 575–581
Rustad LE, Campbell JL, Marion GM, Norby RJ, Mitchell MJ, Hartley AE, Cornelissen JHC, Gurevitch J (2001) A meta-analysis of the response of soil respiration, net nitrogen mineralization, and aboveground plant growth to experimental ecosystem warming. Oecologia 126:543–562
Shaver GR, Johnson LC, Cades DH, Murray G, Laundre JA, Rastetter EB, Nadelhoffer KJ, Giblin AE (1998) Biomass accumulation and CO_2 flux in three Alaskan wet sedge tundras: responses to nutrients, temperature, and light. Ecol Monogr 68:75–99
Shaver GR, Canadell J, Chapin FS, Gurevitch J, Harte J, Henry G, Ineson P, Jonasson S, Melillo J, Pitelka L, Rustad L (2000) Global warming and terrestrial ecosystems: a conceptual framework for analysis. BioScience 50:871–881
Shaw MR, Zavaleta ES, Chiariello NR, Cleland EE, Mooney HA, Field CB (2002) Grassland responses to global environmental changes suppressed by elevated CO_2. Science 298:1987–1990
Solomon AM (1986) Transient-response of forests to CO_2-induced climate change – simulation modeling experiments in eastern North-America. Oecologia 68:567–579
Tingey DT, McVeety BD, Waschmann R, Johnson MG, Phillips DL, Rygiewicz PT, Olszyk DM (1996) A versatile sun-lit controlled-environment facility for studying plant and soil processes. J Environ Qual 25:615–62
Uresk D, Sims PL, Dodd JL (1996) NPP Grassland: Central Plains Experimental Range (SGS) USA, 1970–1975. Data set. Available on-line [http://www.daac.ornl.gov] from Oak Ridge National Laboratory Distributed Active Archive Center, Oak Ridge, Tennessee, USA
Wan S, Norby RJ, Pregitzer KS, Ledford J, O'Neill EG (2004) CO_2 enrichment and warming of the atmosphere enhance both productivity and mortality of maple tree fine roots. New Phytol 162:437–446
Warren-Wilson J (1957) Arctic plant growth. Adv Sci 13:383–388
Williams RS, Norby RJ, Lincoln DE (2000) Effects of elevated CO_2 and temperature-grown red and sugar maple on gypsy moth performance. Global Change Biol 6:685–695
Williams RS, Lincoln DE, Norby RJ (2003) Development of gypsy moth larvae feeding on red maple saplings at elevated CO_2 and temperature. Oecologia 137:114–122
Zavaleta ES, Thomas BD, Chiariello NR, Asner GP, Shaw MR, Field CB (2003a) Plants reverse warming effect on ecosystem water balance. Proc Nat Acad Sci 100:9892–9893
Zavaleta ES, Shaw MR, Chiariello NR, Mooney HA, Field CB (2003b) Additive effects of simulated climate changes, elevated CO_2, and nitrogen deposition on grassland diversity. Proc Nat Acad Sci 100:7650–7654
Zavaleta ES, Shaw MR, Chiariello NR, Thomas BD, Cleland EE, Field CB, Mooney HA (2003c) Responses of a California grassland community to three years of elevated temperature, CO_2, precipitation, and N deposition. Ecol Monog 73:585–604

Chapter 4

Insights from Stable Isotopes on the Role of Terrestrial Ecosystems in the Global Carbon Cycle

DianeE. Pataki · Chun-Ta Lai · Charles D. Keeling · James R. Ehleringer

4.1 Introduction

The use of isotopic tracers in organic matter, water, and atmospheric gases has become an important component of the study of ecology and global change. Physiological and physical processes discriminate against heavy isotopes in predictable ways, so that measurements of isotopes at natural abundance, i.e., naturally occurring concentrations as opposed to artificial labeling experiments, can provide useful information about biological and physical processes at various spatial and temporal scales. Here we review recent progress in the application of naturally occurring stable isotopes in understanding ecosystem physiology and its role in biogeochemistry at ecosystem and global scales, with an emphasis on analyses of isotopes of carbon and oxygen in atmospheric carbon dioxide, a key component of the GCTE core project.

First, it is useful to review the basis for the application of stable isotopes in ecology and to provide some definitions of commonly used terminology. Isotopes of a single element contain differing numbers of neutrons. Some isotopes are stable while others are radioactive. We will focus here on two commonly used stable isotopes in the biological sciences: carbon-13 and oxygen-18. The most abundant isotope of carbon is ^{12}C which constitutes 98.9% of terrestrial carbon atoms; 1.1% of carbon is found as ^{13}C. In oxygen, 99.8% is ^{16}O, and the next most abundant form is ^{18}O constituting 0.2% of oxygen atoms (Fritz and Fontes 1980).

Isotopes are useful in ecology because they influence the physical and chemical properties of molecules. Light isotopes form weaker chemical bonds than heavy isotopes, so chemical reactions with substrates containing light isotopes occur more rapidly than with substrates containing heavy isotopes. This results in a difference in the abundance of heavy isotopes between the substrate and product of the reaction, and is called kinetic fractionation. In an equilibrium reaction, such as equilibration between liquid and gaseous water, the gaseous phase contains more of the light isotope and the liquid phase contains more of the heavy isotope, which results in equilibrium fractionation. There is also diffusive fractionation, which reflects the observation that molecules containing light isotopes diffuse more quickly than their heavier counterparts. A mechanistic understanding of fractionation in biological and physical processes is the basis of using isotopes as tracers of ecosystem processes (Fritz and Fontes 1980; Kendall and McDonnell 1998).

Isotopic measurements are generally expressed as the molar ratios of heavier isotopes to lighter ones (R). Such ratios are difficult to resolve in an absolute sense, and are usually expressed as relative ratios in delta (δ) notation:

$$\delta = (R_{sample} / R_{standard} - 1) \qquad (4.1)$$

where R_{sample} and $R_{standard}$ are the heavy-to-light molar ratios of the sample and of an international standard. If $R_{sample} < R_{standard}$ then δ is negative. For example, plants utilizing the C_3 photosynthetic pathway (the majority of terrestrial plants) typically have values of $\delta^{13}C$ that range from −21 to −35‰, where the symbol ‰ denotes per mil (parts per thousand), while atmospheric CO_2 is currently about −8.0‰ (Keeling et al. 2005) relative to the common standard for carbon stable isotopes, a belemnite from the Pee Dee formation in South Carolina (PDB).

A change in isotopic composition as a result of fractionation can be expressed as a discrimination (Δ):

$$\Delta = (\delta_{source} - \delta_{product}) / (1 + \delta_{product}) \qquad (4.2)$$

where δ_{source} and $\delta_{product}$ are the isotope ratios of the source and product, respectively. With these definitions as a reference we can explore recent progress in the application of stable isotopes to understanding the role of terrestrial ecosystems in the carbon cycle.

4.2 Ecosystem Carbon Cycles

Measurements of the isotopic composition of air sampled near soils and plant canopies can provide information about ecosystem carbon cycles. The isotopic composition of respiration can be applied toward understanding physiological and environmental controls on the flow of carbon through various pools by applying a mechanistic understanding of carbon isotope fractionation. In addition, when different ecosystem pools or fluxes have distinct isotopic signatures, carbon isotopes can be used as a tracer to distinguish between components of the ecosystem carbon cycle.

Currently, a common protocol at a number of sites around the world is to repeatedly sample canopy air at night to obtain estimates of the isotopic composition of ecosystem respiration (see http://basinisotopes.org). We will denote the carbon isotope composition of CO_2 from ecosystem respiration as $\delta^{13}C_R$, and the oxygen isotope ratio of ecosystem respiration as $\delta^{18}O_R$. $\delta^{13}C_R$ is estimated by extracting the isotopic composition of plant- and soil-respired CO_2 from background air with a numerical method derived by Keeling (1958, 1961).

Consider CO_2 inside a forest canopy as a mixture of background CO_2 (not derived from any local sources) and CO_2 released by plant and soil respiration:

$$C_a = C_b + C_R \tag{4.3}$$

where C represents CO_2 concentration (mole fraction of dry air), and subscripts a, b and R represent the total CO_2, the background CO_2 and the ecosystem respired sources, respectively. Based on conservation of mass, we can write:

$$\delta^{13}C_a C_a = \delta^{13}C_b C_b + \delta^{13}C_R C_R \tag{4.4}$$

where $\delta^{13}C$ denotes δ of carbon-13 relative to carbon-12. Substituting $C_R = C_a - C_b$ from Eq. 4.3 to 4.4,

$$\delta^{13}C_a = \frac{C_b(\delta^{13}C_b - \delta^{13}C_R)}{C_a} + \delta^{13}C_R \tag{4.5}$$

Equation 4.5 is a linear function of $1/C_a$, allowing us to estimate $\delta^{13}C_R$ as the intercept of a linear regression with $1/C_a$ as the independent variable. A number of studies have investigated $\delta^{13}C_R$ over a wide range of ecosystems (Buchmann et al. 1997a; Buchmann et al. 1997b; Fessenden and Ehleringer 2002; Flanagan et al. 1996; Harwood et al. 1999; Quay et al. 1989; Sternberg et al. 1989). Pataki et al. (2003b) synthesized measurements from 33 sites and discussed the assumptions inherent in the collection and interpretation of $\delta^{13}C_R$, which include the critical assumption that $\delta^{13}C_b$ and $\delta^{13}C_R$ remain constant during the sampling period. Miller and Tans (2003) also illustrated uncertainties associated with measurement and analytical errors regarding the two-source mixing approach.

Advances in isotope ratio mass spectrometry have increasingly permitted analysis of CO_2 isotopes in smaller air volumes with greater automation, allowing for more frequent sampling (Ehleringer and Cook 1998; Schauer et al. 2003; Tu et al. 2001). Intensive measurements of $\delta^{13}C_R$ over short time-periods, as well as longer-term monitoring on a weekly basis, have led to recent advances in our understanding of ecosystem carbon cycling. Correlations between $\delta^{13}C_R$ and environmental variables have provided evidence that a large proportion of carbon fixed in photosynthesis cycles rapidly through forested ecosystems (Bowling et al. 2002; Ekblad and Högberg 2001; Lai et al. 2005; Ometto et al. 2002). Ekblad and Högberg (2001) observed a correlation between $\delta^{13}C$ of soil-respired CO_2 and atmospheric humidity that lagged by 1–4 days, which they interpreted as an indication of the influence of recent photosynthesis on root respiration. These results were supported by Bowling et al. (2002), who observed a correlation between whole ecosystem $\delta^{13}C_R$ and time-lagged atmospheric vapor pressure deficit (VPD) in several ecosystems in Oregon, USA. VPD strongly influences stomatal conductance, photosynthesis, and the ratio of stomatal conductance to photosynthesis, and therefore $\delta^{13}C$ of assimilate through its influence on C_i/C_a, the ratio of leaf intercellular to ambient CO_2 concentration (see Box 4.1). Scartazza et al. (2004) and Fessenden and Ehleringer (2003) showed that $\delta^{13}C_R$ is also influenced by soil moisture, presumably through the influence of drought on stomatal closure. Although changes in the proportion of above- and belowground respiration and influences of declining soil moisture on the isotopic composition of soil respiration are alternative explanations, Scartazza et al. (2004) show that soil moisture-driven variations in $\delta^{13}C_R$ were related to $\delta^{13}C$ of phloem sugar. Notably, the relationship between $\delta^{13}C_R$ and soil moisture showed a similar slope in two disparate coniferous forests in the eastern and western coastal regions of the United States (Fig. 4.1). Integrated over time, these relationships lead to spatial variability in $\delta^{13}C_R$ as a function of water availability (Fig. 4.2).

In addition to providing a greater mechanistic understanding of ecosystem physiology and its influence on carbon cycling, measurements of the isotopic composi-

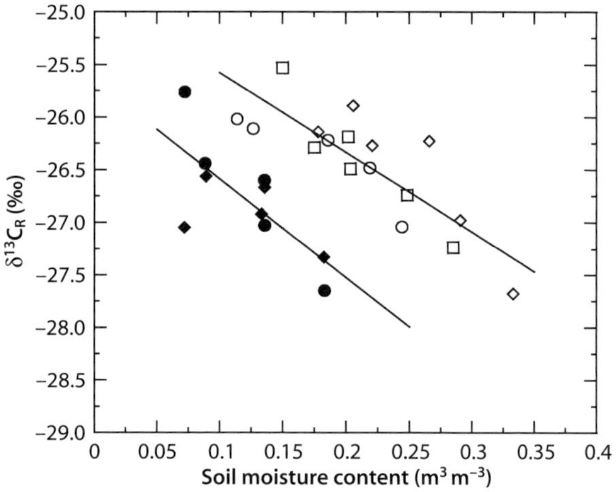

Fig. 4.1. Relationship between the monthly mean carbon isotope composition of ecosystem respiration ($\delta^{13}C_R$) and average soil moisture content during the growing season at two coniferous forests: the Wind River Canopy Crane Facility in Washington, USA (*open symbols*) and Howland Forest in Maine, USA (*closed symbols*). Measurements from different years are indicated by different symbols as *squares* (2001), *circles* (2002) and *diamonds* (2003) (from Lai et al. 2005, reproduced by permission of Blackwell Publishing)

tion of CO_2 can be used to partition ecosystem fluxes into specific component parts (Bowling et al. 1999; Bowling et al. 2001; Lai et al. 2003; Lloyd et al. 1996; Ogée et al. 2003; Yakir and Wang 1996). During daylight hours, Net Ecosystem Exchange (NEE) of CO_2 reflects the balance between photosynthesis (A) and respiration (R):

$$F_{NEE} = F_A + F_R \quad (4.6)$$

where F represents a flux of CO_2. A similar mass balance can be applied to the isotopes of CO_2:

$$\delta^{13}C_{NEE}F_{NEE} = F_\delta = \delta^{13}C_A F_A + \delta^{13}C_R F_R \quad (4.7)$$

where F_δ represents an "isoflux", analogous to net ecosystem exchange of $^{13}CO_2$ (μmol m^{-2} s^{-1} ‰) (Bowling et al. 2001). Values of $\delta^{13}C_{NEE}$, $\delta^{13}C_A$ and $\delta^{13}C_R$ denote the carbon isotope ratios associated with NEE, photosynthetic and respiratory fluxes, respectively.

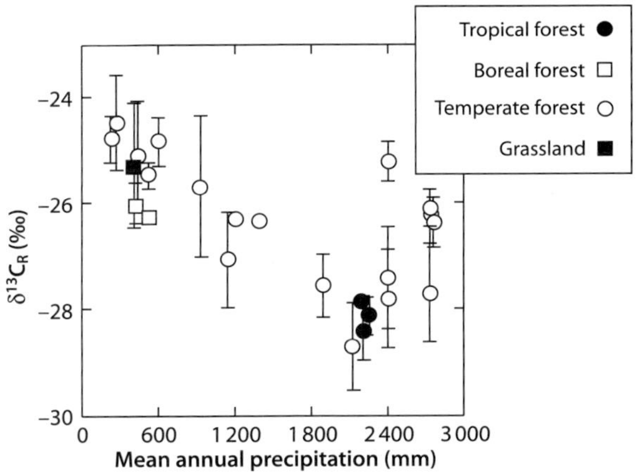

Fig. 4.2. The carbon isotope composition of ecosystem respiration ($\delta^{13}C_R$) at several sites throughout North and South America in relation to mean annual precipitation. *Error bars* show the standard error of all observations for a given site. A correlation is observed across all sites, with the exception of coniferous forests in the northwestern United States, where rainfall exceeds 2 400 mm yr^{-1} (from Pataki et al. 2003b, reproduced by permission of the American Geophysical Union)

F_{NEE} is routinely measured at numerous sites around the world as part of the FLUXNET suite of networks (http://daac.ornl.gov/FLUXNET). To apply Eq. 4.6 and 4.7 to solving for F_A and F_R, methods to measure or estimate the isotopic composition of the individual fluxes are required. Equation 4.5 is commonly used to determine $\delta^{13}C_R$ as discussed earlier. Quantifying $\delta^{13}C_{NEE}$ and $\delta^{13}C_A$, however, is more challenging (Bowling et al. 2003a). Yakir and Wang (1996) used $\delta^{13}C$ and $\delta^{18}O$ measurements to partition F_{NEE} of agricultural fields. They estimated $\delta^{13}C_{NEE}$ and $\delta^{13}C_A$ by measuring the isotopic composition of plants and soil samples, which integrate over seasonal time scales. However, to partition F_{NEE} at half-hourly intervals, capturing diurnal variations in $\delta^{13}C_{NEE}$ and $\delta^{13}C_A$ is necessary (Bowling et al. 2001). Using an aerodynamic approach to estimate canopy conductance and Fick's law of diffusion to estimate C_i, Bowling et al. (2001) calculated $\delta^{13}C_A$ for the whole-canopy:

$$\delta^{13}C_A = \delta^{13}C_a - a - (b - a)\frac{C_i}{C_a} \quad (4.8)$$

which combines Eqs. 4.2 and 4.12. An estimate of canopy-scale $\delta^{13}C_A$ can also be achieved by coupling biochemical photosynthesis models with a stomatal conductance model (Lai et al. 2003). Finally, if another parameter is known such as F_R, which can be modeled as a function of temperature, Eqs. 4.6 and 4.7 can be used to solve for $\delta^{13}C_A$ (Bowling et al. 2003a).

Carbon isotopes are particularly useful for discerning photosynthetic pathways because there is considerable contrast in $\delta^{13}C$ ratios between C_3 and C_4 plants (Box 4.1). In the context of ecosystem respiration, measurements of $\delta^{13}C$ can be used to partition contributions

Box 4.1. Isotopic discrimination in photosynthesis

Discrimination of ^{13}C in photosynthesis occurs in two steps. First, discrimination occurs during diffusion of CO_2 into the stomatal pores, followed by enzymatic discrimination in carboxylation by RuBP carboxylase (Rubisco). For plants utilizing the C_3 photosynthetic pathway, these effects can be represented as:

$$\Delta^{13}C_P = a + (b - a) \cdot C_i / C_a \quad (4.12)$$

where C_i / C_a denotes the ratio of intercellular to ambient CO_2 partial pressure, a the fractionation in diffusion, and b the net fractionation of carboxylation (Farquhar et al. 1982). Because a and b are known (4.4 and 27‰, respectively) photosynthetic discrimination of C_3 plants can be predicted with estimates of C_i / C_a, which is affected by photosynthetic rates and stomatal conductance. It should be noted that this approach neglects resistance to CO_2 transfer from the sub-stomatal cavity to the mesophyll, which may be significant in some species (Ethier and Livingston 2004).

Photosynthetic discrimination by C_4 plants may also be quantified with estimates of C_i / C_a. However, the "leakiness" of the unique C_4 anatomy must also be considered, as it affects the extent to which fractionation of Rubisco influences $\Delta^{13}C_P$, in addition to the C_4 enzyme PEP carboxylase. This is represented as:

$$\Delta^{13}C_P = a + (b_4 + b_3\phi - a) \cdot C_i / C_a \quad (4.13)$$

where b_4 denotes the fractionation factor of PEP carboxylase (−5.7‰), b_3 is the fractionation factor of Rubisco (30‰), and ϕ is the leakiness of bundle sheath cells, where C_4 photosynthesis takes place (Farquhar 1983). Rubisco has a small influence on photosynthetic fractionation in most C_4 plants, and the fractionation of PEP carboxylase is also fairly small. Thus, $\Delta^{13}C_P$ in C_4 plants is significantly lower than in C_3 plants, making these pathways isotopically distinct. Typical values of $\Delta^{13}C_P$ range from 13 to 25‰ in C_3 plants and 2.5 to 5‰ in C_4 plants (Dawson et al. 2002; Farquhar et al. 1989; Lloyd and Farquhar 1994).

of plants utilizing different photosynthetic pathways in C_3-C_4 mixtures (e.g., grassland and savannas), which are mainly distributed in sub- and tropical regions (Still et al. 2003a; Still et al. 2003b). The fraction of C_3 (λ) contributing to total ecosystem respiration can be estimated by:

$$\delta^{13}C_R = \lambda\delta^{13}C_3 + (1-\lambda)\delta^{13}C_4 \quad (4.9)$$

where $\delta^{13}C_3$ and $\delta^{13}C_4$ denote the $\delta^{13}C$ ratios of C_3 and C_4 species, respectively. Still et al. (2003b) used this approach to estimate that C_4 species contributed 60 to nearly 100% of the total respiration between May and September in a tallgrass pasture in central Oklahoma, U.S.A. Lai et al. (2003) found a comparable range of C_4 contributions in a tallgrass prairie with similar species composition and climate. The contribution of C_3 vs. C_4 plants is critical to the application of measurements of $^{13}CO_2$ at a global scale, as we will discuss in the next section.

4.3 The Global Carbon Cycle

There is a high degree of short-term interannual variability associated with the rate of increase in atmospheric CO_2 concentration (Fig. 4.3). We know from statistics of energy use that CO_2 emissions from fossil fuel combustion are increasing at a nearly steady rate from year to year (Andres et al. 2000; IPCC 2001). The principal sources of interannual variability are therefore ocean fluxes and terrestrial ecosystem processes. Isotopes of carbon dioxide are useful to distinguish between these components on regional and global scales.

We have discussed that photosynthesis of terrestrial plants discriminates against $^{13}CO_2$ (Box 4.1), leaving heavier $^{13}CO_2$ behind in the atmosphere, while the process of respiration returns isotopically light carbon to the atmosphere. In contrast, fractionation of CO_2 dissolution into the ocean is small, on the order of 2‰ (Inoue and Sugimura 1985; Wanninkhof 1985; Zhang et al. 1995). Therefore, gradients of $\delta^{13}C$ of CO_2 in the atmosphere may be used to distinguish between terrestrial and oceanic sinks if the magnitude of photosynthetic discrimination and the isotopic composition of respiration can be estimated well enough on a global scale. Because C_3 and C_4 photosynthesis differ greatly in fractionation of $^{13}CO_2$, this requires robust estimates of the proportion of productivity of C_3 vs. C_4 plants globally. However, the information gained from applying these estimates can be significant. Figure 4.3 shows the rate of change of $\delta^{13}C$ of atmospheric CO_2 in addition to the rate of change of atmospheric CO_2 concentration for the last two decades. The close correlation between the two time-series strongly suggests that terrestrial processes have dominated the short-term interannual variability in the growth rate of atmospheric CO_2. By applying mass balance to atmospheric observations of CO_2 and its isotopic composition, sources and sinks of CO_2 can be inferred:

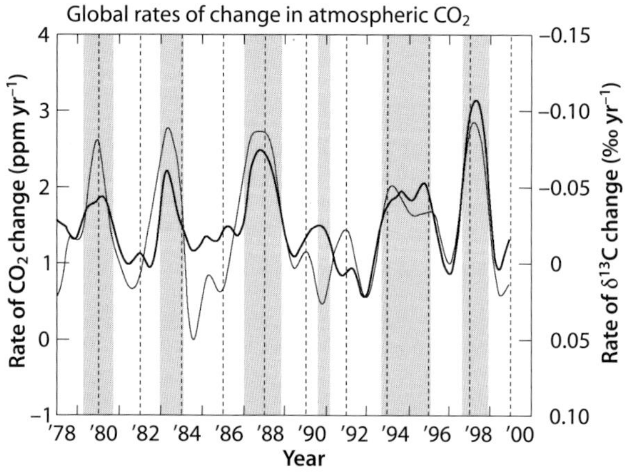

Fig. 4.3. The rate of change in atmospheric CO_2 concentration (*bold line, left axis*) and the rate of change in the carbon isotope composition ($\delta^{13}C$) of atmosphere CO_2 (*right axis*) (from Keeling et al. 2005, used by permission)

$$\frac{\partial}{\partial t}C_a = F_{FF} + F_{DEF} + F_{ao} + F_{oa} + F_{ab} + F_{ba} \quad (4.10)$$

$$\begin{aligned} C_a\frac{\partial}{\partial t}\delta^{13}C_a = &(\delta^{13}C_{FF} - \delta^{13}C_A)F_{FF} \\ &+(\delta^{13}C_{DEF} - \delta^{13}C_A)F_{DEF} \\ &+\varepsilon_{ao}(F_{ao} + F_{oa}) + D_oF_{oa} \\ &+-\Delta_p(F_{ab} + F_{ba}) + D_bF_{ba} \end{aligned} \quad (4.11)$$

where the subscripts 'FF' and 'DEF' denote the fossil fuel- and land-use change-derived releases of CO_2 to the atmosphere, subscripts 'AO' and 'OA' denote the one-way ocean-atmosphere fluxes, and subscripts 'AP' and 'BA' denote the one-way biosphere-atmosphere fluxes (Francey et al. 1995; Fung et al. 1997; Tans et al. 1993). Also included in order to solve Eqs. 4.10 and 4.11 for the oceanic and terrestrial fluxes are symbols to denote CO_2 emissions from fossil fuel combustion, $\delta^{13}C_{FF}$; CO_2 emissions from land-use change, $\delta^{13}C_{DEF}$; fractionation factors associated with the net oceanic, ε_{ao}; and terrestrial discrimination, Δ_P; and two "disequilibrium" terms, D_o and D_b. These latter terms express the extent to which older carbon, dissolved in ocean waters or fixed in photosynthesis in the past and later released, has a more enriched $\delta^{13}C$ than current carbon due to the dilution of atmospheric CO_2 by fossil fuel combustion in the last century.

Mass balances expressed in Eqs. 4.10 and 4.11 have been applied in a number of global studies of carbon sources and sinks that utilized atmospheric measurements to infer regions of carbon uptake (Battle et al. 2000; Ciais et al. 1995; Francey et al. 1995; Fung et al. 1997; Tans et al. 1993). A limitation to this approach is that the parameterization of Δ_P, the discrimination associated with net carbon uptake by the terrestrial biosphere, is difficult to validate with measurements. Ecosystem studies commonly measure the isotopic composition of the one-way, nighttime flux, $\delta^{13}C_R$, but estimating the signature of photosynthetic discrimination from direct measurements is difficult at whole ecosystem scales. For lack of data to the contrary, Δ_P has been generally fixed as a constant value in global mass balance studies (Battle et al. 2000; Francey et al. 1995; Tans et al. 1993). However Randerson et al. (2002) demonstrated that if large-scale regional changes in Δ_P covary with changes in gross fluxes, as could occur during climatic anomalies such as the El Niño Southern Oscilliation (ENSO), applying constant values of Δ_P to Eq. 4.11 could introduce significant errors into global mass balance calculations. Modifying the mass balance approach to explicitly consider gross fluxes and their time varying anomalies could improve top-down estimates of carbon sinks from atmospheric measurements (Randerson et al. 2002).

It has been noted that the peaks in atmospheric CO_2 growth rate have generally been associated with the occurrence of ENSO events (Francey et al. 1995; Keeling et al. 2005; Keeling et al. 1995), which brings warm, dry conditions to highly productive tropical regions (Dai et al. 1997; Ropelewski and Halpert 1987). One hypothesis for the underlying correlation between ENSO conditions and terrestrial sources and sinks of carbon has been that drought conditions, and in particular dry atmospheric conditions, in tropical regions have resulted in reduced productivity and subsequently smaller terrestrial sinks in tropical forests during ENSO years. Measurements of $\delta^{13}C$ of atmospheric CO_2 have generally supported this hypothesis (Fig. 4.3), which would also be expected based on the data in Fig. 4.1 and 4.2. However, recent work utilizing carbon monoxide and methane tracers in addition to measurements of CO_2 concentration and $\delta^{13}C$ of CO_2 suggests that an increased occurrence of biomass burning and wildfire during ENSO years, also associated with drought, may also be an important mechanism (Van der Werf et al. 2004).

The oxygen isotopic composition of CO_2 also has great potential to distinguish between components of the carbon cycle, namely, between assimilation and respiration of the terrestrial biosphere. The oxygen isotopic signature of assimilation and respiration are generally very different – more so than the carbon isotope signatures in ecosystems that contain only C_3 plants. However, oxygen isotopes have been applied less frequently in global carbon studies, largely because of the complexity of modeling oxygen fractionation effects at the global scale.

CO_2 and water in the liquid phase equilibrate isotopically so that the isotopic composition of CO_2 is strongly influenced by the isotopic composition of water where water is abundant, as in plants and soils (Amundson et al. 1998; Farquhar and Lloyd 1993; Farquhar et al. 1993; Hesterberg and Siegenthaler 1991). Assimilation and respiration have different isotopic signatures because leaf water is enriched in oxygen-18 relative to soil water due to the evaporative effects of transpiration. Approximately two-thirds of the CO_2 that diffuses into leaves is not fixed by photosynthesis, but this CO_2 does equilibrate with leaf water and diffuses back into the atmosphere with an isotopically enriched signature that is distinct from that of respired CO_2 which has equilibrated with soil water (Farquhar and Lloyd 1993; Flanagan et al. 1997; Francey and Tans 1987). Quantifying these processes in global models requires hydrologic models that predict the isotopic composition of regional precipitation, as well as ecological models and models of atmospheric transport.

Despite this complexity, there has been recent progress in quantifying the oxygen isotopic composition of CO_2 on a global basis for carbon cycle applications. Ciais et al. (1997a) combined a global biospheric model with global climate model simulations to estimate monthly global surface fluxes of ^{18}O in CO_2 from the terrestrial biosphere. Integrated with an atmospheric tracer transport model, this approach yielded results that agreed well with atmospheric observations, and led to the conclusion that the temporal and spatial patterns of $\delta^{18}O$ of CO_2 in the atmosphere were largely attributable to terrestrial ecosystem processes (Ciais et al. 1997b; Peylin et al. 1999). Cuntz et al. (2003a,b) expanded global simulations of ^{18}O fluxes with a well-integrated suite of model components that provided estimates on the diurnal time-scale. These advances will likely provide a basis for utilizing observations of $\delta^{18}O$ of atmospheric CO_2 to constrain carbon sources and sinks and distinguish between assimilation and respiration components of terrestrial fluxes.

Further research is needed on an interesting aspect of the atmospheric CO_2 record, which showed a decline in $\delta^{18}O$ of CO_2 of about 0.5‰ in the mid-1990s (Gillon and Yakir 2001; Ishizawa et al. 2002). Gillon and Yakir (2001) found variations in the activity of the enzyme of carbonic anhydrase, which catalyzes oxygen isotope exchange between CO_2 and water, suggesting that full equilibration is not reached in many species, particularly C_4 grasses. The authors concluded that large-scale conversions of forested ecosystems to pastures dominated by C_4 plants during the 1990s could have led to a global decline in $\delta^{18}O$ of atmospheric CO_2 of 0.02‰ yr^{-1}. Ishizawa et al. (2002) pointed out that such conversions can only partially explain the observed decline in $\delta^{18}O$ of CO_2, and that an additional mechanism could be an increase in both photosynthesis and respiration of the terrestrial biosphere. Advances in global modeling of $\delta^{18}O$ of CO_2 and additional observations of mechanisms of ecosystem fractionation and the isotopic composition of atmospheric CO_2 will likely lead to further interpretation and application of variations in $\delta^{13}C$ and $\delta^{18}O$ of CO_2 for understanding the role of the terrestrial biosphere in the carbon cycle.

4.4 Future Directions

Research on the application of stable isotopes in ecosystem and global carbon cycles was synthesized and coordinated by the Biosphere-Atmosphere Stable Isotope Network (BASIN) of GCTE beginning with an initial workshop in 1997. The BASIN network (http://basinisotopes.org) is carrying forward with additional coordination and synthesis of traditional aspects of data collection, such as the isotopic composition of ecosystem respiration, as well as other applications of stable isotopes in understanding the role of terrestrial ecosystems in biogeochemistry. Data are increasingly available on the isotopic composition of other gases present in the atmosphere in lower concentrations than CO_2, and therefore more difficult to measure. Stable isotopes of methane (Snover and Quay 2000), N_2O (Perez et al. 2000; Perez et al. 2001), and hydrogen (Rahn et al. 2002), for example, are providing additional information on the role of ecosystem physiology in land-atmosphere interactions. Also promising are new studies on the isotopic composition of non-methane hydrocarbons such as isoprene (Affek and Yakir 2003; Goldstein and Shaw 2003), and the highly abundant atmospheric components, water vapor (Moreira et al. 1997; Yepez et al. 2003) and diatomic oxygen (Angert et al. 2003; Stern et al. 2001), which are strongly influenced by terrestrial processes.

Monitoring of a wider variety of ecosystem types will also provide a broader information base for understanding the influence of the land surface on the atmosphere and scaling from ecosystems to regional and global scales. Data from tropical ecosystems, semi-arid systems, and human-dominated ecosystems have been sparsely represented in the BASIN network, but new datasets in these areas are increasingly available (Ometto et al. 2002; Pataki et al. 2003a; Yepez et al. 2003). New technologies also promise to increase spatial and temporal coverage of the isotopic composition of the atmosphere. Automated samplers for collecting atmospheric samples have recently been deployed in a number of ecosystems and have demonstrated the dynamic influence of ecosystems processes on the atmosphere and the utility of long-term monitoring (Lai et al. 2005; Lai et al. 2003; Schauer et al. 2003). These monitoring programs will be greatly extended by the application of optical measurements of rare isotopes with instruments such as Tunable Diode Lasers (TDL) that measure $^{13}CO_2$ concentrations in real time. TDL and similar technologies will increase analytical capabilities severalfold in terms of sample number, allowing large numbers of samples to be collected over short time-periods in a number of locations throughout plant canopies (Bowling et al. 2003b).

Results from atmospheric monitoring programs in ecosystems are being incorporated into a growing number of models of isotopic fractionation and its influence on the atmosphere (Baldocchi and Bowling 2003; Riley et al. 2003; Suits et al. 2005). These linkages have been largely indirect, that is, as isotopic measurements improved our basic understanding of fractionation and its physical and physiological basis, new theoretical and empirical relationships have been incorporated into ecosystem-atmosphere models. However, direct linkages between isotopic data-streams and ecological models are emerging as data assimilation or "data-model fusion" techniques are becoming common in ecological and biogeochemical applications. Similar to approaches already common in global scale studies of the carbon cycle, these quantitative methods optimize the solution for a given set of equations, e.g., a mass balance of ecosystem carbon sources and sinks, using available datasets as constraints. This method has been used to model ecosystems fluxes (Raupach 2001; Styles et al. 2002), estimate the mean residence time of carbon pools (Barrett 2002; Luo et al. 2003), and partition fluxes into their component parts (Luo et al. 2004). The increasing temporal and spatial coverage of isotopic measurements in various ecosystems will likely be directly incorporated into improved estimates of ecosystem-scale physiological and biogeochemical parameters with these methods.

Measurements of the isotopic composition of trace gases in the atmosphere and within and above terrestrial ecosystems have provided new insights into ecosystem functioning and the role of the terrestrial biosphere in the carbon cycle. These insights have included partitioning carbon cycle components such as terrestrial and oceanic carbon sinks, photosynthetic and respiratory components of NEE, and C_3 and C_4 contributions to ecosystem respiration. Improvements in both measurement techniques and ecosystem- to global-scale models are rapidly resulting in refined estimates of these aspects of the carbon cycle and the underlying mechanisms for their spatiotemporal variations. This is a rapidly evolving area of research, and one in which the GCTE core project made an important and lasting contribution through its support of the BASIN network.

Acknowledgments

We wish to thank all the participants of BASIN and its associated GCTE activities for their effort and support, and the U.S. National Science Foundation for Research Coordination in Biological Networks (RCN) funding of the BASIN activity.

In Memoriam

We gratefully acknowledge the many contributions of our co-author C. D. Keeling, who passed away before this book went to press. Without his pioneering efforts much of the work cited here would not have been possible.

References

Affek HP, Yakir D (2003) Natural abundance carbon isotope composition of isoprene reflects incomplete coupling between isoprene synthesis and photosynthetic carbon flow. Plant Physiol 131:1727–1736

Amundson R, Stern L, Baisen T, Wang Y (1998) The isotopic composition of soil and soil-respired CO_2. Geoderma 82:83–114

Andres RJ, Marland G, Boden T, Bischof S (2000) Carbon dioxide emissions from fossil fuel consumption and cement manufacture, 1751–1991, and an estimate of their isotopic composition and latitudinal distribution. In: Wigley TML, Schimel DS (eds) The carbon cycle. Cambridge University Press, Cambridge, pp 53–62

Angert A, Barkan E, Barnett B, Brugnoli E, Davidson EA, Fessenden JE, Maneepong S, Panapitukkul N, Randerson JT, Savage K, Yakir D, Luz B (2003) Contribution of soil respiration in tropical temperate and boreal forests to the ^{18}O enrichment of atmospheric O_2. Global Biogeochemical Cycles 17:Art. No. 1089

Baldocchi DD, Bowling DR (2003) Modelling the discrimination of $^{13}CO_2$ above and within a temperate broad-leaved forest canopy on hourly to seasonal time scales. Plant, Cell and Environment 26:231–244

Barrett D (2002) Steady state turnover time of carbon in the Australian terrestrial biosphere. Global Biogeochemical Cycles 16:1108

Battle M, Bender ML, Tans PP, White JWC, Ellis JT, Conway T, Francey RJ (2000) Global carbon sinks and their variability inferred from atmospheric O_2 and $\delta^{13}C$. Science:2467–2470

Bowling DR, Baldocchi DD, Monson RK (1999) Dynamics of isotopic exchange of carbon dioxide in a Tennessee deciduous forest. Global Biogeochemical Cycles 13:903–922

Bowling DR, Tans PP, Monson RK (2001) Partitioning net ecosystem carbon exchange with isotopic fluxes of CO_2. Global Change Biology 7:127–145

Bowling DR, McDowell NG, Bond BJ, Law BE, Ehleringer JR (2002) ^{13}C content of ecosystem respiration is linked to precipitation and vapor pressure deficit. Oecologia 131:113–124

Bowling DR, Pataki DE, Ehleringer JR (2003a) Critical evaluation of micrometeorological methods for measuring ecosystem-atmosphere isotopic exchange of CO_2. Agricult Forest Meteorol 116:159–179

Bowling DR, Sargent SD, Tanner BD, Ehleringer JR (2003b) Tunable diode laser absorption spectroscopy for stable isotope studies of ecosystem-atmosphere CO_2 exchange. Agricult Forest Meteorol 118:1–19

Buchmann N, Cuehl J-M, Barigah TS, Ehleringer JR (1997a) Interseasonal comparison of CO_2 concentrations, isotopic composition, and carbon dynamics in an Amazonian rainforest (French Guiana). Oecologia 110:120–131

Buchmann N, Kao W-Y, Ehleringer J (1997b) Influence of stand structure on carbon-13 of vegetation, soils, and canopy air with-in deciduous and evergreen forests in Utah, United States. Oecologia 110:109–119

Ciais P, Tans P, Trolier M, White J, Francey R (1995) A large northern hemisphere terrestrial CO_2 sink indicated by the $^{13}C/^{12}C$ ratio of atmospheric CO_2. Science 269:10981102

Ciais P, Denning AS, Tans PP, Berry JA, Randall DA, Collatz GJ, Sellers PJ, White JWC, Trolier M, Meijer HAJ, Francey RJ, Monfray P, Heimann M (1997a) A three-dimensional synthesis study of $\delta^{18}O$ in atmospheric CO_2. 1. Surface fluxes. J Geophys Res 102:5857–5872

Ciais P, Tans PP, Denning AS, Francey RJ, Trolier M, Meijer HAJ, White JWC, Berry JA, Randall DA, Collatz GJ, Sellers PJ, Monfray P, Heimann M (1997b) A three-dimensional synthesis study of $\delta^{18}O$ in atmospheric CO_2. 2. Simulations with TM2 transport model. J Geophys Res 102:5873–5883

Cuntz M, Ciais P, Hoffman G, Allison CE, Francey RJ, Knorr W, Tans PP, White JWC, Levin I (2003a) A comprehensive global three-dimensional model of $\delta^{18}O$ in atmospheric CO_2: 2. Mapping the atmospheric signal. J Geophys Res 108:doi:10.1029/2002JD003154

Cuntz M, Ciais P, Hoffman G, Knorr W (2003b) A comprehensive global three-dimensional model of $\delta^{18}O$ in atmospheric CO_2: 1. Validation of surface processes. J Geophys Res 108:doi: 10.1029/2002JD003153

Dai A, Fung IY, Del Genio AD (1997) Surface observed global land precipitation variations during 1900–1988. Journal of Climate 10:2943–2962

Dawson TE, Mambelli S, Planboeck AH, Templer PH, Tu KP (2002) Stable isotopes in plant ecology. Annual Review of Ecology and Systematics 33:507–559

Ehleringer JR, Cook CS (1998) Carbon and oxygen isotope ratios of ecosystem respiration along an Oregon conifer transect: preliminary observations based on small-flask samping. Tree Physiology 18:513–519

Ekblad A, Högberg P (2001) Natural abundance of ^{13}C in CO_2 respired from forest soils reveals speed of link between tree photosynthesis and root respiration. Oecologia 127:305–308

Ethier GJ, Livingston NJ (2004) On the need to incorporate sensitivity to CO_2 transfer conductance into the Farquhar-von Caemmerer-Berry leaf photosynthesis model. Plant, Cell and Environment 27:137–153

Farquhar GD, Lloyd J (1993) Carbon and oxygen isotope effects in the exchange of carbon dioxide between plants and the atmosphere. In: Ehleringer JR, Hall AE, Farquhar GD (eds) Stable isotopes and plant carbon-water relations. Academic Press, New York, pp 47–70

Farquhar GD, O'Leary MH, Berry JA (1982) On the relationship between carbon isotope discrimination and the intercellular carbon dioxide concentration in leaves. Aust J Plant Physiol 9:121–137

Farquhar GD, Ehleringer JR, Hubick KT (1989) Carbon isotope discrimination and photosynthesis. Annual Review of Plant Physiology and Plant Molecular Biology 40:503–537

Farquhar GD, Lloyd J, Taylor JA, Flanagan LB, Syvertsen JP, Hubick KT, Wong SC, Ehleringer JR (1993) Vegetation effects on the isotope composition of oxygen in atmospheric CO_2. Nature 363:439–443

Fessenden JE, Ehleringer JR (2002) Age-dependent variations in $\delta^{13}C$ of ecosystem respiration across a coniferous forest chronosequence in the Pacific Northwest. Tree Physiology 22:159–167

Fessenden JE, Ehleringer JR (2003) Temporal variation in $\delta^{13}C$ of ecosystem respiration in the Pacific Northwest: links to moisture stress. Oecologia 136:129–139

Flanagan LB, Brooks J, Varney GT, Berry SC, Ehleringer JR (1996) Carbon isotope discrimination during photosynthesis and the isotope ratio of respired CO_2 in boreal forest ecosystems. Global Biogeochemical Cycles 10:629–640

Flanagan LB, Brooks JR, Varney GT, Ehleringer JR (1997) Discrimination against $C^{18}O^{16}O$ during photosynthesis and the oxygen isotope ratio of respired CO_2 in boreal forest ecosystems. Global Biogeochemical Cycles 11:83–98

Francey RJ, Tans PP (1987) Latitudinal variation in oxygen-18 of atmospheric CO_2. Nature 327:495–497

Francey RJ, Tans PP, Allison CE, Enting IG, White JWC, Troller M (1995) Changes in oceanic and terrestrial carbon uptake since 1982. Nature 373:326–330

Fritz P, Fontes JC (1980) Handbook of environmental isotope geochemistry, Vol I, The terrestrial environment. Elsevier, Amsterdam

Fung I, Field CB, Berry JA, Thompson MV, Randerson JT, Malmström CM, Vitousek PM, Collatz GJ, Sellers PJ, Randall DA, Denning AS, Badeck F, John J (1997) Carbon-13 exchanges between the atmosphere and biosphere. Global Biogeochemical Cycles 11:507–533

Gillon J, Yakir D (2001) Influence of carbonic anhydrase activity in terrestrial vegetation on the ^{18}O content of atmospheric CO_2. Science 291:2584–2587

Goldstein AH, Shaw SL (2003) Isotopes of volatile organic compounds: An emerging approach for studying atmospheric budgets and chemistry. Chemical Reviews 103:5025–5048

Harwood KG, Gillon JS, Roberts A, Griffiths H (1999) Determinants of isotopic coupling of CO_2 and water vapour within a *Quercus petraea* forest canopy. Oecolgia 119:109–119

Hesterberg R, Siegenthaler U (1991) Production and stable isotopic composition of CO_2 in a soil near Bern, Switzerland. Tellus 43B:197–205

Inoue H, Sugimura Y (1985) Carbon isotopic fractionation during the CO_2 exchange process between air and sea water under equilibrium and kinetic conditions. Geochemica et Cosmochimica Acta 49:2453–2460

IPCC (2001) Climate Change 2001: The scientific basis. Contrilbution of Working Group 1 to the Third Assessment Report of the Intergovernmental Panel on Climatic Change. Cambridge University Press, Cambridge

Ishizawa M, Nakazawa T, Higuchi K (2002) A multi-box model study of the role of the biospheric metabolism in the recent decline of $\delta^{18}O$ in atmospheric CO_2. Tellus 54B:307–324

Keeling CD (1958) The concentration and isotopic abundances of atmospheric carbon dioxide in rural areas. Geochim Cosmochim Acta 13:322–334

Keeling CD (1961) The concentration and isotopic abundance of carbon dioxide in rural and marine air. Geochim Cosmochim Acta 24:277–298

Keeling CD, Whorf TP, Wahlen M, van der Plicht J (1995) Interannual extremes in the rate of rise of atmospheric carbon dioxide since 1980. Nature 375:666–670

Keeling CD, Piper SC, Bacastow RB, Wahlen M, Whorf TP, Heimann M, Meijer HAJ (2005) Atmospheric CO_2 and $^{13}CO_2$ exchange with the terrestrial biosphere and oceans from 1978 to 2000: Observations and carbon cycle implications. In: Ehleringer JR, Cerling TE, Dearing MD (eds) A history of atmospheric CO_2 and its implications for plants, animals, and ecosystems. Springer-Verlag, New York, pp 83–113

Kendall C, McDonnell JJ (eds) (1998) Isotope tracers in catchment hydrology. Elsevier, Amsterdam

Lai C-T, Schauer AJ, Owensby CE, Ham JM, Ehleringer JR (2003) Isotopic air sampling in a tallgrass prairie to partition net ecosystem CO_2 exchange. Journal of Geophysical Research 108:doi:10.1029/2002JD003369

Lai C-T, Ehleringer JR, Schauer AJ, Tans PP, Hollinger DY, Paw UKT, Munger JW, Wofsy SC (2005) Canopy-scale $\delta^{13}C$ of photosynthetic and respiratory CO_2 fluxes: observations in forest biomes across the United States. Global Change Biology 11:633–643

Lloyd J, Farquhar GD (1994) ^{13}C discrimination during CO_2 assimilation by the terrestrial biosphere. Oecologia 99:201–215

Lloyd J, Kruijt B, Hollinger DY, Grace J, Francey RJ, Wong SC, Kelliher FM, Miranda AC, Farquhar GD, Gash JHC, Vygodskaya NN, Wright IR, Miranda HS, Schulze E-D (1996) Vegetation effects on the isotopic composition of atmopsheric CO_2 at local and regional scales: theoretical aspects and a comparison between a rain forest in Amazonia and a boreal forest in Siberia. Australian Journal of Plant Physiology 23:371–399

Luo YQ, White LW, Canadell JG, DeLucia EH, Ellsworth DS, Finzi AC, Lichter J, Schlesinger WH (2003) Sustainability of terrestrial carbon sequestration: A case study in Duke Forest with inversion approach. Global Biogeochemical Cycles 17:Art. No. 1021

Miller JB, Tans PP (2003) Calculating isotopic fractionation from atmospheric measurements at various scales. Tellus 55B: 207–214

Moreira M, Sternberg L, Martinelli L, Reynaldo L, Barbosa E, Bonates L, Nepstad D (1997) Contribution of transpiration to forest ambient vapour based on isotopic measurements. Global Change Biology 3:439–450

Ogée J, Peylin P, Ciais P, Bariac T, Brunet Y, Berbigier P, Roche C, Richard P, Bardoux G, Bonnefond J-M (2003) Partitioning net ecosystem carbon exchange into net assimilation and respiration using $\delta^{13}CO_2$ measurements: A cost-effective sampling strategy. Global Biogeochemical Cycles 17:doi: 10.1029/2002GB001995

Ometto PHB, Flanagan LB, Martinelli LA, Moreira MZ, Higuchi N, Ehleringer JR (2002) Carbon isotope discrimination in forest and pasture ecosystems of the Amazon Basin, Brazil. Global Biogeochemical Cycles 16:1109

Pataki DE, Bowling DR, Ehleringer JR (2003a) The seasonal cycle of carbon dioxide and its isotopic composition in an urban atmosphere: anthropogenic and biogenic effects. Journal of Geophysical Research 108:4735

Pataki DE, Ehleringer JR, Flanagan LB, Yakir D, Bowling D, Still CJ, Buchmann N, Kaplan JO, Berry JA (2003b) The application and interpretation of Keeling plots in terrestrial carbon cycle research. Global Biogeochemical Cycles 17

Perez T, Trumbore SE, Tyler SC, Davidson EA, Keller M, de Camargo PB (2000) Isotopic variability of N_2O emissions from tropical forest soils. Global Biogeochemical Cycles 14:525–535

Perez T, Trumbore SE, Tyler SC, Matson PA, Ortiz-Monasterio I, Rahn T, Griffith DWT (2001) Identifying the agricultural imprint on the global N_2O budget using stable isotopes. Journal of Geophysical Research 106A:9869–9878

Peylin P, Ciais P, Denning AS, Tans PP, Berry JA, White JWC (1999) A 3-dimensional study of $\delta^{18}O$ in atmospheric CO_2: contribution of different land ecosystems. Tellus 51B:642–667

Quay PD, King SL, Wilbur D, Wofsy S, Richey J (1989) $^{13}C/^{12}C$ of atmospheric CO_2 in the Amazon Basin; forest and river sources. Journal of Geophysical Research 94:18,327–318,336

Rahn T, Eiler JM, Kitchen N, Fessenden JE, Randerson JT (2002) Concentration and δD of molecular hydrogen in boreal forests: Ecosystem-scale systematics of atmospheric H_2. Geophysical Research Letters 29:Art. No. 1888

Randerson JT, Collatz GJ, Fessenden JE, Munoz AD, Still CJ, Berry JA, Fung iY, Suits N, Denning AS (2002) A possible global covariance between terrestrial gross primary production and ^{13}C discrimination: Consequences for the atmospheric ^{13}C budget and its response to ENSO. Global Biogeochem Cycles 16:doi:10.1029/2001GB001845

Raupach MR (2001) Inferring biogeochemical sources and sinks from atmospheric concentrations: General considerations and applications in vegetation canopies. In: Schulze E-D, et al. (eds) Global biogeochemical cycles in the climate system. Academic Press, San Diego, pp 41–60

Riley WM, Still CJ, Helliker BR, Ribas-Carbo M, Berry JA (2003) ^{18}O composition of CO_2 and H_2O ecosystem pools and fluxes in a tallgrass prairie: Simulations and comparisons to measurements. Global Change Biology 9:1567–1581

Ropelewski CF, Halpert MS (1987) Global and regional scale precipitation patterns associated with the El Niño/Southern Oscillation. Monthly Weather Review 115:1606–1626

Scartazza A, Mata C, Matteucci G, Yakir D, Moscatello S, Brugnoli E (2004) Comparisons of $\delta^{13}C$ of photosynthetic products and ecosystem respiratory CO_2 and their responses to seasonal climate variability. Oecologia 140:340–351

Schauer AJ, Lai C, Bowling DR, Ehleringer JR (2003) An automated sampler for collection of atmospheric trace gas samples for stable isotope analyses. Agricultural and Forest Meteorology 118:113–124

Snover AK, Quay PD (2000) Hydrogen and carbon kinetic isotope effects during soil uptake of atmospheric methane. Global Biogeochemical Cycles 14:25–39

Stern LA, Amundson R, Baisden WT (2001) Influence of soils on oxygen isotope ratio of atmospheric CO_2. Global Biogeochemical Cycles 15:753–759

Sternberg LL, Mulkey SS, Wright JS (1989) Ecological interpretation of leaf carbon isotope ratios: influence of respired carbon dioxide. Ecology 70:1317–1324

Still CJ, Berry JA, Collatz GJ, DeFries RS (2003a) Global distribution of C_3 and C_4 vegetation: Carbon cycle implications. Global Biogeochemical Cycles 17:doi: 10.1020/2001GB001807

Still CJ, Berry JA, Ribas-Carbo M, Helliker BR (2003b) The contribution of C-3 and C-4 plants to the carbon cycle of a tallgrass prairie: an isotope approach. Oecologia 136:347–359

Styles JM, Raupach MR, Farquhar GD, Kolle O, Lawton KA, Brand WA, Werner RA, Jordan A, Schulze E-D, Shibistova O, Lloyd J (2002) Soil and canopy CO_2, $^{13}CO_2$, H_2O and sensible heat flux partitions in a forest canopy inferred from concentration measurements. Tellus 54B:655–676

Suits NS, Denning AS, Berry JA, Still CJ, Kaduk J, Miller JB (2005) Simulation of carbon isotope discrimination of the terrestrial biosphere. Global Biogeochem Cycles 19:doi:10.1029/2003GB002141

Tans PP, Berry JA, Keeling RF (1993) Oceanic $^{13}C/^{12}C$ observations: A new window on Ocean CO_2 uptake. Global Biogeochemical Cycles 7:353–368

Tu KP, Brooks PD, Dawson TE (2001) Using septum-capped vials with continous-flow isotope ratio mass spectrometric analysis of atmospheric CO_2 for Keeling plot applications. Rapid Communications in Mass Spectrometry 15:952–956

Van der Werf GR, Randerson JT, Collatz GJ, Giglio L, Kasibhatla PS, Arellano AF, Olsen SC, Kasischke ES (2004) Continental-scale partitioning of fire emissions during the 1997 to 2001 El Niño/La Niño period. Science 303:73–76

Wanninkhof R (1985) Kinetic fractionation of the carbon isotopes ^{13}C and ^{12}C during transfer of CO_2 from air to seawater. Tellus 37B

Yakir D, Wang XF (1996) Fluxes of CO_2 and water between terrestrial vegetation and the atmosphere estimated from isotope measurements. Nature 380:515–517

Yepez EA, Williams DG, Scott RL (2003) Partitioning overstory and understory evapotranspiration in a semiarid savanna woodland from the isotopic composition of water vapor. Agricultural and Forest Meteorology 119:53–68

Zhang J, Quay PD, Wilbur DO (1995) Carbon-isotope fractionation during gas-water exchange and dissolution of CO_2. Geochemica et Cosmochimica Acta 59:107–114

Chapter 5

Effects of Urban Land-Use Change on Biogeochemical Cycles

Richard V. Pouyat · Diane E. Pataki · Kenneth T. Belt · Peter M. Groffman · John Hom · Lawrence E. Band

5.1 Introduction

Urban land-use change, the conversion of agricultural and natural ecosystems to human settlements, has become an important component of global change. Virtually all of the projected increase in the world's population is expected to occur in cities so that by the year 2007 more than half of the global population is expected to live in urban areas (United Nations 2004). Yet, urban settlements and surrounding areas are complex ecological systems that have only recently been studied from a rigorous ecological perspective (Pickett et al. 2001). Clearly, urban areas represent ecosystems with modified biogeochemical cycles such that fluxes and pools of matter, energy, and organisms differ greatly from the previous ecosystem (Collins et al. 2000). Our knowledge of the magnitude and extent of these biogeochemical changes at local, regional, and global scales is a major area of uncertainty in our understanding of global change.

In the ecological sciences, the ecosystem approach has been used successfully in studying the effects of disturbance and exogenous inputs of nutrients and pollutants on biogeochemical cycles (Likens 1992). To study biogeochemical cycles in urban areas, Grimm et al. (2003) recommended the ecosystem approach where: (1) urban ecosystems are viewed as spatially homogenous, i.e., "well-mixed reactors," or (2) urban ecosystems are considered a heterogeneous assemblage of parts or "patches." Indeed, cities have been viewed as well-mixed reactors in answering questions about how they affect biogeochemical cycles at regional and global scales (Fig. 5.1). In so doing, mass balances have been constructed for entire cities to assess their "ecological footprint," that is, the amount of land resources acquired from other ecosystems needed to sustain the urban population (Rees 2003). Moreover, the mass balance approach allows investigations into the movement of nutrients and pollutants across ecosystem boundaries and the cycling of these materials within the system itself. For example, Baker et al. (2001) constructed a nitrogen balance for the Phoenix metropolitan area and found that inputs (human and

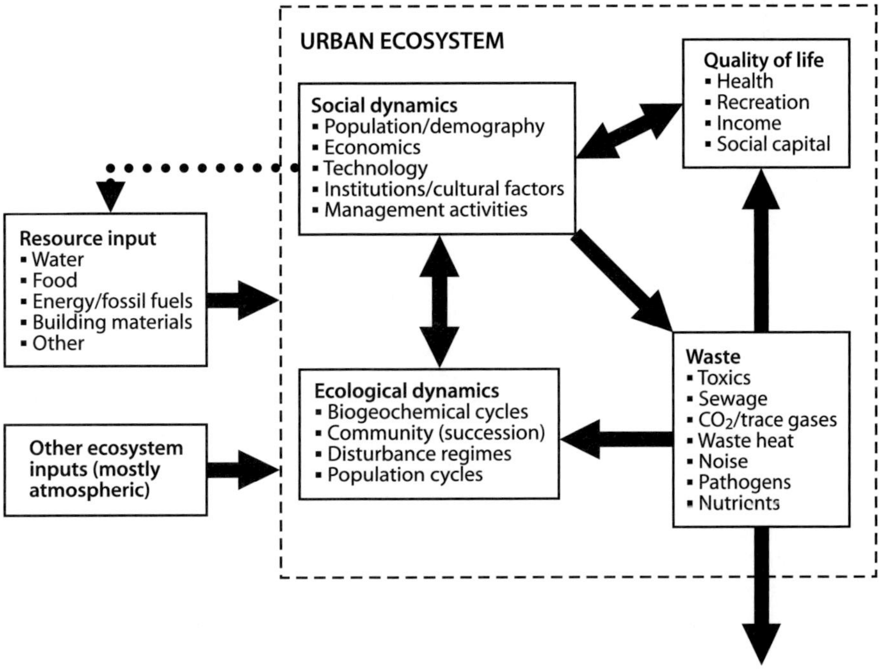

Fig. 5.1. Conceptual diagram of an urban ecosystem showing interrelationships between human and biophysical components. *Dashed line* indicates ecosystem boundary if city considered a "well-mixed reactor" such that inputs and outputs into and out of the ecosystem can be measured. Model includes feedback loop from the human social component of the system to resource acquisition, i.e., demand for resources dependent on population density, quality of life issues, and the ability of ecosystem to assimilate waste

natural) to the urban ecosystem were an order of magnitude higher than inputs to the surrounding desert areas. Furthermore, human hydrologic modifications within the urban area promoted the accumulation of N within the ecosystem.

Urban ecosystems maintain a high state of resiliency due to the socioeconomic influence of their inhabitants. This enables them to continually obtain resources from other areas. However, this dependence disconnects urban populations (and their social systems) from locally occurring ecological constraints (Luck et al. 2001). As a result, the ecological impact of cities can extend to regional and global scales without negative feedbacks directly affecting the system itself (Rees 2003). Moreover, urban areas can affect adjacent non-urban ecosystems through the production and transport of waste products and by altering regional weather patterns (e.g., Cerveny and Balling 1998; Shepherd and Burian 2003; Gregg et al. 2003).

An alternative to the well-mixed reactor approach is to delineate and recognize the parts or patches of urban ecosystems (Collins et al. 2000; Pickett et al. 2001). With this approach, ecosystems and the patches that comprise them are treated as well-mixed reactors at some scales and heterogeneous systems at others (Grimm et al. 2003). In this way input-output budgets can be developed not only for entire cities but also for individual functional parts of the urban ecosystem. This allows investigations of the effects of urban development on biogeochemical processes of a patch of lawn or remnant forest, which themselves may have cycling and fluxes of resources that interact with neighboring patches. Pouyat et al. (in press) suggested that researchers should exploit the heterogeneity of urban landscapes using the diverse array of patches as surrogates for field manipulations, or "natural experiments."

In this chapter we address the question: what is the net effect of urban land-use change on biogeochemical cycles at local and metropolitan area scales? Our use of the term *urban* is inclusive of landscapes having >2500 persons km^{-2} or densities at which human populations cannot be supported by local agricultural production (Ellis 2004). We use the phrase *urban land-use change* to describe coarse scale conversion of natural or agricultural ecosystems to urban land uses, as well as fine scale or local changes in land use that may occur in post-urban development. We will focus on the local scale because it is the most relevant one in making mechanistic assessments of urban biogeochemical processes. Our discussion relies on data available in the literature (mostly from temperate systems of North America) and from current research at the Long Term Ecological Research (LTER) site in Baltimore, MD USA (Baltimore Ecosystem Study). We devote most of our discussion to carbon (C) and nitrogen (N) pools and fluxes, both of which are important at multiple scales (Vitousek et al. 1997; Schlesinger and Andrews 2000) and may be indicators of urban stress and disturbance (Carreiro et al. in press). While heavy metals and organic compounds are important constituents in urban environments, their effects on plants, animals, and human health are beyond the scope of this chapter. Nor do we discuss the far-reaching effects of natural resource importation by cities, i.e., the aforementioned urban ecological footprints. The ability of urban inhabitants to exploit land resources at great distances has important anthropogenic impact on biogeochemical cycles at regional and global scales. But we believe that these impacts should be weighed against the effect of alternative settlement patterns on a per-capita basis – a comparison that requires extensive discussion. Excellent discussions of urban ecological footprints are included in Folke et al. (1997) and Rees (2003).

5.2 Urban Land-Use Change

In developed nations during the last half century, urban growth has occurred in a dispersed pattern relative to older and more densely populated cities that developed before and during the Industrial Revolution. Between 1980 and 2000 alone, land devoted to urban uses grew by more than 34% in the United States (USDA Natural Resource Conservation Service 2001). By contrast, the population grew by only 24% during the same period (USDC Census Bureau 2001). This dispersed growth pattern, or suburbanization, has occurred at the expense of agricultural and forested land (Imhoff et al. 1997). A recent analysis based on satellite imagery suggests that the land area covered by impervious surfaces (such as pavement and buildings) is larger than the surface area covered by herbaceous wetlands in the conterminous United States, or roughly the size of Ohio USA (Elvidge et al. 2004).

Although biogeochemical effects of agricultural land conversions and their recovery after abandonment have been relatively well studied, conversions to urban land uses have received little attention (Pouyat et al. 2002). We know that agricultural uses lead to drastic changes in biogeochemical cycling (Matson et al. 1997). However, once these lands are abandoned, many of the biogeochemical functions of the preagricultural ecosystem are recovered. For example, forest regrowth on agriculturally abandoned land has resulted in a gradual recovery of above- and below-ground C pools in the eastern and central United States (Houghton et al. 1999; Caspersen et al. 2000; Post and Kwon 2000). In fact, what are now considered "pristine" forest habitats at one time were some of the most densely populated regions of the world, for example, the Central American Mayan forests (Turner et al. 2003) and the Amazon River Basin (Heckenberger et al. 2003).

Converting agriculture, forest, and grasslands to urban and suburban land uses entails a complex array of land and ecosystem alterations. Dense human inhabi-

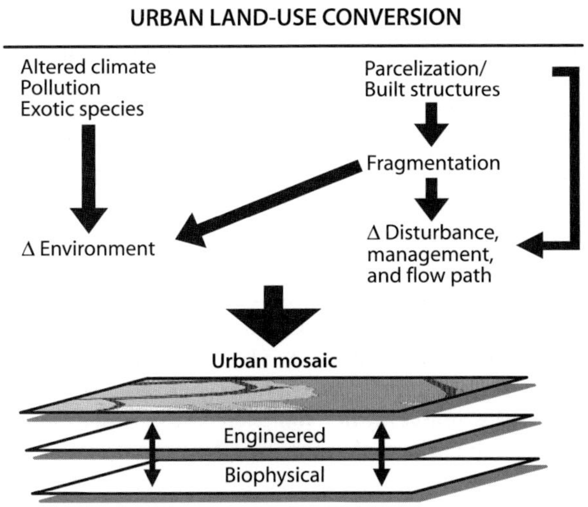

Fig. 5.2. Conceptual diagram of the effect of urban land-use conversions on native ecosystems. As landscapes are urbanized, natural habitats are increasingly fragmented as parcels of land ownership become smaller. During this process humans introduce novel disturbance and management regimes and flow paths to the landscape (*arrows on right*). Concurrently there is a change in the environmental conditions in which the ecosystem functions (*arrow on left*). See text for detailed explanation (modified from Pouyat et al. 2003)

tation along with urban land-use change necessitates the construction of various built structures, e.g., roads, buildings, civil infrastructure, as well as the introduction of human activities. The spatial pattern of these human features and activities are largely the result of "parcelization," or the subdivision of land by property boundaries, as landscapes are developed for human settlement. Parcelization typically leads to habitat loss and fragmentation as landscapes are continually subdivided into ever smaller patches, each with an individual land owner. Overlaid on the patch mosaic are various environmental factors that typically are associated with urban areas. The result is a patchwork or landscape mosaic of remnant ecosystems, managed parcels, and a built environment that overlay the preexisting ecosystem's biophysical features (Fig. 5.2).

5.3 Urban Environmental Factors

A complex of environmental factors altered by urbanization potentially affects biogeochemical cycling: atmospheric, soil, and water pollution; CO_2 emissions; micro- and meso-climates; and introductions of exotic plant and animal species (Pouyat et al. 1995). In some metropolitan areas, the net effect of such factors is analogous to predictions of global environmental change (Carreiro and Tripler 2005), e.g., increased temperatures and rising atmospheric concentrations of CO_2 (Ziska et al. 2004). In the following sections we discuss environmental factors that could affect biogeochemical cycles in urban and suburban landscapes.

5.3.1 Climate and Atmospheric Composition

Urban environments are characterized by localized increases in temperature known as the "heat island" effect (Oke 1990). Urban heat islands occur when vegetation cover is replaced by built structures. This reduces evapotransporation rates while the introduction of building materials increases the absorption and storage of solar energy that later is released as sensible heat (Oke 1990). The combined effect of these changes is to increase minimum (and to a lesser degree maximum) temperatures such that the difference between rural and urban temperatures typically is greatest several hours after sunset. This difference has been as much as 3 to 5 °C (Brazel et al. 2000). While cities embedded within desert ecosystems have warmer temperatures during the night, they typically have lower temperatures during the day, i.e., the "oasis effect" (Brazel et al. 2000).

How higher urban temperatures affect plant growth depends partly on the interactive effect of other environmental factors that may enhance plant growth such as N deposition and concentrations of atmospheric CO_2, the latter of which can reach 520 ppm in urban areas (Idso et al. 2002; Körner and Klopatek 2002; Pataki et al. 2003), or nearly double the preindustrial level of 280 ppm. Still other urban environmental factors may negatively affect plant growth such as tropospheric ozone (Chappelka and Samuelson 1998; Skärby and Ro-Poulsen 1998; Gregg et al. 2003).

The interactive effect of these multiple atmospheric and climate factors on plant productivity is a major uncertainty in global change research due to the difficulty of implementing controlled, factorial experiments in most ecosystems (Norby and Luo 2004). Hence, ecological studies in urban ecosystems have been proposed as a means of improving our understanding of multiple effects of global change (Pouyat et al. 1995; Grimm et al. 2000; Carreiro and Tripler 2005). Although this work is at an early stage, some results have been reported. Ziska et al. (2004) found plant productivity increases of 60 to 115% along an urban-rural gradient in Maryland, USA that corresponded to a 21% increase in CO_2 and a 3.3 °C temperature increase at the urban end of the gradient. Gregg et al. (2003) found that the growth of cottonwood seedlings was highest in the urban portion of an urban-rural gradient in the New York City area because of high ozone concentrations in rural areas downwind of urban pollution. Outside of the Los Angeles area, a large body of research in the San Bernardino Mountains Gradient Study has shown that foliar injury and chlorotic mottling is common at polluted sites due to ozone exposure, concurrent with decreased root biomass and increased aboveground biomass due to high N deposition (Arbaugh et al. 2003; Grulke et al. 1998; Grulke and Balduman 1999).

Additional work is necessary to determine whether these trends are robust in a large number of urban ecosystems.

Differences in air temperature also have been noted for soil temperatures. Mean annual temperatures in highly disturbed soils (0–10 cm depth) on a playground in New York's Central Park were more than 3 °C warmer than in soils in an adjacent wooded area (Mount et al. 1999). In comparing soil heat flux of several urban surfaces, Montague and Kjelgren (2004) found that heat fluxes were greatest beneath asphalt and concrete and least under pine bark mulch. For woodland soils in the New York City metropolitan area, surface temperatures (2 cm depth) differed by as much as 3 °C between urban and rural forest patches (Pouyat et al. 2003). Assuming that these soil temperatures are representative of those in urban areas, the difference between highly disturbed soils and natural forested soils in a metropolitan region can be as high as 6 °C. This difference in temperature can have significant effects on microbial activity and N availability in soil (Carreiro et al. in press).

Higher air temperatures and heat loading by impervious surfaces in urban areas also can affect the temperature of streams. First- and second-order streams in urban areas may have less shade than their rural counterparts and receive considerable runoff from impervious surfaces, which typically store a great amount of heat energy from solar radiation. In headwater streams, the major effect of urban-heat loading occurs early in the storm event as the first flush of runoff reaches the stream. The amount of temperature change is related to the percentage of impervious area in the watershed. For example, stream-temperature monitoring in the Baltimore Ecosystem Study showed that storm events in headwater streams resulted in a rapid increase in temperature. The increase was related directly to the housing density and proportion of impervious cover in the watershed (Fig. 5.3a). Similarly, in a study of headwater streams in Wisconsin and Minnesota USA, the percentage of impervious area in the watershed had a positive, linear relationship with water temperature during dry weather flows (Wang et al. 2003). Stream invertebrates and leaf breakdown processes are highly sensitive to elevated stream temperatures (Wang and Kanehl 2003; Webster and Benfield 1986).

Besides temperature, soil-moisture regimes can be modified by urban environments. Disturbed soils in urban areas typically have limited infiltration capacities due to hydrophobic surfaces, crust formation, and soil compaction (Craul 1992; Pouyat and Effland 1999). Compounding slow infiltration rate is the potential of impervious surfaces to drain concentrated flow into pervious areas. This can prevent water infiltration even during small storm events. Soil-moisture levels also can be reduced in urban areas due to higher air temperatures, which generate higher evaporative and transpirative demand on soil water resources. In contrast to factors that reduce soil moisture levels, soils in urban areas often are irrigated and have abrupt textural and structural interfaces that can restrict drainage resulting in higher soil-moisture content (Craul 1992). Moreover, below-ground infrastructures such as pressurized potable water distribution systems can leak water into adjacent soils by as much as 20 to 30% (Law et al. in press).

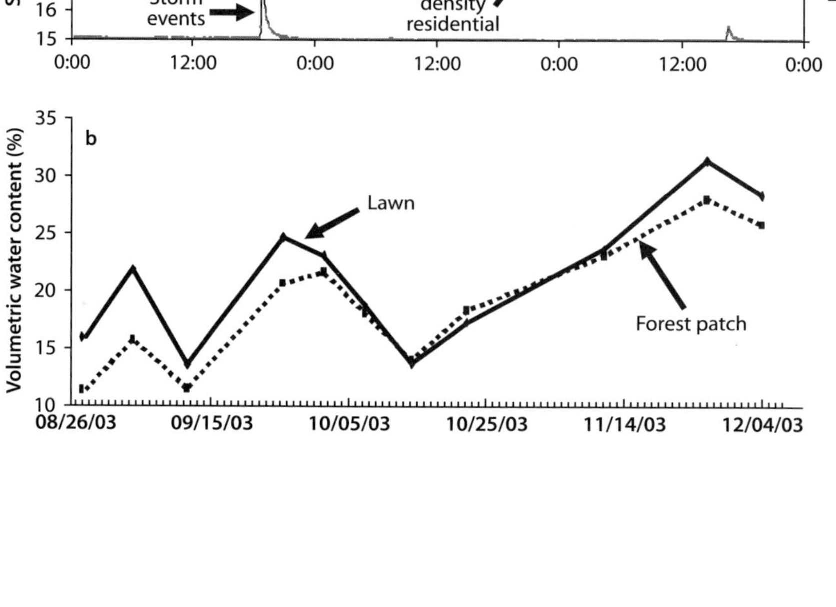

Fig. 5.3.
a Stream temperatures, May 25 to 27, 2004, for high density residential and lower density residential small headwater catchments. Stage at one of watersheds (*right axis legend*) shows two storm events (0.63 and 0.15 inches of rain). **b** Soil moisture (% by volume) of soils sampled at weekly intervals (depth of 0 to 10 cm) at Cub Hill eddy flux tower site of Baltimore Ecosystem Study. In forest patch, measurements taken along three transects perpendicular to slope ($n = 30$). Lawn measurements taken in two parcels along 30-m transects ($n = 24$)

Currently, we lack long-term continuous data with which to make comprehensive comparisons of soil moisture between urban and rural forest patches or between urban forest soils and disturbed soils. In the Baltimore Ecosystem Study, synoptic measurements of soil moisture (0–5 cm depth) in a medium-density residential area and adjacent deciduous woodland indicate that differences in soil moisture occur between land-use and cover types in urban landscapes. Residential lawns within a 150 m radius of the forest patch showed that un-irrigated turf soils had higher moisture levels than forest soils during the summer. There were no differences between the two patch types after leaf drop (end of October) (Fig. 5.3b). Presumably, differences in the summer were due to higher transpiration rates of the broad-leaved trees. In late fall and early winter, the moisture level in turf soil increased more rapidly than in forest soil, possibly reflecting concentrated runoff from roof gutters or snow melt (Fig. 5.3b). How these differences affect C and N cycling is being studied.

5.3.2 Atmospheric and Soil Pollution

Urban environments usually have higher concentrations and depositional fluxes of atmospheric chemicals than rural environments (Gatz 1991). Most urban atmospheric pollutants originate from the combustion of fossil fuels and industrial emissions. These include nitrogen oxides, sulfur oxides, heavy metals, and various organic chemicals. The pollutants are emitted locally into a restricted geographic area, particularly relative to the area from which the resources were derived, resulting in high atmospheric concentrations (Wong et al. 2004). High pollutant concentrations in the atmosphere combined with numerous temperature inversions increase the depositional rates of chemicals into urban areas (Seinfeld 1989). Since the capacity of ecosystems to assimilate atmospheric chemicals is correlated with the amount of leaf area of plants and soil biological activity, highly altered urban ecosystems have greatly diminished capacities to assimilate chemicals, especially when inputs are high.

Not surprisingly, studies of forest ecosystems along urban-rural gradients suggest that urban forest remnants receive relatively high amounts of heavy metals, organic compounds, N, and sulfur (S) in wet and dry atmospheric deposition (Pouyat and McDonnell 1991; Lovett et al. 2000; Wong et al. 2004). Lovett et al. (2000) quantified atmospheric N inputs over two growing seasons in oak forest stands along an urban-rural gradient in the New York City metropolitan area. They found that N entering urban forests in throughfall was 50 to 100% greater than the N flux into rural and suburban forests. These inputs fell off in the suburban stands 45 km from the city (New York's Central Park), which Lovett et al. suggested was due to the reaction of acidic anions with larger alkaline dust particles (Ca^{2+} and Mg^{2+}) of 2 μm or less in diameter that precipitated closer to the city. They hypothesized these particles (urban dust) originated from construction and demolition activity within the city. Similar results were found for the city of Louisville, KY USA and the San Bernardino Mountains in the Los Angeles metropolitan area where both N and base cation deposition rates into urban forest patches have been found to be higher than in rural forest patches (Bytnerowicz et al. 1999; Carreiro et al. in press; Fenn and Bytnerowicz 1993).

Evidence for a similar depositional pattern was found for heavy metals along the New York City transect. Pouyat and McDonnell (1991) found two to threefold increase in contents of lead (Pb), copper (Cu), and nickel (Ni) in urban than in suburban and rural forest soils. A similar pattern but with greater differences was found by Inman and Parker (1978) in the Chicago, IL USA, metropolitan area, where levels of heavy metals were more than five times higher in urban than in rural forest patches. Other urban-rural gradient studies have shown the same pattern though for smaller cities differences between urban and rural stands were narrower (Carreiro et al. in press; Pavao-Zuckerman and Coleman in press). Finally, Wong et al. (2004) found a large gradient of Polycyclic Aromatic Hydrocarbons (PAH) concentrations in forest soils in the Toronto (Canada) metropolitan area, with concentrations decreasing with distance from the urban center to surrounding rural areas by a factor of at least 60.

How these pollutants affect biogeochemical fluxes is uncertain, but preliminary results suggest that the effects are variable and depend on various urban environmental factors (Carreiro et al. in press). For example, Inman and Parker (1978) found slower leaf litter decay rates in urban stands that were highly contaminated with Cu (75.7 mg kg^{-1}) and Pb (399.9 mg kg^{-1}) compared to unpolluted rural stands. Apparently, the heavy metal contamination negatively affected the activity of decomposers in these urban stands. By contrast, where heavy metal contamination of soil was moderate to low relative to other atmospherically deposited pollutants such as N, rates of decomposition and soil N-transformation increased in forest patches near or within major metropolitan areas in southern California (Fenn and Dunn 1989; Fenn 1991), Ohio (Kuperman 1999) and southeastern New York (Carreiro et al. in press).

5.3.3 Introductions of Exotic Species

The expansion of urban areas has resulted in some of the greatest local extinction rates observed in the world (McKinney 2002). Cities also have become epicenters of many of introductions of nonnative species (Steinberg et al. 1997; McKinney 2002), some of which have become important pathogens or insect pests, e.g., Dutch elm disease and the Asian long-horned beetle. Local extinctions of native species and the invasions by urban-adapted nonnative species have resulted in a pattern where the species richness of nonnative species increases from outlying rural areas to urban centers while native species decrease (Blair 2001; Hope et al. 2003).

While native plant and animal species richness may be depressed, plant species richness apparently is greater in urban than in rural environments. Urban plant communities tend to have higher species richness and diversity than natural forests because of planting choices of multiple land owners (Nowak 2000). For instance, Nowak et al. (2002) found 81 and 57 tree species with Shannon-Weiner diversity index values of 3.6 and 3.4 in Baltimore MD and Brooklyn NY, USA, respectively. These values are higher than those found for eastern deciduous forests (1.9 to 3.1) of the United States (Barbour et al. 1980). In Phoenix, USA, overall species richness was greater in the urban area than in the surrounding desert. Within the urban area, species diversity was positively correlated to median family income, such that the highest biodiversity was associated with the wealthiest neighborhoods, a phenomenon termed the "luxury effect" (Hope et al. 2003).

Invasive species can play a disproportionate role in controlling C and N cycles in terrestrial ecosystems (Bohlen et al. 2004; Ehrenfeld 2003). Therefore, the relationship between invasive species abundances and urban land-use change has important implications for biogeochemical cycling of C and N (Pouyat et al. in press). For example, in the northeastern and mid-Atlantic United States where native earthworm species are rare or absent, urban areas are important foci of nonnative earthworm introductions, especially Asian species from the genus *Amynthas*, which are expanding toward outlying forested areas (Steinberg et al. 1997; Groffman and Bohlen 1999; Szlavecz et al. 2006). These invasions into forests have resulted in highly altered C and N cycling processes (Bohlen et al. 2004; Carreiro et al. in press). Other examples of species invasions in urban areas that have altered C and N cycles include shrub (*Berberis thunbergii*), tree (*Rhamnus cathartica*), and grass (*Microstegium vimineum*) species (Ehrenfeld et al. 2001; Heneghan et al. 2002; Kourtev et al. 2002).

5.4 Disturbance and Management Effects

For most urban landscapes, human-caused disturbance is more pronounced during rather than after the land-development process. Urban development of land typically includes the clearing of existing vegetation, massive movements of soil, and the building of structures. The extent and magnitude of these initial disturbances is dependent on infrastructure requirements (e.g., storm water retention ponds), topography, and other site limiting factors. As an example, a topographic change analysis of 30 development projects in Baltimore County USA showed that the total volumetric change of soil per development was positively correlated with mean slope of the site ($r = 0.54$, $p = 0.002$) (McGuire 2004). In addition, these development projects resulted in massive soil disturbances with the potential to have large effects on soil organic C. Using data from this study, we estimated that the amount of soil organic C that was disturbed during a development project covering 2600 m^2 was roughly 2.7×10^4 kg. This assumes that the original soil had a soil organic C density of a deciduous forest (10.3 kg m^{-2}) to a depth of 3 m, i.e., the average depth of the soil disturbance. How much soil organic C actually is lost during the development process is unknown and depends partly on the ultimate fate of the surface soil layers (topsoil).

The previous development scenario typically predates residential, institutional, and commercial land uses, for which turfgrass is the resultant plant-cover type. Nowak et al. (1996) found that residential, institutional, and commercial land uses made up on the average 39.0, 5.7, and 13.6% (total 58.3%) of the land cover in the major U.S. metropolitan areas, respectively. The total estimated lawn area for the conterminous USA is 16.38 ±3.58 million ha (Milesi et al. 2005). Management of this vast area of turfgrass typically includes adding pesticides and supplements of water and nutrients as well as being regularly clipped during the growing season. In the following sections we discuss how these lawn and horticultural management practices may affect biogeochemical cycles.

5.4.1 Lawn and Horticultural Management

To manage turfgrasses in lawns, homeowners and institutional land mangers in the USA apply about 16 million kg of pesticides each year (Aspelin 1997) as well as fertilizers at rates similar to or exceeding those of cropland systems (Talbot 1990). In addition, lawns typically are clipped on a regular basis during the growing season. This management scenario may or may not result in increases in plant productivity. In comparing results from studies of mowed lawns in Wisconsin, California, and Maryland, Falk (1980) estimated that the range for net primary productivity in temperate climates was about 1.0 to 1.7 kg ha yr^{-1}, most of which is below ground. Although above-ground productivity increased with N fertilization and irrigation, total productivity (above and below ground) did not differ. Other studies have shown somewhat lower productivity rates for lawns (0.6 to 0.7 kg ha yr^{-1}) (Blanco-Montero et al. 1995; Jo and McPherson 1995). Still others have reported direct relationships between management and productivity (Heckman 2000; Kopp and Guillard 2002; Golubiewski 2006). As a comparison, net primary production in temperate grasslands ranges from 0.1 to 1.5 kg ha yr^{-1} (Leith 1975).

While total turfgrass productivity may or may not increase with management inputs, soil organic C sequestration appears positively related to supplemental inputs of fertilizer and water. In comparing surface soils of 15 golf courses, Qian and Follett (2002) found that total C sequestration rates ranged from 0.9 to 1.0 t ha^{-1} yr^{-1}. This is similar to rates obtained for the recovery of perennial grasslands following cultivation (1.1 t ha^{-1} yr^{-1}) (Gebhart et al. 1994), but much higher than unmanaged grasslands (0.33 t ha^{-1} yr^{-1}) (Post and Kwon 2000). Similarly, Higby and Bell (1999) found that soil organic matter was higher in fertilized golf course fairways than in adjacent unfertilized areas. Using long-term simulations of the CENTURY model for lawn ecosystems, Qian et al. (2003) showed that N fertilization coupled with a management regime in which grass clippings were left on site increased soil organic C accumulations by up to 59% in comparison to sites that were not fertilized and clippings were removed. How much C is sequestered in soils under turfgrass depends on the differential effects of management practices on above- and below-ground productivity and the rate of decomposition in the surface soil (Qian and Follett 2002).

While lawn care management practices input chemicals at rates comparable to cropland systems, they are potentially less disruptive of biogeochemical cycles. Cropland systems have a greater magnitude and frequency of soil disturbances and generally remove a greater proportion of the standing crop (Asner et al. 1997; Pouyat et al. 2003). As a result, cropland systems lose substantial amounts of soil organic C and N (Matson et al. 1997). The net effect of an initial soil disturbance and subsequent lawn care management on C and N dynamics is less certain. In natural ecosystems, C sequestration increases with N additions until N is no longer limiting (Asner et al. 1997), i.e., a condition of N saturation (Aber et al. 1989). Nitrogen limitation is determined by biological demand and the ability of soil to accumulate N, which, in turn, is attributed partly to site history, climate, soil fertility, and vegetation type (Aber et al. 1998). In cultivated ecosystems, N additions over the long term can increase C storage, but due to annual disturbances of soil and the removal of plant biomass, these gains are less than those for grassland and forested ecosystems. On the other hand, turfgrass ecosystems can accumulate soil organic C at rates similar to those for grasslands and some forests due to the absence of annual soil disturbances (Qian and Follett 2002; Pouyat et al. 2003).

Carbon sequestration is an important regulator of the net effect of ecosystems on atmospheric carbon dioxide levels, which influence the atmospheric "greenhouse effect." This effect is also influenced by other soil-atmosphere gas exchanges, especially nitrous oxide and methane that are influenced by urban land-use change. Kaye et al. (2004) found that lawns in Colorado had reduced methane uptake and increased nitrous oxide fluxes relative to native shortgrass steppe, with fluxes similar to irrigated corn. Goldman et al. (1995) found reduced methane uptake in forests in urban areas. In both cases, these changes were linked to nitrogen enrichment associated with urban land-use.

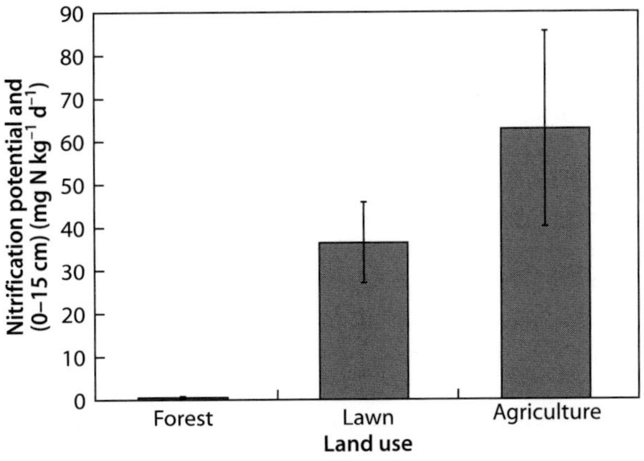

Fig. 5.4. Mean (±S.E.) potential net nitrification rates (mg NO_3^- kg d^{-1}) of 0–15 cm mineral soil samples after 14-d incubation period of 0.04 ha plots in the Baltimore LTER study. *Bars* represent comparison of forest, lawn, and agriculture land-use types ($n = 14, 10,$ and 10, respectively) (data from C. Williams, unpubl.)

The potential of turfgrass ecosystems to sequester soil organic C and thus maintain or increase microbial demand for N suggests a high capacity of these systems to retain N. Preliminary data from comparisons of lawn, forest, and cultivated patch types in the Baltimore Ecosystem Study suggest that lawns fall between temperate forests and cultivated land in C sequestration, N cycling, and N retention (Fig. 5.4). Currently, we are investigating the interactive effects of high N inputs, C pool dynamics, biomass removal, and altered hydrologic pathways in lawn ecosystems.

5.4.2 Management Effort

Perhaps the most distinctive characteristic of urban landscapes is the number and diversity of potential land managers on a per area basis. Parcels of land generally tend to become smaller from rural to suburban and urban residential areas, with a slight increase at the urban core (Pouyat et al. in press). This pattern potentially results in greater variation in management practices in high-density residential areas than in larger rural parcels because of the greater number of potential land managers (Collins et al. 2000). Parcel size also may affect the amount of effort of individual land owners. For example, in comparing fertilizer application rates in two small residential watersheds in Baltimore County, MD USA, Law et al. (2004) found that the application rate per unit of lawn area was higher for the watershed with smaller lot sizes, suggesting that homeowner input is inversely related to lot size. They also found that soil bulk density, an indicator of soil compaction, was positively related to the average annual application rate of fertilizer. The relationship between human impact and lot size also was evident at a coarser scale. In Baltimore, mean bulk densities of surface soils (0–5 cm) in high- and medium-density residential land-uses were 1.30 and 1.17 Mg m^{-3}, respectively. Bulk density was inversely related ($r = 0.52$, $P < 0.001$) to soil organic matter (Pouyat et al. 2002).

In addition to parcel size, management effort varies with socioeconomic factors in residential neighborhoods. In a survey conducted in the Columbus, OH USA, metropolitan area, Robbins et al. (2001) found that residents with houses valued at more than $250 000 were much more likely to apply chemicals on their lawns than owners of houses valued at less than $80 000. However, Law et al. (2004) and Osmond and Platt (2000) found that households with intermediate value applied the most lawn fertilizer. Whatever the relationships between socioeconomic factors and management effort, decisions involving horticultural management in much of the urban landscape are largely independent of government regulations (Robbins et al. 2001).

Management effort also should be related to the magnitude of the natural constraints on the system. For example, irrigation rates for lawns in dry land areas should be higher than for lawns in more temperate climates. The net result of these relationships is that while natural constraints on biological systems in urban landscapes are important, human desires and efforts to overcome these constraints result in vegetation structure and soil characteristics that are remarkably similar across urban settlements on a global scale (McDonnell et al. in press). Therefore, urban landscapes might be similar in vegetation structure and soil characteristics at coarse scales (regional and global) but highly variable at finer scales (Pouyat et al. 2003).

5.5 Effects of Built Environment

To fully grasp the effects of urban land-use change on biogeochemical cycles, we need to understand the ways the built environment affects energy and material flows in urban ecosystems. Our conceptual framework of urban land-use change (Fig. 5.5) incorporates the importance of built structures and the effect of impervious surfaces on ecosystem processing by differentiating between natural and engineered templates, either of which may constrain ecosystem processes. For example, both soil drainage and irrigation infrastructure can partially constrain distributions of plant species and trace gas fluxes, as well as the movement of nutrients and contaminants in urban ecosystems. Perhaps more interestingly, the natural and engineered templates may intersect; the proportion and "connectivity" of each determining the importance of each template in constraining biogeochemical processes. For example, hydrologists use the concept of "effective imperviousness" to more accurately model runoff volumes, that is, the connectivity between the natural and engineered templates. Effective impervious areas (EIA) drain immediately to storm-drain pipes and are equal to the total impervious area (TIA) minus the impervious surfaces draining to pervious areas where infiltration is possible (Sutherland 1995).

If measurements of EIA and TIA were made at the scale of a small watershed (<100 ha) for a given metropolitan area we suggest they would form a continuum of watersheds ranging from totally pervious cover to those with an almost entirely engineered template. Thus, the continuum would represent an array of small watersheds varying in area and connectivity of the engineered and natural templates, at least for hydrologic flow paths. Measurements of connectivity between engineered and natural templates also can be made for other flow paths (e.g., organic-matter transport or atmospheric deposition onto built and biological surfaces) and for other ecosystem delineations at various scales.

These relationships are important in determining the degree to which the engineered system mediates flows of energy and material. A totally engineered system has pathways that entirely separate or disconnect material flows from natural elements of the ecosystem, though over time, all human built structures fail to function to some degree, resulting in exchanges between engineered and natural templates. This disconnect "short circuits" the biological system, which diminishes the system's overall ability to buffer changes in water, nutrient, and contaminant inputs. As a result, the system's capacity to retain or process these materials is altered. For totally engineered systems, atmospherically derived contaminant and nutrient inputs can accumulate on impervious surfaces and be washed off repeatedly by small rainfall events into the engineered system. Thus, there is little chance of being accumulated in soil or biologically processed in the terrestrial system. In addition, gaseous exchanges between the atmosphere and the soil-plant continuum will be diminished, again short circuiting the ability of the biological system to assimilate C or gas-phase contaminants. We propose that the point in the continuum at which the constraints of the engineered and natural templates shift in importance from one to the other may serve as a useful delineation between urban and non-urban-dominated ecosystems.

At some point, all engineered pathways flow to natural systems, for example, sewage treatment outflows, power plant emissions, storm water runoff, and effluent from septic systems. All of these systems have important effects on ecosystems receiving their flows. In most cases, flows from engineered pathways concentrate materials collected from a larger area. The functional size of the area is dependent on the extent and connectivity of the infrastructure. For example, septic systems collect waste from a single detached house and release these flows into a relatively small volume of soil, while sewage treatment

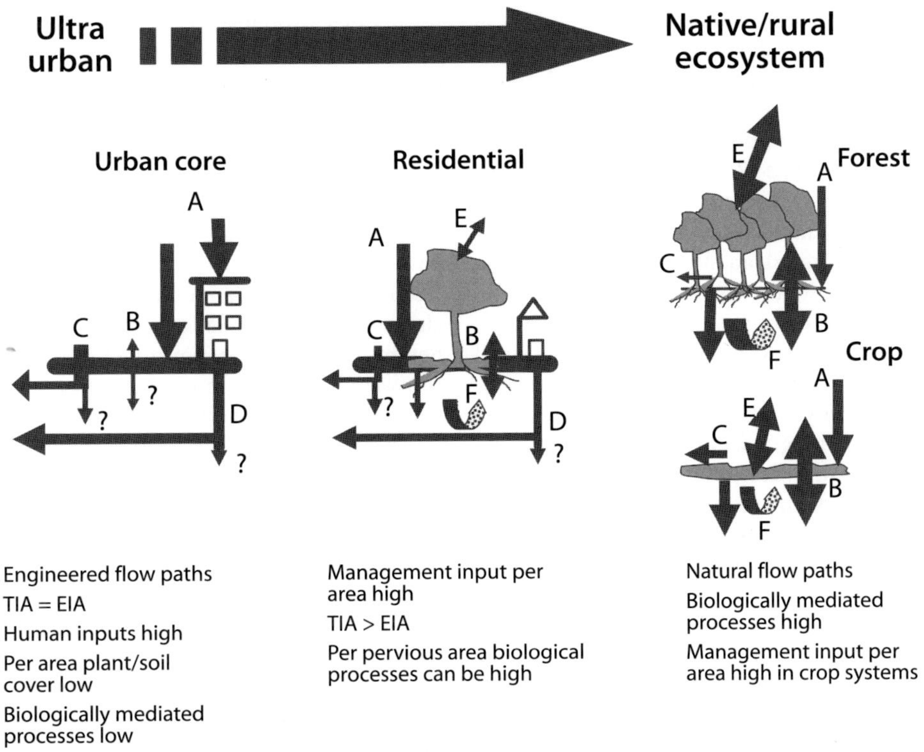

Fig. 5.5. Conceptual framework where anthropogenic and biophysical ecosystem characteristics form a continuum from highly urban (almost entirely human-made) to native or rural ecosystem types (those with the least human modification). Agricultural ecosystems also are depicted. The framework incorporates the importance of built structures and the effect of impervious surfaces on ecosystem processing. At the urban core, engineered flow paths (*C* and *D*) disconnect material inputs and flows (*A*) from natural processing that occurs in native/rural ecosystems. In these areas, the connectivity between the engineered and biophysical templates is low (*B*). In residential areas, the connectivity between engineered and biophysical templates can be relatively high depending on the spatial relationship of impervious and pervious surfaces. Management and environmental inputs in residential areas can be high (*A* and *E*) on a per-unit pervious area. However, depending on site history, soil type, and the concentration of flows these areas can have surprisingly high cycling rates (*F*) for processing or storing these inputs (adopted from Kay et al. 2006)

plants receive waste flows from tens and hundreds of thousands of residences and release flows to surface waters with varying abilities to assimilate the waste. As mentioned previously, the waste flows in both cases are byproducts of resources originating outside the urban ecosystems – often from great distances. Compounding the concentration of flow is the ability of built surfaces to accumulate various atmospherically derived contaminants through the development of organic films that increase the capture efficiency of impervious surfaces (Law and Diamond 1998). These films ultimately increase the movement of contaminants from the atmosphere to surface waters following wash-off (Diamond et al. 2000).

A clear example of the effects of concentrated flows is the impact of impervious cover on stormwater flows and the resulting impact on urban-stream ecosystems. Schueler (1994) and Morse et al. (2003) concluded from reviewing data in the literature that as little as 6% impervious cover in a watershed can result in measurable responses in stream-channel morphology and invertebrate populations. Moreover, changes in soil-drainage patterns associated with urban development and altered stream morphology can have marked effects on soil organic C and N processing, particularly in riparian soils (e.g., Groffman et al. 2002). An important question in urban hydrology is whether responses in stream ecology and riparian function differ between catchments with the same total impervious area but with different measures of effective impervious areas (Walsh 2004).

The tendency of the built environment and human activity to concentrate flow paths and chemical inputs can result in the development of "hot spots" in the landscape. Hotspots are areas or patches that show disproportionately high reaction rates relative to the surrounding area or matrix (McClain et al. 2003). The concept of hotspots developed from studies of N processing in soil cores (Parkin 1987) and riparian zones that showed that anoxic microsites with high C content were zones of elevated denitrification rates. Generally, hotspots are sites where reactants for specific biogeochemical reactions coincide in an environment conducive for the reaction to take place (McClain et al. 2003). Human activities and the introduction of built structures provide such conditions in urban landscapes at various scales. Examples include septic systems, golf greens, stormwater retention basins, and compost piles.

The importance of hotspots in the biogeochemistry of urban ecosystems is largely unknown. In urban fringe areas, there is concern about nitrate contamination of ground water from septic systems, which when functioning properly can be thought of as purposely engineered hot spots for processing human waste (Band et al. 2004). Hotspots in urban landscapes also can be sinks for contaminants, nutrients, or C. For example, the conversion of what was predominately prairie grassland into an urban landscape resulted in the development of C sequestration hotspots in the Boulder, CO USA, metropolitan area (Golubiewski 2003). These hotspots were composed of relatively small parcels with highly managed turfgrass and woody vegetation whose productivity rates and C storage were much higher than in the native grasslands they replaced. Similarly, detention basins designed to capture urban stormwater and protect streams may be hotspots of denitrification, replacing functions that occurred in riparian areas before urbanization (Groffman and Crawford 2003).

5.6 Assessing Biogeochemical Effects – the Importance of Scale

A critical task in using the ecosystem approach is the ability to make practical and meaningful delineations of ecosystem boundaries (Likens 1992). In nonhuman-dominated ecosystems, this task is challenging since many environmental factors vary as a continuum on the landscape. Ecologists and biogeochemists have addressed this challenge by using discontinuities of biophysical processes in the landscape as ecosystem boundaries, e.g., a watershed. Determining boundaries for urban ecosystems is an even greater challenge as there are no generally accepted ecological definitions of "urban" (Mcintyre et al. 2000). Nor do we know how urban environmental factors vary spatially (Pouyat et al. in press). This is especially problematic in viewing cities as well-mixed reactors because urban areas often are expanding in area, forming roughly concentric circles of development (McDonnell and Pickett 1990). For landscapes where development patterns are less dispersed, e.g., desert ecosystems, the boundary of urban expansion is more apparent (e.g., Baker et al. 2001).

To address the difficulty in delineating urban ecosystem boundaries for investigations of biogeochemical cycles, we propose a two-pronged approach: (1) conduct investigations for parts or patches of the urban landscape that have more readily recognizable boundaries, and (2) assess the interrelationship of these patches at multiple scales: one at the scale of individual parcels (<km) each with a distinctive landscape context, site history, land manager, and natural and engineered template, and the other at coarser scales, e.g., watershed or well-defined area of the landscape (e.g., Ellis 2004), to assess the interrelationships among parcels and investigate the net change in biogeochemical cycles due to urban-land transformations. At coarse scales, we suggest comparisons on both a per-unit pervious cover and per-unit total area basis. The latter includes built surfaces and considers urban landscapes as a whole. To distinguish between the two approaches, in the first case, productivity on a pervious-area basis can be greater than the native ecosystem replaced (or nonurbanized areas at the rural end of the continuum); in the second case, a city generally would have lower overall productivity than the native ecosystem it replaced. When human use of fossil fuels are included in these calculations, cities would be considered heterotrophic (primary productivity < total ecosystem respiration) rather than natural ecosystems, which are autotrophic (Grimm et al. 2003).

These relationships suggest a paradox of urban ecosystems: the engineered template reduces the ability of an ecosystem to assimilate or process energy and materials on a per-unit ecosystem basis such that when compared to natural systems urban ecosystems exhibit less biological activity. At the same time, the engineered template and human activity concentrate energy and matter into smaller areas (volumes) such that fluxes and biological activity will be higher on a per-unit pervious area basis than in the native ecosystem replaced. The latter characteristic is significant since the ability of biological systems to assimilate energy and material is determined at the scale at which organisms are using the resource. This can occur at very fine scales, e.g., hotspots. Thus, assessments at coarse scales that fail to distinguish between the built and pervious component of urban ecosystems will miss the scale at which biological processes are being constrained.

A comparison of N budgets of residential watersheds in the Baltimore Ecosystem Study illustrates the importance of scale. Law et al. (2004) calculated N input from fertilizer into a low- and medium-density residential watershed at three spatial aggregations: by the area of the watershed, subdivision, and residential lawns. For the more sparsely populated watershed, application rates were 9.5, 27.8, and 37.1 kg N ha^{-1} yr^{-1} on a watershed, subdivision, and residential lawn area, respectively. For the more densely populated watershed, rates were 12.5, 26.7, and 83.5 kg N ha^{-1} yr^{-1}. Therefore, the greatest distinction between the two watersheds is on a lawn-area basis, or the scale at which added N will be processed by the soil community and management decisions will be made by individual land owners. In addition, the watershed with lower housing densities had an aggregated septic load that was significantly lower than N fertilizer inputs and roughly equal to atmospheric deposition inputs (Band et al. 2004). However, this septic load was concentrated in 1 to 2% of the watershed area (leach fields), resulting in a loading rate that was an order of magnitude higher than N loading per unit of watershed area. Indeed, stream sampling in the Baltimore Ecosystem Study revealed that nitrate concentrations were higher in watersheds at the suburban fringe that are septic-serviced than in central city streams or sanitary sewer-serviced suburbs with similar lawn fertilization rates (Band et al. 2004; Groffman et al. 2004).

As another example, Nowak (2000) estimated the amount of C sequestered by trees in Baltimore to be 0.71 t C $ha^{-1} yr^{-1}$. His estimate was for the entire land area of the city. Estimates for pervious areas in Baltimore are not available, though Jo and McPherson (1995) measured carbon budgets on a per-pervious-area basis for two residential neighborhoods in Chicago, IL USA. They found that C sequestered by woody vegetation was as high as 3.29 t C $ha^{-1} yr^{-1}$. Assuming that C budgets of residential areas are similar across urban ecosystems, comparison of these rates to gross sequestration rates for a 25 yr old loblolly plantation (2.6 t C $ha^{-1} yr^{-1}$) and a naturally regenerating spruce-fir forest (1.0 t C $ha^{-1} yr^{-1}$) (Birdsey 1996) suggests the urban values are somewhat low when the entire landscape is considered (thus the net change from converting from a natural ecosystem to an urban ecosystem), but high when only pervious surfaces are considered. This suggests that on average, urban trees have higher growth rates than trees growing in nonirrigated and nonfertilized stands under rural environmental conditions. This finding is consistent with measurements of the productivity of urban trees (McPherson 2000), urban-rural comparisons of containerized plants (Gregg et al. 2003), and eddy flux tower measurements above a residential area (Hom et al. 2003).

On the basis of these comparisons, we propose that the net effect of urban land-use change on biogeochemical cycles depends on the overall constraints of the native ecosystem replaced. For example, converting a temperate deciduous forest ecosystem to an urban landscape results in a decrease (per land area) in plant productivity, but potentially higher rates on a pervious area basis. By contrast, converting a desert ecosystem to an urban landscape should result in both a land-area and per-pervious-area increase in primary productivity. Differences in the direction of the ecosystem response are the result of the relatively severe limitations in soil-water availability in desert vs. temperate environments. Thus, the technical ability and desire of humans to manage for specific types of plant communities irrespective of natural limiting factors have resulted in a convergence of urban vegetation structure and ecosystem function on a global basis (Pouyat et al. 2003).

5.7 Summary and Conclusions

Urban areas represent ecosystems with modified biogeochemical cycles such that fluxes and pools of matter, energy, and organisms differ greatly from the previous unaltered ecosystem. The conversion of agriculture, forest, and grasslands to urban and suburban land uses entails a complex array of land and ecosystem alterations. These include the construction of various built structures, e.g., roads, buildings, and civil infrastructure, as well as the introduction of horticultural management practices. In addition to land altering activities, a complex of urban environmental factors can potentially affect biogeochemical cycling including atmospheric and soil pollution, CO_2 emissions, micro- and meso-climates, and introductions of exotic plant and animal species. In some metropolitan areas, the net effect of such factors is analogous to predictions of global environmental change, which we feel represents an opportunity to study ecosystem responses to such factors at a scale not practical in controlled field experiments.

To fully understand the effects of urban land-use change on biogeochemical cycles, the effect of the built environment must be considered. A conceptual framework of urban land-use change was presented that incorporates the importance of built structures and the effect of impervious surfaces on ecosystem processing. This framework differentiates between natural and engineered systems, either of which may constrain biogeochemical processes. A totally engineered system has pathways that entirely separate or disconnect material flows from natural elements of the ecosystem, though all human built structures fail to function to some degree, resulting in exchanges between engineered and natural templates. This disconnect "short circuits" the biological system, which diminishes the system's overall ability to buffer changes in water, nutrient, and contaminant inputs.

To assess the overall impact of urban land-use change on biogeochemical cycles, we propose comparisons at: (1) individual parcels each with a distinctive set of characteristics, and (2) coarser scales to assess the interrelationships among parcels and investigate the net change in biogeochemical cycles due to urban land transformations. At coarse scales, we suggest comparisons on both per-unit pervious cover and per-unit land area (ecosystem) basis. The latter includes built surfaces and considers urban landscapes as a whole. Preliminary analyses of the available data suggest that process rates on a pervious-area basis can be greater than those of the native ecosystem replaced, while on an ecosystem basis urban areas tend to have lower overall process rates than to the native ecosystem replaced. The former comparison suggests the continued importance of biological processes in urban areas, which we believe results from human efforts to overcome environmental constraints on biological processes and the tendency for horticultural management activities and urban infrastructure to concentrate flows of energy and matter into the remaining biologically active areas in the urban landscape.

We conclude that global and regional comparisons of urban ecosystems on an ecosystem and pervious-area basis will provide an analysis of the net effect of urban land conversion on biogeochemical cycles, e.g., soil C pools and fluxes, as well as insight into the mechanisms causing those changes. We hypothesize that the net effect of urban land-use conversion will depend partly on the characteristics of the native or rural ecosystem replaced. For arid regions, the net overall effect of urbanization may be higher productivity rates and thus the potential to actually increase C sequestration; in more humid environments, the net effect will be a reduction in C sequestration rates. Therefore, we suggest that cross-system comparisons on regional and global scales will be necessary to determine the net effect of urbanization on biogeochemical cycles.

Acknowledgments

Funding support came from the USDA Forest Service, Northern Global Change Program and Research Work Unit (NE-4952), Syracuse, NY; Baltimore Ecosystem Study's Long Term Ecological Research grant from the National Science Foundation (DEB 97-14835); and the Center for Urban Environmental Research and Education, University of Maryland Baltimore County.

References

Aber JD, Nadelhoffer KJ, Steudler P, Melillo JM (1989) Nitrogen saturation in northern forest ecosystems. Bioscience 39:378–386
Aber JD, McDowell WH, Nadelhoffer KJ, Magill AH, Berntson G, Kamekea M, McNulty SG, Currie WS, Rustad LE, Fernandez IJ (1998) Nitrogen saturation in temperate forest ecosystems: hypothesis revisited. BioScience 48:921–934
Arbaugh M, Bytnerowicz A, Grulke N, Fenn M, Poth M, Temple P, Miller P (2003) Photochemical smog effects in mixed conifer forests along a natural gradient of ozone and nitrogen deposition in the San Bernadino Mountains. Environment International 29:401–406
Asner GP, Seastedt TR, Townsend AR (1997) The decoupling of terrestrial carbon and nitrogen cycles. BioScience 47:226–234
Aspelin AL (1997) Pesticide industry sales and usage: 1994 and 1995 market estimates. US Environmental Protection Agency, Biological and Economic Analysis Division, Office of Pesticide Programs, Washington, DC
Baker LA, Hope D, Xu Y, Edmonds J, Lauver L (2001) Nitrogen balance for the central Arizona-Phoenix (CAP) ecosystem. Ecosystems 4:582–602
Band LE, Cadenasso ML, Grimmond S, Grove JM, Pickett STA (2004) Heterogeneity in urban ecosystems: Patterns and process. In: Lovett GM, Jones CG, Turner M, Weathers KC (eds). Ecosystem Function in Heterogeneous Landscapes. Springer-Verlag, New York
Barbour MG, Burk JH, Pitts WD (1980) Terrestrial plant ecology. Benjamin/Cummings Publishing Co., Menlo Park, CA
Birdsey RA (1996) Carbon storage for major forest types and regions in the conterminous United States. In: Sampson RL, Hair D (eds) Forests and global change, Volume 2: Forest management opportunities for mitigating carbon emissions. American Forests, pp 1–26
Blair RB (2001) Birds and butterflies along urban gradients in two ecoregions of the United States: Is urbanization creating a homogenous fauna? In Lockwood JL, Mckinney ML (eds) Biotic homogenization. Kluwer Academic/Plenum Publishers, New York, pp 33–56
Blanco-Montero CA, Bennett TB, Neville P, Crawford CS, Milne BT (1995) Potential environmental and economic impacts of turfgrass in Albuquerque, NM (USA). Landscape Ecology 10:121–128
Bohlen PJ, Groffman PM, Fahey TJ, Fisk MC, Suarez E, Pelletier DM, Fahey RT (2004) Ecosystem consequences of exotic earthworm invasion of north temperate forests. Ecosystems 7:1–12
Brazel A, Selover N, Vose R, Heisler G (2000) The tale of two climates – Baltimore and Phoenix urban LTER sites. Climate Research 15:123–135
Bytnerowicz A, Fenn ME, Miller PR, Arbaugh MJ (1999) Wet and dry pollutant deposition to the mixed conifer forest. In: Miller PR, McBride JR (eds). Oxidant air pollution impacts in the montane forests of Southern California: A case study of the San Bernardino Mountains. Springer-Verlag, New York, pp 235–269
Carreiro MM, Tripler CE (2005) Forest remnants along urban-rural gradients: examining their potential for global change research. Ecosystems 8:568–582
Carreiro MM, Pouyat RV, Tripler CE, Zhu W (in press) Carbon and nitrogen cycling in soils of remnant forests along urban-rural gradients: Case studies in New York City and Louisville, Kentucky. In: McDonnell MJ, Hahs A, Breuste J (eds) Comparative ecology of cities and towns. Cambridge University Press, New York
Caspersen JP, Pacala SW, Jenkins JC, Hurtt GC, Moorcroft PR, Birdsey RA (2000) Contributions of land-use history to carbon accumulation in U.S. forests. Science 290:1148–1151
Cerveny RS, Balling RC (1998) Weekly cycles of air pollutants, precipitation and tropical cyclones in the coastal NW Atlantic region. Nature 394:561–563
Chappelka AH, Samuelson LJ (1998) Ambient ozone effects on forest trees of the eastern United States: A review. New Phytologist 139:91–108
Collins JP, Kinzig AP, Grimm NB, Fagan WF, Hope D, Wu J, Borer ET (2000) A new urban ecology. American Scientist 88:416–425
Craul PJ (1992) Urban soil in landscape design. John Wiley & Sons, New York
Diamond ML, Priemer DA, Law NL (2001) Developing a multimedia model of chemical dynamics in an urban area. Chemosphere 44:1655–1667
Ehrenfeld JG (2003) Effects of exotic plant invasions on soil nutrient cycling processes. Ecosystems 6:503–523
Ehrenfeld JG, Kourtev P, Huang W (2001) Changes in soil functions following invasions of exotic understory plants in deciduous forests. Ecological Applications 11:1278–1300
Ellis EC (2004) Long-term ecological changes in the densely populated rural landscapes of China. In: Asner GP, DeFries RS, Houghton RH (eds) Ecosystems and land use change. Geophysical Monographs Vol. 153. American Geophysical Union, Washington, DC, pp 303–320
Elvidge CD, Milesi C, Dietz JB, Tuttle BT, Sutton PC, Nemani R, Vogelmann JE (2004) U.S. constructed area approaches the size of Ohio. EOS, Transaction of the American Geophysical Union 85:233–234
Falk JH (1980) The primary productivity of lawns in a temperate environment. Journal of Applied Ecology 17:689–696
Fenn ME (1991) Increased site fertility and litter decomposition rate in high pollution sites in the San Bernardino Mountains. Forest Science 37:1163–1181
Fenn ME, Bytnerowicz A (1993) Dry deposition of nitrogen and sulfur to ponderosa and Jeffrey pine in the San Bernardino National Forest in southern California. Environmental Pollution 81:277–285
Fenn ME, Dunn PH (1989) Litter decomposition across an air-pollution gradient in the San Bernardino Mountains. Soil Science Society of America Journal 53:1560–1567
Folke CA, Jansson J, Costanza R (1997) Ecosystem appropriation by cities. Ambio 26:167–172
Gatz DF (1991) Urban precipitation chemistry: A review and synthesis. Atmospheric Environment 25:1–15
Gebhart DL, Johnson HB, Mayeux HS, Polley HW (1994) The CRP increases soil organic carbon. Journal of Soil and Water Conservation 49:488–492
Goldman MB, Groffman PM, Pouyat RV, McDonnel MJ, Pickett SJA (1995) CH_4 uptake and N availability in forest soils along an urban to rural gradient. Soil Biology & Biochemistry 27:281–286
Golubiewski NE (2003) Carbon in conurbations: Afforestation and carbon storage as consequences of urban sprawl in Colorado's Front Range. Ph.D. Dissertation, University of Colorado, Boulder, Colorado

Golubiewski NE (2006) Urbanization increases grassland carbon pools: affects of landscaping in Colorado's Front Range. Ecological Applications 16:555–571

Gregg JW, Jones CG, Dawson TE (2003) Urbanization effects on tree growth in the vicinity of New York City. Nature 424:183–187

Grimm NB, Grove JM, Pickett STA, Redman CL (2000) Integrated approaches to longterm studies of urban ecological systems. BioScience 50:571–84

Grimm NB, Baker LJ, Hope D (2003) An ecosystem approach to understanding cities: familiar foundations and uncharted frontiers. In: Berkowitz AR, Nilon CH, Hollweg KS (eds) Understanding urban ecosystems: A new frontier for science and education. Springer-Verlag, New York, pp 95–114

Groffman PM, Bohlen PJ (1999) Soil and sediment biodiversity: cross-system comparisons and large-scale effects. BioScience 49:139–148

Groffman PM, Crawford MK (2003) Denitrification potential in urban riparian zones. Journal of Environmental Quality 32: 1144–1149

Groffman PM, Crawford MK, Zipperer WC, Pouyat RV, Band LE, Colosimo MF (2002) Soil nitrogen cycle processes in urban riparian zones. Environmental Science and Technology 36: 4547–4552

Groffman PM, Law NL, Belt KT, Band LE, Fisher GT (2004) Nitrogen fluxes and retention in urban watershed ecosystems. Ecosystems 7:393–403

Grulke NE, Balduman L (1999) Deciduous conifers: High N deposition and O_3 exposure effects on growth and biomass allocation in Ponderosa Pine. Water, Air and Soil Pollution 116:235–248

Grulke NE, Andersen CP, Fenn ME, Miller PR (1998) Ozone exposure and nitrogen deposition lowers root biomass of ponderosa pine in the San Bernadino Mountains, California. Environmental Pollution 103:63–73

Heckenberger MJ, Kuikuro A, Kuikuro UT (2003) Amazonia 1492: Pristine forest or cultural parkland? Science 301:1710–17–14

Heckman JR, Liu H, Hill W, Demilia M, Anastasia L (2000) Kentucky bluegrass responses to mowing practice and nitrogen fertility management. Journal Sustainable Agriculture 15: 25–33

Heneghan L, Clay C, Brundage C (2002) Observations on the initial decomposition rates and faunal colonization of native and exotic plant species in an urban forest fragment. Ecological Restoration 20:108–111

Higby JR, Bell PF (1999) Low soil nitrate levels from golf course fairways related to organic matter sink for nitrogen. Communication in Soil Science and Plant Analysis 30:573–588

Hom J, Grimmond S, Golub D, Offerle B, Nowak D, Heisler G, Pouyat R, Zipperer W (2003) Studies on carbon dioxide concentrations and carbon flux in a forested region in suburban Baltimore USA. Baltimore Ecosystem Study Annual Meeting, Baltimore, MD. October 2002. Abstract available at: http://www.beslter.org/frame7-page_7q.html

Hope D, Gries C, Zhu W, Fagan W, Redman CL, Grimm NB, Nelson AL, Martin C, Kinzig A (2003) Socioeconomics drive urban plant diversity. Proceedings of the National Academy of Sciences 100: 8788–8792

Houghton RA, Hackler JL, Lawrence KT (1999) The U.S. carbon budget: Contributions from land-use change. Science 285: 574–8

Idso SB, Idso CD, Balling RC (2002) Seasonal and diurnal variations of near-surface atmospheric CO_2 concentration within a residential sector of the urban CO_2 dome of Phoenix, AZ, USA. Atmospheric Environment 36:1655–1660

Imhoff ML, Lawrence WT, Elvidge CD, Paul T, Levine E, Privalsky MV, Brown V (1997) Using nighttime DMSP/OLS images of city lights to estimate the impact of urban land use on soil resources in the United States. Remote Sensing and Environment 59:105–117

Inman JC, Parker GR (1978) Decomposition and heavy metal dynamics of forest litter in northwestern Indiana. Environmental Pollution 17:34–51

Jo H, McPherson EG (1995) Carbon storage and flux in urban residential greenspace. Journal of Environmental Management 45: 109–133

Kay JP, Burke IC, Mosier AR, Guerschman JP (2004) Methane and nitrous oxide fluxes from urban soils to the atmosphere. Ecological Applications 14:975–981

Kay JP, Groffman PM, Grimm NB, Baker LA, Pouyat RV (2006) A distinct urban biogeochemistry? Trends in Ecology and Evolution 21:192–199

Koerner B, Klopatek J (2002) Anthropogenic and natural CO_2 emission sources in an arid urban environment. Environmental Pollution 116:S45–S51

Kopp KL, Guillard K (2002) Clipping management and nitrogen fertilization of turfgrass: growth, nitrogen utilization, and quality. Crop Science 42:1225–1231

Kourtev PS, Ehrenfeld JG, Huang W (2002) Exotic plant species alter microbial structure and function in the soil. Ecology 85:3152–3166

Kuperman RG (1999) Litter decomposition and nutrient dynamics in oak-hickory forests along a historic gradient of nitrogen and sulfur deposition. Soil Biology and Biochemistry 31:237–244

Law NL, Diamond ML (1998) The role of organic films and the effect on hydrophobic organic compounds in urban areas: an hypothesis. Chemosphere 36:2607–2620

Law NL, Band LE, Groffman PM, Belt KT, Fisher GT (in press) Water quality trends and determinants in developing urban-suburban watersheds. Hydrological Processes

Law NL, Band LE, Grove JM (2004) Nutrient input from residential lawn care practices. Journal of Environmental Management 47:737–755

Leith H (1975) Primary productivity of the major vegetational units of the world. In: Whittaker RH (ed) Primary productivity of the biosphere. Springer-Verlag, New York. pp 203–216

Likens G (1992) The ecosystem approach: its use and abuse. Ecology Institute, Oldendorf/Luhe (Germany), Excellence in ecology, book 3. pp 159

Lovett GM, Traynor MM, Pouyat RV, Carreiro MM, Zhu W, Baxter JW (2000) Atmospheric deposition to oak forests along an urban-rural gradient. Environmental Science and Technology 34:4294–4300

Luck MA, Jenerette GD, Wu J, Grimm NB (2001) The urban funnel model and the spatially heterogeneous ecological footprint. Ecosystems 4:782–796

Mcintyre NE, Kowles-Yánez K, Hope D (2000) Urban ecology as an interdisciplinary field: differences in the use of "urban" between the social and natural sciences. Urban Ecosystems 4:5–24

Matson PA, Parton WJ, Power AG, Swift MJ (1997) Agricultural intensification and ecosystem properties. Science 277:504–509

McClain ME, Boyer EW, Dent CL, Gergel SE, Grimm NB, Groffman PM, Hart SC, Harvey JW, Johnston CA, Mayorga E, McDowell WH, Pinay G (2003) Biogeochemical hot spots and hot moments at the interface of terrestrial and aquatic ecosystems. Ecosystems 6:301–312

McDonnell MJ, Pickett STA (1990) Ecosystem structure and function along urban-rural gradients: an unexploited opportunity for ecology. Ecology 71:1232–1237

McDonnell MJ, Hahs A, Breuste J (eds) (in press) Comparative ecology of cities and towns. Cambridge University Press, New York

McGuire M (2004) Using DTM and LiDAR data to analyze human induced topographic change. In: Proceedings of ASPRS 2004 fall conference, September 12–16, 2004, Kansas City, MO. PDF available at: http://eserv.asprs.org/eseries/source/Orders/

McKinney ML (2002) Urbanization, biodiversity, and conservation. BioScience 52:883–890

McPherson EG (2000) Urban forests and climate change. In: Abdollahi KK, Ning ZH, Appeaning A (eds) Global climate change and the urban forest. Franklin Press, Baton Rouge, LA, pp 17–30

Milesi C, Running SW, Elvidge CD, Dietz JB, Tuttle BT, Nemani RR (2005) Mapping and modeling the biogeochemical cycling of turf grasses in the United States. Environmental Management 36:426–438

Montague T, Kjelgren R (2004) Energy balance of six common landscape surfaces and the influence of surface properties on gas exchange of four containerized tree species. Sciencia Horticulturae 100:229–249

Morse CC, Huryn AD, Cronan C (2003) Impervious surface area as a predictor of the effects of urbanization on stream insect communities in Maine, U.S.A. Environmental Monitoring and Assessment 89:95–127

Mount H, Hernandez L, Goddard T, Indrick S (1999) Temperature signatures for anthropogenic soils in New York City. In: Kimble JM, Ahrens RJ, Bryant RB (eds.) Classification, correlation, and management of anthropogenic soils, Proceedings – Nevada and California, September 21-October 2, 1998. USDA Natural Resource Conservation Service, National Survey Center, Lincoln, Nebraska, pp 147–140

Norby RJ, Luo Y (2004) Evaluating ecosystem responses to rising CO_2 and global warming in a multi-factor world. New Phytologist 162:281–293

Nowak DJ (2000) The interactions between urban forests and global climate change. In: Abdollahi KK, Ning ZH, Appeaning A (eds) Global climate change and the urban forest. Franklin Press, Baton Rouge, LA, pp 31–44

Nowak DJ, Rowntree RA, McPherson EG, Sisinni SM, Kerkmann ER, Stevens JC (1996) Measuring and analyzing urban tree cover. Landscape and Urban Planning 36:49–57

Nowak DJ, Crane DE, Stevens JC, Ibarra M (2002) Brooklyn's urban forest. USDA Forest Service General Technical Report NE-290

Oke TR (1990) The micrometeorology of the urban forest. Quarterly Journal of Royal Meteorological Society 324:335–349

Osmond DL, Platt J (2000) Characterization of suburban nitrogen fertilizer and water use on residential turf in Cary, North Carolina. Horticultural Technology, 10, April–June

Parkin TB (1987) Soil microsites as a source of denitrification variability. Soil Science Society of America Journal 51:1194–1199

Pataki DE, Bowling DR, Ehleringer JR (2003) Seasonal cycle of carbon dioxide and its isotopic composition in an urban atmosphere: anthropogenic and biogenic effects. Journal of Geophysical Research – Atmospheres 108(D23), 4735

Pavao-Zuckerman MA, Coleman DC (in press) Physical, chemical and biological properties of urban soils in Asheville, NC. Urban Ecosystems

Pickett STA, Cadenasso ML, Grove JM, Nilon CH, Pouyat RV, Zipperer WC, Costanza R (2001) Urban ecological systems: linking terrestrial ecology, physical, and socio-economic components of metropolitan areas. Annual Review of Ecology and Systematics 32:127–157

Post WM, Kwon KC (2000) Soil carbon sequestration and land-use change: Processes and potential. Global Change Biology 6:317–327

Pouyat RV, Effland WR (1999) The investigation and classification of humanly modified soils in the Baltimore Ecosystem Study. In: Kimble JM, Ahrens RJ, Bryant RB (eds) Classification, correlation, and management of anthropogenic soils, Proceedings – Nevada and California, September 21-October 2, 1998. USDA Natural Resource Conservation Service, National Survey Center, Lincoln, Nebraska, pp 141–154

Pouyat RV, McDonnell MJ (1991) Heavy metal accumulation in forest soils along an urban-rural gradient in southeastern New York. Water, Air, and Soil Pollution 57–58:797–807

Pouyat RV, McDonnell MJ, Pickett STA, Groffman PM, Carreiro MM, Parmelee RW, Medley KE, Zipperer WC (1995) Carbon and nitrogen dynamics in oak stands along an urban-rural gradient. In: Kelly JM, McFee WW (eds) Carbon forms and functions in forest soils. Soil Science Society of America, Madison, Wisconsin, pp 569–587

Pouyat RV, Groffman P, Yesilonis I, Hernandez L (2002) Soil carbon pools and fluxes in urban ecosystems. Environmental Pollution 116:S107–S118

Pouyat RV, Russell-Anelli J, Yesilonis ID, Groffman PM (2003) Soil carbon in urban forest ecosystems. In: Kimble JM, Heath LS, Birdsey RA, Lal R (eds) The potential of U.S. forest soils to sequester carbon and mitigate the greenhouse effect. CRC Press, Boca Raton, FL, pp 347–362

Pouyat RV, Carreiro MM, Groffman PM, Pavao-Zuckerman M (in press) Investigative approaches of urban biogeochemical cycles: New York and Baltimore Cities as case studies. In: McDonnell MJ, Hahs A, Breuste J (eds) Comparative ecology of cities and towns. Cambridge University Press, New York

Qian YL, Follett RF (2002) Assessing soil carbon sequestration in turfgrass systems using long-term soil testing data. Agronomy Journal 94:930–935

Qian YL, Bandaranayake W, Parton WJ, Mecham B, Harivandi MA, Mosier AR (2003) Long-term effects of clipping and nitrogen management in turfgrass on soil organic carbon and nitrogen dynamics: The CENTURY model simulation. Journal of Environmental Quality 32:1695–1700

Rees WE (2003) Understanding urban ecosystems: an ecological economics perspective. In: Berkowitz AR, Nilon CH, Hollweg KS (eds) Understanding urban ecosystems: A new frontier for science and education. Springer-Verlag, New York, pp 115–136

Robbins P, Polderman A, Birdkenholtz T (2001) Lawns and toxins: An ecology of the city. Cities 18:369–380

Schlesinger WH, Andrews JA (2000) Soil respiration and the global carbon cycle. Biogeochemistry 48:7–20

Schueler T (1994) The importance of imperviousness. Watershed Protection Techniques 1:100–111

Seinfeld JH (1989) Urban air pollution: State of the science. Science 243:745–752

Skärby L, Ro-Poulsen H (1998) Impact of ozone of forests: a European perspective. New Phytologist 139:109–122

Shepherd JM, Burian SJ (2003) Detection of urban-induced rainfall anomalies in a major coastal city. Earth Interactions 7:1–14

Steinberg DA, Pouyat RV, Parmelee RW, Groffman PM (1997) Earthworm abundance and nitrogen mineralization rates along an urban-rural land use gradient. Soil Biology and Biochemistry 29:427–430

Sutherland RC (1995) Methods for estimating the effective impervious area of urban watersheds. Watershed Protection Techniques 2:282–284

Szlavecz K, Placella S, Pouyat RV, Groffman PM, Csuzdi C, Yesilonis ID (2006) Invasive earthworms and N-cycling in remnant forest patches. Applied Soil Ecology 32:54–62

Talbot M (1990) Ecological lawn care. Mother Earth News 123:60–73

Turner BLI, Matson PA, McCarthy J (2003) Illustrating the coupled human-environment system for vulnerability analysis: three case studies. Proceedings of National Academy of Sciences 100: 8080–8085

United Nations (2004) World urbanization prospects: The 2003 Revision. United Nations Department of Economic and Social Affairs, Population Division Report no. E.04.XIII.6, United Nations, New York, NY

US Department of Agriculture, Natural Resource Conservation Service (2001) Summary report: 1997 National Resources Inventory (revised December 2001). US Department of Agriculture, Natural Resource Conservation Service, Washington, DC

US Department of Commerce, Census Bureau (2001) Statistical Abstract of the United States, 2001. Table 1046. available at http://www.census.gov/prod/www/statistical-abstract-us.html

Vitousek PM, D'Antonio CM, Loope LL, Rejmanek M, Westbrooks R (1997) A significant component of human-caused global change. New Zealand Journal of Ecology 21:1–16

Walsh CJ (2004) Protection of in-stream biota from urban impacts: Minimize catchment imperviousness or improve drainage design? Marine and Freshwater Research 55:317–326

Wang LH, Kanehl P (2003) Influences of watershed urbanization and instream habitat on macroinvertebrates in cold water streams. Journal of the American Water Resources Association 39:1181–1196

Wang LH, Lyons J, Kanehl P (2003) Impacts of urban land cover on trout streams in Wisconsin and Minnesota. Transactions of the American Fisheries Society 132:825–839

Webster JR, Benfield EF (1986) Vascular plant breakdown in freshwater ecosystems. Annual Review of Ecological Systematics 17:567–594

Wong F, Harner T, Liu Q, Diamond ML (2004) Using experimental and forest soils to investigate the uptake of polycyclic aromatic hydrocarbons (PAHs) along an urban-rural gradient. Environmental Pollution 129:387–398

Ziska LH, Bunce JA, Goins EW (2004) Characterization of an urban-rural CO_2/temperature gradient and associated changes in initial plant productivity during secondary succession. Oecologia 139:454–458

Chapter 6

Saturation of the Terrestrial Carbon Sink

Josep G. Canadell · Diane E. Pataki · Roger Gifford · Richard A. Houghton · Yiqi Luo · Michael R. Raupach
Pete Smith · Will Steffen

6.1 Introduction

There is strong evidence that the terrestrial biosphere has acted as a net carbon (C) sink over the last two and half decades. Its strength is highly variable year-to-year ranging from 0.3 to 5.0 Pg C yr^{-1}; an amount of significant magnitude compared to the emission of about 7 Pg C yr^{-1} from fossil fuel burning (Prentice et al. 2001; Schimel et al. 2001; Sabine et al. 2004). Uncertainties associated with C emissions from land-use change are large. On average, the terrestrial C sink is responsible for removing from the atmosphere approximately one third of the CO_2 emitted from fossil fuel combustion, thereby slowing the build-up of atmospheric CO_2. The ocean sink is of similar magnitude (Sabine et al. 2004). Given the international efforts to stabilize atmospheric CO_2 concentration and climate (i.e., Kyoto Protocol, C trading markets), the terrestrial C sink can be viewed as a subsidy to our global economy worth trillions of dollars. Because many aspects of the terrestrial C sink are amenable to purposeful management, its basis and dynamics need to be well understood.

There is a sparse understanding of the multiple processes responsible for the terrestrial C sink, the relative importance of each of them, and their future dynamics. Such information is needed to predict the future strength and distribution of biospheric terrestrial sinks, and therefore critical for the design of future pathways of decarbonization.

Terrestrial biospheric models and fully carbon-coupled General Circulation Models (GCMs) are progressively incorporating more C processes thought to drive significant carbon-climate feedbacks. However, being limited by poor understanding and the slow process of building ever more complex models, they still represent a partial set of processes. This limits the ability to produce robust projections of the biosphere's role in the future atmospheric CO_2 growth and its impacts on climate change.

In this paper we demonstrate that the underlying ecology of terrestrial biospheric CO_2 sinks suggests that, despite having the potential for increased C sink owing to atmospheric and climate change over the next decades, most of the biological sinks will eventually level-off and subsequently declined to zero (hereafter referred as "sink saturation") whereby no further C will be removed from the atmosphere.

Coupled with this sink decline, global warming and deforestation have the potential to destabilize large biospheric C pools (hereafter referred as "vulnerable C pools") which would add CO_2 to the atmosphere. This C source component will further diminish the net gains of C sinks and could even diminish the sink strength beyond zero, thereby moving from being a C sink to a source during this century. Such an eventuality would put further pressure on society to select higher targets of CO_2 emission reductions from fossil fuel burning.

In this paper, we first introduce briefly the current understanding of the global distribution of C sources and sinks. Second, we discuss processes thought to be responsible for the current terrestrial C sink and their likely future dynamics highlighting their potential contribution to saturation of the terrestrial C sink. Third, we discuss the dynamics of the future terrestrial C balance based on results from stand-alone and fully coupled carbon-climate models, and their accuracy in representing various physiological and ecological sink/source processes. Recommendations for future research are made at the end of the chapter.

6.2 Location of the Current Terrestrial Carbon Sinks

The dual constraint of the measured global atmospheric CO_2 concentration growth rate and the relatively weak north-south CO_2 concentration gradient indicate a net sink of about 2 Pg C yr^{-1} in the northern extratropics (Tans et al. 1990). These observations are corroborated by additional information from $\delta^{13}C$ and O_2/N_2, as well as increasingly sophisticated inverse modeling with augmented flask data (Rayner et al. 1999; Kaminiski et al. 1999; Gurney et al. 2002).

Despite the robustness of the global and latitudinal average estimates, inverse modeling calculations are poorly constrained at the continental scale, particularly with regard to the longitudinal partition of the sink. Model intercomparisons suggest that the Northern Hemisphere terrestrial sink is relatively evenly distributed across North America, Europe and Asia, with boreal North America being C neutral (Gurney et al. 2002; Yuen et al. 2005). Large uncertainties still remain for northern Africa, tropical America, temperate Asia and boreal Asia. Interpretations of global distributions of atmospheric O_2/N_2 ratio (Keeling et al. 1996) and inverse analyses (Rayner et al. 1999; Gurney et al. 2002) conclude that tropical ecosystems are not a strong net sink, and most likely they are C neutral. Given that tropical forests are well documented as current large CO_2 emitters via deforestation, this implies that there must be a strong balancing tropical sink of equivalent magnitude to that of emissions from tropical deforestation. During the 1980s–1990s this was estimated to be 0.6 to 2.2 Pg C yr^{-1} (DeFries et al. 2002; Archard et al. 2002; Houghton 2003).

Atmospheric data provides limited information about the processes driving terrestrial sinks. Such information requires the addition of ground-based observations, experiments, and modeling (Canadell et al. 2000; Global Carbon Project 2003). Processes responsible for the sinks in the tropical, mid-, and high-latitudes need to be equally understood since the saturation of any one can have large impacts on the growth rate of atmospheric CO_2.

6.3 Dynamics of Processes that Contribute to Carbon Sink Saturation

A decrease in the long term terrestrial net CO_2 sink can arise from a decrease in the C uptake component or increase in the emission component (see Box 6.1). Key sink processes are CO_2 fertilization of photosynthesis, N fertilization of net primary production, woody vegetation thickening and encroachment, forest regrowth in abandoned cropland, and afforestation/resforestation. Key C emission processes are soil respiration in response to warming, permafrost thawing and subsequent decomposition of organic matter, fires, deforestation, and peatland drainage. Sink saturation occurs when the increase in efflux becomes equal to the increase in uptake for a given period of time producing no increase in net uptake.

When assessing whether and when C sinks might saturate in the future, there are two types of dynamics to be considered: *(i)* Saturation of C uptake processes which refers to the diminishing sensitivity of a sink process to an external forcing or stimulus (e.g., CO_2 fertilization effect on photosynthesis as atmospheric CO_2 increases); and *(ii)* increasing C emission driven by external forcing (e.g., warming effects on heterotrophic respiration).

We refer to C pool saturation when the annual emission itself becomes equal to the annual uptake which involves two types of processes: *(i)* the "fast response" C uptake (e.g., net photosynthesis) is matched by the "fast response" C emission (e.g., heterotrophic respiration), or *(ii)* long timescale, "slow response" processes cause pools to reach maximum size (e.g., tree death rate matches new tree growth rate or, for soil, the capacity for organic matter stabilization by clay surfaces is exhausted).

Box 6.1. Description of GPP, NPP, NEP and NBP

Plants take up carbon dioxide via photosynthesis. The sum of photosynthesis over a year is termed Gross Primary Productivity (GPP). The actual C fixed into plants, the Net Primary Productivity (NPP) of an ecosystem or growth of an individual plant, is the balance between GPP and the C lost through plant respiration (i.e., the construction and maintenance cost). It is not until C losses by microbial respiration (heterotrophic respiration, Rh) in litter and soil, and by herbivory are accounted for that we obtain the net C balance of an ecosystem. This net balance is termed Net Ecosystem Productivity (NEP). However, when considering long periods of time and large regions (or the whole terrestrial biosphere for that matter) we need to include other processes that contribute to loss of C such as, fires, harvest, erosion, and export of C in river flow. The ecosystem C balance measured over the full disturbance/recovery cycle is often termed the Net Biome Productivity (NBP) which for a system in equilibrium is zero. Continuous enrichment of the resource base on which the biome depends by CO_2 enrichment or N-deposition could lead to a positive NBP. Continuous global warming might lead to a positive NBP initially followed by a negative NBP later.

6.4 Processes Contributing to Terrestrial Carbon Sink Saturation

Our current understanding of the global net C sinks requires invoking three types of processes (Table 6.1): *(i)* processes driven by changes in atmospheric composition (e.g., CO_2 fertilization on photosynthesis, nitrogen (N) deposition fertilization, pollution damage); *(ii)* processes driven by climate change (e.g., temperature and precipitation effects on gross primary production and heterotrophic respiration); and *(iii)* processes driven by land-use change or land management (e.g., deforestation, forest regrowth, woody encroachment, forest thickening due to fire suppression, afforestation and reforestation, changes in soil C pools under cultivation and grazing). The relative strengths of each of these processes contributing to the current sink and its spatial distribution remain largely to be evaluated (Houghton 2002; Smith 2005).

6.4.1 Processes Driven by Atmospheric Composition Change

6.4.1.1 *CO_2 Fertilization Effect*

Short term responses of leaf net photosynthetic rate to CO_2 concentration saturate at around 800–1 000 ppm. The response shows a compensation point (zero net rate) of about 50 ppm for C_3 species, followed by a near-linear increase up to around present ambient atmospheric CO_2

Table 6.1.
Sink and source processes driving the anthropogenic perturbation of the net C exchange between land and atmosphere

Processes	Sink, source
Processes driven by atmospheric composition change	
CO_2 fertilization of photosynthesis	Sink
Nitrogen deposition fertilization on net primary production	Sink
Air pollution effects on net primary production	Reduced sink
Processes driven by climate change	
Temperature and precipitation effects on net primary production	Sink/reduced sink
Temperature and precipitation effects on heterotrophic respiration	Source
Permafrost thawing	Source
Shifts in natural disturbances (wildfires and insect attacks)	Source/sink
Shifts in vegetation types	Sink/source
Processes driven by land-use change and land management	
Afforestation and reforestation	Sink
Forest regrowth in abandoned cropland	Sink
Vegetation thickening and encroachment	Sink
Shifts in fire regimes and other disturbances	Source/reduced source
Soil erosion and C burial	Source/sink
Crop management	Source/sink
Deforestation	Source
Peatland drainage	Source

concentration which is determined by the abundance of the CO_2-fixing enzyme RuBP carboxlyase (Rubisco) (Farquhar et al. 1980; Farquhar and Sharkey 1982). That is, when the activity of this enzyme is limiting, overall photosynthetic rate rises proportionally with CO_2 concentration. At higher concentrations, the photosynthetic rate becomes limited by the regeneration capacity of the CO_2 acceptor molecule, ribulose-1,5-biphosphate (RuBP), resulting in a curvilinear response to increasing CO_2 concentration (Fig. 6.1a). From this leaf-level saturation response, we can infer that in the absence of any other limiting factor (e.g., light, nutrients, water) net primary productivity will not increase with increasing CO_2 beyond 800–1000 ppm (Fig. 6.1b). When other limiting factors co-occur with CO_2, the response and saturation point is different. In the short term, water limitation increases the saturation point with respect to atmospheric CO_2 concentration because water stress reduces stomatal conductance thereby decreasing the CO_2 concentration inside of the leaf. Nutrient limitations, however, can significantly decrease the saturating concentration.

Plants growing in a CO_2 richer atmosphere with sub-optimal resources acclimate to the greater C availability by lowering photosynthetic capacity below that expected for a given CO_2 concentration (Long et al. 2004). This down regulation of C fixation appears to express an optimization of the deployment of limited C and nutrient resources, particularly N (Xu et al. 1994; Wolfe et al. 1998).

Herbaceous species growing in sub-optimal conditions with multiple levels of atmospheric CO_2 (including sub-ambient CO_2 concentrations) showed that the biomass of C_3 plants is linearly related to CO_2 at CO_2 concentrations corresponding to pre-industrial (280 ppm) to present (380 ppm) (Fig. 6.1c; Polley et al. 1992; Polley et al. 1993; Polley et al. 1994; Dippery et al. 1995). At future CO_2 concentrations greater than 450–550 ppm most species enter the saturating responses zone. Woody species have similar responses with a consistent linear increase in plant biomass from pre-industrial to ambient concentrations, and often a saturating growth response in the same range as for herbaceous species (450–550 ppm) (Fig. 6.1d). Hättenschwiler and Körner (2000) found the same behavior for six European temperate tree species growing in the low light regime of a forest understory.

It is well established that N limitation prevents the full expression of the CO_2 fertilization effect on plant growth, and that a progressive N limitation is likely to be the principal factor driving the above saturation curves (Luo et al. 2004; Reich et al. 2006; Groenigen et al. 2006). That is, N becomes increasingly limiting (at least in the period up to a few years typical of elevated CO_2 experiments) as both N and C are immobilized in new plant and soil organic matter fractions. Other nutrients, particularly phosphorus, also limit plant growth responses to increasing atmospheric CO_2 (Barrett and Gifford 1995; Dukes et al. 2005; Edwards et al. 2005).

There are systems that already show saturation at present CO_2 concentrations due to nutrient or other environmental constraints (Körner et al. 2007, Chap. 3 of this volume). This is the case for: *(i)* a mature temperate trees

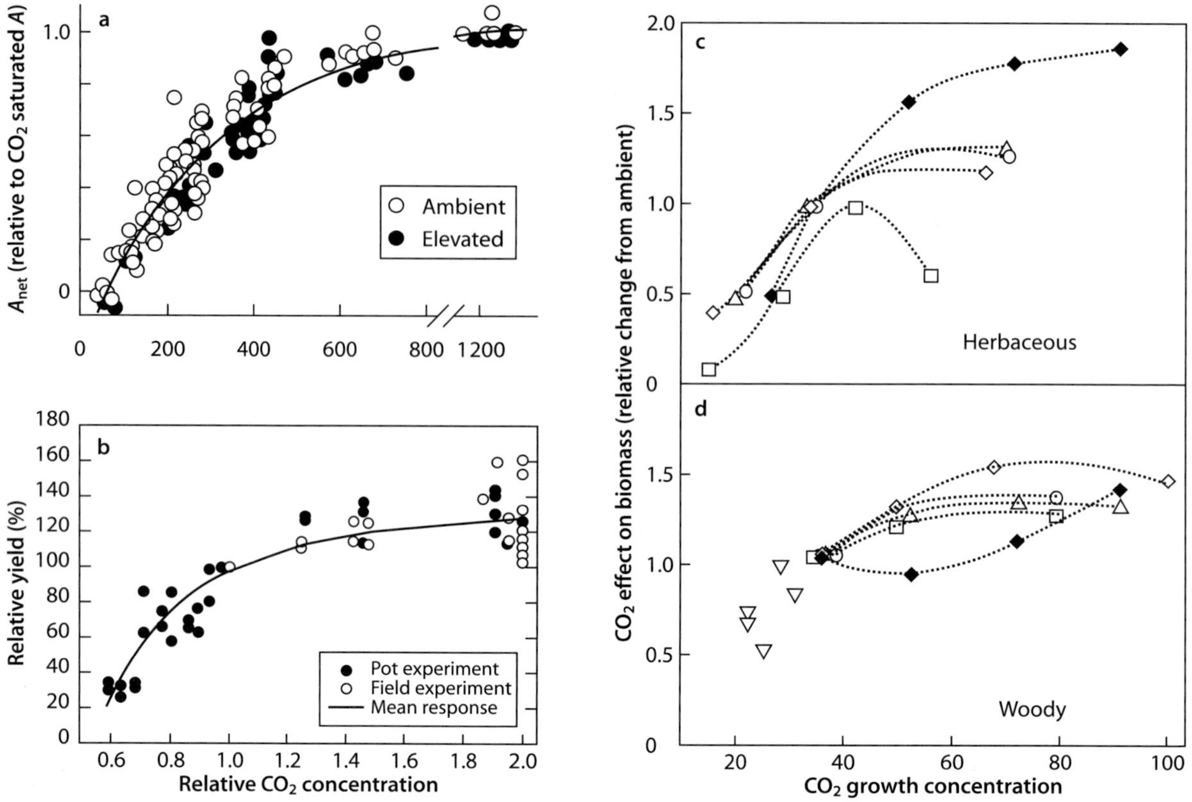

Fig. 6.1. Plant responses with saturation dynamics to increasing levels of atmospheric CO_2. **a** Typical A/C_i curve: Leaf net C assimilation at increasing leaf internal CO_2 concentrations (CO_2 concentration at the site of carboxylation in the chloroplasts). **b** Wheat responses to increasing CO_2 levels (data from field experiments (*open symbols*) and pot/glasshouse experiments (*solid symbols*, Olesen and Bindi 2002). **c** Data for the herbaceous species *Abutilon theophrasti* (Dippery et al. 1995), *Homogyne alpina* (Hättenschwiler and Körner 1996), *Arabidopsis thaliana* (Ward and Strain 1997), and *Glycine max* (*closed symbols*, Rogers et al. 1983; *open symbols*, Allen et al. 1991). **d** Data for the woody species *Quercus alba* (Norby and O'Neill 1989), *Liriodendron tulipifera* (Norby and O'Neill 1991), *Pinus taeda* (Rogers et al. 1983), and *Liquidambar styraciflua* (*closed symbols*, Rogers et al. 1983; *open symbols*, Tolley and Strain 1984). For *Prosopis glandulosa*, data were available for subambient concentrations only, and the *dashed line* is a linear regression ($R^2 = 0.55$, $P = 0.09$, Polley et al. 1994)

in Switzerland (100 years, 30–40 m tall) which after three years of CO_2 fumigation showed no growth response (Körner et al. 2005); *(ii)* trees growing nearby natural CO_2 springs (Tognetti et al. 2000); *(iii)* 30 yr old *Pinus* spp. growing at the high elevation tree-line in the Swiss Alps (Handa et al. 2005); and *(iv)* grassland in California after 5 years of elevated CO_2 exposure (Dukes et al. 2005). A number of other studies showed no responses to increased atmospheric CO_2 (Mooney et al. 1999; Luo et al. 2006).

Eleven Free-Air CO_2 Enrichment (FACE) experiments encompassing bogs, grasslands, desert, and young temperate tree stands report an average increased NPP of 12% at 550 ppm when compared to ambient CO_2 (Nowak et al. 2004). Four FACE studies on forest stands showed a 23% median increased NPP, an expectedly high response for stands made up of young trees and saplings (Norby et al. 2005). A meta-analysis of over a hundred studies shows about ⅔ of the experiments responding positively to increased CO_2 (Luo et al. 2006). This analysis also shows that ecosystems under elevated CO_2 can accumulate N, supporting the hypothesis that terrestrial ecosystems have certain capacity to gradually acquire additional N required to continue accumulating C under increasing CO_2 concentrations (Gifford 1992, 1994; but see Reich et al. 2006; Groenigen et al. 2006).

Just as increased leaf-level photosynthesis does not necessarily translate into whole plant growth, increased NPP does not necessarily translate into increased Net Ecosystem Productivity (NEP) (i.e., net C sequestration). Ecosystem-level net C fixation is constrained by water, temperature, light, nutrients, and a host of biotic interactions between individuals and species that involve CO_2 emissions from ecosystems. There are few elevated CO_2 experiments that have been able to measure annual net C exchange owing to intrinsic methodological difficulties. Over a three-year period, NEP of a nutrient-limited tundra ecosystem exposed to a step increase in CO_2 in Alaska exhibited a pulse-increase producing a temporary C sink (Oechel 1994). The positive NEP declined over time and completely disappeared after three years. More recently a Mojave Desert shrubland that was still accumulating C (i.e., positive NEP) exposed to 8 years of elevated CO_2 using FACE technology showed in its 8th year a decreased annual net C uptake at elevated CO_2 compared to ambient levels (Jasoni et al. 2005). These examples illustrate that a variety of responses of NEP to CO_2 can occur in different ecosystems.

In summary, ecosystem responses to large step increases of CO_2 concentration can show a strong saturation behavior driven by resource limitation. Plant growth is most stimulated under pre-industrial to present atmospheric CO_2 concentrations with some systems showing saturation at present CO_2 concentrations. For a large portion of systems studied, CO_2 exposure to higher CO_2 concentrations does stimulate plant growth. In these cases, increasing CO_2 concentrations results in a saturation response at around 500–600 ppm, a much lower level than that expected based on solely leaf-level physiological grounds. It is unclear to what extent over decades of gradual annually increasing CO_2 concentration the acquisition of atmospheric N by symbiotic and free-living N fixation (and by increased N capture from N deposition and N-cycle turnover) will be enhanced by elevated CO_2 concentration.

The CO_2 fertilization effect on C accumulation is not solely the result of photosynthesis fertilization, but also of increased water use efficiency by plant canopies owing to increased photosynthesis at lower stomatal conductances (Morison et al. 1985; see review by Mooney et al. 1999). The result can be a slower rate of depletion of soil moisture, so that higher rates of photosynthesis may be maintained for longer fractions of soil wetting/drying cycles under high CO_2, particularly in arid and semiarid ecosystems. This effect has been consistently reported for water-limited systems such as deserts (Pataki et al. 2000), Mediterranean grassland (Field et al. 1996) and savanna ecosystems (Owensby et al. 1997). The implication is that "well-watered" ecosystems, including the most productive forest regions of the tropics and boreal zone (likely to play the largest role in C sequestration), take the least advantage from the beneficial effects of reduced stomatal conductance. Therefore, they may be the first systems to approach an upper limit for the physiological effects of increasing CO_2.

6.4.1.2 *Fertilization by Nitrogen Deposition*

Given that N often limits plant growth in terrestrial ecosystems, particularly temperate regions (Vitousek and Howarth 1991), it has long been speculated that increased plant growth from N deposition is likely to be an important C sink process. Much of this reactive atmospheric N results from N fertilizer use in agriculture and the combustion of fossil fuel in industrial activities. The first process produces reactive N, predominantly as NH_3 (and NO_3); the second produces NO_x ($NO + NO_2$), which eventually deposit on vegetation and soils as dry and wet deposition (Galloway and Cowling 2002). However, it is known from fertilization trials and long term N deposition studies that there is an upper limit to which additions of N into the system will results in increased NPP. Past this threshold, pollutants associated with N deposition have a negative effect on ecosystems particularly tropospheric ozone (Aber et al. 1998; see Sect. 6.4.1.3).

Early modeling studies showed that N deposition could have a significant impact on enhancing C sequestration in forests, due to their high C/N ratio, particularly at mid-latitude forests in the Northern Hemisphere, where N deposition is highest and N-limitation is common (Holland et al. 1997; Townsend et al. 1996). These models suggested that N deposition could result in increased C sequestration at a rate of 0.1 to 2.3 Pg C yr^{-1}. At its upper plausible limit, N fertilization could account for the entire net terrestrial C sink. Subsequent field experimentation suggests 0.25 Pg C yr^{-1} as the upper realistic limit of C stimulation by N deposition (Nadelhoffer et al. 1999). More recent analyses using the Biome-BGC model and new N deposition data reinforce this lower end value with an additional C uptake due to N deposition of 0.19 Pg C yr^{-1} and 0.26 Pg C yr^{-1} for the decades of 1980s and 1990s respectively (Galina Churkina, in preparation). That analysis attributed 8 Pg C of the terrestrial sink during the period 1950–2000 to fertilization by N deposition in Northern Hemisphere forests.

Global analyses of N deposition impacts continue to be poorly constrained by lack of high quality spatially-explicit datasets of N deposition (but see Holland et al. 2005 for the US and Europe) and an incomplete knowledge of the various pathways by which deposited N is incorporated into ecosystem components, including N absorption directly by plant canopies.

Taking into account future projections of increased N deposition over the next 100 years, largely in tropical Americas, Southern Africa, and China-India (Galloway and Cowling 2002; Fig. 6.2), it is unlikely that N deposition will create any major additional C sinks. Field experiments using N additions in phosphorus-limited

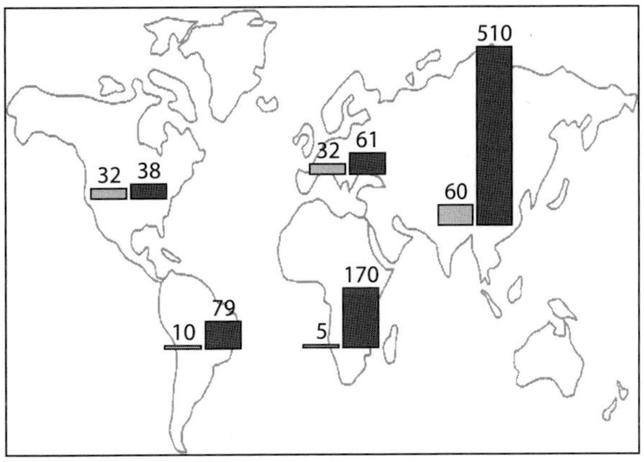

Fig. 6.2. Current and future projections of reactive nitrogen N creation rates in different regions of the world (Galloway and Cowling 2002). *Left bar* indicates the reactive N in 1996 (Tg N yr^{-1}), *right bar* indicates maximum reactive N at the time of projected peak population in 2100

tropical forest show higher emissions of nitrogen oxides than in N-limited forests, both after single pulse and chronic N additions (Hall and Matson 1999). In fact, anthropogenic N inputs may reduce productivity due to the indirect effects on acidity and the availability of phosphorus and cations (Matson et al. 1999). In China and India, most of the N deposition increases in will occur on agricultural lands with little or no sink potential. Therefore the fertilization effect of N deposition on C uptake at present is probably close to the saturation level if it has not already been reached.

There are, however, two related processes by which N may be able to stimulate further C uptake. First, N-deposition concurrent with increasing atmospheric CO_2 concentration may be acting synergistically to progressively increase the standing C pool in terrestrial ecosystems by partially reducing the limitation that its availability imposes on C uptake (Gifford et al. 1996). Second, increased soil respiration in a warmer world, provided water is not limiting, will release additional soil N which can stimulate C uptake to some degree (see Sect. 6.4.2.2).

6.4.1.3 *Air Pollution*

Air pollution, particularly tropospheric ozone (O_3) background concentrations, has increased since the beginning of the industrial revolution. Tropospheric O_3 is formed in photochemical reactions between carbon monoxide, methane (CH_4) and other hydrocarbons, and NO_x. All precursors are largely the result of fossil fuel combustion (Felzer et al. 2004 and references within). Unlike other more localized pollutants, tropospheric ozone is a major regional issue with hot spots in Eastern US, Eastern Europe and China, and with teleconnections to other regions making tropospheric O_3 a global pollutant (Sanz and Millán 1999; Derwent et al. 2004).

Plants exposed to tropospheric O_3 reduce C uptake because O_3 damage cellular tissue which in turn reduces photosynthesis and stomatal conductance. Historical analyses of the possible negative effects of O_3 damage on terrestrial carbon sequestration show a reduced sink between 0.1 and 0.3 Pg C yr^{-1} for the period 1950–1995 (Felzer et al. 2005). The suppression of the sink although small is very significant over the 45-year cumulative period, and the relative change was several times larger for hot spot regions such as US, Europe and China.

Physiologically stressed vegetation is often more susceptible to insect and fungal attacks. Karnosky et al. (2002) showed that aspen (*Populus tremuloides*) developed 3–4-fold increases in levels of rust infection (*Melampsora medusae*) following long exposures to O_3. This further exacerbated the reduction in C update.

Background O_3 concentrations and peak levels are predicted to stay high or increase in the future (Fowler et al. 1999).

6.4.2 Processes Driven by Climate Change

6.4.2.1 *Precipitation and Temperature Effects on Net Primary Production*

Increased length of the growing season, increased precipitation, increased temperature, and decreased cloud cover are factors thought to be responsible for an estimated 6% increased terrestrial NPP (3.4 Pg C) over the last two decades of the 20th century (Nemani et al. 2003); this suggests a possible increase in C sequestration. IPCC (2001) reported that the net sink in the Northern Hemisphere increased from 0.2 Pg C yr^{-1} during the decade of the 1980s to 1.4 Pg C yr^{-1} during the 1990s with associated large uncertainties.

These are the processes and trends, along with CO_2 fertilization, that support model projections suggesting a likely increase in the terrestrial C sink over the next half a century. The same model projections suggest that soil respiration will surpass levels of C uptake later in this century, and that the tropical sink will be ultimately destabilized due to drought (Cao and Woodward 1998; Cramer et al. 2001; Cox et al. 2000; Fung et al. 2005; Friedlingstein et al. 2006). This is projected to lead to an overall decline of the net terrestrial sink which becomes a source by the end of the century in some of the model projections. A key model assumption here is that soil respiration will be as highly temperature dependent in the long term as is found in short term experiments (months to a few years).

Interestingly, new observations are already showing a dampening of regional C sinks strength due to increased climate variability (particularly due to hotter droughts), and thus challenging the expectation of C sink enhancement over the next few decades. An analysis of the entire Northern Hemisphere shows that since 1994 the well known acceleration of C uptake during early spring was offset by decreased uptake during summer, most likely due to drier summers in mid and high latitudes (Angert et al. 2005). The heatwave in 2003 alone reduced the gross primary productivity of European ecosystems by 30% resulting in a net CO_2 source to the atmosphere of 0.5 Pg C yr^{-1} (Ciais et al. 2005), equivalent to about 4 years of C accumulation in these systems lost in a single extreme event of just a few weeks duration.

In the tropics, drought can enhance NEP in the short term because it suppresses heterotrophic respiration (Saleska et al. 2003). However, over the longer term, drought and associated fire (Werf et al. 2004) are likely to diminish the Net Biome Productivity (NBP) in regions that experience an increasingly dry environment.

Thus, increases in intra- and inter-annual variability (frequency and/or intensity) as predicted by some climate models under higher greenhouse gas (GHGs) concentrations (Giorgi et al. 2004; Meehl and Tebaldi 2004) could depress the net C sink of the Northern Hemisphere and even push it to become a CO_2 source. If the increase in climate variability (e.g., frequency of extreme events) that has occurred in the past decade were to persist as part of climate change, the net terrestrial C sink could soon be declining, well before model predictions that show a possible sink saturation between the middle and end of this century (Cox et al. 2000; Cramer et al. 2001; Friedlingstein et al. 2006).

6.4.2.2 *Temperature and Water Effects on Heterotrophic Respiration*

The sensitivity of soil C pools to global warming is the single biggest uncertainty in the C cycle to determine future growth rates of atmospheric CO_2. Terrestrial ecosystem models and coupled C cycle GCMs suggest large positive feedbacks to climate due to an assumed high sensitivity of soil heterotrophic respiration to warming (Cramer et al. 2001; Cox et al. 2000; Friedlingstein et al. 2006). This assumption is supported by some short term soil warming field experiments and laboratory soil incubations which consistently show an increase in soil respiration when exposed to elevated temperatures. A meta-analysis of thirty-two field studies showed a 20% increase in soil respiration when soils were experimentally heated 0.5 °C to 5 °C for periods ranging from 2 to 9 years (Rustad et al. 2001; Fig. 6.3). Studies with laboratory soil incubations at various temperatures have also shown that temperature sensitivity is usually high at low temperatures (Q_{10} as high as 8 close to 0 °C) and low at higher temperatures (Q_{10} at around 2 for temperatures above 20 °C) (Kirschbaum 1995).

Other studies show either no correlation between respiration fluxes and temperature or enhanced soil respiration that disappears after some time (Giardina and Ryan 2000; Valentini et al. 2000; Jarvis and Linder 2000; Luo et al. 2001). The latter experiments suggest that once the more labile C fraction has been respired, the remaining more recalcitrant and stable pools are not sensitive to temperature.

More recently, modeling and experimental studies have shed light on this apparent paradox by showing that more recalcitrant pools of Soil Organic Carbon (SOC) do not show lower sensitivity to temperature than labile pools (Knorr et al. 2005; Fang et al. 2005; Fierer et al. 2005). In fact the studies show that slow turnover pools, the largest and therefore dominant component, are equally (Fang et al. 2005) or more (Knorr et al. 2005) sensitive to warming than fast pools.

It is important to note that concurrent with organic C mineralisation to CO_2 during heterotrophic respiration, there is N mineralisation to ammonium and nitrate that could feedback onto productivity. The meta-analysis of soil warming experiments found that associated with an average 20% increase of heterotrophic respiration with warming there was an average 46% increase in net mineralisation of soil N (Rustad et al. 2001). Given the C cycle is tightly coupled with the N cycle (Gifford 1994) some of this increased production of mineral N can be taken up by plants and converted to biomass; this would compensate to some extent for soil C lost under warmer conditions.

Thus it is not well established as to the extent global warming will decrease soil C pools. Certainly the N-cycle feedback can reduce the magnitude of the net long term decline in soil C due to increased temperature. In some ecosystems the compensation can be significant whereas in ecosystems with largest soil C pools (e.g., tundra and taiga with frozen soils and peatlands which are highly sensitive to temperature) the compensation will be smaller. A more detail analysis of C emissions from frozen soils and peatlands are described in Sect. 6.4.2.3 and 6.4.3.7.

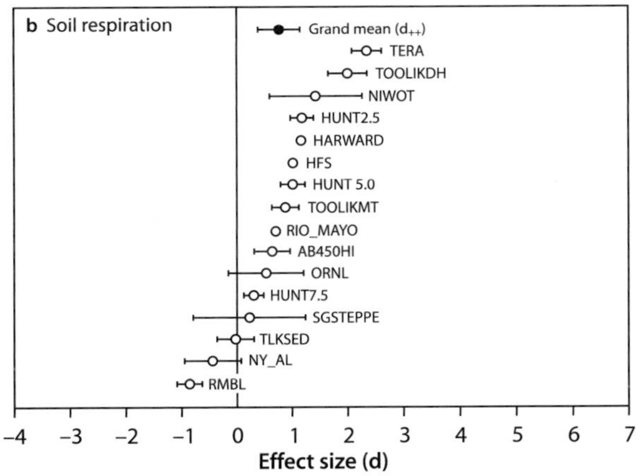

Fig. 6.3. Effects of field experimental warming on soil respiration in 16 studies (Rustad et al. 2001). Mean effect sizes and 95% confidence interval for individual experiments. Soils exposed to elevated temperatures of 0.3 to 6.0 °C for periods of 2 to 9 years

6.4.2.3 *Permafrost Thawing*

Frozen soils hold over 400 Pg C which have been accumulated over thousands of years (Sabine et al. 2004). Approximately 54% of frozen soils occur in Eurasia, largely in Russia, and 46% in North America, largely in Canada (Tarnocai et al. 2003). An additional 500 Pg C exist in frozen loess that have been accumulated in glacial times in the north plains of Siberia (Zimov et al. 2006). With the rapid warming of the northern latitudes well above the global averages (IPCC 2001), C in frozen ground can be exposed to warmer aerobic conditions which would result in increased decomposition of organic matter and thus C emissions. In fact, widespread observations exist of permafrost thawing leading to the development of thermokarst and lake expansion, followed by lake drainage as the permanent permafrost further degrades (Camill 2005; Smith et al. 2005).

Preliminary estimates show that permafrost area could shrink by up to 25% with a mean global warming of 2 °C (Anisimov et al. 1999). More recent estimates using a fully coupled GCM show that of the 10 million km^2 of present-day permafrost as little as 1 million km^2 near surface permafrost will remain by the end of this century (Lawrence et al. 2005). Melting permafrost will increase CO_2 and CH_4 emissions, and it is estimated for the Canadian permafrost alone that up to 48 Pg C could be sensitive to oxidation under a 4 °C warming scenario (Tarnocai 1999).

In addition to C exchange between land and atmosphere, lateral transport of DOC from thawing permafrost is another process by which C is lost from peatlands. Frozen permafrost watersheds in Western Siberia release little DOC to the rivers while permafrost-free watersheds show considerably higher amounts that are proportional to the extent of the peatland area (Frey and Smith 2005). For Western Siberia, climate models predict a doubling of the area above a mean average temperature of −2 °C (which coincides with permafrost distribution) and an associated 700% increase in DOC concentrations in streams. Part of the DOC reaching the Arctic Ocean will oxidize and return to the atmosphere.

A preliminary global estimate suggests that up to 5 Pg C could be released from permafrost over the next 20 years and up to 100 Pg C in the next 100 years if it is assumed that 25% of the C locked in frozen soils could be oxidized (Gruber et al. 2004; Raupach and Canadell 2006). This amount could increase to as much as 200 Pg C if we include the loess frozen sediments which have been recently reported. Whether this C is released initially as CO_2 or as CH_4 depends on the local hydrological conditions. Although studies suggest an increased gross C uptake by the newly established vegetation (Payette et al. 2004), long term warming and fertilization studies show increased losses of C from deep soil layers (Mack et al. 2004). These losses were sufficient to counteract increased plant biomass C, thereby contributing a net C flux to the atmosphere.

There are compensating feedbacks to consider. The northwards movement of the permafrost boundary is accompanied by northward movement of the vegetation that fixes C and thus compensating to some extent the loss of soil C (see Sect. 6.4.2.4). However, thawing will expose deep deposits of C to aerobic conditions and the losses of the integrated soil profile (including deep soil layers) will offset the possible C gains from increased productivity.

6.4.2.4 *Shifts in Vegetation Types*

The distribution of the world's vegetation has changed throughout glacial and interglacial periods largely forced by climate and to lesser extent atmospheric CO_2. Consequently it is expected that the distribution of vegetation will change commensurately to the extent of future climate change with associated changes in the size of carbon pools and fluxes (VEMAP members 1995).

Vegetation shifts will change significantly regional C balances, but two of the most important ones at the global scale are: *(i)* woody encroachment in high latitudes due to warmer temperatures (e.g., boreal summer green trees moving into herbaceous systems) (see Sect. 6.4.3.3 for other types of woody encroachment), and *(ii)* the more controversial die-back of Amazonia and subsequent savannization driven by increased water stress (Cox et al. 2004; Schaphoff et al. 2006). In the first case, warmer temperatures are already allowing for longer growing seasons and a substantial increase of C pools in vegetation which will likely result in an additional C sink (but see Schaphoff et al. 2006 who argue otherwise due to the concomitant increased soil respiration; see Sect. 6.4.2.1). In the second case, savannization of Amazonia would result in substantial emissions to the atmosphere which is the largest contributor to a positive carbon-climate feedback projected to increase global warming by 0.1–1.5 °C by the end of this century (Friedlingstein et al. 2006).

Some of these vegetation shifts may have transient dynamics with abrupt releases of C emissions, often associated with disturbances (e.g., fire), several times bigger than the slower increased of C pools in biomass and soils in other regions of the biosphere (Smith and Shugart 1993).

6.4.3 Processes Driven by Land-Use Change and Land Management

6.4.3.1 *Afforestation and Reforestation*

Globally, the upper "biological" boundary of C sequestration by land management could be estimated as the amount of C lost historically from land-use change, about 180–200 Pg C (DeFries et al. 1999). This could decrease atmospheric CO_2 concentration at the end of this century by 40 ppm to 70 ppm (House et al. 2002). Taking account of expanding demands for food, fiber, energy and urbanization, the potential of forest C sequestration is reduced to an achievable capacity of only 10 to 50 Pg C (Cannell 2003) or 5% to 25% of the upper boundary for C sequestration estimated based only on biophysical considerations.

In addition, there is a whole fabric of interlinked environmental and socio-economic constraints and opportunities which will ultimately determine the achievable capacity to sequester C such as costs of land for sequestration and for maintaining C stores, environmental requirements for other resources, environmental constraints, social factors, economical feasibility, institutional factors, and demographics (Raupach et al. 2004).

Thus, despite a significant potential for creating C sinks through afforestation and reforestation, the ultimate impact on atmospheric CO_2 growth will be relatively modest and limited in time because globally there will be a balance between intensification (and reduced requirements of land) and an expanding population largely in less developing world with a large demand for new land to meet food and energy requirements. However, a larger potential exist for future enhancement of the C sink if there is greater adoption of biofuels, which could provide the incentive for the establishment of additional plantations.

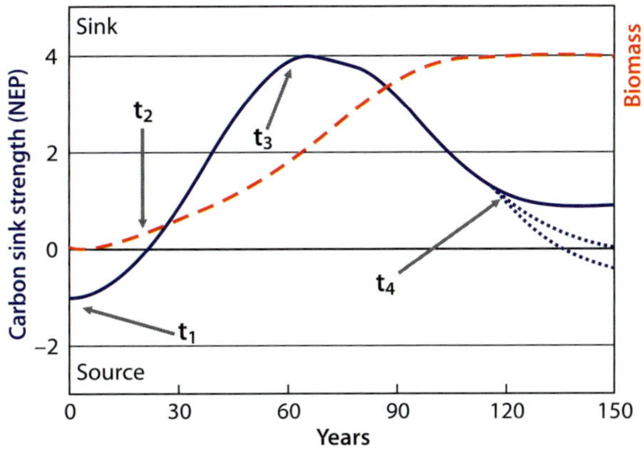

Fig. 6.4. Diagrammatic representation of C sink strength (i.e., net primary productivity, kg C yr^{-1}) and biomass (kg C) accumulation over time of an hypothetical forest stand. t_1: initial state (e.g., after clear cut); t_2: flip point from C source to sink; t_3: maximum net C uptake; t_4: slow down of the C sink, potentially becoming neutral or even flipping from C sink to source. *Dotted lines* indicate alternative trajectories at maturity

6.4.3.2 *Forest Regrowth on Abandoned Cropland*

Forest regrowth on abandoned agricultural land has been identified as one of the most significant process to explain the net C sink in the Northern Hemisphere (USA: Houghton et al. 2000; Pacala et al. 2001; Europe: Janssens et al. 2005). The western movement of agricultural lands from the forests of the east to the prairies of the midwest has been the main driver of land abandonment and subsequent regrowth in eastern North America since the beginning of the last century, while agricultural intensification was the principal driver in Europe (Kauppi et al. 1992; see in Ramankutty and Foley 1999; Goldewijk 2001).

This agricultural shift has resulted in a substantial expansion of relatively young forests with fast growth rates in abandoned cropland, and therefore, with high C sink capacity. Forest cover in Europe and North America has increased over this period and is considered to be an important, if not the dominant, component of their current C sink. Caspersen et al. (2000) showed that for Eastern USA, where the largest amount of cropland abandonment has taken place, that regrowth explains 98% of the sink that is driven by the dynamics of forest demographics (i.e., age structure). Only 2% of forest growth was attributable to increased CO_2 and N deposition.

Stand NEP declines with stand age as measured in a variety of forests in the temperate and boreal regions of Europe, Siberia, and North America (Law et al. 2003; Desai et al. 2005; Fig. 6.4). As forests reach maturity (maximum C pool size) they cease to be sinks (Bond-Lamberty et al. 2004) and even later may become C sources (Goulden et al. 1998; Schulze et al. 1999; Carey et al. 2001).

From a regional and global perspective, persistence of the C sink due to regrowth can only be sustained by continuous expansion of new forests, something that is unlikely to happen in the light of increasing pressures on existing land for multiple uses (urbanization, industrialization, recreation in the industrial world) including fossil fuel substitution with energy crops. In fact, crop abandonment has largely ceased in North America and Europe, while in other parts of the world conversion to cropland is still the primary means by which to meet the food demands of an increasing human population. Forest management can maintain a sustained C sink over the years but at a much lower strength because wood extraction generally returns C to the atmosphere. Old growth forest may not be important globally for the removal of C from the atmosphere, but they hold large C pools that cannot be matched by other land uses (Knohl et al. 2003; Field and Kaduk 2004).

6.4.3.3 *Vegetation Thickening and Encroachment*

Vegetation thickening in forests and woodland, and woody encroachment in savannas and grasslands increase the amount of C stored on land. Many processes are responsible, but fire suppression as a means to control wildfire damage and fuel reduction by grazing may be the most important. Other factors may be also important (for a review see Gifford and Howden 2001).

Managed fire exclusion during the twentieth century has resulted in an increase of biomass in forest and woodlands in many regions (Mouillot and Field 2005; Luger and Moll 1993; Houghton et al. 2000), and the potential exists for further accumulation. For instance, for the USA, forest and woodland biomass in 1990 was 75% of what it was in 1700 and 63% of what it would be in the absence of disturbance; that is, if all forests had the biomass estimated for undisturbed forests lacking wildfire. In the unrealistic scenario of excluding fire from all areas currently lacking trees, the estimated upper limit of C sequestration by woody encroachment would be about 2 Pg C yr^{-1} (Scholes and Hall 1996), although it would not be sustainable for too long.

Woody thickening and encroachment is an important C sink process estimated to account for a substantial proportion of the sink in various parts of the world although it is recognized as the single least well constrained of all sink estimates. For Australia, the introduction of livestock grazing and fire suppression has shifted the dominance from natural open woodlands to tree-grass dominated systems with an estimated 0.035 Pg C yr^{-1} sink in northeastern Australia alone, equivalent to 25% of the total national net emissions (Burrows et al. 2002). Forest thickening and woody encroachment in the US accounts for 0.12–0.13 Pg C yr^{-1} or 22–40% of the apparent total US terrestrial sink (Pacala et al. 2001).

The suppression of wildfire has allowed C accumulation, a process which is not sustainable in the long term as the accumulated C can eventually return to the atmosphere when fire occurs (see Sect. 6.4.3.4). Forests under routine control-burn management will hold less C than the maximum possible, but over the long term they may hold more than forests experiencing occasional catastrophic fires. The fact that the burned area in the extra tropics has increased in a number of regions (Boreal North America and Europe: Mouillot and Field 2005; Fig. 6.5) may indicate that fire exclusion has already reached, or exceeded, a maximum C storage, unless attempts to prevent fire are intensified. Furthermore, this increase in wildfires is also enhanced by the interactions with a changing climate.

In savannas, woody thickening also tends to saturate. As the global terms of trade for ruminant products continues its long term decline, the extent of woody encroachment onto grazed tropical grasslands may continue for decades. However, C accumulation does not always occur. Little or no net C accumulation was found in wet grasslands invaded by *Prosopis*, *Larrea*, and *Juniperus* (Jackson et al. 2002) and drier grasslands invaded by *Juniperus virginiana* (Smith and Johnson 2003). Thus the C sink estimates attributed to woody encroachment may not be great.

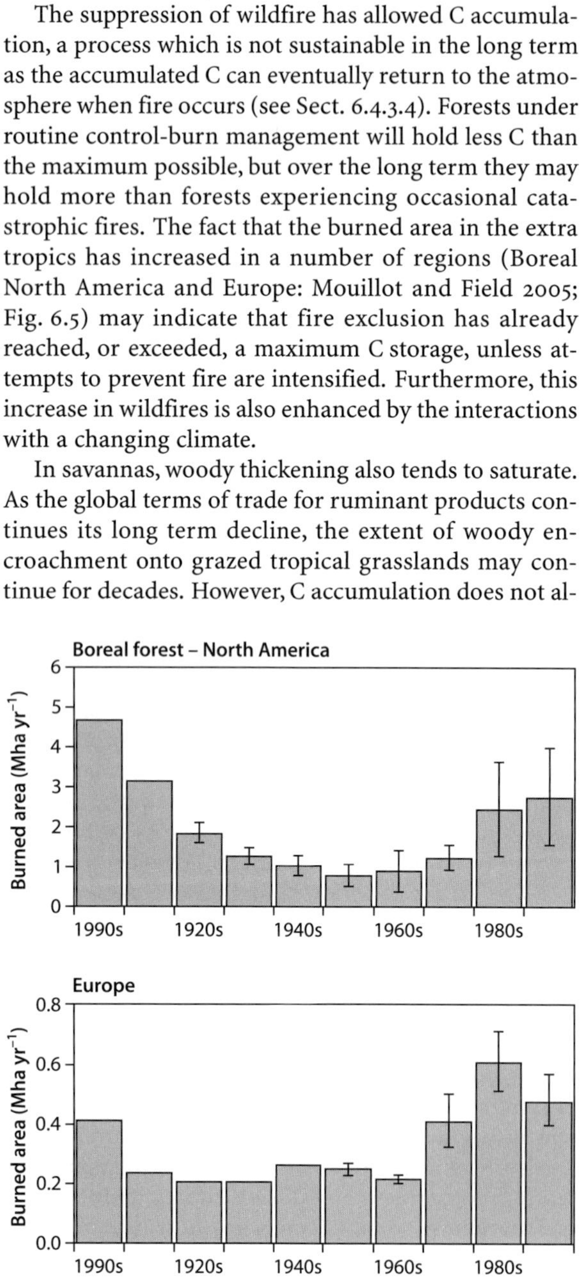

Fig. 6.5. Temporal trend of burned areas (in Mha yr^{-1}) for the 20th century for Europe and Boreal forest in North America (Mouillot and Field 2005)

6.4.3.4 *Shifts in Fire Regimes and Other Disturbances*

Disturbances to vegetation can have dramatic effects on short-term C dynamics, although they are neutral in the long term C cycle. That is, the often quick C release during a disturbance event (e.g., wildfire) will be balanced by C uptake of regrowing vegetation during the following decades (fire as a tool to clear land is considered in Sect. 6.4.3.7 on Deforestation). However, changing climate, atmospheric CO_2 concentrations, and land use and management can alter the frequency and intensity of disturbance regimes and the net contribution to the atmospheric CO_2 growth. It is during the transient dynamics in moving from one fire regime to another that a C imbalance is created: a sink when reducing the frequency of disturbances frequency and a source when increasing the frequency of disturbances.

The annual gross C flux to the atmosphere from global savanna and forest fires (excluding biomass burning for fuel and land clearing) is estimated to be in the range of 1.7 to 4.1 Pg C (Mack et al. 1996). Thus, the potential for fire to alter the terrestrial C sink is significant. One of the most striking examples of fire sensitivity to climate (e.g., drought) was the fires in tropical regions during El Niño in 1997–1998. Fire emissions contributed 2.1 ±0.8 Pg C or the equivalent of 66% of the CO_2 growth rate anomaly during that period (Werf et al. 2004). Although most of the fires ignited by human activities, the area burned and intensity were largely the result of drought.

Despite the sensitivity of fire frequency and intensity to climate change and variability, the long term trends over the past century were driven primarily by the implementation of fire suppression policies in temperate regions and increased use of fire to clear forest in tropical regions.

Fire frequency in the boreal forests of Canada and Russia decreased during most of the last century but has increased during the last two decades as a result of changes in management practices perhaps reinforced by strong climate warming (Dixon and Krankina 1993; Kasischke and Stocks 2000). As a result, modeled C emissions from fires in Canada have consistently increased over the last 40 years, and changed the net C balance of the Canadian forest from being a sink to a small source (Fig. 6.6; Kurtz and Apps 1999).

Future fire activity will continue to be dominated to some extent by fire suppression policies and the use of fire for forest clearing (see Sect. 6.4.3.3). However, greenhouse induced climate change, especially if accompanied by increased climate variability and particularly drought events, will affect overall fire extent and intensity. Models predict increased fire activity with a warmer climate for North America, Russia and Europe (Stocks et al. 1998; Flannigan et al. 1998; Flannigan et al. 2000).

In the tropics, El Niño-drought events have an overriding effect on fire intensity, and therefore are likely to determine the future strength of the terrestrial C sink in that part of the world to a large extent (along with deforestation). Future dynamics of El Niño/La Niña cycles are still uncertain but one view is that there will be more frequent or more severe El Niño-like conditions in a warmer climate (Timmermann et al. 1999).

In addition, the interactive nature between management practices (e.g., fire suppression in temperate forests, deforestation) and climate change largely determine the future fire frequencies. One example concerns how the unnaturally high biomass accumulation in forests due to fire exclusion may respond to warming and increased drought frequency in some parts of the world. Grissino-Mayer and Swetnam (2000) have suggested that excessive biomass accumulation in the southwest of the USA is responsible for an increase in the areas burned in recent years. A second example is the evidence that increased landscape fragmentation in tropical regions as a result of higher pressure for multiple land uses favors the rapid spread of fire (Nepstad et al. 1999; Laurance 2000).

There are other sources of disturbances linked to climate change that are important for the net carbon balance. One of the most notable examples is insect damage linked to warmer conditions in boreal forests which has been as extensive as fire during some years (Kurz and Apps 1999; Carroll et al. 2004). Shifts in the regimes of hurricanes, tornadoes, and windstorms are also linked to global warming although their effects on the future net carbon balance are more difficult to be determined.

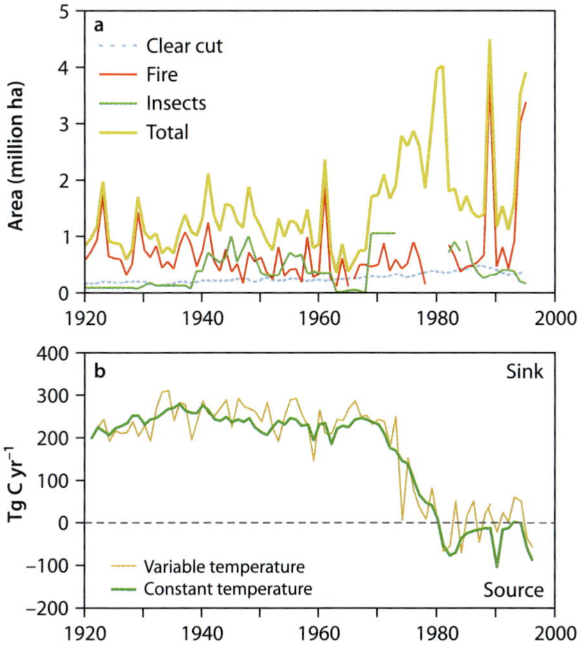

Fig. 6.6. Disturbances and its effects on the net carbon balance of Canada's forests (1920–1995): **a** Area in Mha yr^{-1} disturbed by fire and insect damage, and **b** net ecosystem C fluxes (after Kurz and Apps 1999)

6.4.3.5 *Soil erosion and Carbon Burial*

Changes in land cover and use are estimated to contribute 10–100 times the natural background levels of soil erosion, stimulating the increase of sediment load into the world's rivers by 2.3 Pg yr^{-1} since pre-agricultural times (Syvitski et al. 2005). Of the total C sediment in rivers, about 0.4 Pg C yr^{-1} is total organic C and 0.4 Pg C yr^{-1} is dissolved inorganic C (Richey 2004). Part of this C is intercepted by dams (estimated to be 1–3% of total river sediment (Syvitski et al. 2005)) and wetlands and the rest reaches the coastal zones. For both pathways only a fraction of that C is stored in long term pools and the rest is mineralized releasing CO_2 back to the atmosphere. Recent estimates on the CO_2 outgassing of rivers shows that a large portion of riverine C is mineralized during transport and quickly returned to the atmosphere as CO_2 (Richey et al. 2002). Although this C redistribution may have important effects on regional C balances, the fact that C previously stored in soils may end up under different sequestration characteristics (e.g., anaerobic conditions at the bottom of dams or in ocean sediments) may not necessarily result in a net global greenhouse gas sink. This is only the case if the C residence time in sediments is so much longer than that in the upper soil horizons from which soil was eroded; this would also need to overcompensate for the outgassing during river transport and mineralization at the final destination (C in sediments may be emitted as CH_4 under anaerobic conditions). Therefore, claims for net C sinks due to sediment burial as big as 1 Pg C yr^{-1} globally (Smith et al. 2001; Stallard 1998) are unlikely to be true and require further investigation. Likewise, little is know on the future of this C sink.

6.4.3.6 *Crop Management*

Cultivation has resulted in a loss of 55 Pg C from soils worldwide (Cole et al. 1996), and improved management practices can restore some of this soil C. For example, the introduction of conservation tillage in the USA has increased soil organic matter pools by about 1.4 Pg over the past 30 years (Donigian et al. 1994), with the potential to store a further 5 Pg over the next 50 years (Kern and Johnson 1993; Lal et al. 1998). If best management practices could be implemented at the global scale, agricultural soils could mitigate up to ⅓ of the current annual atmospheric CO_2 increased. However, this would be effective for only 20–50 years (Smith 2004) because soils, as with many other biological C sinks, have an upper limit above which no additional C can be stored. This upper limit could change to a lower level from that of pre-agriculture (e.g., if major soil erosion has occurred) or to a higher level if future climate change increased productivity (e.g., at higher atmospheric CO_2 concentrations).

Over the next 50–100 years, if one includes all available management practices, the potential exists for agricultural management to sequester 0.075 to 0.208 Pg C yr^{-1} in USA arable land (Lal et al. 1998 Metting et al. 1999). In the USA, full adoption of best management practices would likely restore SOC to about 75–90% of its pre-cultivation total (Donigian et al. 1994). Similarly in Europe, it is estimated that 0.113 Pg C yr^{-1} could be offset over the next 100 years (including C offsets from bioenergy crops planted on surplus arable land) (Smith et al. 2000). Changes in management practices in Europe and the USA combined may restore ⅓ to ½ of SOC lost through agriculture globally.

In addition to the limited effectiveness of this process given the ultimate saturation of the pool, the biophysical potentials described above are probably unachievable when socio-economic and other environmental constraints are taking into account. Recent estimates for soil C sequestration potential suggest a global technical potential of about 1.9 Pg C yr^{-1}, whereas the realistically achievable potential is estimated to be 0.2–0.4 Pg C yr^{-1}, with the level of implementation of mitigation measures determined by the market price of CO_2 equivalents (Smith et al. 2006). For Europe, using projections in actual management trends suggest negligible actual sequestration between 1990 and 2000, and negligible sequestration expected by 2010 due to lack of policy incentives to encourage C sequestration (Smith et al. 2005).

6.4.3.7 *Deforestation*

Land-cover shifts from one type to another are responsible for large C fluxes in and out of the terrestrial biosphere. Human activities have altered land cover for thousands of years but the intensity and magnitude have accelerated during the last centuries. Historically, between 3 253 and 3 470 × 10^6 ha have been converted from natural vegetation, approximately 10% of the total land surface. This has reduced global NPP by about 5% and released 182–199 Pg C to the atmosphere (DeFries et al. 1999). Overall, changes in land use and cover since 1850 are responsible for 33% of the increased concentrations of CO_2 observed in the atmosphere (Houghton 1998), 68% of which was due to cropland establishment (Houghton 1999). Some of these values have been challenged on the basis of lower deforestation rate obtained using remote sensing measurements (Archard et al. 2002; Defries et al. 2002) but the overall values are likely to remain high (Fearnside and Laurance 2004). Recent high resolution remote sensing analyses in the Brazilian Amazon show that unaccounted selective logging increased previously reported deforestation areas by 60% to 120% (Asner et al. 2005).

Future conversion of forest and grasslands to pastures and croplands will without doubt increase atmospheric CO_2 concentration. With increasing population, increased consumption, and apparent shifts in diets, either land must become more productive or agricultural area must be expanded. In the most developed world the former is being achieved through technological intensification, but major expansion of agricultural land must occur in many parts of the world, particularly in subtropical and tropical countries where the technology or capital for intensification may not be available. Using the UN intermediate population estimates about ⅓ of the Earth's land cover will change in the next hundred years (Alcamo et al. 1996a,b), with the largest changes expected within the next few decades (Walker et al. 1999).

In a compilation of dozens of scenarios of C emissions from anthropogenic land-use change, the IPCC-SRES (2000) predicts that the largest emissions will result from deforestation in tropical Africa, Asia and Latin America. Deforestation in all these three major regions is expected to decrease towards the end of this century to a small fraction of the levels in 1990. It is worth noting that the declines in Asia and Africa are driven by the depletion of their forests, while Latin America shows the highest uncertainty given the extent of forest resources. Carbon emissions from tropical deforestation may release between 85–130 Pg C by 2100 (Houghton 2005).

In addition to the primary effect of increased C emissions resulting from deforestation, there is a secondary (or amplifier) effect on increasing atmospheric CO_2 due to the reduced sink capacity owing to the shorter residence time of C pools in the newly established pastures or croplands (Gitz and Ciais 2004). Using the IPCC-SRES A2 future scenario, the amplifier effect is responsible for an extra 61 ppm of atmospheric CO_2 by the end of the century, after subtracting the additional C uptake by the oceans and remaining vegetation. The magnitude of this effect, which is not accounted for in climate model projections (IPCC 2001), is of a similar magnitude to the carbon-climate feedbacks reported recently (Jones et al. 2005; Friedlingstein et al. 2006).

It is difficult to know whether the current size of the deforestation source will remain steady for a number of decades before declining or whether it will accelerate C emissions due to the implementation of new land-use practices (Carvalho et al. 2001) followed by a step decline. The role of deforestation in terrestrial sink saturation is, therefore, complex but large given the size of this emission term in the global C budget.

6.4.3.8 *Peatland Drainage*

Water-logged peatlands are subject to CO_2 emissions from drying out and from fire. A drop of the water table brought about by land use and/or climate change, can expose large quantities of C to decomposition that otherwise are locked under anaerobic conditions. Interestingly, recent observations suggest that increased atmospheric CO_2 concentrations alone could also lead to a loss of peatland C through larger fluxes of Dissolved Organic Carbon, DOC (Freeman et al. 2004).

High-latitude peatlands have been a net C sink since early in the Holocene (Smith et al. 2005 for Russia's West Siberian lowland peatlands) but there has been a recent slowdown or cessation in net C accumulation (Botch et al. 1995; Peteet et al. 1998; Turunen et al. 2001; Turetsky et al. 2002; Friborg et al. 2003). There exists a balance in wetlands between increased CO_2 emissions when water tables fall or temperature goes up, and CH_4 emissions when water tables rise. In cold regions, changes in precipitation and temperature will largely determine the net balance between CO_2 and CH_4 emissions. Friborg et al. (2003) illustrates this complexity by showing that a Siberian wetland, despite being a net C sink, was an important source of radiative forcing owing to emissions of CH_4 (which has a warming potential 21 times larger than that of CO_2 over 100 years).

Lowland tropical peatlands, accounting for about 10% of global peatland extent, contain up to 70 Pg C in deposits as deep as 20 m (Page et al. 2002; Page et al. 2004 and references therein). Indonesia, Malaysia, Brunei and Thailand hold the largest tropical peatlands in the world which have been an overall net C sink since late Pleistocene (Page et al. 2004). However, over the last decade a combination of intense draining for agriculture and increasing climate variability in the form of more intense droughts (associated or not to El Niño events) have destabilized this millennia-long C sink (Page et al. 2004; Aldhous 2004; Murdiyarso and Lebel 2007, Chap. 21 of this volume).

During El Niño 1997–1998 events in Indonesia, burning of peat and vegetation resulted in an estimated loss of C between 0.81 and 2.57 Pg in 1997 equivalent to 13% to 40% of the mean annual global C emissions from fossil fuels (Page et al. 2002).

With large tracts of drained swamp forests and new peatland conversion projects to agriculture, tropical peatlands will continue contributing to increased C emissions over the coming decades. With some projections of decreased precipitation over the tropics during the dry season (Wenhong Li, in preparation) and the possibility of El Niño-like events becoming more intense or frequent under a warmer climate, C emissions from peatlands have increased and will continue to diminish the strength of the terrestrial net C sink.

Together, high latitude and tropical peatlands account for over 450 Pg C (Sabine et al. 2004). A preliminary estimate suggests that up to 100 Pg C of CO_2 equivalents could be released to the atmosphere from wetlands and peatlands over the next 100 years (Gruber et al. 2004).

6.5 Integration and Model Predictions

So far it has been difficult to integrate the above disparate knowledge under a common framework that allows the various processes and top-down measurements to constrain one another, and informs estimates of the relatively contribution and spatial distribution of the multiple processes driving the net terrestrial C sink. Suggestive of how little constrained the results are is that, added together, the sink estimates from each individual sink process described in Sect. 6.4 amount to a larger terrestrial net C sink than can be accounted for in the global C budget or the observed growth of atmospheric CO_2. This creates a paradoxical problem of an apparent missing large C source.

The limited understanding of some of the processes and pools has also prevented models from representing the necessary dynamics and underlying processes. The result is that there is no single model that includes all phenomena described above. Global climate models used to predict the effect of increasing GHGs on future climate (GCMs), and specifically the ones with coupled C cycle models (i.e., in the C4MIP family of models; Friedlingstein et al. 2006) deal just with simplified versions of photosynthesis and respiration.

Nevertheless, there have been important global and regional analyses. McGuire et al. (2001) attempted a global analysis to partition the historical global C sink from 1920 to 1992 including climate and atmospheric CO_2 change, land conversion to crops and crop abandonment, and forest harvests. They reported that from 1920 to 1957 terrestrial ecosystems were responsible for a net release of 8.8 Pg C, largely due to cropland establishment. During the period 1958 to 1992 the terrestrial biosphere stored 14.3 Pg C largely due to the physiological effects of rapidly rising atmospheric CO_2, although cropland establishment continued to release C. This is one of the best global mechanistic analyses to date but it only accounted for a portion of the sink processes believed to be important.

A sectoral approach has been used by some nations to attribute national sinks and sources to specific activities of their economy and ecosystems. Such studies do not provide information on the ultimate causes of the various sink processes but partition the individual fluxes contributing to a regional C sink. For the conterminous USA (1980–1989) eight atmosphere-land fluxes contributing to the overall C sink were reported (Pacala et al. 2001): (1) forest tree growth, (2) other forest components such as nonliving organic matter, (3) agricultural soils, (4) non-forest systems including woody encroachment, (5) wood products both in use and in landfills, (6) sediments of reservoirs and rivers, (7) atmospheric C fixed by U.S. ecosystems and then exported by rivers, or (8) exported commercially (food and fiber). The analysis showed that only 30% of the sink was the result of sectors susceptible to the influences of CO_2 fertilization (i.e., (1), (3), and (4)), which was until recent years thought to be the primary driver of the entire terrestrial sink. A similar analysis has been also presented for Europe (Janssens et al. 2003, 2005).

Several attempts have been made to assess the future terrestrial C sink using different types of model: *(i)* physiological and biogeochemical models without dynamic vegetation (do not allow for carbon-climate feedbacks), *(ii)* Dynamic Global Vegetation Models (DGVMs) (do not allow for carbon-climate feedbacks) (Fig. 6.7a), and *(iii)* the C4MIP GCM family (provide coupled carbon-climate projections and allow for carbon-climate feedbacks, but most of them do not have yet dynamic vegetation) (Fig. 6.7b).

Despite the large differences in complexity of the various models, all projections show varying degrees of increased NPP and heterotrophic respiration during the next century due to increasing atmospheric CO_2 concentration (i.e., CO_2 fertilization effects) and associated climate changes. This results in increased global NEP during the next few decades and a subsequent decline of the terrestrial C sink strength. The models under categories *(i)* and *(ii)* that emphasised N limitation without increased N fixation showed the fastest saturation and subsequent decline of NEP. The timing and magnitude of the downturn depends critically on the sensitivity of soil respiration to warming and how the response of biological N fixation is represented (Gifford et al. 1996; Cao and Woodward 1998; Kicklighter et al. 1999; Cramer et al. 2001; Fig. 6.7a).

More recently GCMs that have been further developed to take into account carbon-climate feedbacks after Cox et al. (2000) showed that the feedback resulted in an additional 200 ppm in the atmosphere by the end of this century (model type iii). There are now 11 GCMs with such capacity and they were recently evaluated for the magnitude of the carbon-climate feedback (Friedlingstein et al. 2006). By the end of the century, the C feedback is modeled to account for an additional CO_2 of 20 to 200 ppm with 8 of the models falling between 50 and 100 ppm (Fig. 6.7b). This could lead to an additional 0.1 to 1.5 °C in surface warming.

A few of the models in categories *(i)* and *(ii)* account for additional processes such as N limitation and fire; models in category *(iii)* are driven exclusively by simple representations of photosynthesis and soil respiration as the overall controls of C exchange between the terrestrial biosphere and the atmosphere. The exclusion of some of the sink processes and vulnerable C pools (Fig. 6.8) tend to overestimate the CO_2 fertilization effect and underestimate biospheric C emissions, thus underestimating what we think is a larger magnitude of carbon-climate feedbacks than currently estimated. The omission of four of these processes/pools will illustrate this point:

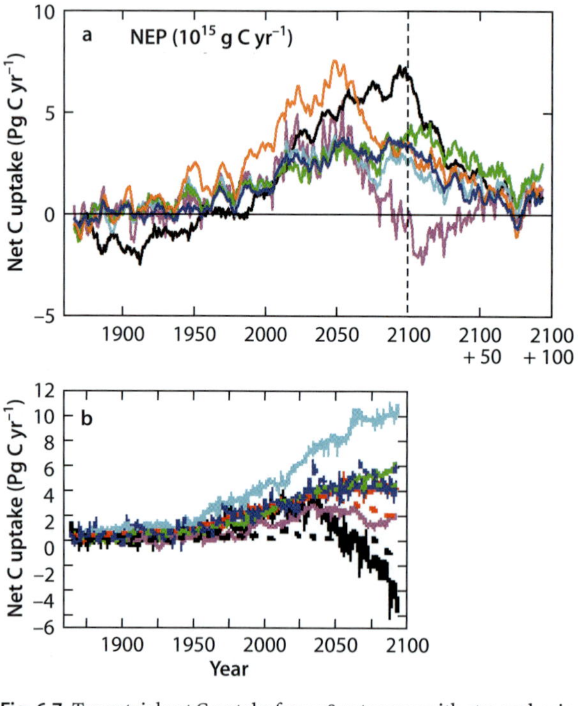

Fig. 6.7. Terrestrial net C uptake from 1850 to 2100 with atmospheric CO_2 and climate change: **a** six dynamic global vegetation models (Cramer et al. 2001); and **b** ten global circulation models with coupled carbon cycle model (Friedlingstein et al. 2006)

The legacy of land-use change. We discussed in Sect. 6.4.3.2 and 6.4.3.3 that there is strong evidence that a significant component of the current northern hemisphere net terrestrial C sink may be the result of past land practices, particularly from forest regrowth in abandoned cropland and vegetation thickening due to fire exclusion. If that is the case and given the saturation nature of these processes, we should expect a rather quick sink decline within a few decades as forests mature and its sink strength diminishes accordingly (Fig. 6.9). Our current estimates of the future strength of the terrestrial sink (IPCC 2001; Friedlingstein et al. 2006) falsely attribute any C sink that is due to vegetation recovery from past disturbances to the CO_2 fertilization effect with the consequent overestimation of the sensitivity of CO_2 for future sinks.

Nitrogen limitation. Hungate et al. (2003) calculated the amount of N required to accumulate between 260 and 890 Pg C over the 21st century as predicted by terrestrial ecosystem models used in the IPCC Third Assessment (IPCC 2001). Despite accounting for large increases in atmospheric N deposition and biological fixation, and decreased C/N ratios, Hungate et al. were only able to account for 1.2 to 6.1 Pg of N by the end of the 21st century, not sufficient to meet the models demands in all but two simulations for the high end N production (6.1 Pg N),

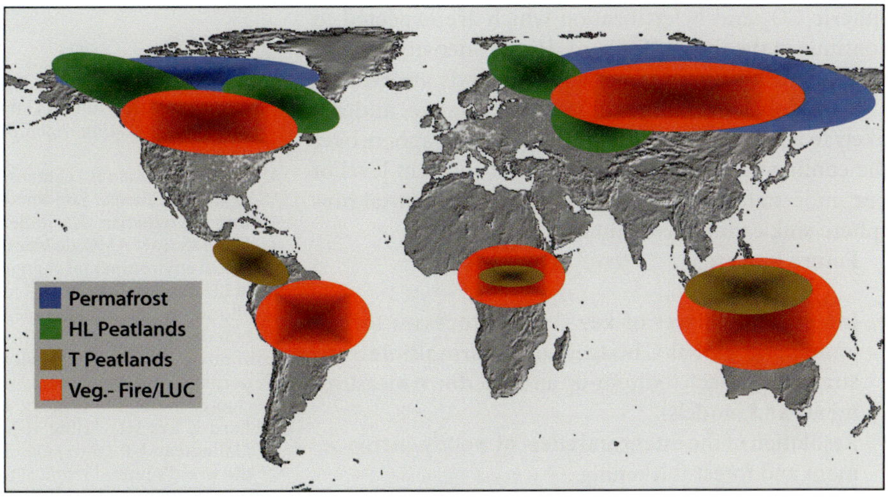

Fig. 6.8.
C pools vulnerable to global warming and land-use change. They include: *(i)* frozen ground, *(ii)* tropical and high-latitude peatlands, and *(iii)* vegetation susceptible to land use and fire

and in all simulations for the low end (1.2 Pg C). Therefore, failure to account for what is a well established constraint on terrestrial NPP (P is also important in the tropics) will result in a gross overestimation of the CO_2 fertilization effect.

Carbon oxidation by fire. Currently there are no coupled carbon-climate models that distinguish soil respiration emissions from fire emissions. Very warm years such as El Niño in 1997–1998 were used to test the sensitivity of soil respiration to temperature which was then used to forecast the soil C sensitivity to future warming (Cox et al. 2000) without accounting for the fact that those same years had high fire emissions. The unintended consequence of this false attribution is to overestimate C emissions from soil respiration. The total C emissions are not necessarily overestimated but omitting fire as a process results in a more smooth dynamics of changes in the sink capacity over time while changes brought about by high fire activity can result in major vegetation shifts which would diminish the strength of the terrestrial sink more abruptly.

Carbon in frozen ground. Soil C databases used in global modeling are still incomplete. Particularly C stored in frozen ground, both in soils and sediments, is poorly known and often estimated only for to the first 30 cm of the soil profile. Sect. 6.4.2.3 reported new findings on the depth and an extension of frozen carbon, particularly with the finding of 500 Pg C in loess sediments in Siberia not previously accounted for (Fig. 6.8). An underestimation of the size of C pools will lead to the underestimation of the potential carbon-climate feedback and therefore overestimation of the potential terrestrial C sink.

6.6 Summary and Conclusions

The current measured net C sink in the Northern Hemisphere and the inferred biosphere sink over the tropics

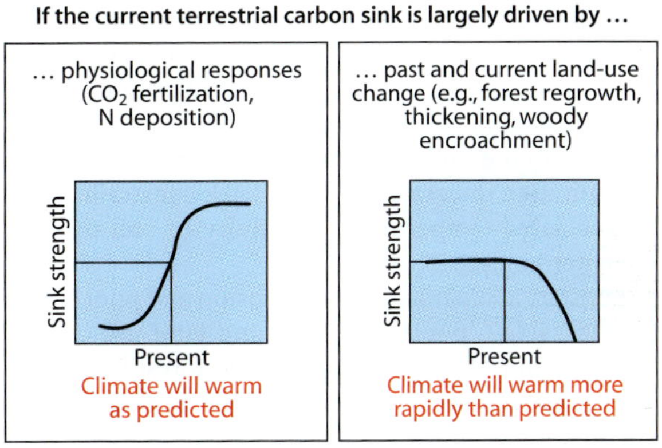

Fig. 6.9. Two alternative scenarios on future dynamics of the terrestrial net biospheric C sink

are not permanent features of the terrestrial biosphere but the result of an imbalance driven by past and present human activities. The sink strength shows high interannual variability and there is evidence that the sink cannot be sustained indefinitely. It is conceivable that by the end of this century or earlier the terrestrial sink may significantly decrease or disappear. Uncertainties in partitioning the historical and current C sinks into various possible processes limits our capacity to predict the future sink strength. Plausible processes that can explain the current sink are land-use change, including forest regrowth on abandoned croplands and fire exclusion, and responses of ecosystem C storage to the related increases in atmospheric CO_2, N deposition and climate change, among others acting individually or synergistically/antagonistically. Some of the sink processes have response curves that saturate at higher forcing levels expected to be reached during the next few decades to a century, while others produce inherently temporary sinks that disappear over time. The saturation of the sink will occur sooner if the prevalent sink processes are related to legacies of past land-use change, and slower, if the prevailing processes are driven by physiological effects of atmospheric CO_2 and N fertilization which are expected to continue in the future (Fig. 6.9). High C-density regions on land such as permafrost and peatlands are vulnerable to global warming and land-use change, and are likely to add large amounts of C into the atmosphere over the coming decades (Fig. 6.8). With the present level of our understanding, it is certain that the terrestrial biosphere sink cannot be maintained indefinitely.

Future Research:

- Integrated analyses of key C sink processes for regional and global C budgets using a multiple constraints approach (bottom-up and top-down measurements and models).
- Resolution of the extent and effect of woody encroachment and forest thickening.
- Integration of the legacy effects of past land-use change (e.g., forest age structure) in biogeochemical models.
- Persistence of new sinks created by afforestation and reforestation.
- Temperature and soil water effects on heterotrophic respiration integrated with feedback constraints, and associated temperature sensitivity of soil organic matter pools.
- Accurate assessment of the extension and pool size of vulnerable C pools to climate and land-use change (e.g., C in frozen ground, peatlands, forest biomass).
- Increase sophistication of C models in coupled carbon-climate GCMs (i.e., Earth System Models) with the inclusion of key terrestrial C-cycle and land surface processes (e.g., vegetation shifts, anthropogenic and wild-fires, forest-age structure, nutrient feedbacks) and vulnerable C pools (e.g., C in frozen soils and sediments, C in cold and tropical peatlands).

Acknowledgments

This paper is the result of over a decade of research and discussions in relation to the future of the terrestrial C sink and its contribution to the atmospheric CO_2 growth. The issue of the possible saturation of the terrestrial C sink was first identified by the Global Change and Terrestrial Ecosystems (GCTE) core project of the International Geosphere-Biosphere Program (Walker et al. 1999) and later on further developed by the integrative efforts of the Global Carbon Project (Global Carbon Project 2003) of the Earth System Science Partnership. We thank Chris B. Field for providing insightful comments in early versions of this manuscript.

Josep G. Canadell and Michael Raupach acknowledge and thank the Australian Greenhouse Office and CSIRO-Marine and Atmospheric Research for supporting the GCP International Project Office in Canberra (and previously the GCTE office).

References

Aber JD, McDowell WH, Nadelhoffer KJ, Magill AH, Berntson G, Kamakea M, McNulty S, Currie W, Rustad L, Fernandex I (1998) Nitrogen saturation in temperate forest ecosystems. Bioscience 48:921–934

Alcamo J, Kreileman E, Leemans R (ed) (1996a) Integrated Scenarios of Global Change. London: Pergamon Press

Alcamo J, Kreileman GJJ, Bollen JC, van den Born GJ, Gerlagh R, Krol MS, Toet AMC, de Vries HJM (1996b) Baseline scenarios of global environmental change. Global Environ Change 6:261–303

Allen LH, Bisbal EC, Boote KJ, Jones PH (1991). Soybean dry matter allocation under subambient and superambient levels of carbon dioxide. Agron J 83:875–883

Anisimov OA, Nelson FE, Pavlov AV (1999) Predictive scenarios of permafrost development under conditions of global climate change in the XXI century. Earth Cryology 3:15–25

Archard F, Eva HD, Stibig H-J, Mayaux P, Gallego J, Richards T, Malingreau J-P (2002) Determination of deforestation rates of the world's humid tropical forests. Science 297:999–1002

Asner GP, Knapp DE, Broadbent EN, Oliveira PJC, Keller M, Silva JN (2005) Selective logging in the Brazilian Amazon. Science 310:480–482

Barrett DJ, Gifford RM (1995) Acclimation of photosynthesis and growth by cotton to elevated CO_2: Interactions with severe phosphate deficiency and restricted rooting volume. Australian J Plant Physiology 22:955–963

Bond-Lamberty B, Wang C, Gower ST (2004) Net primary production and net ecosystem production of a boreal black spruce wildfire chronosequence. Global Change Biol 10:473–487

Botch MS, Kobak KI, Vinson TS, Kolchugina TP (1995) Carbon pools and accumulation in peatlands of the former Soviet Union. Global Biogeochem Cycles 9:37–46

Burrows WH, Henry BK, Back PV, Hoffmann MB, Tit LJ, Anderson ER, Menke N, Danaher T, Carter JO, McKeon GM (2002) Growth and carbon stock change in eucalypt woodlands in northeast Australia: ecological and greenhouse sink implications. Global Change Biol 8:769–784

Camill P (2005) Permafrost thaw accelerates in boreal peatlands during late-20th century climate warming. Climatic Change 68:135–152

Canadell JG, Mooney HA, Baldocchi DD, Berry JA, Ehleringer JR, Field CB, Gower ST, Hollinger DY, Hunt JE, Jackson RB, Running SW, Shaver GR, Steffen W, Trumbore SE, Valentini R, Bond BY (2000). Carbon Metabolism of the Terrestrial Biosphere: a multi-technique approach for improved understanding. Ecosystems 3:115–130

Cannell MGR (2003) Carbon sequestration and biomass energy offset: theoretical, potential and achievable capacities globally, in Europe and the UK. Biomass Bioenergy 24:97–116

Cao M, Woodward FI (1998) Dynamic responses of terrestrial ecosystem carbon cycling to global climate change. Nature 393:249–252

Carey EV, Sala A, Keane R, Callaway RM (2001) Are old forests underestimated as global carbon sinks? Global Change Biol 7:339–344

Carroll AL, Taylor SW, Régnière J, Safranyik L (2004) Effects of Climate Change on Range Expansion by the Mountain Pine Beetle in British Columbia. In: Mountain Pine Beetle Symposium: Challenges and Solutions, pp 223–232, October 30–31, 2003, Kelowna, British Columbia. TL Shore, JE Brooks, and JE Stone (eds). Natural Resources Canada, Canadian Forest Service, Pacific Forestry Centre, Information Report BC-X-399, Victoria, BC. 298 p

Carvalho G, Barros AC, Moutinho PRS, Nepstad DC (2001) Sensitive development could protect Amazonia instead of destroying it. Nature 409: 131

Caspersen JP, Pacala SW, Jenkins JC, Hurtt GC, Moorcroft PR, Birdsey RA (2000) Contributions of land use history to carbon accumulation in U.S. forests. Science 290:1148–1151

Ciais P, Reichstein M, Viovy N, Granier A, Ogee J, Allard V, Aubinet M, Buchmann N, Bernhofer C, Carrara A, Chevallier F, De Noblet N, Friend AD, Friedlingstein P, Grunwald T, Heinesch B, Keronen P, Knohl A, Krinner G, Loustau D, Manca G, Matteucci G, Miglietta F, Ourcival JM, Papale D, Pilegaard K, Rambal S, Seufert G, Soussana JF, Sanz MJ, Schulze ED, Vesala T, Valentini R (2005) Europe-wide reduction in primary productivity caused by the heat and drought in 2003. Nature 437:529–533

Cole V, et al. (1996) Agricultural options for mitigation of greenhouse gas emissions. In: Climate Change 1995. Impacts, Adaptations and mitigation of climate change: Scientific-Technical Analyses, Watson RT, Zinyowera MC, Moss RH, Dokken DJ (eds). Cambridge University Press, New York, pp 745–771

Cox PM, Betts RA, Jones CD, Spall SA, Totterdell IJ (2000) Acceleration of global warming due to carbon-cycle feedbacks in a coupled climate model. Nature 408:184–187

Cox PM, Betts RA, Collins M, Harris PP, Huntingford C, Jones CD (2004) Amazonian forest dieback under climate-carbon cycle projections for the 21st Century. Theo Appl Climatol 78:137–156

Cramer W, Bondeau A, Woodward FI, Prentice IC, Betts RA, Brovkin V, Cox PM, Fisher V, Foley J, Friend AD, Kucharik C, Lomas MR, Ramankutty N, Sitch S, Smith B, White A, Young-Molling C (2001) Global responses of terrestrial ecosystems structure and function to CO_2 and climate change: results from six dynamic global vegetation models. Global Change Biol 7:357–373

DeFries RS, Field CB, Fung I, Collatz GJ, Bounoua L (1999) Combining satellite data and biogeochemical models to estimate global effects of human-induced land cover change on carbon emissions and primary productivity. Global Biogeochem Cycles 13:803–815

DeFries RS, Houghton RA, Hansen MC, Field CB, Skole D, Townshend J (2002) Carbon emissions from tropical deforestation and regrowth based on satellite observations for the 1980s and 1990s. PNAS 99: 14256–14261

Derwent RG, Stevenson DS, Collins WJ, Johnson CE (2004) Intercontinental transport and the origins of the ozone observed at surface sites in Europe. Atmospheric Environment 38:1891–1901

Desai AR, Bolstad PV, Cook BD, Davis KJ, Carey EV (2005) Comparing net ecosystem exchange of carbon dioxide between an old-growth and mature forest in the upper Midwest, USA. Agriculture, Forest Meteorol 128:33–35

Dippery JK, Tissue DT, Thomas RB, Strain BR (1995) Effects of low and elevated CO_2 on C_3 and C_4 annuals. I. Growth and biomass accumulation. Oecologia 101:13–20

Dixon RK, Krankina ON (1993) Forest fires in Russia: carbon dioxide emissions to the atmosphere. Canadian J For Res 23: 700–705

Donigian AS Jr, Barnwell TO Jr, Jackson RB IV, Patwardhan AS, Weinrich KB, Rowell AL, Chinnaswamy RV, Cole CV (1994) Assessment of Alternative Management Practices and Policies Affecting Soil Carbon in Agroecosystems of the Central United States. U.S. EPA Report EPA/600/R-94/067, Athens, 194 pp

Dukes JS, Chiariello NR, Cleland EE, Moore LA, Shaw MR, Thayer S, Tobeck T, Mooney HA, Field CB (2005) Responses of Grassland Production to Single and Multiple Global Environmental Changes. PLoS Biology 3, e319 DOI: 10.1371/journal.pbio.0030319

Edwards EJ, McCaffery S, Evans JR (2005) Phosphorus status determines biomass response to elevated CO_2 in a legume: C_4 grass community. Global Change Biol 11:1968–1981

Fang C, Smith P, Moncrieff JB, Smith JU (2005) Similar response of labile and resistant soil organic matter pools to changes in temperature. Nature 433:57–59

Farquhar GD, Sharkey TD (1982) Stomatal conductance and photosynthesis. Ann Rev Plant Physiol 33:317–345

Farquhar GD, von Caemmerer S, Berry J (1980) A biochemical model of photosynthetic CO_2 assimilation in leaves of C_3 plants. Planta 149:78–90

Fearnside PM, Laurance WF (2004) Tropical deforestation and greenhouse-gas emissions. Ecological Appl 14:982–986

Felzer B, Kicklighter D, Melillo J, Wang C, Zhuang Q, Prinn R (2004) Effects of ozone on net primary production and carbon sequestration in the conterminous United States using a biogeochemistry model. Tellus 56B: 230–248

Felzer B, Reilly J, Mellillo J, Kicklighter D, Sarofim M, Wang C, Prinn R, Zhuang Q (2005) Future effects of ozone on carbon sequestration and climate change policy using a global biogeochemical model. Climatic Change 73:345–373

Field CB, Kaduk J (2004) The carbon balance of an old-growth forest: building across approaches. Ecosystems 7:525–533

Field CB, Chapin FS, III, Chiariello NR, Holland EA, Mooney HA (1996) The Jasper Ridge CO_2 experiment: design and motivation. In: Carbon Dioxide and Terrestrial Ecosystems, Koch G, Mooney HA (eds). Academic Press, San Diego, pp 121–145

Fierer N, Craine JM, McLauchlan K, Schimel JP (2005) Litter quality and the temperature sensitivity of decomposition. Ecology 86:320–326

Flannigan MD, Bergeron Y, Engelmark O, Wotton BM (1998) Future wildfire in circumboreal forests in relation to global warming. J Veg Sci 9:469–476

Flannigan MD, Stocks BJ, Wotton BM (2000) Climate change and forest fires. Science of the total environment 262:221–229

Fowler D, Cape JN, Coyle M, Flechard C, Kuylenstienra J, Hicks K, Derwent D, Johnson C, Stevenson D (1999) The global exposure of forests to air pollutants. Water Air Soil Pollut 116:5–32

Freeman C, Fenner N, Ostle NJ, Kang H, Dowrick DJ, Reynolds B, Lock MA, Sleep D, Hughes S, Hudson J (2004) Export of dissolved organic carbon from peatlands under elevated carbon dioxide levels. Nature 430:195–198

Frey KE, Smith LC (2005) Amplified carbon release from vast West Siberian peatlands by 2100. Geophysical Res Lett 32: No. L09401

Friborg T, Soegaard H, Christensen TR, Lloyd CR, Panikov NS (2003) Siberian wetlands: where a sink is a source. Geophysical Res Lett 30:2129

Friedlingstein P, Cox P, Betts R, Bopp L, von Bloh W, Brovkin V, Doney VS, Eby M, I. Fung I, Govindasamy B, John J, C. Jones C, Joos F, Kato T, Kawamiya M, Knorr W, Lindsay K, Matthews HD, Raddatz T, Rayner P, Reick C, Roeckner E, Schnitzler KG, Schnur R, Strassmann K, Weaver AJ, Yoshikawa C, Zeng N (2006) Climate-carbon cycle feedback analysis. Results from the C4MIP model intercomparison. J Clim 19:3337–3353

Fung I (2005) Drier summers cancel out the CO_2 uptake enhancement induced by warmer springs. PNAS 102:10823–10827

Galloway JN, Cowling EB (2002) Reactive nitrogen and the world: 200 years of change. Ambio 31:64–71

Giardina CP, Ryan MG (2000) Evidence that decomposition rates of orgnaic carbon in mineral sil do not vary with temperature. Nature 404: 861

Gifford RM (1992) Interaction of carbon dioxide with growth-limiting environmental factors in vegetation productivity: implications for the global carbon cycle. Advances Bioclimatol 1:25–58

Gifford RM (1994) The global carbon cycle: a view point on the missing sink. Australian J Plant Physiol 21:1–15

Gifford RM, Howden M (2001) Vegetation thickening in an ecological perspective: significance to national greenhouse gas inventories. Environ Sci Policy 4:59–72

Gifford RM, Lutze JL, Barrett DJ (1996) Global atmospheric change effects on terrestrial carbon sequestration: Exploration with a global C- and N-cycle model (CQUESTN). Plant and Soil 187:369–387

Giorgi F, Bi X, Jeremy Pal J (2004) Mean, interannual variability and trends in a regional climate change experiment over Europe. II: climate change scenarios (2071–2100) Climate Dynamics 23: 839–858

Gitz V, Ciais P (2004) Future expansion of agriculture and pasture acts to amplify atmospheric CO_2 levels in response to fossil-fuel and land-use change emissions. Climatic Change 67:161–184

Global Carbon Project (2003) Science framework and implementation. Canadell JG, Dickinson R, Hibbard K, Raupach M, Young O (eds). Earth System Science Partnership Report No. 1; GCP Report No. 1, 69 pp, Canberra

Goldewijk KK (2001) Estimating global land-use change over the past 300 years: The HYDE Database. Global Biogeochem Cycles. 15:417–433

Goulden ML, Wofsy SC, Harden JW, Trumbore SE, Crill PM, Gower ST, FriesT, Daube BC, Fan SM, Sutton DJ, Bazzaz A, Munger JW (1998) Sensitivity of Boreal forest carbon balance to soil thaw. Science 279:214–217

Grissino-Mayer HD, Swetnam TW (2000) Century-scale climate forcing of fire regimes in the American Southwest. Holocene 10: 213–220

Groenigen van K-J, Six J, Hungate BA, de Graaff M-A, van Breemen N, van Kessel C (2006) Element interactions limit soil carbon storage. PNAS 103:6571–6574

Gruber N, Friedlingstein P, Field CB, Valentini R, Heimann M, Richey JF, Romero P, Schulze E-D, Chen A (2004) The vulnerability of the carbon cycle in the 21st Century: An assessment of carbon-climate-human interactions. In: Global Carbon Cycle, integrating human, climate, and the natural world, Field CB, Raupach M (eds). Island Press, Washington, DC., pp 45–76

Gurney KR, Law RM, Denning AS, Rayner PJ, Baker D, Bousquet P, Brhwiler L, Chen Y-H, Ciais P, Fan S, Fung IY, Gloor M, Heimann M, Higuchi K, John J, Maki T, Kaksyutov S, Masarie K, Peylin P, Prather M, Pak BC, Randerson J, Sarmiento J, Toguchi S, Takahashi T, Yuen C-W (2002) Towards robust regional estimates of CO_2 sources and sinks using atmospheric transport models. Science 415:626–630

Hall SJ, Matson PA (1999) Nitrogen oxide emissions after nitrogen additions in tropical forests. Nature 400:152–155

Handa IT, Körner C, Hattenschwiler S (2005) A test of the treeline carbon limitation hypothesis by in situ CO_2 enrichment and defoliation. Ecology 86:1288–1300

Hättenschwiler S, Körner C (1996) Effects of elevated CO_2 and increased nitrogen deposition on photosynthesis and growth of understory plants in spruce model ecosystems. Oecologia 106:172–180

Hättenschwiler S, Körner C (2000) Tree seedling response to in situ CO_2-enrichment differs among species and depend on understorey light availability. Global Change Biol 6:213–226

Holland EA, Braswell BH, Lamarque J-F, Townsend A, Sulzman J, Muller J-F, Dentener F, Brasseur G, Levy II H, Penner JE, Roelofs G-J (1997) Variations in the predicted spatial distribution of atmospheric nitrogen deposition and their impact on carbon uptake by terrestrial ecosystems. J Geophys Res 102: 15849–15886

Houghton RA (1998) Historic role of forests in the global carbon cycle. In: Carbon Dioxide Mitigation in Forestry and Wood Industry, Kohlmaier GH, Weber M, Houghton RA (eds). Springer-Verlag, Berlin, pp 1–24

Houghton RA (1999) The annual net flux of carbon to the atmosphere from changes in land use 1850–1900. Tellus Ser. B 51B: 298–313

Houghton RA (2002) Magnitude, distribution and causes of terrestrial carbon sinks and some implications for policy. Climate Policy 2:71–88

Houghton RA (2003) Why are estimates of the terrestrial carbon balance so different? Global Change Biol 9:500–509

Houghton RA (2005) Tropical deforestation as a source of greenhouse gas emissions. In: Tropical Deforestation and Climate Change, P. Moutinho P, Schwartzman S (eds). Instituto de Pesquisa Ambiental da Amazonia, Belem, and Environmental Defence, Washington DC, pp 13–21

Houghton RA, Hackler JL, Lawrence KT (1999) The US carbon budget: contributions from land-use change. Science 285:574–578

Houghton RA, Hackler JL, Lawrence KT (2000) changes in terrestrial carbon storage in the United States. 2: The role of fire and fire management. Global Ecol Biogeogr 9:145–170

House JI, Prentice IC, Le Quere C (2002) Maximum impacts of future reforestation or deforestation on atmospheric CO_2. Global Change Biol 8:1047–1052

Hungate BA, Dukes JS, Shaw MR, Luo Y, Field CB (2003) Nitrogen and Climate Change. Science 302:1512–1513

IPCC (2001) Climate Change 2001. The scientific basis. Houghton JT, Ding Y, Griggs DJ, Noguer M, van der Linder PJ, Dai X, Maskell K, Johnson CA (eds). Cambridge University Press, Cambridge

IPCC-SRES (2000) Special Report on Emissions Scenarios. Nakicenovic N, Swart S (eds). Cambridge University Press, Cambridge

Jackson RB, Banner JL, Jobbagy E.G., Pockman WT, Wall DH (2002) Ecosystem carbon loss with woody plant invasion of grasslands. Nature 418:623–626

Janssens IA, Freibauer A, Ciais F, Smith P, Nabuurs G-J, Folberth G, Schalamadinger B, Hutjes RWA, Ceulemans R, Schulze E-D, Valentini R, Dolman AJ (2003) Europe's terrestrial biosphere absorbs 7 to 12% of European anthropogenic CO_2 emissions. Nature 300:1538–1542

Janssens IA, Freibauer A, Schlamadinger B, Ceulemans R, Ciais P, Dolman AJ, Heimann M, Nabuurs G-J, Smith P, Valentini R, Schulze E-D (2005) The carbon budget of terrestrial ecosystems at country-scale – a European case study. Biogeosciences 2:15–26

Jarvis P, Linder S (2000) Constraints to growth of boreal forests. Nature 405:904–905

Jasoni RL, Smith SD, Arnone III, JA (2005) Net ecosystem CO_2 exchange in Mojave Desert shrublands during the eight year of exposure to elevated CO_2. Global Change Biol 11:749–756

Jones C, McConnell C, Coleman K, Cox P, Falloon P, Jenkinson D, Powlson D (2005) Global climate change and soil carbon stocks; predictions from two contrasting models for the turnover of organic carbon in soil. Global Change Biol 11:154–166

Karnosky DF, Percy KE, Xiang B, Callan B, Noormets A, Mankvska B, Hopkin A, Sober J, Jones W, Dickson RE, Isebrands JG (2002) Interacting elevated CO_2 and tropospheric O_3 predisposes aspen (*Populus tremuloides* Michx.) to infection by rust (*Melampsora medusae* f. sp. *Tremuloidae*). Global Change Biol 8:329–338

Kasischke ES, Stocks BJ (2000) Introduction. In: Fire, climate change, and carbon cycling in the boreal forest, Kasischke ES, Stocks BJ (eds). Springer-Verlag, New York, pp 1–5

Kauppi PE, Mielikäinen K, Kuusela K (1992) Biomass and carbon budget of European forests, 1971 to 1990. Science 256:70–74

Keeling RF, Piper SC, Heimann M (1996) Global and hemispheric CO_2 sinks deduced from changes in atmospheric O_2 concentration. Nature 381:218–221

Kern JS, Johnson MG (1993) Conservation tillage impacts on national soil and atmospheric carbon levels. Soil Sci Soc Amer J 57:200–210

Kicklighter DW, Bruno M, Dönges S, Esser G, heimann M, Helfrich J, Ift F, Joos F, Kaduk J, Kohlmaier GH, McGuire AD, Melillo JM, Meyer R, Moore III B, Nadler A, Prentice IC, Sauf W, Schloss AL, Sitch S, Witternberg U, Würth G (1999) A first-order analysis of the potential role of CO_2 fertilization to affect the global carbon budget: a comparison of four terrestrial biopshere models. Tellus 51 B: 343–366

Kirschbaum MUF (1995) The temperature dependence of soil organic decomposition and the effect of global warming on soil organic carbon storage. Soil Biol. Biochem. 27:753–760

Knohl A, Schulze E-D, Kolle O, Buchmann N (2003) Large carbon uptake by an unmanaged 250-year-old deciduous forest in Central Germany. Agricultural Forest Meteorol 118:151–167

Knorr W, Prentice IC, House JI, Holland EA (2005) Long-term sensitivity of soil carbon turnover to warming. Nature: 298–301

Körner C (1996) The response of complex multispecies systems to elevated CO_2. In: Global Change and Terrestrial Ecosystems. Walker B, Steffen W. Cambridge, Cambridge University Press, pp 20–42

Körner C, Asshoff R, Bignucolo O, Hattenschwiler S, Keel SG, Pelaez-Riedl S, Pepin S, Siegwolf RTW, Zotz G (2005) Carbon flux and growth in mature deciduous forest trees exposed to elevated CO_2. Science 309:1360–1362

Körner C, Morgan J, Norby R (2007) CO_2 fertilization: when, where, how much? In: Terrestrial Ecosystems in a Changing World, Canadell J, Pataki D, Pitelka L (eds). The IGBP Series. Springer-Verlag, Berlin

Kurz WA, Apps MJ (1999) A 70-yr retrospective analysis of carbon fluxes in the Canadian forest sector. Ecol Appl 9:526–547

Lal R, Kimble JM, Follet RF, Cole CV (1998) The potential of U.S. cropland to sequester carbon and mitigate the greenhouse effect. Ann Arbor Press, Chelsea, MI. pp 128

Lamarque JF (2005) Nitrogen deposition onto the United States and western Europe: Synthesis of observations and models. Ecological Appl 15:38–57

Laurance WF (2000) Mega-development trends in the Amazon: Implications for global change Environ Monit Assess 61: 113–122

Law BE, Sun OJ, Campbell J, Van tuyl S, Thorton PE (2003) Changes in carbon storage and fluxes in a chronosequence of ponderosa pine. Global Change Biol 9:510–524

Long SP, Ainsworth EA, Rogers A, Ort DR (2004) Rising atmospheric carbon dioxide: plants face the future. Annual Review Plant Biology 55:591–628

Luger AD, Moll EJ (1993) Fire protection and afromontane forest expansion in cape fynbos. Biolog Conserv 64:51–56

Luo Y, Wan S, Hui D, Wallace LL (2001) Acclimatization of soil respiration to warming in a tall grass prairie. Nature 413:622–625

Luo Y, Currie WS, Dukes JS, Finzi A, Hartwig U, Hungate B, McMurtrie RE, Oren R, Parton WJ, Pataki DE, Shaw MR, Zak DR, Field CB (2004) Progressive nitrogen limitation of ecosystem responses to rising atmospheric carbon dioxide. Bioscience 54:731–739

Luo Y, Hui D, Zhang D (2006) Elevated CO_2 stimulates net accumulations of carbon and nitrogen in land ecosystems: a meta-analysis. Ecology 87:53–63

Mack F, Hoffstadt J, Esser G, Goldammer JG (1996) Modeling the influence of vegetation fires on the global carbon cycle. In: Biomass burning and global change Vol. In Levine JS (ed), MIT Press, Cambridge, MA

Mack MC, Schuur EAG, Bret-Harte MS, Shaver GR, Chapin III FS (2004) Ecosystem carbon storage in arctic tundra reduced by long-term nutrient fertilization. Nature 431:440–443

Matson PA, McDowell WH, Townsend AR, Vitousek PM (1999) The globalization of N deposition: ecosystem consequences in tropical environments. Biogeochemistry 46:67–83

McGuire AD, Sitch S, Clein JS, Dargaville R, Esser G, Foley J, Heimann M, Joos F, Kaplan J, Kicklighter DW, Meier RA, Melillo JM, Moore III B, Prentice IC, Ramankutty N, Reichenau T, Schloss A, Tian H, Williams LJ, Wittenberg U (2001) Carbon balance of the terrestrial biosphere in the twentieth century: Analyses of CO_2, climate and land-use effects with four process-based ecosystem models. Global Biogeochem Cycles 15:183–206

Meehl GA, Tebaldi C (2004) More Intense, More Frequent, and Longer Lasting Heat Waves in the 21st Century. Science 305:994–997

Metting FB, Smith JL, Amthor JS (1999) Science needs and new technology for soil carbon sequestration. In: Carbon sequestration in soils: Science, monitoring and beyond, Rosenberg NJ, Izaurralde RC, Malone EL (eds). Battelle Press, Columbus, Ohio, pp 1–34

Mooney H, Canadell J, Chapin FS, Ehleringer J, Körner Ch, McMurtrie R, Parton W, Pitelka L, Schulze D-E (1999) Ecosystem Physiology Responses to Global Change. In: The Terrestrial Biosphere and Global Change: Implications for Natural and Managed Ecosystems, Walker BH, Steffen WL, Canadell J, Ingram JSI (eds). Cambridge University Press, London, pp 141–189

Morison JIL (1985). Sensitivity of stomata and water use efficiency to high CO_2. Plant Cell Environ 8:467–474

Mouillot F, Field CB (2005) Fire history and the global carbon budget: a 1° × 1° fire history reconstruction for the 20th century. Global Change Biol 11:398–420

Murdiyarso D, Lebel L (2007) Southeast Asian fire regimes and land development policy. In: Terrestrial Ecosystems in a Changing World, Canadell J, Pataki D, Pitelka L (eds). The IGBP Series. Springer-Verlag, Berlin

Nabuurs G-J (2004) Current consequences of past actions: how to separate direct from indirect. In: The Global Carbon Cycle: Integrating Humans, Climate and the Natural World, Field C, Raupach M (eds). Island Press, Washington D.C., pp 317–326

Nadelhoffer KJ, Emmett BA, Gundersen P, Kjonaas OJ, Koopmans CJ, Schleppi P, Tietema A, Wright RF (1999) Nitrogen deposition makes a minor contribution to carbon sequestration in temperate forests. Nature 398:145–148

Nemani RR, Keeling CD, Hashimoto H, Jolly WM, Piper SC, Tucker CJ, Mynemi RB, Running SW (2003) Climate-driven increases in global terrestrial net primary production from 1938 to 1999. Science 300:1560–1563

Nepstad DC, Verissimo A, Alencar A, Nobres C, Lima E, Lefebvre P, Schlesinger P, Potter C, Mounthiho P, Mendoza E, Cochrane M, Brooks V (1999) Large-scale impoverishment of Amazonian forests by logging and fire. Nature 398:505–508

Norby RJ, O'Neill EG (1989) Growth dynamics and water use of seedlings of *Quercus alba* L. in CO_2-enriched atmospheres. New Phytol 111:491–500

Norby RJ, O'Neill EG (1991). Leaf area compensation and nutrient interactions in CO_2-enriched seedlings of yellow-poplar (Liriodendron tulipifera L.). New Phytol 117:515–528

Norby RJ, DeLucia EH, Gielend B, Calfapietra C, Giardina CP, King JS, Ledford J, McCarthy HR, Moore DJP, Ceulemans R, De Angelis P, Finzi AC, Karnosky DF, Kubiskel ME, Luka M, Pregitzer KS, Scarascia-Mugnozza GE, Schlesinger WH, Oren R (2005) Forest response to elevated CO_2 is conserved across a broad range of productivity. PNAS 102: 18052–18056

Nowak RS, Ellsworth DS, Smith SD (2004) Functional responses of plants to elevated atmospheric CO_2 – do photosynthetic and productivity data from FACE experiments support early predictions? New Phytol 162:253–280

Oechel WC, Cowles S, Grulke N, Hastings SJ, Lawrence B, Prudhomme T, Riechers G, Strain B, Tissue D, Vourlitis G (1994) Transient nature of CO_2 fertilization in Arctic tundra. Nature 371:500–503

Olesen JE, Bindi M (2002) Consequences of climate change for European agricultural productivity, land use and policy. European J Agron 16:239–262

Owensby CE, Ham JM, Knap AK, Bremer D, Auen LM (1997) Water vapor fluxes and their impact under elevated CO_2 in a C_4-tallgrass prairie. Global Change Biol 3:189–195

Pacala SW, Hurtt GC, Baker D, Peylin P, Houghton RA, Birdsey RA, Heath L, Sundquist ET, Stallard RF, Ciais P, Moorcroft P, Caspersen JP, Shevliakova E, Moore B, Kohlmaier G, Holland E, Gloor M, Harmon ME, Fan S-M, Sarmiento JL, Goodale CL, Schimel D, Field CB (2001) Consistent land- and atmosphere-based U.S. carbon sink estimates. Science 292:2316–2320

Page SE, Siegert F, Rieley JO, Boehm H-D V, Jaya A, Limin S (2002) The amount of carbon released from peat and forest fires in Indonesia during 1997. Nature 420:61–65

Page SE, Wust RAJ, Weiss D, Rieley JO, Shotyk W, Limin SH (2004) A record of late Pleistocene and Holocene carbon accumulation and climate change from an equatorial peat bog (Kalimantan, Indonesia): implications for past, present and future carbon dynamics. Journal of Quaternary Science 19:625–635

Pataki DE, Huxman TE, Jordan DN, Zitzer SF, Coleman JS, Smith SD, Nowak RS, Seemann JR (2000) Water use of two Mojave Desert shrubs under elevated CO_2. Global Change Biol 6:889–897

Payette S, Delwaide A, Caccianiga M, Beauchemin M (2004) Accelerated thawing of subartic peatland permafrost over the last 50 years. Geophysical Research Letters 31: L18208

Peteet D, Andreev A, Bardeen W, Mistretta F (1998) Long-term Arctic peatland dynamics, vegetation and climate history of the Pur-Taz region, Western Siberia. Boreas 27:115–126

Polley HW, Johnson HB, Mayeux HS (1992) Growth and gas exchange of oats (*Avena sativa*) and wild mustard (*Brassica kaber*) at subambient CO_2 concentrations. Int J Plant Sci 153:453–461

Polley HW, Johnson HB, Marino BD, Mayeux HS (1993) Increase in C_3 plant water-use efficiency and biomass over glacial to present CO_2 concentrations. Nature 361:61–64

Polley HW, Johnson HB, Mayeus HS (1994) Increasing CO_2: comparative responses of the C_4 grass *Schizachyrium* and grassland invader *Prosopis*. Ecology 75:976–988

Prentice IC (2001) The carbon cycle and atmospheric carbon dioxide. In: IPCC – Climate Change 2001. The scientific basis, Houghton JT, Ding Y, Griggs DJ, Noguer M, van der Linder PJ, Dai X, Maskell K, Johnson CA (eds). Cambridge University Press, Cambridge, pp 182–237

Ramankutty N, Foley JA (1999) Estimating historical changes in global land cover: croplands from 1700 to 1992. Global Biogeochem Cycles 13:997–1027

Raupach M, Canadell JG, Bakker D, Ciais P, Sanz M-J, Fang JY, Melillo J, Romero-Lankao P, Sathaye J, Schulze D, Smith P, Tschirley J (2004) Interactions between CO_2 stabilisation pathways and requirements for a sustainable Earth System. In: Global Carbon Cycle, Integrating Humans, Climate and the Natural World, Field C, Raupach M (eds), Island Press, Washington, D.C., pp 45–76

Raupach MR, Canadell JG (2006) Observing a vulnerable carbon cycle. In: Observing the continental scale greenhouse gas galance of Europe, Dolman H, Valentini R, Freibauer A (eds). Springer-Verlag, Berlin (in press)

Rayner PJ, Enting IG, Francey RJ, Langenfelds R (1999) Reconstructing the recent carbon cycle from atmospheric CO_2, delta ^{13}C and O_2/N_2 observations. Tellus B 51:213–232

Reich PB, Hobbie SE, Lee T, Ellsworth DS, West JB, Tilman D, Knops JMH, Naeem S, Trost J (2006) Nitrogen limitation constrains sustainability of ecosystem response to CO_2. Nature 440: 922–925

Richey JE (2004) Pathways of atmospheric CO_2 through fluvial systems. In: The Global Carbon Cycle: Integrating Humans, Climate and the Natural World, Field C, Raupach M (eds). Island Press, Washington D.C., pp 329–340

Richey JE, Melack JM, Aufdenkampe AAK, Balleter VM, Hess L (2002) Outgassing from Amazonia Rivers and wetlands as a large tropical source of atmospheric CO_2. Nature 416:617–620

Rogers HH, Bingham GE, Cure JD, Smith JM, Surano KA (1983) Responses of selected plant species to elevated carbon dioxide in the field. J Environ Quality 12:569–574
Rustad LE, Campbell J, Marion GM, Norby RJ, Mitchell MJ, Hartley AE, Cornelissen JHC, Gurevitch J, GCTE-NEWS (2001) A meta-analyses of the response of soil respiration, net N mineralization, and aboveground plant growth to experimental ecosystem warming. Oecologia 126:543–562
Sabine CL, Heimann M, Artaxo P, Bakker DCE, Chen C-TA, Field CB, Gruber N, Le Quere C, Prinn RG, Richey JE, Romero P, Sathaye JA, Valentini R (2004) Current status of past trends of the global carbon cycle. In: Global Carbon Cycle, Integrating Humans, Climate and the Natural World, Field C, Raupach M (eds). Island Press, Washington, D.C, pp 17–44
Saleska SR, Miller SD, Matross DM, Goulden ML, Wofsy SC, Rocha HR, Camargo PB, Crill P, Daube BC, Freitas HC, Hutyra L, Keller M, Kirchhoff V, Menton M, Munger JW, Pyle EH, Rice AH, Silva H (2003) Carbon in amazon forests: unexpected seasonal fluxes and disturbance-induced losses. Science 302: 1554–1557
Sanz MJ, Millán MM (1999) Ozone in the Mediterranean Region: Evidence of Injury to Vegetation. In: Forest Dynamics in Heavily Polluted Regions, Innes, Oleksyn (eds). CABI Publishing, pp 165–192
Schaphoff S, Lucht W, gerten D, Sitch S, Cramer W, Prentice C (2006) Terrestrial biosphere carbon storage under alternative climate projections. Climatic Change 74:97–122
Schimel DS, House JI, Hibbard KA, Bousquet P, Ciais P, Peylin P, Braswell BH, Apps MA, Baker D, Bondeau A, Canadell J, Churkina G, Cramer W, Denning AS, Field CB, Friedlingstein P, Goodale C, Heimann M, Houghton RA, Melillo JM, Moore III B, Murdiyarso D, Noble I, Pacala SW, Prentice IC, Raupach MR, Rayner PJ, Scholes RJ, Steffen WL, Wirth C (2001) Recent patterns and mechanisms of carbon exchange by terrestrial ecosystems. Nature 414:169–172
Schlesinger WH, Andrews JA (2000) Soil respiration and the global carbon cycle. Biogeochemistry 48:7–20
Scholes RJ, Hall DO (1996) The carbon budget of tropical savannas, woodlands and grasslands. In: Global Change: Effects on coniferous forests and grasslands. Breymeyer AI, Hall DO, Melillo JM, Agren GI (eds). SCOPE 56, John Wiley, Chichester, pp 70–100
Schulze ED, Lloyd J, Kelliher FM, Wirth C, Rebmann C, Luhker B, Mund M, Knohl A, Milyukova IM, Schulze W, Ziegler W, Varlagin AB, Sogachev AF, Valentini R, Dore S, Grigoriev S, Kolle O, Panfyorov MI, Tchebakova N, Vygodskaya NN (1999) Productivity of forests in the Eurosiberian boreal region and their potential to act as a carbon sink – a synthesis. Global Change Biol 5:703–722
Smith P (2004) Soils as carbon sinks: the global context. Soil Use Manag 20:212–218
Smith P (2005) An overview of the permanence of soil organic carbon stocks: influence of direct human-induced, indirect and natural effects. European J Soil Science 56:673–680
Smith DL, Johnson LC (2003) Expansion of Juniperus virginiana L. in the Great Plains: changes in soil carbon dynamics. Global Biogeochem Cycles 17, doi: 10.1029/2002GB001990
Smith TM, Shugart HH (1993) The transient response of terrestrial carbon storage to a perturbed climate. Nature 361: 523–526
Smith P, Powlson DS, Smith JU, Falloon P, Coleman K (2000) Meeting Europe's climate change commitments: Quantitative estimates of the potential for carbon mitigation by agriculture. Global Change Biol 6:525–539
Smith SV, Renwick WH, Buddemeier RW, Crossland CJ (2001) Budgets of soil erosion and deposition for sediments and sedimentary organic C across the conterminous United States. Global Biogeochem Cycles 15:697–707
Smith P, Andrén O, Karlsson T, Perälä P, Regina K, Rounsevell M, van Wesemael B (2005a) Carbon sequestration potential in European croplands has been overestimated. Global Change Biol 11:2153–2163
Smith LC, Sheng Y, MacDonald GM, Hinzman LD (2005b) Disappearing Artic lakes. Science 308: 1429
Smith P, Martino D, Cai Z, Gwary D, Janzen HH, Kumar P, McCarl B, O'Mara F, Rice C, Scholes RJ, Sirotenko O, Howden M, McAllister T, Ogle S, Pan G, Romanenkov V, Schneider U, Towprayoon S, Wattenbach M, Smith JU (2006) Greenhouse gas mitigation in agriculture. Phil Trans Royal Soc, B. (submitted)
Stallard RF (1998) Terrestrial sedimentation and the carbon cycle: coupling weathering and erosion to carbon burial. Global Biogeochem Cycles 12:231–257
Stocks BJ, Fosberg MA, Lynham TJ, Mearns L, Wotton BM, Yang Q, Jin J-Z, Lawrence K, Hartley GR, Mason JA, McKenney DW (1998) Climate change and forest fire potential in Russian and Canadian boreal forests. Clim Change 38:1–13
Syvitski JPM, Vorosmarty CH, Kettner AJ, Green P (2005) Impact of humans on the flux of terrestrial sediment to the global coastal ocean. Science 308:376–380
Tans PP, Fung IY, Takahashi T (1990) Observational constraints on the global CO_2 budget. Science 247:1431–1438
Tarnocai C, Kimble J, Broll G (2003) Determining carbon stocks in cryosols using the Northern and Mid Latitudes soil database. In: Permafrost, Phillips, Springman, Arenson (eds). Swets and Zeitlinger, Lisse, pp 1129–1134
Timmermann A, Oberhuber J, Bacher A, Esch M, Latif M, Roeckner E (1999) Increased El Niño frequency in a climate model forced by future greenhouse warming. Nature 398:694–697
Tognetti R, Cherubini P, Innes JL (2000) Comparative stem-growth rates of Mediterranean trees under background and naturally enhanced ambient CO_2 concentrations. New Phytol 146:59–74
Townsend AR, Braswell BH, Holland EA, Penner JE (1996) Spatial and temporal patterns in terrestrial carbon storage due to deposition of fossil fuel nitrogen. Ecol Appl 6:806–814
Turetsky M, Wieder K, Halsey L, Vitt D (2002) Current disturbance and the diminishing peatland carbon sink. Geophysical Research Letters 29: 10.1029/2001Gl014000
Turunen J, Tahvanainen T, Tolonen K, Pitkanen A (2001) Carbon accumulation in West Siberian mires, Russia. Global Biogeochem Cycles 15:285–296
Valentini R, Matteucci G, Dolman AJ, Schulze ED, Rebmann C, Moors EJ, Granier A, Gross P, Jensen NO, Pilegaard K, Lindroth A, Grelle A, Bernhofer C, Grunwald T, Aubinet M, Ceulemans R, Kowalski AS, Vesala T, Rannik U, Berbigier P, Loustau D, Guomundsson J, Thorgeirsson H, Ibrom A, Morgenstern K, Clement R, Moncrieff J, Montagnani L, Minerbi S, Jarvis PG (2000) Respiration as the main determinant of carbon balance in European forests. Nature 404:861–865
VEMAP members (1995) Vegetation/ecosystem modeling and analysis project: Comparing biogeography and biogeochemistry models in a continental-scale study of terrestrial ecosystem responses to climate change and CO_2 doubling. Global Biogeochem Cycles 9:407–438
Vitousek PM, Howarth RW (1991) Nitrogen limitation on land and in the sea: How can it occur? Biogeochemistry 13:87–115
Walker BH, Steffen WL, Langridge J (1999) Interactive and integrated effects of global change on terrestrial ecosystems. In: The Terrestrial Biosphere and Global Change. Implications for Natural and Managed Ecosystems, Walker BH, Steffen WL, Canadell J, Ingram JSI (eds). Cambridge University Press, Cambridge, pp 329–375
Ward JK, Strain BR (1997) Effects of low and elevated CO_2 partial pressure on growth and reproduction of Arabidopsis thaliana from different elevations. Plant Cell Envir 20:254–260
Werf GR, Randerson JT, Collatz GJ, Giglio L, Prasad S. Kasibhatla PS, Arellano Jr AF, Olsen SC, Kasischke ES (2004) Continental-scale partitioning of fire emissions during the 1997 to 2001 El Niño/La Niña Period. Science 303:73–76
Wolfe DW, Gifford RM, Hilbert D, Luo Y (1998) Integration of photosynthetic acclimation to CO_2 at the whole plant level. Global Change Biol 4:879–893
Xu D-Q, Gifford RM, Chow WS (1994) Photosynthetic acclimation in pea and soybean to high atmospheric CO_2 partial pressure. Plant Physiology 106:661–671
Yuen CW, Higuchi K, TRANSCOM-3 Modellers (2005) Impact of Fraserdale CO_2 observations on annual flux inversion of the North American boreal region. Tellus 57B: 203–209
Zimov SA, Schuur EAG, Chapin III FS (2006) Permafrost and the global carbon budget. Science 312:1612–1613

Part B

Changing Biodiversity and Ecosystem Functioning

Chapter 7

Functional Diversity – at the Crossroads between Ecosystem Functioning and Environmental Filters

Sandra Díaz · Sandra Lavorel · F. Stuart Chapin III · Paula A. Tecco · Diego E. Gurvich · Karl Grigulis

7.1 Introduction

Functional diversity, i.e., the *kind*, *range*, and *relative abundance* of functional traits present in a given community, is one of the major factors influencing ecosystem functioning (Chapin et al. 2000b; Díaz and Cabido 2001; Loreau et al. 2001; Hooper et al. 2005). Through its effects on ecosystem functioning, plant functional diversity (hereafter FD) is both affected by and affects all the major global change drivers identified by the international scientific community (Fig. 7.1, path 1). Changes in climate, atmospheric composition, land-use and disturbance regime, and biotic exchanges (deliberate or accidental introduction of organisms to an ecosystem) all have non-random effects on FD, i.e., they select for or against species bearing certain traits. For example, the more drought- and/or frost-sensitive species are eliminated first under an increased frequency of climatic extreme events (Díaz et al. 1999). Slow-growing species are eliminated when ecosystems experience nitrogen loading (Thompson 1994). Non-resprouting species are strongly reduced by increased fire return intervals (Russell-Smith et al. 2002; Johnson and Cochrane 2003). At the same time, through its effect on ecosystem processes, FD influences climate, carbon exchange with the atmosphere,

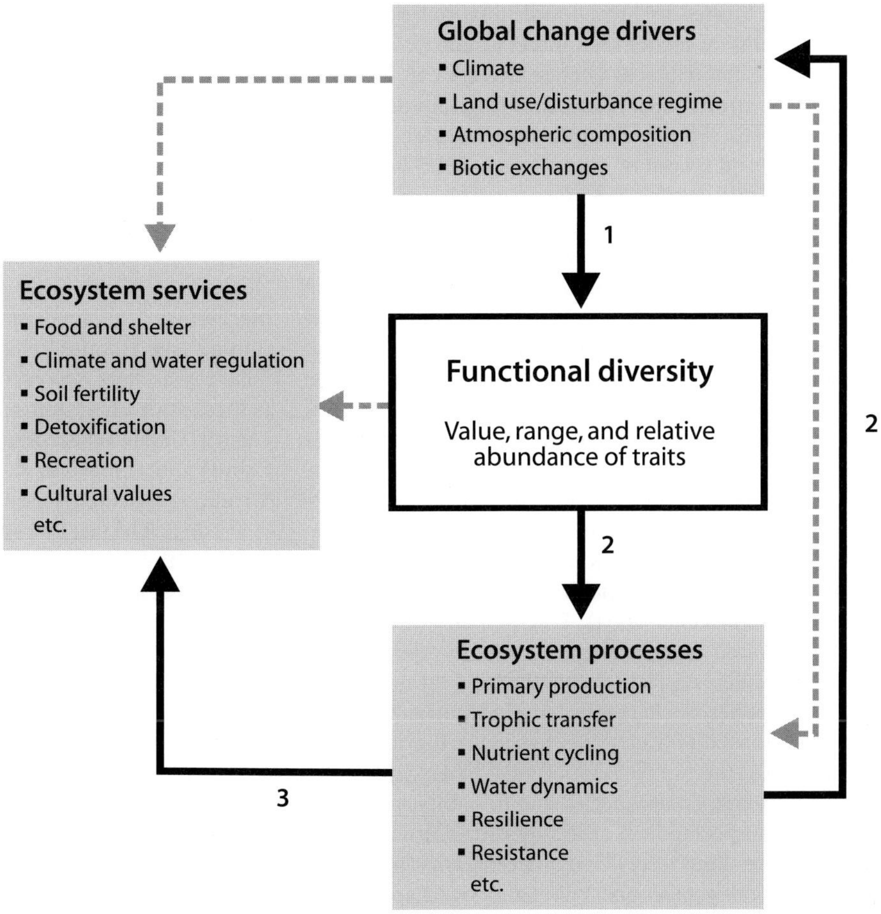

Fig. 7.1.
Functional diversity is both a response variable modified by, and a factor modifying global change drivers. *Solid arrows* indicate relationships addressed in this chapter. See text for description of numbered links

disturbance regime, and the susceptibility to, and consequences of biotic exchanges (Fig. 7.1, path 2). In this chapter we focus on the processes of mutual influence and feedbacks between global change drivers and FD, and discuss the empirical evidence supporting the effect of different components of FD.

7.2 Environmental Filters Affect FD

FD at any site is the result of the action of environmental filters on the regional species pool. Environmental filters are non-random factors that narrow the range of functional traits in a local community. Filtering operates primarily at the level of ecological sorting (i.e., changing the proportions of different genotypes represented in the established community, but not altering the genotypes themselves), but can lead to stabilizing or directional natural selection over evolutionary time. Climate, disturbance regime, some aspects of atmospheric composition, and biotic interactions are major environmental filters (Keddy 1992; Woodward and Diament 1991; Díaz et al. 1998, 1999), and strongly determine which traits and functions can survive at any particular site. It has been suggested that, at a given site, species richness is limited by the regional species pool, whereas functional diversity is limited by local conditions that determine the availability of niches (Schmid et al. 2002).

The four major global change drivers can be interpreted as environmental filters: they filter out organisms bearing certain sets of traits and allow the establishment, persistence and spread of others (see Lavorel et al. 2007, Chap. 13 of this volume, for more examples of environmental filtering of plant functional traits). Therefore, the dramatic biodiversity loss now experienced at the global scale has two aspects: random extinction as a result of the reduction of inhabitable area (i.e., as expected on the basis of island biogeography theory; MacArthur and Wilson 1967), and non-random global or local extinction as a result of the filtering effect of global change drivers. The latter leads to a biota that not only is taxonomically poorer, but also represents a *biased* subset of the range of traits initially available. Recent assessments have shown strong asymmetries in the extinction risk of species with certain traits (McKinney and Lockwood 1999), or belonging to different habitats (Brook et al. 2003) or biomes (Thomas et al. 2004). Fragmentation because of land use, as well as increasing the random probability of extinction, differentially increases the local extinction risk of certain species, by altering water, wind, nutrient dynamics and flammability (Cochrane et al. 1999). Biotic homogenization, i.e., the replacement of local biotas with widespread non-indigenous species due to habitat modification and transportation of exotic species, is also a process where 'winners' and 'losers' bear different traits (McKinney and Lockwood 1999; see also Vilà et al. 2007, Chap. 8 of this volume). Non-random extinctions and functional shifts in vegetation cover as a consequence of the filtering effects of global change drivers are at least as alarming as global extinctions due to reduction in habitat area. This is because their potential effects on ecosystem processes and services (Fig. 7.1, paths 1, 3), and feedbacks to global environmental drivers (Fig. 7.1, path 2) can be dramatic well before species disappear from the face of the Earth.

7.3 FD effects on Global Change Drivers

Although the evidence of effects of FD on global change drivers and ecosystem services mediated by altered ecosystem functioning is accumulating fast, it is not equally strong for the different components of FD. Theory suggests that the kind, relative abundance, and range of traits present in a community should affect ecosystem processes and thus ecosystem services and global change drivers. However, at present there is much more empirical support for the role of the traits of the dominants (*kind* and *abundance* of traits) than for that of the *range* of traits present. On the other hand, although many (and sometimes dramatic) examples suggest that indirect interactions are crucially important in triggering some ecosystem-level feedbacks, there is still no integrated theoretical framework explicitly linking FD with these feedbacks, or predicting the consequences of indirect interactions in systems with different degrees of FD. In this section, we summarize some basic theoretical issues and the empirical evidence related to each of these components.

7.3.1 The Traits of the Dominants

Despite some remaining controversy on precisely which mechanisms explain diversity effects on ecosystem functioning (see Hooper et al. 2005 for an updated review), most researchers now recognize that FD is more relevant to local-scale ecosystem functioning than taxonomic diversity (Grime 1998, 2002; Chapin et al. 2000b; Díaz and Cabido 2001; Loreau et al. 2001; Tilman 2001; Naeem and Wright 2003; Petchey 2004; Petchey et al. 2004; Hooper et al. 2005). The fact that species show different attributes, and thus their contributions to ecosystem processes are not equal or interchangeable, is at the heart of the mechanisms proposed to explain how biodiversity should affect ecosystem processes. However, most studies of the effects of diversity on local-scale ecosystem functioning have considered only species richness, or have taken a very coarse approach to FD. For example, they consider a small number of life forms or taxonomic groups (e.g., grasses, legumes, composites), all of them with similar abundances (see Box 7.1 for a discussion of links between species richness and FD, and issues involved in FD measurement).

Box 7.1. Measuring FD

Despite the widely recognized importance of FD in determining ecosystem processes, species richness (number of species present in a given community) continues to be the most used and manipulated measurement of diversity (Díaz and Cabido 2001; Petchey 2004). This is partly because of practical reasons. First, it is much easier to count species than to weigh them by their abundance. However, this means in practice ignoring the fact that the effect of a species in an ecosystem is largely a function of its biomass (see Sect. 7.3.1). Giving the same 'weight' to extremely abundant and extremely rare species, as happens in species richness, can substantially obscure links between diversity and ecosystem processes. Second, the objective quantification of FD is far from straightforward. The most common measurement of FD is functional type richness, but the results strongly depend of the grouping into functional types (Díaz and Cabido 2001; Petchey 2004), which may or may not reflect true functional differences among species and in any case makes comparisons among different studies extremely difficult. The most promising methods of quantifying FD thus appear to be those based trait distances between species (Díaz and Cabido 2001; Naeem and Wright 2003; Petchey 2004). These require the measurement (or literature compilation) of the attributes of many species, which is time- and resource-consuming. Also, there is so far no ideal FD index, since none of the available metrics can simultaneously account for functional distances between species independently from species richness and at the same time deal with multidimensional- attribute spaces (see Mason et al. 2005, Petchey et al. 2004 and Moulliot et al. 2005 for thorough comparison of indices). Moreover, even an index with all the desirable properties (e.g., those enumerated by Mason et al. 2005) would be of little use if it is not based on functionally important traits (see Table 7.1 and Lavorel et al. 2007, Chap. 13 of this volume, for examples of 'relevant' traits for different ecosystem processes).

Beyond practical considerations, the use of species richness as a surrogate for FD is based on assumptions, which are unrealistic for most natural ecosystems (Díaz and Cabido 2001; Naeem and Wright 2003). FD is often assumed to vary in direct proportion to species diversity (Lawton et al. 1998; Tilman 1999; Naeem 2002). Species richness should be an adequate surrogate for functional richness only if there is a linear increase in niche space occupation as species richness increases. Theoretically, this can happen only when there is random (Fig. 7.2a) or uniform (Fig. 7.2b) occupation of niche space (Díaz and Cabido 2001), and when all species are equally different (addition of any species to a community and the contribution of each species to functional diversity is independent of species richness (Naeem and Wright 2003). However, these cases are uncommon in nature, as compared to aggregate occupation of niche space (Holling et al. 1996). This is because in nature random assemblage and extinction are much less common than non-random ones (Zobel 1997; Díaz and Cabido 2001; see also Sect. 7.2). When aggregate occupation of niche space is related to strong convergence of different species into contrasting functional types [e.g., annual graminoids, succulents and dwarf shrubs in deserts, Fig. 7.2c], species richness overestimates functional richness. When aggregate occupation of niche space is due to strong differentiation in niche space among different genotypes or phenotypes within a single species (e.g., herbivore-tolerant and herbivore-intolerant grass genotypes in prairie-dog dominated landscapes, Jaramillo and Detling 1988; intraspecific differentiation of the tree *Metrosideros polymorpha* in the Hawaiian islands, Cordell et al. 1998), species richness underestimates functional richness. These mismatches between taxonomic and functional richness can have important practical consequences. Only in cases in which extinction is happening at random should we expect a linear decrease in functional richness (and presumably in associated ecosystem functioning) with species richness loss. In other cases, the consequences of extinctions will likely be non-linear. For example, in the case illustrated in Fig. 7.2d, an assessment of species number loss could grossly underestimate functional richness loss. Biodiversity assessments based only on species richness and checklists will tend to overlook processes of functional extinction, that is, the extinction of local genotypes or phenotypes without immediate risk of global (Estes et al. 1989; Luck et al. 2003), Functional extinction often leads to ecosystems that retain the same richness and taxonomic composition of species, but cannot sustain necessary functions for their long-term persistence (e.g., the 'half-empty forest' of Redford and Feinsinger 2001).

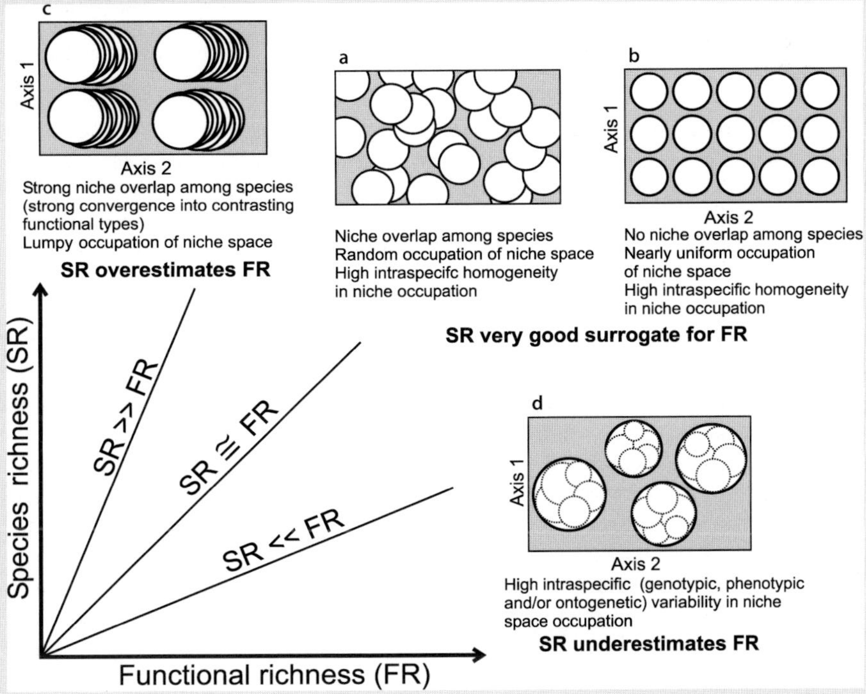

Fig. 7.2. Extreme cases of links between plant species richness and functional richness. *Axes 1* and *2* in the *shaded rectangles* represent different resource or disturbance axes. *Solid-line circles* represent the fundamental niches of different species; *dashed-line circles* represent the niche of different genotypes, phenotypes or ontogenetic stages within a single species (reprinted from Díaz and Cabido 2001, with permission from Elsevier)

In an analysis of synthetic-community and field experiments involving plants, Díaz and Cabido (2001) found that rates and magnitudes of ecosystem processes, such as plant and soil biomass production, decomposition, mineralization or various aspects of stability, were more consistently associated with functional composition (presence of certain *kinds* of plant functional traits; e.g., species with nitrogen-fixing symbionts, or with particularly fast growth rate, or evergreen leaves) and functional richness (number of different plant functional types) than with species richness. Functional composition was associated with ecosystem processes more often than was functional richness, even though most of these experiments kept the relative abundance among species unnaturally even (but see Petchey's 2004 considerations on the assignment of species to functional groups). This can hardly come as a surprise. The fact that the morpho-functional traits of the dominant plant species strongly determine ecosystem properties has been recognized since the earliest days of ecology, and has often been referred to as the 'mass-ratio' hypothesis (Grime 1998). Locally abundant plant traits determine the rate and magnitude (Grime 1977; Hobbie 1992; Chapin et al. 1993; Aerts 1995; Aerts and Chapin 2000; Herbert et al. 1999; Lavorel and Garnier 2002; Díaz et al. 2004; Garnier et al. 2004), and stability (Lepš et al. 1982; McGillivray et al. 1995; Grime et al. 2000) of major ecosystem processes. The existence of recurrent suites of plant traits (e.g., 'acquisitive' vs. 'conservative' syndromes) with predictable controlling effects on ecosystem processes such as primary production, trophic transfer, carbon storage, or nutrient cycling has been proposed (Grime 1977; Chapin 1980; Chapin et al. 1993), and empirically supported at the local (e.g., Chapin et al. 1996; Grime et al. 1997; Garnier et al. 2004) and trans-regional levels (e.g., Díaz et al. 2004).

By affecting ecosystem processes, the traits of the dominant plant species (i.e., *kind* and *relative abundance* of traits) have considerable impact on ecosystem services and therefore human well-being (Fig. 7.1, path 3; Table 7.1). On the other hand, global change drivers affect, and to a certain degree are affected by, locally dominant plant traits (Fig. 7.1, paths 1 and 2). The FD approach thus offers the interesting perspective of tracing some causal relationships and mutual feedbacks between global change drivers, locally dominant plant traits, ecosystem properties, and ecosystem services.

Table 7.1. Causal connections between some individual plant traits, ecosystem processes, and ecosystem services. *ES* = ecosystem service. ES classification follows MEA (2003)

Individual traits	Ecosystem processes	Ecosystem services
Leaf nutrient concentration Leaf toughness Specific leaf area Leaf turnover rate	Net primary productivity Relative productivity rate Consumption by herbivores ungulates (trophic transfer) Decomposition and net mineralization	Soil formation and fertility Food and fiber production Carbon sequestration Resistance of production of goods and services over time Cultural and aesthetic ES (e.g. autumn colours, traditional ungulate-dependent landscapes)
Canopy size Leaf phenology Leaf distribution over stems Growth form Canopy architecture	Evapotranspiration, heat exchange Albedo, roughness	Climate regulation
Shoot height Wood density Lifespan Leaf nutrient concentration	Accumulation of standing biomass	Carbon sequestration Climate regulation (indirectly by carbon sequestration)
Tissue water content Canopy architecture Leaf nutrient and resine content Texture and architecture of litter	Vegetation flammability Fuel load (undecomposed litter) Soil insulation from air temperature and moisture	Nutrient cycling Soil formation, fertility and stability Climate regulation Carbon sequestration Resilience of ES over time
Seed shape and mass	Persistence in the soil seed bank Seed transport by animals	Resilience of production of goods and services over time Maintenance of food-provisioning and cultural ES mediated by animals
Root depth Root architecture	Evapotranspiration Soil exploration by root systems	Climate regulation Water runoff Soil stability

7.3.1.1 *FD and Climate*

FD both responds to and affects the climate system. Local FD is strongly determined by the climatic filter (Fig. 7.1, path 1; Lavorel et al. 2007, Chap. 13 of this volume). In addition, FD affects climate regulation, which is one of the most essential services that natural ecosystems provide to humans (MAE 2003; Fig. 7.1, path 3). FD affects climate indirectly, through changes in the carbon sequestration capacity of the biota (i.e., the more carbon biological systems can sequester, the less will be contributed to the greenhouse effect in the atmosphere), and also directly, through changes in the biophysical properties of the land surface. Not surprisingly, these effects are most pronounced where the ecological changes occur over large areas. For example, in northeast Siberia, widespread fires and other disturbances lead to a replacement of mosses and evergreens, which have a long season (but low rates) of photosynthetic activity, with deciduous plants that have a shorter season (and higher rates) of photosynthesis. Mosses also insulate the soil, so the depth of soil thaw is shallower and soil respiration is less, particularly in late summer and early winter. The net effect of disturbance-induced changes in FD is therefore to increase the magnitude of both summer carbon gain and winter carbon loss. In this way recent increases in disturbance contribute to the increased seasonal amplitude of atmospheric carbon dioxide concentration at high latitudes (Zimov et al. 1999). The feedback from increased carbon dioxide to climate occurs at the global scale because of the rapid mixing of the atmosphere.

FD also interacts with the climate system through changes in water and energy exchange (Foley et al. 2003). Complex canopies often absorb more energy (have a lower albedo) because canopy complexity increases the probability that other surfaces will absorb reflected radiation before it is lost to space. For this reason there is a decrease in albedo from conifer forests to deciduous forests to grasslands. Particularly in open-canopied ecosystems, which account for 70% of the ice-free terrestrial surface (Graetz 1991), species in the ecosystem will likely influence stand-level albedo, with their importance depending on abundance, height, and structural properties. In addition, diversity often contributes to the structural complexity of leaf arrangement in canopies and therefore the efficiency with which incoming radiation is trapped. For example, the increase in shrub density in arctic tundra in response to regional warming (Sturm et al. 2001) has reduced regional albedo and increased regional heating (Chapin et al. 2000a). Structural diversity of the canopy also increases the efficiency with which heat and moisture are transferred to the atmosphere. Even a low density of trees (e.g., <100 ha^{-1}) in a savanna or woodland substantially increases turbulent exchange with the atmosphere (Thompson et al. 2004).

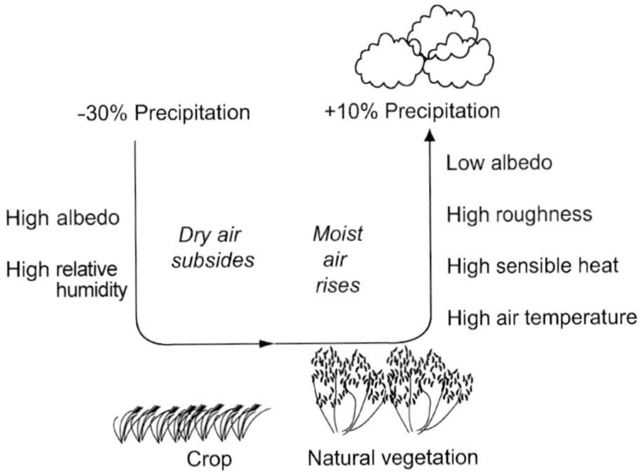

Fig. 7.3. Effects on regional climate of conversion of heathland to croplands in southwestern Australia (Chambers 1998) (reproduced from Chapin et al. 2002, with permission from Springer-Verlag)

The diversity of patches on a landscape exerts an additional impact on biophysical coupling between land and atmosphere. Patches that are larger in size than the depth of the planetary boundary layer (i.e., their smallest dimension is at least 10 km) and which differ in albedo or surface temperature from neighboring patches create convection cells, in which air rises above the warm patch; this air is replaced by cooler moister air that flows laterally from adjacent patches (Pielke 2001). In Australia, for example, the replacement of native heath vegetation by croplands increased regional albedo. As a result, air tended to rise over the dark heathland, drawing moist air from the croplands to the heathlands. The net effect was a 10% increase in precipitation over heathlands and a 30% decrease in precipitation over croplands (Fig. 7.3) (Chambers 1998).

7.3.1.2 *Fd and Land-Use/Disturbance Regime*

Changes in FD brought about by shifting climatic or land-use conditions can in turn produce changes in an ecosystem's disturbance regime and/or resource basis, and thus its ecosystem-service value and suitability for certain land uses. These include, for example, changes in carrying capacity for livestock, capacity for water retention and flood regulation, control of human and domestic animal disease, or amenity values. These processes sometimes start with sharp increases in abundance of native species that were previously non-dominant (e.g., woody encroachment; Scholes and Archer 1997) or the spread over the landscape of exotic species. The latter is the result of the explosive increase in biotic exchanges during the 19th and 20th centuries, often in combination with changes in climate and land use (Fig. 7.1, path 1; see Vilà et al. 2007, Chap. 8 of this volume). Once established, some exotics become invasive and alter ecosystem productivity, resource dynamics, stability, and disturbance regime, often irreversibly (Fig. 7.1, paths 2 and 3). A review of over 150 studies to evaluate the mechanisms underlying the impacts of exotic plant invasions on plant and animal community structure, nutrient cycling, hydrology, and fire regimes revealed that studies examining effects of invaders on ecosystem processes often tend to attribute impacts to differences in important functional traits of the invader as compared to the resident species (Levine et al. 2003). The majority of examples where significant effects on ecosystem processes were reported concerned invasions by a life form previously absent from a native ecosystem (e.g., grasses) and species with traits allowing them to tap into unexploited resources (e.g., nitrogen-fixing, deep rooting). However, life forms readily present in an ecosystem could also have impacts, e.g., on the nitrogen cycle or fuel accumulation for fire, when invaders had attributes conferring them greater efficiency in resource use than the natives.

A recent study illustrates how climate and land-use change and biotic exchanges can alter, and be altered by FD, in a feed-forward, highly irreversible process. The northern Mediterranean basin agricultural land abandonment over the last century (Moreno et al. 1998) and ongoing climate change (Pausas 2004) have resulted in increasing frequencies of very large, intense fires. In Catalonia (NE Spain), higher fire frequency seems to promote the expansion of the large, evergreen, resprouting tussock grass *Ampelodesmos mauritanica* (Vilà et al. 2001). Near-monocultures of *A. mauritanica* now occupy original shrubland sites. Grigulis et al. (2005) tested the hypothesis of a positive feedback between *A. mauritanica* abundance and changing fire regimes by combining measurements across a natural gradient of density of *A. mauritanica* near Barcelona, Spain with model simulations of fire and vegetation dynamics using the landscape modeling platform LAMOS (Lavorel et al. 2000; Cousins et al. 2003). The invasion of shrublands by *A. mauritanica* produced a series of spectacular modifications in community structure and ecosystem properties, which translated to changes in vegetation and fire regimes at the landscape-scale (Fig. 7.4):

1. Aboveground biomass nearly doubled in plots with high vs. low density of *A. mauritanica* (Fig. 7.4a). This occurred because *A. mauritanica* replaced native shrubs, but also some grasses and herbs, and also because the standing biomass of *A. mauritanica* increased significantly. Therefore the morphological/functional composition of the community was considerably modified by the invasion. The quantity of litter also increased significantly, leading to much higher total fuel loads in high invasion plots. The specific aboveground net primary productivity (Garnier et al. 2004) of all components of the community decreased significantly with increasing density of *A. mauritanica*.

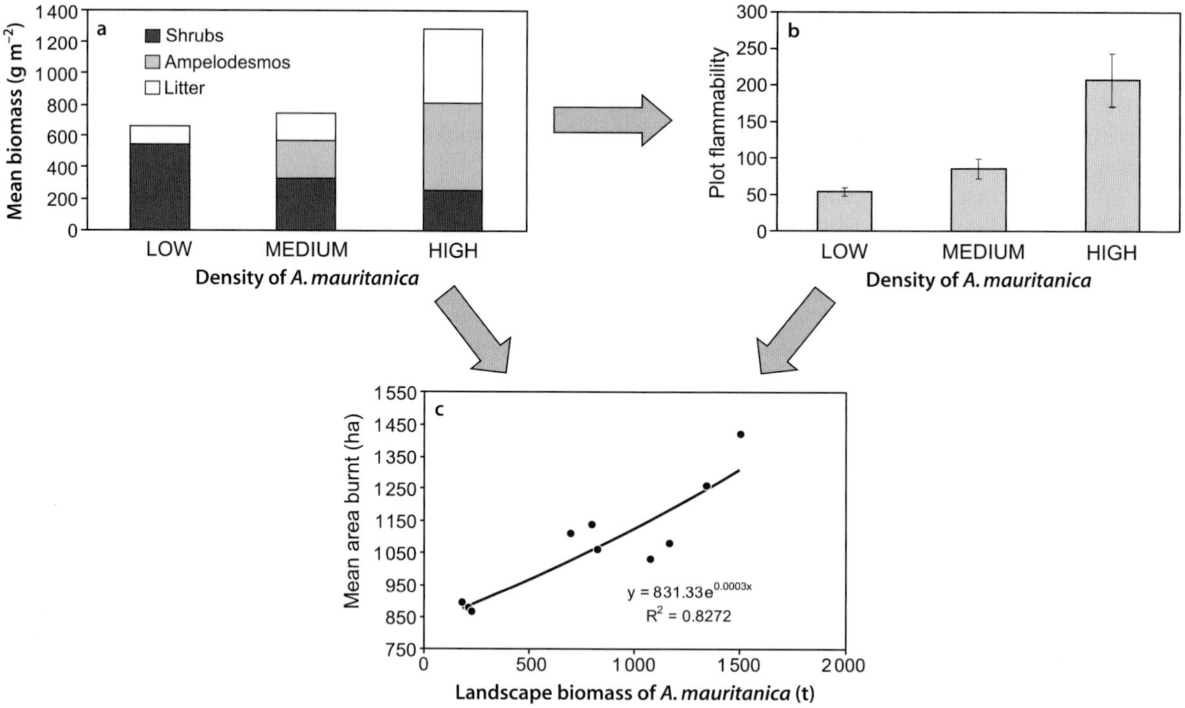

Fig. 7.4. Positive feedback between fire regimes and invasion by *Ampelodesmos mauritanica* in Catalonia. **a** Functional composition of communities with increasing density of *A. mauritanica*. **b** Plot-level flammability calculated from individual flammabilities weighed by the relative biomass contributions of species. **c** Effect of total biomass of *A. mauritanica* on mean area burnt during fire years (reprinted from Grigulis et al. 2005, with permission from Blackwell)

2. The considerable litter accumulation observed with increasing density of *A. mauritanica* was explained not only by the greater production of litter, but also by the fact that its litter decomposed at a 30% slower rate than that of the resident shrubs.
3. Under standardised conditions, *A. mauritanica* ignited much more rapidly than any of the shrub species, especially its dry litter, and once burning produced a more intense flame, again especially as litter. Calculated plot flammability taking into account changes in functional composition and specific flammabilities showed a >40-fold increase from low- to high-density plots (Fig. 7.4b).
4. Simulations of coupled vegetation- fire dynamics parameterised from these measurements of ecosystem properties and life history characteristics of *A. mauritanica* showed that invasion success and contribution to community biomass of *A. mauritanica* increased abruptly with decreasing fire return intervals. Total area burned in the landscape during each fire year was positively and exponentially related to the total biomass of *A. mauritanica* present in the landscape (Fig. 7.4c). Landscapes can hence switch from regimes of small localized to extensive fires as a result of the spread of *A. mauritanica* under decreasing fire return intervals. These patterns were explained by the increasing connectivity of highly flammable vegetation as *A. mauritanica* becomes more frequent in the landscape, coupled with the direct positive effects of fire on the demography of this grass.
5. Hence invasion of Catalan shrublands by *A. mauritanica* threatens a range of services that these natural ecosystems provide to rural and urban populations. Greatest of these is exposure to wildfire, especially in suburban areas that surround the city of Barcelona. Effects on native biodiversity were not yet detectable for plants, but may occur at later stages of invasion, in particular via the notable changes in fire regimes. Finally, invasion by *A. mauritanica* is perceived by the local population and managers as a threat to landscape amenity value.

7.3.2 The Role of Interactions

We have shown above that three components of FD – the *kind*, *relative abundance* and *range* of traits – are important in explaining biodiversity effects on ecosystem processes. In Sect. 7.3.1 we presented strong evidence of the role of the first two components: at any particular location, and within the envelope defined by climatic and disturbance conditions, ecosystem functioning is largely driven by the attributes of the dominant species, which are in turn strongly determined by the action of environmental filters. We know much less, however, about the functional role of the *range* of traits present in any given ecosystem, i.e., does it matter whether the organisms present are very similar or very different in terms of functional traits? The overwhelming evidence in favor of the mass-ratio hypothesis, which states that instantaneous functioning of ecosystems is determined to a large extent by the trait values of the dominant contributors to plant biomass; Grime (1998), might lead to the conclusion that the ecosystem role of subordinate species can be safely considered negligible. However, both theoretical knowledge of interactions and empirical examples tell us otherwise. Subordinate species are known – or expected – to play a significant role in long-term ecosystem stability (see e.g., Grime 1998; Dukes 2001; Lyons and Schwartz 2001; Lyons et al. 2005). Interactions between dominant and subordinate species within a given trophic level, and across trophic levels, are at the core of some of the mechanisms by which biodiversity affects ecosystem functioning (Chapin et al. 2000b). Two mechanisms involving species coexistence within the trophic level of primary producers are based on trait differences (*range* of traits) among species. These are resource use complementarity and facilitation.

Resource use complementarity is based on temporal and spatial niche partitioning (e.g., different rooting depths, phenologies, establishment mechanisms in regeneration gaps, forms of nitrogen that can be taken up from the soil), which reduces interspecific competition (MacArthur and Levins 1967). As a mechanism by which biodiversity influences local-scale ecosystem functioning, resource use complementarity assumes that the larger the functional trait distance between plants, the stronger the complementarity effect, and the more complete the total resource use by the community is expected to be (Trenbath 1974; Ewel 1986; Vandermeer 1989; see Hooper et al. 2005 for a detailed review). Complementarity in resource use is indeed a common coexistence mechanism in plant communities (e.g., Silvertown et al. 1999; McKane et al. 2002; Hooper and Dukes 2004). However, it is still uncertain whether this higher FD consistently leads to higher ecosystem 'performance' (e.g., higher productivity, nutrient retention, stability) beyond combinations of a very small number of species, such as those found in agriculture and forestry. Many cases in which species differences lead to higher biomass production involve nitrogen-fixing legumes, and thus can be interpreted as facilitation (Fridley 2001, see below).

Complementarity in resource use leading to higher ecosystem 'performance' in the absence of nitrogen-fixing legumes has been empirically demonstrated in some cases (e.g., van Ruijven and Berendse 2003), but seems far from universal, and its importance likely depends on the environmental context (Fridley 2003; Dimitrakopoulos and Schmid 2004; Hooper and Dukes 2004; Petchey 2004; Hooper et al. 2005). Grime (2002) has argued that, while resource use complementarity is definitely important as a coexistence mechanism, there is no reason based on natural selection why it should lead to 'enhanced' ecosystem functioning. The admission and persistence of organisms in plant communities and ecosystems depend upon their individual fitness rather than their contribution to community- or ecosystem-level properties, and coexistence and contribution to ecosystem processes often depend on different plant traits. Whereas complementarity facilitates the entry and persistence of species in the community it cannot be assumed that this will necessarily lead to complementary roles in sustaining the ecosystem. For example, Hooper and Dukes (2004), using synthetic serpentine-grassland communities, found that functional trait differences among species lead to strong complementarity in resource use among plants, but these were not sufficient to cause consistent increases in community productivity with increasing number of species. Dimitrakopoulos and Schmid (2004) measured biodiversity effects on plant biomass in synthetic calcareous-grassland communities planted on soil of different depths, and found that the importance of resource use complementarity increased as soil depth increased, suggesting that the biodiversity effect on biomass production may be strongly constrained in shallow soils.

The second interaction-based mechanism by which trait differences between species within the same trophic level can influence local-scale ecosystem functioning is facilitation. Probably the best example of this is the frequent stimulation of total production of species mixtures in the presence of nitrogen-fixing species. This is well known from the agricultural and forestry literature (Trenbath 1974; Vandermeer 1989; Cannell et al. 1992) and has been one of the most consistent effects of FD in synthetic-community experiments on the role of diversity in local-scale ecosystem functioning. Other plant-plant facilitation examples are the nurse effect by which larger plants ameliorate harsh climatic conditions or herbivore pressure for seedlings of other species (see Callaway 1992 and Bruno et al. 2003 for examples and detailed review of other facilitation mechanisms). Despite the increased interest in facilitative interactions in the past few years, there are not many documented examples of effects at the level of ecosystem functioning beyond biomass production (often positive effects, see above) and invasion resistance (both positive and negative effects; see Marler et al. 1999; Stampe and Daehler 2003). The literature on the effects of facilitation or resource use complementarity on ecosystem ser-

Box 7.2. Indirect interactions challenge the FD approach

Indirect interactions are those in which a species, through direct interaction with another species or modification of resources, alters the abundance of a third species with which it is not directly interacting. Such interactions challenge the FD approach, since there is no way to predict the ecosystem-level consequences of species additions or loss on the basis of the kind, range or relative abundance of traits.

Example 1
Facilitation among Woody Invaders in Temperate and Subtropical South America

The glossy privet (*Ligustrum lucidum*) is an evergreen, bird-dispersed and shade-tolerant Asiatic tree that invades seminatural woodlands and grasslands in southern South America. Originally introduced as an ornamental plant, this species apparently does not grow faster than native species, nor can it use soil resources not taken up by the resident communities (Gurvich et al. 2005). Its seedlings are susceptible to grazing by livestock and its recruitment has been reported to be strongly limited by the availability of recruitment micro-sites and by the interference from herbaceous vegetation (Mazia et al. 2001). On these bases, one would hardly predict its aggressive expansion over the landscape in the last decades. In central Argentina open woodlands its success is the result of indirect interactions with the native fauna and flora, mediated by the presence of another woody invader, the fire hawthorn (*Pyracantha angustifolia*). Unlike native species, both exotic shrubs produce fleshy fruits in autumn and winter, thus offering an extra food source to native frugivorous birds during a period of scarcity (Tecco et al. 2006). The branching architecture of *P. angustifolia* is attractive to perching birds, thus facilitating the dispersal of *L. lucidum* seeds under its canopy. Once germinated, *L. lucidum* seedlings grow slowly under *P. angusti-folia*, but as a shade tolerant, survives much better than the herbaceous dicots and grasses which outcompete it in the open (Tecco et al. in press). Also, the thorny, dense, umbrella-like canopy of *P. angustifolia* protects the seedlings from browsing and trampling by livestock. Once established, *L. lucidum* forms dense, shady, almost impenetrable thickets with an extremely poor understorey, both in terms of biomass and species richness (Lichstein et al. 2004). This, in turn, is expected to have important effects on ecosystem processes, services, and the land-use value of the land. For example, the total standing biomass and water uptake of *L. lucidum* near-monocultures are likely to be higher than those of stands dominated by semideciduous native vegetation. Carbon and nutrient cycling may also change by alteration of decomposition processes through altered litter quality and quantity and by a cooler, moister microhabitat. The effect of the winter extra supply of food for frugivorous birds by these two invasive species can have ripple effects on bird communities and the fleshy-fruited species dispersed by them.

Example 2
Gophers and Goatgrass in Californian Grasslands

The expansion of the exotic barbed goatgrass (*Aegilops triuncialis*) over the Californian landscape is also the consequence of several indirect interactions between three trophic levels that can hardly be predicted on the basis of the traits of the dominant species or the range of traits present in the community. According to Eviner and Chapin (2003), gophers are both attracted to patches dominated by goatgrass and negatively affect the latter. Apparently, the root architecture of goatgrass stabilizes the soil and makes it more efficient for gophers to burrow in patches dominated by this species. The negative effects of gophers on goatgrass stem from two processes. First, their burrowing activity buries goatgrass plants and thus slows down its expansion over the landscape. Second, they reduce the infection of goatgrass seedheads by the fungus *Ulocladium*, thus retarding germination and putting this species in a disadvantaged position with respect to native grasses. On the other hand, the presence of gophers decreases with heavy grazing. The expansion of goatgrass in heavily grazed paddocks decreases both the pastoral quality (seedheads can cause injury to livestock and its leaves are very poor nutritionally) and the amenity value (less native biodiversity, spiky seedheads are uncomfortable to hikers) of these Californian landscapes.

vices is scarce (see Díaz et al. 2005), and, to our knowledge, there is no published study explicitly addressing the effects of these mechanisms on global change drivers.

Resource use complementarity and facilitation are based on trait differences between species, i.e., their effects are expected to be maximal when species are very different from each other. Therefore, in those systems where these mechanisms indeed lead to enhanced ecosystem functioning, one should reasonably predict increased or decreased ecosystem 'performance' on the basis of an expanding or shrinking range of plant traits (Hooper 1998; Petchey 2003). However, there are other types of interactions that are both common and play an essential role in the preservation – or disruption – of ecosystem processes and services, and yet are not directly linked to the range of traits present in the local community. Major examples are interactions involving keystone species (Power et al. 1996), interactions involving ecosystem engineers (Jones et al. 1994), and indirect interactions (Box 7.2). 'Ecological surprises' in which the removal or introduction of species has triggered dramatic changes in ecosystem processes and services, often with very high cultural and economic costs, usually involve one or more of these types of interactions (see Díaz et al. 2005 for review and examples of 'ecological surprises'). In all these cases, changes in ecosystem processes and services, and feedbacks to global change drivers, are not easily predictable from species richness or from any of FD components (i.e., neither from the kind, or the relative abundance, or the range of traits present). The functional *identity* (*kind* of traits) of the introduced or removed species is of prime importance in determining these unexpected (and often negative) ecosystem-level impacts. Unlike in the mass-ratio hypothesis however, there is no easy a priori link between the traits of the species in question and putative ecosystem processes or feedbacks onto global change drivers. This represents one of the toughest challenges to the FD approach, and an extremely promising venue for new research.

7.4 Summary and Conclusions

Functional diversity (FD) comprises the *kind*, *range*, and *relative abundance* of functional traits present in a given community. Major global change drivers, including changes in climate, atmospheric composition, land-use/disturbance regime, and biotic exchanges, affect and are affected by FD in a non-random, and often predictable way. There is overwhelming evidence that the *kind* of traits that are present in higher abundance are major drivers of short-term ecosystem processes and their feedbacks onto global change drivers. Both the response of plant traits to environmental filters and their effects on biogeochemical and biophysical ecosystem processes (including simple, direct interactions with herbivores and decomposers) can be predicted with acceptable accuracy, although present theory cannot account for the mutually-neutralizing or synergistic effects of combinations of traits.

The role of the *range* of traits in influencing ecosystem functioning by resource use complementarity and facilitation is less clear, and fewer empirical examples are available, although both theory and some experimental evidence suggest that the role of subordinate species is important in maintaining long-term stability. Finally, there are other biodiversity-related effects, which are the result of indirect interactions, and/or interactions involving keystone species and/or ecosystem engineers, that cannot be predicted on the basis of the kind, range or relative abundance of traits, and remain a major challenge for the FD approach.

Non-linearities that are triggered by changes in climate, land-use regime and/or biotic interactions that involve altered FD, and in turn affect global change drivers and ecosystem services, represent a major threat to the integrity of the life-support systems. There is therefore an urgent need for more theoretical and empirical studies that would lead to better understanding and anticipating their consequences.

The quantification of functional diversity is not free from difficulties, and it is certainly less straightforward than counting the number of species, at least in areas with well-described floras. However, it provides a much stronger insight into the links between community structure, ecosystem functioning, and global change drivers and ecosystem functioning than does the consideration of species richness alone. Major venues for future research in FD are (1) the identification of those functional traits or functional trait combinations that are more likely to trigger ecosystem and landscape-level non-linearities; (2) continued efforts to measure functional traits under a wide range of biomes and regions, following standardized lists and protocols (see Lavorel et al. 2007, Chap. 13 of this volume); (3) more empirical investigation on how important are the roles of resource use complementarity and facilitation in determining ecosystem functioning under different degrees of environmental filtering; (4) more empirical and theoretical work on how FD effects on ecosystem processes translate into changes in global change drivers and ecosystem services; and (5) the design of easily-workable FD indices that could be applied to a wide range of natural situations, and thus broaden the possibilities of empirically testing the functional role of biodiversity.

Acknowledgments

We are grateful to members the GCTE Networks on Plant Functional Types and Removal Experiments on the Role of Biodiversity in Ecosystem Functioning for insightful discussions that contributed to the development of the ideas presented here. We thank D. Abal-Solís for drawing some of the figures. S. Díaz is supported by CONICET, FONCyT and Agencia Córdoba Ciencia S.E.

References

Aerts R (1995) The advantages of being evergreen. Trends Ecol Evol 10:402–407

Aerts R, Chapin FSI (2000) The mineral nutrition of wild plants revisited: a re-evaluation of processes and patterns. Adv Ecol Res 30:1–67

Brook BW, Sodhi NS, Ng PKL (2003) Catastrophic extinctions follow deforestation in Singapore. Nature 424:420–423

Bruno JF, Stachowicz JJ, Bertness MD (2003) Inclusion of facilitation into ecological theory. Trends Ecol Evol 18:119–125

Callaway RM (1992) Effects of shrubs on recruitment of *Quercus douglasii* and *Quercus lobata* in California. Ecology 73:2118–2128

Cannell MGR, Malcolm DC, Robertson PA (1992) The ecology of mixed species stands. Blackwell Scientific Publications, Oxford

Chambers S (1998) Short- and Long-Term Effects of Clearing Native Vegetation for Agricultural Purposes. Ph.D. Dissertation Flinders University of South Australia

Chapin FS III (1980) The mineral nutrition of wild plants. Annu Rev Ecol Syst 11:233–260

Chapin FSI, Autumn K, Pugnaire F (1993) Evolution of suites of traits in response to environmental stress. Am Nat 142:S78–S92

Chapin FSI, Bret-Harte MS, Hobbie SE, Zhong H (1996) Plant functional types as predictors of transient responses of arctic vegetation to global change. J Veg Sci 7:347–358

Chapin FS III, Eugster W, McFadden JP, Lynch AH, Walker DA (2000a) Summer differences among arctic ecosystems in regional climate forcing. J Climate 13:2002–2010

Chapin FS III, Zavaleta ES, Eviner VT, Naylor RL, Vitousek PM, Reynolds HL, Hooper DY, Lavorel S, Sala OE, Hobbie SE, Mack MC, Díaz S (2000b) Functional and societal consequences of changing biotic diversity. Nature 405:234–242

Chapin FS III, Matson PA, Mooney HA (2002) Principles of terrestrial ecosystem ecology. Springer, New York

Cochrane MA, Alencar A, Schulze MD, Souza CM, Nepstad DC, Lefebvre P, Davidson EA (1999) Positive feedbacks in the fire dynamic of closed canopy tropical forests. Science 284:1832–1835

Cordell S, Goldstein G, Mueller-Dombois D, Webb D, Vitousek PM (1998) Physiological and morphological variation in *Metrosideros polymorpha*, a dominant Hawaiian tree species, along an altitudinal gradient: the role of phenotypic plasticity. Oecologia 113:188–196

Cousins SAO, Lavorel S, Davies I (2003) Modelling the effects of landscape pattern and grazing regimes on the persistence of plant species with high conservation value in grasslands in south-eastern Sweden. Lands Ecol 18:315–352

Díaz S, Cabido M (2001) Vive la différence: plant functional diversity matters to ecosystem processes. Trends Ecol Evol 16:646–655

Díaz S, Cabido M, Casanoves F (1998) Plant functional types and environmental filters at a regional scale. J Veg Sci 9:113–122

Díaz S, Cabido M, Casanoves F (1999) Functional implications of trait-environment linkages in plant communities. In: Weiher E, Keddy PA (eds) Ecological assembly rules: Perspectives, advances, retreats pp 338–362. Cambridge University Press, Cambrige

Díaz S, Hodgson JG, Thompson K, Cabido M, Cornelissen JHC, Jalili A, Montserrat-Martí G, Grime JP, Zarrinkamar F, Asri Y, Band SR, Basconcelo S, Castro-Díez P, Funes G, Hamzehee B, Khoshnevi M, Pérez-Harguindeguy N, Pérez-Rontomé MC, Shirvany FA, Vendramini F, Yazdani S, Abbas-Azimi R, Bogaard A, Boustani S, Charles M, Dehghan M, de Torres-Espuny L, Falczuk V, Guerrero-Campo J, Hynd A, Jones G, Kowsary E, Kazemi-Saeed F, Maestro-Martínez M, Romo-Díez A, Shaw S, Siavash B, Villar-Salvador P, Zak M (2004) The plant traits that drive ecosystems: Evidence from three continents. J Veg Sci 15:295–304

Díaz S, Tilman D, Fargione J, Chapin III FS, Dirzo R, Kitzberger T Gemmill B, Zobel M, Vilà M, Mitchell C, Wilby A, Daily GC, Galetti M, Laurance WF, Pretty J, Naylor R, Power A, Harvell D (2005) Biodiversity regulation of ecosystem services. In: Hassan R, Scholes R, Ash N (eds) Ecosystems and human wellbeing: current state and trends: Findings of the Condition and Trends Working Group. Washington (D.C.) Island Press, pp 297–329

Dimitrakopoulos PG, Schmid B (2004). Biodiversity effects increase linearly with biotope space. Ecol Lett 7:574–583

Dukes JS (2001) Biodiversity and invasibility in grassland microcosms. Ecology 126:563–568

Estes JA, Duggins DO, Rathbun GB (1989) The ecology of extinctions on kelp forest communities. ConsBiol 3:251–264

Eviner VT, Chapin FS III (2003) Gopher-plant-fungal interactions affect establishment of an invasive grass. Ecology 84:120–128

Ewel JJ (1986) Designing agroecosystems for the humid tropics. Annu Rev Ecol Syst 17:245–271

Foley JA, Costa MH, Delire C, Ramankutty N, Snyder P (2003) Green surprises? How terrestrial ecosystems could affect Earth's climate. Frontiers Ecol Env 1:38–44

Fridley JD (2001) The influence of species diversity on ecosystem productivity: how, where, and why? Oikos 93:514–526

Fridley JD (2003) Diversity effects on production in different light and fertility environments: an experiment with communities of annual plants. J Ecol 91:396–406

Garnier E, Cortez J, Billès G, Navas M-L, Roumet C, Debussche M, Laurent G, Blanchard A, Aubry D, Bellmann A, Neill C, Toussaint J-P (2004) Plant functional markers capture ecosystem properties during secondary succession. Ecology 85:2630–2637

Graetz RD (1991) The nature and significance of the feedback of change in terrestrial vegetation on global atmospheric and climatic change. Climatic Change 18:147–173

Grigulis K, Lavorel S, Davies ID, Dossantos A, Lloret F, Vilà M (2005) Landscape-scale positive feedbacks between fire and expansion of the large tussock grass, *Ameplodesmos mauritanica*, in Catalan shrublands. Global Change Biology 11:1042–1053

Grime JP (1977) Evidence for existence of three primary strategies in plants and its relevance to ecological and evolutionary theory. Am Nat 111:1169–1194

Grime JP (1998) Benefits of plant diversity to ecosystems: immediate, filter and founder effects. J Ecol 86:902–910

Grime JP (2002) Declining plant diversity: empty niches or functional shifts? J Veg Sci 13:457–460

Grime JP, Thompson K, Hunt R, Hodgson JG, Cornelissen JHC, Rorison IH, Hendry GAF, Ashenden TW, Askew AP, Band SR, Booth RE, Bossard CC, Campbell BD, Cooper JEL, Davison AW, Gupta PL, Hall W, Hand DW, Hannah MA, Hillier SH, Hodkinson DJ, Jalili A, Liu Z, Mackey JML, Matthews N, Mowforth MA, Neal AM, Reader RJ, Reiling K, Ross-Fraser W, Spencer RE, Sutton F, Tasker DE, Thorpe PC, Whitehouse J (1997) Integrated screening validates primary axes of specialisation in plants. Oikos 79:259–281

Grime JP, Brown VK, Thompson K, Masters GJ, Hillier SH, Clarke IP, Askew AP, Corker D, Kielty JP (2000) The response of two contrasting limestone grasslands to simulated climate change. Science 289:762–765

Gurvich DE, Tecco PA, Díaz S (2005) Plant invasion in undisturbed ecosystems: The triggering attribute approach. J Veg Sci 16:723–728

Herbert DA, Fownes JH, Vitousek PM (1999) Hurricane damage to a Hawaiian forest: nutrient supply rate affects resistance and resilience. Ecology 80:908–920

Hobbie SE (1992) Effects of plant species on nutrient cycling. Trends Ecol Evol 7:336–339

Holling CS, Peterson G, Marples P, Sendzimir J, Redford K, Gunderson L, Lambert D (1996) Self-organization in ecosystems: lumpy geometries, periodicities and morphologies. In: Walker B, Steffen W (eds) Global Change and Terrestrial Ecosystems. Cambridge University Press, Cambridge pp 346–384

Hooper DU (1998) The role of complementarity and competition in ecosystem responses to variation in plant diversity. Ecology 79:704–719

Hooper DU, Dukes JS (2004) Overyielding among plant functional groups in a long-term experiment. Ecol Lett 7:95–105

Hooper DU, Chapin FS III, Ewel JJ, Hector A, Inchausti P, Lavorel S, Lawton JH, Lodge DM, Loreau M, Naeem S, Schmid B, Setälä H, Symstad AJ, Vandermeer J, Wardle DA (2005) Effects of biodiversity on ecosystem functioning: a consensus of current knowledge and needs for future research. Ecol Monogr 75:3–35

Jaramillo VJ, Detling JK (1988) Grazing history, defoliation, and competition: Effects on shortgrass production and nitrogen accumulation. Ecology 69:1599–1608

Johnson EA, Cochrane MA (2003) Disturbance regime interactions In: Lovejoy TE, Hannah L (eds) Climate Change and Biodiversity: Synergistic Impacts. Yale University Press, pp 39–44

Jones CG, Lawton JH, Shachak M (1994) Organisms as ecosystem engineers. Oikos 69:373–386

Keddy PA (1992) Assembly and response rules: two goals for predictive community ecology. J Veg Sci 3:157–164
Lavorel S, Garnier E (2002) Predicting changes in community composition and ecosystem function from plant traits: revisiting the Holy Grail. Func Ecol 16:545–556
Lavorel S, Davies I, Noble I (2000) LAMOS: a landscape modelling shell. In Hawkes BC, Flannigan MD (eds) Landscape Fire Modeling Workshop, pp 25–28
Lawton JH, Naeem S, Thompson LJ, Hector A, Crawley MJ (1998) Biodiversity and ecosystem function: getting the Ecotron experiment in its correct context. Funct Ecol 12:848–852
Lepš J, Osbornová-Kosinová J, Rejmánek M (1982) Community stability, complexity and species life history strategies. Vegetatio 50:53–63
Levine JM, Vilà M, D'Antonio CM, Dukes JS, Grigulis K, Lavorel S (2003) Mechanisms underlying the impacts of exotic plant invasions. Proceedings of the Royal Society of London, Series B 270:775–781
Lichstein JW, Grau HR, Aragón R (2004) Recruitment limitation in secondary forests dominated by an exotic tree. J Veg Sci 15:721–728
Loreau M, Naeem S, Inchausti P, Bengtsson J, Grime JP, Hector A, Hooper DU, Huston MA, Raffaelli D, Schmid B, Tilman D, Wardle DA (2001) Biodiversity and ecosystem functioning: Current knowledge and future challenges. Science 294:804–808
Luck G, Daily G, Ehrlich P (2003) Population diversity and ecosystem services. Trends Ecol Evol 18:331–336
Lyons KG, Schwartz MW (2001) Rare species loss alters ecosystem function – invasion resistance. Ecol Lett 4:358–365
Lyons, KG, Brigham CA, Traut BH, Schwartz MW (2005) Rare species and ecosystem functioning. Cons Biol 19:1019–1024
MacArthur RH, Levins R (1967) The limiting similarity, convergence, and divergence of coexisting species. Am Nat 101:377–385
MacArthur RH, Wilson EO (1967) The theory of island biogeography. Princeton University Press, Princeton
MacGillivray CW, Grime JP, ISP Team (1995) Testing predictions of the resistance and resilience of the vegetation subjected to extreme events. Func Ecol 9:640–649
MAE (Millennium Ecosystem Assessment) (2003) Ecosystem and Human Well-Being. Island Press, Washington
Marler MJ, Zabinski CA, Callaway R (1999) Mycorrhizae indirectly affect competitive effects of an invasive forb on a native bunchgrass. Ecology 80:1180–1186
Mason NWH, MacGillivray K, Steel JB, Wilson JB (2003) An index of functional diversity. Journal of Vegetation Science 14:571–578
Mason NWH, Mouillot M, Lee WG, Wilson JB (2005) Functional richness, functional evenness and functional divergence: the primary components of functional diversity. Oikos, 111:112–118
Mazia CN, Chaneton EJ, Ghersa CM, León RJ (2001) Limits to tree species invasion in pampean grassland and forest communities. Oecologia 128:594–602
McKane RB, Johnson LC, Shaver GR, Nadelhoffer KJ, Rastetter EB, Fry B, Giblin AE, Kielland K, Kwiatkowski BL, Laundre JA, Murray G (2002) Resource-based niches provide a basis for plant species diversity and dominance in artic tundra. Nature 415:68–71
McKinney M, Lockwood J (1999) Biotic homogenization: a few winners replacing many losers in the next mass extinction. Trends Ecol Evol 14:450–453
Moreno JM, Vásquez A, Velez R (1998) Recent history of forest fires in Spain. In: Moreno JM (ed) Large forest fires. Backhuys Publishers, Leiden, pp 159–186
Mouillot D, Mason WHN, Dumay O, Wilson JB (2005) Functional regularity: a neglected aspect of functional diversity. Oecologia 142:353–359
Naeem S (2002) Disentangling the impacts of diversity on ecosystem functioning in combinatorial experiments. Ecology 83:2925–2935
Naeem S, Wright JP (2003) Disentangling biodiversity effects on ecosystem functioning: deriving solutions to a seemingly insurmountable problem. Ecol Lett 6:567–679
Pausas JG (2004) Changes in fire and climate in the eastern Iberian Peninsula (Mediterranean basin). Climatic Change 63:337–350
Petchey OL (2003) Integrating methods that investigate how complementarity influences ecosystem functioning. Oikos 101: 323–330
Petchey OL (2004) On the statistical significance of functional diversity effects. Func Ecol 18:297–303
Petchey OL, Hector A, Gaston KJ (2004) How do different measures of functional diversity perform?. Ecology 85(3):847–857
Pielke RA (2001) Influence of the spatial distribution of vegetation and soils on the prediction of cumulus convective rainfall. Rev Geoph 39:151–177
Power ME, Tilman D, Estes J, Menge BA, Bond WJ, Mills LS, Daily G, Castilla JC, Lubchenco J, Paine RT (1996) Challenges in the quest for keystones. BioScience 46:609–620
Redford KH, Feinsinger P (2000). The half-empty forest: Sustainable use and the ecology of interactions. In: Reynolds J, Mace GM, Redford KH, Robinson JG (eds) Conservation of Exploited Populations. Cambridge University Press, Cambridge, pp 370–400
Russell-Smith J, Stanton P (2002) Fire regimes and fire management of rainforest communities across northern Australia. In: Bradstock RA, Williams JE, Gill AM (eds) Flammable Australia. The fire regimes and biodiversity of a continent. Cambridge University Press, Cambridge, pp 329–350
Schmid B, Joshi J, Schläpfer F (2002). Empirical evidence for biodiversity-ecosystem functioning relationships. In: Kinzig AP, Pacala SW, Tilman D (eds) The functional consequences of biodiversity: empirical progress and theoretical extensions. Monographs in Population Biology 33, Princeton University Press, Princeton and Oxford, pp 120–150
Scholes R, Archer S (1997) Tree-grass interactions in savannas. Annu Rev Ecol Syst 28:517–544
Silvertown J, Dodd ME, Gowing DJG, Mountford JO (1999) Hydrologically defined niches reveal a basis for species richness in plant communities. Nature 400:61–63
Stampe D, Daehler CC (2003) Mycorrhizal identity affects plant community structure and invasion: a microcosm study. Oikos 100: 362–372
Sturm M, Racine C, Tape K (2001) Increasing shrub abundance in the artic. Nature 411:546–547
Tecco PA, Gurvich DE, Díaz S, Perez-Harguindeguy N, Cabido M (2006) Positive interaction between invasive plants: the influence of *Pyracantha angustifolia* on the recruitment of native and exotic woody species. Aust Ecol 31:293–300
Tecco PA, Díaz S, Gurvich DE, Pérez-Harguindeguy N, Cabido M, Bertone GA (2007) Experimental evidence of positive association between exotic woody species: Facilitation by *Pyracantha angustifolia* on *Ligustrum lucidum* sapling survival. App Veg Sci, in press
Thomas CD, Cameron A, Green RE, Bakkenes M, Beaumont LJ, Collingham YC, Erasmus BFN, Ferreira de Siqueira M, Grainger A, Hannah L, Hughes L, Huntley B, van Jaarsveld AS, Midgley GF, Miles L, Ortega-Huerta MA, Townsend Peterson A, Phillips OL, Williams SE (2004) Extinction risk from climate change. Nature 427:145–148
Thompson K (1994) Predicting the fate of temperate species in response to human disturbance and global change. In: Boyle TJB, Boyle CEB (eds) Biodiversity, temperate ecosystems and global change. Springer-Verlag, Berlin, pp 61–76
Thompson C, Beringer J, Chapin FS III, McGuire AD (2004) Structural complexity and land-surface energy exchange along a vegetation gradient from artic tundra to boreal forest. J Veg Sci 15:397–406
Tilman D (1999) The ecological consequences of changes in biodiversity: a search for general principles. Ecology 80:1455–1474
Tilman D (2001) Commentary – An evolutionary approach to ecosystem functioning. Proc Nat Acad Sci USA 98:10979–10980
Trenbath BR (1974) Biomass productivity of mixtures. Adv Agron 26:177–210
van Ruijven J, Berendse F (2003) Positive effects of plant species diversity on productivity in the absence of legumes. Ecol Lett 6:170–175
Vandermeer JH (1989) The ecology of intercropping. Cambridge University Press, Cambridge
Vilà M, Lloret F, Ogheri E, Terradas J (2001) Positive grass-fire feedback in Mediterranean Basin woodlands. For Ecol Manag 147:3–14
Woodward F, Diament AD (1991) Functional approaches to predicting the ecological effects of global change. Func Ecol 5:202–212
Zimov SA, Davidov SP, Zimova GM, Davidova AI, Chapin FS III, Chapin MC, Reynolds JF (1999) Contribution of disturbance to high-latitude amplification of atmospheric CO_2. Science 284:1973–1976
Zobel M (1997) The relative role of species pools in determining plant species richness: an alternative explanation of species coexistence? Trends Ecol Evol 12:266–269

Chapter 8

Linking Plant Invasions to Global Environmental Change

Montserrat Vilà · Jeffrey D. Corbin · Jeffrey S. Dukes · Joan Pino · Stanley D. Smith

8.1 Introduction

Biotic invasions have been recognized as an important element of global change (Vitousek 1994). Introductions of alien species into novel habitats have increased in tandem with travel and international trade (McNeely 2001). Many species have been introduced accidentally (e.g., in water ballast, in soil, or as crop seed "contaminants"), but some have been intentionally introduced as ornamentals, food, or fiber products.

The introduction of alien species can have many ecological impacts, and contribute to the homogenization of biological systems worldwide (Lockwood and McKinney 2001). Biotic invasions, along with alterations in land-use patterns and disturbance regimes, are among the major causes of biodiversity loss worldwide (Soulé 1991). In cases where alien species have quantitative or qualitative trait differences from native species, invasions can also alter ecosystem processes such as nutrient cycling dynamics and disturbance regimes (Vitousek et al. 1987; D'Antonio and Vitousek 1992; D'Antonio and Corbin 2003; Levine et al. 2003). There are some well known cases of devastating effects of invasive plants on ecosystems such as the invasion of annual grasses in western U.S. (Mack 1981) or the invasion of pines in South-African shrublands (Le Maitre et al. 1996). These dramatic invasions emphasize that invaders often parallel environmental changes that are taking place at the regional scale. Therefore, research on the links between invasions and environmental changes is urgent and timely.

Biotic invasions are capable of interacting with other anthropogenic changes in the environment to alter biodiversity and ecosystem processes in invaded habitats. For example, there is evidence from a variety of ecosystems that N inputs favor alien plant species (Huenneke et al. 1990; Vinton and Burke 1995; Maron and Connors 1996). Furthermore, land-use changes such as clearing for agriculture, road building (Gelbard and Belnap 2003), or alteration of disturbance regimes (Mack and D'Antonio 1998; D'Antonio et al. 1999) have been shown to facilitate plant invasions.

The aim of this chapter is to present evidence of the interactions between several components of global environmental change and plant invasions. We focus on the effects of increasing atmospheric CO_2 concentrations ($[CO_2]$), climate change, terrestrial eutrophication, and changes in land-use/cover on the distribution and performance of plant invasions. We focus on plant invasions because in terrestrial ecosystems plants interact most dramatically with environmental and landscape changes, often reverberating to higher tropic levels. We do not consider the interactions between changes in disturbance regimes "per se", such as wildfires and invasions, because they have been extensively reviewed elsewhere (Mack and D'Antonio 1998; D'Antonio et al. 1999).

8.2 Plant Invasions and Elevated CO_2

Atmospheric CO_2 concentration has increased ca. 35% since pre-industrial times, and is predicted to rise to double the preindustrial concentration by the end of this century. A great deal of research has shown that plant photosynthesis and growth are stimulated by elevated $[CO_2]$. Although invasive and native plant species responded similarly across a wide range of controlled-environment experiments, i.e., the average invasive species does not respond more strongly than the average native, several problematic invasive species are nonetheless strong responders (Dukes 2000). Indeed, recent research has highlighted several cases in which invasive species may be favored by the $[CO_2]$ increase (Moore 2004).

Results from several controlled-environment studies found growth of invasive species to be more strongly stimulated by elevated $[CO_2]$ than growth of native species that were either closely related or sympatric in the same habitat. The average E/A ratio (ratio of growth or final biomass at elevated/ambient $[CO_2]$) for an experimental doubling of $[CO_2]$ was 1.82 in non-native species vs. 1.28 in native species – 2.35 in the non-native vine *Lonicera japonica* vs. 1.40 in native sympatric vines (Sasek and Strain 1991); 1.56 in the non-native *Rhododendron ponticum* vs. 1.12 in native understory plants in a Swiss forest (Hättenschwiler and Körner 2003); and 1.54 in *Bromus tectorum* vs. 1.31 in native Great Basin grasses (Smith et al. 1987).

Most elevated [CO_2] research has examined responses of plants to projected future concentrations, but it is important to note that the present [CO_2] already represents a significant increase over pre-industrial times. Ziska (2003) found an average stimulation of biomass of six invasive species of 46% in response to a doubling of [CO_2]. However, the growth response among these species to the increase in [CO_2] during the 20th century (from ca. 280 to 370 μmol mol^{-1}, a 30% increase) was significantly higher, averaging 110%, and was much higher than in native species. This indicates that many plant invasions that have occurred in the 20th century have probably been stimulated by rising levels of [CO_2]. In addition to increased branching and leaf area, Ziska and colleagues also observed significantly higher pollen production in common ragweed (Ziska and Caulfied 2000) and higher spine production in Canada thistle (Ziska 2002), both in the past century and projected into the next century. This indicates that elevated [CO_2] may increase pollen counts from non-native weeds to the detriment of human health, and that weeds such as Canada thistle may become even more invasive at elevated [CO_2].

In a community context, Dukes (2002) found *Centaurea solstitialis* (yellow starthistle) to increase in biomass and seed production by ca. 70% in response to a doubling of [CO_2] in both monoculture and polyculture, whereas total biomass of the polyculture (made up primarily of native species) increased by less than 30%. This suggests a differential success of *Centaurea* in a community dominated by native species. Although the response of a species in a whole community is less straightforward than the response of individual pot-grown plants, the response of the monocultures and polycultures as a whole were similar in the Dukes (2002) experiment. In a similar vein, *Lonicera japonica* (Japanese honeysuckle, a C_3 vine) increased ANPP 2.9-fold and cover 2.2-fold at the Oak Ridge FACE site, whereas the total understory community did not show increased ANPP and only a 1.3-fold increase in cover (Belote et al. 2003).

At the Nevada Desert Free-air CO_2 Enrichment (FACE) Facility, an annual invasive grass (*Bromus madritensis* ssp. *rubens*; red brome) exhibited higher differential photosynthesis, productivity and seed production than native annuals at elevated CO_2 during an extremely wet (El Niño) year (Smith et al. 2000). *Bromus madritensis* is closely related to *B. tectorum* (cheatgrass), an annual Eurasian grass that has invaded most of the intermountain west of the U.S.A. (Mack 1981). *Bromus tectorum*, whose growth has also been shown to be differentially stimulated over native species at elevated [CO_2] (Smith et al. 1987), is well known to stimulate large grassland fires in this region (Knapp 1998), which in turn has converted large areas of diverse shrublands to annual grasslands (Young and Evans 1978). Similarly, the non-native *B. madritensis* has been shown to stimulate wildfires in the Mojave Desert (Brooks 1999), a region with no evolutionary history of wildfire on a recurring basis. This led Smith et al. (2000) to conclude that elevated [CO_2], through a differential stimulation of growth in *B. madritensis*, could potentially convert large areas of desert scrub to an alien-dominated annual grassland.

Why are invasive species more responsive to elevated [CO_2] than are natives in some of these recent studies? Three trends have been observed. First, there are potential plant architectural differences. Ziska (2003) examined the underpinnings of increased growth, namely the production of leaf area and net assimilation rate (NAR, a growth-related index of photosynthetic CO_2 assimilation) in six non-native species. Interestingly, increases in leaf area were consistently responsible for the stimulation of growth in these non-native species rather than increases in NAR. Similarly, Huxman and Smith (2001) found that elevated CO_2 stimulated growth of a Mojave Desert invasive grass much more than that of a native forb, even though both species had similar increases in photosynthesis. In a study of a non-native and native vines, Sasek and Strain (1991) observed that the differentially increased growth in the non-native vine at elevated [CO_2] was due to accelerated branching (up to 3-fold higher) that in turn resulted in a greater increase in leaf area. The precise mechanism for such a response is not presently known.

Second, there may be differences in plant construction cost between invasive and native plants. Several studies have shown invasive species to have lower construction costs than sympatric natives (Baruch and Goldstein 1999; Nagel and Griffin 2001), and elevated [CO_2] exacerbates this difference. In *Bromus madritensis*, plant construction cost is significantly lower at elevated [CO_2] than at ambient [CO_2], a response that was not observed in a native annual grass, *Vulpia octoflora* (Nagel et al. 2004). Lower construction cost allows *Bromus* seedlings to grow more quickly and assume larger size and greater seed rain than the native *Vulpia*, with *Bromus* exhibiting differentially greater growth than *Vulpia* at elevated [CO_2] (Table 8.1); this should give *Bromus* a strong competitive advantage during the short time window that soil resources are available in this desert ecosystem. Similar to *Bromus*, lower construction cost of the wetland macrophyte *Lythrum salicaria* (purple loosestrife) at elevated [CO_2] resulted in over a 200% increase in carbon fixation per unit of energy invested in leaf biomass, a much higher rate than in sympatric native species (Nagel and Griffin 2004). This dramatic increase in carbon assimilation efficiency occurred in *Lythrum* despite only modest increases in photosynthesis and no change in leaf respiration per unit leaf area.

Third, elevated [CO_2] may bring about changes in reproductive allocation, specifically investment of carbon and nitrogen in seeds. *Bromus madritensis* makes more seeds per unit carbon and nitrogen investment at elevated [CO_2] than in ambient [CO_2] (Huxman et al. 1998, 1999),

Table 8.1. Construction cost, relative growth rate (*RGR*), final plant biomass and total seed number in the native annual grass *Vulpia octoflora* and the invasive grass *Bromus madritensis* growing during a wet year at the Nevada Desert FACE Facility in the Mojave Desert. Elevated $[CO_2] = 550\ \mu mol\ mol^{-1}$. Also included is the ratio of the invasive-to-native species (*I*/*N*) for each parameter at each $[CO_2]$ (data are from Nagel et al. (2004))

Parameter	$[CO_2]$	*Vulpia octoflora*	*Bromus madritensis*	*I*/*N*
Construction cost ($g\ g^{-1}\ d^{-1}$)	Ambient	1.95	1.75	0.9
	Elevated	1.78	1.55	0.87
RGR ($g\ g^{-1}$ biomass d^{-1})	Ambient	0.054	0.068	1.26
	Elevated	0.062	0.108	1.74
Biomass (g)	Ambient	0.08	0.51	6.4
	Elevated	0.11	0.89	8.1
Seed number $plant^{-1}$	Ambient	128	231	1.8
	Elevated	226	533	2.36

thus significantly increasing their differential abundance in community seed rain (Smith et al. 2000).

Our current knowledge concerning the effects of elevated $[CO_2]$ on invasive plant species has mainly focused on plant performance and plant ecophysiology. However, we are still poorly prepared to predict the consequences for communities and ecosystems. This is because most studies that have examined the effects of elevated $[CO_2]$ on plants have been conducted on potted plants grown in glasshouses, a problem that has been alleviated in recent years by the advent of various technologies (e.g., FACE and Open-Top Chambers (OTC's)) that have allowed natural communities of plants to be continuously exposed to elevated $[CO_2]$. Additionally, relatively few studies have examined the CO_2 responses of invasive species grown in competition with the communities that they invade. Responses of individual species do not consistently predict the success of species grown in competition. Future research efforts need to prioritize (1) the mechanistic underpinnings for the differential success of invasive species to elevated $[CO_2]$, and (2) responses of invasive species grown under "realistic" circumstances.

8.3 Plant Invasions and Climatic Change

As fossil fuel burning and tropical forest clearing increasingly change the composition of the atmosphere, the Earth's climate is expected to change in tandem (IPCC 2001). Model predictions suggest that the future climate will be warmer, and that precipitation patterns will change. Although the rate and extent of these changes will depend largely on future human actions, even the mildest of expected changes would likely lead to range shifts among plant and animal species, along with changes in plant and animal phenology (IPCC 2001). Indeed, many species are already responding to changes in temperatures and growing season lengths (Peñuelas et al. 2002; Parmesan and Yohe 2003; Root et al. 2003). Will some invasive plant species be favored by these climate changes? If so, which ones?

Clearly, invasive species whose native ranges are warmer than their introduced ranges would be at an advantage. Two mechanisms would favor these species: first, the species should withstand the increasingly hot extreme temperatures better than natives, and second, they should experience less mortality due to extreme cold events. Examples of the first mechanism can be found among the C_4 species. The higher optimum temperature for photosynthesis found in C_4 species would provide an advantage over C_3 plants in a warmer environment (Ehleringer et al. 1997). Increases in warm extremes could help native C_4 grasses invade C_3 grasslands in New Zealand (White et al. 2000, 2001). Patterson (1995) provided examples of several tropical and warm-temperate agricultural weeds that are likely to become more problematic in the United States as temperatures increase. In an example of the second mechanism, warming may already have benefited cold-sensitive (i.e., frost-sensitive) invasive plants in Swiss forests. There, as the frequency of frost return intervals has declined over the past century, the abundance and diversity of alien evergreen broadleaved species have increased (Walther 2002).

Species interactions may be as sensitive to changes in precipitation patterns as to warming. In regions where precipitation increases, ornamental species that had been restricted to gardens by water limitation could become more problematic. In western North America, a wetter climate could cause several invasive grasses to become denser, with negative consequences for native plant and animal species (Dukes and Mooney 1999). On the other hand, a drier climate in this same general region could increase the prevalence and impacts of shrubby *Tamarix* species that invade riparian zones (Zavaleta and Royval 2002).

Certain properties are likely to confer advantages under any type of climate change. These properties include the ability to tolerate a wide range of climates, the ability to shift ranges rapidly, and the lack of dependence on other organisms for pollination and seed dispersal. Invasive species tend to have all of these favorable properties (Dukes and Mooney 1999).

Several studies of herbaceous plants have shown that the native ranges of invasive species span greater latitudinal distances or geographic ranges than the native ranges of non-invasive species (Rejmánek 2000). While debate continues about the cause for this pattern, a consequence is that many successful invaders tolerate a wide range of climates, and should be relatively insensitive to climate shifts over most of their ranges. Malcolm et al. (2002) presented evidence that the rate at which climatic zones shift under projected warming may, in some locations, exceed maximum rates of plant migration observed in post-glacial time periods, which suggests that the most mobile species will be favored by rapid warming. Indeed, several traits favoring long-distance dispersal are commonly found in invasive species (Rejmánek 1996), suggesting that these species will be among the fastest to migrate to newly suitable habitats. Finally, the ranges of invasive species are rarely limited to the range in which coevolved pollinators or seed dispersers exist. In fact, invasive species seem particularly adept at forming new mutualisms (Richardson et al. 2000a). Therefore, while some native species may be disadvantaged if mutualistic relationships are disrupted by climate change, most invasive species will not.

A handful of case studies have examined how climate change could alter the success of particular invasive species by combining climate model predictions with inferred environmental constraints. Based on two future climate scenarios, Beerling et al. (1995) predicted northward shifts of the European range of *Fallopia japonica*, which is native to Asia. In both scenarios, the new potential distribution moved from central Europe into Scandinavia, and into Iceland. This model was purely based on climate suitability, and did not limit ranges based on soil types or interactions with currently established species. In a climate envelope exercise carried out by Richardson et al. (2000b), potential ranges of five of South Africa's more noxious invasive plants decreased to 63–92% of their starting range under a ~2 °C warming. In this analysis, plant ranges were restricted to suitable soil types. Kritikos et al. (2003a,b) predicted potential ranges of a tropical woody vine, *Cryptostegia grandiflora*, and a woody legume, *Acacia nilotica*, under several future climate scenarios. Although the two invaders had not yet filled their current potential ranges in Australia, the studies suggest both species' ranges would expand under warming, and especially with decreased drought stress.

8.4 Plant Invasions and Land Eutrophication

Though N is the most abundant component of the Earth's atmosphere, it is unavailable for most vegetation unless it is "fixed", or converted from inert atmospheric N_2 into other forms of N, through such processes as lightning or biological fixation. As a result, it is frequently the nutrient most limiting primary productivity in the temperate region (Vitousek and Howarth 1991). Human alteration of the N cycle, however, has increased the rate of N fixation to such an extent that human-derived N now exceeds natural processes (Vitousek et al. 1997). Today, ecosystems in eastern North America, northern Europe, and localized habitats downwind of large sources of N such as urban areas receive orders of magnitude more N in wet and dry deposition than they would receive from natural sources (Aber et al. 1989, 1998; Galloway et al. 1995, 2003; Fenn et al. 1998).

Increasing N inputs have had important consequences for species composition and diversity of plant communities (e.g., Kellner 1993; Wedin and Tilman 1996). A number of European communities have changed dramatically as a result of increased N deposition. For example, the diversity of vascular plants, bryophytes, and lichens has declined in British acid grasslands (Stevens et al. 2004) and Dutch chalk grasslands (Bobbink 1991) as a result of decades of increased N inputs. Species composition has also shifted toward species characteristic of N-rich habitats in forests in France (Thimonier et al. 1992), the Netherlands (van Breemen and van Dijk 1988; DeVries et al. 1995), and Sweden, Norway and Finland (Falkengren-Grerup 1986; Hogbom and Hogberg 1991; Rosen et al. 1992).

Against the general background of changes in species composition and diversity following atmospheric N deposition, increased availability of N may facilitate the establishment and success of alien species (Jefferies and Maron 1997; Fenn et al. 2003b). Enhancement of N availability has been shown to favor establishment and success of fast-growing invasive species in a variety of habitats (e.g., Huenneke et al. 1990; Vinton and Burke 1995; Maron and Connors 1996). Therefore, the potential impact of N deposition would be expected to be greatest in relatively infertile habitats, by permitting the invasion of species characteristic of high-N conditions into formerly low- or moderate-N habitats.

The role of elevated N inputs in facilitating species invasions has been shown most clearly in infertile desert, grassland, and shrubland habitats of western and southwestern North America. Urban centers in the Southwest, including Phoenix and Tucson, AZ, Las Vegas, NV, and Los Angeles, CA have contributed to N deposition rates as high as 29 kg $ha^{-1}\,yr^{-1}$ (Fenn et al. 2003a). Alien annual grasses, such as *Bromus madritensis* ssp. *rubens* and *Schismus* spp., and an alien forb, *Erodium cicutarium*, have invaded *Larrea tridentata* (Creosote bush) shrublands in the Mojave and Sonoran deserts. Brooks (2003) found that N and NPK fertilization (32 kg N $ha^{-1}\,yr^{-1}$) increased dominance of alien species in the Mojave Desert. While alien biomass increased in fertilized plots by 52–56% relative to controls, the density, biomass and diversity of native annual species decreased in fertilized plots. In contrast, fertilization (10 kg N ha^{-1} yr^{-1}) of a Mojave Desert annual plant community before the invasion of the alien grasses and forbs led to an increase in the biomass of native annual species (Romney et al. 1978). Brooks (2003) hypothesized that increased competition with alien species for soil water and other nutrients led to the decline in native species. Thus, native desert annuals are apparently able to take advantage of N inputs when aliens are scarce, but not when aliens are abundant (Brooks 2003).

Alien annual grasses have also invaded coastal sage scrub communities and serpentine grasslands in California. The increasing abundance of grass species such as *Avena* spp., *Bromus* spp., and *Vulpia* spp. in southern California coastal sage scrub ecosystems has coincided with increasing N deposition rates to >45 kg N ha^{-1} yr^{-1} (Allen 2003). In order to tie grass invasion to N inputs, Allen (2003) added N (60 kg N ha^{-1} yr^{-1} for 10 years) to a shrubland receiving low deposition rates. Annual grasses responded positively to N fertilization, though there was no effect on shrub growth or mortality, suggesting that changes in shrub cover requires many years to take place. The invasion of annual grasses such as *Lolium multiflorum*, *Avena* spp., and *Bromus mollis* into serpentine grasslands in the San Francisco Bay Area has also coincided with elevated N deposition (>10–15 kg N ha^{-1} yr^{-1}) (Huenneke et al. 1990; Weiss 1999). Serpentine soils in California have served as refuges for native grass and forb species because their low levels of N, P and K, coupled with high Mg/Ca ratios and high heavy metal content, have historically acted as barriers to alien species invasions (Murphy and Ehrlich 1989). Responses of serpentine vegetation to experimental N fertilization indicate that increasing N inputs are a driving factor in the invasion (Huenneke et al. 1990). Without active management to reduce the competitive effects of fast-growing alien species, native grasses and annual forbs, including rare serpentine endemics, may be lost from these communities with cascading effects to other trophic levels (Weiss 1999).

Elevated atmospheric N deposition may also increase the susceptibility and/or impacts of introduced pathogens or disease in forests (Brown 1995; Bobbink et al. 1998; Throop and Lerdau 2004). In eastern North America, introduced pests and pathogens such as the gypsy moth (*Lymantria dispar*), beech bark disease (*Cryptococcus fagisuga* and *Nectria* spp.) and the hemlock woolly adelgid (*Adelges tsugae*) have had substantial impacts on the community composition and ecosystem functioning of forests (Liebhold et al. 1995; Jenkins et al. 1999). The role that N inputs play in these invasions is not well-established, but there are indications that elevated foliar and bark N concentrations in trees experiencing N deposition have led to increased susceptibility to attack (Latty et al. 2003). Future research should consider the interactions between atmospheric N deposition, pathogen invasion and forest health.

8.5 Plant Invasions and Changes in Land Use/Cover

Man-induced transformation of natural systems is leading to a major biodiversity crisis in recent times (Vane-Wright et al. 1991; Reynolds et al. 2007, Chap. 8 of this volume). In the last century, land transformations have been extensive, rapid, and in most cases irreversible. Changes in land-use/cover (LUC hereafter) can be defined as land transformations from one ecosystem "state" to another in terms of ecosystem structure, composition and function (Hobbs 2000). Because plant invasions occur across landscapes, their dispersal and spread depends on landscape structure and dynamics (With 2002). For this reason, it is surprising that few studies have examined interactions between invasion patterns and changes in LUC.

There is a general presumption among ecologists and managers that changes in LUC directly increase biological invasions. Human population density has been used as a surrogate of LUC transformation, and it has been shown that there is a positive relationship between human population density and regional plant species density (Vilà and Pujadas 2001) or frequency distribution of species (Pysek et al. 1998). Moreover, observational analyses of the current relationship between invasions and LUC types almost always conclude that the most invaded habitats are the ones which directly or indirectly suffer major environmental fluctuations or changes and/or are "less natural" (e.g., urban, agricultural). Although these approaches support the notion that changes in LUC are drivers of invasion, we should emphasize that changes in LUC include a broad range of ecosystem degradation and ecosystem recovery changes (Hobbs 2000; Reynolds et al. 2007, Chap. 20 of this volume). Therefore, land transformation does not necessarily have a sole directional effect on plant invasions. Furthermore, differences in the distribution or abundance of an alien plant within a LUC type are not amenable to the static analysis of contemporary LUC if differences are due to differences in LUC history. For example, in Catalonia (NE Spain), shrublands and woodlands invaded by prickly-pear cacti, *Opuntia* spp., are the ones that were croplands 30 years ago (Vilà et al. 2003).

There are two main reasons why, in general, changes in LUC can influence invasions. First, LUC changes are frequently accompanied by disturbances. Disturbances are unanimously acknowledged to favor plant invasions because they provide "open windows" to the entrance of propagules and better conditions (i.e., more light, space and nutrients per capita) for seedling establishment and growth (Hobbs and Huenneke 1992). Regional surveys and flora screening for alien species often find higher alien species richness in altered ecosystems (e.g., ruderal and anthropogenic ecosystems) than in unaltered ones (e.g., forests, shrublands). For example, edges of highly used roads enhance dispersal and increase light levels and disturbance, which promote establishment of alien herbs (Parendes and Jones 2000).

Second, changes in LUC include modifications in landscape composition and structure, such as fragmentation, which increases the length of edges and the prevalence of corridors that can facilitate invasions. There is clear evidence that fragmentation creates more edge, which facilitates the spread of invasive species to habitat remnants within a man-modified land matrix. For example, in Quebec (Canada), forest fragments surrounded by agricultural fields that have been intensively farmed with the use of commercial fertilizers and herbicides contain a higher percentage of alien weeds than forests surrounded by low intensity farms (Boutin and Jobin 1998). In the same way, transportation corridors increase the spread and the opportunity of establishment of alien plants. In European countries, the density of alien plant species is positively correlated to the length of transport networks (Vilà and Pujadas 2001). The influence of transportation corridors on invasions is positively related to their improvement and frequency of use. For instance, in Utah there is higher cover and richness of alien plant species in semiarid grasslands and woodlands adjacent to paved roads than in graded or 4WD tracks (Gelbard and Belnap 2003). Likewise, in Glacier National Park (Montana-US), grasslands adjacent to primary and secondary roads have higher species richness than along back trails (Tyser and Worley 1992).

To properly test if changes in LUC directly increase invasions, we must relate spatially and temporally explicit knowledge of LUC changes to the presence, abundance and population performance of alien species. We propose the following LUC attributes as land transformation parameters for analysis: direction, intensity, number of stages and number of stage types (Table 8.2). Direction refers to the trajectory of the change, which can be to a more degraded or to a more recovered or restored state than the initial LUC; intensity means the magnitude of the change; number of stages means the number of steps analysed in the whole time series and; finally, the number of stage types refer to the identities of the LUC types that have characterized a given location through time. We expect invasions to be more frequent where the direction of LUC changes leads to greater degradation, where intensity is higher, and where more types of LUC changes have taken place. For example, we might expect higher abundance and diversity of alien species in oldfields than in mature woodland, in areas where the LUC changes have been large in extent or very extreme, and where repeated LUC changes have offered multiple openings for invasion to occur. To our knowledge, there are very few studies aimed to test this hypothesis. A positive relationship between intense habitat degradation over the last 50 years and high diversity of alien plants has been observed in several wetland and dune communities around the Mediterranean Basin (Pino et al. 2006).

Overall, despite the wide interest in biological invasions at a variety of scales, studies exploring invasion patterns and their main landscape correlates from a geographically explicit point of view are scarce. Moreover, interactions among plant invasions and the type, magnitude, trajectory, and direction of landscape changes remain to be explored.

8.6 Multiple Interactions

Most research on the effects of global environmental changes on biological invasions has examined effects of single environmental factors. Although different elements of global change act at different spatial scales, and their impacts are different depending on the biome considered, environmental changes interact in complex ways, complicating predictions of ecosystem responses (Norby et al. 2007, Chap. 3 of this volume). For example, the success of broad-leaved alien trees in Switzerland can be related to both recent climate shifts and elevated atmospheric [CO_2] (Hättenschwiler and Körner 2003).

Ongoing changes on land and in the atmosphere decrease our capacity to predict which introduced species are most likely to become invaders and which ecosystems are most vulnerable to invasion. To overcome this uncertainty, regional-scale studies from a geographically explicit point of view appear to be relevant to identify invasion hot spots and to compare the relative role of climatic, edaphic, landscape and socioeconomic factors on invasion patterns. Preliminary studies in Catalonia, a high diversity ecosystem region of 32 000 km^2 in NE Spain, point out that mean annual temperature influences plant invader diversity, which is also favoured by landscape diversity and regional disturbances (Pino et al. 2005). Ecological niche modeling, a technique in which georeferenced occurrences of alien species are linked to geographically explicit environmental datasets (Peterson

Table 8.2. Proposed attributes of changes in land-use/cover (LUC) and their influence on plant invasions. *S* represents stages of LUC changing thought time from more (S_0) to less natural (S_n)

LUC attributes	Increasing invasion scenario		Decreasing invasion scenario	
Direction	Degradation:	($S_0 \rightarrow S_n$)	Restoration:	($S_n \rightarrow S_0$)
Intensity	More intense:	($S_0 \rightarrow S_n$)	Less intense:	($S_0 \rightarrow S_{n-1}$)
Number of stages	3 stages:	($S_0 \rightarrow S_{n-1} \rightarrow S_n$)	2 stages:	($S_0 \rightarrow S_n$)
Number of stage types	3 stage types:	($S_0 \rightarrow S_{n-1} \rightarrow S_n$)	2 stage types:	($S_0 \rightarrow S_{n-1} \rightarrow S_0$)

et al. 2003), is another recent, reliable approach used to predict the potential distribution of invasions on a species-by-species basis. In essence, the ecological niches are modeled by combining occurrence points and environmental information of species in their native distribution area using a selected set of environmental datasets and diverse algorithms. These models are then projected onto potential receptor areas to identify geographic regions showing ecological conditions inside and outside of the species' niche. A number of recent examples including vertebrates, plants, and insects are available in the literature (Peterson and Vieglais 2001; Peterson et al. 2003; Roura-Pascual et al. 2004). Both approaches offer the hope that realistic predictions of future plant invasion patterns are possible.

Moreover, as changes in land-use cause vegetation changes and alter ecosystem biogeochemistry (Jackson et al. 2007, Chap. 19 of this volume; Reynolds et al. 2007, Chap. 20 of this volume), plant invasions transform vegetation structure, and can directly or indirectly affect ecosystem functioning (Ehrenfeld 2003; Levine et al. 2003), and consequently can exacerbate changes in land-use or environmental changes. Research on the feedbacks of plant invasions on environmental change has been scarce, and has mainly focused on how certain fire prone invasive plant species increase fire risk (D'Antonio et al. 1999), or whether invaders that expand their populations in tandem with land degradation limit restoration efforts back to former community composition (Corbin and D'Antonio 2004). There are many research areas which deserve greater attention. For example, it remains to be explored if plant invaders, responding with fast growth and high fitness to elevated [CO_2] and N deposition, increase soil respiration and N availability. Factorial field experiments comparing the responses of native and alien species to multiple environmental changes could provide answers to these questions.

8.7 Summary and Conclusions

Invasion by alien species is one of the most conspicuous human driven ecological changes, often with undesirable effects on biodiversity and ecosystem functioning. However, few studies have addressed how environmental change (i.e., elevated [CO_2], climate change, N deposition and changes in LUC) can influence the success of plant invaders, invasion impacts and feedbacks. Because global change components are interacting in multiple ways, increasing the uncertainty of species and ecosystem responses (Norby et al. 2007, Chap. 3 of this volume), there is an urgent need to increase research in this area.

While studies of invasive species' responses to elevated [CO_2] have mainly focused on ecophysiology and growth, studies examining responses to changes in climate, N deposition and land-use have examined patterns of plant distribution and abundance in the context of community structure. When comparing invasive with closely related or coexisting native species, the invasives often perform better than the natives or the alien but non-invasives. The mechanisms through which aspects of global change benefit invasive species and harm native species have begun to be explored. A particularly promising direction is to consider whether specific life-history traits, such as those that confer fast growth, broad ecological niches, high seed production or long-distance dispersal, offer hope in the development of predictions of future invasions (Grotkopp et al. 2002). We suggest that the exploration of plant invasions in the context of plant functional types (Lavorel et al. 2007, Chap. 13 of this volume) deserves further consideration. We also propose that exploring the multifactor effects of global change on plant invaders would increase our predictive power on habitat vulnerability to invasion.

Acknowledgments

We thank M. Rejmánek for comments on an earlier version of this chapter. Partial funding was provided by the Spanish "Ministerio de Ciencia y Tecnología" (REN2000-0361/GLO), and the VI European Union Framework Programme Research Project ALARM (Assessing large-scale environmental risks with tested methods- Contract no. 506675) to MV and JP, and the US National Science Foundation and the Department of Energy to SDS.

References

Aber JD, McDowell WH, Nadelhoffer KJ, Magill A, Berntson G, Kamakea M, McNulty SG, Currie W, Rustad L, Fernandez I (1998) Nitrogen saturation in temperate forest ecosystems: Hypotheses revisited. BioScience 48:921–934

Aber JD, Nadelhoffer KJ, Steduler P, Melillo JM (1989) Nitrogen saturation in northern forest ecosystems. BioScience 39:378–386

Allen EB (2003) Restoration of *Artemisia* shrublands invaded by exotic annual *Bromus*: A comparison between southern California and the Intermountain region. In: Hild AL, Shaw NL, Meyer SE, Schupp EW, Booth T (Compilers) Seed and soil dynamics in shrubland ecosystems: Proceedings; August 12–16, 2002, Laramie, Wyoming, Proceedings RMRS-P-31. U.S. Department of Agriculture Forest Service, Rocky Mountain Research Station, Ogden, Utah, pp 9–17

Baruch Z, Goldstein G (1999) Leaf construction cost, nutrient concentration and net CO_2 assimilation of native and invasive species in Hawaii. Oecologia 121:183–192

Beerling DJ, Huntley B, Bailey JP (1995) Climate and the distribution of *Fallopia japonica*: use of an introduced species to test the predictive capacity of response surfaces. Journal of Vegetation Science 6:269–282

Belote RT, Weltzin JF, Norby RJ (2003) Response of an understory plant community to elevated [CO_2] depends on differential responses of dominant invasive species and is mediated by soil water availability. New Phytologist 161:827–835

Bobbink R (1991) Effects of nutrient enrichment in Dutch chalk grasslands. Journal of Applied Ecology 28:28–41

Bobbink R, Hornung M, Roelofs JGM (1998) The effects of air-borne nitrogen pollutants on species diversity in natural and semi-natural European vegetation. Journal of Ecology 86:717–738

Boutin C, Jobin B (1998) Intensity of agricultural practices and effects on adjacent habitats. Ecological Applications 8:544–557
Brooks ML (1999) Alien annual grasses and fire in the Mojave Desert. Madroño 46:13–19
Brooks ML (2003) Effects of increased soil nitrogen on the dominance of alien annual plants in the Mojave Desert. Journal of Applied Ecology 40:344–353
Brown VC (1995) Insect herbivores and gaseous air pollutants: current knowledge and predictions. In: Harrington R, Stork NE (eds) Insects in a changing environment. Academic Press, London, pp 22–29
Corbin JD, D'Antonio CM (2004) Effects of invasive species on soil nitrogen cycling: Implications for restoration. Weed Technology 18:1464–1467
D'Antonio CM, Corbin JD (2003) Effects of plant invaders on nutrient cycling: Using models to explore the link between invasion and development of species effects. In: Canham CD, Cole JC, Lauenroth WK (eds) Models in ecosystem science. Princeton University Press, Princeton, pp 363–384
D'Antonio CM, Vitousek PM (1992) Biological invasions by exotic grasses, the grass/fire cycle, and global change. Annual Review of Ecology and Systematics 23:63–87
D'Antonio CM, Dudley TL, Mack MC (1999) Disturbance and biological invasions: direct effects and feedbacks. In: Walker LR (ed) Ecosystems of disturbed ground. Elsevier, New York, New York, USA, pp 413–452
DeVries W, Leeters EM, Hendricks CA (1995) Effects of acid deposition on Dutch forest ecosystems. Water, Air and Soil Pollution 85:1063–1068
Dukes JS (2000) Will the increasing atmospheric CO_2 concentration affect the success of invasive species? In: Mooney HA, Hobbs RJ (eds) Invasive Species in a Changing World. Island Press, Washington, DC, pp 95–113
Dukes JS (2002) Comparison of the effect of elevated CO_2 on an invasive species (*Centaurea solstitialis*) in monoculture and community settings. Plant Ecology 160:225–234
Dukes JS, Mooney HA (1999) Does global change increase the success of biological invaders? Trends in Ecology and Evolution 14:135–139
Ehleringer JR, Cerling TE, Helliker BR (1997) C_4 photosynthesis, atmospheric CO_2 and climate. Oecologia 112:285–299
Ehrenfeld J (2003) Effects of exotic plant invasions of ecosystem nutrient cycling processes. Ecosystems 6:503–523
Falkengren-Grerup U (1986) Soil acidification and vegetation changes in deciduous forests in southern Sweden. Oecologia 70:339–347
Fenn ME, Poth MA, Aber JD, Baron JS, Bormann BT, Johnson DW, Lemly AD, McNulty SG, Ryan DF, Stottlemyer R (1998) Nitrogen excess in North American ecosystems: Predisposing factors, ecosystem responses, and management strategies. Ecological Applications 8:706–733
Fenn ME, Haeuber R, Tonnesen GS, Baron JS, Grossman-Clarke S, Hope D, Jaffe DA, Copeland S, Geisher L, Rueth HM, Sickman JO (2003a) Nitrogen emissions, deposition, and monitoring in the western United States. BioScience 53:391–403
Fenn ME, Baron JS, Allen EB, Rueth HM, Nydick KR, Geiser L, Bowman WD, Sickman JO, Meixner T, Johnson DW, Neitlich P (2003b) Ecological effects of nitrogen deposition in the western United States. BioScience 53:404–420
Galloway JN, Schlesinger WH, Levy H II, Michaels A, Schnoor JL (1995) Nitrogen fixation: Anthropogenic enhancement – environmental response. Global Biogeochemical Cycles 9:235–252
Galloway JN, Aber JD, Erisman JW, Seitzinger SP, Howarth RW, Cowling EB, Cosby BJ (2003) The nitrogen cascade. BioScience 53:341–356
Gelbard JL, Belnap J (2003) Roads as conduits for exotic plant invasions in a semiarid landscape. Conservation Biology 17:420–432
Grotkopp E, Rejmánek M, Rost TL (2002) Toward a causal explanation of plant invasiveness: seedling growth and life-history strategies of 29 pine (*Pinus*) species. The American Naturalist 159: 396–419
Hättenschwiler S, Körner C (2003) Does elevated CO_2 facilitate naturalization of the non-indigenous *Prunus laurocerasus* in Swiss temperate forests? Functional Ecology 17:778–785
Hobbs RJ (2000) Land-use Changes and Invasions. In: Mooney HA, Hobbs RJ (eds) Invasive species in a changing world. Island Press, Washington, pp 55–64
Hobbs RJ, Huenneke LF (1992) Disturbance, diversity, and invasion: implications for conservation. Conservation Biology 6:324–337
Hogbom L, Hogberg P (1991) Nitrate nutrition of *Deschampsia flexuosa* (L.) Trin. in relation to nitrogen deposition in Sweden. Oecologia 87:488–494
Huenneke LF, Hamburg SP, Koide R, Mooney HA, Vitousek PM (1990) Effects of soil resources on plant invasion and community structure in California serpentine grassland. Ecology 71:478–491
Huxman TE, Smith SD (2001) Photosynthesis in an invasive grass and native forb at elevated CO_2 during an El Niño year in the Mojave Desert. Oecologia 128:193–201
Huxman TE, Hamerlynck EP, Jordan DN, Salsman KJ, Smith SD (1998) The effects of parental CO_2 environment on seed quality and subsequent seedling performance in *Bromus rubens*. Oecologia 114:202–208
Huxman TE, Hamerlynck EP, Smith SD (1999) Reproductive allocation and seed production in *Bromus madritensis* ssp. *rubens* at elevated atmospheric CO_2. Functional Ecology 13:769–777
IPCC (2001) Climate Change 2001: Impacts, Adaptation, and Vulnerability. Contribution of Working Group II to the Third Assessment Report of the Intergovernmental Panel on Climate Change. Cambridge University Press, Cambridge, United Kingdom
Jackson R, Jobbagy E (2007) Vegetation changes and carbon and water dynamics. In: Canadell J, Pataki D, Pitelka L (eds) Terrestrial ecosystems in a changing world. Springer-Verlag, Berlin
Jefferies RL, Maron JL (1997) The embarrassment of riches: atmospheric deposition of nitrogen and community and ecosystem processes. Trends in Ecology and Evolution 12:74–78
Jenkins JC, Aber JD, Canham (1999) Hemlock woolly adelgid impacts on community structure and N cycling rates in eastern hemlock forests. Canadian Journal of Forest Research 29:630–645
Kellner O (1993) Effects on associated flora of sylvicultural nitrogen fertilization repeated at long intervals. Journal of Applied Ecology 30:563–574
Knapp PA (1998) Spatio-temporal patterns of large grassland fires in the Intermountain West, U.S.A. Global Ecol Biogeogr 7:259–272
Kriticos DJ, Sutherst RW, Brown JR, Adkins SW, Maywald GF (2003a) Climate change and biotic invasions: a case history of a tropical woody vine. Biological Invasions 5:145–165
Kriticos DJ, Sutherst RW, Brown JR, Adkins SW, Maywald GF (2003b) Climate change and the potential distribution of an invasive alien plant: *Acacia nilotica* ssp *indica* in Australia. Journal of Applied Ecology 40:111–124
Latty EF, Canham CD, Marks PL (2003) Beech bark disease in northern hardwood forests: the importance of nitrogen dynamics and forest history for disease severity. Canadian Journal of Forest Research 33:257–268
Lavorel S, Díaz S, Pausas J, Garnier E, Neilson R (2007) Plant functional types: are we getting any closer to the Holy Grail? In: Canadell J, Pataki D, Pitelka L (eds) Terrestrial ecosystems in a changing world. Springer-Verlag, Berlin
Le Maitre DC, Wilgen BV, Chapman R, McKelly D (1996) Invasive plants and water resources in the Eastern Cape Province, South Africa: modelling the consequences of a lack of management. Journal of Applied Ecology 33:161–172
Levine JM, Vilà M, D'Antonio CM, Dukes JS, Grigulis K, Lavorel S (2003) Mechanisms underlying the impact of exotic plant invasions. Philosophical Transactions of the Royal Society of London B 270:775–781
Liebhold AM, MacDonald WL, Bergdahl D, Mastro VC (1995) Invasion by exotic forest pests: A threat to forest ecosystems. Forest Science Monographs 30:1–49
Lockwood JL, Mckinney ML (2001) Biotic homogenization. Kluwer Academic. Plenum Publishers, New York
Mack RN (1981) Invasion of *Bromus tectorum* L. into western North America: an ecological chronicle. Agro-Ecosystems 7:145–165
Mack MC, D'Antonio CM (1998) Impacts of biological invasions on disturbance regimes. Trends in Ecology and Evolution 13:195–198
Malcolm JR, Markham A, Neilson RP, Garaci M (2002) Estimated migration rates under scenarios of global climate change. Journal of Biogeography 29:835–849

Maron JL, Connors PG (1996) A native nitrogen-fixing shrub facilitates weed invasion. Oecologia 105:302–312

McNeely JA (2001) The great reshuffling. Human dimensions of invasive alien species. IUCN, Cambridge

Moore PD (2004) Favoured aliens for the future. Nature 427:594

Murphy DD, Ehrlich PR (1989) Conservation biology of California's remnant native grasslands. In: Huenneke LF, Mooney HA (eds) Grassland structure and function: The California annual grassland. Klewer Academic Publishers, The Netherlands, pp 201–211

Nagel JM, Griffin KL (2001) Construction cost and invasive potential: comparing *Lythrum salicaria* (Lythraceae) with co-occurring native species along pond banks. American Journal of Botany 88:2252–2258

Nagel JM, Griffin KL (2004) Can gas-exchange characteristics help explain the invasive success of *Lythrum salicaria*? Biological Invasions 6:101–111

Nagel JM, Huxman TE, Griffin KL, Smith SD (2004) CO_2 enrichment reduces the energetic cost of biomass construction in an invasive desert grass. Ecology 85:100–106

Norby R, Rustad L, Beier C, Zavaleta E, Parton B, McMurtrie R (2007) Multiple factor interactions on ecosystem function (CO_2, temperature, nutrients, water (TERACC)). In: Canadell J, Pataki D, Pitelka L (eds) Terrestrial ecosystems in a changing world. Springer-Verlag, Berlin

Parendes LA, Jones JA (1999) Role of light availability and dispersal in exotic plant invasion along roads and streams in the HJ Andrews Experimental Forest, Oregon. Conservation Biology 14:64–75

Parmesan C, Yohe G (2003) A globally coherent fingerprint of climate change impacts across natural systems. Nature 421:37–42

Patterson DT (1995) Weeds in a changing climate. Weed Science 43:685–701

Peñuelas J, Filella I, Comas P (2002) Changed plant and animal life cycles from 1952 to 2000 in the Mediterranean region. Global Change Biology 8:531–544

Peterson AT, Vieglais DA (2001) Predicting the species invasions using ecological niche modeling: New approaches from bioinformatics attack a pressing problem. BioScience 51:363–371

Peterson AT, Papes M, Kluza DA (2003) Predicting the potential invasive distribution of four alien plant species in North America. Weed Science 51:863–868

Pino J, Font X, Carbó J, Jové M, Pallarés L (2005) Large-scale correlates of alien plant invasion in Catalonia (NE Spain). Biological Conservation 122:339–350

Pino J, Seguí JM, Alvarez N (2006) Invasibility of four plant communities in the Llobregat delta (Catalonia, NE of Spain) in relation to their historical stability. Hydrobiologia 570:257–263

Pysek P, Prach K, Mandák B (1998) Invasions of alien plants into habitats of central european landscape: an historical pattern. In: Starfinger U, Edwards K, Kowarik I, Williamson M, editors. Plant invasions: ecological mechanisms and human responses. Backhuys Publishers, Leiden, pp 23–32

Rejmánek M (1996) A theory of seed plant invasiveness: the first sketch. Biological Conservation 78:171–181

Rejmánek M (2000) Invasive plants: approaches and predictions. Austral Ecology 25:497–506

Reynolds JF, Maestre F, Lambin EF, Stafford-Smith DM, Valentin C (2007) Natural and human dimensions of land degradation: causes and consequences. In: Canadell J, Pataki D, Pitelka L (eds) Terrestrial ecosystems in a changing world. Springer-Verlag, Berlin

Richardson DM, Allsopp N, D'Antonio CM, Milton SJ, Rejmánek M (2000a) Plant invasions – the role of mutualisms. Biological Reviews 75:65–93

Richardson DM, Bond WJ, Dean RJ, Higgins SI, Midgley GF, Milton SJ, Powrie LW, Rutherford MC, Samways MJ, Schulze RE (2000b) In: Mooney HA, Hobbs RJ (eds) Invasive alien species and global change: a South African perspective. Invasive species in a changing world. Island Press, Washington, pp 303–349

Romney EM, Wallance A, Hunter RB (1978) Plant response to nitrogen fertilization in the Northern Mojave Desert and its relationship to water manipulation. In: West NE, Skujins J (eds) Nitrogen in desert ecosystems. Dowden, Hutchinson & Ross Inc., Stroudsburg, pp 232–243

Root TL, Price JT, Hall KR, Schneider SH, Rosenzweig C, Pounds JA (2003) Fingerprints of global warming on wild animals and plants. Nature 421:57–60

Rosen K, Gunderson P, Tegnhammer L, Johansson M, Frogner T (1992) Nitrogen enrichment of Nordic forests: the concept of critical loads. Ambio 21:364–368

Roura-Pascual N, Suarez AV, Gómez C, Pons P, Touyama Y, Wild AL, Peterson AT (2004) Geographical potential of Argentine ants (*Linepithema humile* Mayr) in the face of global climate change. Proceedings of the Royal Society of London 271: 2527–2534

Sasek TW, Strain BR (1991) Effects of CO_2 enrichment on the growth and morphology of a native and an introduced honeysuckle vine. American Journal of Botany 78:69–75

Smith SD, Strain BR, Sharkey TD (1987) Effects of CO_2 enrichment on four Great Basin grasses. Functional Ecology 1:139–143

Smith SD, Huxman TE, Zitzer SF, Charlet TN, Housman DC, Coleman JS, Fenstermaker LK, Seemann JR, Nowak RS (2000) Elevated CO_2 increases productivity and invasive species success in an arid ecosystem. Nature 408:79–82

Soulé ME (1991) Conservation: tactics for a constant crisis. Science 253:744–750

Stevens CJ, Dise NB, Mountford JO, Gowing DJ (2004) Impact of nitrogen deposition on the species richness of grasslands. Science 303:1876–1879

Thimoneir A, Dupouey JL, Timbal J (1992) Floristic changes in the herb-layer vegetation of a deciduous forest in the Lorraine Plain under the influence of atmospheric deposition. Forest Ecology and Management 55:149–167

Throop HL, Lerdau MT (2004) Effects of nitrogen deposition on insect herbivory: Implications for community and ecosystem processes. Ecosystems 7:109–133

Tyser RW, Worley CA (1992) Alien flora in grasslands adjacent to road and trail corridors in Glacier National Park, Montana (U.S.A.). Conservation Biology 6:253–262

van Breemen N, van Dijk HFG (1988) Ecosystem effects of atmospheric deposition on nitrogen in The Netherlands. Environmental Pollution 54:249–274

Vane-Wright RI, Humphries CJ, Williams PH (1991) What to protect? Systematics and the agony of choice. Biological Conservation 55:235–254

Vilà M, Pujadas J (2001) Land-use and socio-economic correlates of plant invasions in European and North African countries. Biological Conservation 100:397–401

Vilà M, Burriel JA, Pino J, Chamizo J, Llach E, Porterias M, Vives M (2003) Association between *Opuntia* spp. invasion and changes in land-cover in the Mediterranean region. Global Change Biology 9:1234–1239

Vinton MA, Burke IC (1995) Interactions between individual plant species and soil nutrient status in shortgrass steppe. Ecology 76:1116–1133

Vitousek P (1994) Beyond global warming: ecology and global change. Ecology 75:1861–1876

Vitousek PM, Howarth RW (1991) Nitrogen limitation of land and sea: How can it occur? Biogeochemistry 13:87–115

Vitousek PM, Walker LR, Whiteaker LD, Mueller-Dombois D, Matson PA (1987) Biological invasion by *Myrica faya* alters ecosystem development in Hawaii. Science 238:802–804

Vitousek PM, Howarth RW, Likens GE, Matson PA, Schindler D, Schlesinger WH, Tilman GD (1997) Human alteration of the global nitrogen cycle: Causes and consequences. Issues in Ecology 1:1–17

Walther GR (2002) Weakening of climatic constraints with global warming and its consequences for evergreen broad-leaved species. Folia Geobotanica 37:129–139

Wedin DA, Tilman D (1996) Influence of nitrogen loading and species composition on the carbon balance of grasslands. Science 274:1720–1723

Weiss SB (1999) Cars, cows, and checkerspot butterflies: Nitrogen deposition and management of nutrient-poor grasslands for a threatened species. Conservation Biology 13:1476–1486

White TA, Campbell BD, Kemp PD, Hunt CL (2000) Sensitivity of three grassland communities to simulated extreme temperature and rainfall events. Global Change Biology 6:671–684

White TA, Campbell BD, Kemp PD, Hunt CL (2001) Impacts of extreme climatic events on competition during grassland invasions. Global Change Biology 7:1–13

With KA (2002) The landscape ecology of invasive spread. Conservation Biology 16:1192–1203

Young JA, Evans RA (1978) Population dynamics after wildfires in sagebrush grasslands. Journal of Range Management 31:283–289

Zavaleta ES, Royval JL (2002) Climate change and the susceptibility of U.S. ecosystems to biological invasions: two cases of expected range expansion. In: Root TL (ed) Wildlife Responses to Climate Change. Island Press, Washington, pp 277–341

Ziska LH (2002) Influence of rising atmospheric CO_2 since 1900 on early growth and photosynthetic response of a noxious invasive weed, Canada thistle (*Cirsium arvense*). Functional Plant Biology 29:1387–1392

Ziska LH (2003) Evaluation of the growth response of six invasive species to past, present and future atmospheric carbon dioxide. Journal of Experimental Botany 54:395–404

Ziska LH, Caulfield FA (2000) Rising CO_2 and pollen production of common ragweed (*Ambrosia artemisiifolia*), a known allergy-inducing species: implications for public health. Australian Journal of Plant Physiology 27:893–898

Chapter 9

Plant Biodiversity and Responses to Elevated Carbon Dioxide

Catherine Potvin · F. Stuart Chapin III · Andrew Gonzalez · Paul Leadley · Peter Reich · Jacques Roy

9.1 Ten Years of GCTE Research: Apprehending Complexity

The general assumption, in the first phase of elevated carbon dioxide (CO_2) research, was that increasing atmospheric CO_2 concentration would act as a fertilizer for plant systems. Under this "large-plant-scenario" (Fig. 9.1), increased photosynthesis at elevated CO_2 would lead to higher plant biomass which would induce a negative atmospheric feedback loop that would offset anthropogenic carbon (C) emissions.

A second general assumption in early research was that the effects of elevated CO_2 on plant community structure could be predicated using a C-based perspective with clear "winners" and "losers" based on species' ability to take advantage of the increased atmospheric CO_2 concentration (e.g., plant with C_3 vs. C_4 photosynthetic pathways). The reality unveiled by a decade of research proved more complex and the "large plant scenario" largely untenable.

9.1.1 Effects of CO_2 on Plant Diversity Through Alterations of the Physical Environment

A large number of studies on plants growing in pots, monoliths and *in situ*, now yield a broad perspective of responses to elevated CO_2 (e.g., Bazzaz 1990; Poorter 1993; Körner 2001; Navas 1998; Woodward 2002; Poorter and Navas 2003). Reviews have used meta-analysis to identify statistically significant trends (Curtis and Wang 1998; Wand et al. 1999; Kerstiens 2001). Pot studies have generally confirmed the C-based perspective of the differential response of functional groups of plant species to elevated CO_2: plants with the C_3 photosynthetic pathway generally respond more that C_4 plants (due to greater stimulation of net C fixation), fast-growing plants more than slow-growing plants (due to stronger C sinks), and N-fixing plants more than non-fixers (because the high C cost of N fixation is overcome). However, an important revelation of *in situ* experiments, both within and outside of the GCTE network, is that this C-based perspective gives an incomplete understanding of how plants respond to elevated CO_2 in multi-species communities. Apparently C-based functional groupings of plant response to CO_2 break down (Poorter and Navas 2003) while the effects of CO_2 on ecosystem water balance and nutrient cycling often play a larger role on plant community structure than previously suspected (Fig. 9.1).

Elevated CO_2 can modify soil nutrient availability, nutrient uptake capacity, and nutrient use efficiency (Bassirirad et al. 1996; Hungate 1999; Körner 2001) all of which are likely to modify plant community structure in

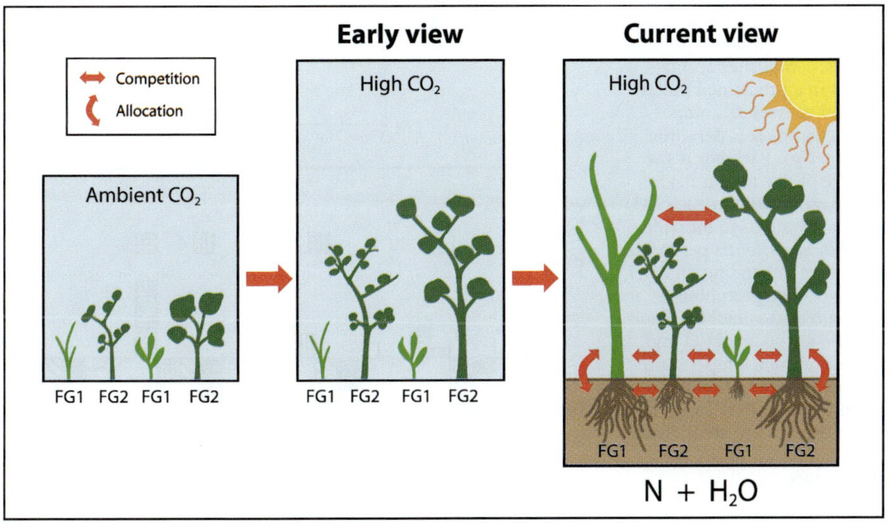

Fig. 9.1.
Simple cartoon illustrating the early and current views of the effect of elevated CO_2 on plant ecosystems. Current knowledge suggest that elevated CO_2 is likely to affect, minimally, photosynthetic carbon uptake, plant growth and allocation, N and water uptake leading to differential above- and belowground competition. Moreover response to elevated CO_2 is not consistent within functional groups. Here *FG1* and *FG2* refer to two hypothetic functional groups

nutrient limited ecosystems. Yet, the effects of CO_2 on plant community structure acting indirectly through nutrient availability are poorly studied. On the other hand, several studies indicate that soil water status is often key in determining species-specific responses to elevated CO_2 *in situ* (e.g., Owensby et al. 1993; Volk et al. 2000; Morgan et al. 2004), although there are also a number of studies where this is not the case (e.g., Arp et al. 1993; Zavaleta et al. 2003a,b; Nowak et al. 2004). The effects of CO_2 on soil water balance may help explain why many of the expected direct CO_2 effects on diversity, such as the encroachment of C_3 woody plants into C_4 savannas (Walker 2001) or the increase in competitive ability of C_3 species compared to C_4 species in grasslands, are not consistently observed (Morgan et al. 2004). Species-specific responses in calcareous grasslands may also largely depend on CO_2 effects on water balance. In study of CO_2 effects on calcareous grassland, Leadley et al. (1999) found that *Carex flacca* had a greater growth response to elevated CO_2 than the dominant grass *Bromus erectus*. Volk et al. (2000) provided compelling evidence that the large differences in response of *Carex* and *Bromus* to elevated CO_2 may be explained by the effects of elevated CO_2 on soil water content. More generally, Stocklin and Körner (1999) identified the functional group of mesophytic species as the most responsive to increased CO_2 in monoliths of these calcareous grassland communities. Apparently, reductions in stomatal conductance at elevated CO_2 that lead to modifications in soil water status can play a key role altering plant community structure.

Climate apparently also modulates community responses to elevated CO_2. Roy and collaborators (unpublished) compared the CO_2 response of Mediterranean annual herbaceous species over several growing seasons. Their experiment documented large inter-annual variability in the response of individual species to CO_2. The response of most annual species to elevated CO_2 switched from being highly negative in some years to being highly positive in other years. For example, in the community composed of 2 grasses, 2 composites and 2 legumes, the biomass response of the annual grass *Bromus madritensis* (Fig. 9.2) changed from a 70% reduction to a 300% increase depending on the years. Because the planting was repeated similarly every year for 5 years in greenhouses tracking outside temperature and precipitation, the results suggest that climatic conditions might drive the outcome of this elevated CO_2 experiment. Similarly, the presence of an interaction between climatic variability and response to CO_2 for perennial herbaceous communities was reported by Owensby et al. (1996) for a Kansas tallgrass prairie. On wet years, there was no significant biomass response to CO_2, while on dry years, the C_4 species *Andropogon gerardii* was stimulated and drove a positive community biomass response.

Roy (unpublished) further noted that CO_2 interacted with the diversity level of the community: year to year variability was less in the communities with the higher number of species. Several experiments both in the field and in microcosms suggest that increasing diversity increases the stability of ecosystem properties (Hooper et al. 2005). Tilman and Downing (1994) reported for example that, in the Minnesota grassland, plots with a high diversity resisted drought better then plots with low diversity. An issue that remains largely unanswered is the extent to which elevated CO_2 and diversity would interact to either increase or decrease the stability of eco-

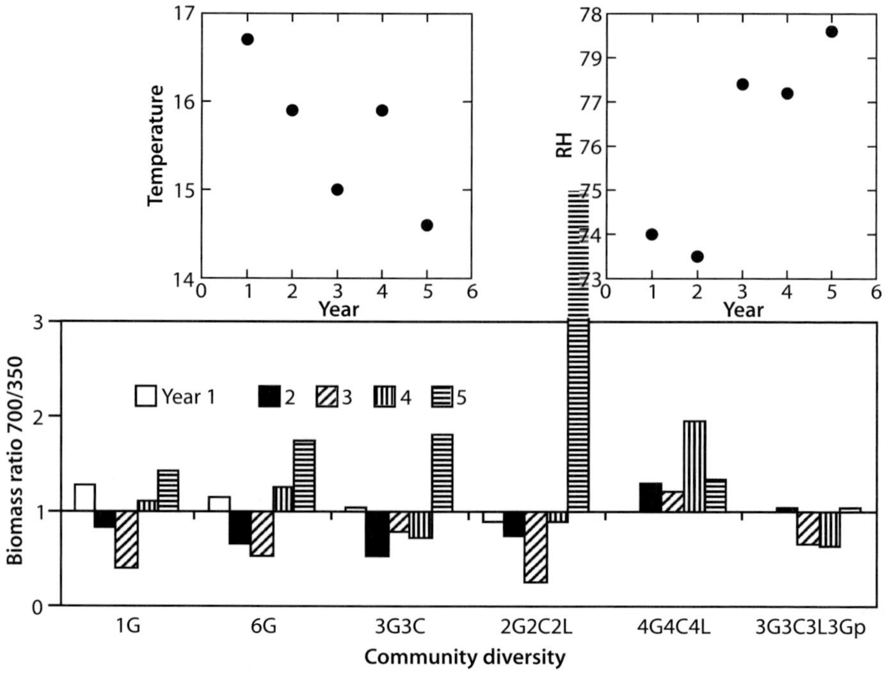

Fig. 9.2.
CO_2 response of the annual grass *Bromus madritensis* (total aboveground biomass at senescence) when grown for 5 successive years in communities of various complexity established by planting young seedlings of annual or perenial grasses, composites and legumes in a predefined pattern (annual grasses: *G*, composites: *C*, legumes: *L*, perennial grasses: *Gp*). Each *bar* is the average of one to four 0.5 m^2 model ecosystems at each CO_2 level, with 360 to 30 individuals of *B. madritensis* per model ecosystem depending on diversity level. Two greenhouses were run at 350 µmol CO_2 mol^{-1} and two at 700 µmol CO_2 mol^{-1}. Average annual temperature and relative humidity in the greenhouses for the five years of the experiment are given in the *small panels* above the main figure

system properties. Clearly responses to elevated CO_2 are modulated by climate, water and nutrient availability with different feedback loops:

- Changes in atmospheric CO_2 alter the relative availability of resources (C, water, and nutrients), which in turn alters competitive interactions among plants potentially leading to changes in diversity or dominance patterns.
- Environmental variability affects species-specific responses to elevated CO_2 while community diversity and composition in turn influences both the inter-annual variability and the direction of the response to elevated CO_2.

9.2 Temporal Variation and Response to Elevated CO_2

9.2.1 Reproductive and Evolutionary Aspects of the Response to Elevated CO_2

Adding to complexity, researchers recognized that the response of vegetative biomass to elevated CO_2 is not sufficient to understand long-term community responses. CO_2 induced differences in number and quality of seeds should have strong impact on the diversity of plant communities over the long-term. In a meta-analysis covering 79 species, Jablonski et al. (2002) reported a stimulation of the number of reproductive organs or their weight by about 20%. Crop species were more responsive (+25%) than wild species (+4%) despite a similar stimulation of vegetative biomass. Variability is very high between genera (–26% in *Ipomea* sp to +30% in *Brassica* sp.) as well as within genera (from +2% to +60% in *Raphanus* sp.). Individual seed size was marginally increased but seed nitrogen content was reduced by about 14% in both wild species and crops. In a single experiment in the field with 17 species, Thürig et al. (2003) found a significant increase in the number of flowering shoots (+24%) and seeds (+29%) with very large responses in some species (flowering shoot +236% in *Trifolium pratense* and +1 000% in *Carex flacca*). In perhaps the first field experiment to compare tree species reproductive output, Stiling et al. (2004) found a four-fold increase in acorn density for two oak species and no change in the third one.

Furthermore, rising atmospheric CO_2 concentrations may alter the genetic structure of plant populations through directional selection; i.e., within a population, those individuals that have the greatest response to CO_2 in terms of growth and reproduction will tend to exclude those individuals of the same species with weaker responses. If such evolutionary responses occur responses of plant species observed in short-term experiments may not be representative of the response of plants to long-term, slowly rising atmospheric CO_2 concentrations. Several studies have demonstrated genetic variation in responses to CO_2 concentration (Wang et al. 1994; Stewart and Potvin 1996; Schmid et al. 1996; Roumet et al. 2002; Spinnler et al. 2003; Mohan et al. 2004; Bidart-Bouzat 2004; Ward and Kelly 2004) suggesting that elevated CO_2 may drive rapid directional selection. However only a few studies experimentally tested the evolutionary responses to controlled selection under various CO_2 levels. *Brassicae juncea* did not show genetic adaptation to a combined CO_2 and temperature increase, possibly due to inbreeding depression in a stressful environment (Potvin and Toussignant 1996). Derner et al. (2004) found that the increase in CO_2 responsiveness from the first to the third generation of wheat was not driven by genetic changes but by CO_2-induced changes in seed quality. Andalo et al. (1996) reported a decrease in seed germination of *Arabidopsis thaliana* when grown in two successive generations under elevated CO_2. Ward et al. (2000), also on *A. thaliana*, is the only report of a selective effect of CO_2 but genetic adaptation was found only under selection at low Pleistocene-like CO_2 concentration. Evolutionary responses at elevated CO_2 resulted in a shortened life cycle that did not allow for increased biomass accumulation. It appears that:

- Variations in the response of growth, allocation, phenology, or ontogeny (Kinugasa et al. 2003; Ward et al. 2000; LaDeau and Clark 2001) can explain the observed large interspecific differences in reproductive responses.
- In principle, it is critical to incorporate the effects of evolutionary processes in our predictions of ecosystems responses to global change (Ward and Kelly 2004). However, too few studies exist to give us a clear idea of the evolutionary potential in the face of elevated CO_2.

9.2.2 Communities at Equilibrium Versus Dynamic Systems

Community dynamics also has to be taken into account to determine whole systems response to elevated CO_2. Changes in the relative contribution of individual species under elevated CO_2 could alter community production with consequent changes in community composition (Körner 2001; Potvin and Vasseur 1997; Reynolds et al. 2001; Chapin et al. 2000). Several experiments have focused on understanding the impact of elevated CO_2 on community composition. We use three long-term CO_2 enrichment studies conducted on grasslands to illustrate this point: the calcaeous grassland experiment of Switzerland (Leadley et al. 1999; Niklaus and Körner 2004), the grazed pasture of Southern Québec, Canada (Potvin and Vasseur 1997; Vasseur and Potvin 1997) and the annual grassland of the Jasper Ridge Biological Preserve in California (Zavaleta et al. 2003a,b). Floristically the richest of those grassland sites was the Swiss grassland with a total of 70 vascular species (Leadley et al. 1999), followed by the Californian site with a total of 43 species (Zavaleta et al. 2003a) then the Québec site with about 10 species (Potvin and Vasseur 1997). CO_2 enrichment lasted six years in Switzerland, five years in Québec while Zavaleta et al. (2003a,b) reported three years of CO_2 enrichment. These three experiments provide a unique opportunity to examine the extent to which elevated CO_2 might alter species composition with a possible feed-back loop on community biomass accumulation.

In the Swiss grassland, the number of species present in the plots increased through time, 1996 vs. 1999, but exposure to elevated CO_2 had no significant effect on species richness. In contrast elevated CO_2 significantly enhanced evenness estimated as the modified Hill ratio E^* (Niklaus and Körner 2004). At the same time, community productivity increased under elevated CO_2 and the increase in aboveground biomass could be predicted by precipitation. The highest biomass enhancement was observed after five years of enrichment (+31%) and the effect seemed to decrease thereafter (+18% enhancement on year 6). At the Québec field site, between 1992 and 1996, total species number was reduced by 10% under elevated CO_2, 47% in the ambient open-tops and 31% in the control plots (Potvin unpublished). Through time evenness, measured as Simpson's index, decreased under ambient CO_2 while it increased under elevated CO_2 (Potvin and Vasseur 1997). Destructive harvest, carried-out at the end of the five years of CO_2 enrichment, showed that biomass accumulation was not statistically different between the high (360 g m^{-2} ±125) and the ambient (486 g m^{-2} ±152) open-tops nor the control plots (435 g m^{-2} ±118) (Potvin unpublished). Three years of elevated CO_2 over California grassland led to a reduction in species number of 8% (Zavaleta et al. 2003a) however no information was given on evenness. The Californian study examined the effect of N deposition, precipitation and warming on diversity. Interestingly the effect of these different global change drivers on diversity was purely additive. The two enrichments (N and CO_2) reduced diversity, while precipitation increased diversity and warming had no effect. Norby et al. (2007, Chap. 3 of this volume) provides an exhaustive analysis of the results from the Jasper Ridge Global Change Experiment (JRGCE). In the JRGCE, elevated CO_2 led to increased plant biomass when acting alone but to a reduction in NPP when applied in conjunction with other global change drivers: temperature, N deposition and precipitation.

Together these results suggest that elevated CO_2 will elicit changes in biodiversity, whether negative or positive, and that these changes may or may not be associated with changes in community biomass. This observation is supported by other studies. An experiment combining elevated CO_2 with sward management of temperate grassland monoliths (Teyssonneyre et al. 2002) showed that cutting frequency modified the biodiversity response of the monoliths. At low cutting frequency the decline in species diversity was partly alleviated by elevated CO_2, a result with much resemblance to the Québec experiment. Furthermore, elevated CO_2 may interact with a number of other components of global change including N deposition, changes in precipitation, and temperature. CO_2 is thought to stimulate production and C storage primarily in areas of nitrogen deposition (Schimel 1995; Nadelhoffer et al. 1999). In these nutrient-rich environments, enhanced productivity is often associated with reduced species diversity (Tilman and Downing 1994). However, in the JRGCE, Zaveletta et al. (2003a,b) found no interactions between CO_2, N, precipitation and temperature on plant diversity. The diversity responses of any combination of treatments could be predicted from the sum of the responses to individual treatments. Additive responses of this type would make prediction of CO_2 responses in the future much more tractable, but additional multifactor experiments will be essential for testing the generality of this pattern.

The three field experiments discussed above differ fundamentally from those using growth chambers, in which a fixed set of species is exposed to elevated CO_2, because the communities under study were moving targets where species richness, identity and dominance were

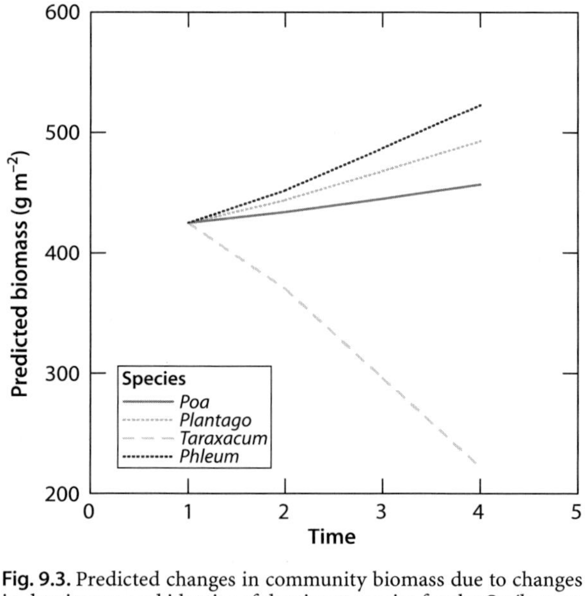

Fig. 9.3. Predicted changes in community biomass due to changes in dominance and identity of dominant species for the Québec experiment. The model assumes an arbitrary time scale leading to changes in dominance in the community. Predicted biomass was calculated as $PB = S_a b_a + S_b(b_i + b_j + b_k)$ where S_a is the area occupied by the dominant species, S_b is the area occupied by each of the other species, b_a is the biomass per area of the dominant species and b_{i-k} the biomass per area of the three companion species. At time 1, community biomass was modeled for even stands where Simpson's index, $D = 4$. Dominance increased through time to a maximum, $D = 1.53$, when 80% of the area was occupied by a single species. The model was run four times with either *Phleum pratense*, *Plantago major*, *Taraxacum officinale*, or *Poa pratensis* as dominant species. The scenario in which *Poa* becomes dominant is analog to observations from the Québec elevated CO_2 experiments

constantly changing. As such they might provide a more realistic appraisal of the effect of elevated CO_2. The study of dynamic communities raises the possibility that biomass response to elevated CO_2 could be hard to detect in early successional sites if large changes in biomass associated with a shift in species composition are taking place. A simple model was built to examine the effect of shifts in dominance of the four most common species on community biomass at the Québec site. The simulation shows that both changes in dominance pattern and in the identity of the dominant species strongly affect community biomass (Fig. 9.3). Community biomass decreased by 47%, when *Taraxacum* became dominant, while it increased by 23%, when *Phleum* became dominant. Biomass in the high CO_2 chambers was 25% lower then that of the ambient CO_2 chambers after five years of enrichment. The simulation shows that succession-mediated changes in biomass could either mask or amplify changes in biomass due to CO_2 enrichment. Leadley et al. (1999) reported vast differences in responsiveness to elevated CO_2 between *Lotus corniculatus, Bromus erectus* and *Carex flacca* and the other species during the first three years of results for the Swiss grassland experiment. In fact, the authors explain the observed lag in the response of aboveground biomass by the growth rate of these species; implicitly suggesting that successional pattern is important for an understanding of biomass responses to CO_2. Bolker et al. (1995), using an individual-based forest model, similarly reported that changes caused by successional sequence were larger than direct CO_2 responses. In their simulation, they identify red oak as a species driving many of the patterns observed. Two years of CO_2 enrichment in the Nevada Desert FACE Facility (NDFF) reported a 42% reduction in density for the native annual species coupled with a 40% increase in aboveground biomass. *Bromus madritensis*, an invasive exotic species, became a higher proportion of total plant density at elevated CO_2 (Smith et al. 2000). Those changes in community biomass associated with a shift in composition illustrate the challenge facing ecologists in predicting the effect of elevated CO_2 on dynamic communities (Díaz 2001).

An important legacy of the last ten years of elevated CO_2 research is the recognition that natural communities must been studied as dynamic systems. Studies of the dynamic responses of communities to elevated CO_2 indicate that:

- Elevated CO_2 might elicit changes in community diversity both in term of species richness or evenness. In dynamic communities, species identity and dominance patterns can drive the response to CO_2.
- Community biomass can change regardless of the effect of CO_2 due to succession alone. As both effects can be of similar magnitude, it may be difficult to partition the effect of CO_2 and succession on transient communities.
- Although the effects of elevated CO_2 on biodiversity may be less pronounced than those of other global change drivers such as land use, climate, N deposition, and biotic exchange, elevated CO_2 will likely interact with these drivers in ways that differ among biomes (Sala et al. 2000).

9.3 Biodiversity Loss and Response to Elevated CO_2

9.3.1 Species Diversity and Response to Elevated CO_2

The studies reported in the previous section suggest that elevated CO_2 has the potential, at least under certain circumstances, to alter biodiversity. The interaction between CO_2 and diversity must therefore be examined in the context of the diversity-productivity relationship. The GCTE decade has witnessed an enormous effort by ecologists to try and disentangle the links existing between diversity and ecosystem functioning (e.g., Chapin et al. 2000; Tilman et al. 2001; Spehn et al. 2005). As pointed out by Hooper et al. (2005) biodiversity has often been confounded with species richness although other components of biodiversity such as evenness, composition and presence/absence of key species all exert strong effects on ecosystem properties. Loreau and Hector (2001) suggested that a diversity effect might be due to either sampling or complementarity. Sampling refers to the likelihood that a key species be present in a community while complementarity describes interactions among species. Research on elevated CO_2 suggests that the response to elevated CO_2 is largely mediated by species identity and competitive interactions (see Körner et al. this volum;, Hanley et al. 2004; Navas et al. 2002; Ramsier et al. 2005). It is thus likely that elevated CO_2 could modify the diversity/productivity relationship. Three projects explicitly addressed this question by manipulating CO_2 concentration and plant diversity: the Swiss calcareous grassland study (Stocker et al. 1999; Niklaus et al. 2001), Roy's annual herbaceous communities (see Sect. 9.2.1) and the Minnesota, USA, BioCON project (Reich et al. 2001a,b, 2004) (Fig. 9.4).

To answer the question: "How does species loss influence biomass and its response to elevated CO_2?" the BioCON project relied on plots planted with 16, 9, 4 and 1 species while the Swiss experiment involved plots of 31, 15 and 5 species. In the BioCON project, total biomass of plots with 9, 4, and 1 species was, respectively 2, 13, and 50% less than that of plots with 16 species (Reich et al. 2001a,b; Reich et al. 2004). The enhancement of biomass resulting from elevated CO_2 decreased with declining diversity. In ambient plots, the average stimulation of total biomass in response to elevated CO_2 was 22%, 18%, 10% and 7% in, respectively, the 16-, 9-, 4- and 1-species plots (Reich et al. 2001a,b). Differences in diversity accounted for a five-fold difference in the effect of CO_2 enrichment Whereas elevated CO_2 increased biomass by

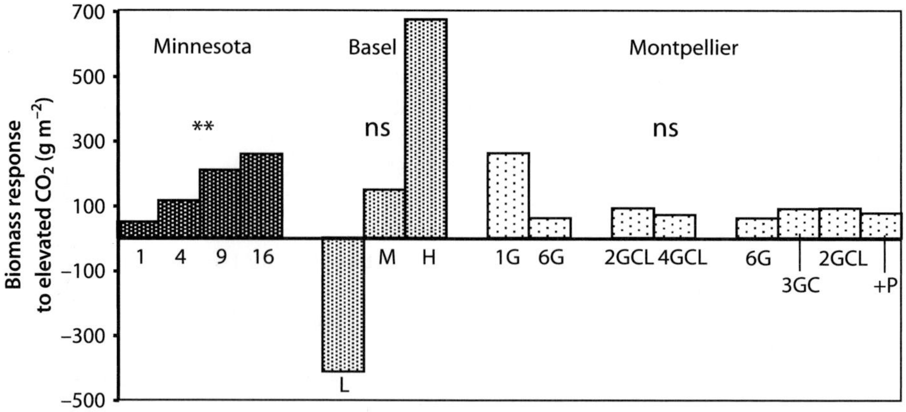

Fig. 9.4. Comparison of three experiments studying the interactions between CO_2 and plant diversity. In all cases, the CO_2 effect is expressed as the absolute change in total plant biomass at elevated vs. ambient CO_2 concentrations. *Minnesota:* Effects of diversity on total plant biomass response averaged over two years of CO_2 exposure in the field using FACE rings (Reich et al. 2001). Communities were planted with 1, 4, 9, or 16 plant species. *Basel:* Biomass response of calcareous grassland communities after five years of CO_2 exposure in the field using screen-aided CO_2 exposure (Niklaus et al. 2001). Communities were planted with 5 (*L* = low diversity), 12 (*M* = medium diversity) and 31 (*H* = high diversity) species, but species richness declined over the duration of the experiment in all treatments. *Montpellier:* CO_2 response of planted Mediterranean grassland communities of various complexity (*G* = annual grasses, *C* = composites, *L* = legumes, *P* = perennial grasses, where the number indicates the number of species within each functional group). Each bar is the average of four 0.5 m² model ecosystems grown in climate-controlled greenhouses (J. Roy, unpublished data)

258 g m^{-2} in the most diverse plots, it increased biomass by only 47 g m^{-2} in the monocultures. Likewise, the Swiss experiment found a significant effect of diversity on biomass; this effect, however, declined with time (Niklaus et al. 2001). Elevated CO_2 had a significant but highly variable effect on the biomass of high-diversity plots but no such effect on the medium and low-diversity communities. Conversely, Roy and collaborators found a significant increase in plant biomass production at elevated CO_2 in Mediterranean grassland communities, but the addition of species within functional groups did not alter this response.

Two complementary BioCON experiments allow further partitioning of the effect of functional groups and species within functional groups on CO_2 responses. These experiments include plots with *(i)* 1 or 4 species from single functional groups (C_4 grasses, C_3 grasses, N-fixing legumes and non-N-fixing herbaceous species) and *(ii)* four species varying from 4 to 1 in functional groups. All these plots were replicated across four CO_2 and N levels. Biomass increment in response to elevated CO_2, enriched N, or their combination was greater in 4-species than in 1-species plots (Fig. 9.5). Furthermore, plots planted with each forb species grown in monoculture had, on average, 44% less total biomass than plots with all four forbs species. Across all four functional groups, single functional group plots with one instead of four species had 29% less biomass on average (Reich et al. 2004). Decreasing the number of groups from 3 or 4 to 1 or 2 resulted in 20% lower total biomass on average, even with species richness constant at 4. Total biomass responses to CO_2 and N reflect effects on fine root biomass (Fig. 9.5). For instance, on average, in plots with four functional groups, elevated CO_2 increased fine root biomass by 191 g m^{-2}, but as the number of groups decreased to 3, 2, and 1, elevated CO_2 increased fine root biomass by 91, 90, and only 14 g m^{-2}, respectively.

In the BioCON project, functional group diversity was not a prerequisite for species richness to have significant effects and this effect was seen in disparate functional groups. This contrasts with the Mediterranean grassland study that did not find evidence that reductions in the number of functional groups altered the growth response to elevated CO_2 In the Swiss experiment, species responding positively to CO_2 enrichment did not follow common functional type classification. Thus while elevated CO_2 might well influence community composition, it is difficult to predict changes. Together, the BioCON, Swiss grassland, and Mediterranean grassland experiments provide insights on how a species poor world would respond to elevated CO_2:

- Decreases in diversity can lead to lower total plant biomass and to reduced responsiveness to elevated CO_2. This diversity effect is highly variable ranging from very pronounced (BioCON), relatively small (Swiss grassland) to nil (Mediterranean grassland).
- In the BioCON project, holding species richness constant, plots with decreasing functional group richness had less biomass and smaller enhancements of biomass in response to CO_2, when compared with plots with greater functional group richness. This was not the case in the Mediterranean grassland study.

9.3.2 Ecosystem C Fluxes in a Species-Poor World

The previous section suggests that species-poor communities could accumulate less biomass and be less responsive to elevated CO_2 with the magnitude of responses

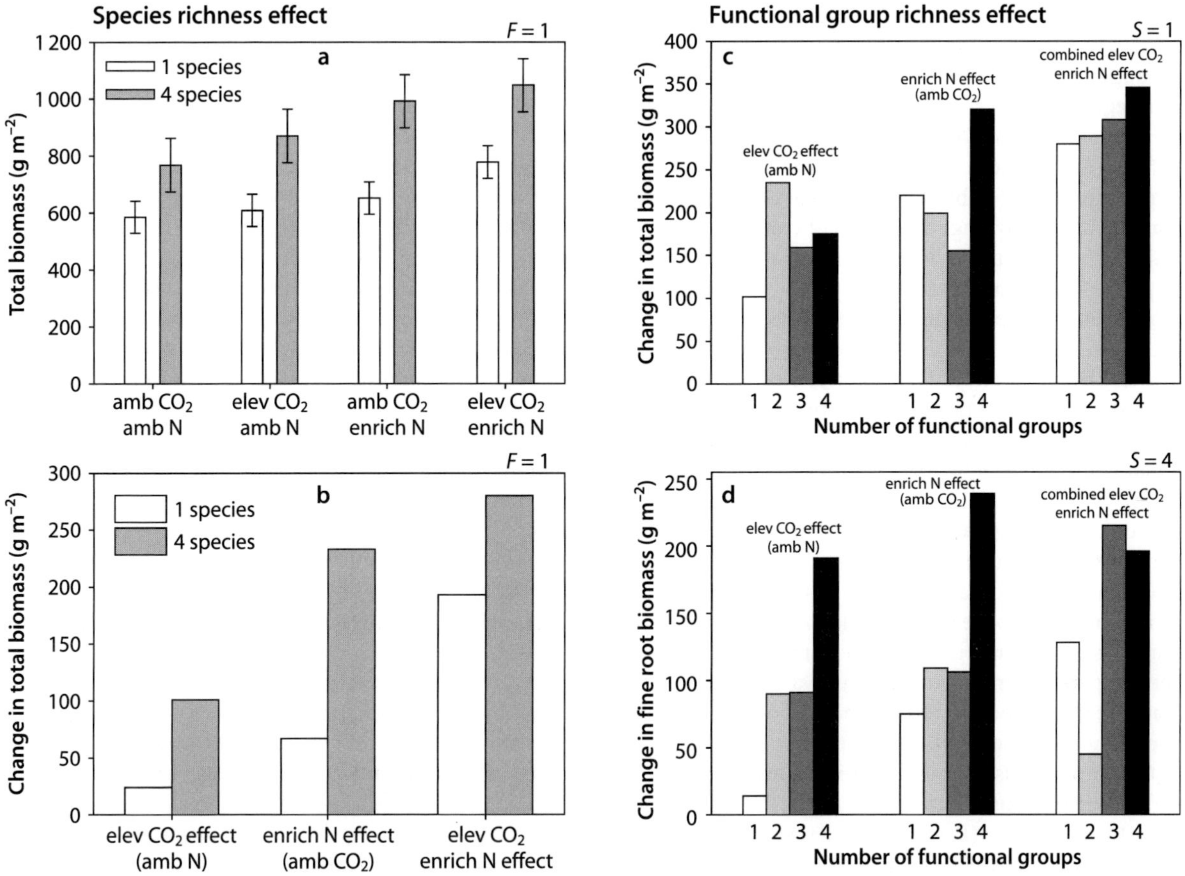

Fig. 9.5. Effects of species richness and functional group richness on biomass accumulation and response to elevated CO_2. **a** Total biomass and **b** change in total biomass (above- and below-ground, 0–20 cm depth) (+1 S.E.) for plots planted with one functional group and either 1 or 4 species, grown at four combinations of ambient (368 µmol mol^{-1}) and elevated (560 µmol mol^{-1}) CO_2, and ambient or enriched N (4 g N m^{-2} yr^{-1}). Data averaged over two harvests in each year from 1998 to 2001. **c** Change in total biomass and **d** in fine root biomass in plots with 4 species, and either 1, 2, 3 or 4 functional groups, grown at four combinations of ambient and elevated CO_2, and ambient and enriched N

modulated by resource availability. This raises a question: Would a species-poor world become a less efficient C sink? If this were the case, the initial assumption of elevated CO_2 research, coined earlier as "the large-plant-scenario", would be invalidated. In fact, we know little about the effects of species loss on the capacity of ecosystems to take up C in the long term. So, where does the future of research on elevated CO_2 and diversity lie?

First, if we want to understand the consequences of species loss on the ability of ecosystem to sequester C, we need to develop better ways to estimate C cycling. Although it is widely acknowledged that C sequestration and net primary production (NPP) are not equivalent (Körner 1995; Karnosky 2003) most experiments examining the diversity/productivity relationship have used either NPP or biomass accumulation to report on ecosystem function. Catovsky et al. (2002) emphasize the urgent need to move beyond estimates of NPP toward better estimations of net ecosystem exchange (NEP). It is important to understand the diversity effects on plant biomass, on litter decomposition as well as on soil respiration (both autotrophic and heterotrophic) in order to estimate appropriately C fluxes in and out of the system. In the only published measurements of the effects of plant diversity on NEP, Stocker et al. (1999) found that reductions in plant diversity did not alter whole ecosystem CO_2 fluxes in calcareous grasslands at elevated or ambient CO_2 concentrations. At the ecosystem level, Chambers et al. (2004) modeled the response to elevated CO_2 of tree biomass and wood litter for a Central Amazon forest. An important conclusion of their study is that coarse litter is important in determining the forest carbon balance and should be considered as a C flux. They argue that under elevated CO_2, the gains in carbon due to increased growth would be roughly balanced by the increased stock of coarse litter. Using microcosms, Van Voris et al. (1980) proposed that the complexity of CO_2 efflux time series, i.e., the number of cyclic components required to compose a time series, might provide useful information regarding ecosystem C cycling. In laboratory microcosms of an old-field ecosystem, they demonstrated that the complexity of CO_2 flux time series was negatively related to the amount of calcium lost in the leachate, an indicator of system disturbance in their experiment. This result suggests that the analysis of CO_2 fluxes may be a useful integrative measure of the effects of disturbance (e.g., species loss) on the C cycle.

Second, we need to better understand the impact on community disassembly on C cycling. Habitat destruction and climate change are and will continue to be the major drivers of species loss and subsequent ecosystem responses (e.g., Lawton and May 1995; Thomas et al. 2004a). For example, using an area-based extrapolation of climate envelopes, Thomas et al. (2004b) suggested that climate change might lead to up to 35% of the taxa being "committed to extinction" in their study regions over the next 50 years. To date, experiments addressing the functional consequences of species loss have typically been performed over small spatial and temporal scales relative to the size, generation times and possible extinction rates of the organisms involved (e.g., Sect. 9.3.1, Tilman et al. 1997; Hector et al. 1999; Lepš et al. 2001). Because extinction in response to habitat destruction and climate change is a non-random process dominated by non-equilibrium population dynamics (Kareiva and Wennergren 1995), we need to better understand the long-term effects of community disassembly.

Community disassembly and its affects on C uptake and on the response to elevated CO_2 have received relatively little attention (Lawton 1994; Sala et al. 1996; Grime 1998; Gonzalez and Chaneton 2002). For example, the identity of species extinctions (e.g., rare vs. dominant; Grime 1998) and the rate at which such extinctions occur could generate non-linear ecosystem responses (Lamont 1995). Furthermore, long-term temporal changes in the abundances and biomass of remnant populations (e.g., Laurance et al. 1997; Didham et al. 1998) have been largely missing from biodiversity-C cycling studies conducted to date. The first Ecotron experiment (Naeem et al. 1994) reported that higher diversity communities consumed more CO_2 than lower diversity communities. It was suggested, however, that the experimental design could not separate diversity and composition effects (Huston 1997), but an additional shortcoming was the limited duration of the experiment. McGrady-Steed et al. (1997) studied microbial communities that exhibited greater inter-community variation in CO_2 flux at low diversity than at high diversity. However, this effect seemed to be due to differences in species composition among low diversity communities (Morin and McGrady-Steed 2004), and not to diversity effects per se.

The results obtained to date do not demonstrate whether a species-poor world will become a less efficient C sink or not. We might however be able to flip the question around and propose management strategies to increase C storage at the landscape level. GCTE is coming to an end at the same time as the Kyoto Protocol is coming into effect. The Nations of the world are now seeking ways to promote the uptake of C through land-use changes while ensuring the stability of these sinks. Hooper et al. (2005) summarized our knowledge to date by indicating that community composition is at least as important as species richness to determine ecosystem properties. Therefore management strategies that incorporate biodiversity will have to pay attention to the identity of species and their interactions (Hiremath and Ewel 2001). Furthermore estimates of the potential for C sequestration need to incorporate a precise assessment not only of C stocks but also of C losses. For example, C accounting methodologies proposed for afforestation and reforestation projects under the Clean Development Mechanism (www.unfccc.org) need to inventory coarse woody debris as well as living biomass.

Many questions remain for the next generation of global change research:

- How is the relationship between net primary productivity and net ecosystem productivity affected by biodiversity, elevated CO_2 and other global change drivers?
- Can we predict which species are most likely to be lost by habitat fragmentation or climate change?
- How will the process of species loss, driven by habitat fragmentation and climate change, affect CO_2 fluxes and the response of ecosystems to elevated CO_2?
- In what ways, if any, will future species loss alter the C source and sink capacities of different ecosystems?

9.4 Summary and Conclusions

In situ studies of communities suggest that the interactions between biodiversity, atmospheric CO_2 concentration, and C cycling are very complex. Rising CO_2 will probably act on plant diversity through a number of very indirect pathways; e.g., by altering the relative availability of resources such as water and nutrients thereby altering competitive interactions among plants. Second, changes in plant community structure due to rising CO_2 concentrations, particularly in highly dynamic systems, may be as or more important in determining biomass responses than the direct effect of elevated CO_2. Finally, species loss resulting from global change might reduce productivity as well as responsiveness to elevated CO_2, although the response will largely depend on species identity.

Over all, simple scenarios invoking a clear and simple biospheric loop between atmospheric CO_2 and terrestrial biomass due to CO_2 fertilization need to be abandoned and replaced by an understanding of a complex system involving both biotic and abiotic feed-back loops. Our ability to either manage or predict the relationship between diversity and ecosystem function is still limited.

References

Andalo C, Godelle B, Lefranc M, Mousseau M-A, Hill-Bottraud I (1996) Elevated CO_2 decreases seed germination in *Arabidopsis thaliana*. Glob Ch Biol 2:129–135

Arp WJ, Drake BG, Pockman WT, Curtis PS, Whigham DF (1993) Interactions between C_3 and C_4 saltmarsh plant species during four years of exposure to elevated CO_2. Vegetatio 104/105:133–144

Bassirirad H, Thomas RB, Reynolds JF, Strain BR (1996) Differential responses of root uptake kinetics of NH_4^+ and NO_3^- to enriched atmospheric CO_2 concentration in field-grown loblolly pine. Plant, Cell and Environment 19:367–371

Bazzaz FA (1990) The response of natural ecosystems to rising global CO_2 levels. Ann Rev Ecol Syst 21:167–196

Bidart-Bouzat MG (2004) Herbivory modifies the lifetime fitness response of *Arabidopsis thaliana* to elevated CO_2. Ecology 85(2):297–303

Bolker BM, Pacala SW, Bazzaz FA, Canham CD, Levin SA (1995) Species-diversity and ecosystem response to carbon dioxide fertilization – Conclusions from a temperate forest model. Glob Ch Biol 1(5):373–381

Catovsky S, Bradford MA, Hector A (2002) Biodiversity and ecosystem productivity: implications for carbon storage. Oikos 97(3):443–448

Chambers JQ, Higuchi N, Texeira LM, dos Santos J, Laurance SG, Trumbore SE (2004) Response of tree biomass and wood litter to disturbance in Central Amazon forest. Oecologia 141(40): 596–611

Chapin III FS, Zavaleta ES, Eviner VT, Naylor RL, Vitousek PM, Reynolds HL, Hooper DU, Lavorel S, Sala OE, Hobie SE, Mack MC, Díaz S (2000). Consequences of changing biodiversity. Nature 405 (11):234–242

Curtis PS, Wang X (1998) A meta-analysis of elevated CO_2 effects on woody plant mass, form, and physiology. Oecologia 113:299–313

Derner JD, Tischler CR, Polley HW, Johnson HB (2004) Intergenerational above- and belowground responses of spring wheat (*Triticum aestivum* L.) to elevated CO_2. Basic Appl Ecol 5(2):145–152

Díaz S (2001) Complex interactions between plant diversity, succession and elevated CO_2. TREE 16(12) 667

Didham RK, Hammond PM, Lawton JH, Eggleton P, Stork, NE (1998) Beetle species responses to tropical forest fragmentation. Ecol Monog 68:295–323

Gonzalez A, Chaneton E (2002) Heterotroph species extinction, abundance and biomass dynamics in a fragmented microecosystem. J Anim Ecol 71:594–602

Grime JP (1998) Benefits of plant diversity to ecosystems: immediate, filter and founder effects. J Ecol 86:902–910

Hanley ME, Tropimov S, Taylor G (2004) Species-level effects more important than functional group-level responses to elevated CO_2: evidence from simulated curves. Functional ecology 18:304–313

Hector A, Schmid B, Beierkuhnlein C (1999) Plant diversity and productivity experiments in European grasslands. Science 286: 1123–1127

Hiremath AJ, Ewel JJ (2001) Ecosystem nutrient use efficiency, productivity, and nutrient accrual in model tropical communities Ecosystems 4:669–682

Hooper DU, Chapin III FS, Ewel JJ, Hector A, Inchausti P, Lavorel S, Lawton JH, Lodge DM, Loreau M, Naeem S, Schmid B, Setala H, Symstad AJ, Vandermeer J, Wardle DA (2005) Effects of biodiversity on ecosystem functionaing: a consensus of current knowledge. Ecological Monographs 7591) 3–35

Hungate BA (1999) Ecosystem responses to rising atmospheric CO_2: feedbacks through the nitrogen cycle. In: Luo Y, Mooney HA, eds. Carbon Dioxide and Environmental Stress. San Diego, USA: Academic Press, 265–285

Huston MA (1997) Hidden treatments in ecological experiments: re-evaluating the ecosystem function of biodiversity. Oecologia 110:449–460

Jablonski LM, Wang XZ, Curtis PS (2002) Plant reproduction under elevated CO_2 conditions: a meta-analysis of reports on 79 crop and wild species. New Phytol 156(1):9–26

Kareiva P, Wennergren U (1995) Connecting landscape patterns to ecosystem and population processes. Nature 373:299–302

Karnosky DF (2003) Impacts of elevated atmospheric CO_2 on forest trees and forest ecosystems: Knowledge gaps. Environment international 29:161–169

Kerstiens G (2001) Meta-analysis of the interaction between shade-tolerance, light environment and growth response of woody species to elevated CO_2. Act Oecol 22:61–69

Kinugasa T, Hikosaka K, Hirose T (2003) Reproductive allocation of an annual, *Xanthium canadense*, at an elevated carbon dioxide concentration. Oecologia 137(1):1–9

Körner Ch (1995) Toward a better experimental basis for upscaling plant responses to elevated CO_2 and climate warming. Plant Cell and Environment 18:1101–1110

Körner C (2001) Biosphere responses to CO_2 enrichment. Ecol Appl 10:1590–1619

Körner C, Morgan J, Norby R (2007) CO_2 fertilization: when, where, and how much. In: Canadell J, Pataki D, Pitelka L (eds) Global change and terrestrial ecosystems. Springer-Verlag, Heidelberg, this volume

LaDeau SL, Clark JS (2001) Rising CO_2 levels and the fecundity of forest trees. Science 292(5514):95–98

Lamont BB (1995) Testing the effect of ecosystem composition structure on its functioning. Oikos 74:283–295

Laurance WF, Laurance SG, Ferreira LV, Rankin-de Merona JM, Gascon C, Lovejoy TE (1997) Biomass Collapse in Amazonian Forest Fragments. Science 278:1117–1118

Lawton JH (1994) What do species do in ecosystems? Oikos 71:367–374

Lawton JH, May R (1995) Extinction Rates. Oxford University Press, Oxford

Leadley PW, Niklaus PA, Stocker R, Körner C (1999) A field study of the effects of elevated CO_2 on plant biomass and community structure in a calcareous grassland. Oecologia 118(1):39–49

Lepš J, Brown VK, Díaz Len TA, Gormsen D, Hedlund K, Kailová J, Korthals GW, Mortimer SR, Rodriguez-Barrueco C, Roy J, Santa Regina I, van Dijk C van der Putten WH (2001) Separating the chance effect from other diversity effects in the functioning of plant communities. Oikos 92:123–134

Loreau M, Hector A (2001) Partitioning selection and complementarity in biodiversity experiments. Nature 412:72–76

McGrady-Steed J, Harris PM, Morin PJ (1997) Biodiversity regulates ecosystem predictability. Nature 390:162–164

Mohan JE, Clark JS, Schlesinger WH (2004) Genetic variation in germination, growth, and survivorship of red maple in response to subambient through elevated atmospheric CO_2. Glob Ch Biol 10(2):233–247

Morgan JA, Mosier AR, Milchunas DG, LeCain DR, Nelson JA, Parton WJ (2004) CO_2 enhances productivity, alters species composition, and reduces digestibility of short grass steppe vegetation. Ecol Appl 14:208–219

Morin PJ, McGrady-Steed J (2004) Biodiversity and ecosystem functioning in aquatic microbial systems: a new analysis of temporal variation and species richness-predictability relations. Oikos 104:458–466

Nadelhoffer KJ, Emmertt BA, Gundersen P, Kjonaas OJ, Koopmans CJ, Schleppi P, Tietema A, Wright RF (1999) Nitrogen deposition makes a minor contribution to carbon sequestration in temperate forests. Nature 398:145–148

Naeem S, Thompson LJ, Lawler SP, Lawton JH, Woodfin RM (1994) Declining biodiversity can alter the performance of ecosystems. Nature 368:734–736

Navas ML (1998) Individual species performance and response of multi-specific communities to elevated CO_2: a review. Funct Ecol 12:721–727

Navas ML, Garnier E, Austin MP, Viaud A, Gifford RM (2002) Seeking a sound index of competitive intensity: Application to the study of biomass production under elevated CO_2 along a nitrogen gradient. Austral Ecology 27:463–473

Niklaus PA, Körner Ch (2004) Synthesis of a six-year study of calcareous grassland responses to in situ CO_2 enrichment. Ecological Monographs 74(3) 491–511

Niklaus PA, Leadley PW, Schmid B, Körner C (2001) A long-term field study on biodiversity X elevated CO_2 interactions in grassland. Ecol Monog 7:341–356

Nowak RS, Zitzer SF, Babcock D, Smith-Longozo V, Charlet TN, Coleman JS, Seemann JR, Smith SD (2004) Elevated atmospheric CO_2 does not conserve soil water in the Mojave desert. Ecology 85:93–99

Owensby CE, Coyne PI, Ham JM, Auen LM (1993) Biomass production in a tallgrass prairie ecosystem exposed to ambient and elevated levels of CO_2. Ecol Appl 3:644–653
Owensby C, Ham JM, Knapp AK, Rice CW, Coyne PI, Auen LM (1996) Ecosystem-level responses of tallgrass prairie to elevated CO_2. Carbon dioxide and terrestrial ecosystems. GW. Koch and HA Mooney. San Diego, Academic Press, pp 147–162
Poorter H (1993) Interspecific variation in the growth response of plants to an elevated ambient CO_2 concentration. Vegetatio 104/105:77–97
Poorter H, et al. (1998) Do slow-growing species and nutrient-stressed plants respond relatively strongly to elevated CO_2? Glob Ch Biol 4:693–697
Poorter H, Navas ML (2003). Plant growth and competition at elevated CO_2: on winners, losers and functional groups. New Phytol 157(2):175–198
Potvin C, Tousignant D (1996) Evolutionary consequences of simulated global change: Genetic adaptation or adaptive phenotypic plasticity. Oecologia 108(4):683–693
Potvin C, Vasseur L (1997) Long-term CO_2 enrichment of a pasture community: Species diversity and dominance pattern. Ecology 78:666–667
Ramseir D, Connolly J, Bazzaz FA (2005) Carbon dioxide regime, species identity and influence of species initial abundance as determinants of change in stand biomass composition in five-species communitites; an investigation using a simplex design and RGRD analysis. Journal of Ecology 93:502–511
Reich PB, Knops J, Tilman D, Craine J, Ellsworth D, Tjoelker M, Lee T, Naeem S, Wedin D, Bahauddin D, Hendrey G, Jose G, Wrage K, Goth J, Bengston W (2001a) Plant diversity enhances ecosystem responses to elevated CO_2 and nitrogen deposition. Nature 410:809–812
Reich PB, Tilman D, Craine J, Ellsworth D, Tjoelker M, Knops J, Wedin D, Naeem S, Bahauddin D, Goth J, Bengston W, Lee T (2001b) Do species and functional groups differ in acquisition and use of C, N and water under varying atmospheric CO_2 and N deposition regimes? A field test with 16 grassland species. New Phyt 150:435–448
Reich PB, Tilman D, Naeem S, Ellsworth D, Knops J, Craine J, Wedin D, Trost J (2004) Species and functional diversity independently influence biomass accumulation and its response to CO_2 and N. Proc Nat Acad Sci USA 101 (27) 10101–10106
Reynolds JF, Bugmann H, Pitelka LF (2001) How much physiology is needed in forest gap models for stimulating long-term vegetation response to global changeÉ Challenges, limitations, and potentials. Climatic Change 51:541–557
Roumet C, Laurent G, Canivenc G, Roy J (2002) Genotypic variation in the response of two perennial grass species to elevated carbon dioxide. Oecologia 133(3):342–348
Sala OE, Lauenroth WK, McNaughton SJ, Rusch G, Zhang X (1996) Biodiversity and ecosystem function in grasslands. Functional Roles of Biodiversity: a Global Perspective. In: Mooney HA, Cushman JH, Medina E, Sala OE, Schulze E-D (eds) pp 129–149, John Wiley and Sons, London
Sala OE, Chapin FS, III, Armesto JJ, Berlow E, Bloomfield J, Dirzo R, Huber-Sanwald E, Huenneke LF, Jackson RB, Kinzig A, Leemans R, Lodge DM, Mooney HA, Oesterheld M, Poff NL, Sykes MT, Walker BH, Walker M, Wall DH (2000) Global Biodiversity Scenarios for the Year 2100. Science 287:1770–1774
Schimel D S (1995) Terrestrial Ecosystems and the Carbon-Cycle. Glob Ch Biol 1(1):77–91
Schmid B, Birrer A, Lavigne C (1996) Genetic variation in the response of plant populations to elevated CO_2 in a nutrient-poor, calcareous grassland. In: Körner C, Bazzaz FA, (eds) Carbon dioxide, populations and communities. Physiological Ecology Series. San Diego, CA, USA: Academic Press, pp 31–50
Smith SD, Huxman TE, Zitzer SF, Charlet TN, Housman DC, Coleman JS, Fenstermaker LK, Seemann JR, Nowak RS (2000) Elevated CO_2 increase productivity and invasive species success in an arid ecosystem. Nature 408:79–82
Spehn EM, Hector A, Joshi J, Scherer-Lorenzen M, Schmid B, Bazeley-White E, Beirkuhnlein C, Caldeira MC, Diemers M, Dimitrakopoulos PG, Finn JA, Freitas H, Giller PS, Good J, Harris R, Hogberg P, Huss-Danell K, Jumpponen A, Koricheva J, Leadley PW, Loreau M, Minns A, Mulder CPH, O'Donovan, Ottawy SJ, Palmborg C, Pereira JS, Pfisterer AB, Prinz A, Read DJ, Schulze ED, Siamantziouras A-S D, Terry AC, Troumbis AY, Woodward EI, Yachl SY, Lawton JH (2005) Ecosystem effects of biodiversity manipulations in european grasslands. Ecological Monographs 7591) 37–63
Spinnler D, Egli P, Körner C (2003) Provenance effects and allometry in beech and spruce under elevated CO_2 and nitrogen on two different forest soils. Basic Appl Ecol 4(5):467–478
Stewart J, Potvin C (1996) Effects of elevated CO_2 on an artificial grassland community: Competition, invasion and neighborhood growth. Funct Ecol 10:157–166
Stiling P, Moon D, Hymus G, Drake B (2004) Differential effects of elevated CO_2 on acorn density, weight, germination, and predation among three oak species in a scrub-oak forest. Glob Ch Biol 10(2):228–232
Stocker R, Körner C, Schmid B, Niklaus PA, Leadley PW (1999) A field study of the effects of elevated CO_2 and plant species diversity on ecosystem-level gas exchange in a planted calcareous grassland. Glob Ch Biol 5(1):95–105
Stocklin J, Körner C (1999) Interactive effects of elevated CO_2, P availability and legume presence on calcareous grassland: results of a glasshouse experiment. Funct Ecol 13(2):200–209
Teyssonneyre F, Picon-Cochard C, Soussana JF (2002) How can we predict the effects of elevated CO_2 on the balance between perennial C_3 grass species competing for light? New Phytol 154(1): 53–64
Thomas CD, Cameron A, Green RE, Bakkenes M, Beaumont LJ, Collingham YC, Barend F, Erasmus N, Ferreira de Siqueira M, Grainger A, Hannah L, Hughes L, Huntley B, van Jaarsveld AS, Midgley GF, Miles L, Ortega-Huerta MA, Peterson AT, Phillips OL, Williams SE (2004a) Extinction risk from climate change. Nature 427:145–148
Thomas JA, Telfer MG, Roy DB, Preston CD, Greenwood JJD, Asher J, Fox R, Clarke RT, Lawton JH (2004b) Comparative losses of British butterflies, birds, and plants and the global extinction crisis. Science 303:1879–1881
Thürig B, Körner C, Stocklin J (2003) Seed production and seed quality in a calcareous grassland in elevated CO_2. Glob Ch Biol 9(6):873–884
Tilman D, Downing JA (1994) Biodiversity and stability in grasslands. Nature 367:363–365
Tilman D, Knops J, Wedin D (1997) The influence of functional diversity and composition on ecosystem processes. Science 277:1300–1302
Tilman D, Reich PB, Knops J, Wedin D, Mielke T, Lehman C (2001) Diversity and productivity in the long-term grassland experiment. Science 294:843–845
Van Voris P, O'Neill RV, Emanuel WR, Shugart Jr HH (1980) Functional complexity and ecosystem stability. Ecology 61:1352–1360
Vasseur L, Potvin C (1997) Natural pasture community response to enriched CO_2 atmosphere. Plant Ecol 135:31–41
Volk M, Niklaus PA, Körner C (2000) Soil moisture effects determine CO_2 responses of grassland species. Oecologia 125(3):380–388
Walker BH (2001) Tropical savanna. In: Chapin III FS, Sala OE, Huber-Sannwald E (eds) Global Biodiversity in a Changing Environment: Scenarios for the 21st Century. Springer-Verlag, New York, pp 139–156
Wand SJE, Midgley GF, Jones MH, Curtis PS (1999) Responses of wild C_4 and C_3 (Poaceae) species to elevated atmospheric CO_2 concentration: a meta-analytic test of current theories and perceptions. Glob Ch Biol 5:723–741
Wang ZM, Lechowicz MJ, Potvin C (1994) Early selection of black spruce seedlings and global change: Which genotypes should we favor? Ecol Appl 4:604–616
Ward JK, Kelly JK (2004) Scaling up evolutionary responses to elevated CO_2: lessons from *Arabidopsis*. Ecol Lett 7(5):427–440
Ward JK, Antonovics J, Thomas RB, Strain BR (2000) Is atmospheric CO_2 a selective agent on model C_3 annuals? Oecologia 123(3):330–341
Woodward FI (2002) Potential impacts of global elevated CO_2 concentrations on plants. Curr Opin Plant Biol 5:207–211
Zavaleta ES, Shaw MR, Chiariello NR, Mooney HA, Field CB (2003a) Additive effects of simulated climate changes, elevated CO_2, and nitrogen deposition on grassland diversity. PNAS 100(13):7650–7654
Zavaleta ES, Shaw MR, Chiariello NR, Thomas BD, Cleland EE, Field CB, Mooney HA (2003b) Grassland responses to three years of elevated temperature, CO_2, precipitation, and N deposition. Ecol Monog 73(4):585–604

Chapter 10

Predicting the Ecosystem Consequences of Biodiversity Loss: the Biomerge Framework

Shahid Naeem · Robert Colwell · Sandra Díaz · Jennifer Hughes · Claire Jouseau · Sandra Lavorel · Peter Morin Owen Petchey · Justin Wright · BioMERGE

10.1 Biodiversity and Ecosystem Functioning: a Synthesis

10.1.1 Why Biodiversity Matters to Global Change Ecology

Global change ecology has become increasingly concerned with understanding how linkages between changes in biological diversity (henceforth, *biodiversity*) affect ecosystem functioning (Chapin et al. 2000; Loreau et al. 2001; Hooper et al. 2005). By biodiversity we mean, "the variability among living organisms from all sources including, *inter alia*, terrestrial, marine and other aquatic ecosystems and the ecological complexes of which they are part; this includes diversity within species, between species and of ecosystems." This definition, because it has been adopted by the Convention on Biological Diversity and the Millennium Assessment probably has the broadest subscription in comparison to the many other definitions proposed for biodiversity (Duraiappah and Naeem 2005). It clearly defines biodiversity as more than just species or taxonomic diversity ("variability of living organisms from all sources"). It includes variability in biotic interactions ("ecological complexes") and geography (diversity within and among populations, species, and ecosystems).

The relationship between biodiversity and ecosystem functioning is important in global change ecology for several reasons. First, biodiversity is experiencing unprecedented declines in virtually all ecosystems due predominantly to anthropogenic processes such as the conversion of natural ecosystems to less diverse managed systems (e.g., farms, plantations, aquaculture), biological invasions, enhanced rates of N deposition, habitat fragmentation, and climate change (Ehrlich 1988; Wilson 1988; Soulé 1991; Groombridge 1992; Hawksworth 1995; Heywood 1995; Wilcove et al. 1998; Sala et al. 2000; Balmford et al. 2002; Raven 2002). Several studies have shown that biodiversity loss can significantly lower primary production, which is critical to all other ecosystem functions (e.g., Naeem et al.1994; Tilman et al. 1996; Hector et al. 1999).

Second, biosphere-geosphere feedbacks in Earth System processes are a critical component of climate change biology (Díaz et al. 1993; Mooney et al. 1993; Walker and Steffan 1996; Woodwell 1995; Andreae 2002), thus understanding what roles biodiversity may play in these feedbacks is important given widespread declines in biodiversity. For example, carbon sequestration by terrestrial plants represents an important feedback in climate change, and CO_2 impacts on production are affected by plant species composition and diversity (e.g., Bolker et al. 1995; Zavaleta et al. 2003; Reich et al. 2004).

Third, the Convention on Biological Diversity (CBD), signed and ratified in the early 1990s by the vast majority of the world's nations, declared that biodiversity underpins Earth's ecosystems and that biodiversity conservation is fundamental to achieving environmental sustainability and sustainable development (http://www.biodiv.org/). Establishing a scientific basis for the relationship between biodiversity and ecosystem functioning would facilitate the development and implementation of science-based policy for the CBD to achieve its goals of biodiversity conservation and, ultimately, sustainable development.

An additional consideration of the importance of biodiversity in global change ecology is its relationship to ecosystem services, most of which are governed by ecosystem functions. Ecosystem services are defined as the benefits people obtain from ecosystems (Daily et al. 1997; Costanza et al. 1997; Naeem 2001a). The importance of biodiversity to ecosystem services is highlighted in the Millennium Assessment (MA), a consensus of nearly 1 400 experts from 95 countries, by the following statement:

> Biodiversity represents the foundation of ecosystems that, through the services they provide, affect human well-being. These include provisioning services such as food, water, timber, and fiber; regulating services such as the regulation of climate, floods, disease, wastes, and water quality; cultural services such as recreation, aesthetic enjoyment, and spiritual fulfillment; and supporting services such as soil formation, photosynthesis, and nutrient cycling. The MA considers human well-being to consist of five main components: the basic material needs for a good life, health, good social relations, security, and freedom and choice and action. Human well-being is the result of many factors, many directly or indirectly linked to biodiversity and ecosystem services while others are independent of these.

10.1.2 Linking Change in Biodiversity with Change in Ecosystem Functioning

In spite of the logical inseparability of biodiversity and ecosystem functioning, in that neither exists in the absence of the other, little theoretical or empirical work prior to 1992 had explicitly addressed whether changes in biodiversity could have predictable impacts on ecosystem function. Traditionally, biodiversity was largely considered an epiphenomenon or dependent function of abiotic factors, such as climate, greography, or edaphic factors (Naeem 2002; Naeem et al. 2002). Intercropping studies (e.g., Vandermeer 1989; Vandermeer et al. 2002), the study of keystone species (e.g., Power et al. 1996, and studies of ecosystem engineers (e.g., Jones et al. 1994; Wright et al. 2002), had demonstrated that species could potentially have large impacts on ecosystem function (Statzner and Moss 2004), but these were viewed as local or secondary effects. Over the last decade, however, this view has changed. Theoretical and small-scale (e.g., $<10\ m^2$) experimental studies have introduced the idea that biodiversity can regulate abiotic factors through its control over ecosystem function, although whether these findings apply at larger scales ($>1\ km^2$) typical of natural systems remains unclear (Naeem 2002; Loreau et al. 2002; Hooper et al. 2005).

That biodiversity was seldom linked to ecosystem functioning prior to the 1990s is not surprising in light of the fact that the study of biodiversity and the study of ecosystem functioning have been, and continue to be, largely independent lines of ecological research (McIntosh 1985; Likens 1992; Schulze and Mooney 1993; Grimm 1995; Díaz and Cabido 2001). To link biodiversity with ecosystem functioning, in particular making the latter a dependent function of the former, required a synthesis that would bring together community and ecosystem ecology.

10.1.3 Lessons Learned from Early Debates

Like syntheses in most scientific disciplines (Kuhn 1962), considerable debate surrounded early findings (Naeem et al. 2002). Three important lessons emerged from this debate, arising mostly from sources of confusion. First, the idea that ecosystem function is controlled, at least in part, by biodiversity does not imply that taxonomic or species richness, in and of itself, somehow controls ecosystem function. Biodiversity refers to both taxonomic (e.g., species or genetic) *and* ecological (e.g., life history or functional) diversity over all spatial and temporal scales and of these components, it is functional diversity that matters. Functional diversity refers to the diversity of species traits rather than taxonomic diversity, where traits are those that either govern how a species responds to environmental change or how it effects ecosystem processes. For example, a grassland may have 200 species of herbaceous plant species, but with respect to nitrogen fixation, it may be considered to have only two functional types – legumes (N fixers) and non-legumes (see Chap. 7 and 13). Thus, biodiversity can control ecosystem function when changes in biodiversity involve changes in functional diversity (number and relative abundance of functional types) over space or time (Smith et al. 1997; Hooper and Vitousek 1998; Díaz and Cabido 2001; Lavorel and Garnier 2002; Petchey and Gaston 2002; Naeem and Wright 2003; Symstad and Tilman 2001). Change in taxonomic diversity affects ecosystem functioning only in so much as it correlates with change in functional diversity.

Second, because biodiversity and biomass are often related, in the sense that larger expanses of ecosystems often contain more individuals and more species than smaller expanses, there is an additional conflation of biodiversity loss with biomass loss. For example, removal of a forest by clear cutting will significantly alter watershed discharge, but it is not the loss of biodiversity *per se* that is responsible for this change in ecosystem function – it is the loss of vegetation, or more simply, the loss of biomass. If, however, one replaced a functionally diverse forest with an equivalent biomass of a monoculture plantation and watershed discharge was affected, then in this case, the change in biodiversity led to the change in ecosystem function.

Third, the associations between biodiversity and ecosystem functioning derived from large-scale observational studies are often different both in magnitude and sign from those derived from small-scale experimental studies. Many of the debates have concerned seeming conflicts between findings from within-ecosystem experiments and those of cross-ecosystem correlational studies (e.g., Levine 2000; Naeem 2001b). Such contrasts in findings are not surprising given that larger scale studies include variation in ecosystem function due not only to variation in biodiversity, but variation in many other factors, such as climate, nutrient supply rate, and history, that are held constant in experimental systems (Troumbis et al. 2000; Troumbis and Memtas 2000; Chase and Leibold 2002; Fridley 2003). The lesson learned is that if one wants to predict ecosystem response to change in biodiversity across large scales, one has to understand how other factors (controlled in experiments and uncontrolled in observational studies) affect both biodiversity and ecosystem function.

These sources of confusion that fueled the early debates have since become active research areas in the study of how biodiversity and ecosystem functioning are linked. Indeed, the framework we discuss below is an outcome of research in theses areas.

10.1.4 What We Have Learned about the Relationship between Biodiversity and Ecosystem Function

Although the history of research on biodiversity and ecosystem functioning (BEF) is one of point-counterpoint debate, a consensus has emerged (Loreau and Hector 2001) and syntheses and directions for future research identified (Loreau et al. 2002; Hooper et al. 2005). Numerous experiments and much theory have accumulated over the last decade and BEF research continues to be one of the most active areas in ecological research. Space does not permit reviewing this enormous literature, but much of it is covered in several reviews (Chapin III et al. 2000; Loreau et al. 2002; Hooper et al. 2005). Briefely, however, BEF research has demonstrated that

- change in biodiversity has significantly altered ecosystem functioning and the services it provides in many well-documented studies,
- species' functional characteristics strongly influence ecosystem functioning,
- the mechanisms and pathways by which biodiversity loss impacts ecosystem functioning vary among systems,
- species vary in their impact on ecosystem functioning from slight to relatively subtantial,
- and, whereas a small number of species may provide levels of ecosystem functioning similar to species-rich communities, more species are needed to insure continued provisioning of ecosystem function over longer periods of time (Hoppe et al. 2005).

For this chapter, it is important to note that while theory (e.g., Hughes et al. 2002; Loreau 2004; Petchey et al. 2004), grassland experiments (e.g., Fridley 2003; Foster and Dickson 2004; Zavaleta and Hulvey 2004; Roscher et al. 2005; Spehn et al. 2005), microcosm research (e.g., Fox 2004; Heemsbergen et al. 2004; Wohl et al. 2004), and marine/aquatic studies (e.g., Duffy et al. 2001; Emmerson et al. 2001; Stachowicz et al. 2002; Solan et al. 2004), have all contributed significantly to advancing the frontiers of BEF research, only a few have considered the implications of applying BEF to the scales typical of most ecosystems (e.g., Emmerson and Raffaelli 2000; Troumbis and Memtas 2000; Bai et al. 2004; Cardinale et al. 2004; Statzner and Moss 2004). The purpose of this chapter is to establish a framework by which BEF research can expand to larger scale investigations.

10.1.5 The Scientific Framework for Linking Biodiversity and Ecosystem Functioning

The initial BEF framework was developed over a decade ago, but has stimulated a tremendous amount of research. Vitousek and Hooper (1993) developed a graphical framework for linking biodiversity and ecosystem functioning that captured the heart of the synthesis, but was insufficiently detailed to avoid the controversy that would eventually emerge. They plotted biodiversity on the x-axis and ecosystem functioning on the y-axis then postulated several possible relationships between the two. As simple as this might seem, it represented a radical departure from traditional ecology. By plotting biodiversity on the x-axis, it portrayed biodiversity as the *independent* factor while portraying ecosystem functioning as the *dependent* factor, traditionally the y-axis. Almost always, in ecology, one plots biodiversity as a dependent function, usually of abiotic factors such as temperature, precipitation, or productivity (e.g., Gaston 2000). A wide variety of relationships between biodiversity and ecosystem functioning have been proposed in accordance with the Vitousek-Hooper framework (Martinez 1996; Schläpfer and Schmid 1999; Naeem 2002).

Graphical approaches such as the Vitousek-Hooper framework are heuristic devices that motivate and facilitate scientific discourse, but adapting such frameworks to the real world is seldom straightforward (Naeem et al. 2002). In particular, the most significant challenge to adapting abstract frameworks to the real world requires dealing with scale (Naeem 2001b; Symstad et al. 2003). Both biodiversity and ecosystem functioning are complex, scale-dependent terms that are not readily quantified by single variables and therefore not readily related to each other by simple bivariate plots. "Ecosystem function" refers to a virtually limitless set of processes associated with ecosystems ranging from global (e.g., global sequestration of atmospheric CO_2) to local processes (e.g., grassland productivity). Likewise, "biodiversity" refers to an equally limitless set of ecological measures that quantify biodiversity from large-scale (e.g., biome) to small-scale (e.g., genotypic variability within a population for the expression of a single gene) properties of ecological systems. Consider Fig. 10.1 in which the scales of biodiversity and ecosystem functioning are presented as a bivariate plot. Plotting ecosystem function against biodiversity across such a plot would yield an enigmatic figure that is almost impossible to interpret mechanistically even though conceptually it captures the central idea.

When experimental tests of the Vitousek-Hooper framework were conducted, in reality, only a small region of the bivariate plot shown in Fig. 10.1 was explored, but interpretations of these studies were seldom constrained to the region explored. Much of this experimental research crossed out "biodiversity" and wrote in "plant species richness," crossed out "ecosystem functioning" and wrote in "primary production," and then plotted data collected from combinatorial experiments whose units (pots and plots) were seldom more than a few meters on a side and seldom conducted for more than one or a few years (Naeem et al. 1994; Naeem and Li 1997; Hooper and Vitousek 1997; Tilman et al. 1996; Tilman et al. 1997;

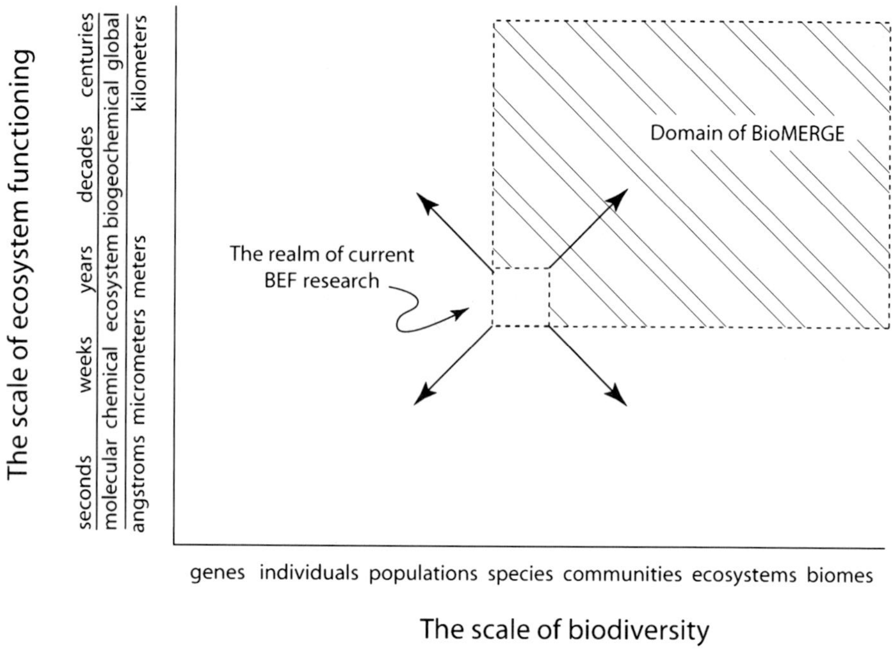

Fig. 10.1.
The realm of current research in Biodiversity and Ecosystem Functioning (BEF) and the realm of BioMERGE. Each *axis* abstractly and approximately represents the scales of biodiversity and ecosystem functioning, illustrating the latter as a dependent function of the former. The *center box* illustrates the average domain of current BEF research, which has focused on relating species diversity to ecosystem functioning at small spatial (in the realm of square meters) and temporal scales (in the realm of weeks to years). The *arrows* indicate directions of new research necessary for the expansion of BEF research. The *larger box* encompassing the *upper right quadrant* represents the domain of BioMERGE, which seeks to extend BEF research to larger scales

Wardle et al. 1997; Symstad et al. 1998; Hector et al. 1999; Mulder et al. 1999; Sankaran and McNaughton 1999; Spehn et al. 2000; Symstad 2000; Symstad and Tilman 2001; Engelhardt and Ritchie 2001; Hooper et al. 2002; Petchey et al. 2002; Schmid 2002; Schmid et al. 2002; Fukami and Morin 2003). Microcosm studies (e.g., Naeem and Li 1997; McGrady-Steed and Morin 2000; Petchey et al. 2002; Fukami and Morin 2003) were often conducted for longer periods of time in terms of numbers of generations, but they were still small in spatial scale (small with respect to the typical size of aquatic ecosystems, though in terms of organism size, the spatial scale was probably larger than most experimental studies).

In spite of the limited scope of this research, because a positive, saturating or asymptotic (or linear in a semilogarithmic plot) relationship between biodiversity and ecosystem functioning was often found (e.g., Naeem et al. 1994; Tilman et al. 1996; Hector et al. 1999), the claim was made that biodiversity enhances ecosystem functioning. This claim ignited a strong debate (Guterman 2000; Kaiser 2000). Even though the findings themselves were not questioned, the mechanisms were unclear, the use of species richness as a proxy for biodiversity and primary production as a proxy for ecosystem functioning seemed questionable, and the findings often appeared to be the inverse of what was known from correlative studies of biodiversity and ecosystem functioning conducted at larger scales (e.g., Wardle et al. 2000). Finally, these early experiments lacked considerations of functional diversity and trophic structure. Although later experiments would examine these issues, the initial lack of treatment of these issues offered only limited applicability of such findings to worlds that were functionally (not just taxonomically) diverse, multitrophic, and much larger in scale.

The new framework we discuss here brings biogeography into the picture, and by so doing, allows for the expansion of BEF research beyond its currently limited domain of small-scale studies to large-scale evaluations of ecosystems and environmental consequences of biodiversity loss (Fig. 10.1). Biogeography, broadly defined, is the study of past and present distributions of species and populations. As such, it encompasses many topics including evolutionary ecology, community assembly, species saturation, species area curves, ecogeographic rules, zonation, endemism, environmental heterogeneity, allometry and scaling rules, historical factors and paleoecology, and much more. In this framework we use the term *biogeography* in a narrow sense to specifically refer to geographic patterns of distribution and abundance, and we use the term macroecology (*sensu* Brown 1995) to refer more specifically to allometry and scaling rules, though the biogeography and macroecology are closely linked (Blackburn and Gaston 2003).

This chapter reviews the most recent conceptual framework for BEF research. The framework we discuss is being developed by a research coordinating network (RCN) whose acronym is BioMERGE (Biotic Mechanisms of Ecosystem Regulation in the Global Environment). For convenience, we simply refer to this framework as the BioMERGE framework.

It is important to note at the outset that *BioMERGE is not a research agenda*, such as those periodically established by the Ecological Society of America (Lubchenco et al. 1991; Palmer et al. 2004) or by international agencies such as GCTE or DIVERSITAS. Rather, it is a synthetic research framework that develops and evolves through participation by any and all interested individuals.

10.2 The BioMERGE Framework

The BioMERGE framework consists of three sub-frameworks. The first framework is structural, one that links species (taxonomic diversity), trophic groups, and functional diversity with ecosystem functioning. The second is an expansion of the Vitousek-Hooper BEF framework to include biogeographic principles. The third is a research implementation framework that serves as a guide for conducting an effective large-scale BEF research program. These three sub-frameworks reflect a summary of the previous decade of BEF research, identify existing bodies of ecological knowledge that can be brought into the BEF synthesis, and point to needed research and future directions that will enable us to better understand the ecosystem consequences of biodiversity loss.

10.2.1 The BioMERGE Structural Sub-Framework

The structural framework for BioMERGE outlines how species and ecosystems are linked and provides the ecological foundation for the conceptual and empirical

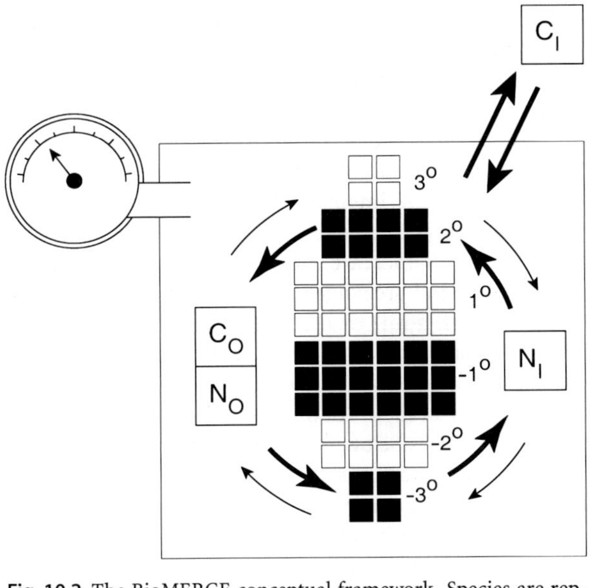

Fig. 10.2. The BioMERGE conceptual framework. Species are represented by *small squares* that are either autotrophs (here limited to photo-autotrophs or 1° producers), and either autotrophically derived heterotrophs (2° and 3° consumers) or microbial decomposers (–1°) or consumers of microbial decomposers (–2° and –3°). The autotrophs draw down inorganic carbon (C_I) (e.g., atmospheric CO_2 for terrestrial plants) and uptake inorganic nutrients to produce organic material that is distributed throughout the trophic groups of species. All contribute through mortality to a pool of dead organic material (C_O and N_O) that is decomposed into its inorganic constituents. There are many other routes (*thinner arrows*), but the overall pattern consists of internal cycling and carbon exchange that tracks energy acquisition through photosynthesis. Ecosystem functioning therefore involves any of the *broad arrows* while biodiversity represents taxonomic diversity, trophic structure, and functional diversity (not shown) within trophic groups

frameworks discussed below. This structural framework differs little from previous frameworks and serves only to focus attention on the key elements of BEF research. An ecosystem is considered to be a collection of species representing four trophic groups (autotrophic species, autotrophically-derived species, decomposer, or decomposer-derived species, see Fig. 10.2) that cycle material between organic and inorganic forms. Omnivory or other complex trophic interactions are acknowledged, but left out of the graphic for simplicity.

Graphically, this framework is illustrated by portraying an ecosystem as a box within which an assemblage of species with a specific trophic structure resides. Within this box, nutrients are represented as two organic (living species and dead organic matter) and one inorganic pool. The box (ecosystem) draws down and returns inorganic carbon (CO_2) to the atmosphere, but much of the nutrient cycling is considered to be confined to the ecosystem itself. To this system, a meter or measuring device that monitors ecosystem function (e.g., net primary production or total ecosystem respiration) is connected. This framework illustrates BEF research as the study of how ecosystem function, or the attached meter, responds to species loss. Traditional ecosystem ecology would not pay much attention to the species within the box, but would instead focus primarily on how changes in cycles and fluxes affected the meter, whereas traditional community ecology would pay little attention to the cycles, fluxes, or the meter associated with the system.

The main points of this structural framework are,

- it draws significant attention to the decomposer community, which is frequently ignored,
- traditionally defined trophic groups, such as primary consumers, herbivores, and carnivores, are not considered functional groups,
- ecosystem function refers only to organic-inorganic matter transformation (i.e., not invasion resistance or aesthetic properties).

10.2.2 The BioMERGE BEF Sub-Framework: an Expansion of the Vitousek-Hooper Framework

The Vitousek-Hooper BEF framework presented a number of alternative trajectories that ecosystem function could trace (e.g., linear, asymptotic, or logistic) as biodiversity declines. In contrast, the BioMERGE framework assumes an asymptotic trajectory that represents an average for a range of possible trajectories that ecosystem function could follow as biodiversity declines (Fig. 10.3). The asymptotic curve reflects the perspective that the loss of species does not, on average, dramatically impact ecosystem function until functional (F) and taxonomic diversity (S) are equal (i.e., species are singular rather

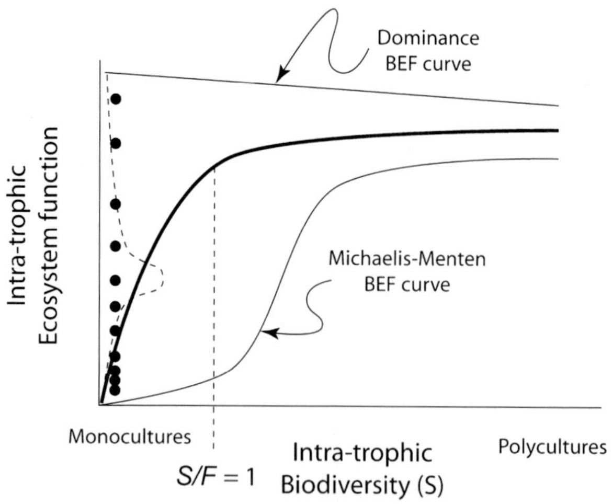

Fig. 10.3. The BioMERGE expansion of the Vitousek and Hooper framework. At the monoculture end, the majority of species (each represented as a *closed circle*) exhibit small levels of functioning whereas a few exhibit high levels following a general log-normal like pattern. The *x*-axis plots biodiversity as intra-trophic species richness (*S*), such as species richness of plants or species richness of herbivores. The maximum species richness (*right-most portion of curve*) represents an ecosystem in which species richness is stable and high. The asymptotic, *solid line*, represents the general BEF relationship based on the assumption that a significant portion of the species in an ecosystem are redundant with respect to specific ecosystem functions. That is, once at least one species is present per functional group ($S/F = 1$, where F = the number of functional groups), additional species have significantly smaller impacts on ecosystem functioning. The *dashed lines* indicate boundaries of combinatorial variants in species composition. The lower boundary is based on a Michaelis-Menten like model of ecosystem function, which assumes that biodiversity acts like a catalyst for biogeochemical processes. The upper boundary is based on a dominance model in which no species can achieve higher ecosystem functioning than that of the species with the dominant level of functioning among the monocultures. The *line* is illustrated as a slightly negative slope to allow for the fact that in multi-species assemblages, at least some of the resources are controlled by other species, no matter how rare they are. These boundaries represent the range of possible expressions of ecosystem functioning for a specific level of species richness. Note that many alternative boundaries are possible

than redundant)(Naeem 1998). The "elbow" in the curve therefore occurs when $S/F = 1$.

The rationale for using a range of trajectories rather than a single trajectory is to acknowledge the fact that different sequences of biodiversity loss will result in different patterns of change in ecosystem functioning. A community will trace a specific path through the diversity-by-function bivariate space for a specific sequence of biodiversity loss, but there are many different paths possible – that is, history, or the pattern of disassembly (or assembly) matters (Fukami and Morin 2003).

The multiplicity of pathways a trajectory can take in biodiversity-function space therefore describes two distinct, but important properties of biodiversity change. The first property is the combinatorial nature of change. A step towards decreasing (or increasing) biodiversity implies an enormous number of possible changes in community structure and composition. The second property is temporal contingency or the historical component of disassembly (or assembly). One can get to any community composition or structure from another by an enormous number of possible pathways and how one gets there affects ecosystem function. Consider, for example, that a BEF experiment that constructs replicate communities of 1, 8, and 16 species to determine the relationship between biodiversity and ecosystem function, there are 26 292 possible species combinations between a 16 species community and an 8 species community and 26 316 between an 8 species community and a monoculture. Thus, the problems are not simply the inability to replicate all 8 species communities, of which there are 12 870 different kinds, but the inability to know how the ecosystem functioning observed for 8 species communities might have been affected if they were the result of a staggering number of possible disassembly (or assembly) pathways. Obviously, for practical reasons, unless the community itself is very small (<10 spp.), experimental studies can only explore such an enormous number of possible communities in a very limited way.

The BioMERGE framework also considers that among the possible single species (or monoculture) treatments only a few species exhibit high levels of ecosystem function while the rest show lower levels (Fig. 10.3). That is, species with higher abundance (in terms of biomass) are assumed to make larger contributions to ecosystem functioning. This portion of the framework derives from the common, lognormal-like pattern of distribution and abundance observed at the large spatial scales more appropriate for communities than what has been typically employed in experimental research (Preston 1962; May 1975; Hughes 1986; Adler 2004). While debate continues concerning whether the canonical lognormal (Preston 1962), sequential breakage (Sugihara 1980), neutral (Bell 2000; Hubbell 2001) or other models best explain biogeographical patterns in relative abundance at larger scales, the fact remains that lognormal-like abundance patterns are common. An important part of this exercise, which we discuss below, is to use species traits to derive relative abundance. It is at this point that biogeography, in the form of incorporating patterns of commonness and rarity, enters the BioMERGE framework. There are many other aspects of biogeography that could be included, but they are beyond the scope of our chapter.

Using a lognormal-like distribution of abundance to describe the distribution of monoculture ecosystem functions leads to the null expectation that only a few species will dominate ecosystem function while many others (the rare species) will make lesser contributions. For example, in the case of plants, the null expectation would be that dominant species in polyculture will be the most productive species in monoculture and rare species in polyculture would have the smallest abundance as monocultures. In the absence of experimentation, of course, one cannot know for certain if a rare species in polyculture is rare because of suppression by competitive dominants, but could be quite abundant if released from suppression by extinction of the dominants. For example, while it is difficult to imagine that a rare orchid in a tropical rainforest would completely replace the biomass of the rainforest if it were the only species left, it is less clear whether any one of 900 species of trees in an Amazonian rainforest would replace the bulk of the biomass of that forest in the absence of the other 899.

The issues surrounding dominance, relative growth rates, and the relationship between monoculture and polyculture production are many and complex and have been a central part of the debates surrounding BEF literature (Garnier et al. 1997; Hodgson et al. 1998; Lepš et al. 2001; Loreau and Hector 2001; Adler and Bradford 2002; Schmid 2002). We will not review these issues here, but they do point to the need for a contemporary BEF framework to explicitly include dominance.

The main points of this expanded Vitousek-Hooper BEF framework are,

- an emphasis on a *range* of possible pathways ecosystem function can take in the face of biodiversity loss rather than a single pathway
- the inclusion of both functional and taxonomic diversity reflected by an average asymptotic trajectory, with the asymptote occurring where ecological redundancy is lost
- dominance (commonness and rarity) is integrated into the framework by the assumption of a lognormal-like distribution of function among the monocultures

Like previous BEF research, this framework assumes that environmental influences over community composition and structure (e.g., gradients, frequency of disturbance, seasonality, and so forth) are constant.

10.2.3 The BioMERGE Research Implementation Sub-Framework

The third BioMERGE sub-framework is one in which we have outlined the necessary steps in an evaluation of the ecosystem consequences of biodiversity loss at scales much larger than experimental work generally encompasses. There are three steps. First, establish an operational biota using what is known to set the stage and using ecological principles to estimate what is unknown (i.e., fill in the missing species, their abundance, and their function). Second, establish diversity-function transition matrices for estimating changes in ecosystem function given changes in diversity. Third, select an extinction scenario to determine the realm of possible expressions of ecosystem function that could occur during the course of biodiversity loss. These steps are detailed below.

Step 1: Establish an Operational Biota

Virtually no biota for any ecosystem is fully known nor is likely to be known in the near future if for no other reason than the fact that microbial diversity remains, in spite of many advances in molecular methods, still impossible to determine fully (Øvreås 2000; Torsvik et al. 2002). Given a location on Earth, however, one can establish an *operational* biota, or a biota derived inventories, surveys, and from ecological principles such as biogeographic theories of relative abundance solely for the purposes of BEF analyses. Like climate modeling to understand ecosystem or Earth System responses to elevated CO_2, understanding the range of responses of ecosystems or Earth Systems to biodiversity loss provides valuable insights into the potential consequences of such change. We propose several steps.

Step 1.1. Estimate species richness for major taxa in the community. Global patterns in biodiversity are not random so, at least in principle, there are some predictable elements for species richness and relative abundance given the geographical position of the ecosystem. For example, even in the absence of taxonomic information, such as floral and faunal surveys, we can estimate species richness for various taxa (plants and vertebrates) using the environmental correlates of species richness such as potential evapotranspiration, elevation, latitude, and so forth). This information provides an approximate magnitude for the biodiversity axis on the BEF curve.

Step 1.2. Establish relative abundance. This step requires employing a lognormal-like distribution, as discussed earlier, to derive the expected relative abundance for the number of species in the ecosystems. For example, to derive the relative abundance of an ecosystem containing S species (N_i, where N equals the maximum number of individuals of the ith species supported by the ecosystem) we draw S numbers from a log normal distribution with its shape determining coefficients (mean and variance) based on what is known for the region, what is known from similar regions, or from the estimated S from step 1.1.

To derive the relative biomass of species based on abundance we could use the allometric scaling formula,

$$M_i = 9\,191\, N_i^{-1.341}$$

(Enquist et al. 1998), where M_i is the biomass and N_i the density of the ith species. This is not necessarily a universal law, but it is a strong, empirically robust relationship that can be used to approximate relative biomass.

While use of such methods has long been a tradition in ecology (Horn 2004; Statzner and Moss 2004), they are not without its controversies (e.g., Cyr and Walker

2004; Li et al. 2004; Marquet et al. 2004; Tilman et al. 2004). For example, allometric coefficients from the metabolic theory of ecology (MTE) (Brown et al. 2004), such as that used for M_i above, are often derived across enormous taxonomic (e.g., cyanobacteria to giant redwood trees), spatial (across biomes), and temporal scales (data collected over decades). Not surprisingly, the predictive power of scaling rules based on MTE declines rapidly as one reduces the taxonomic, spatial, and temporal scales of investigation. Indeed, they may be practically useless for predicting function for a small system typical for most ecological research (Tilman et al. 2004). The BioMERGE approach is at an intermediate scale – providing predictions at the ecosystem level for multitrophic communities over moderate time scales – thus such caution in applying MTE is appropriate.

Step 1.3. Establish species relationships to ecosystem functioning. Step 1.1 and 1.2 provide species richness and relative abundances, but response- and effect-functional traits (*sensu* Lavorel and Garnier 2002) are needed to convert changes in the estimated biota to changes in ecosystem functioning. Response traits are species traits that govern ecosystem response to environmental change (e.g., drought tolerance traits), whereas effect traits are traits that govern change in ecosystem function as a consequence of change in species abundance or metabolic activities (e.g., hosting symbiotic N-fixing prokaryotes). The study of functional traits and functional diversity is a rapidly growing field (Smith et al. 1997; Hooper and Vitousek 1998; Díaz and Cabido 2001; Symstad and Tilman 2001; Lavorel and Garnier 2002; Petchey and Gaston 2002; Naeem and Wright 2003), but acquisition of this information is still difficult. Even in the absence of any information, an extensive and rapidly growing list of ecological patterns may be usable for providing, at least to a first approximation, trait information for estimating BEF relationships. For example, leaf, stem, and root biomass are strongly correlated with one another, often with regression r^2 values >0.80 (Enquist and Niklas 2002). This means one can estimate biomass allocations for species even if they have not been determined empirically.

Another example of the utility of allometric traits comes from Wright et al.(2004) who found that leaf mass per area, photosynthetic assimilation rates, leaf nitrogen, leaf phosphorus, dark respiration rate, and leaf lifespan all showed strong patterns of associations for plants from an enormous range of habitats covering most of the range of conditions experienced in terrestrial ecosystems. Once again, even if the only information available is from herbarium sheets, one might be able to estimate important plant traits such as those studied by Wright et al. (2004).

One final example is the estimation of biomass production (primary or secondary) from growth rates using the following allometric scaling formula:

$$G_i = M_i^{0.763}$$

(Niklas and Enquist 2002), where G is the growth rate of the ith species in kg dry matter yr^{-1}. There are many other examples for which physiological or functional traits of species may be estimated from density or biomass data.

While these examples are illustrative, we repeat a caution we raised earlier. The scales over which scaling coefficients are derived should match the scales over which predictions are being made. If they do not, then prediction errors associated with such mismatches (e.g., Tilman et al. 2004) should be stated.

Step 1.4. Establish patterns for trophic structure. Again, for many ecosystems, trophic structure is unlikely to be known, but even so, some efforts can be taken to include this important aspect of an ecosystem's biota. Given autotrophic biomass, one can use empirically derived relationships between the biomasses of different trophic groups or, in the absence of any empirical information, use approximate efficiencies of energy transfer across trophic groups. In many ecosystems, for example, carnivore diversity and even abundance are likely to be known, and Carbone and Gittleman (2002) have shown that about 10 000 kg of prey supports about 90 kg of a given species of carnivore for a wide range of species. The purpose of this step is to determine how much autotrophic biomass is being consumed by herbivores and how much herbivores are regulated by carnivores, an important aspect of ecosystems studied since the 1960s (e.g., Hairston et al. 1960; Polis and Strong 1996).

Below-ground food webs are much tougher to deal with, especially microbes, which may represent the dominant biomass and diversity in many ecosystems (Torsvik et al. 2002). Some studies argue that soil organisms occur in functionally redundant assemblages (Andren et al. 1995; Beare et al. 1995; Liiri et al. 2002), while others argue that redundancy may not be prevalent in stressed systems (e.g., Freckman and Virginia 1997) or in microbial communities (Cavigelli and Robertson 2000). Experimental studies argue for idiosyncratic responses or no response to biodiversity loss in soil communities (Mikola and Setälä 1998) while studies of mycorrhizal fungi argue the opposite (Van der Heijden et al. 1998). The linkages between plant diversity and soil communities and processes are also complex (Setälä and Huhta 1991; Harte and Kinzig 1993; Wardle and Nicholson 1996; De Mazancourt et al. 1998; Bardgett and Shine 1999; Naeem et al. 2000; Mikola et al. 2000; Spehn et al. 2000; Joffre and Ågren 2001; Mikola et al. 2001; Stephan et al. 2001; Mulder et al. 2003; Wardle et al. 2004). In the BioMERGE framework, we assume that the linkage is positive – that soil communities positively affect plants by mineralizing nutrients and that their diversity-function relationships follow the same pattern for other trophic levels (i.e., a positive, asymptotic BEF curve). This is an untested assumption with important consequences and requires considerable caution in interpreting results. One can relax this assumption by using several different BEF curves, but perhaps the most important point to make here is the need to find out if biodiversity in other trophic groups follows patterns similar to what has been observed for plants.

Step 2: Establish Diversity-Function Transition Matrices

A common approach to relating species to ecosystem functioning is to classify species into functional groups, but the wide variety and somewhat eclectic classification schemes make them difficult to use in BEF research (Lavorel and Garnier 2002; Naeem and Wright 2003). Rather than using functional groups, we propose directly using the functional traits of species that are the basis for forming functional groups in the first place (Lavorel and Garnier 2002; Naeem and Wright 2003).

Every organism possesses traits that characterize its response to environmental change and its effect on ecosystem processes. For example, in the face of a drought, plants with traits such as tap roots, persistent dormant seeds, thick cuticles, and greater water-use efficiency are more likely to persist than species that lack these traits. With respect to nitrogen cycling, plants with traits that facilitate microbial fixation of atmospheric nitrogen, either by harboring *Rhizobia* in nodules or supplying carbon to N-fixing bacteria in soils, impact soil N and N cycling in a variety of ways. If legumes lacked drought-tolerant traits in an ecosystem, then, based on response-traits, we could predict that a drought might negatively impact net N fixation.

To translate changes in biodiversity to changes in ecosystem functioning at large spatial scales requires four steps, each of which is illustrated in Fig. 10.4. First, a matrix of environmental variables and a matrix of species' abundances is constructed that reflects conditions prior to the application of a driver. Second, formulae are derived for adjusting each species' density or biomass according to its response functional trait. Third, using the effect functional traits, changes in ecosystem functioning are derived. Fourth, each trophic group is treated separately and the influence of one trophic group upon another is applied (see discussion above). These trophically adjusted, post-biotic-change abundances are then used to estimate what the resulting impacts on ecosystem function would be. Steps 2–4 yield the post biotic-change environmental matrix.

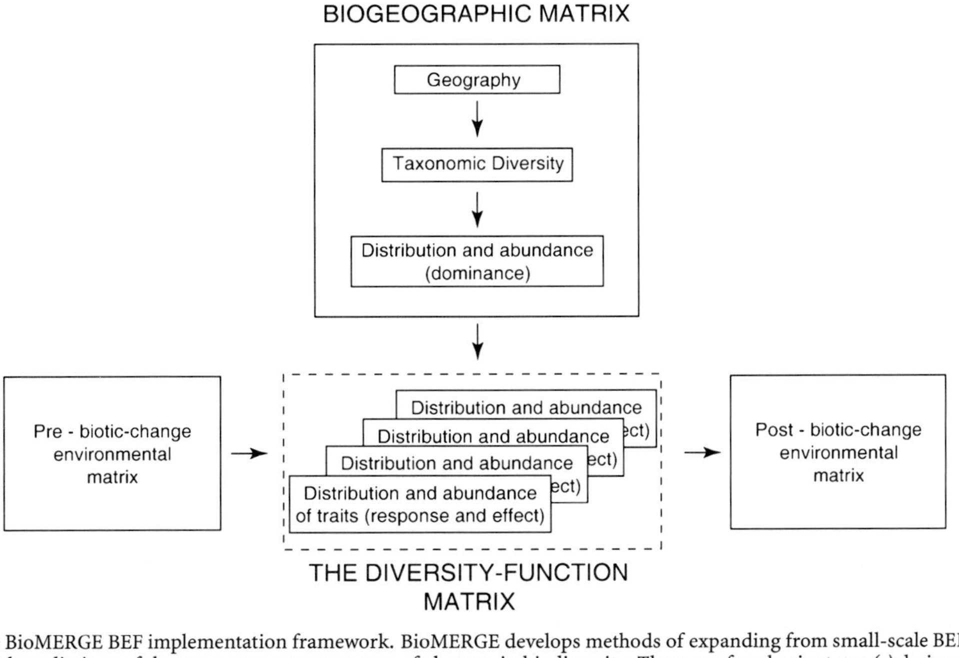

Fig. 10.4. The BioMERGE BEF implementation framework. BioMERGE develops methods of expanding from small-scale BEF research to landscape level predictions of the ecosystem consequences of changes in biodiversity. There are four basic steps; *(1)* designation or construction of a biota, *(2)* development of diversity-function transformation matrices, *(3)* application of driver scenarios for biodiversity change (most often local-extinction scenarios), and *(4)* estimation of the post-driver consequences of the selected driver(s). Constructing a biota (*top most box*) consists of using biodiversity information (i.e., species richness) developed for specific geographic locations. If distribution and abundance patterns (i.e., dominance patterns) are unavailable, then biogeographic principles are used to estimate likely patterns of distribution and abundance. Under the assumption that dominance patterns apply to all trophic levels, diversity-function matrices are developed (*boxes within dashed box on bottom*). These matrices consist of response- and effect-functional trait transformation matrices developed for each trophic group. The right most-box consists of the driver selected for investigation of ecosystem response to biodiversity change (e.g., drought, fire, increasing habitat fragmentation, elevated levels of N deposition, etc.) while the right-most box consists of the predictions for the outcome of the driver applied to the model system

Step 3: Select a Trait-Based Extinction Scenario for Change in Biodiversity

Once one has the model for converting changes in the environment to changes in ecosystem functioning (Fig. 10.4), one can explore how this model behaves when biodiversity is lost, but one has to chose a scenario for how biodiversity loss will occur. One can choose two routes for exploring biodiversity loss. First, as is done in experimental and theoretical BEF research, one can employ a neutral scenario, a combinatorial approach in which all possible compositions due to all possible sequences of species loss are examined. This neutral approach treats extinction probabilities for all species as a random variable.

The second is to employ trait-based extinction scenarios. By this we mean selecting an extinction driver (e.g., climate change, habitat fragmentation, or invasion) and varying extinction probabilities based on species' traits. For example, if the driver is habitat fragmentation, higher probabilities of extinction might be applied to rare species, species with small range sizes, species with small body size, or species in higher trophic levels. Extinction scenarios (based on expert opinion) were used by Sala et al. (2000) to derive predictions for the future of global biodiversity. This framework both extends and narrows this approach by using field-derived data rather than expert opinion on species' traits and drivers to predict specific changes in biodiversity for specific ecosystems rather than the globe. Note that it is at this point that environmental change, when treated as a driver, is incorporated into the analyses.

The main points of this implementation BEF framework are,

- instead of functional groups, response and effect traits are used
- biodiversity is as an explicit part of the conversion of changes in environmental factors to changes in ecosystem functioning through trait-based extinction scenarios

10.3 Discussion: Towards a Large Scale BEF

Initial work in biodiversity and ecosystem functioning focused on simply demonstrating that changes in biodiversity, however defined by researchers, could, in fact, change ecosystem functioning even if all other conditions were held constant. Confirmation of the hypothesis, formalized by the Vitousek-Hooper framework, was a critical first step in bringing biodiversity into our considerations of global change ecology. The next phase, to go beyond the confirmation of the Vitiousek-Hooper framework to one in which predictive applications of BEF principles can be employed in environmental problem solving, is even more critical. The BioMERGE framework is a suggestion for how to pursue this second phase. It does so by embedding much more ecology into its frameworks. By "more ecology," we mean that the BioMERGE framework

- uses patterns from biogeography (i.e., log-normal like projections) to estimate to estimate relative abundance,
- uses allometric scaling and macroecological principles to ascribe function to species,
- uses material and energy-efficiency transfer rates and trophic dynamics to estimate the influence of trophic structure on ecosystem function
- and uses trait-based extinction scenarios to link environmental change with changes in biodiversity.

Rather than supplying specific predictions of ecosystem response to biodiversity loss, it provides boundaries and central tendencies of ecosystem response to biodiversity loss. It also allows for different scenarios of biodiversity loss.

Like other synthetic frameworks, BioMERGE brings with it both the strengths and the limitations of the individual frameworks it incorporates. BioMERGE follows in the footsteps of many other synthetic frameworks, such as the Vitousek-Hooper framework, ecological stoichiometry (Elser and Sterner 2002), the unified neutral theory of biodiversity and biogeography (Hubbell 2001), the metabolic theory of ecology (Brown et al. 2004), island biogeography (MacArthur and Wilson 1967), each of which have tremendous predictive power, but whose limitations have made them each sources of much controversy. For example, in plant ecology, the continuum concept (Austin 1985) is a synthetic framework that combines autecology and biogeography to predict ecosystem function from biodiversity (Austin 1985, 1999). This framework, however, stands in conflict with frameworks that emphasize biotic interactions and system complexity (e.g., trophic-level dynamics, competition, facilitation and indirect and higher-order interactions) (Austin 1985; Callaway 1997). The inclusion of interactions in empirical work based on the continuum concept, however, is readily achieved (e.g., Garnier et al. 1997; Navas et al. 1999; Groves et al. 2003). The way forward is to embrace more complexity when necessary, but resist it when the gains in predictive power outweigh the costs of the added complexity. The BioMERGE framework goes beyond the continuum concept by the inclusion of intra- and intertrophic interactions, metabolic theory, and biodiversity and ecosystem functioning, thereby incorporating the strengths of the continuum concept and other frameworks but at the cost of increased complexity.

For the reader, and to be honest, for the authors as well, the BioMERGE frameworks require us to conduct research that seems at first glance to be broad, imprecise, and far afield from traditional ecology. We urge, however, that ecologists not be deterred. Consider that early attempts to model global climate and the global carbon cycle began with equally coarse approximations, but the desire as well as the need to understand global warming led to concerted efforts by researchers to model climate change as best they could. Over time, data, models, and methods improved, and while climate change research is still not a precise science, it provides valuable insights into the potentials for anthropogenic forcing of climate processes to lead to significant and detrimental climate change. Given this history for climate change research, certainly understanding "biodiversity change" warrants equal attention.

The experimental and theoretical BEF research spurred by the growth of global change ecology led to a robust set of results that strongly suggest that within-trophic level changes in diversity are associated with changes in the magnitude and variability of ecosystem processes (Loreau et al. 2002; Loreau and Hector 2001). To be sure, many questions remain unanswered. To what extent are experimental results due to facilitation, niche complementarity, or sampling? Which components of biodiversity (is it functional diversity, the identity of species, or dominance) are the most important in determining the impacts of biodiversity loss? Compared to abiotic controls, do biodiversity controls over ecosystem functions matter? Is stability (low variability in ecosystem function in high diversity treatments) due to insurance (Yachi and Loreau 1999), statistical averaging (Doak et al. 1998), or other causes? These questions are not readily addressable by the small-scale experimental approaches that have typified BEF research, but large-scale experiments are impractical. Perhaps answers to these questions may come using observational studies and the BioMERGE frameworks.

Lessons learned from the controversies surrounding early phases of BEF research warn against indiscriminate employment of the BioMERGE framework. The BioMERGE framework has been partially and successfully employed by Solan et al. (2004) to predict ecosystem function (specifically, biogenic mixing) in a marine (estuarine) ecosystem. Similar studies are under way for North American and European grasslands and New World tropical rainforests. Clearly, BEF is scaling up. Perhaps the next step should be to combine a traditional combinatorial BEF experiment with a BioMERGE study and compare their findings.

Acknowledgments

We thank Louis Pitelka and Pep Canadell for instigating this contribution and the members of BioMERGE who attended its second workshop and provided much input. We especially thank GCTE's fostering of an international research program on biodiversity and ecosystem functioning. Through GCTE's activities, in particular Pep Canadell, M. Loreau, P. Inchausti, L. Lebel, and many participants of BioMERGE, this National Science Foundation (USA) funded Research Coordinating Network was established. We thank P. Canadell, S. Tjossem, and two anonymous reviewers for critical evaluation of the manuscript. This work was supported by NSF DEB-04-35178.

References

Adler PB (2004) Neutral models fail to reproduce observed species-area relationships in Kansas grasslands. Ecology 85:1265–1272

Adler PB, Bradford JB (2002) Compensation: an alternative method for analyzing diversity-productivity experiments. Oikos 96:411–420

Andreae MO (2002) Humanity: Passenger or pilot on Spaceship Earth? In: Global Change Newsletter. International Geosphere Biosphere Program, Stockholm, pp 2–6

Andren O, Bengtsson J, Clarholm M (1995) Biodiversity and species redundancy among litter decomposers. In: Collins HP, Robertson GP, Klug MJ (eds) The significance and regualtion of soil biodiversity. Kluwer Academic Publishers, Amsterdam, Netherlands, pp 141–151

Austin MP (1982) Use of a relative physiological performance value in the prediction of performance in multi species mixtures from monoculture performance. Journal of Ecology 70:559–570

Austin MP (1985) Continuum concept, ordination methods, and niche theory. Annual Review of Ecology, Evolution, snd Systematics 16:39–61

Austin MP (1999) The potential contribution of vegetation ecology to biodiversity research. Ecography 22:465–484

Bai Y, Han X, Wu J, Chen Z, Li L (2004) Ecosystem stability and compensatory effects in the Inner Mongolia grassland. Nature 431:181–184

Balmford A, Bruner A, Cooper P, Costanza R, Farber S, Green RE, Jenkins M, Jefferiss P, Jessamy V, Madden J, Munro K, Myers N, Naeem S, Paavola J, Rayment M, Rosendo S Roughgarden J, Trumper K, Turner KR (2002) Economic reasons for conserving wild nature. Science 297:950–953

Bardgett RD, Shine A (1999) Linkages between plant litter diversity, soil microbial biomass and ecosystem function in temperate grasslands. Soil Biology and Biochemistry 31:317–321

Beare MH, Coleman DC, Crossley DA, Hendrix PF, Odum EP (1995) A hierarchical approach to evaluating the significance of soil biodiversity to biogeochemical cycling. Plant and Soil 170:5–122

Bell G (2000) The distribution of abundance in neutral communities. American Naturalist 155:606–617

Blackburn TM, Gaston KJ (eds) (2003) Macroecology: Concepts and Consequences. Blackwell Publishing, Malden, Massachusetts

Bolker BM, Pacala SW, Bazzaz FA, Canham CD, Levin SA (1995) Species diversity and ecosystem response to carbon dioxide fertizliation: conclusions form a temperate forest model. Global Change Biology 1:373–381

Brown JH (1995) Macroecology. University of Chicago Press, Chicago

Brown JH, Gillooly JF, Allen AP, Savage VM, West GB (2004a) Toward a metabolic theory of ecology. Ecology 85:1771–1789

Brown JH, Gillooly JE, Allen AP, Savage VM, West GB (2004b) Response to forum commentary on "toward a metabolic theory of ecology". Ecology 85:1818–1821

Callaway RM (1997) Positive interactions in plant communities and the individualistic-continuum concept. Oecologia 112:143–149

Carbone C, Gittleman JL (2002) A common rule for the scaling of carnivore density. Science 295:2273–2276

Cardinale BJ, Ives AR, Inchausti P (2004) Effects of species diversity on the primary productivity of ecosystems: extending our spatial and temporal scales of inference. Oikos 104:437–450

Cavigelli MA, Robertson GP (2000) The functional significance of denitrifier community composition in a terrestrial ecosystem. Ecology 81:1402–1414

Chapin III FS, Zavaleta ES, Eviner VT, Naylor RL, Vitousek PM, Reynolds HL, Hooper DU, Lavorel S, Sala OE, Hobbie SE, Mack MC, Díaz S (2000) Consequences of changing biodiversity. Nature 405:234–242

Chase JM, Leibold MA (2002) Spatial scale dictates the productivity-biodiversity relationship. Nature 416:427–430

Costanza R, R. d' Arge, Groot Rd, Farber S, Grasso M, B. Hannon, Limburg K, Naeem S, O'Neill RV, Paruelo J, Raskin RG, Sutton P, Belt. MVd (1997) The value of the world's ecosystem services and natural capital. Nature 357:253–260

Cyr H, Walker SC (2004) An illusion of mechanistic understanding. Ecology 85:1802–1804

Daily GC, Alexander S, Ehrlich PR, Goulder L, Lubchenco J, Matson PA, Mooney HA, Postel S, Schneider SH, Tilman D, Woodwell GM (1997) Ecosystem services: Benefits supplied to human societies by natural ecosystems. Issues in Ecology 2:1–18

De Mazancourt C, Loreau M, Abbadie L (1998) Grazing optimization and nutrient cycling: when do herbivores enhance plant production? Ecology 79:2242–2252

Díaz S, Cabido M (2001) Vive la différence: plant functional diversity matters to ecosystem processes. Trends in Ecology and Evoluition 16:646–655

Díaz S, Grime JP, Harris J, McPherson E (1993) Evidence of a feedback mechanism limiting plant response to elevated carbon dioxide. Nature 364:616–617

Doak DF, Bigger D, Harding-Smith E, Marvier MA, O'Malley R, Thomson D (1998) The statistical inevitability of stability-diversity relationships in community ecology. American Naturalist 151:264–276

Duffy JE, McDonald SK, Rhode JM, Parker JD (2001) Grazer diversity, functional redundancy, and productivity in seagress beds: An experimental test. Ecology 82:2417–2434

Duraiappah AK, Naeem S (2005) Synthesis Report on Biodiversity. M. Assessment, Island Press

Ehrlich PR (1988) The loss of biodiversity: causes and consequences. In: Wilson EO (ed) Biodiversity. National Academy Press, Washington, DC, pp 21–27

Elser JJ, Sterner R (2002) Ecological Stoichiometry: The Biology of Elements from Molecules to the Biosphere. Princeton University Press, Princeton

Emmerson MC, Raffaelli DG (2000) Detecting the effects of diversity on measures of ecosystem functioning: experimental design, null models and empirical observations. Oikos 91:195–203

Emmerson MC, Solan M, Emes C, Paterson DM, Raffaelli DG (2001) Consistent patterns and the idiosyncratic effects of biodiversity in marine ecosystems. Nature 411:73–77

Engelhardt KAM, Ritchie ME (2001) Effects of macrophyte species richness on wetland ecosystem functionng and services. Nature 411:687–689

Enquist BJ, Niklas KJ (2002) Global allocation rules for patterns of biomass partitioning in seed plants. Science 295:1517–1520

Enquist BJ, Brown JH, West GB (1998) Allometric scaling of plant energetics and population density. Nature 395:163–165

Foster BL, Dickson TL (2004) Grassland diversity and productivity: The interplay of resource availability and propagule pools. Ecology 85:1541–1547

Fox JW (2004) Effects of algal and herbivore diversity on the partitioning of biomass within and among trophic levels. Ecology 85:549–559

Freckman DW, Virginia RA (1997) Low-diversity antarctic soil nematode communities: distribution and response to disturbance. Ecology 78:363–369

Fridley JD (2003) Diversity effects on production in different light and fertility environments: an experiment with communities of annual plants. J Ecology 91:396–406

Fukami T, Morin PJ (2003) Productivity–biodiversity relationships depend on the history of community assembly. Nature 424:423–426

Garnier E, Navas M-L, Austin MP, Lilley JM, Gifford RM (1997) A problem for biodiversity-productivity studies: how to compare the productivity of multispecific plant mixtures to that of monocultures? Acta Oecologica 18:657–670

Gaston K (2000) Global patterns in biodiversity. Nature 405: 220–227

Grimm NB (1995) Why link species and ecosystems? A perspective from ecosystem ecology. In: Jones CG, Lawton JH (eds) Linking Species and Ecosystems. International Thomson Publishing, New York, pp 5–15

Groombridge B (ed) (1992) Global biodiversity. Chapman and Hall, London, United Kingdom

Groves RH, Austin MP, Kaye PE (2003) Competition between Australian native and introduced grasses along a nutrient gradient. Austral Ecology 28:491–498

Guterman L (2000) Have ecologists oversold biodiversity? Some scientists question experiments on how numerous species help ecosystems. The Chronicle of Higher Education 47:A24–26

Hairston NGS, Smith FE, Slobodkin LB (1960) Community structure, population control, and competition. American Naturalist 106:249–257

Harte J, Kinzig AP (1993) Mutualism and competition between plants and decomposers: implicatins for nutrient allocation in ecosystems. American Naturalist 141:829–846

Hawksworth DL (ed) (1995) Biodiversity measurement and estimation. Chapman and Hall, London

Hector A, Schmid B, Beierkuhnlein C, Caldeira MC, Diemer M, Dimitrakopoulos PG, Finn JA, Freitas H, Giller PS, Good J, Harris R, Högberg P, Huss-Danell K, Joshi J, Jumpponen A, Körner C, Leadley PW, Loreau M, Minns A, Mulder PH, Donovan GO, Otway SJ, Pereira JS, Prinz A, Read DJ, Scherer-Lorenzen M, Schulze MD, Siamantziouras ASD, Spehn EM, Terry AC, Troumbis AY, Woodward FI, Yachi S, Lawton JH (1999) Plant diversity and productivity experiments in European grasslands. Science 286:1123–1127

Heemsbergen DA, Berg MP, Loreau M, van Hal JR, Faber JH, Verhoef HA (2004) Biodiversity effects on soil processes explained by interspecific functional dissimilarity. Science 306:1019–1020

Heywood VH (ed) (1995) Global biodiversity assessment. UNEP and Cambridge University Press, Cambridge

Hodgson JG, Thompson K, Bogaard A, Wilson PJ (1998) Does biodiversity determine ecosystem function? The Ecotron experiment reconsidered. Functional Ecology 12:843–848

Hooper DU, Vitousek PM (1997) The effects of plant composition and diversity on ecosystem processes. Science 277:1302–1305

Hooper DU, Vitousek PM (1998) Effects of plant composition and diversity on nturient cycling. Ecological Monographs 68: 121–149

Hooper DU, Solan M, Symstad A, Díaz S, Gessner MO, Buchmann N, Degrange V, Grime P, Hulot F, Mermillod-Blondin F, Roy J, Spehn E, van Peer L (2002) Species diversity, functional diversity and ecosystem functioning. In: Loreau M, Naeem S, Inchausti P (eds) Biodiversity and Ecosystem Functioning. Oxford University Press, Oxford, pp 195–208

Hooper DU, Chapin III FS, Ewel JJ, Hector A, Inchausti P, Lavorel S, Lawton JH, Lodge DM, Loreau M, Naeem S, Schmid B, Setälä H, Symstad AJ, Vandermeer J, Wardle DA (2005) Effects of biodiversity on ecosystem functioning: a consensus of current knowledge and needs for future research. Ecological Applications 75:3–35

Horn HS (2004) Commentary on Brown et al.'s "toward a metabolic theory of ecology". Ecology 85:1816–1818

Hubbell SP (2001) The Unified Neutral Theory of Biodiversity and Biogeography. Princeton University Press, Princeton

Hughes RG (1986) Theories and models of species abundance. The American Naturalist 128:879–899

Hughes JB, Ives AR, Norberg J (2002) Do species interactios buffer environmental variation (in theory)? In: Loreau M, Naeem S, Inchausti P (eds) Biodiversity and Ecosystem Functioning: Synthesis and perspectives. Oxford University Press, Oxford, pp 92–101

Joffre R, Ågren GI (2001) From plant to soil: Litter production and decomposition. In: Roy J, Saugier B, Mooney HA (eds) Terrestral Productivity. Academic Press, San Diego, CA, USA, pp 83–99

Jones CG, Lawton JH, Shachak M (1994) Organisms as ecosystem engineers. Oikos 69:373–386

Kaiser J (2000) Rift over biodiversity divides ecologists. Science 289:1282–1283

Kuhn TS (1962) The structure of scientific revolutions. The University of Chicago Press, Chicago
Lavorel S, Garnier E (2002) Predicting changes in community composition and ecosystem functioning from plant traits: revisiting the Holy Grail. Functional Ecology 16:545–556
Lepš J, Brown VK, Díaz Len TA, Gormsen D, Hedlund K, Kailová J, Korthals GW, Mortimer SR, Rodriguez-Barrueco C, Roy J, Santa Regina I, van Dijk C,van der Putten WH (2001) Separating the chance effect from other diversity effects in the functioning of plant communities. Oikos 92:123–134
Levine JM (2000) Complex interactions in a streamside plant community. Ecology 81:3431–3444
Li B-a, Gorshkov VG, Makarieva AM (2004) Energy partitioning between different-sized organisms and ecosystem stability. Ecology 85:1811–1813
Liiri M, Setälä H, Haimi J, Pennanen T, Fritze H (2002) Relationship between soil microarthropod species diversity and plant growth does not change when the system is disturbed. Oikos 96:137–149
Likens GE (1992) The Ecosystem Approach: Its Use and Abuse. Ecology Institute, Oldendorf/Luhe, Germany
Loreau M (2004) Does functional redundancy exist? Oikos 104:606–611
Loreau M, Hector A (2001) Partitioning selection and complementarity in biodiversity experiments. Nature 412:72–76
Loreau M, Naeem S, Inchausti P, Bengtsson J, Grime JP, Hector A, Hooper DU, Huston MA, Raffaelli D, Schmid B, Tilman D, Wardle DA (2001) Biodiversity and ecosystem functioning: Current knowledge and future challenges. Science 294:806–808
Loreau M, Naeem S, Inchausti P (eds) (2002) Biodiversity and Ecosystem Functioning: Synthesis and Perspectives. Oxford University Press, Oxford
Lubchenco J, Olson AM, Brubacker LB, Carpenter SR, Holland MM, Hubbell SP, Levin SA, MacMahon JA, Matson PA, Melillo JM, Mooney HA, Pulliam HR, Real LA, Regal PJ, Risser PG (1991) The sustainable biosphere initiative: an ecological research agenda. Ecology 72:371–412
MacArthur RH, Wilson EO (1967) The theory of island biogeography. Princeton University Press, Princeton, N.J
Marquet PA, Labra FA, Maurer BA (2004) Metabolic ecology: Linking individuals to ecosystems. Ecology 85:1794–1796
Martinez ND (1996) Defining and measuring functional aspects of biodiversity. In: Gaston K (ed) Biodiversity: A biology of numbers and difference. Blackwell Science Ltd., Oxford, pp 114–148
May RM (1975) Patterns of species abundance and diversity. In: Cody ML, Diamond JM (eds) Ecology and evolution of communities. Belknap Press of Harvard University, Cambridge, MA, pp 81–120
McGrady-Steed J, Morin PJ (2000) Biodiversity, density compensation, and the dynamics of populations and functional groups. Ecology 81:361–373
McIntosh RP (1985) The background of ecology. Cambridge University Press, Cambridge, United Kingdom
Mikola J, Setälä H (1998) Relating species diversity to ecosystem functioning: mechanistic backgrounds and experimental approach with a decomposer food web. Oikos 83
Mikola J, Barker GM, Wardle DA (2000) Linking above-ground and below-ground effects in autotrophic microcosms: effects of shading and defoliation on plant and soil properties. Oikos 89:577–587
Mikola J, Yeates GW, Barker GM, Wardle DA, Bonnre KI (2001) Effects of defoliation intensity on soil food-web proerties in an experimental grassland community. Oikos 92:333–343
Mooney HA, Fuentes ER, Kronberg BI (eds) (1993) Earth System responses to global change: contrasts between North and South America. Academic Press, San Diego
Mulder CPH, Koricheva J, Huss-Danell K, Högberg P, Joshi J (1999) Insects affect relationships between plant species richness and ecosystem processes. Ecology Letters 2:237–246
Mulder C, Breure AM, Joosten JHJ (2003) Fungal functional diversity inferred along Ellenberg's abiotic gradients: Palynological evidence from different soil microbiota. Granna 42:55–64
Naeem S (1998) Species redundancy and ecosystem reliability. Conservation Biology 12:39–45
Naeem S (2001a) How changes in biodiversity may affect the provision of ecosystem services. In: Hollowell VC (ed) Managing Human Dominated Ecosystems. Missouri Botanical Garden Press, St. Louis, pp 3–33
Naeem S (2001b) Experimental validity and ecological scale as tools for evaluating research programs. In: Gardner RH, Kemp WM, Kennedy VS, Petersen JE (eds) Scaling relationships in experimental ecology. Columbia University Press, New York, pp 223–250
Naeem S (2002) Ecosystem consequences of biodiversity loss: The evolution of a paradigm. Ecology 83:1537–1552
Naeem S, Li S (1997) Biodiversity enhances ecosystem reliability. Nature 390:507–509
Naeem S, Wright JP (2003) Disentangling biodiversity effects on ecosystem functioning: Deriving solutions to a seemingly insurmountable problem. Ecology Letters 6:567–579
Naeem S, Thompson LJ, Lawler SP, Lawton JH, Woodfin RM (1994) Declining biodiversity can alter the performance of ecosystems. Nature 368:734–737
Naeem S, Hahn D, Schuurman G (2000) Producer-decomposer codependency modulates biodiversity effects. Nature 403: 762–764
Naeem S, Loreau M, Inchausti P (2002) Biodiversity and Ecosystem Functioning: The emergence of a synthetic ecological framework. In: Loreau M, Naeem S, Inchausti P (eds) Biodiversity and ecosystem functioning:synthesis and perspectives. Oxford University Press, Oxford, pp 3–11
Navas M-L, Garnier E, Austin MP, Gifford RM (1999) Effect of competition on the responses of grasses and legumes to elevated atmospheric CO_2 along a nitrogen gradient: differences between isolated plants, monocultures and multi-species mixtures. New Phytologist 143:323–331
Niklas KJ, Enquist BJ (2002) Invariant scaling relationships for interspecific plant biomass production rates and body size. Proceedings of the National Academy of Sciences of the United States of America 98:2922–2927
Øvreås L (2000) Population and community level approaches for analysing microbial diversity in natural environments. Ecology Letters 3:236–251
Palmer M, Bernhardt E, Chornesky E, Collins S, Dobson A, Duke C, Gold B, Jacobson R, Kingsland S, Kranz R, Mappin M, Martinez ML, Micheli F, Morse J, Pace M, Pascual M, Palumbi S, Reichman OJ, Simons A, Townsend A, Turner M (2004) Ecology for a Crowded Planet. Science 304:1251–1252
Petchey OL, Gaston K (2002) Functional diversity (FD), species richness and community composition. Ecology Letters 5: 402–411
Petchey OL, Morin PJ, Hulot F, Loreau M, McGrady-Steed J, Lacroix G, Naeem S (2002) Contributions of aquatic model systems to our understanding of biodiversity and ecosystem functioning. In: Loreau M, Naeem S, Inchausti P (eds). Oxford University Press, UK, Oxford, pp 127–138
Petchey OL, Downing AL, Mittelbach GG, Persson L, Steiner CF, Warren PH, Woodward G (2004) Species loss and the structure and functioning of multitrophic aquatic systems. Oikos 104:467–478
Polis GA, Strong DR (1996) Food web complexity and community dynamics. American Naturalist 147:813–846
Power ME (1996) Challenges in the quest for keystones. BioScience 46:609–620
Preston FW (1962) The canonical distribution of commonness and rarity. Ecology 43:185–215, 410–432
Raven PH (2002) Science, sustainability, and the human prospect. Science 297:954–958
Reich PB, Tilman D, Naeem S, Ellsworth DS, Knops J, Craine J, Wedin D, Trost J (2004) Species and functional group diversity independently influence biomass accumulation and its response to CO_2 and N. Proceedings of the National Academy of Sciences of the United States of America 101:10101–10106
Roscher C, Temperton VM, Scherer-Lorenzen M, Schmit M, Schumacher J, Schmid B, Buchmann N, Weisser WW, Schulze ED (2005) Overyielding in experimental grassland communities – irrespective of species pool or spatial scale. Ecol Letters 8:419–429

Sala OE, Chapin FS, III, Armesto JJ, Berlow E, Bloomfield J, Dirzo R, Huber-Sanwald E, Huenneke LF, Jackson RB, Kinzig A, Leemans R, Lodge DM, Mooney HA, Oesterheld M, Poff NL, Sykes MT, Walker BH, Walker M, Wall DH (2000) Global Biodiversity Scenarios for the Year 2100. Science 287:1770–1774

Sankaran M, McNaughton SJ (1999) Determinants of biodiversity regulate compositional stabitliy of communities. Nature 401: 691–693

Schläpfer F, Schmid B (1999) Ecosystem effects of biodiversity: A classification of hypotheses and exploration of empirical results. Ecological Applications 9:893–912

Schmid B (2002) The species richness-productivity controversy. Trends in Ecology and Evolution 17:113–114

Schmid B, Joshi J, Schläpfer F (2002) Empirical evidence for biodiversity-ecosystem functioning relationships. In: Kinzig A, Pacala SW, Tilman D (eds) The functional consequences of biodiversity. Princeton University Press, Princeton, New Jersey

Schulze ED, Mooney HA (eds) (1993) Biodiversity and Ecosystem Function. Springer-Verlag, New York

Setälä H, Huhta V (1991) Soil fauna increase *Betula pendula* growth: laboratory experiments with coniferous forest floor. Ecology 72:665–671

Smith TM, Shugart HH, Woodward FI (eds) (1997) Plant functional types. Cambridge University Press, Cambridge

Solan M, Cardinale BJ, Downing AL, Engelhardt KAM, Ruesink JL, Srivastava DS (2004) Extinction and ecosystem function in the marine benthos. Science 306:1177–1180

Soulé ME (1991) Conservation: tactics for a constant crisis. Science 253:744–750

Spehn EM, Joshi J, Schmid B, Alphei J, Körner C (2000) Plant diversity effects on soil heterotrophic activity in experimental grassland ecosystems. Plant and Soil 224:217–230

Spehn EM, Hector A, Joshi J, Scherer-Lorenzen M, Schmid B, Bazeley-White E, Beierkuhnlein C, Caldeira MC, Diemer M, Dimitrakopoulos PG, Finn JA, Freitas H, Giller PS, Good J, Harris R, Hogberg P, Huss-Danell K, Jumpponen A, Koricheva J, Leadley PW, Loreau M, Minns A, Mulder CPH, O'Donovan G, Otway SJ, Palmborg C, Pereira JS, Pfisterer AB, Prinz A, Read DJ, Schulze ED, Siamantziouras ASD, Terry AC, Troumbis AY, Woodward FI, Yachi S, Lawton JH (2005) Ecosystem effects of biodiversity manipulations in European grasslands. Ecological Monographs 75:37–63

Stachowicz JJ, Fried H, Osman RW, Whitlatch RB (2002) Biodiversity, invasion resistance, and marine ecosystem function. Reconciling pattern and process. Ecology 83:2575–2590

Statzner B, Moss B (2004) Linking ecological function, biodiversity and habitat: a mini-review focusing on older ecological literature. Basic & Applied Ecology 5:97–106

Stephan A, Meyer A, Schmid B (2001) Plant diversity affects culturable soil bacteria in experimental grassland communities. Journal of Ecology 88:988–998

Sugihara G (1980) Minimal community structure: an explanation of species abundance patterns. American Naturalist 116:770–787

Symstad AJ (2000) A test of the effects of functional group richness and composition on grassland invasibility. Ecology 81:99–109

Symstad AJ, Tilman D (2001) Diversity loss, recruitment limitation, and ecosystem functioning: Lessons learned from a removal experiment. Oikos 92:424–435

Symstad AJ, Tilman D, Wilson J, Knops J (1998) Species loss and ecosystem functioning: effects of species identity and community composition. Oikos 81:389–397

Symstad AJ, Chapin III FS, Wall DH, Gross KL, Huenneke LF, Mittelbach GG, Peters DPC, Tilman D (2003) Long-term and large-scale perspectives on the relationship between biodiversity and ecosystem functioning. Bioscience 53:89–98

Tilman D, Wedin D, Knops J (1996) Productivity and sustainability influenced by biodiversity in grassland ecosystems. Nature 379: 718–720

Tilman D, Knops J, Wedin D, Reich P, Ritchie M, Sieman E (1997) The influence of functional diversity and composition on ecosystem processes. Science 277:1300–1302

Tilman D, HillerRisLambers J, Harpole S, Dybzinski R, Fargione J, Clark C, Lehman C (2004) Does metabolic theory apply to community ecology? It's a matter of scale. Ecology 85:1797–1799

Torsvik V, Øvreås L, Thingstad TF (2002) Prokaryotic diversity – magnitude, dynamics, and controlling factors. Science 296: 1064–1066

Troumbis AY, Memtas D (2000) Observational evidence that diversity may increase productivity in Mediterranean shrublands. Oecologia 125:101–108

Troumbis AY, Dimitrakopoulous PG, Siamantzioura A-SD, Memtsas D (2000) Hidden diversity and productivity patterns in mixed Mediterranean grasslands. Oikos 90:549–559

Van der Heijden MGA, Klironomos JN, Ursic M, Moutoglis P, Streitwolf-Engel R, Boller T, Wiemken A, Sanders R (1998) Mycorrhizal fungal diversity determines plant biodiversity, ecosystem variability and productivity. Nature 396:69–72

Vandermeer J (1989) The ecology of intercropping. Cambridge University Press, Cambridge, U.K

Vandermeer J, Lawrence D, Symstad A, Hobbie S (2002) Effect of biodiversity on ecosystem functioning in managed ecosystems. In: Naeem S, Loreau M, Inchausti P (eds) Biodiversity and ecosystem functioning:synthesis and perspectives. Oxford University Press, Oxford, pp 221–233

Vitousek PM, Hooper DU (1993) Biological diversity and terrestrial ecosystem biogeochemistry. In: Schulze ED, Mooney HA (eds) Biodiversity and Ecosystem Function. Springer-Verlag, New York, pp 3–14

Walker B, Steffan W (eds) (1996) Global change and terrestrial ecosystems. Cambridge University Press, Cambridge, U.K

Wardle DA, Nicholson S (1996) Synergistic effects of grassland plant species on soil microbial biomass and activity: implications for ecosystem-level effects of enriched plant diversity. Functional Ecology 10:410–416

Wardle DA, Bonner KI, Nicholson KS (1997) Biodiversity and plant litter: experimental evidence which does not support the view that enhanced species richness improves ecosystem function. Oikos 79:247–258

Wardle DA, Huston MA, Grime JP, Berendse F, Garnier E, Laurenroth WK, Setälä H, Wilson SD (2000) Biodiversity and ecosystem function: an issue in ecology. Bulletin of the Ecological Society of America 81:235–239

Wardle DA, Bardgett RD, Klironomos JN, Setala H, van der Putten WH, Wall DH (2004) Ecological Linkages Between Aboveground and Belowground Biota. Science 304:1629–1633

Wilcove DS, Rothstein D, Dubow J, Philips A, Losos E (1998) Quantifying threats to imperiled species in the United States. Bioscience 48:607–615

Wilson EO (1988) The current state of biological diversity. In: Wilson EO (ed) Biodiversity. National Academy Press, Washington, D.C., pp 3–18

Wohl DL, Arora S, Gladstone JR (2004) Functional redundancy supports biodiversity and ecosystem function in a closed and constant environment. Ecology 85:1534–1540

Woodwell GM (1995) Biotic feedbacks from the warming of the Earth. In: Woodwell GM, Mackenzie FT (eds) Biotic Feedbacks in the Global Climate System. Oxford University Press, New York, pp 3–21

Wright J, Jones CG, Flecker AS (2002) An ecosystem engineer, the beaver, increases species richness at the landscape scale. Oecologia 132:96–101

Wright IJ, Reich PB, Westoby M, Ackerly DD, Baruch Z, Bongers F, Cavender-Bares J, Chapin FS, Cornelissen JHC, Diemer M, Flexas J, Garnier E, Groom PK, Gulias J, Hikosaka K, Lamont BB, Lee T, Lee W, Lusk C, Midgley JJ, Navas ML, Niinemets Ü, Oleksyn J, Osada N, Poorter H, Poot P, Prior L, Pyankov VI, Roumet C, Thomas SC, Tjoelker MG, Veneklaas E, Villar R (2004) The worldwide leaf economics spectrum. Nature 428:821–827

Yachi S, Loreau M (1999) Biodiversity and ecosystem functioning in a fluctuating environment: The insurance hypothesis. Proceedings of the National Academy of Sciences of the United States of America 96:1463–1468

Zavaleta ES, Hulvey KB (2004) Realistic species losses disproportionately reduce grassland resistance to biological invaders. Science 306:1175–1177

Zavaleta ES, Shaw MR, Chiariello NR, Mooney HA, Field CB (2003) Additive effects of simulated climate changes, elevated CO_2, and nitrogen deposition on grassland diversity. PNAS 100:7650–7654

Part C

Landscapes under Changing Disturbance Regimes

Chapter 11

Plant Species Migration as a Key Uncertainty in Predicting Future Impacts of Climate Change on Ecosystems: Progress and Challenges

Guy F. Midgley · Wilfried Thuiller · Steven I. Higgins

11.1 Introduction

The prediction of climate change impacts on ecosystems is a challenging problem that has been addressed by both modeling and experimental approaches. The many varied experimental (e.g., Beier et al. 2004) and observational approaches (e.g., Parmesan and Yohe 2003) to investigating ecosystem and species responses to climate change are not addressed here. Modeling approaches also vary widely in approach and scale, from local "patch" scale simulation of system processes and competitive interactions (Shugart et al. 1992; Solomon 1986) to global scale simulation of vegetation functional and structural types (Prentice et al. 1992; Woodward et al. 1995). At "intermediate" geographic scales of landscapes and regions one modeling approach taken has been species-based, with the unit of simulation being the species' geographic range (Guisan and Zimmermann 2000; Huntley et al. 1995; Peterson 2001). Simulation approaches at local and global scales mentioned above are strongly process-oriented. They ignore species range preferences evident at the intermediate landscape and regional scale, and by default assume that a global species or functional pool is available to colonize each site, should environmental conditions be suitable at that site. This has largely concealed a key uncertainty in predicting ecosystem responses to climate change – that which is governed by the underlying spatial rearrangements of species' geographic ranges.

It is widely accepted that climate change impacts on natural species and communities will affect the geographic ranges of species (Parmesan and Yohe 2003; Walther et al. 2002). Indeed, early signs of climate change may be identified by species range shifts (Walther et al. 2001), strongly supporting the idea that range shifts are likely to continue in a larger number of species, and enhancing the possibility for new species invasions (Mooney and Hobbs 2000). What are the implications for predicting vegetation change in response to climate change? Importantly, species in existing plant communities are unlikely to possess similar migration capabilities, and as a result of differential migration and persistence, the communities of the future are unlikely to maintain the composition of those of today with novel communities likely to emerge (Neilson et al. 2005; Pitelka et al. 1997). Clearly, predictions of changing ecosystem function need to account for the species-specific range shifts that may result, apart from the direct conservation need to identify which species are not likely to migrate, and may therefore suffer range reductions and the threat of extinction (Thomas et al. 2004; Thuiller 2004). Considerations of migration and geographic range shift therefore have two direct applications – to modellers of ecosystem function whose models are underpinned by the presence of species with particular functional traits as ecosystem components, and to conservation planners who need to assess threats to biodiversity (Hannah et al. 2002).

The migration capabilities of organisms depend fundamentally both on the distances offspring move from their parents (dispersal) and on the number of offspring a parent produces (Skellam 1951). A great variety of processes is responsible for moving diaspores (Ridley 1930), hence predicting the distances that diaspores move is no trivial task (Higgins et al. 2003; Nathan et al. 2003). Similarly, the diverse range of processes that determines the number of offspring that individuals produce means that forecasting rates of population increase is a complex task (Neubert and Caswell 2000). Research on predicting migration rates has more recently sought to understand the extent to which migration rates are demographically or dispersal limited (Clark et al. 2003).

It is clear that the structure and composition of the world's vegetation can be broadly correlated with climate regime, but it is far more complex to derive from this observation the mechanisms that may drive range shifts as climate changes. This is for three main reasons: First, it is not known what proportion of the worlds species have current geographic ranges which are directly controlled by climate and not by other determinants such as disturbance regime (Bond et al. 2003), human land-use (Hobbs 2000), or strong inter-species interactions (either mutualistic or antagonistic) (Davis et al. 1998). Such factors are at most only indirectly linked to climate. Second, even if we knew precisely how climate controlled species ranges, it would remain a substantial challenge to predict how ranges might change dynamically due to lags in adult mortality and range contractions, which might delay range responses as climate changes. Third, human impacts that have fragmented the landscape and altered populations of biological dispersal vectors (Janzen and Martin 1982) have introduced barriers and filters to the dispersal and establishment of propagules. Many of these factors will retard migration rates (Collingham et al. 1996; Higgins et al. 2003; Schwartz et al. 2001). Other human activities, such as human mediated dispersal (Hodkinson and Thompson 1997) may accelerate spread rates. Plant species migration therefore represents a major uncertainty in the prediction of vegetation response to climate change (Higgins et al. 2003).

In spite of the problems identified above, there has been significant progress in modeling plant species' ability to migrate, driven by the need to project threats both of invasions by alien species and climate change to biodiversity (Clark et al. 2003). In this chapter, we review this progress in broad terms, and identify some key challenges and opportunities that remain.

11.2 Will Migration Be Necessary for Species Persistence?

Four broad responses may follow when populations of sessile organisms are confronted with a change in environmental conditions that compromise their physiological performance, namely local extinction, *in situ* persistence, *in situ* adaptation over generations (also termed an orthoselective response), and migration (Barnosky 1987). These responses are not all mutually exclusive, but may occur concurrently. For example, genetic patterns documented in several tree species reveal the synergistic effect of apparently recent (post-glacial) concurrent migration and selection in spatially separated populations (Cwynar et al. 1987; Davis and Shaw 2001).

In situ persistence is essentially a first line of defense for sessile organisms that can be accomplished either through high levels of tolerance and a long life span of the vegetative stage, or longevity and dormancy of propagules. Many species of extreme environments have evolved significant abilities to persist under adverse or unpredictable conditions, such as the propagules of desert annuals (van Rheede van Oudtshoorn and van Rooyen 1999), or long-lived trees (Bond and Midgley 2001). Persistence may allow many species of extreme environments to maintain a presence in the landscape long after suitable climate conditions have changed. Indeed, it now seems as though populations situated at the so-called "rear edge" of a species' geographic range undergoing a spatial shift are particularly important for the persistence of genetic diversity and evolution of species (Hampe and Petit 2005).

Genetic selection and adaptation is possibly an underappreciated control of species response to climate change (Davis and Shaw 2001), bearing in mind that rapid genetic change via selection under future anthropogenic climate change is possibly only relevant for the shortest-lived plant species. Nonetheless, selection by past climate change may determine current-day spatial variation in physiological optima of broad-ranging species in ways which affect predictions of species responses to climate change. For example, the response of the evergreen gymnosperm *Pinus contorta* to climate change and atmospheric CO_2 rise reverses from being positive throughout its range (if a single optimum genotype is assumed to inhabit the entire species range), to becoming slightly to strongly negative given the current distribution of varying genotypes (Rehfeldt et al. 1999). Nowak et al. (1994) illustrate a range of migration and adaptive responses to Pleistocene-Holocene change in woody plant species of the northwestern Great Basin, and demonstrate the clear existence of "orthoselective" species. These are species that possess high levels of genetic variation that permit their *in situ* persistence during periods of climate change through local selection of pre-adapted genotypes. The factors that determine adaptive plasticity are likely complex and species-specific, and it is too early to be able to generalize about how this response will emerge as anthropogenic climate change continues, but the topic deserves more attention in the context of species persistence.

There is currently no analysis available of what proportion of local, regional or global floras may persist or adapt to climate change *in situ*, but it is widely recognised that migration to keep pace with suitable climate conditions is a necessary response in organisms that lack extreme stress tolerance, longevity or genetic plasticity. Indeed, the rate and amount of climate change may ultimately exceed the capacity of either of these *in situ* responses. Studies of paleo-records strongly suggest that migration has been a widespread response to climate change in Pleistocene floras (e.g., Davis 1976; Davis 1983; Huntley 1990; Huntley and Birks 1983; Webb 1981; Webb 1992) and faunas (Graham 1992) and early monitoring efforts have identified incipient range shifts already occurring in response to recent climatic changes (Parmesan et al. 1999; Parmesan and Yohe 2003).

Spatially distinct genetic patterns in lodgepole pine in western Canada have been ascribed to migration accompanied by selection during post-glacial times (Cwynar et al. 1987), and Davis and Shaw (2001) illustrate several examples of concurrent genetic adaptation and migration, most conclusively for Scots Pine which has migrated extensively in northern Europe since the Last Glacial Maximum, and now shows physiological responses to climate cues that vary depending on the geographic provenance of the species. Thus, a combination of migration and concurrent selection has interacted to allow this species to occupy a range that is potentially wider than if the species had remained genetically homogenous across its range.

It is also apparent that the relative migration capabilities of species are reflected in the global patterns of biodiversity rich regions (e.g., Myers et al. 2000). It seems likely that Pleistocene climatic oscillations between glacial and interglacial climatic conditions have repeatedly sifted local biota according to their migration capabilities, and those regions which were exposed to relatively low rates and amounts of climate change have retained higher numbers of species as endemics (Jansson 2003). This pattern strongly suggests that migration has been a key determinant of the current global distribution of biodiversity, and that high rates of migration across the board are unlikely. Indeed, in the mid-latitude western Cape region of southern Africa, the key characteristics of an endemic include the likelihood of its being ant-dispersed (McDonald and Cowling 1995).

Given expected rates of anthropogenic climate change, what migration rates are required to allow species or species assemblages to keep pace? Two approaches have been taken to address this question. The first is derived from modeling global vegetation types using Dynamic Global Vegetation Models (DGVMs), and the second from modeling individual species ranges.

11.2.1 Vegetation-Type Models

Enhanced DGVMs represent one of the key products of the GCTE project (Cramer et al. 2001), and were designed to model ecosystem processes as determined by interacting vegetation and biogeochemical cycles (e.g., Bachelet et al. 2003; Box 1996; Foley et al. 1998; Holoyac 2000; Kirilenko et al. 2000; Sitch et al. 2003; Smith et al. 2001). Because of their complexity and the resultant computational demands in running them, DGVMs that simulate global vegetation structure and function commonly run at coarse spatial scales – typically more than 2 × 2 degrees of latitude and longitude (although they may be run at a rather fine temporal scale of days to hours and even minutes for some key processes such as photosynthesis). Given that propagule dispersal by plants only approaches distances appropriate to this coarse scale under extremely rare circumstances, DGVMs were clearly never designed to estimate migration requirements under climate change scenarios, but nonetheless have been used to do so (e.g., Malcolm et al. 2002).

Analyses of the required migration rate for ten biome types (representing between 60 and 80% of the Earth's surface) simulated by two vegetation models was estimated to be predominantly in the <300 m yr^{-1} range (Malcolm et al. 2002). For a minority of biome types (representing between 10 and 20% of the Earth's surface) required migration rates of between 1 000 and 10 000 m yr^{-1} were estimated (Fig. 11.1), not exceeding by far rates of migration recorded in the paleo-record (e.g., MacDonald 1993). The lowest required migration rates were found in tropical latitudes, but this increased strongly at latitudes of 40° and higher, such that temperate and circumpolar biomes had higher required rates than did tropical and subtropical biomes.

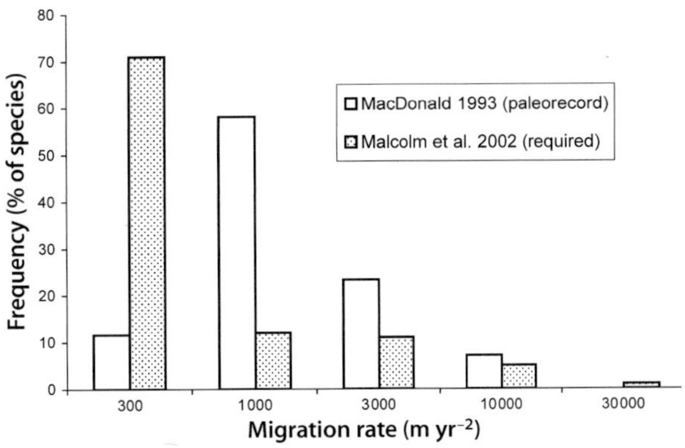

Fig. 11.1. Frequency distribution of post-glacial migration rates (from MacDonald 1993) in relation to required migration rates (from Malcolm et al. 2002). *x*-axis labels represent upper bounds of the migration rate categories

The concept behind Malcolm et al.'s (2002) analysis has been justifiably criticised (Loehle and LeBlanc 1996). Indeed, plant assemblages have never been stable through past variations in climate (Prentice 1986; Webb 1981). Many modern species assemblages do not appear to have long histories (Birks 1993; Davis 1983) and therefore communities or biomes are thought unlikely to move as an entity under changing environment conditions (Birks 1986; Huntley and Webb 1988).

Given the limitation of modeling vegetation types or biomes as whole entities, we analysed, for the purposes of this chapter, the required migration rates of individual plant functional types that define the world's biomes in the Sheffield DGVM (SDGVM, Woodward et al. 1995). In this DGVM, the geographic limits of individual functional types are defined by their physiological tolerances and performance relative to competing functional types. Biomes are then mapped using the relative composition of functional types. This approach is more justifiable than that of Malcolm et al. (2002) for a number of reasons. Most important is the fact that biomes may share functional types (and even species) with adjacent biomes, and the Malcolm et al. (2002) approach therefore overestimates required migration rates of shared functional types.

Our results for SDGVM are presented at a coarse spatial scale (Fig. 11.2) but nevertheless show that all functional types modeled show some migration to new ranges (Table 11.1), and that the most significant required migration rates are for temperate, subtropical and tropical functional types (C_3 shrub/grass, C_4 grass and deciduous broad-leaved tree functional types). Required migration rates are evenly spread with latitude, in contrast to the strong latitudinal patterning predicted by (Malcolm et al. 2002). This lack of latitudinal trend in our analysis is due partly to modeling relatively few functional types, which increases the likelihood of types being shared by

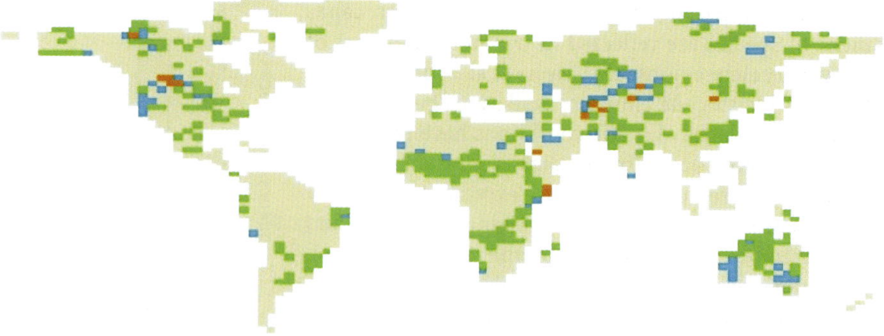

Fig. 11.2.
Global map of 3.75 × 2.5° pixels which experience the arrival of a novel plant functional type as projected by the SDGVM (Woodward 1995) using a climate change scenario for ~2050 according to the GCM HadCM3 (different *colours* represent the number of novel PFTs migrating into a cell, *green* = 1 PFT, *blue* = 2 PFT, *red* = 3 PFT)

Table 11.1.
Migration rates of 6 major plant functional types in terms of numbers of 2.5° pixels (median shift, mean shift and standard error), as projected by the SDGVM (Woodward 1995) using a climate change scenario for ~2050 according to the GCM HadCM3

PFT	C3	C4	Dcbl	Evbl	Dcnl	Evnl
# pixels	251	56	29	67	55	15
Median shift	4.5	3.75	3.75	2.5	2.5	2.5
Mean shift	8.33	4.50	4.93	3.47	3.77	4.37
Std error	0.47	0.37	0.68	0.20	0.29	0.81

adjacent geographic pixels. As such, these results are not directly comparable with those of (Malcolm et al. 2002). However, despite differences in approach, both analyses underline the central message that biogeochemically-based modeling approaches of vegetation predict a substantial requirement for migration of biomes and functional types to new geographic ranges.

The need to incorporate dispersal limitations into DGVM forecasts has been recognised for some time (e.g., Pitelka et al. 1997), but the challenges are substantial. A recent treatment (Neilson et al. 2005) has reiterated that, apart from the challenge of simulating dispersal realistically at such coarse spatial scale, it is difficult to assign a distribution of potential migration rates to plant functional types. This is because plant functional types group species based on functional or vegetative attributes (Smith et al. 1997), and there are not yet well known relationships between these attributes and attributes of fecundity and dispersal. Several issues need to be considered when trying to generalize species-specific knowledge about dispersal and fecundity. For example, how does dispersal ability relate to traits that define either the response to environmental factors or ecosystem functioning (Lavorel and Garnier 2002), and on what basis should functional types be split to distinguish between poor and good dispersers?

It is clear that by incorporating greater scientific understanding of plant functional types and a necessary increase in computational power, future DGVMs will resolve functional types more finely, and operate at finer spatial scale. It may, therefore, become feasible to assign dispersal and fecundity characteristics to functional types that more closely represent groups of species and even individual species, and model their migration more credibly. This step will be critical to allow DGVMs to account for lags in ecosystem response to climate change induced by migration limitations (Neilson et al. 2005), which currently introduce is ignored as an uncertainty in DGVM simulations.

11.2.2 Species-Based Models

Species-based approaches (also called niche-based models) have received increasing interest in the past decade because of their apparent ability to project the potential geographic range responses of multiple species, finally allowing some quantification of species' range shifts in response to climate change. Several recent studies have applied this class of models to generate and estimate risk of species extinctions (Thomas et al. 2004), as well as species' range shifts, species turnover, and variation in species richness (Iverson and Prasad 2002; Bakkenes et al. 2002; Thuiller 2003; Schwartz et al. 2001; Peterson 2003), under global climate change. Recent steps have included the use of species-based models to select areas for species persistence (Araújo and Williams 2000) and to assess the persistence of species in existing reserve networks in the face climate change (Araújo et al. 2004).

Despite considerable research on statistical models of species-climate relationships and their application to climate change studies, the uncertainties in forecasts of such models arising from ecological and methodological considerations have not so far been estimated. Species-based approaches have primarily been derived by developing statistical relationships between known species ranges and environmental variables taken as surrogates for physiologically relevant variables (Austin et al. 1994; Austin and Smith 1989; Guisan and Zimmermann 2000). The approach relies heavily on the concept of Hutchinson's realized niche, recently clarified by Leibold (1995) and Pulliam (2000). In this regard, the most frequent over-simplification found in the literature is the statement that, due to the observed distributions being constrained by biotic interactions, species-based approaches *de facto* quantify the realized niche of species, and never the fundamental niche (Pulliam 2000). As a direct consequence, projecting these models into climatically changed future conditions is, at least according to theory, likely to generate mistakes (Davis et al. 1998), but see Pearson and Dawson (2003) and reply from Hampe (2004).

The degree of error in predictions of niche-based models should thus be related in some way to the competitive status of the species for a given resource (weak vs. strong competitor), which in turn should allow the prediction of whether the species occupies its full fundamental niche or only part of it. Recent analyses have shown that it may be possible to include inter-specific competition in these models (Leathwick and Austin 2001; Anderson et al. 2002) but the competitive status of species can also be expected to change with a changing climate (Davis et al. 1998; Hughes 2000).

A further source of uncertainty is due to the fact that these models are static in space and time and are conceptually unable to cope with non-equilibrium situations, since they do not distinguish between the transient and equilibrium responses of species to a stochastically and dynamically changing environment (Guisan and Zimmermann 2000). As a direct consequence, they cannot theoretically include migration in a dynamic way. To cope with this limitation, recent risk assessment analyses for future climate change scenarios have commonly used two crude assumptions about migration, namely no migration beyond a species present site ("zero migration" or "null migration") and perfect migration to all sites, or "full migration" (Peterson et al. 2002; Thomas et al. 2004; Thuiller 2004). These approximations bracket the most pessimistic and optimistic estimates of future species range size, but have so far incorporated only the simplest migration limitations (i.e., migration into contiguous elements of the modeled landscape (Peterson et al. 2002)) to refine these estimates. The projected impacts of climate change on species turnover (an index of community compositional change and probably linked with ecosystem function) are very strongly influenced by the contrasting migration assumptions of zero and universal migration (Fig. 11.3), and this clearly demonstrates the current limitation in forecasting climate change impacts on biodiversity, and ecosystem structure and function, induced by uncertainties in migration rate.

11.3 Measurements and Models of Migration Rates

Much of our understanding of plant migration is derived from paleo-botanical studies of northern hemisphere regions (Davis 1976; Huntley and Birks 1983; Webb 1992). These studies indicated that several tree genera appeared to migrate rapidly from equatorward "refugial" populations of the Last Glacial Maximum (LGM), some 20 000 years ago, to occupy their current ranges several degrees of latitude poleward. Although some of these

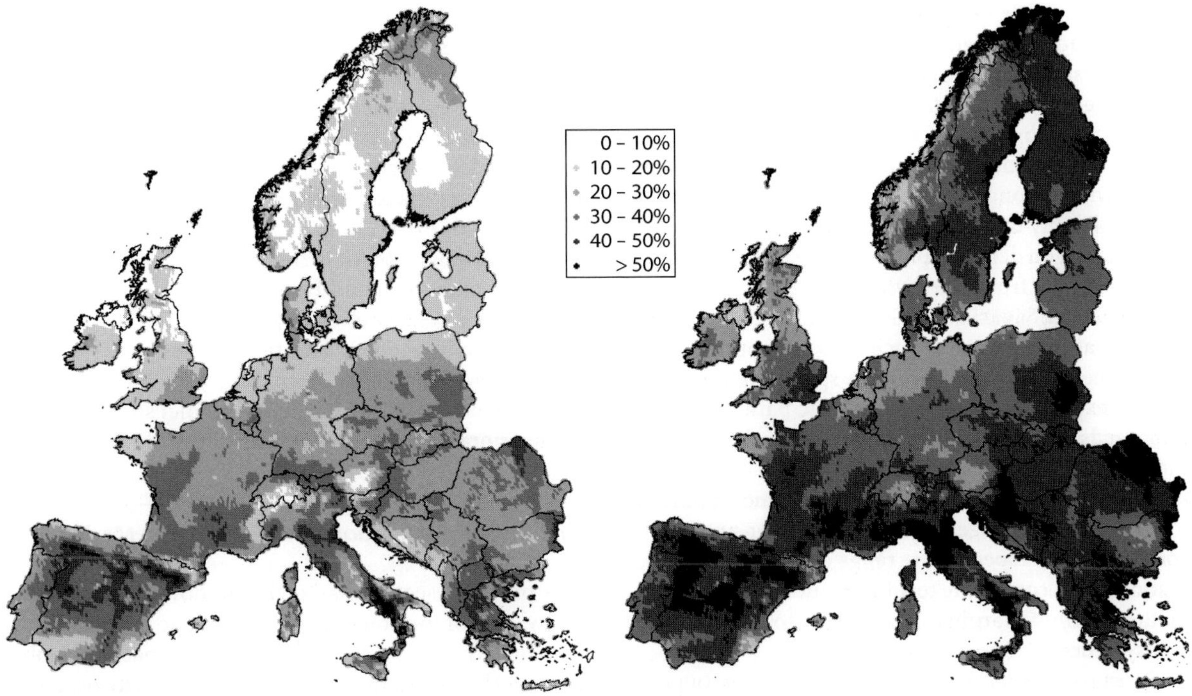

Fig. 11.3. Rate of plant species turnover in Europe for a climate scenario simulated by the GCM HadCM3 under the A1 SRES assumptions (Nakicenovic and Swart 2000), assuming either no species dispersal (*left hand panel*), or universal dispersal (*right hand panel*)

paleo-botanical studies have recently been questioned (e.g., McClachlan and Clark 2004), the balance of evidence suggests that rapid spread (>100 m yr^{-1}) is possible, at least for some species. This apparent empirical evidence for rapid migration cannot be reconciled with models of plant spread that assumed that dispersal should be approximated as a diffusion process (Skellam 1951). This conundrum (termed "Reid's paradox" by Clark et al. 1998) was resolved by re-examining the simple diffusion assumption that lies at the heart of classical models of spread. Relaxing the assumption that the distribution of seed dispersal distances is Gaussian was found to be consistent with empirical data on the distribution of seed dispersal distances. Moreover, the resulting spread rates were consistent with the rates estimated from the paleo-ecological evidence (Clark 1998; Higgins and Richardson 1999). Hence a new class of spread models have emerged that essentially all treat dispersal as a stratified process (Shigesada et al. 1995). That is, they accommodate the possibility that most diaspores move relatively short distances and that a few diaspores move relatively long distances.

The "resolution" of Reid's paradox has shifted attention away from dispersal and focused it on the other component of spread models, namely the demographic component (Clark et al. 2003). Our empirical knowledge of demographic rates suggest that most populations are neither increasing nor decreasing, that is they have population growth rates close to zero (Franco and Silvertown 2004). A population with a zero growth rate cannot spread, hence most populations for which we have empirical data on demographic rates would be predicted not to spread. However, growth rates could become positive as a result of disturbances caused by climate change, due to the opening up of new geographic areas suitable for occupation by species. Very little is known about how rapidly population growth at the margins of species' ranges will respond to an "improving" climate, and the new availability of space in adjacent areas.

Most data on demographic rates are, for practical reasons, collected from sites where the study species are common (e.g., Sagarin and Gaines 2002), but the demographic rates estimated at such sites are not representative of the conditions to which a spreading population may be exposed. Ideally demographic parameter variation needs to be estimated from the so-called "core" to the edge of species ranges – seldom done, such an approach may yield useful insights into demographic factors controlling range limits (e.g., Brewer and Gaston 2003). It seems especially at range edges that this information could contribute to a fuller understanding of the vulnerability of marginal populations to temporal variability (Vucetich and Waite 2003).

Spreading alien invasive species provide an opportunity to estimate the demographic rates that are appropriate for spread models. For instance, Higgins (2001) estimated demographic and dispersal parameters for spreading invasive plants. While much can be learned from the spread of invasive species about the potential spread of native species under climate change, several factors mean that the knowledge gained is not directly transferable. Notably, spreading alien species are often demographic super-organisms because of predator and competitor release (Keane and Crawley 2002). This suggests that to forecast the spread rate of native species under climate change we need to estimate the likelihood that predators and competitors will accompany the target species, which is clearly a complex task. Nonetheless, studies of invasive plants do reveal the complexities of predicting whether a species that appears to be physiologically suited to a site will indeed invade that site. This subject has both entertained and frustrated invasion biologists for decades, and although significant progress has been made, experience from invasions suggests that our predictive power remains modest (Rejmanek et al. 2004). Studies of spreading invasives also inform us that alien species will migrate rapidly; simple calculations based on the differences in demographic rates between native and alien species with similar life histories suggest that aliens will indeed migrate orders of magnitude faster than native species (Richardson et al. 2000).

A great difference for plants between past episodes of climate change and the current anthropogenic warming is the human alteration of landscapes, involving fragmentation of pristine habitat into patches of varying sizes and connectivity, and the creation of barriers to dispersal. Theoretical models of spread in landscapes influenced by fragmentation and habitat loss show that there may be critical thresholds in landscape connectivity beyond which migration processes cease (Higgins et al. 2003), but the data needed to test these models are far from complete, and will be very challenging to collect. More detailed models of spread that include occasional long distance dispersal have demonstrated that although thresholds may not exist, linear decreases in spread rate with increasing fragmentation and habitat loss are to be expected (Collingham and Huntley 2000; Higgins and Richardson 1999; Lavorel et al. 1995; Malanson 2003). These studies emphasize that additional dispersal traits that define how dispersal will interact with human-transformed landscapes, are needed to predict spread rates in contemporary landscapes (Higgins et al. 2003).

11.4 Linking Migration and Niche Based Models

The previous section has reviewed how an incomplete knowledge of the potential migration rate of species limits our current capacity to predict the impacts of global climate change on the future geographic distribution of species, their range sizes and even potential vulnerability to extinction. Yet very few studies have combined mechanistic models of migration with spatially-explicit models of geographic range shifts. Niche-based models offer a potentially useful tool to predict species range shifts for indigenous species, and even areas susceptible to invasion by alien species (albeit these correlation approaches can only interpolate, not extrapolate into the future). Recent studies (Hoffman 2001; Peterson 2003; Peterson et al. 2003; Peterson and Vieglais 2001; Welk et al. 2002) have shown that this approach could provide a robust alternative to more complex, process-based models (Sutherst et al. 2000). However, despite the usefulness of such niche-based models in drawing potential maps of risk, species migration is not routinely incorporated in this approach.

Some preliminary attempts to integrate niche-based models and mechanistic dispersal models have been carried out. For instance, Iverson, Schwartz and colleagues (Iverson et al. 2004; Schwartz et al. 2001) have recently developed interesting analyses to predict the shift of trees due to climate changing in eastern US, based on the coupling between a niche-based model simulating the future potential suitable area for a species, and a cellular automaton simulating seed dispersal into a fragmented landscape. Such "hybrid" approaches combining different models are promising because they allow inclusion of transient responses of species to changing environments (see also Collingham (2000) and Malanson (2003)). These approaches are also useful in that they encompass a spatial scale within which migration can be simulated realistically, especially by differentiating between long- and short-distance dispersal of propagules, and assessing the relative importance of uncertainty inherent in their estimation. More studies of this kind are needed in order to develop better forecasts of future species range size, distribution and vulnerability to extinction under a wide range of migration rate assumptions (Higgins et al. 2003).

11.5 Summary and Conclusions

As we show above, a failure to incorporate migration limitations into models of vegetation response to climate change greatly compromises their predictive capability, and the uncertainty due to migration is therefore substantial. Species range shifts have been a ubiquitous response by plant species during Pleistocene climate change, and early signs of this response are evident in modern assemblages. Recent work has increased our understanding of the dispersal limitations to migration rate, but there has been far less focus on the issues which govern population establishment and growth rate, especially at the edge of species' ranges.

An overall understanding of community responses to climate change would also benefit from better understanding of *in situ* adaptive responses, as these appear to be significant in some species. Much has been learned from reconstructions of past migration patterns in the paleo-record, and from studies of alien invasive plants, but these "natural experiments" are limited in that they represent special cases where species migration occurs over landscapes unfragmented by human activities (paleo-record), or are experiencing release from predators and pathogens (alien species).

Finally, promising approaches are being developed that address the issue of how human transformation of landscapes will modify migration rates, and that combine mechanistic migration models with spatially explicit models of species geographic ranges at spatial scales relevant to simulating plant propagule dispersal and demographic behavior. These approaches will provide useful insights into biodiversity change under climate and land-use change scenarios. However, the potential increase in spatial resolution of DGVM simulations, and their increasing capacity to simulate more finely defined plant functional types, will allow them to provide an independent alternative assessment of the role of migration in determining the future structure and function of the ecosystems of the Earth.

Acknowledgments

GFM acknowledges the Henry Luce Foundation and the Center for Applied Biodiversity Science for financial support during the course of this study, and is grateful to Ian Woodward and Mark Lomas for the provision of data in Figure 11.2 and table 11.1.

References

Anderson RP, Gomez-Laverde M, Peterson AT (2002) Geographical distributions of spiny pocket mice in South America: insights from predictive models. Global Ecology and Biogeography 11:131–141

Araújo MB, Williams P (2000) Selecting areas for species persistence using occurrence data. Biological Conservation 96:331–345

Araújo MB, Cabeza M, Thuiller W, Hannah L, Williams PH (2004) Would climate change drive species out of reserves? An assessment of existing reserve-selection methods. Global Change Biology 10:1618–1626

Austin MP, Smith TM (1989) A new model for the continuum concept. Vegetatio 83:35–47

Austin MP, Nicholls AO, Doherty MD, Meyers JA (1994) Determining species response functions to an environmental gradient by means of a beta-function. Journal of Vegetation Science 5:215–228

Bachelet D, et al. (2003) Simulating past and future dynamics of natural ecosystems in the United States. Global Biogeochemical Cycles 17(2), 1045, doi:10.1029/2001GB001508

Bakkenes M, Alkemade JRM, Ihle F, Leemans R, Latour JB (2002) Assessing effects of forecasted climate change on the diversity and distribution of European higher plants for 2050. Global Change Biology 8:390–407

Barnosky CW (1987) Response of vegetation to climate changes of different duration in the late Neogene. Trends in Ecology and Evolution 2:247–250

Beier C, et al. (2004) Novel approaches to study climate change effects on terrestrial ecosystems in the field: Drought and passive nighttime warming. Ecosystems 7:583–597

Birks HJB (1986) Late-Quaternary biotic changes in terrestrial and lacustrine environments, with particular reference to north-west Europe. In: Berglund BE (ed) Handbook of Holocene Paleoecology and Paleohydrology. Wiley, Chichester

Birks HJB (1993) Quaternary paleoecology and vegetation science – current contributions and possible future developments. Rev. Paleobot. Palynol. 79:153–177

Bond WJ, Midgley JJ (2001) Ecology of sprouting in woody plants: the persistence niche. TREE 16:45–51

Bond WJ, Midgley GF, Woodward FI (2003) The importance of low atmospheric CO_2 and fire in promoting the spread of grasslands and savannas. Global Change Biology 9:973–982

Box EO (1996) Plant functional types and climate at the global scale. Journal of Vegetation Science 7:309–320

Brewer AM, Gaston KJ (2003) The geographical range structure of the holly leaf-miner. II. Demographic rates. Journal of Animal Ecology 72:82–93

Clark JS (1998) Why trees migrate so fast: confronting theory with dispersal biology and the paleo-record. American Naturalist 152:204–224

Clark JS, et al. (1998) Reid's paradox of rapid plant migration: dispersal theory and interpretation of palaeoecological records. BioScience 48:13–24

Clark JS, Lewis M, McLachlan JS, HilleRisLambers J (2003) Estimating population spread: What can we forecast and how well? Ecology 84:1979–1988

Collingham YC, Huntley B (2000) Impacts of habitat fragmentation and patch size upon migration rates. Ecological Applications 10:131–144

Collingham YC, Hill MO, Huntley B (1996) The migration of sessile organisms: a simulation model with measurable parameters. Journal of Vegetation Science 7:831–846

Cramer W, et al. (2001) Global response of terrestrial ecosystem structure and function to CO_2 and climate change: results from six dynamic global vegetation models. Global Change Biology 7:357–373

Cwynar LC, MacDonald GM (1987) Geographical variation of lodgepole pine in relation to population history. American Naturalist 129:463–469

Davis MB (1976) Pleistocene biogeography of temperate deciduous forests. Geoscience and Man 13:13–26

Davis MB (1983) Holocene vegetational history of the eastern United States. In: Wright HE (ed) Late-Quaternary environments of the United States. University of Minneapolis, Minneapolis, MN, USA

Davis MB, Shaw RG (2001) Range shifts and adaptative responses to quaternary climate change. Science 292:673–679

Davis AJ, Jenkinson LS, Lawton JH, Shorrocks B, Wood S (1998a) Making mistakes when predicting shifts in species range in response to global warming. Nature 391:783–786

Davis AJ, Lawton JH, Shorrocks B, Jenkinson LS (1998b) Individualistic species responses invalidate simple physiological models of community dynamics under global environmental change. Journal of Animal Ecology 67:600–612

Foley JA, Levis S, Prentice IC, Pollard D, Thompsons SL (1998) Coupling dynamics models of climate and vegetation. Global Change Biology 4:561–579

Franco M, Silvertown J (2004) A comparative demography of plants based upon elasticities of vital rates. Ecology 85:531–538

Graham RW (1992) Late pleistocene faunal changes as a guide to understanding effects of greenhous warming on the mammalian fauna of North America. In: Peters RL, Lovejoy T (eds) Global warming and biological diversity. Yale University, New Haven and London

Guisan A, Zimmermann NE (2000) Predictive habitat distribution models in ecology. Ecological Modelling 135:147–186

Hampe A (2004) Bioclimate envelope models: what they detect and waht they hide. Global Ecology & Biogeography 13:469–476

Hampe A, Petit RJ (2005) Conserving biodiversity under climate change: the rear edge matters. Ecology Letters 8(5):461–467

Hannah L, Midgley GF, Millar D (2002) Climate change integrated conservation strategies. Global Ecology and Biogeography 11:485–495

Higgins SI, et al. (2003) Forecasting plant migration rates: managing uncertainty for risk assessment. Journal of Ecology 91:341–347

Higgins SI, Richardson DM (1999) Predicting plant migration rates in a changing world: The role of long-distance dispersal. American Naturalist 153:464–475

Higgins SI, Richardson DM, Cowling RM (2001) Validation of a spatial simulation model of a spreading alien plant population. Journal of Applied Ecology 38:571–584

Higgins SI, Nathan R, Cain ML (2003) Are long-distance dispersal events in plants usually caused by nonstandard means of dispersal? Ecology 84(8):1945–1956

Hobbs RJ (2000) Land-use changes and invasions. In: Mooney HA, Hobbs HA (eds) The impact of global change on alien species. Island Press, Washington D.C

Hodkinson DJ, Thompson K (1997) Plant dispersal: the role of man. Journal of Applied Ecology 34:1481–1496

Hoffman MH (2001) The distribution of *Senecio vulgaris*: capacity of climate change models for predicting adventitious ranges. Flora 196:395–403

Holoyac M (2000) Spatiotemporal dynamic models of plant populations and communities. The American Naturalist 156:378–389

Hughes L (2000) Biological consequences of global warming: is the signal already apparent? TREE 15:56–61

Huntley B (1990) European post-glacial forests: compositional changes in response to climate change. Journal of Vegetation Science 1:507–518

Huntley B, Birks HJB (1983) An atlas of past and present pollen maps for Europe 0–13 000 years ago. Cambridge University Press, Cambridge

Huntley B, Webb TI (1988) Vegetation history. Kluwer, Dordrecht

Huntley B, Berry PM, Cramer W, McDonald AP (1995) Modelling present and potential future ranges of some European higher plants using climate response. Journal of Biogeography 22:967–1001

Iverson LR, Prasad AM (2002) Potential redistribution of tree species habitat under five climate change scenarios in the eastern US. Forest Ecology and Management 155:205–222

Iverson LR, Schwartz MW, Prasad A (2004) How fast and far might tree species migrate in the eastern United States due to climate change? Global Ecology & Biogeography 13:209–219

Jansson R (2003) Global patterns in endemism explained by past climatic change. Proceedings of the Royal Society London B 270:583–590

Janzen DH, Martin PS (1982) Neotropical anachronisms: the fruits the gomphotheres ate. Science 215:19–27

Keane R, Crawley MJ (2002) Exotoc plant invasions and the enemy release hypothesis. Trends in Ecology and Evolution 17:164–170

Kirilenko AP, Belotelov NV, Bogatyrev BG (2000) Global model of vegetation migration: incorporation of climatic variability. Ecological Modelling 132:125–133

Lavorel S, Garnier E (2002) Predicting changes in community composition and ecosystem functioning from plant trait: revisiting the Holy Grail. Functional Ecology 16:545–556

Lavorel S, Gardner RH, O'Neill RV (1995) Dispersal of annual plants in hierarchically structured landscapes. Landscape Ecology 10: 277–289

Leathwick JR, Austin MP (2001) Competitive interactions between tree species in New Zealand's old-growth indigenous forests. Ecology 82:2560–2573

Leibold MA (1995) The niche concept revisited: mechanistic models and community context. Ecology 76:1371–1382

Loehle C, LeBlanc D (1996) Model-based assessments of climate change effects on forests: A critical review. Ecological Modelling 90:1–31

MacDonald GM (1993) Fossil pollen analysis and the re-construction of plant invasions. Advances in Ecological Research 24:67–110

Malanson GP (2003) Dispersal across continuous and binary representations of landscapes. Ecological Modelling 169:17–24

Malcolm JR, Markham A, Neilson RP, Garaci M (2002) Estimated migration rates under scenarios of global climate change. Journal of Biogeography 29:835–849

McClachlan JS, Clark JS (2004) Reconstructing historical ranges with fossil data at continental scales. Forest Ecology and Management 197:139–147

McDonald DJ, Cowling RM (1995) Towards a profile of an endemic mountain fynbos flora: implications for conservation. Biological Conservation 72:1–12

Mooney HA, Hobbs RJ (2000) Invasive species in a changing world. Island Press, Washington, D.C, California

Myers N, Mittermeier RA, Mittermeier CG, Da Fonseca GAB, Kent J (2000) Biodiversity hotspots for conservation priorities. Nature 403:853–858

Nakicenovic N, Swart R (eds) (2000) Emissions Scenarios: A Special Report of Working Group III of the Intergovernmental Panel on Climate Change. Cambridge University Press, Cambridge

Nathan R, Perry G, Cronin JT, Strand AE, Cain ML (2003) Methods for estimating long-distance dispersal. Oikos 103:261–273

Neilson RP, Pitelka LF, Solomon AM, Nathan R, Midgley GF, Fragoso JMV, Lischke H, Thompson K (2005) Forecasting regional to global plant migration in response to climate change. BioScience 55:749–759

Neubert M, Caswell H (2000) Demography and dispersal: Calculation and sensitivity analysis of invasion speed for structured populations. Ecology 81:1613–1628

Nowak CL, Nowak RS, Tausch RJ, Wigand PE (1994) Tree and shrub dynamics in northwestern Great Basin woodland and shrub steppe during the late-Pleistocene and Holocene. American Journal of Botany 81:265–277

Parmesan C, Yohe G (2003) A globally coherent fingerprint of climate change impacts across natural systems. Nature 421: 37–42

Parmesan C, et al. (1999) Poleward shifts in geographical ranges of butterfly species associated with regional warming. Nature 399:578–583

Pearson RG, Dawson TE (2003) Predicting the impacts of climate change on the distribution of species: are bioclimate envelope models useful? Global Ecology & Biogeography 12: 361–372

Peterson AT (2001) Predicting species' geographic distributions based on ecological niche modeling. The Condor 103:599–605

Peterson AT (2003) Predicting the geogaphy of species' invasions via ecological niche modellingg. The Quarterly Review of Biology 78:419–433

Peterson AT, Vieglais DA (2001) Predicting species invasions using ecological niche modeling: new approaches from bioinformatics attack a pressing problem. BioScience 51:363–371

Peterson AT, et al. (2002) Future projections for Mexican faunas under global climate change scenarios. Nature 416:626–629

Peterson AT, Papes M, Kluza DA (2003) Predicting the potential invasive distributions of four alien plant species in North America. Weed Science 51:863–868

Pitelka LF, et al. (1997) Plant migration and climate change. American Scientist 85:464–501

Prentice IC (1986) Vegetation response to past climatic variation. Vegetatio 67:131–141

Prentice IC, Cramer W, Harrison SP, Leemans R, Monserud RA, Solomon AM (1992) A global biome model based on plant physiology and dominance, soil properties and climate. Journal of Biogeography 19:117–134

Pulliam HR (2000) On the relationship between niche and distribution. Ecology Letters 3:349–361

Rehfeldt GE, Ying CC, Spittlehouse DL, Hamilton jr. DA (1999) Genetic responses to climate in *Pinus contorta*: Niche breadth, climate change, and reforestation. Ecological Monographs 69: 375–407

Rejmanek M, Richardon DM, Higgins SI, Pitcairn MJ, Grotkopp E (2004) Ecology of invasive plants: state of the art. In: Mooney HA, McNeely JA, Neville L, Schei PJ, Waage J (eds) Invasive alien species: searching for solutions, Island, Washington, D.C

Richardson DM, et al. (2000) Invasive alien organisms and global change: A South African Perspective. In: Mooney HA, Hobbs HA (eds) The impact of global change on alien species. Island Press, Washington D.C

Ridley HN (1930) The dispersal of plants throughout the world. Reeve, Ashford, UK

Sagarin RD, Gaines SD (2002) The 'abundant centre' distribution: to what extent is it a biogeographical rule? Ecology Letters 5:137–147

Schwartz MW, Iverson LR, Prasad AM (2001) Predicting the potential future distribution of four tree species in Ohio using current habitat availability and climatic forcing. Ecosystems 4:568–581

Shigesada N, Kawasaki K, Takeda, Y (1995) Modelling stratified diffusion in biological invasions. American Naturalist 146:229–251

Shugart HH, Smith TM, Post WM (1992) The application of individual-based simulation models for assessing the effects of global change. Annual Review of Ecology and Systematics 23:15–38

Sitch S, et al. (2003) Evaluation of ecosystem dynamics, plant geography and terrestrial carbon cycling in the LPJ Dynamic Global Vegetation Model. Global Change Biology 9:161–185

Skellam JG (1951) Random dispersal in theoretical populations. Biometrika 38:196–218

Smith TM, Shugart HH, Woodward FI (1997) Plant Functional Types, Their Relevance to Ecosystem Properties and Global Change. Cambridge Univ. Press, Cambridge

Smith B, Prentice IC, Sykes MT (2001) Representation of vegetation dynamics in modelling of European ecosystems: comparison of two contrasting approaches. Global Ecology & Biogeography 10 (ETEMA special issue):621–637

Solomon AM (1986) Transient response of forests to CO_2-induced climate change: simulation modeling experiments in eastern North America. Oecologia 68:567–579

Sutherst RW, Maywald HF, Russel BL (2000) Estimating vulnerability under global change: modular modelling of pests. Agriculture, Ecosystems and Environment 82:303–319

Thomas CD, et al. (2004) Extinction risk from climate change. Nature 427:145–148

Thuiller W (2003) BIOMOD: Optimising predictions of species distributions and projecting potential future shifts under global change. Global Change Biology 9:1353–1362

Thuiller W (2004) Patterns and uncertainties of species' range shifts under climate change. Global Change Biology 10:2020–2027

van Rheede van Oudtshoorn K, van Rooyen MW (1999) Dispersal biology of desert plants. Springer-Verlag, Berlin

Vucetich JA, Waite TA (2003) Spatial patterns of demography and genetic processes across the species' range: Null hypotheses for landscape conservation genetics. Conservation Genetics 4:639–645

Walther G-R, Burga CA, Edwards PJ (2001) Fingerprints of climate change – adapted behaviour and shifting species ranges. Kluwer Academic/Plenum Publishers, New York

Walther G-R, et al. (2002) Ecological responses to recent climate change. Nature 416:389–395

Webb TI (1981) The past 11,000 years of vegetational change in eastern North America. BioScience 31:501–506

Webb TI (1992) Past changes in vegetation and climate: lessons for the future. In: Peters RL, Lovejoy TE (eds) Global warming and biological diversity. Yale University, New Haven and London

Welk E, Schubert K, Hoffman MH (2002) Present and potential distribution of invasive garlic mustard (*Allaria petiolata*) in North America. Diversity and Distribution 8:219–233

Woodward FI, Smith TM, Emanuel WR (1995) A global land primary productivity and phytogeography model. Global Biogeochemical Cycles 9:471–490

Chapter 12

Understanding Global Fire Dynamics by Classifying and Comparing Spatial Models of Vegetation and Fire

Robert E. Keane · Geoffrey J. Cary · Ian D. Davies · Michael D. Flannigan · Robert H. Gardner · Sandra Lavorel James M. Lenihan · Chao Li · T. Scott Rupp

12.1 Introduction

Wildland fire is a major disturbance in most ecosystems worldwide (Crutzen and Goldammer 1993). The interaction of fire with climate and vegetation over long time spans, often referred to as the fire regime (Agee 1993; Clark 1993; Swetnam and Baisan 1996; Swetnam 1997), has major effects on dominant vegetation, ecosystem carbon budget, and biodiversity (Gardner et al. 1996; Lenihan et al. 1998; Ryan 1991; Starfield and Chapin 1996). These effects include cycling nutrients, regulating succession, maintaining diversity, reducing biomass, controlling insect and disease populations, triggering interactions between vegetation and animals, and maintaining important biological and biogeochemical processes (Johnson 1992; Agee 1993; Crutzen and Goldammer 1993; DeBano et al. 1998). Carbon cycles are intimately linked to fire regimes because of the accumulation and combustion of fuel and post-fire vegetation development (Kasischke et al. 1995; Olsen 1981). Fire has sculpted landscapes for millennia by dictating pattern dynamics and community composition (Knight 1987; Swanson et al. 1997).

Modification of the fire regime due to climate warming (e.g., Cary and Banks 1999) may overwhelm many of the commonly investigated ecosystem responses to climate change, including ecosystem productivity, plant migration, and species extinction (Weber and Flannigan 1997; Ryan 1991; Keane et al. 1998). Large nonlinear changes in vegetation composition and structure are likely to occur in response to climate warming and human land use because of the overriding influence of the fire regime (Flannigan and Van Wagner 1991; Crutzen and Goldammer 1993). Carbon storage in some forest ecosystems may significantly decline if fire increases due to climate warming (Kasischke et al. 1995). Because of the inherent complexity of the interactions of fire, climate, and vegetation across multiple scales, simulation modeling is a critical tool for exploring and understanding climate change implications.

A major problem in projecting ecological change and understanding its mechanisms is the lack of non-equilibrium dynamics in many ecological models. The inclusion of disturbance, especially fire, is essential for dynamic vegetation models to simulate transient changes in vegetation composition and structure. Understanding landscape dynamics in relation to fire and how these dynamics may be altered by climate and land-use changes is critical. Additionally, understanding human impacts on the fire regime is essential for projecting vegetation change in human-modified landscapes, which now occupy large proportions of the globe. Using our current understanding of fire behavior, fire ecology and fire-weather, a set of dynamic fire-climate-vegetation models could be developed to simulate fire effects at temporal and spatial scales relevant to vegetation change.

One approach to addressing the effects of changing climate on fire and vegetation dynamics is simulation modeling. At the global scale there appear to be two options: (1) a suite of landscape models that spatially simulate succession and fire dynamics at local scales, or (2) coarse scale vegetation models, called Dynamic Global Vegetation Models or DGVM, that have been developed to predict vegetation response to climate change at global scales (Neilson and Running 1996; Lenihan et al. 1998; Thonicke et al. 2001). However, there are drawbacks to both. The available landscape models are quite different in design and application, developed for different ecosystems and fire regimes, and as such cannot be brought together for comprehensive global simulations. Second, the incorporation of fire into DGVMs to predict these ecosystem responses at global scales has been problematic because of confounding environmental complexities that govern global fire regimes and the general lack of understanding of which fire regime characteristics are important at coarse scales.

This chapter summarizes an effort by the Landscape Fire Working Group, formed under the aegis of the Global Change and Terrestrial Ecosystems Project (GCTE – Task 2.2.2) to identify those fire processes that are essential to dynamically model landscape vegetation changes. This was accomplished by classifying and comparing existing fine-scale, spatial, non-equilibrium process-based ecological models that simulate the dynamic interactions between climate, fire and vegetation, which we called Landscape Fire Succession Models or LFSMs. By classifying all existing LFSMs, we were able to inventory the diverse methods currently used to simulate vegetation, climate, and fire interactions in a spatial domain. We then selected a set of models from a suite of classification categories to perform an extensive LFSM comparison on neutral landscapes to identify the relative importance and sensitivity of simulated fire to terrain, fuel pattern, climate, and weather. Once the important processes were identified, we developed a set of recommendations for including fire dynamics in DGVMs.

12.2 Background

We define LFSMs as models that simulate the linked processes of fire and succession (i.e., vegetation development) in a spatial domain. Although the complexity of spatial relationships of vegetation and fire dynamics may vary from model to model, all LFSMs, by definition, produce time-dependent, georeferenced results in the form of digital maps or GIS layers. Additional processes may have been incorporated into the LFSM simulation, such as timber harvesting and biogeochemical modeling (explicit simulation of the flow of energy, carbon, water, and other elements within an ecosystem or landscape), but the minimum requirement for a LFSM is the explicit linkage between fire and succession. Climatic processes need not be explicitly incorporated into the LFSM, but, because of our interests in climate change, special attention was given to those models that consider the direct effect of weather on fire occurrence and vegetation change.

A series of six GCTE sponsored workshops attended by a wide variety of international ecological modelers and ecologists was held from 1999 to 2003 to synthesize current landscape fire modeling into an organized framework (see Hawkes and Flannigan 2000). One product of these workshops was an objective, quantitative protocol for comparing LFSM simulations across neutral landscapes (i.e., artificial landscapes where topography and vegetation characteristics are controlled) and regional climates to determine the relative sensitivity of predictions to model structure and complexity (Cary et al. 2006). A standardized set of model descriptive elements (MDE) was developed to qualitatively contrast and compare LFSMs, and the values for these elements were estimated by the modelers and entered into a database (Rupp et al. 2001). Information included in the MDE data base included initial purpose of the model, the ecosystem type being simulated, nature of the vegetation being represented and the method of succession, climate variables and drivers, the temporal and spatial scales of predictions, and computing constraints (Keane et al. 2004).

Using workshop findings and the MDE database information, we identified four essential components in LFSMs that represent the primary processes governing the landscape simulation of vegetation and fire: (1) vegetation succession, (2) fire ignition, (3) fire spread, and (4) fire effects. Any LFSM needed to contain all of these components. We assumed any other ecosystem and landscape process simulated by an LFSM, such as harvesting and insect epidemics, could be added as other components or incorporated into one or more of these four primary components. For example, fuel accumulation would be considered part of the vegetation succession component. We debated whether fire extinguishments – when a spreading fire actually goes out – was another component, but decided it should be part of the spread component (i.e., extinguishment is the lack of spread) for simplicity.

Each component for every LFSM was described by the approach, scale, and strategy by the modelers or authors, and this information was also entered into the MDE database (Reinhardt et al. 2001; Keane and Finney 2003; Keane et al. 2004). The approach defines the general design of the model as probabilistic (based on stochastic processes), empirical (based on relationships described by data), or physical (based on fundamental physical processes). Spatial scales are either regional (1 000s of km^2), landscape (10s of km^2), forest stand (<1 ha), or at the level of the individual plant (~m^2). The strategy describes the algorithms, tools, or techniques used to represent a simulation component. Many LFSM components were developed by merging multiple approaches, scales, and strategies.

12.3 Model Classification

The classification effort provided the framework to interpret differences between LFSMs and for stratifying LFSMs for the subsequent comparisons (see Keane et al. 2004 for full details). The classification was based on the four primary processes that influence fire and vegetation dynamics – fire ignition, fire spread, fire effects, and vegetation succession – and the approaches used to represent these processes in the models. The classification space for each component was described in three dimensions by the gradients of stochasticity, complexity, and mechanism that best characterize the simulation of that component. Stochasticity was defined as the amount of randomness inherent in the component design, or the degree at which probabilistic functions influence the simulation of that component. Complexity was defined as the inherent detail incorporated into the design of a simulated component. Models with low complexity have modest sophistication in simulation detail and those with a large number of parameters are considered quite complex. Mechanism was the degree to which fundamental physical or chemical processes are represented in the simulation of a LFSM component. This resulted in 12 evaluation elements (4 components by 3 gradients) for each model. Together, these elements represented a formal description that can be used to objectively compare other models. There is some unavoidable ambiguity in these gradients, but overall we felt these 12 evaluation elements provided a standardized, comprehensive, and somewhat objective context in which to assess LFSMs. In addition, this method can be used to classify and compare other types of models by modifying, eliminating, or adding gradients and components.

We conducted an inventory of existing LFSMs using information gained from workshop participants, a review of the literature, and correspondence with modelers. Only LFSMs that were published in some form were considered, and this search found 44 LFSMs that were used in the classification effort. We contacted the developers of these models and asked them to rate the simulation of the four components (succession, fire ignition, fire spread, and fire effects) by the three evaluation gradients (stochasticity, complexity, and mechanism) using a scale from zero to 10 (zero meant that it was not modeled or applicable and 10 represented the highest level of stochasticity, mechanism, or complexity). Some modelers did not reply, so we assigned our own ratings based on a thorough review of publications on the model. The values assigned to each evaluation element were compiled into a database and then analyzed to identify groups of similar models. To ensure consistency across modeler evaluations of their own models, we created another database with our own assignments of evaluation elements based on published literature and our knowledge of the model.

To identify natural clusters or groups, the evaluation element data were ordinated using principal components analysis (PCA) and clustered using TWINSPAN techniques in the PC-ORD package (McCune and Mefford 1999). The See5 statistical software (Quinlan 2003) was also used to cluster the evaluation element data using Ward's minimum variance hierarchical clustering, which is a divisive clustering technique. The ordination and clustering results alone were not sufficient for developing a comprehensive classification because of the high variance in evaluation elements across models. Therefore, we revised the MDE database created from the workshops so that keywords were used to describe various explanatory categories such as approach, strategy, scale and other descriptive attributes by LFSM component. We then compared the frequency of keywords for each category across all LFSMs to qualitatively identify similar characteristics and natural clusters.

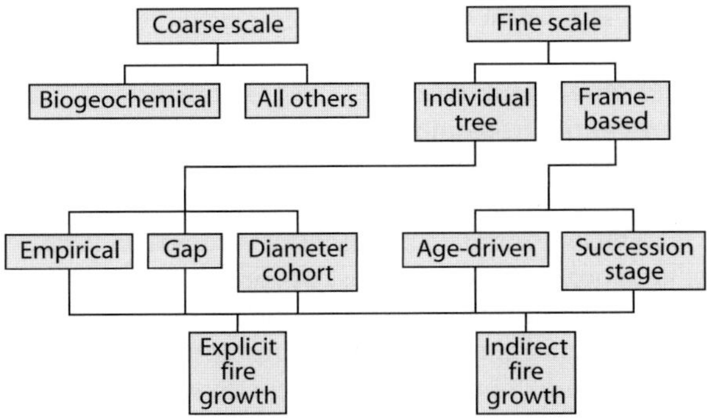

Fig. 12.1. Final set of categories for landscape fire succession models classified using the degree of stochasticity, complexity, and mechanism in the design of the model components of succession, fire ignition, fire spread and fire effects

A general LFSM classification was developed from the fusion of the ordination, clustering, and keyword comparison results (Fig. 12.1). This classification has 12 hierarchically nested classes that are distinguished by their scale of application (coarse vs. fine), representation of vegetation (individual plant cohorts vs. framed-based community), simulation of succession (age, empirical, gap-phase, or successional pathway) and the explicit or implicit simulation of fire spread. The locations of all LFSMs and six classification categories are shown in ordination space in Fig. 12.2. A dichotomous key for the classification was constructed from the MDE database using See5 analysis. We then related common keywords to the dichotomous key to name and identify important branches in the dichotomy.

12.4 Model Comparison

The model comparison was undertaken to identify the relative influence of environmental processes on fire dynamics. It is comprised of three phases of which only results from the first phase are presented in this chapter. In the first phase, we evaluated the sensitivity of fire spread and ignition components without the influence of fire effects and subsequent vegetation development (Phase 2 and 3, respectively). We compared five LFSMs that covered three classification categories for this exercise: LANDSUM, FIRESCAPE, EMBYR, SEMLAND, and a special application of the LAMOS modeling shell called LAMOS-DS (Table 12.1). These models are quite different in many aspects including a wide diversity in their approaches for simulating fire spread and ignition, representation of vegetation, and the complexity of climate and fire linkages (see Fig. 12.2, Table 12.1).

12.4.1 The Models

EMBYR is an event-driven, grid-based simulation model of fire ignition and spread designed to represent the landscapes and fire regimes of Yellowstone National Park (Hargrove et al. 2000). The pattern of forest succession of lodgepole pine forests is simulated by a Markov model, with fuels sufficient to sustain crown fires developing as a function of forest stand age. Fire spread is simulated by examining each burning site and determining spread to the eight neighboring sites as a function of fuel type, fuel moisture, wind speed and direction, and slope. A qualitative index of fire severity of each burned site, estimated as a function of fuel type, fuel moisture, wind speed and burn rate, is used to determine when fire intensity is sufficient to result in a stand-replacing fire.

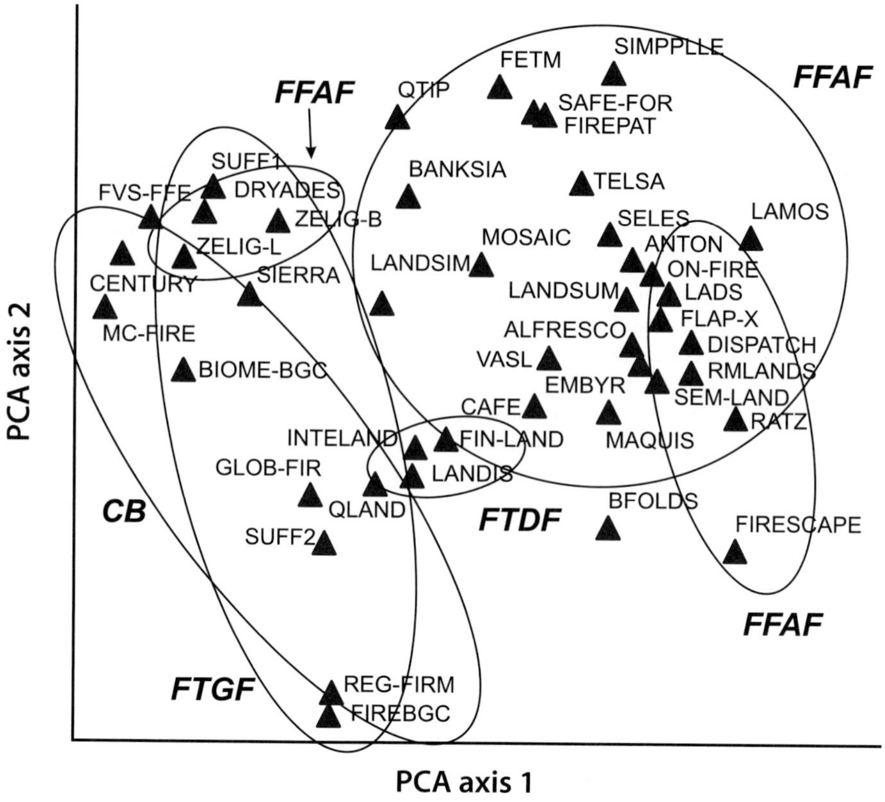

Fig. 12.2.
Delineation of six model classes in ordination space for the developed classification using ratings assigned by the modelers. This shows the relative position between classes and the similarity of models within a class

Table 12.1. Landscape fire succession models included in the model comparison effort with a general description of its application

Model name	References	Ecosystem	Geographic area	Weather database
EMBYR	(Gardner et al. 1996; Hargrove et al. 2000)	Lodgepole pine forests	Central Rockies, USA	Glacier National Park
FIRESCAPE	(Cary and Banks 1999; Cary 2002)	Eucalypts forest	Southeastern Australia	Australian Capital Territory
LAMOS(DS)	(Lavorel et al. 2000)	Any	Portugal	Corsica, Italy
LANDSUM	(Keane et al. 1997; Keane et al. 2002)	Any	Northern Rockies, USA	Glacier National Park
SEM-LAND	(Li 2000, 2001)	Spruce-fir forests	Canada	Edmonton, Alberta

EMBYR propagates fire by user-defined probabilities of fire spread. The probabilities are dependent on the age of the forest stand and the fuel moisture conditions (i.e., probabilities of spread increase with increasing age and decreasing fuel moisture) and are adjusted during the simulation by the local topography, fuel moisture, wind speed and direction.

FIRESCAPE generates spatial patterns of fire regime (Gill 1975) for *Eucalyptus* dominated landscapes in southeastern Australia (Cary and Banks 1999). It operates on a daily time step that changes to hourly whenever a fire ignites. Daily weather is generated using a type stochastic climate generator (Richardson 1981; Matalas 1967; Cary and Gallant 1997; McCarthy and Cary 2002) and ignition locations are generated from an empirical model of lightning strike locations (Cary 1998). The spread of fire from cells to immediate neighbors is a function of elliptical fire spread (Van Wagner 1969) and Huygens' Principle (Anderson et al. 1982). Head fire rate of spread is determined from the equation form of McArthur's Forest Fire Danger Meter (McArthur 1967; Noble et al. 1980) and fuel loads are modeled using Olson's (1963) model of biomass accumulation which has been parameterized for a range of Australian systems (Fox et al. 1979; Raison et al. 1986). Fire line intensity ($kW\,m^{-1}$) is calculated for the spread of fire from one cell to the next for characterizing this aspect of the fire regime and for determining the extinction of the individual fire events.

LAMOS-DS is an implementation of LAMOS (Lavorel et al. 2000) with a contagious fire spread model working on a daily time step. It is a simple model, sensible to daily minimum and maximum temperature, precipitation, fuel amount and slope. LAMOS-DS contains two principle functions; one to estimate pan evaporation (Bristow and Campbell 1984; Roderick 1999) which, together with precipitation, produces a moisture budget, and a second equation to modify spread probabilities as a function of slope (Li 2000) and intensity. Fire intensity is the product of three linear functions: fuel load (0–1 kg m^{-2}), moisture (0–200 mm) and temperature (5–25 °C). Temperature during the course of the fire is interpolated between the daily minimum and maximum by a symmetrical sine function. Fires are assumed to begin when temperature is at the daily maximum. Fuel is consumed in proportion to the resulting fireline intensity.

The LANDscape SUccession Model (LANDSUM) is a spatially explicit vegetation dynamics simulation program wherein succession is treated as a deterministic process using a pathway or frame-based community sequence approach, and disturbances (e.g., fire, insects, and disease) are treated as stochastic processes with all but fire occurring at the polygon scale (Keane et al. 1997, 2002). LANDSUM simulates fire spread from user-specified wind speed and direction, slope, and fuel type (burn, no-burn) determined from the succession stage using equations from Rothermel (1972) and Albini (1976).

The SEM-LAND model (Spatially Explicit Model for LANdscape Dynamics) simulates fire regimes and associated forest landscape dynamics resulting from long-term interactions among forest fire events, landscape structures, and weather conditions. A fire process is simulated in two stages: initiation and spread (Li 2000, 2001). The fire initiation stage continues from the presence of a fire ignition source in a forest stand until most trees in that stand have been burned. Once most trees are burned, the fire has the potential to spread to its surrounding cells, and whether a neighboring cell would be burned is a function of the fire spread probability. The fire spread probability is determined by not only fuel and weather conditions, but also by slope in landscape topography. The Canadian Forestry Fire Weather Index system (FWI) (Van Wagner 1987) and the Canadian Forest Fire Behavior Prediction system (FBP) (Forest Canada Fire Danger Group 1992; Hirsh 1996) were used to drive weather, fire spread, and fuel moisture interactions in the model simulation.

12.4.2 The Comparison Design

The comparison involved evaluating the effects of variation in terrain, fuel pattern (i.e., vegetation), climate, and weather on the simulated annual number of fire ignitions and area burned using a set of neutral landscapes and native simulation parameters (i.e., simulation parameters quantified for the region for which the model was developed). The simulation landscapes were represented by a $1\,000 \times 1\,000$ array of square 50 m pixels. Variation in terrain was introduced by generating three landscapes with a systematic pattern of valleys and peaks – flat, undulating, and mountainous – characterized by maximum slope values of 0°, 15° and 30° and relief of 0 m, 1 250 m and 2 500 m respectively (Fig. 12.3a) created by a two-dimensional sine function with periodicity of 16.67 km (333.3 pixels).

Two patterns of fuel treatment – finely clumped and coarsely clumped spatial pattern – were evaluated (Fig. 12.3b,c). We created 10 replicates of finely and coarsely clumped fuel maps by randomly allocating values from the series 0.1, 0.2, 0.3, …, 1.0 (inclusive) to either 50×50 pixel (625 ha) clumps (coarsely clumped) or 10×10 pixel (25 ha) clumps (finely clumped) so that values were evenly represented across landscapes. Fuel maps were transformed differently for each model to produce either fuel or age related maps that were meaningful to individual models (Cary et al. 2006).

Simulations were performed using weather relevant to the location where the model had been previously parameterized and tested, although we attempted to standardize the amount of variability represented across the weather years chosen for each model. Ten, year-long se-

a Terrain b Finely clumped c Coarsely clumped

Fig. 12.3. Neutral landscapes used as inputs in the model comparison effort. **a** topography is represented by an "egg carton" shape with lowest and highest elevation indicated by *lightest* and *darkest color* respectively, **b** the finely clumped fuel landscape was created by randomly assigning fuel types or associated vegetation types to each pixel on the simulation landscape in 25 ha patches, and **c** the coarsely clumped fuel pattern was created by randomly assigning fuel types within 625 ha patches

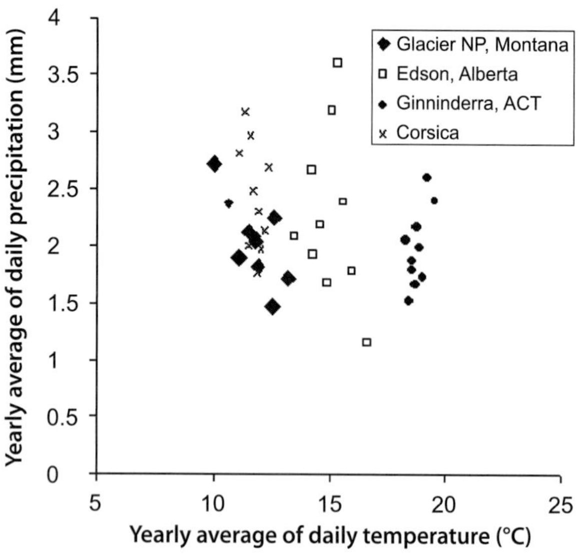

Fig. 12.4. Comparison of average daily temperature and precipitation from Glacier National Park (Montana), Edson (Alberta), Ginninderra (Australian Capital Territory) and Corsica (Italy) weather databases. The temperature data for Glacier National Park and Ginninderra are daily maximum temperatures. The temperature for Corsica is daily average temperature and for Edson it is observed at 1200 LST

quences of daily weather were selected from available data for each model so that the distributions of average annual temperature and precipitation in the selected weather years (Fig. 12.4) matched the corresponding distributions in the available data sets. The differences among the weather set for a particular model represents the weather treatment in our design.

Our design included three climates – observed, warmer/wetter and warmer/drier. The observed climate was comprised of the weather sequences outlined above. The other climates were derived by adding 3.6 °C (mid-range of projected global average temperature increase of 1.4 to 5.8 °C) (IPCC 2001) to daily temperature and by incrementing daily rainfall amounts by –20% (warmer/drier) and +20% (warmer/wetter).

A total of 1800 year long simulations were run for each model from the 18 unique combinations of terrain (flat, undulating, mountainous), fuel pattern (finely and coarsely clumped), climate (observed, warmer/wetter, warmer/drier), and weather (ten one-year replicates), given that there were ten replicate maps of each fuel pattern. Fires affected fuel load/age within each simulation but, since simulations were for only a single year, no vegetation succession algorithms were invoked. The sensitivity of area burned to terrain, fuel pattern, climate, and weather was assessed from the variance explained (r^2) by each variable and their interactions determined from a fully factorial ANOVA performed in the SAS statistical package. Prior to our analyses, we transformed area burned by the natural logarithm to remove highly skewed residuals that resulted from untransformed data.

12.5 Results and Discussion

12.5.1 Model Classification

Results from the model classification effort illustrate the great diversity in model approaches used to simulate vegetation and fire dynamics. This diversity will always preclude a perfect classification because modellers will always emphasize those processes that are important to the local ecosystem or landscape for which the model was developed. For example, many models simulated fire effects as the simple killing of all trees because most fires were stand-replacing for their native landscape. The fire effects component had the least influence in the classification of LSFMs because of the low diversity and low complexity in fire effects simulation across models. In contrast, succession and fire spread were the components that most heavily influenced the clustering and ordination results because of the wide variety of simulation techniques represented. Succession was most important because the three evaluation gradients (complexity, stochasticity, and mechanism) were not closely related in succession simulations. Classification design is nearly always governed by its intended application, so there will never be the ideal LFSM classification for all purposes.

The three gradients used in the classification were not perfectly orthogonal. Highly complex models tend to include many mechanistic functions that tend to have a low degree of stochasticity (Gardner et al. 1999), and they often were built specifically to remove the stochasticity, so that these approaches could be mutually exclusive and the gradients correlated. Some highly complex systems, such as lightning dynamics, must be represented by stochastic functions because of insufficient knowledge and computer resources. We only found significant correlation between the complexity and mechanism gradients ($R^2 = 0.71$, $p < 0.0132$).

The classification can be used for many purposes by providing (1) a common language for communication between managers, modelers, and research scientists, (2) a means to quickly and efficiently characterize or describe a model relative to others, (3) a starting point for managers to select the most appropriate model to implement for their areas of interest, (4) a method for scientists and other modelers to select the most appropriate models for particular situations, and (5) the context to evaluate or compare simulation approaches for each component to build new models or refine old ones. The classification contains 12 categories based on gradients of complexity, stochasticity, and mechanism so it is only useful if these gradients are important to selecting, evaluating, or comparing models. This classification approach can be used for other applications by selecting different evaluation gradients or different components.

12.5.2 Model Comparison

The primary finding of the comparison effort is the overriding influence of climate and weather on fire dynamics (Table 12.2). The natural log-transformed area burned was most sensitive to climate (4 models), weather (4 models), and their interaction (3 models) (Table 12.2). Only FIRESCAPE showed any sensitivity to terrain and only EMBYR showed sensitivity to fuel pattern. The variance explained by weather year was much greater than that explained by the climate treatment for EMBYR, LANDSUM and SEM-LAND simulations. The converse was observed for FIRESCAPE and LAMOS-DS, perhaps because the inter-annual variation between the weather replicates for these locations was lower than for other weather stations despite efforts to standardize this across all models. Differences in variance in area burned explained by the weather replicate treatment may result from differences in inter-annual variability in weather between sites, not differences in model formulation *per se*. Nevertheless, the finding of the importance of inter-annual variability in weather (compared with climate) to fire regime has important implications for the inclusion of fire into higher order models, like DGVMs, because our results indicate increased variability in daily weather may result in greater simulated burned area than large shifts in mean temperature and precipitation.

There are a number of important differences between the models used in this comparison. First, fire regime is an input in LANDSUM whereas it is an emergent property for the other models. Ordinarily, the area burned in LANDSUM would not vary among the climate treatments, however for this comparison the probability of ignition success was made sensitive to the Keetch-Byram Drought Index. For other models, climate affects either the area burned from the same number of ignitions (EMBYR, SEM-LAND) or both the number of ignitions and the area burned resulting from them (FIRESCAPE, LAMOS-DS). Second, there are differences in representation of processes associated with fire ignition and spread. Using the gradient scores from the classification effort, we found that the variance in area burned appeared to be primarily related to the fire ignition module. It was highest for models with complex, mechanistic, deterministic ignition modules (Fig. 12.5).

12.6 Summary and Conclusions

It was clear from the results of the comparison that the inclusion of both climate and daily weather variation into fire models is essential for simulating fire regime. This is especially important if the effects of climate change on fire dynamics are important. Fires burn the most area in warm, dry years, especially when coupled with antecedent drought, and the frequency of these droughty years

Table 12.2. Relative sums of squares attributed to different sources of variation in the comparison of sensitivity of ln-transformed area burned to terrain (terrain), fuel pattern (fuel), climate (climate) and weather replicate (weather), and their interactions. Treatments and interactions considered unimportant (explaining less than 0.05 and 0.025 of total variance respectively are blank). Significant treatments and interactions ($P < 0.05$) are indicated by an asterisk (*). Note that not all significant sources are considered important

Source	DF	Model				
		EMBYR	FIRESCAPE	LAMOS	LANDSUM	SEMLAND
Terrain	2		0.293*			
Fuel	1	0.217*	*		*	*
Terrain × fuel	2		*			
Climate	2	*	0.418*	0.278*	0.178*	0.370*
Terrain × climate	4		*			
Fuel × climate	2	*				*
Terrain × fuel × climate	4		*			
Weather	9	0.329*	0.087*	*	0.333*	0.542*
Terrain × weather	18		0.025*		*	
Fuel × weather	9	0.031*	*			*
Terrain × fuel × weather	18	*				
Climate × weather	18	0.096*	*	*	0.224*	0.046*
Terrain × climate × weath	36		0.025*			
Fuel × climate × weather	18	*				
Terr × fuel × clim × weath	36					
Model	179	0.744	0.905	0.401	0.766	0.971

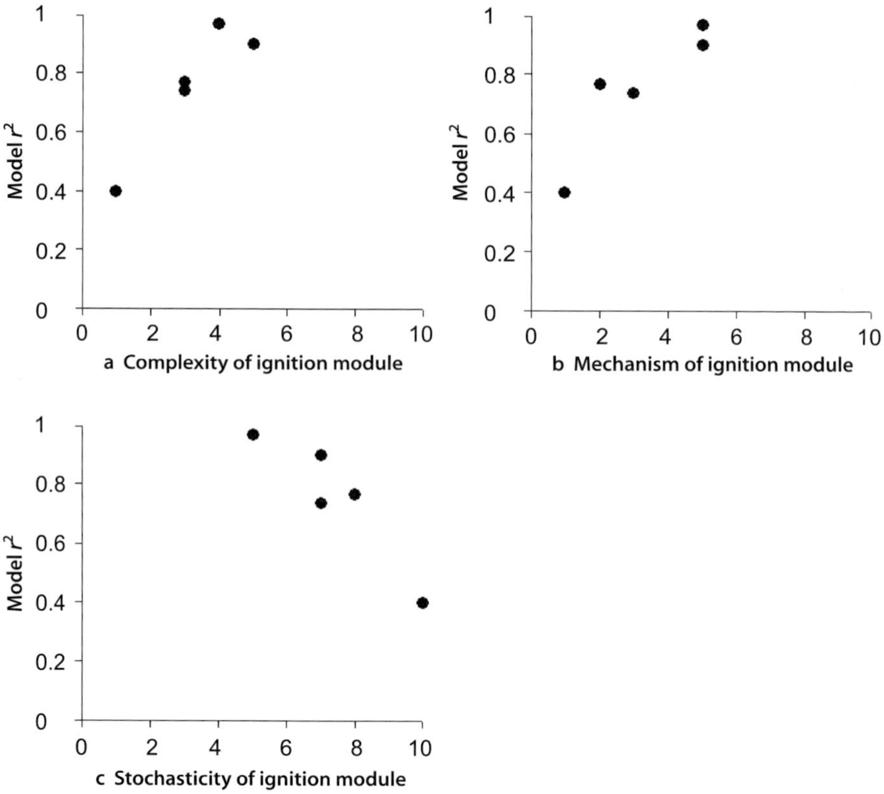

Fig. 12.5. Relationship between level of **a** complexity, **b** stochasticity, and **c** mechanism of fire model ignition modules assessed from the classification effort and model r^2 computed from the comparison effort

eventually defines the fire regime. Furthermore, variation in weather must be adequately captured to effectively simulate climate change responses to fire and vegetation. A surprising result was the relatively low sensitivity of fuel pattern in influencing fire regime. This may mean that coarse scale DGVMs need not be concerned with the pattern of vegetation or the spread of fire within large simulation pixels. The relative insignificance of terrain in determining fire ignition and spread dynamics may also mean that coarse scale models need not account for topographic complexity with the simulation unit when estimating burned area.

The results from the classification can be applied to compare other simulation experiments. Subsequent phases ot he the model comparison need to evaluate ignition and area burned over longer simulation years (100 years) to assess the sensitivity of succession and fire effects on the simulation of fire regime. This should provide a foundation to dynamically represent fire and vegetation in coarse scale DGVMs or LFSMs by taking into account fire plant functional types (Díaz and Cabido 1997). Next, there will be a need to modify the comparison method to evaluate the effect of fuel treatments and fire suppression on fire dynamics by simulating various fuel treatment patterns and management ignition patterns with the weather scenarios for current climate. Results from this effort could provide valuable information on the effectiveness of fuel treatments under severe weather conditions.

Acknowledgments

This work was mostly supported by the National Center for Ecological Analysis and Synthesis (NCEAS), a Center funded by the National Science Foundation (Grant #DEB-0072909), the University of California, and the Santa Barbara campus. We also thank all participants in the four NCEAS workshops especially Andrew Fall, Florent Mouillot, Carol Miller, Don McKenzie, Dan Fagre, and Mike Wotton. We thank Russ Parsons, USDA Forest Service Fire Sciences Laboratory for technical assistance and technical review and CSIRO Tropical Savannas Research Centre, Darwin, Australia for support. This work was also partially funded by a grant (NS-7327) from NASA's Earth Science Applications Division.

References

Agee JK (1993) Fire ecology of Pacific Northwest Forests. Island Press, Washington DC USA

Albini FA (1976) Estimating wildfire behavior and effects. USDA Forest Service General Technical Report INT-30. Intermountain Research Station, Ogden, UT, USA. 45 pages

Anderson DG, Catchpole EA, DeMestre NJ, Parkes E (1982) Modeling the spread of grass fires. J. Austral. Math. Soc. 23:451–466

Bristow KL; Campbell GS (1984) On the relationship between incoming solar radiation and daily maximum and minimum temperature. Agricultural and Forest Meteorology 31:159–66

Cary GJ (1998) Predicting fire regimes and their ecological effects in spatially complex landscapes. Doctoral dissertation. Australian National University, Canberra, AU

Cary GJ (2002) Importance of a changing climate for fire regimes in Australia. In: Bradstock RA, Gill AM, Williams JE (eds) Flammable Australia: the fire regimes and biodiversity of a continent. Cambridge University Press, Cambridge, UK, pp 26–46

Cary GJ, Banks JCG (1999) Fire regime sensitivity to global climate change: An Australia perspective. *in* J L Innes, M Beniston, and MM Verstraete, Editors. Advances in Global Change Research: Biomass burning and its inter-relationships with the climate system. Kluwer Academic Publishers, London, UK

Cary GJ, Gallant JC (1997) Application of a stochastic climate generator for fire danger modelling. Proceedings of the Bushfire '97 Australasian Bushfire Conference, July 1997, Darwin, Northern Territory, Australia

Cary GJ, Keane RE, Gardner RH, Lavorel S, Flannigan MD, Davies ID, Li C, Lenihan JM, Rupp TS, Mouillot F (2006) Comparison of the sensitivity of landscape-fire-succession models to variation in terrain, fuel pattern and climate. Landscape Ecology 21: 121–137

Clark JS (1988) Effect of climate change on fire regimes in northwestern Minnesota. Nature 334:233–235

Clark JS (1993) Fire, climate change, and forest processes during the past 2000 years. Geological Society of America, Special Paper 276:295–308

Crutzen PJ, Goldammer JG (1993) Fire in the Environment: The ecological, atmospheric and climatic importance of vegetation fires. John Wiley and Sons, New York, NY, USA

DeBano LF, Neary DG, Folliott PF (1998) Fire's Effect on Ecosystems. John Wiley and Sons, New York, New York USA. 254 pages

Díaz S, Cabido M (1997) Plant functional types and ecosystem function in relation to global change. Journal of Vegetation Science 8:121–133

Flannigan MD, Van Wagner CE (1991) Climate change and wildfire in Canada. Canadian Journal of Forest Research 21:66–72

Forestry Canada Fire Danger Group (1992) Development and structure of the Canadian Forest Fire Behaviour Prediction Systems. Information Report ST-X-3. Ottawa: Forestry Canada, Science and Sustainable Development Directorate. 63 pages

Fox BJ, Fox MD, McKay GM (1979) Litter accumulation after fire in a eucalypt forest. Australian Journal of Botany 27:157–65

Gardner RH, Hargrove WW, Turner MG, Romme WH (1996) Climate change, disturbances and landscape dynamics. In: Walker BH, Steffen WL (eds) Global change and terrestrial ecosystems. Cambridge University Press, Cambridge, MA, USA, pp 149–172

Gardner RH, Romme WH, Turner MG (1999) Predicting forest fire effects at landscape scales. Pages 163–185 *in* DJ Mladenoff and WL Baker Editors. Spatial modeling of forest landscape change: approaches and applications. Cambridge University Press, Cambridge, United Kingdom

Gill AM (1975) Fire and the Australian flora: a review. Australian Forestry 38, 4–25

Hargrove WW, Gardner RH, Turner MG, Romme WH, Despain DG (2000) Simulating fire patterns in heterogeneous landscapes. Ecological Modelling 135:243–263

Hawkes BC; Flannigan MD, Editors (2000) Landscape fire modeling-challenges and opportunitites. Northern Forestry Centre Information Report NOR-X-371, Canadian Forestry Service, Victoria, British Columbia. 68 pages

Hirsch KG (1996) Canadian Forest Fire Behavior Prediction (FBP) System: User's Guide. Natural Resources Canada, Canadian Forest Service, Northwest Region, Northern Forest Centre, Edmonton, Alberta, Special Report Number 7. 12 pages

IPCC (2001) Climate Change 2001: The Scientific Basis. IPCC Third Assessment Report: Summaries for Policymakers WG I Climate Change 2001

Johnson EA (1992) Fire and vegetation dynamics: studies from the North American boreal forest. Cambridge University Press, Cambridge

Kasischke ES, Christensen JL, Stocks BJ (1995) Fire, global warming, and the carbon balance of boreal forests. Ecological Applications 5:437–451

Keane RE, Finney MA (2003) The simulation of landscape fire, climate, and ecosystem dynamics. In: Veblen TT, Baker WL, Montenegro GT, Swetnam TW, editors. Fire and Global Change intemperate Ecosystems of the Western Americas. Springer-Verlag, New York, USA

Keane RE, Long D, Basford D, Levesque BA (1997) Simulating vegetation dynamics across multiple scales to assess alternative management strategies. Pages 310–315 *in* Conference Proceedings – GIS 97, 11th Annual symposium on Geographic Information Systems – Integrating spatial information technologies for tomorrow. GIS World, INC., Vancouver, British Columbia, Canada

Keane RE, Ryan KC, Finney MA (1998) Simulating the consequences of altered fire regimes on a complex landscape in Glacier National Park, USA. Tall Timbers Fire Ecology Conference 20:310–324

Keane RE, Parsons R, Hessburg P (2002) Estimating historical range and variation of landscape patch dynamics: Limitations of the simulation approach. Ecological Modelling 151:29–49

Keane RE, Cary G, Davies ID, Flannigan MD, Gardner RH, Lavorel S, Lenihan JM, Li C, Rupp TS (2004) A classification of landscape fire succession models: spatially explicit models of fire and vegetation dynamics. Ecological Modelling 179(1):3–27

Knight DH (1987) Parasites, lightning, and the vegetation mosaic in wilderness landscapes. In: Turner MG (ed) Landscape heterogeneity and disturbance. Springer-Verlag, New York, USA, pp 59–83

Lavorel S, Davies ID, Nobel IR (2000) LAMOS: a LAndscape MOdelling Shell. Pages 25–28 *in* B. Hawkes and MD Flannigan, editors. Landscape fire modeling-challenges and opportunities. Natural Resources Canada, Canadian Forest Service, Vancouver, BC, Canada

Lenihan JM, Daly C, Bachelet D, Neilson RP (1998) Simulating broad scale fire severity in a dynamic global vegetation model. Northwest Science 72:91–103

Li C (2000) Reconstruction of natural fire regimes through ecological modelling. Ecological Modelling 134:129–144

Li C (2001) Fire disturbance patterns and forest age structure. Natural Resource Modeling 14:495–521

Matalas NC (1967) Mathematical assessment of synthetic hydrology. Water Resources Research 3, 937–45

McArthur AG (1967) Fire behavior in eucalypt forests. Leaflet Number 107 Commonwealth of Australia Forestry and Timber Bureau. 16 p

McCarthy MA, Cary GJ (2002) Fire regimes in landscapes: models and realities. Pages 77–94 *in* R Bradstock, J Williams, AM Gill, editors. Flammable Australia: the fire regimes and biodiversity of a continent. Cambridge University Press, Cambridge, United Kingdom

McCune B, Mefford MJ (1999) PC-ORD, Multivariate analysis of ecological data. Version 4. MjM Software Design, Gleneden Beach, Oregon, USA

Neilson RP, Running SW (1996) Global dynamic vegetation modelling: coupling biogeochemistry and biogeography models. Pages 451–465 *in* Global change and terrestrial ecosystems. Cambridge University Press, New York, New York, USA

Noble IR, Bary GAV, Gill AM (1980) McArthur's fire-danger meters expressed as equations. Australian Journal of Ecology 5:201–203

Olsen J (1981) Carbon balance in relation to fire regimes. Pages 327–378 *in* HA Mooney, Bonnicksen TM, Christensen NL, Lotan JE, Reiners WA (Technical Coordinators), editor. Proceedings of the Conference Fire Regimes and Ecosystem Properties. USDA Forest Service. Washington Office Report WO-3

Pausas JG, Austin MP, Noble IR (1997) A forest simulation model for predicting eucalypt dynamics and habitat quality for arboreal marsupials. Ecological Applications 7:921–933

Quinlan JR (2003) Data mining tools See5 and C5.0. RULEQUEST RESEARCH, St. Ives, NSW, Australia. www.rulequest.com/see5-info.html

Raison RJ, Woods PV, Khanna PK (1986) Decomposition and accumulation of litter after fire in sub-alpine eucalypt forests. Australian Journal of Ecology 11, 9–19

Reinhardt ED, Keane RE, Brown JK (2001) Modeling fire effects. International Journal of Wildland Fire 10:373–380

Richardson CW (1981) Stochastic simulation of daily precipitation, temperature, and solar radiation. Water Resources Research 17:182–90

Roderick ML (1999) Estimating the diffuse component from daily and monthly measurements of global radiation. Agric For Meteorol 95:169–185
Rothermel RC (1972) A mathematical model for predicting fire spread in wildland fuels. Research Paper INT-115, United States Department of Agriculture, Forest Service, Intermountain Forest and Range Experiment Station, Ogden, Utah. 56 pages
Rupp TS, Keane RE, Lavorel S, Flannigan MD, Cary GJ (2001) Towards a classification of landscape fire succession models. GCTE News 17:1–4
Ryan KC (1991) Vegetation and wildland fire: implications of global climate change. Environment International 17:169–178
Starfield AM, Chapin FS (1996) Model of transient changes in arctic and boreal vegetation in response to climate and land use change. Ecological Applications 6:842–864
Swanson FJ, Franklin JF, Sedell JR (1997) Landscape patterns, disturbance, and management in the Pacific Northwest, USA. In: Zonnneveld IS, Forman RTT (eds) Changing Landscapes: An Ecological Perspective. Springer-Verlag, New York, pp 191–213
Swetnam TW (1997) Mesoscale disturbance and ecological response to decadal climatic variability in the American Southwest. Journal of Climate 11:3128–3147
Swetnam TW, Baisan CH (1996) Historical fire regime patterns in the soutwestern United States since AD 1700. Pages 11–32 *in* CD Allen, editor. Fire effects in southwestern forests, Proceedings of the 2nd La Mesa Fire Symposium. Rocky Mountain Forest and Range Experiment Station, Forest Service, USDA
Thonicke KS, Venevski S, Sitch S, Cramer W (2001) The role of fire disturbance for global vegetation dynamics: coupling fire into a Dynamic Global Vegetation Model. Global Ecology and Biogeography Letters 10:661–678
Van Wagner CE (1987) Development and structure of the Canadian Forest Fire Weather Index System. Forestry Technical Report No. 35. Ottawa: Canadian Forestry Service. 36 p
Weber MG, Flannigan MD (1997) Canadian boreal forest ecosystem structure and function in a changing climate: impact on fire regimes. Environmental Review 5:145–156

Chapter 13

Plant Functional Types: Are We Getting Any Closer to the Holy Grail?

Sandra Lavorel · Sandra Díaz · J. Hans C. Cornelissen · Eric Garnier · Sandy P. Harrison · Sue McIntyre · Juli G. Pausas Natalia Pérez-Harguindeguy · Catherine Roumet · Carlos Urcelay

13.1 In Search of the Holy Grail

Functional classifications have been seen as a necessary tool for the simplification of floristic complexity in global vegetation models (Neilson et al. 1992; Prentice et al. 1992; Foley et al. 1996; Woodward and Cramer 1996), for mapping vegetation patterns at key times in the past (Prentice and Webb 1998; Prentice et al. 2000), and for monitoring effects of global change or management on vegetation distribution and ecosystem processes (Díaz et al. 2002a; Cruz et al. 2002). Plant functional classifications were first designed by grouping plants a priori based on knowledge of their function, or based on observed correlations among their morphological, physiological, biochemical, reproductive or demographic characteristics (Woodward and Cramer 1996; Smith et al. 1997). It was assumed that these classifications would allow to predict changes in ecosystem processes directly from projected changes in plant species composition in response to global change. This idea was challenged by the recognition that functional effect groups (species with a similar effect on one or several ecosystem functions; e.g., primary productivity, nutrient cycling, Gitay and Noble 1997; Walker et al. 1999) and functional response groups (groups of species with a similar response to a particular environmental factor; e.g., resource availability, disturbance or CO_2; Gitay and Noble 1997; Lavorel et al. 1997) do not necessarily coincide. Although there have been sustained efforts to refine plant functional type (PFT) concepts and terminology (Gitay and Noble 1997; Lavorel et al. 1997; Lavorel and Garnier 2002), the search for a single, functionally comprehensive yet relatively parsimonious, plant functional classification has remained an elusive Holy Grail.

The Holy Grail requires to focus on functional traits of terrestrial vascular plants that (1) can together represent the key responses and effects of vegetation at various scales from ecosystems to landscapes, biomes, and continents; (2) are suitable for relatively easy, inexpensive and standardised measurement over the world; and (3) can hence be used to devise a satisfactory functional classification for global-scale modeling and mapping of the biosphere.

A large amount of research has been initiated on plant furnctional traits and PFTs since the early days of GCTE, where the Holy Grail was formulated largely by modellers. In this chapter we first summarise theoretical and empirical progress on the understanding of the response traits that are relevant to different aspects of environmental change. Recently numerous empirical studies have made considerable progress in elucidating how plant traits can be related to plant function in relation to main environmental constraints (see Sect. 13.2), and how these same traits are then relevant to the distribution of species along gradients of climate, nutrient availability, and disturbance. Additional theoretical (Ackerly 2003) and empirical (Ackerly 2004a) work has also analysed the role of phenotypic plasticity, ecological sorting and natural selection in determining present patterns of association between plant traits and environmental gradients. The correlational approach, which has formed the bulk of that research, has been quite successful in detecting significant associations between particular plant traits and environmental factors (reviewed in Sect. 13.3). Understanding how these traits may then influence biotic interactions, and eventually shape local vegetation through community assembly has however appeared to be a non-trivial scaling exercise, and we review current state of the art in Sect. 13.4. Likewise, scaling from individual plant traits that are affected by environmental changes to ecosystem effects has proved more challenging than initially anticipated because the traits that determine the response to specific environmental factors overlap directly, indirecly, or not at all with the traits that determine ecosystem functions such as biogeochemical cycling or flammability (Lavorel and Garnier 2002; reviewed in Sect. 13.5). Finally, we return to asking how these results have helped modellers working at larger scales, and what key challenges remain.

13.2 Individual Plant Structure and Function

Among all possible traits measurable on an individual plant, those of interest to global syntheses and modeling must fill four conditions: *(i)* bear some relationship to plant function; *(ii)* be relatively easy to observe and quick

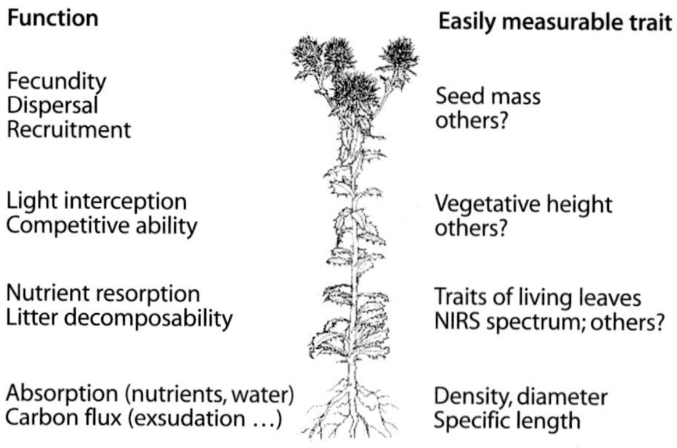

Fig. 13.1. Examples of soft traits and associated functions

to quantify ('soft' traits; Hodgson et al. 1999), *(iii)* using measurements that can be standardized across a wide range of species and growing conditions (Cornelissen et al. 2003b); *(iv)* have a consistent ranking – not necessarily constant absolute values – across species when environmental conditions vary (Garnier et al. 2001; Cornelissen et al. 2003a; Shipley and Almeida-Cortez 2003). Such traits are called 'functional traits' or 'functional markers' (Garnier et al. 2004), and can be morphological, ecophysiological, biochemical, demographical or phenomenological (Fig. 13.1). At this stage, the interest lies in the role of these traits for main plant functions, while their role in ecosystem dynamics is not specified, and they may turn to be response and/or effect traits. Information on soft traits is available for large numbers of species from all types of vegetation (see Díaz et al. 2004; Wright et al. 2004). Over the years, a consensus has been growing about which soft traits are best related to key plant functions such as resource economy or recruitment and how these then become response traits to climate and disturbances (Grime et al. 1997; Westoby et al. 2002; Cornelissen et al. 2003b) (Table 13.1). Comparability of data and global coverage must be ensured by using standardised lists of traits, which are a consensus about which traits are critical for the ecological challenges ahead (Table 13.1), and measuring them with standard methodologies (McIntyre et al. 1999; Díaz et al. 2002b; Cornelissen et al. 2003b), now freely available (Cornelissen et al. 2003b[1]).

Plant functional traits are considered as reflecting adaptations to variation in the physical environment and trade-offs (ecophysiological and/or evolutionary) among different functions within a plant. Co-varying sets of traits have been associated to 'major axes of plant specialization', that are consistent across environments, biogeographic regions and major plant taxa (Grime 1977; Chapin et al. 1993; Díaz et al. 2004). More generally, the analysis of plant functional trait responses to environmental variation, and of their effects on ecosystem function has been guided by the recognition that plants are constrained for performing alternative functions simultaneously, such as resource capture and conservation (Grime 1979; Chapin et al. 1993; Poorter and Garnier 1999), acquisition of several different resources (light and water, Smith and Huston 1989; light and nutrients, Tilman 1988), or growth and reproduction (Silvertown et al. 1993; Solbrig 1993).

A synthesis of empirical and theoretical studies proposed that at least four axes of plant specialization should be considered (Westoby et al. 2002). The first and best understood axis is represented by the specific leaf area (SLA) – leaf life span trade-off and is associated with turnover time of plant parts (including through herbivory), nutrient residence times and rate of response to favourable growth conditions. The global relevance of this axis was confirmed by an analysis across four floras from different biomes and biogeographic regions (Díaz et al. 2004). The second axis, representing the trade-off between fecundity and seed mass addresses establishment opportunities and success in the face of hazards, respectively. Seed mass and fecundity are negatively correlated, even after correcting for plant size. The third axis represented by potential plant height, carries several trade-offs and adaptive elements, and captures multiple constraints such as the density and height of shading competitors, water economy, and response to disturbance. The global relevance of plant height was confirmed by cross-continental analysis (Díaz et al. 2004). Finally, a fourth axis representing the coupled variation between twig size and leaf size determines the texture of canopies. Although a decrease in leaf size is common in dry, high light or cold conditions, the costs and benefits of small vs. large leaves remain to be formalized. Still, identifying an axis relevant to temperature variation, and especially response to extremes, stands as a challenge to ecophysiologists and ecologists.

However, the previous syntheses focused on aboveground traits, which are easily accessible. Recent studies have endeavoured to provide the same kind of information belowground, focusing on the functions of root traits, trade-offs among them, and how they relate to aboveground traits that could then be used as proxies for belowground function. Suites of correlated traits appear to also exist in roots, and represent a trade-off between root longevity and growth rate. Generally, low specific root length (SRL: ratio between root length and root biomass) is associated with thick, dense roots with low nitrogen and high lignin concentrations (Comas and Eissenstat 2004; Craine et al. 2002; Craine et al. 2003), a syndrome usually found in slow growing species or plants grown in limiting conditions (Reich et al. 1998a; Ryser 1998; Comas et al. 2002; Comas and Eissenstat 2004). Available data linking root morphology and plant functioning, and analogies to leaf traits suggest that this suite of traits reflects root longevity (Eissenstat 1991; Eissenstat

[1] http://www.publish.csiro.au/nid/65/paper/BT02124.htm.

Table 13.1. Association of plant functional traits with (1) plant responses to four classes of environmental change (i.e., 'environmental filters'), (2) plant competitive strength and plant 'defense' against herbivores and pathogens (i.e., 'biological filters'), and (3) plant effects on biogeochemical cycles and disturbance regimes. See also Chapin et al. (1993), Díaz et al. (1999), Weiher et al. (1999), Lavorel (2002) and Lavorel and Garnier (2002) for details, including 'hard traits' corresponding with the soft traits given here. Soil resources include water and nutrient availability. Disturbance includes any process that destroys major plant biomass (e.g., fire, storm, floods, extreme temperatures, ploughing, landslides, severe herbivory or disease). Note that effects on disturbance regime may also result in effects on climate or atmospheric CO_2 concentration, for instance fire promotion traits may be linked with large-scale fire regimes, which in turn may affect regional climates

	Climate response	CO_2 response	Response to soil resources	Response to disturbance	Competitive strength	Plant defence/ protection	Effects on biogeochemical cycles	Effects on disturbance regime
Whole-plant traits								
Growth form	*	*	*	*	*	*	*	*
Life form	*	*	*	*	*		*	*
Plant height	*	*	*	*	*	*	*	*
Clonality	*	?	*	*	*			?
Spinescence	*	?			*	*		?
Flammability		?			*	?	*	*
Leaf traits								
Specific leaf area	*	*	*		*	*	*	
Leaf size	*	?	*		*	*	*	
Leaf dry matter content	*	?	*			*	*	*
Leaf N and P concentration	*	*	*	*	*	*	*	
Physical strength of leaves	*	?	*	*		*	*	
Leaf life-span	*	*	*	*	*	*	*	*
Leaf phenology	*		*		*		*	*
Photosynthetic pathway	*	*			*			
Leaf frost resistance	*				*	*		
Stem and below-ground traits								
Stem specific density	*	?	?	*		*	*	*
Twig dry matter content	*	?	?	?		*	*	*
Twig drying time	*	?	?				?	*
Bark thickness			*	*		*		?
Specific root length	*	?	*		*	*		?
Diameter of fine root	*	?	*					
Distribution of rooting depth	*	*	*	*	*		*	*
95% rooting depth	*	?	*		*			*
Nutrient uptake strategy	*	*	*	*	*		*	
Regenerative traits								
Dispersal mode				*				
Dispersule shape and size				*				
Seed mass			*	*	*	*		
Resprouting capacity								

et al. 2000; Ryser 1996), greater nutrient use efficiency and resistance to herbivore and physical damage (Craine et al. 2001). These traits are thought to minimize nutrient losses, allowing plants to grow larger at low nutrient supply rates. Opposite traits tend to maximize root surface area and length per unit biomass, thereby allowing quick exploration of soil resources and rapid growth. However, root traits are definitely not easily measured. Therefore, there are a number of challenges still to be resolved: *(i)* to identify traits that are closely related to key root functions such as nutrient acquisition, anchoring, rhizospheric activity, decomposition rate; *(ii)* to normalize root traits measurements for broad comparisons; *(iii)* to test relationships between leaf and root traits for later use of leaf traits as easier proxies. Close relationships between leaf and root nitrogen concentration have already been shown at a global scale (Craine and Lee 2003; Craine et al. 2005).

As recognised by early work (Theophrastus ca. 300 BC; Raunkiaer 1934), growth forms are one expression of trade-offs among traits and adaptation to different types of environments. As such, they provide comprehensive links between key traits, plant response to their environment and their effecs on ecosystem function (Chapin 1993), and this is why large scale dynamic models have used them as the basis for their plant functional classifications. However, there is also an important range of variation in trait values and detailed trade-offs among traits within each life form. For instance, the trade-off between leaf lifespan (and underlying protective traits) and traits that promote leaf productivity (e.g., SLA) operates both within and between life forms (Díaz et al. 2004; Wright et al. 2004), but relationships between seed dispersal syndromes and seed size depend on seed size (Westoby et al. 1990).

13.3 Traits and Environmental Gradients

The functional meaning of plant traits for response to environmental variation has been identified through observations of the variations in trait values across environments differing for one or several factors, targeted experiments and modeling. Most of our knowledge is derived from the first approach, and therefore directly applies to spatial variation. Space-for-time substitution is then used to assume that the same changes in trait values would be observed following an environmental change of the same magnitude. Growing numbers of experiments have confirmed this assumption for responses to single environmental factors such as nutrient availability (Dyer et al. 2001) or grazing (Bullock et al. 2001), as do models of community or ecosystem dynamics (Pausas 1999; Colasanti et al. 2001; Ackerly 2003; Boer and Stafford Smith 2003; Hickler et al. 2004).

This combination of approaches has advanced the understanding of the adaptive significance of traits, or combinations of traits, to predict the response of organisms to climate, nutrients and disturbance (Table 13.1). Globally, response to climate, considered in terms of response to extreme low or high temperatures, and to gradients of moisture availability, is associated with variations in life form, leaf traits, rooting depth and lateral root spread, and genome size (McGillivray 1995; Díaz and Cabido 1997; Pavón et al. 2000; Niinemets 2001; Schenk and Jackson 2002, 2005). Specific leaf area (SLA), leaf dry matter content (LDMC) and leaf chemical composition co-vary with soil resource availability (Cunningham et al. 1999; Poorter 1999) as do specific root length, tissue density and diameter (Reich et al. 2003; Robinson and Rorison 1988). Responses to CO_2 depend on life cycle, relative growth rate (RGR), photosynthetic pathway, and stochiometric relationships (Poorter et al. 1996). Response to water availability is associated with variation in SRL, root diameter and root architecture (Fitter 1991; Wright and Westoby 1999; Nicotra et al. 2002). RGR, leaf and root morphology, and seed mass determine response to shading (Leishman and Westoby 1994; Reich et al. 1998b). Response to disturbance is associated with variation in life cycle, plant height, architecture, resprouting and seed traits (McIntyre and Lavorel 2001; Bond and Migdley 2001; Pausas et al. 2004; Díaz et al. in press). Following, we summarise recent progress in the identification of plant traits associated with response to the two simple factors that vary most significantly across communities within a landscape: resource availability and disturbance.

13.3.1 Plant Functional Response to Mineral Resource Availability

Early syntheses on changes in species traits along nutrient gradients (Grime 1977; Chapin 1980) recognised that species from nutrient-rich habitats tend to be inherently fast-growing. This goes along with rapid resource capture and fast turn-over of organs leading to poor internal conservation of resources, while the reverse is true for species from nutrient-poor habitats (Tables 13.1, 13.2). More recently, a series of quantitative traits has been associated with this fundamental trade-off in plant functioning (Reich et al. 1992; Grime et al. 1997; Poorter and Garnier 1999; Díaz et al. 2004; Wright et al. 2004). Fast-growing species from nutrient-rich habitats usually have a combination of high SLA, high SRL and relatively more fine roots, high tissue nutrient (in particular nitrogen) concentration, low tissue density and cell wall content, high rates of carbon and nutrient uptake, and short-lived leaves and roots. Opposite traits characterize species from nutrient-poor habitats, in which the mean residence time of nutrients tends to be maximized through longer-lived organs and/or higher resorption of nutrients from senescing organs (Ryser 1996; Garnier and Aronson 1998; Boot 1989; Aerts and Chapin 2000; Westoby et al. 2002).

13.3.2 Plant Functional Response to Disturbance

Disturbances are defined here as natural or land use related events that remove biomass or individuals (Grime 1977). Although disturbances must be considered for relevant plant functional classifications (Lavorel et al. 1997; Lavorel and Cramer 1999; Pausas et al. 2003), a theoretical framework comparable to that developed for resource economy remains elusive. Moving beyond the well-

Table 13.2. Summary of traits associated with contrasted environmental conditions and their feed forward effects on community structure and ecosystem processes in perennial grasslands

Level of organization	High fertility High disturbance	Low fertility Low disturbance
Individual traits	High SLA, low LDMC, low leaf C:N; allocation to leaves and stems; high plasticity	Large size, long-lived, low SLA, high LDMC, high leaf C:N; allocation to roots; low plasticity
	Numerous, persistent small seeds with high dispersal; selfing	Few larger seeds with low dispersal and no seed bank; outcrossing
Species interactions	Exploitative competitors: rapid depletion of resources; symmetric interactions	Conservative competitors: tolerance to low resource levels; asymmetric interactions
	Predominance of competition	Predominance of facilitation and allelopathy
	High herbivory	High mycorhizal associations
Community	Abundance of forbs and some stoloniferous grasses	Abundance of cespitous grasses
Ecosystem processes	Fast rates of biogeochemical cycling; rapid litter decomposition; high NPP	Slow rates of biogeochemical cycling; slow litter decomposition; low NPP
	High palatability	High fuel loads, high flammability

known ruderal syndrome (Grime 1977) requires better understanding of regeneration traits. There is good evidence that seed size has a fundamental evolutionary and ecological significance for post-disturbance colonization, competitive response and tolerance to abiotic stress (Venable et al. 1988; Westoby et al. 2002). However the role of this trait for seed persistence, dispersal or seedling growth and survival is debated (Marañon and Grubb 1993; Thompson et al. 1993; Hughes et al. 1994). Overall, traits determining population persistence have so far received limited attention in functional trait analyses (Eriksson and Ehrlén 2001). In an analysis of the sensitivity of population growth rate to species demographic parameters, Silvertown et al. (1993) found a correspondence between longevity and resource-rich environments, survival and resource shortage, and fecundity and disturbance. These patterns still need to be matched with variation in soft traits through meta-analyses of large demographic and trait data bases.

Syntheses targeted at specific disturbances organised under the banner of GCTE have highlighted recurrent patterns of plant specialization in relation to soil disturbance, grazing and fire, as well as the nuances that need to be applied to them.

Soil disturbance consistently favours plants with a suite of traits additional to the ruderal syndrome (Lavorel and McIntyre 1999c). Plant species tolerant to soil disturbance by ploughing or mammal digging are characterised by: a short and prostrate stature, with either a stoloniferous architecture in perennial grasses, or flat rosettes in forbs, high fecundity and a small dormant seed pool. Intolerant species are typically larger tussock grasses or dicotyledons, with low fecundity and no seed dormancy mechanisms, and with low plasticity in their morphology. Finally, a group of indifferent species has an architecture characterised by leafy stems, with high morphological plasticity, and high seed dormancy (Lavorel et al. 1998, 1999a,b).

There are few empirically tested generalizations about which plant traits are positively or negatively associated with ungulate grazing, and the validity of some widely recognized trait responses to grazing has remained mostly untested at the global scale. There have been suggestions in the literature, based on pair-wise regional comparisons (Díaz et al. 2001; Adler et al. 2004, 2005), that evolutionary history of grazing by ungulates, as well as habitat productivity, determine what plant traits are favoured by grazing. Díaz et al. (in press) have asked what plant traits are consistently associated with grazing at the global scale, and whether these traits varied with precipitation (a surrogate for resource availability) and evolutionary history of grazing. A quantitative analysis of 195 studies from all over the world confirmed that overall grazing favoured annuals over perennials, short-statured over tall-statured plants, prostrate over erect plants, and stoloniferous or rosette over tussock architecture. This analysis demonstrated for the first time that some of the response patterns disappeared or were substantially stronger or weaker under particular combinations of precipitation and evolutionary history of herbivory. For example, in dry regions with long evolutionary history of ungulate herbivory, grazing did not favor annual plants over perennial plants.

Pausas et al. (2004) analysed regeneration strategies worldwide for woody plant species from ecosystems that are subject to stand-replacement (crown) fires. It is commonly assumed that the main traits allowing persistence after stand-replacement fires are resprouting capacity and the ability to retain a persistent seed bank (termed 'propagule persistence capacity'). Different combinations of these two traits have been preferentially selected in floras with different evolutionary histories. Although all four possible binary combinations appear in most fire-prone ecosystems, the relative proportions of each type (and the dominant type) differ. In Australian heathlands, the proportion of resprouters and non-resprouters is relatively even, compared with other fire-prone ecosystems, though post-fire obligate resprouters (resprouters without a seed bank) are almost absent. In the Mediterranean basin, most resprouters are obligate, while in California shrubs resprouters are evenly segregated among those having propagules that persist after fire (facultative species) and those without propagule persistence capacity (obligate resprouters). Species with neither persistence mechanism are rare in most fire-prone shrublands. Although data was limited, the review also highlighted some clear trade-offs with other traits (e.g., height), as well as the importance of considering the phylogenetic relatedness for a proper understanding of functional traits and trade-offs. For instance, in the Mediterranean basin flora, most resprouters have fleshy fruits and most non-resprouters have dry fruits. However, this pattern is not due to an ecological trade-off, but to a common lineage, as demonstrated by a phylogenetically controlled analysis (Pausas and Verdú 2005).

The data syntheses relating response traits to fire and grazing have still only dealt with individual disturbances. A further challenge lies in the understanding of disturbance regimes. For example, in many grasslands of the world disturbance regimes combine fire, grazing and/or mowing, fertilization and soil disturbance. Because regimes represent fixed combinations of disturbances that recur on the landscape, they can mask the interactions between the individual disturbances within that structure plant communities (Collins 1987). Woodlands and savannas are often shaped by the combination of grazing by wild and domestic herbivores, together with intentional or unintentional fire (Bond et al. 2003). Additional disturbances associated with farming and forestry can co-occur with grazing and fire to structure woodlands and forests (McIntyre and Martin 2001; McIntyre et al. 2002; Dale et al. 2001). Grazing tolerance of a number of species from Australian subtropical grasslands was found to covary with other disturbances (McIntyre et al. 2003), with the grazing tolerance of native species tending to decline in the presence of soil disturbance or water enrichment. Novel combinations of human-induced and natural disturbances are already widespread and are expected to be features of the future. Their effects can range from changes in dominance of different PFTs to dramatic shifts associated to plant invasions (D'Antonio and Vitousek 1992).

Pausas and Lavorel (2003) proposed a unifying framework that may apply to a variety of disturbance types. This framework is based on the recognition that plants can have persistence strategies at different levels of organisation and provides an understanding of the order in which different mechanisms act on plant persistence in disturbed systems. The main parameters to determine persistence in chronically disturbed ecosystems are those related to: Individual-persistence capacity, Propagule-persistence capacity (persistence at the population level), Competitive capacity (persistence at the community level) and Dispersal capacity (persistence at the landscape level). The approach by Pausas et al. (2004) is a special case of this approach. Much remains to be done to identify traits relevant to different levels of response, and how these may vary depending on context. Nevertheless, sixteen possible functional types could be obtained by assuming a simple binary classification of the four levels of persistence.

13.3.3 Projecting Changes in Plant Functional Traits in Response to Global Change

Natural gradients usually combine variations in climate, resource availability, and disturbance regimes. These underlying simple gradients may be explicit for analyses of plant response to altitude (Pavón et al. 2000) or agricultural disturbance (Kleyer 1999), though again these combinations are likely to not be stable in the future. In many other instances however, the nature and amount of environmental variation underlying complex gradients (*sensu* Austin and Smith 1989) along which traits are studied has not been quantified or formalized. For instance, this is the case of successional gradients (Bazzaz 1996; Prach et al. 1997; Garnier et al. 2004; Richardson et al. 2005) or ecotones (Dodd et al. 2002) which are prime opportunities to obtain time series of plant functional traits.

Interpretation and projections of plant distributions along complex gradients are problematic because traits associated with different factors, such as water and nutrient stress, can overlap (Díaz et al. 2004) or be independent (Cunningham et al. 1999; Dyer et al. 2001). Non-overlapping trait responses are commonly observed when one environmental factor involves adult traits (e.g., nutrient availability) and other regeneration traits (e.g., disturbance) (Shipley et al. 1989; Leishman and Westoby 1992; Thompson et al. 1996; Lavorel et al. 1999b). In addition, the role of genetically-based plasticity of traits in response to environmental changes has largely been ignored from analyses, though it may contribute to population persistence (Strand and Weisner 2004; Stanton et al. 2000). Attempts to understand linkages, trade-offs and dependency among traits based on basic ecophysiological and evolutionary mechanisms, as reviewed in Sect. 13.2, are the way forward to address these complex responses (Ackerly et al. 2000; Westoby et al. 2002). In the context of global change, our ability to generate reliable projections of future vegetation is further limited by the fact that different combinations of climatic factors could occur in the future, as they have in the past (Jackson and Williams 2004), as will novel combinations of atmospheric CO_2 concentration, climate and disturbance regimes. Our ability to apply plant functional responses to generate future vegetation projections must therefore rely on more experimental and modeling work. One remaining challenge for modeling, however, lies in the construction of PFTs from analyses of continuous plant traits. This methodological challenge can be solved with statistical approaches (Pillar and Sosinsky 2003; Nygaard and Erjnaes 2004), but is also a more fundamental issue because of the assumption that future responses to multiple factors will be stable within groups.

13.4 Scaling from Individual Plants to Communities: from Response Traits to Community Assembly

Having learned how populations of individual species respond to environmental variation across landscapes, understanding how communities assemble remains a significant challenge (Weiher and Keddy 1999; Ackerly 2003; Suding et al. 2003). This challenge is particularly significant when novel environmental conditions and landscape fragmentation by land use may lead to entirely new assemblages, as have different environmental conditions in the past (Jackson and Williams 2004). Effects of changes in climate or land use might be modeled as changes in the strength of different abiotic (climatic, atmospheric CO_2 concentration, resource availability, disturbance) and biotic (competition, predation, mutualisms) filters that successively constrain which species and traits, from a regionally available pool, can persist at a site (Woodward and Diament 1991; Díaz et al. 1999; Naeem and Wright 2003; see also Díaz et al. 2007, Chap. 7 of this volume). We should then be able to predict the trait composition of communities by combining knowledge of (1) the regional species pool, (2) the nature and strength of different filters, (3) the response traits associated with each filter and (4) the rules that shape assembly (Campbell et al. 1999).

Attempts to predict interspecific competition from plant traits have had mixed success (Keddy et al. 1998; Wardle et al. 1998), in particular because species competitive rankings are sensitive to nutrient availability (Keddy et al. 2000), disturbance (Suding and Goldberg 2001), and to mycorrhizal associations (van der Heijden et al. 1998; Wardle et al. 1998). A reasonable consensus has been reached regarding tolerance of competition by neighbours (competitive response) where plant height and seed mass are positively associated with greater tolerance (Goldberg 1991). However, the attributes that determine competitive effects of plants upon their neighbours, and how these may vary with resources and disturbance, remain to be elucidated. Grime's competitive syndrome (1977, 2001) includes attributes such as tall stature, wide spread, nutrient monopolization and abundant litter production. Tilman (1988) also emphasised the importance of height in productive conditions, but in nutrient-poor environments leaf and root attributes associated with nutrient uptake are those expected to confer competitive effects (Craine et al. 2001, 2005). Further, Goldberg and Novoplansky (1997) proposed that competitive effects of species may depend on whether nutrients or water is most limiting. Relating this idea to the strategy model by Ackerly (2004b) in the case of nutrient limitation, a 'conservative competitor' strategy, with leaf traits promoting resource conservation, such as a low SLA, high tissue density and long life span, is expected. Otherwise an 'exploitative competitor' strategy, with opposite leaf traits, is expected (Michalet 2001; Liancourt et al. 2005). In addition, other types of plant-plant interactions such as facilitation and allelopathy may gain importance in either resource-poor and physically stressful, or resource-rich and physically benign environments (Bertness and Callaway 1994; Pellissier 1998; Bruno et al. 2003). Which plant traits are conducive to these other mechanisms remains to be elucidated, but are most likely related to the nutrient and water acquisition vs. conservation syndromes (Liancourt et al. 2005), including their effects on herbivory.

Another fundamental interaction structuring communities is herbivory (Crawley 1992). There is good evidence that structural and chemical traits known to be associated with nutrient or climatic gradients influence patterns of herbivory by generalist invertebrates (Grime et al. 1996; Wardle et al. 1998; Cornelissen et al. 1999; Pérez-Harguindeguy et al. 2003) and vertebrates (McKey et al. 1978; Bryant et al. 1991). Nutrient and/or water limitation tends to select for conservative strategies which result in low attractiveness to these herbivores due for example, to leaf high tensile strength (toughness) and low nutritive value (e.g., high C/N ratio) (Bryant et al. 1983; Coley et al. 1985; Cebrían 1998; Díaz et al. 2004). The opposite applies to resource-rich environments. How antiherbivore defences may then feedback to ecosystem productivity *via* the soil is discussed in Sect. 13.5.

Finally, the strength of interactions among plants is also expected to co-vary with other key traits along environmental gradients. This is because the nature and the quantity of the production of secondary biochemical compounds involved in anti-herbivore defense, litter decomposition, or allelopathy (Pellissier and Souto 1999) can also be affected by microclimate and resource availability (Herms and Mattson 1992; Hartley and Jones 1997).

Other types of trophic interactions that influence community structure and ecosystem processes are associations with soil microbes, for instance mycorrhizae (van der Heijden et al. 1998; Klironomos et al. 2000; Langley and Hungate 2003; Read et al. 2004; Rillig 2004). Cornelissen et al. (2001) revealed consistent large and significant differences in inherent relative growth rate (RGR), foliar chemistry and leaf litter decomposability among plants with mycorrhizal association strategies. This results in slow carbon cycling in ericoid and ectomycorrhizal plant species from temperate ecosystems with low pH, vs. low nitrogen availability and fast carbon cycling in arbuscular and non-mycorrhizal species, found in more nitrogen-rich ecosystems with higher pH. The relative abundance of dependent and non-dependent species in a community will determine the importance of this plant-fungus interaction (Urcelay and Díaz 2003). All together, these findings support Read's (1991) hypothesis that mycorrhizal type is an important component of a plant's strategy in the context of nutrient availability. However, at a global scale we still know little about links between plant-associated microbes and plant traits, or about belowground plant traits in general. Recent efforts in this field are promising (Jackson et al. 1996; Ryser 1996; Craine et al. 2003; Craine and Lee 2003; Wardle 2002).

In order to better capture the complexities of community assembly, and how these may link to individual plant traits, Suding et al. (2003) proposed that trade-offs among species traits (e.g., ability to capture and cycle resources quickly vs. leaf toughness) determine community structure through the nature and intensity of competition and other interspecific interactions depending on environmental conditions. Our current knowledge, as summarised above, highlights a number of these linkages. For instance, we expect that nutrient-poor environments will select for species with leaf traits promoting resource conservation, such as low SLA, high tissue density and long life span. As a consequence of these primary traits, predominant plants in such environments will compete with their neighbours by sequestering nutrients – possibly with the help of mycorrhizal associations, by accumulating poorly degradable litter (Berendse 1994), and sometimes through allelopathy or nutrient immobilization (Michelsen et al. 1995; Hättenschwiller and Vitousek 2000); while also facilitating subordinates by herbivore protection. The converse would be expected in nutrient-rich environments (Table 13.2).

13.5 Scaling from Communities to Ecosystems: from Response Traits to Effect Traits

The 'Holy Grail' hypothesis states that environmental changes will lead to changes in community composition and thus in plant traits, and these in turn will affect ecosystem functioning. This hypothesis was first approached by matching lists of response attributes with known effects of some of these attributes (or their correlates) on ecosystem processes (Díaz et al. 1999; Walker et al. 1999; Eviner and Chapin 2003; see also Díaz et al. 2007, Chap. 7 of this volume; see Tables 13.1 and 13.2 and Table 1 *in* Lavorel and Garnier 2002). Inspections of these lists have revealed that the resource axis has maximum overlap between response and effect traits, whereas overlaps are few in the case of disturbance. One essential step to understanding the causes of these differing degrees of overlap has been the analysis of the specific functions of the traits involved in either response or effect (Lavorel and Garnier 2002; see Table 13.1). For the resource axis it showed that responses to resource availability and effects on biogeochemistry are jointly constrained by the trade-off between acquisition and conservation strategies, and their characteristic traits (Chapin et al. 1993; Grime 2001). For example, plants growing in adverse environments have low specific leaf area, high C/N ratio, and high tensile strength. These traits make them less palatable to generalist herbivores (see above), and persist in litter, thereby strongly influencing decomposition (Wardle et al. 1998; Cornelissen et al. 1999; Pérez Harguindeguy et al. 2000). This way, herbivore-induced changes in the balance of palatable and unpalatable species lead to changes in the net litter quality and therefore in decomposer activity (Wardle et al. 1998; Cebrían et al. 1998; Wardle 2002). Association with N_2 fixing bacteria is another trait that provides feedback on ecosystem productivity *via* N-rich litter. This mechanism is in particular a recurrent one underlying the impacts of invasive species on nutrient cycling (D'Antonio and Corbin 2003). On the other hand, regeneration and demographic traits associated with response to disturbance (e.g., fire, grazing) are known to have little direct relation with adult ecophysiological traits, and would therefore be of little relevance to biogeochemistry.

The ultimate goal of response-effect analyses should be the formulation of parsimonious quantitative relationships expressing the different components of each ecosystem function in relation to particular traits (Lavorel and Garnier 2002; Eviner and Chapin 2003). These relationships would make it possible to use traits to scale from individual plants and the communities they form to the ecosystem level (Dawson and Chapin 1993). Such formulations have been proposed for aboveground primary productivity (Chapin 1993 and further modifications by Lavorel and Garnier 2002 and Garnier et al. 2004). Specific annual net primary productivity (SANPP; "ecosystem efficiency", Reich et al. 1997) expresses ANPP per gram of green biomass, and can be written as:

$$\mathrm{SANPP} = \frac{\log_e \left[\sum_{i=1}^{n\,\mathrm{species}} p_i x e^{\mathrm{RGR}_i x (tf-to)_i} \right]}{\Delta T}$$

where p_i is the relative contribution of species i to the biomass of the community, RGR_i and $(tf-to)_i$ are the aboveground relative growth rate and period of active growth of species i, respectively, and ΔT is the period over which SANPP is assessed. Garnier et al. (2004)

tested this relationship against independent measurements of leaf traits and productivity and showed that specific leaf area, LDMC, and leaf nitrogen concentration are indeed correlated with SANPP due to the well-established links between these traits and RGR (Reich et al. 1992; Poorter and Garnier 1999). In this equation, contributions of individual species to ecosystem function are proportional to their abundance in the community. Díaz et al. (this volume) further discuss the role of dominant species in functional diversity – ecosystem function relationships.

Correlations have also been established between rates of litter decomposition and leaf traits (SLA, LDMC and leaf tensile strength) of individual species (Cornelissen et al. 1999) and over communities (Garnier et al. 2004). However, in contrast to the case of ANPP there is no mechanistic model available. Empirical relationships between response and effects have been proposed for several other environmental factors and ecosystem functions (Chapin 2003; Eviner and Chapin 2003; Diaz et al. 2007, Chap. 7 of this volume), but developing formal models must be the next step.

Our understanding remains even more limited when attempting to link disturbance response and effect traits. Lavorel and Garnier (2002) showed that the list of traits relevant to ecosystem flammability has in fact minimal *direct* overlap with traits relating to fire response. When functional linkages between traits promoting fire tolerance and those involved in ecosystem flammability exist (Bond and Midgley 1995), even once phylogeny has been accounted for (Schwilk and Ackerly 2001), they result from associations or trade-offs between fire response traits and the actual traits that determine flammability, and are therefore *indirect*. For instance, a high growth rate is required to increase the success of seed regeneration after fire, and is also often associated with canopy architectures with many thin stems and high surface/volume ratios, which promote fire. Large underground structures allow resprouting, which increases fire survival, and drought tolerance, which allows low water potential and hence increases flammability. Closer investigations using phylogenetically independent analyses across floras evolved in high vs. low fire regimes, or sites with high vs. low resources, are needed to explore this issue further. Further knowledge will also be gained by analysing trait and ecosystem processes for cases of invasions that trigger positive fire feedback loops (D'Antonio 2000; Grigulis et al. 2005). Likewise, formal analyses and models linking grazing response strategies and palatability along resource gradients are still needed to build on the abundant but dispersed grazing literature (Landsberg et al. 1999; Adler et al. 2004). Pasture agronomists have shown that patterns of pastoral value (i.e., biomass quantity and quality over the growing season) along gradients of nutrient availability can be related to response traits such as LDMC (Cruz et al. 2002; Duru et al. 2004).

13.6 So, Are We Getting Closer to the Holy Grail? Scaling beyond Ecosystems

13.6.1 Plant Functional Traits and Landscape Dynamics

Plant functional types have been for a long time 'building blocks' of models of patch and landscape dynamics. At the landscape scale their most widespread use has been in models that couple successional dynamics as represented by applications of the Vital Attributes (VA) model (Noble and Slatyer 1980) with sub-models of lateral processes such as seed dispersal and disturbance propagation. These applications have been particularly successful in the case of 'Landscape Fire Succession Models' (*sensu* Keane et al. 2007, Chap. 12 of this volume), probably because the VA scheme was first designed to capture plant response to fire, and represents an intermediate level of complexity that is amenable to sophisitication by addition of new processes. Examples include applications to Australian wet sclerophyllous forest (Noble and Gitay 1996), Mediterranean shrublands (Pausas 2003) and of coniferous-dominated forests of northwestern America (Roberts and Betz 1997).

More complex versions of VA, such as the FATE model (Moore and Noble 1990) have been adapted to model landscape-fire interactions (LAMOS, Grigulis et al. 2005; FATELAND, Pausas and Ramos 2005) and responses to other disturbances such as grazing (Cousins et al. 2003) by including a series of semi-quantitative traits (e.g., for shade tolerance, recruitment). The latest developments of VA-based landscape models can take into account continuous traits and processes e.g., for biomass production and the dispersal phase (Grigulis et al. 2005). Current challenges lie in including biogeochemical processes, as captured by plant effect traits into the current models that simulate responses to changing soil resources, disturbances and their modifications by climate. Such models will make it possible to better capture important non-linearities associated with landscape dynamics (Reynolds et al. 1997; Boer and Stafford Smith 2003).

13.6.2 Regional to Global Models – Revisiting the Early Functional Classifications

Dynamic Global Vegetation Models (DGVMs: Foley et al. 1996; Friend et al. 1997; Woodward et al. 1998; Potter and Klooster 1999; Sitch et al. 2003; Gerten et al. 2004) simulate natural vegetation distribution and terrestrial carbon cycling in response to atmospheric and soil environment, disturbance and their interactions. The properties of the vegetation system arise out of competition between a limited number of PFTs whose behavior is, in turn, determined by basic physiological constraints.

Aerts R, Chapin III FS (2000) The mineral nutrition of wild plants revisited: a re-evaluation of processes and patterns. Advances in Ecological Research 30:1–67
Arora VK, Boer GJ (2005) A parameterization of leaf phenology for the terrestrial ecosystem component of climate models. Global Change Biology 11:39–59
Austin MP, Smith TM (1989) A new model for the continuum concept. Vegetatio 83:35–47
Bazzaz FA (1996) Plants in Changing Environments – Linking physiological, population and community ecology. Cambridge University Press, Cambridge
Berendse F (1994) Litter decomposability – a neglected component of plant fitness. Journal of Ecology 82:187–190
Berry SL, Roderick ML (2002a) CO_2 and land-use effects on Australian vegetation over the last two centuries. Australian Journal of Botany 50:511–531
Berry SL, Roderick ML (2002b) Estimating mixtures of leaf functional types using continental-scale satellite and climate data. Global Ecology and Biogeography 11:23–39
Berry SL, Roderick ML (2004) Gross primary productivity and transpiration flux of the Australian vegetation from 1788 to 1988 AD: effects of CO_2 and land use change. Global Change Biology 10:1884–1898
Bertness MD, Callaway R (1994) Positive interactions in communities. Trends in Ecology and Evolution 9:191–193
Bigelow NH, Brubaker LB, Edwards ME, Harrison SP, Prentice IC, Anderson PM, Andreev AA, Bartlein PJ, Christensen TR, Cramer W, Kaplan JO, Lozhkin AV, Matveyeva NV, Murray DF, McGuire AD, Razzhivin VY, Rtchie JC, Smith B, Walker DA, Gajewski K, Wolf V, Holmqvist BH, Igarashi Y, Kremenetskii K, Paus A, Pisaric MFJ, Volkova VS (2003) Climate change and Arctic ecosystems: 1. Vegetation changes north of 55°N between the last glacial maximum, mid-Holocene and present. Journal of Geophysical Research 108:8170, doi: 8110.1029/2002JD002558
Boer M, Stafford Smith M (2003) A plant functional approach to the prediction of changes in Australian rangeland vegetation under grazing and fire. Journal of Vegetation Science 14: 333–344
Bond WJ, Midgley JJ (1995) Kill thy neighbour: an individualistic argument for the evolution of flammability. Oikos 73:79–85
Bond WJ, Midgley JJ (2001) Ecology of sprouting in woody plants: the persistence niche. Trends in Ecology and Evolution 16: 45–51
Bond WJ, Midgley GF, Woodward FI (2003) The importance of low atmospheric CO_2 and fire in promoting the spread of grasslands and savannas. Global Change Biology 9:973–982
Boot RGA (1989) The significance of size and morphology of root systems for nutrient acquisition and competition. In: Lambers H, Cambridge ML, Konings H, Pons TL (eds) Causes ans consequences of variation in growth rate and productivity of higher plants. SPB Academic Publishing, The Hague, pp 299–311
Box EO (1981) Macroclimate and Plant Forms:An Introduction to Predictive Modelling in Phytogeography. Junk, The Hague
Brant SP, Chapin III FS, Klein DR (1983) Carbon/nutrient balance of boreal plants in relation to vertebrate herbivory. Oikos 40: 357–368
Bruno JF, Stachowicz JJ, Bertness MD (2003) Inclusion of facilitation into ecological theory. Trends in Ecology and Evolution 18: 119–125
Bryant JP, Provenza FD, Pastor J, Reichardt PB, Clausen TP, du Toit JT (1991) Interactions between woody plants and browsing mammals mediated by secondary metabolites. Annual Review of Ecology and Systematics 22:431–446
Bullock JM, Franklin J, Stevenson MJ, Silvertown J, Coulson SJ, Gregory SJ, Tofts R (2001) A plant trait analysis of responses to grazing in a long-term experiment. Journal of Applied Ecology 38:253–267
Campbell C, Stafford Smith DM, Ash JE (1999) A rule-based model for the functional analysis of vegetation change in Australasian grasslands. Journal of Vegetation Science 10:723–730
Cebrian J, Williams M, McClelland J, Valiela I (1998) The dependence of heterotrophic consumption and C accumulation on autotrophic nutrient content in ecosystems. Ecology Letters 1:165–170
Chapin III FS (1993) Functional role of growth forms in ecosystem and global processes. In: Ehrlinger JR, Field CB (eds) Scalling physiological processes: leaf to globe. Academic press, San Diego, pp 287–312
Chapin III FS (2003) Effects of plant traits on ecosystem and regional processes: a conceptual framework for predicting the consequences of global change. Annals of Botany 91:1–9
Chapin III STA, Autumn K, Pugnaire F (1993) Evolution of suites of traits in response to environmental stress. The American Naturalist 142:578–592
Colasanti RL, Hunt R, Askew AP (2001) A self-assembling model of resource dynamics and plant growth incorporating plant functional types. Functional Ecology 15:676–687
Coley PD, Bryant JP, Chapin III S (1985) Resource availability and plant antiherbivore defense. Science 230:895–899
Collins SC (1987) Interaction of disturbances in a tallgrass prairie: a field experiment. Ecology 68:1243–1250
Comas LH, Eissenstat DM (2002) Linking root traits to potential growth rate in six temperate tree species. Oecologia 132:34–43
Comas LH, Eissenstat DM (2004) Linking fine root traits to maximum potential growth rate among 11 mature temperate tree species. Functional Ecology 18:388–397
Cornelissen JHC, Pérez-Harguindeguy N, Díaz S, Grime JP, Marzano B, Cabido M, Vendramini F, Cerabolini B (1999) Leaf structure and defence control litter decomposition rate across species and life forms in regional flora on two continents. New Phytologist 143:191–200
Cornelissen HJC, Aerts R, Cerabolini B, Werger MJA, van der Heijden MGA (2001) Carbon cycling traits of plant species are linkedwith mycorrhizal strategy. Oecologia 129:611–619
Cornelissen JHC, Cerabolini B, Castro-Díez P, Villar-Salvador P, Montserrat-Martí G, Puyravaud JP, Maestro M, Werger MJA, Aerts R (2003a) Functional traits of woody plants: correspondence of species rankings between field adults and laboratory-grown seedlings? Journal of Vegetation Science 14:311–322
Cornelissen JHC, Lavorel S, Garnier E, Díaz S, Buchmann N, Gurvich DE, Reich PB, ter Steege H, Morgan HD, van der Heijden MGA, Pausas JG, Poorter H (2003b) Handbook of protocols for standardised and easy measurement of plant functional traits worldwide. Australian Journal of Botany 51:335–380
Cousins S, Lavorel S, Davies ID (2003) Modelling the effects of landscape pattern and grazing regimes on the persistence of plant species with high conservation value in grasslands in south-eastern Sweden. Landscape Ecology 18:315–332
Craine JM, Lee WG (2003) Covariation in leaf and root traits for native and non-native grasses along an altitudinal gradient in New Zealand. Oecologia 134:471–478
Craine JM, Froehle J, Tilman DG, Wedin DA, Chapin III FS (2001) The relationships among root and leaf traits of 76 grassland species and relative abundance along fertility and disturbance gradients. Oikos 93:274–285
Craine JM, Tilman D, Wedin D, Reich PB, Tjoelker MG, Knops J (2002) Functional traits, productivity and effects on nitrogen cycling in 33 grassland species. Functional Ecology 16:585–595
Craine JM, Wedin DA, Chapin FSI, Reich PB (2003) Relationship between the structure of root systems and resource use for 11 North American grassland plants. Plant Ecology 165:85–100
Craine JM, Lee WG, Bond WJ, Williams RJ, Johnson LC (2005) Environmental constraints on a global relationship among leaf and root traits of grasses. Ecology 86:12–19
Cramer W, Bondeau A, Woodward FI, Prentice IC, Betts RA, Brovkin V, Cox PM, Fisher V, Foley J, Friend AD, Kucharik C, Lomas MR, Ramankutty N, Sitch S, Smith B, White A, Young-Molling C (2001) Dynamic responses of global terrestrial ecosystems to changes in CO_2 and climate. Global. Change Biology 7:357–374
Crawley MJ (ed) (1992) Natural Enemies. Blackwell Scientific Publications, Oxford
Cruz P, Duru M, Therond O, Theau JP, Ducourtieux C, Jouany C, Al Haj Khaled R, Ansquer P (2002) Une nouvelle approche pour caractériser la valeur d'usage des prairies naturelles et seminaturelles. Fourrages 172:335–354
Cunningham SA, Summerhayes B, Westoby M (1999) Evolutionary divergences of leaf structure and chemistry, comparing rainfall and soil nutrient gradients. Ecological Monographs 69:569–588

D'Antonio CM (2000) Fire, plant invasions, and global changes. In: Mooney HA, Hobbs RJ (eds) Invasive species in a changing world. Island Press, Covela, pp 65–94

D'Antonio CM, Corbin JD (2003) Effects of plant invaders on nutrient cycling – Exploring the link between invasion and development of species effects. In: Canham C, Cole JJ, Lauenroth WK (eds) Models in Ecosystem Ecology. Princeton University Press, Princeton, pp 363–384

D'Antonio CM, Vitousek PM (1992) Biological invasions by exotic grasses, the grass/fire cycle, and global change. Annual Review of Ecology and Systematics 23:63–87

Dale VH, Joyce LA, McNulty S, Neilson RP, Ayres MP, Flannigan MD, Hanson PJ, Irland LC, Lugo AE, Peterson CJ, Simberloff D, Swanson FJ, Stocks BJ, Wotton BM (2001) Climate change and forest disturbances. Bioscience 51:723–734

Danell K, Bergström R (2002) Mammalian herbivory in terretrial environments. In: Herrera CMP, O. (ed) Plant-animal interactions. Blackwell Publishing, Oxford, pp 107–131

Dawson TE, Chapin III FS (1993) Grouping plants by their form – Function characteristics as an avenue for simplification in scaling between leaves and landscapes. In: Field CB, Ehleringer JR (eds) Scaling physiological processes from the leaf to the globe. Academic Press, pp 313–319

Díaz S, Cabido M (1997) Plant functional types and ecosystem function in relation to global change. Journal of Vegetation Science 8:463–474

Díaz S, Cabido M, Casanoves F (1999) Functional implications of trait-environment linkages in plant communities. In: Weiher E, Keddy P (eds) Ecological assembly rules – Perspectives, advances, retreats. Cambridge University Press, Cambridge, pp 338–362

Díaz S, Noy-Meir I, Cabido M (2001) Can grazing response of herbaceous plants be predicted from simple vegetative traits ? Journal of Applied Ecology 38:497–508

Díaz S, Briske D, McIntyre S (2002a) Range management and plant functional types. In: Hodkingson K, Grice AC (eds) Global Rangelands: Progress and Prospects. CAB International, Wallingford, pp 81–100

Díaz S, McIntyre S, Lavorel S, Pausas J (2002b) Does hairiness matter in Harare? – Global comparisons of plant trait responses to disturbance. New Phytologist 154:7–9

Díaz S, Hodgson JG, Thompson K, Cabido M, Cornelissen JHC, Jalili A, Montserrat-Martí G, Grime JP, Zarrinkamar F, Asri Y, Band SR, Basconcelo S, Castro-Diez P, Funes G, Hamzehee B, Khoshnevi M, Pérez-Harguindeguy N, Pérez-Rontomé MC, Shirvany FA, Vendramini F, Yazdani S, Abbas-Azimi R, Bogaard A, Boustani S, Charles M, Dehghan M, De Torres-Espuny L, Falczuk V, Guerrero-Campo J, Hynd A, Jones G, Kowsary E, Kazemi-Saeed F, Maestro-Martínez M, Romo-Díez A, Shaw S, Siavash B, Villar-Salvador P, Zak MR (2004) The plant traits that drive ecosystems: Evidence from three continents. Journal of Vegetation Science 15:295–304

Díaz S, Lavorel S, Chapin III FS, Tecco PA, Gurvich DE, Grigulis K (2007) Functional diversity – at the crossroads between ecosystem functioning and environmental filters. In: Canadell J, Pitelka LF, Pataki D (eds) Terrestrial Ecosystems in a Changing World. Springer-Verlag, this volume

Díaz S, Lavorel S, McIntyre S, Falczuk V, Casanoves F, Milchunas D, Skarpe C, Rusch G, Sternberg M, Noy-Meir I, Landsberg J, Zangh W, Clark H, Campbell BD (in press) Grazing and plant traits – A global synthesis. Global Change Biology

Dodd MB, Lauenroth WK, Burke IC, Chapman PR (2002) Associations between vegetation patterns and soil texture in the shortgrass steppe. Plant Ecology 158:127–137

Duru M, Cruz P, Magda D (2004) Using plant traits to compare sward structure and composition of grass species across environmental gradients. Applied Vegetation Science 7:11–18

Dyer AR, Goldberg DE, Turkington R, Sayre C (2001) Effects of growing conditions and source habitat on plant traits and functional group definition. Functional Ecology 15:85–95

Eissenstat DM (1991) The ecology of root lifespan. Advances in Ecological Research 27:1–60

Eissenstat DM, Wells CE, Yanai RD, Whitbeck JL (2000) Building roots in a changing environment: implications for root longevity. New Phytologist 147:33–42

Eriksson O, Ehrlén J (2001) Landscape fragmentation and the viability of plant populations. In: Silvertown J, Antonovics J (eds) Integrating ecology and evolution in a spatial context. Blackwell Science, Oxford, pp 157–175

Eviner VT, Chapin III FS (2003) Functional matrix: a conceptual framework for predicting multiple plant effects on ecosystem processes. Annual Review of Ecology and Systematics 34: 455–485

Fitter A, Stickland TR (1991) Architectural analysis of plant root systems. 2. Influence of nutrient supply on architecture in contrasting plant species. New Phytologist 118:383–389

Foley JA, Prentice IC, Ramankutty S, Levis S, Pollard D, Sitch S, Haxeltine A (1996) An integrated biosphere model of land surface processes, terrestrial carbon balance and vegetation dynamics. Global Biogeo-chemical Cycles 10:603–628

Friend AD, Stevens AK, Knox RG, Cannell MGR (1997) A process-based, terrestrial biosphere model of ecosystem dynamics (Hybrid v3.0). Ecological Modelling 95:249–287

Garnier E, Aronson J (1998) Nitrogen-use efficiency from leaf to stand level: clarifying the concept. In: Lambers H, Poorter H, Van Vuuren MMI (eds) Inherent variation in plant growth. Physiological mechanisms and ecological consequences. Backhuys Publishers, Leiden, The Netherlands, pp 515–538

Garnier E, Cortez J, Billès G, Navas M-L, Roumet C, Debussche M, Laurent G, Blanchard A, Aubry D, Bellmann A, Neill C, Toussaint J-P (2004) Plant functional markers capture ecosystem properties during secondary succession. Ecology 85: 2630–2637

Garnier E, Laurent G, Bellmann A, Debain S, Berthelier P, Ducout B, Roumet C, Navas M-L (2001) Consistency of species ranking based on functional leaf traits. New Phytologist 152:69–83

Gerten D, Schaphoff S, Haberlandt U, Lucht W, Sitch S (2004) Terrestrial vegetation and water balance: hydrological evaluation of a dynamic global vegetation model. Journal of Hydrology 286:249–270

Gitay H, Noble IR (1997) What are functional types and how should we seek them? In: Smith TM, Shugart HH, Woodward FI (eds) Plant Functional Types. Their Relavance to Ecosystem Properties and Global Change. Cambridge University Press, Cambridge, pp 3–19

Goldberg DE (1991) Competitive effect and response: hierarchies and correlated traits in the early stages of competition. Journal of Ecology 79:1013–1030

Goldberg DE, Novoplansky A (1997) On the relative importance of competition in unproductive environments. Journal of Ecology 85:409–418

Grassl H (2000) Status and improvements of coupled general circulation models. Science 288:1991–1997

Grigulis K, Lavorel S, Davies ID, Dossantos A, Lloret F, Vilà M (2005) Landscape-scale positive feedbacks between fire and expansion of the large tussock grass, *Ampelodesmos mauritanica*, in Catalan shrublands. Global Change Biology 11:1042–1053

Grime JP (1977) Evidence for the existence of three primary strategies in plants and its relevance to ecological and evolutionary theory. The American Naturalist 111:1169–1194

Grime JP (1979) Plant strategies and vegetation processes. John Wiley and Sons, Chichester

Grime JP (2001) Plant strategies, vegetation processes and ecosystem properties, 2nd edn. John Wiley and Sons, Chichester

Grime JP, Corneliessen HJC, Thompson K, Hodgson JG (1996) Evidence of a causal connection between anti-herbivore defence and the decomposition rate of leaves. Oikos 77:489–494

Grime JP, Thompson K, Hunt R, Hodgons JG, Cornelissen JHC, Rorison IH, Hendry GAF, Ashenden TW, Askew AP, Band SR, Booth RE, Bossard CC, Campbell BD, Cooper JEL, Davison AW, Gupta PL, Hall W, Hand DW, Hannah MA, Hillier SH, Hodkinson DJ, Jalili A, Liu Z, Mackey JML, Matthews N, Mowforth MA, Neal AM, Reader RJ, Reiling K, Ross-Fraser W, Spencer RE, Sutton F, Tasker DE, Thorpe PC, Whitehouse J (1997) Integrated screening validates primary axis of specialisation in plants. Oikos 79:259–281

Hanley M, Lamont B (2002) Relationships between physical and chemical attributes of congeneric seedlings: how important is seedling defence? FUNCTIONAL ECOLOGY 16:216–222

Harrison SP, Foley JA (1995) General circulation models, climate change and ecological impact studies. In: Pernetta J, Leemans R, Elder D, Humphrey S (eds) The Impact of Climate Change on Ecosystems and Species: Environmental Context. IUCN, Gland, Switzerland, pp 9–26
Harrison SP, Prentice IC (2003) Climate and CO_2 controls on global vegetation distribution at the last glacial maximum: analysis based on palaeovegetation data, biome modelling and palaeoclimate simulations. Global Change Biology 9:983–1004
Harrison SP, Yu G, Takahara H, Prentice IC (2001) Plant diversity and palaeovegetation in East Asia. Nature 413:129–130
Hartley SE, Jones CG (1997) Plant chemistry and herbivory, or why the world is green. In: Crawley JM (ed) Plant ecology. Blackwell Science Ltd., Oxford, pp 284–335
Hättenschwiller S, Vitousek PM (2000) The role of polyphenols in terrestrial ecosystem nutrient cycling. Trends in Ecology and Evolution 15:238–243
Herms DA, Mattson WJ (1992) The dilemma of plants: to grow or defend. Quaterly Review of Biology 67:293–335
Hickler T, Smith B, Sykes MT, Davis MB, Sugita S, Walker K (2004) Using a general vegetation model to simulate vegetation dynamics in northeastern U.S.A. Ecology 85:519–530
Hodgson JG, Wilson PJ, Hunt R, Grime JP, Thompson K (1999) Allocating C-S-R plant functional types: a soft approach to a hard problem. Oikos 85:282–294
Hooper DU, Ewel JJ, Hector A, Inchausti P, Lavorel S, Lodge D, Loreau M, Naeem S, Schmid B, Setälä H, Symstad AJ, Vandermeer J, Wardle DA (2005) Effects of biodiversity on ecosystem functioning: a consensus of current knowledge and needs for future research. Ecological Monographs 75:3–35
Hughes L, Dunlop M, French K, Leishman MR, Rice B, Rodgerson L, Westoby M (1994) Predicting dispersal spectra: a minimal set of hypotheses based on plant attributes. Journal of Ecology 82:933–950
Jackson ST, Williams JW (2004) Modern analogs in Quaternary paleoecology: here today, gone yesterday, gone tomorrow? Annual Review of Planetary Science 32:495–530
Jackson RB, Canadell J, Ehleringer JR, Mooney HA, Sala OE, Schulze ED (1996) A global analysis of root distribution for terrestrial biomes. Oecologia 108:389–411
Joussaume S, Taylor KE (2000) The Paleoclimate Modeling Intercomparison Project. In: Braconnot P (ed) Paleoclimate Modelling Intercomparison Project (PMIP). Proceedings of the third PMIP workshop, vol. WCRP-111; WMO/TD-No. 1007, La Huardière, Canada, pp 9–24
Keane RE, Cary GJ, Davies ID, Flannigan MD, Gardner RH, Lavorel S, Lenihan JM, Li C, Rupp TS (2007) Understanding global fire dynamics by classifying and comparing spatial models of vegetation and fire dynamics. In: Canadell J, Pitelka LF, Pataki D (eds) Terrestrial Ecosystems in a Changing World. Springer-Verlag, this volume
Keddy PA (1992) Assembly and response rules: two goals for predictive community ecology. Journal of Vegetation Science 3:157–164
Keddy PA, Fraser LH, Wisheu IC (1998) A comparative approach to examine competitive response of 48 wetland plant species. Journal of Vegetation Science 9:777–786
Keddy P, Gaudet C, Fraser LH (2000) Effects of low and high nutients on the competitive hierarchy of 26 shoreline plants. Journal of Ecology 88:413–423
Kleyer M (1999) Distribution of plant functional types along gradients of disturbance intensity and resource supply in an agricultural landscape. Journal of Vegetation Science 10:697–708
Klironomos JN (2003) Variation in plant response to native and exotic arbuscular mycorrhizal fungi. Ecology 84:2292–2301
Klironomos JN, McCune J, Hart MM, Neville J (2000) The influence of arbuscular mycorrhizae on the relationship between plan biodiversity and productivity. Ecology Letters 3:137–141
Kohfeld KE, Harrison SP (2000) How well can we simulate past climates? Evaluating the models using global palaeoenvironmental datasets. Quaternary Science Review 19:321–346
Landsberg J, O'Connor T, Freudenberger D (1999) The impacts of livestock grazing on biodiversity in natural ecosystems. In: Jung HG, Fahey GC (eds) Nutritional Ecology of Herbivores. Proc. Vth International Symposium on the Nutrition of Herbivores. American Soc. Anim. Sci., Savoy, Illinois, pp 752–777
Langley JA, Hungate B (2003) Mycorrhizal controls on belowground litter quality. Ecology 84:2302–2312
Lavorel S, Cramer W (eds) (1999) Plant functional response to land use and natural disturbance
Lavorel S, Garnier E (2002) Predicting the effects of environmental changes on plant community composition and ecosystem functioning: revisiting the Holy Grail. Functional Ecology 16:545–556
Lavorel S, McIntyre S (1999c) Plant functional types: is the real world too complex ? In: Proceedings of the VIth International Rangelands Congress, Townsville, Australia, pp 905–911
Lavorel S, McIntyre S, Landsberg J, Forbes D (1997) Plant functional classifications: from general groups to specific groups based on response to disturbance. Trends in Ecology and Evolution 12:474–478
Lavorel S, Touzard B, Lebreton J-D, Clément B (1998) Identifying functional groups for response to disturbance in an abandonned pasture. Acta Oecologica 19:227–240
Lavorel S, McIntyre S, Grigulis K (1999a) Plant response to disturbance in a Mediterranean grassland: how many functional groups ? Journal of Vegetation Science 10:661–672
Lavorel S, Rochette C, Lebreton J-D (1999b) Functional groups for response to disturbance in Mediterranean old fields. Oikos 84:480–498
Leishman MR, Westoby M (1992) Classifying plants into groups on the basis of associations of individual traits – evidence from Australian semi-arid woodlands. Journal of Ecology 80:417–424
Leishman MR, Westoby M (1994) The role of seed size in seedling establishment in dry soil conditions-experimental evidence from semi-arid species. Journal of Ecology 82:249–258
Liancourt P, Callaway RM, Michalet R (2005) Stress tolerance and competitive response ability determine the outcome of biotic interactions. Ecology 86:1611–1618
Marañon T, Grubb PJ (1993) Physiological basis and ecological significance of the seed size and relative growth rate relationship in Mediterranean annuals. Functional Ecology 7:591–599
McGillivray CW (1995) Testing predictions of the resistance and resilience of vegetation subjected to extreme events. Functional Ecology 9:640–649
McIntyre S, Lavorel S (2001) Livestock grazing in sub-tropical pastures: steps in the analysis of attribute response and plant functional types. Journal of Ecology 89:209–226
McIntyre S, Martin TG (2001) Biophysical and human influences on plant species richness in grasslands – comparing variegated landscapes in sub-tropical and temperate regions. Austral Ecology 26:233–245
McIntyre S, Martin TG (2002) Managing intensive and extensive land uses to conserve grassland plants in sub-tropical eucalypt woodlands. Biological Conservation 107:241–252
McIntyre S, Lavorel S, Landsberg J, Forbes TDA (1999) Disturbance response in vegetation – towards a global perspective on functional traits. Journal of Vegetation Science 10:621–630
McIntyre S, Heard KM, Martin TG (2002) How grassland plants are distributed over five human-created habitats typical of eucalypt woodlands in a variegated landscape. Pacific Conservation Biology 7:274–285
McIntyre S, Heard KM, Martin TG (2003) The relative importance of cattle grazing in sub-tropical grasslands – does it reduce or enhance plant diversity? Journal of Applied Ecology 40:445–457
McKey D, Waterman PG, Gartlan JS, Struhsacker TT (1978) Phenolic content of vegetation in two African rainforests: ecological implications. Science 202:61–64
Michalet R (2001) Facteurs abiotiques et traits biologiques déterminant les interactions biotiques, la répartition des populations dominantes et la structuration des communautés terrestres: Application à quelques écosystèmes de l'Arc Alpin. In. Université J. Fourier, Grenoble
Michelsen A, Schmidt IK, Dighton J, Jones HE, Callaghan TV (1995) Inhibition of growth and effects on nutrient uptake of arctic graminoids by leaf extracts – allelopathy or resource competition between plants and microbes. Oecologia 103:407–418
Midgley GF, Thuiller W, Higgins SI (2007) Plant species migration as a key uncertainty in predicting future impacts of climate change on ecosystems: progress and challenges. In: Canadell J, Pitelka LF, Pataki D (eds) Terrestrial Ecosystems in a Changing World. Springer-Verlag, this volume

Moore AD, Noble IR (1990) An individualistic model of vegetation stand dynamics. Environmental Management 31:61–81

Naeem S, Wright JP (2003) Disentangling biodiversity effects on ecosystem functioning: deriving solutions to a seemingly insurmontable problem. Ecology Letters 6:567–579

Neilson RP, King GA, DeVelice RL, Lenihan JM (1992) Regional and local vegetation patterns: the responses of vegetation diversity to subcontinental air masses. In: Hansen AJ, di Castri F (eds) Landscape Boundaries. Springer-Verlag, pp 129–149

Nicotra AB, Babicka N, Westoby M (2002) Seedling root anatomy and morphology: an examination of ecological differentiation with rainfall using phyllogenetically independent contrasts. Oecologia 130:136–145

Niinemets Ü (2001) Global-scale climatic controls of leaf dry mass per area, density, and thickness in trees and shrubs. Ecology 82:453–469

Noble IR, Gitay H (1996) A functional classification for predicting the dynamics of landscapes. J Vegetat Sci 7:329–336

Noble IR, Slatyer RO (1980) The use of vital attributes to predict successional changes in plant communities subject to recurrent disturbances. Vegetatio 43:5–21

Norby RJ, Rustad LE, Dukes JS, Ojima DS, Parton WJ, Del Grosso SJ, McMurtrie RE, Pepper DA (2007) Multiple factor interactions on ecosystem function. In: Canadell J, Pitelka LF, Pataki D (eds) Terrestrial Ecosystems in a Changing World. Springer-Verlag, this volume

Nygaard B, Erjnaes R (2004) A shortcut around the fourth corner – a new approach to functional interpretation of vegetation data. Journal of Vegetation Science 15:49–56

Overpeck J, Whitlock C, Huntley B (2003) Terrestrial biosphere dynamics in the climate system: past and future. In: Alverson KD, Bradley RS, Pedersen T (eds) Paleoclimate, Global Change and the Future. Springer-Verlag, Berlin-Heidelberg-New York, pp 81–103

Pausas JG (1999) Response of plant functional types to changes in the fire regime in Mediterranean ecosystems: a simulation approach. Journal of Vegetation Science 10:717–722

Pausas JG (2003) The effect of landscape pattern on Mediterranean vegetation dynamics: A modelling approach using functional types. Journal of Vegetation Science 14:365–374

Pausas J, Lavorel S (2003) A hierarchical deductive approach for functional types in disturbed ecosystems. Journal of Vegetation Science 14:409–416

Pausas JG, Ramos JI (2005) Landscape Analysis and Simulation Shell (LASS). Environmental Modelling and Software in press

Pausas JG, Verdú M (2005) Plant persistence traits in fire-prone ecosystems of the Mediterranean Basin: A phylogenetic approach. Oikos 109:196–202

Pausas JG, Rusch GM, Leps J (eds) (2003) Plant Functional Types in relation to disturbance and land use

Pausas JG, Bradstock RA, Keith DA, Keeley JE, Network tGF (2004) Plant functional traits in relation to fire in crown-fire ecosystems. Ecology 85:1085–1100

Pavón N, Hernández-Trejo H, Rico-Gray V (2000) Distribution of plant life forms along an altitude gradient in the semi-arid valley of Zapoltitlán, Mexico. Journal of Vegetation Science 11: 39–42

Pellissier F (1998) The role of soil community in plant population dynamics: is allelopathy a key component? Trends in Ecology and Evolution 13:407–407

Pellissier F, Souto XC (1999) Allelopathy in northern temperate and boreal semi-natural woodland. Critical Reviews in Plant Sciences 18:637–652

Pérez-Harguindeguy N, Díaz S, Cornelissen JHC, Vendramini F, Cabido M, Castellanos A (2000) Chemistry and toughness predict leaf litter decomposition rates over a wide spectrum of functional types and taxa in central Argentina. Plant and Soil 218:21–30

Pérez-Harguindeguy N, Díaz S, Vendramini F, Cornelissen JHC, Gurvich DE, Cabido M (2003) Leaf traits and herbivore selection in the field and in cafeteria experiments. Austral Ecol 28:642–650

Pickett EJ, Harrison SP, Hope G, Harle K, Dodson JR, Kershaw AP, Prentice IC, Backhouse J, Colhoun EA, D'Costa D, Flenley J, Grindrod J, Haberle S, Hassell C, Kenyon C, Macphail M, Martin H, Martin AH, McKenzie M, Newsome JC, Penny D, Powell J, Raine JI, Southern W, Stevenson J, Sutra JP, Thomas I, van der Kaars S, Ward J (2004) Pollen-based reconstructions of biome distributions for Australia, Southeast Asia and the Pacific (SEAPAC region) at 0, 6 000 and 18 000 ^{14}C yr B.P. Journal of Biogeography 31:1–65

Pillar VD, Sosinsky EJ (2003) An improved method for searching plant functional types by numerical analysis. Journal of Vegetation Science 14:323–332

Poorter L (1999) Growth responses of 15 rain-forest tree species to a light gradient: the relative importance of morphological and physiological traits. Functional Ecology 13:396–410

Poorter H, Garnier E (1999) Ecological significance of inherent variation in relative growth rate and its components. In: Pugnaire FI, Valladares F (eds) Handbook of Functional Plant Ecology. Marcel Dekker INC., New-York, pp 81–120

Poorter H, Roumet C, Campbell BD (1996) Interspecific variation in the growth response of plants to elevated CO_2: a search for functional types. In: Körner C, Bazzaz FA (eds) Carbon dioxyde, populations, and communities. Academic Press, San Diego, California, pp 375–412

Potter CS, Klooster SA (1999) Dynamic global vegetation modelling for prediction of plant functional types and biogenic trace gas fluxes. Global Ecology and Biogeography 8:473–488

Prach K (1997) Changes in species traits during succession: a search for pattern. Oikos 79:201–205

Prentice IC, Cramer W, Harrison SP, Leemans R, Monserud RA, Solomon AM (1992) A global biome model based on plant physiology and dominance, soil properties and climate. Journal of Biogeography 19:117–134

Prentice IC, Jolly D, participants B (1998) Mid-Holocene and glacial-maximum vegetation geography of the northern continents and Africa. Journal of Biogeography 25:997–1005

Prentice IC, Heimann M, Sitch S (2000) The carbon balance of the terrestrial biosphere: Ecosystem models and atmospheric observations. Ecological Applications 10:1553–1573

Prentice IC, Bondeau A, Cramer W, Harrison SP, Hickler T, Lucht W, Sitch S, Smith B, Sykes M (2007) Dynamic global vegetation models: tools to understand the biosphere. In: Canadell J, Pitelka LF, Pataki D (eds) Terrestrial Ecosystems in a Changing World. Springer-Verlag, this volume

Raunkiaer C (1934) The life forms of plants and statistical plant geography. Oxford University Press, Oxford

Read DJ (1991) Mycorrhizas in ecosystems. Experientia 47:376–391

Read DJ, Leake JR, Perez-Moreno J (2004) Mycorrhizal fungi as drivers of ecosystem processes in heathland and boreal forest biomes. Canadian Journal of Botany 82:1243–1263

Reich PB, Walters MB, Ellsworth DS (1992) Leaf lifespan in relation to leaf, plant and stand characteristics among diverse ecosystems. Ecological Monographs 62:356–392

Reich PB, Walters MB, Ellsworth DS (1997) From tropics to tundra: a global convergence in plant functioning. Proceedings of the National Academy of Sciences, USA 94:13730–13734

Reich PB, Tjoelker MG, Walters MB, Vanderklein DW, Buschena C (1998a) Close association of RGR, leaf and root morphology, seed mass and shade tolerance in seedlings of nine boreal tree species grown in high and low light. Functional Ecology 12:337–338

Reich PB, Walters MB, Tjoelker MG, Vanderklein DW, Buschena C (1998b) Photosynthesis and respiration rates depend on leaf and root morphology and nitrogen concentration in nine boreal species differing in relative growth rate. Functional Ecology 12:337–338

Reich PB, Wright IJ, Cavender-Bares J, Craine JM, Oleksyn J, Westoby M, Walters MB (2003) The evolution of plant functional variation: traits, spectra, and strategies. International Journal of Plant Sciences 164:143–164

Richardson SJ, Peltzer DA, Allen RB, McGlone MS (2005) Resorption proficiency along a chronosequence: Responses among communities and within species. Ecology 86:20–25

Rillig MC (2004) Arbuscular mycorrhizae and terrestrial ecosystem processes. Ecology Letters 7:740–754

Roberts DW, Betz DW (1997) Simulating landscape vegetation dynamics of Bryce Canyon National Park with the vital attributes/fuzzy systems model VAFS/LANDSIM. In: Mladenoff DJ, Baker WL (eds) Spatial modeling of forest landscape change: approaches and applications. Cambridge University Press, Cambridge, pp 99–124

Robinson D, Rorison IH (1988) Plasticity in grass species in relation to nitrogen supply. Functional Ecology 2:249–257

Ryser P (1996) The importance of tissue density for growth and life span of leaves and roots: a comparison of five ecologically contrasting grasses. Functional Ecology 10:717–723

Ryser P (1998) Intra- and interspecific variation in root length, root turnover and the underlying parameters. In: Lambers H, Poorter H, Van Vuuren MMI (eds) Inherent variation in plant growth. Physiological mechanisms and ecological consequences. Backhuys Publishers, Leiden, The Netherlands, pp 441–465
Schenk HJ, Jackson RB (2002) Rooting depths, lateral root spreads and below-ground / above-ground allometries of plants in water-limited ecosystems. Journal of Ecology 90:480–494
Schenk HJ, Jackson RB (2005) Mapping the global distribution of deep roots in relation to climate and soil characteristics. Geoderma 126:129–140
Schwilk DW, Ackerly DD (2001) Flammability and serotiny as strategies: correlated evolution in pines. Oikos 94:326–336
Shipley B, Almeida-Cortez J (2003) Interspecific consistency and intraspecific variability of specific leaf area with respect to irradiance and nutrient availability. Ecoscience 10:74–79
Shipley B, Keddy PA, Moore DRJ, Lemky K (1989) Regeneration and establishment strategies of emergent macrophytes. Journal of Ecology 77:1093–1110
Silvertown J, Franco M, Pisanty I, Mendoza A (1993) Comparative plant demography – relative importance of life-cycle components to the finite rate of increase in woody and herbaceous perennials. Journal of Ecology 81:465–476
Sitch S, Smith B, Prentice IC, Arneth A, Bondeau A, Cramer W, Kaplan JO, Levis S, Lucht W, Sykes MT, Thonicke K, Venevsky S (2003) Evaluation of ecosystem dynamics, plant geography and terrestrial carbon cycling in the LPJ dynamic global vegetation model. Global Change Biology 9:161–185
Smith T, Huston M (1989) A theory of the spatial and temporal dynamics of plant communities. Vegetatio 83:49–69
Smith TM, Shugart HH, Woodward FI (eds) (1997) Plant functional types, their relevance to ecosystem properties and global change. Cambridge University Press, Cambridge
Solbrig OT (1993) Plant traits and adaptative strategies: their role in ecosystem function. In: Schulze ED, Mooney HA (eds) Biodiversity and ecosystem function, vol 99. Springer-Verlag, Berlin, pp 97–116
Stanton ML, Roy BA, Thiede DA (2000) Evolution in stressful environments. I. Phenotypic variability, phenotypic selection, and response to selection in five environmental stresses. Evolution 54:93–111
Strand JA, Weisner SEB (2004) Phenotypic plasticity – contrasting species-specific traits induced by identical environmental constraints. New Phytologist 163:449–451
Suding KN, Goldberg DE (2001) Do disturbances alter competitive hierarchies? Mechanisms of change following gap creation. Ecology 82:2133–2149
Suding KN, Goldberg DE, Hartman KM (2003) Relationships among species traits: separating levels of response and identifying linkages to abundance. Ecology 84:1–16
Thompson K, Band SR, Hodgson G (1993) Seed size and shape predict persistence in soil. Functional Ecology 7:236–241
Thompson K, Hillier SH, Grime JP, Bossard CC, Band SR (1996) A functional analysis of a limestone grassland community. Journal of Vegetation Science 7:371–380
Tilman D (1988) Plant strategies and the dynamics and structure of plant communities. Princeton University Press, Princeton
Urcelay C, Díaz S (2003) The mycorrhizal dependence of subordinates determines the effect of arbuscular mycorrhizal fungi on plant diversity. Ecology Letters 6:388–391
van der Heijden MGA, Boller T, Wiemken A, Sanders IR (1998a) Different arbuscular mycorrhizal fungal species are potential determinants of plant community structure. Ecology 79: 2082–2091
van der Heijden MGA, Klironomos JN, Ursic M, Mouroglis P, Striwolf-Engel R, Boller T, Wiemken A, Sanders I (1998b) Mycorrhizal fungal diversity determines plant biodiversity, ecosystem variability and productivity. Nature 396:69–72
Venable DL, Brown JS (1988) The selective interaction of dispersal, dormancy, and seed size as adaptations for reducing risk in variable environments. American Naturalist 131:361–384
Walker B, Kinzig A, Langridge J (1999) Plant attribute diversity, resilience and ecosystem function: the nature and significance of dominant and minor species. Ecosystems 2:95–113
Wardle DA (2002) Communities and ecosystems. Linking the aboveground and belowground components. Princeton University Press, Princeton
Wardle DA, Barker GM, Bonner KI, Nicholson KS (1998) Can comparative approaches based on plant ecophysiological traits predict the nature of biotic interactions and individual plant species effects in ecosystems? J Ecology 86:405–420
Wardle DA, Baker GM, Yeattes GW, Bonner KI, Ghani A (2001) Impacts of introduced browsing mammals in New Zealand forest on decomposer communities,soil biodiversity and ecosystem properties. Ecological Monographs 71:587–614
Wardle DA, Bardgett RD, Klironomos JN, Setälä H (2004) Ecological linkages between aboveground and belowground biota. Science 304:1629–1633
Weiher E, Keddy P (eds) (1999) Ecological assembly rules – Perspectives, advances, retreats. Cambridge University Press, Cambridge
Weiher E, van der Werf A, Thompson K, Roderick M, Garnier E, Eriksson O (1999) Challenging Theophrastus: A common core list of plant traits for functional ecology. Journal of Vegetation Science 10:609–620
Westoby M (1998) A leaf-height-seed (LHS) plant ecology strategy scheme. Plant and Soil 199:213–227
Westoby M, Rice B, Howell J (1990) Seed size and plant growth form as factors in dispersal spectra. Ecology 71:1307–1315
Westoby M, Falster DS, Moles AT, Vesk PA, Wright IJ (2002) Plant ecological strategies: come leading dimensions of variation between species. Annual Review of Ecology and Systematics 33: 125–159
Woodward FI, Cramer W (1996) Plant functional types and climatic changes: introduction. Journal of Vegetation Science 7:306–308
Woodward FI, Lomas MR, Betts AK (1998) Vegetation-climate feedbacks in a greenhouse world. Philisophical Transactions of the Royal Society of London, Series B 353:29–38
Wright IJ, Westoby M (1999) Differences in seedling growth behaviour among species: trait correlations across species, and trait shifts along nutrient compared to rainfall gradients. Journal of Ecology 87:85–97
Wright IJ, Reich PB, Westoby M, Ackerly DD, Baruch Z, Bongers F, Cavender-Bares J, Chapin FSI, Cornelissen JHC, Diemer M, Flexas J, Garnier E, Groom PK, Gulias J, Hikosaka K, Lamont BB, Lee T, Lee W, Lusk C, Midgley JJ, Navas M-L, Niinemets Ü, Oleksyn J, Osada N, Poorter H, Poot P, Prior L, Pyankov VI, Roumet C, Thomas SC, Tjoelker MG, Veneklaas E, Villar R (2004) The worldwide leaf economics spectrum. Nature 428:821–827

Chapter 14

Spatial Nonlinearities: Cascading Effects in the Earth System

Debra P.C. Peters · Roger A. Pielke Sr. · Brandon T. Bestelmeyer · Craig D. Allen · Stuart Munson-McGee · Kris M. Havstad

14.1 Introduction

Nonlinear behavior is prevalent in all aspects of the Earth System, including ecological responses to global change (Gallagher and Appenzeller 1999; Steffen et al. 2004). Nonlinear behavior refers to a large, discontinuous change in response to a small change in a driving variable (Rial et al. 2004). In contrast to linear systems where responses are smooth, well-behaved, continuous functions, nonlinear systems often undergo sharp or discontinuous transitions resulting from the crossing of thresholds. These nonlinear responses can result in surprising behavior that makes forecasting difficult (Kaplan and Glass 1995). Given that many system dynamics are nonlinear, it is imperative that conceptual and quantitative tools be developed to increase our understanding of the processes leading to nonlinear behavior in order to determine if forecasting can be improved under future environmental changes (Clark et al. 2001).

Although most global change studies have examined nonlinear behavior through time (e.g., Pascual and Ellner 2000; Gill et al. 2002; Gerber et al. 2004), it is increasingly recognized that spatial interactions, transport processes, and landscape complexity are important in generating nonlinear behavior through time *and* across space (Aber et al. 1999; Reiners and Driese 2001, 2004). In particular, contagious processes that propagate nonlinearly through time from small to broad spatial extents (i.e., spatial nonlinearities) often generate surprising behavior where dynamics at one scale cannot be easily predicted based on information obtained at finer or broader scales (Holling 1992, 1996; Nieminen 2003). These cascading effects often result in severe consequences for the environment and human welfare (i.e., catastrophes) that are expected to be particularly important under conditions of changes in climate and land use (NRC 2001; Steffen et al. 2004). A key challenge to Earth System science in the face of global change is to understand and predict these cascading effects and their catastrophic consequences (Steffen et al. 2004).

Earlier frameworks have described nonlinear, catastrophic behavior in terms of spatial propagation, positive feedbacks, and thresholds. Holling's (1973) seminal work linked disturbances to alternative stable states and thresholds between them through feedback mechanisms. The concept of self organized criticality introduced the notion that perturbations at critical thresholds may involve self-propagating changes at a variety of spatial and temporal scales (Bak et al. 1988; Chen et al. 1991). Together, these ideas linked spatial and temporal pattern to describe how dynamics may be self-organized. This linkage illustrated the value of recognizing coupled spatial and temporal patterns in ecosystems at multiple scales, and underlies the search for feedback mechanisms (Folke et al. 2004), critical thresholds of connectivity in applications of percolation theory (Davenport et al.1998) and the recognition of patch- to landscape pattern to explain catastrophic shifts in the pattern of vegetation (Rietkerk et al. 2004).

Despite these advances, our understanding of and ability to predict events that propagate across scales (i.e., cascade) and produce catastrophic changes remains limited. For example, this class of ecological problems includes insect outbreaks that spread nonlinearly from fine to broad scales (Swetnam and Lynch 1993). Because the rate and extent of an insect outbreak may be related to climatic patterns (Speer et al. 2001) and interactions with disturbance (Scheller and Mladenoff 2005), changes in climate and land use are expected to have large effects on these dynamics. Similarly, single lightning strikes can initiate broad-scale wildfires as a result of positive feedbacks among weather, fire behavior, land-use patterns, and vegetation interacting across multiple scales. Nonlinear patterns in connectivity of fires are often related to spatial variation in fuel loads (Miller and Urban 2000). Extensive and rapidly expanding fires may be driven by feedbacks with the atmosphere. Although fire (and other) behaviors have been described with respect to the internal forces of self-organized criticality (Drossel and Schwabl 1992), they have not been considered with reference to the role of feedbacks involving a variety of external drivers.

In this chapter, we *first* briefly describe a general conceptual and mathematical framework for understanding and forecasting spatially nonlinear responses to global change in driving variables. The *second* goal of this chapter is to illustrate the utility of our framework in describing the spread of catastrophic events using one historical example (the Dust Bowl) and two current examples (wildfires, invasive species and desertification). *Finally*, we discuss the consequences of applying these ideas to forecasting future dynamics under a changing global environment. Given the continuing challenges associated with global change, our synthetic approach that crosses disciplinary boundaries to include interactions and feedbacks across multiple scales shows great potential to increase our ability to forecast catastrophic events and to develop strategies for minimizing their occurrence and impacts.

14.2 Conceptual Framework

In our framework, we focus on catastrophic events that start small, and propagate nonlinearly to influence broad spatial extents (described in detail in Peters et al. 2004). Mathematically, spatial nonlinearities can be illustrated as:

$$dY/dt = g(I_g, E_g) + f(Y, E_f) + D(Y, E_D) + c(Y, E_c)$$

where each term is associated with one of the four stages in the general model, and is a function of different parameters. All terms except g depend, at least in part, on properties of Y. As the rate of change in Y increases (decreases) through time, it is increasingly governed by terms towards the right (left) [from $g(I_g, E_g)$ to $c(Y, E_c)$] as the amount, connectivity, and spatial extent of Y increases (decreases). Three thresholds occur between the various stages. Stage 14.1 (initiation of Y) is defined by $g(I_g, E_g)$ where (I_g) is internal factors and (E_g) is external factors (e.g., weather) that influence initiation. Stage 14.2 (within patch spread of Y) is defined by $f(Y, E_f)$ where Y is within patch properties and (E_f) is external factors that influence patch heterogeneity (e.g., local weather). Stage 14.3 (spread of Y among patches) is defined by $D(Y, E_D)$ where Y is among patch heterogeneity in Y and E_D is external factors that influence among patch spread (e.g., template heterogeneity). Stage 14.4 (land-atmosphere feedbacks that influence the spread of Y) is defined by $c(Y, E_c)$ where Y is broad scale properties of Y and E_c is broad scale processes or forcing functions.

Our framework includes cross scale interactions, threshold behavior, and feedback mechanisms that generate spatial nonlinearities. There are four key characteristics of our framework: (1) feedback mechanisms are prevalent at a number of scales, (2) thresholds in the dynamic behavior of the system are crossed through time with broader scale consequences, (3) the dominant process controlling system dynamics changes through time and across space, and (4) connectivity along spatial units is important to the generation of cascading dynamics. Although the significance of feedbacks and thresholds is recognized in many disciplines (Elsner and Tsonis 1992; Zeng et al. 1993; Hethcote 2000; Scheffer et al. 2001; Tsonis 2001), the incorporation of processes across spatial scales that cross traditional disciplinary boundaries is required to understand and forecast these events.

14.3 Insights to Global Change Issues

14.3.1 Historical Example: the Dust Bowl of the 1930s

Extreme climatic events have played important roles in ecosystem dynamics historically (Swetnam and Betancourt 1998), and are expected to become increasingly important if the frequency, intensity, and magnitude of extreme events increase with time (Easterling et al. 2000). One historical event that was related to extreme climate interacting with land use was the Dust Bowl that occurred in the 1930s in the United States. This catastrophic event involved localized wind erosion from individual agricultural fields that propagated nonlinearly to generate massive dust storms that were accompanied by mass migrations and economic hardship felt throughout the U.S. Insights gained by examining this historical event within the context of our new cross scale framework can improve our ability to understand, mitigate, and forecast similar catastrophic events.

The Dust Bowl or the "Dirty Thirties" was characterized by a series of years with very low rainfall and high temperatures that generated intense drought conditions throughout the central Great Plains (Worster 1979). Recent analyses indicate that these atmospheric conditions were likely caused by anomalous tropical sea surface temperatures (Schubert et al. 2004). Although these atmospheric conditions were unusual for the 20th century, major droughts have occurred in this region once or twice a century for the past 400 years (Woodhouse and Overpeck 1998). Thus, it is unlikely that extreme climatic conditions alone were sufficient to result in the Dust Bowl, but rather that land-atmosphere interactions increased its severity (Schubert et al. 2004).

Importantly, the Dust Bowl was preceded in the 1920s with government policies that favored cultivation of drought sensitive crops on increasingly marginal land (Hurt 1981). Thus, the landscape consisted of a mosaic of cultivated, drought-susceptible land interspersed with small areas of native grassland. Hot, dry weather combined with strong winds in the 1930s resulted in decreased plant cover and high plant mortality on these cultivated fields (Fig. 14.1a); similar patterns were observed on native grassland, but the effects were not as severe (Weaver and Albertson 1936, 1940; Albertson and Weaver 1942). Low plant cover and strong winds resulted in localized wind erosion and blowing dust at the scale of individual fields (Fig. 14.1b). At the landscape scale, these small dust storms became aggregated among fields (Fig. 14.1c) to generate massive dust storms ("black blizzards") that

Fig. 14.1. **a** Drought in the 1930s resulted in decreased plant cover and high plant mortality at the scale of individual fields (Morton County, Kansas, 1938. Photo courtesy of USDA-ARS-Wind Erosion Research Unit and Kansas State University, http://www.weru.ksu.edu/), **b** wind erosion from within a field in South Dakota, (R. Lord, 1938. Miscellaneous Publication No. 321, U.S. Department of Agriculture. Photos courtesy of the National Oceanic and Atmospheric Administration, http://www.photolib.noaa.gov), **c** dust rising from a landscape of agricultural fields (photo courtesy of USDA-ARS-Wind Erosion Research Unit and Kansas State University, http://www.weru.ksu.edu/), and **d** aggregation of dust storms resulted in Black Sunday April 14, 1935 (photo courtesy of USDA-ARS-Wind Erosion Research Unit and Kansas State University, http://www.weru.ksu.edu/)

spread to the regional and continental scales (Fig. 14.1d). Blowing soil from the Great Plains was documented as far as the East coast, over 1500 km away (Miller 1934). Although it was not documented at the time, it is likely that these large dust storms had important feedbacks to the weather through high atmospheric dust loading and high albedo (Fig. 14.2). These changes would have reduced rainfall (Rosenfeld et al. 2001) and increased temperatures, similar to land-atmosphere feedbacks documented in Saharan Africa (Clausen et al. 1999).

Mass migrations, reduced quality of life, and agricultural depression as a result of these dust storms had economic and social ripples to other parts of the country (Lockeritz 1978). Approximately 35 million acres of formerly cultivated land was destroyed with >200 million acres of cropland with reduced topsoil (Yearbook of Agriculture 1934). In April 1935, the U.S. Congress declared soil erosion "a national menace". Federal drought assistance to farmers has been estimated at $1 billion (in 1930s dollars) (Warrick 1980). In addition, migration of people from the Dust Bowl region created economic strain in other parts of the U.S.

We can understand the sequence of processes leading to the Dust Bowl using our conceptual framework that includes multiple stages, thresholds, and a change in dominant process through time and across space. The Dust Bowl was initiated (Stage 14.1) by two interacting conditions: consecutive years with extreme weather com-

Fig. 14.2. Land-atmosphere interactions can result in regional- to continental-scale dust storms, such as this one from the African Sahara, February 26, 2000 (photo courtesy of SeaWiFS and NASA Goddard Space Flight Center, http://visibleearth.nasa.gov)

bined with individual farmer decisions to cultivate increasingly marginal farmland. Onset of the drought would have resulted in high mortality of plants and an increase in the amount of bare soil on individual fields. Strong winds would have generated soil erosion from individual fields after thresholds in wind velocity and plant cover were crossed (Stage 14.2). Because a number of farmers made the same decisions about crop and field selection, and the drought conditions were widespread, a second threshold would have been crossed related to connectivity among fields such that dust storms developed and spread across the landscape (Stage 14.3). At this stage, the rate and spatial extent of wind erosion would have been determined by the number, size, and spatial arrangement of fields interacting with regional scale weather conditions (Fig. 14.3). As the spatial extent of wind erosion continued to increase, another threshold would have been crossed where land-atmosphere interactions would have become operative to generate positive feedbacks and the creation of massive dust storms at the regional to continental scale (Stage 14.4).

It was recognized during the Dust Bowl period that a complex set of interactions were involved that included weather, vegetation, soils, and human activity (Great Plains Committee 1937). Although the committee stated that "all too frequently what appears to be good to the individual in the long run is not good for the people of the region" (Great Plains Committee 1937), our understanding about how these individual decisions interact with climate and ecological systems has remained limited. For example, conservation practices remain focused on the protection of individual fields or a small collection of fields rather than reducing connectivity among a large number of fields. Our framework suggests that limiting connectivity among fields is key to reducing the severity of these kinds of events.

Drought and other extreme climatic events (e.g., floods, hurricanes) will continue to occur in the Earth System (Easterling et al. 2000). Nonlinear interactions and positive feedbacks among climate, land use, and land cover can result in the propagation of impacts across broad spatial scales in relatively short time periods. However, the spatial extent and impact of these events can be mitigated, and possibly forecast, by understanding how these spatial nonlinearities develop and spread. In some cases, limiting landscape scale connectivity can reduce these impacts.

14.3.2 Wildfire

Wildfires are dominant forces shaping the structure and dynamics of forests, savannas, and grasslands as well as their neighboring or imbedded urban areas worldwide (Scholes et al. 2003). Policies associated with natural areas, particularly in developed countries, have often promoted the development of nearly continuously distrib-

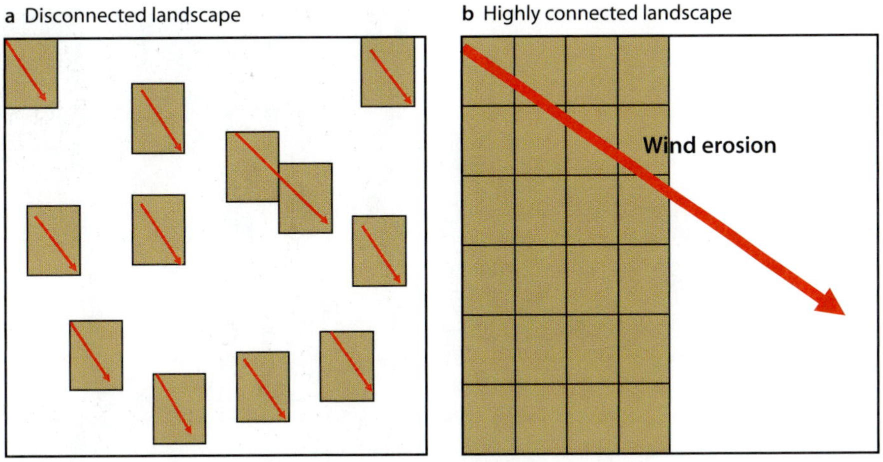

Fig. 14.3. Potential for wind erosion is less for landscapes consisting of a few, widely distributed bare fields (**a**) compared with a landscape that is mostly bare fields (**b**). Farming practices in the 1920s and 1930s in the U.S. predisposed these landscapes to high connectivity resulting from interactions between drought and strong winds, thus leading to the Dust Bowl

uted fuel loads that increasingly allow large, rapidly spreading fires to emerge (USDA 1978). The costs of wildfires are substantial: annual wildfire suppression costs in the U.S. now routinely exceed US$1 billion yr^{-1}. In addition, wildfires can have substantial impacts on atmospheric carbon monoxide (CO) and fine particulates (Scholes et al. 2003). For example, fires in Southeast Asia (2001) resulted in plumes of CO that extended across the Pacific Ocean to the west coast of North America (Steffen et al. 2004). Regional effects are also important: CO emitted by the Hayman fire in Colorado, USA (2003) was at least five times the annual amount produced by industrial sources in that state (Graham 2003). These pulse emissions occurred over a period of days with longer-term impacts on rainfall from the fine particles that extended hundreds of kilometers downwind.

Although it is recognized that short- and long-term weather conditions interacting with the amount, moisture content, and spatial distribution of fuels affect the extent, rate of spread, and severity of wildfires (Graham 2003), the factors that determine if a wildfire can be contained or if it will "blow up" and create catastrophic conditions have not been quantified. For example, 14 firefighters were killed during the Storm King Mountain fire in western Colorado USA (1994) following a sudden wind shift that created surface winds exceeding 55 km h^{-1} and flames 60–90 m high (Butler et al. 1998). Highly connected fuels interacted nonlinearly with the heat of the fire and with a larger scale wind shift to create this rapid change in wind direction and speed. In addition, positive feedbacks among landscape structure, fire, and weather can occur if the climate in the region warms (Laurance and Williamson 2001; McKenzie et al. 2004). Thus, the explosive spread of wildfires across landscapes is not easily forecast based either on fine scale fire behavior or broad scale atmospheric conditions. Our framework that focuses on spatial nonlinearities resulting from connectivity within and among patches of vegetation, and feedbacks to the atmosphere provides new insights into these complex fire dynamics.

We distinguish four major stages of wildfire behavior that are associated with three thresholds or nonlinear changes in the rate of fire spread through time and across space (Peters et al. 2004). This model is supported by published data from two fires in Colorado, USA with similar behavior, yet very different spatial extents (Peters et al. 2004). Similar patterns of fire spread have been documented for other major fires.

In Stage 14.1, fire that is ignited naturally or started by human activities can either spread or stop (Fig. 14.4a). The fire can cross a threshold (T1) to Stage 14.2 by spreading within a patch if local weather conditions, fuel load, and connectivity are sufficient (Fig. 14.4b) (Whelan 1995). For canopy fires to spread, the trees must be sufficiently close for flames to move from one tree to another within a patch. For surface fires to spread, threshold amounts and spatial connectivity of herbaceous fuels must be present. In addition, surface fires can move upward into tree canopies when vertically continuous "ladder" fuels (e.g., small subcanopy trees) connect the burning surface fuels with canopy fuels, generating localized torching.

A second threshold (T2) is crossed when the fire enters Stage 14.3 and burns from one patch to another at varying rates that depend on connectivity and spatial distribution of fuel load as well as interactions with local weather conditions (Fig. 14.4c). Patches that are poorly connected to other patches have low probabilities of the fire spreading whereas fire frequently spreads among patches that are highly connected. Parts of the landscape with low fuel connectivity often burn more slowly and less completely than highly connected parts of the landscape (Turner et al. 2003). As the fire continues to increase in extent and intensity, a third threshold can be crossed (T3) to move the fire into Stage 14.4 that depends on interactions and feedbacks between the fire and the atmosphere. As heated gases from the fire rise into the atmosphere, low air pressure pulls air into the fire, thus creating strong winds. These surface winds drive fire behavior by accelerating fire

Fig. 14.4. The four stages and three thresholds involved in wildfires: **a** a lightning strike can initiate a fire that either goes out or **b** spreads within a patch. Through time, the fire can **c** spread among patches across a landscape and **d** blow up when the heat and intensity of the fire interact with atmospheric conditions to provide a positive feedback to the fire

intensity and rate of spread. These positive feedbacks between fire activity and wind circulation develop rapidly resulting in a "blowup" of the fire with preheating of fuel and "spotting" ahead of the fire front (Fig. 14.4d). Explosive fire activity of this sort can generate large pyrocumulus clouds, particularly when atmospheric conditions are unstable and susceptible to convectional cloud formation (Byram 1954). Under these "blowup" conditions, all parts of the landscape burn, often at higher temperatures, regardless of fuel load or connectivity.

Although general stages of fire behavior have been documented previously and wildfires have been extensively studied at individual scales, we are still missing a clear understanding of the key processes and conditions that lead to catastrophic fire behavior (Steffen et al. 2004). Our approach provides a framework for linking these extensive datasets and models that were developed for specific applications and scales. Understanding cross-scale interactions and feedbacks will improve our ability to identify the key processes generating fire behavior across scales, and to forecast the conditions under which fine scale processes cascade nonlinearly to impact broad spatial extents with consequences for land-atmosphere interactions. Understanding these cross scale interactions will improve our ability to forecast fire behavior under changing weather and land-use regimes that include the continued use of fire as a management tool for many regions of the world (Scholes et al. 2003).

14.3.3 Invasive Species and Desertification

We illustrate the spread of invasive species using woody plant encroachment into perennial grasslands and associated land degradation (i.e., desertification). A wide range of native and exotic species exhibit similar patterns and dynamics (e.g., Hobbs and Humphries 1995; Mack et al. 2000). Desertification is a major problem globally: ca. 40% of the Earth's land surface consists of drylands that are susceptible to desertification and support ca. 20% of the world's human population (Reynolds and Stafford Smith 2002). Conversion of grasslands with homogeneous plant cover to shrublands or woodlands with discontinuous cover interspersed with bare areas often results in local to global consequences, such as increased soil erosion by wind with dust generation to the atmosphere (Schlesinger et al. 1990; Tegen and Fung 1995; Husar et al. 2001). The problem of desertification is complicated by the presence of multiple interacting processes, threshold behavior, and feedbacks with meteorological, hydrological, ecological, and human dimensions (e.g., Rietkerk and van de Koppel 1997; Zeng et al. 1999; Ludwig and Tongway 2000; Reynolds and Stafford Smith 2002; Wilcox et al. 2003). Although numerous studies have been conducted on desertification, a clear consensus is missing regarding the key processes involved and how they produce a variety of responses under apparently similar conditions as well as similar ecological consequences under different conditions (Peters et al. 2006). We propose that cross-scale linkages among three system elements are key determinants of desertification dynamics: (1) local soil and grass degradation associated with large herbivore grazing and other factors, (2) connectivity of erosion processes at multiple scales, and (3) land-atmosphere feedbacks.

Desertification is initiated by the introduction of woody plant seeds into a grass-dominated system (Stage 14.1) as a result of vectors, such as wind, water, and cattle, which transport seeds from woody plant-dominated areas to grasslands (Brown and Archer 1987). In some cases, initiation events fail because microenvironmental conditions are insufficient for establishment. In other cases, particularly when excessive grazing by large herbivores reduces grass cover and competitiveness with a resulting decrease in fire frequency, shrub establishment events succeed and a threshold is crossed (T1) where shrub invasion proceeds (Fig. 14.5). Local spread of a patch of established woody plants within a grassland then occurs (Stage 14.2), either through vegetative expansion or local seed dispersal followed by establishment (Archer 1990). Feedback mechanisms among woody plants and soil properties influence the rate of within-patch expansion. As woody plant size or density increase within a patch, the spatial extent of bare area between woody plants increases, such

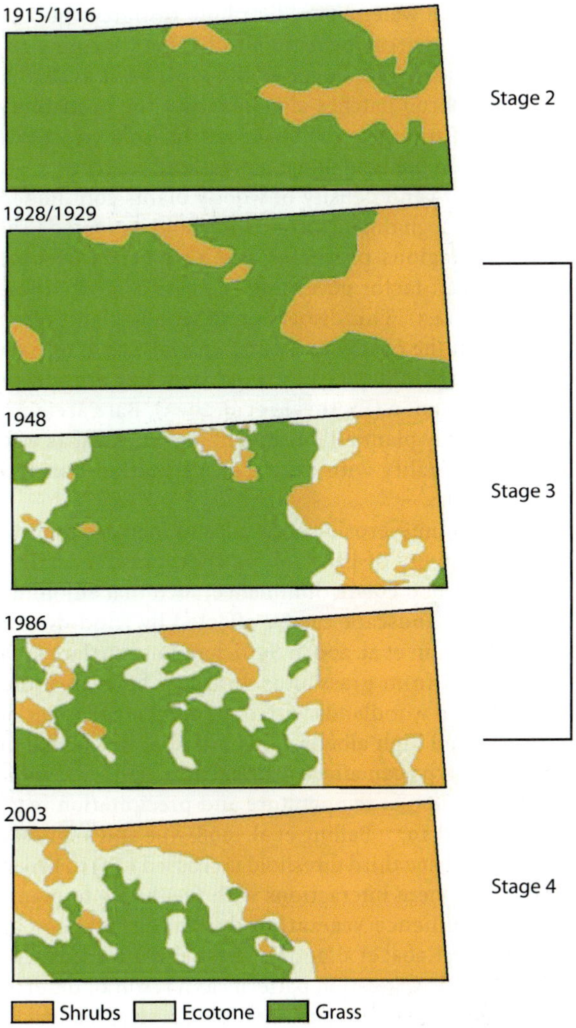

Fig. 14.5. The four stages and three thresholds in desertification. The five panels (942 ha total area) show three classes of vegetation: grasslands, ecotones containing grasses and shrubs, and shrublands through time in the Chihuahuan Desert of southern New Mexico. Field surveys (1915, 1928–1929; Gibbens et al. 2004), black and white (1948) and color infrared photos (1986) and Quickbird satellite images (2003) were scanned at 1200 dpi and geometrically corrected to the satellite image. Boundaries of three classes were manually digitized and areas were calculated in ArcGIS (data shown in Peters et al. 2004). *Stage 14.1:* Desertification begins with the introduction of shrub seeds into a grass-dominated system. In some cases, initiation events fail, and in other cases, they succeed and a threshold is crossed (*T1*) where shrub proliferation (Stage 14.2) proceeds. At this site in NM, this initiation occurred prior to 1915. *Stage 14.2:* Established shrubs proliferate around initial colonizers to form expanding patches (shown in 1915 map). *Stage 14.3:* As the size and density of woody plants continue to increase through time, a second threshold is crossed (*T2*) where contagious processes among patches, in particular wind erosion of bare soil patches, become the dominant factors governing the rate of desertification. *Stage 14.4:* Eventually sufficient land area is converted from grassland (low bare area, low albedo) to a shrubland (high cover of bare area, high albedo) that atmospheric conditions are affected, in particular wind speed, temperature, and precipitation, and a third threshold in crossed (*T3*). Because 92% of the land area of the surrounding research site was shrub dominated in 1998 (Gibbens et al. 2004), we hypothesize that *T3* occurred after 1986

that wind and water can redistribute resources in bare interspaces to areas beneath shrubs, thus forming resource islands (Schlesinger et al. 1990). Seed availability within shrub patches also increases the probability of new initiation events that result in isolated woody plants within the neighboring grassland matrix.

As the size and density of woody plants continue to increase through time, a second threshold is crossed (T2) where contagious processes among patches become the dominant factor governing the rate of desertification (Stage 14.3). This dispersion of woody plants often depends on the connectivity and spatial extent of bare areas that are influenced by wind or water erosion or their combination (Breshears et al. 2003). Bare areas influence woody plant establishment and survival as well as grass mortality with effects on vegetation patterns and dynamics.

Through time, erosion continues to increase the size and connectivity of bare soil patches, grass mortality, and hence woody plant dominance, such that adjoining areas of the landscape can be affected by wind-deposited soil (Youlin et al. 2001). Sufficient land surface area is converted from grasslands with low bare area and low albedo to woodlands with high percentage cover of bare area and high albedo. This broad-scale change in land-cover type can affect regional atmospheric conditions, in particular temperature and precipitation (e.g., Charney et al. 1977; Balling et al. 1998; Xue and Fennessy 2002) such that a third threshold is crossed (T3) in which land-atmosphere interactions with feedbacks to plants strongly influence vegetation dynamics (Stage 14.4) (Pielke 2001; Kabat et al. 2004). These climatic feedbacks to broad-scale vegetation patterns as a result of desertification have been documented: biophysical feedbacks may have resulted in a drier climate and a shift to desert vegetation ca. 5 500 years ago in the Sahara region of Africa (Claussen et al. 1999; deMenocal et al. 2000). Increased wind erosion and deposition of sand off the coast of West Africa may have resulted from this change in climate and shift in vegetation (deMenocal et al. 2000). Dust and other airborne particles can reduce water droplet sizes in clouds to result in reduced rainfall (Rosenfeld et al. 2001) that would then generate feedbacks to the vegetation.

Interactions among climate, land use, and land management will continue to be important drivers in future desertification dynamics. In these water-stressed environments, directional changes in climate or increases in its variability could increase aridity and push these systems to states and dynamics that go beyond current and past experience. For example, an increase in the frequency or severity of drought will likely interact nonlinearly with land management decisions to result in nonlinear increases in the rate and extent of desertification (Squires 2001). By accounting for cross-scale interactions, our approach provides new directions for combating desertification that are currently constrained by observations and models based on particular scales (Peters et al. 2004). Developing site-specific remedies requires quantification of cross-scale interactions among human activities, livestock grazing, drought, and other factors that are coupled by spatial patterns of vegetation at multiple scales. Data collected according to our interdisciplinary framework can be used to identify the key processes governing patterns and rates of desertification as well as thresholds. These data will indicate optimal strategies for manipulating connectivity across scales in order to minimize negative impacts and to capitalize on opportunities for remediation.

14.4 Forecasting Spatial Nonlinearities and Catastrophic Events

Catastrophic events resulting from spatial nonlinearities often result in major changes in ecosystem properties and services as well as loss of life or quality of life, and economic hardship. Our ability to forecast these events, particularly in the presence of environmental and societal changes, is constrained by our limited understanding of the processes, feedbacks, and nonlinear interactions that result in these spatially complex dynamics (Sarewitz et al. 2000). Most of our ecological understanding is based on experiments, observations, and simulation modeling from individual spatial or temporal scales. Experiments are most often conducted at fine scales whereas observations and simulation modeling can include multiple scales from fine to broad. Results from these single or multiple scale studies are often unable to detect cross scale interactions and dynamics that propagate across space (Peters et al. 2004).

We advocate experimental, observation, and modeling networks of studies that explicitly address cross scale interactions as the most fruitful approach to forecasting catastrophic events resulting from spatial nonlinearities. Existing networks of research sites, such as the U.S. National Science Foundation supported Long Term Ecological Research program, provide invaluable site-based information, often with historical and spatial context (Hobbie et al. 2003). However, there is currently insufficient spatial coverage by existing sites as well as insufficient integration, infrastructure, and measures of connectivity across scales, both within and among sites, to assess whether skillful forecasting is possible of how, when, and why small events cascade nonlinearly to result in broad-scale catastrophic impacts. Networks that attempt to address these types of national level, continental scale problems are on the horizon (e.g., National Ecological Observatory Network; http://www.neoninc.org/) that will provide the necessary information to improve our understanding and ability to forecast and mitigate these events.

14.5 Summary and Conclusions

Given the complexity of the issues in Earth System science, future research will need to adopt approaches that cross traditional disciplinary boundaries in order to address cross scale interactions, threshold behavior, and feedback mechanisms leading to catastrophic events (Steffen et al. 2004). Our understanding of broad-scale patterns and dynamics has improved through collaborative efforts among ecosystem ecologists and atmospheric scientists (e.g., Rial et al. 2004). In addition, human dimensions are increasingly recognized by ecologists as integral to explaining system dynamics (e.g., Reynolds and Stafford Smith 2002). However, cross-disciplinary studies are needed to understand and forecast nonlinear dynamics and threshold behavior through time and across space (Peters et al. 2004). The blending of ideas and terms across scientific disciplines by our framework is an important step in this new level of cross-disciplinary research. Future steps include integrated experiments and modeling studies that synthesize technologies and expertise from diverse disciplines. Our ability to show similarities in system dynamics from seemingly disparate disciplines indicates that much is to be gained by cross disciplinary efforts that capitalize on the strengths of each discipline.

Acknowledgments

We thank Andrea Laliberte for assistance in image analyses and Dallas Staley for assistance in manuscript preparation. This work was supported by National Science Foundation grants DEB 0080412 (DPCP, BTB, KMH), DEB 0080529 (DPCP), DEB 0087289 (DPCP, BTB), and DEB 0217631 (RAP), and by the U.S. Long Term Ecological Research Network Office. This work is a contribution of the Global Change in Terrestrial Ecosystems (GCTE), the Land Project, Biologic Aspects of the Hydrologic Cycle (BAHC), and the Integrated Land Ecosystem-Atmosphere Processes Study (iLEAPS) projects of the International Geosphere-Biosphere program (IGBP). This chapter is Sevilleta publication number 364.

References

Aber JD, Burke IC, Acock B, Bugmann HKM, Kabat P, Menuat JC, Noble IR, Reynolds JF, Steffen WL, Wu J (1999) Hydrological and biogeochemical processes in complex landscapes: What is the role of temporal and spatial ecosystem dynamics? Pages 335–355 In: Tenhunen JD, Kabat P (ed.) Integrating hydrology, ecosystem dynamics, and biogeochemistry in complex landscapes. John Wiley & Sons, Chichester, UK

Albertson FW, Weaver JE (1942) History of the native vegetation of western Kansas during seven years of continuous drought, Ecol Monog 12:23–51

Archer S (1990) Development and stability of grass/woody mosaics in a subtropical savanna parkland, Texas, U.S.A. J Biogeo 17:453–462

Bak P, Tang C, Weisenfeld K (1988) Self organized criticality. Physical Review A: 38:364–374

Balling RC Jr., Klopatek JM, Hildebrandt ML, Moritz CK, Watts CJ (1998) Impacts of land degradation on historical temperature records from the Sonoran Desert. Clim Change 40:669–681

Breshears DD., Whicker JJ, Johansen MP, Pinder JE III (2003) Wind and water erosion and transport in semi-arid shrubland, grassland and forest ecosystems: quantifying dominance of horizontal wind-driven transport. Earth Surf Proc Land 28:1189–1209

Brown JR, Archer S (1987) Woody plant seed dispersal and gap formation in a North American subtropical savanna woodland: the role of domestic herbivores. Vegetatio 73:73–80

Butler BW, Bartlette RA, Bradshaw LS, Cohen JD, Andrews PL, Putnam T, Mangan RJ (1998) Fire behavior associated with the 1994 South Canyon Fire on Storm King Mountain, Colorado. USDA-Forest Service, Rocky Mountain Research Station Research Paper RMRS-RP-9

Byram GM (1954) Atmospheric conditions related to blowup fires. USDA-Forest Service, Southeastern Forest Experiment Station Paper 54

Charney JG, Quirk WK, Chow S-H, Kornfield J (1977) A comparative study of the effects of albedo change on drought in semi-arid regions. J Atmos Sci 34:1366–1385

Chen K, Bak P, Obukhov SP (1991) Self-organized criticality in a crack-propagation model of eathquakes. Physical Review A: 43:625–630

Clark JS, Carpenter SR, Barber M, Collins S, Dobson A, Foley JA, Lodge DM, Pascual M, Pielke R Jr., Pizer W, Pringle C, Reid WV, Rose KA, Sala O, Schlesinger WH, Wall DH, Wear D (2001) Ecological forecasts: an emerging imperative. Science 293:657–660

Claussen M, Kubatzki C, Brovkin V, Ganopolski A, Hoelzmann P, Pachur HJ (1999) Simulation of an abrupt change in Saharan vegetation at the end of the mid-Holocene. Geophys Res Lett 24:2037–2040

Davenport DW, Breshears DD, Wilcox BP, Allen CD (1998) Viewpoint: Sustainability of piñon-juniper ecosystems – a unifying perspective of soil erosion thresholds. J. Range Manage 51:229–238

deMenocal PB, Ortiz J, Guilderson T, Adkins J, Sarnthein M, Baker L, Yarusinki M (2000) Abrupt onset and termination of the African Humid Period: Rapid climate response to gradual insolation forcing. Quaternary Sci Rev 19:347–361

Drossel B, Schwabl F (1992) Self-organized critical forest-fire model. Physical Review Letters 69:1629–1632

Easterling DR, Evans JL, Groisman TRK, Kunkel KE, Ambenje (2000) Observed variability and trends in extreme climate events: a brief review. Bull Amer Meteor Soc 81:417–425

Elsner JB, Tsonis AA (1993) Nonlinear prediction, chaos, and noise. Bull Amer Meteor Soc 73:49–60

Folke C, Carpenter S, Walker B, Scheffer M, Elmqvist T, Gunderson L, Holling CS (2004) Regime shifts, resilience, and biodiversity in ecosystem management. Ann Rev Ecol System 35:557–581

Gallagher R, Appenzeller T (1999) Beyond reductionism: introduction to special section on complex systems. Science 284:79–109

Gerber S, Joos F, Prentice IC (2004) Sensitivity of a dynamic global vegetation model to climate and atmospheric CO_2. Global Change Biol 10:1223–1239

Gibbens RP, McNeely RP, Havstad KM, Beck RF, Nolen B (2005) Vegetation changes in the Jornada Basin from 1858 to 1998. J Arid Environ 61:651–668

Gill RA, Polley HW, Johnson HB, Anderson LJ, Maherali H, Jackson RB (2002) Nonlinear grassland responses to past and future atmospheric CO_2. Nature 417:279–282

Graham RT, Technical Editor (2003) Hayman Fire Case Study: Summary. Gen. Tech. Rep. RMRS-GTR-115. U.S. Department of Agriculture, Forest Service, Rocky Mountain Research Station. Ogden, UT USA

Great Plains Committee (1937) The future of the Great Plains. 75th Congress, 1st Session, H.R. Doc. 144

Hethcote HW (2000) The mathematics of infectious diseases. Soc Indus Appl Math Rev 42:599–653

Hobbie JE, Carpenter SR, Grimm NB, Gosz JR, Seastedt TR (2003) The U.S. Long Term Ecological Research (LTER) Program. BioScience 53:21–32

Hobbs RJ, Humphries SE (1995) An integrated approach to the ecology and management of plant invasions. Conserv Biol 9:761–770

Holling CS (1973) Resilience and stability of ecological systems. Annual Review of Ecology and Systematics 4:1–23

Holling CS (1992) Cross-scale morphology, geometry, and dynamics of ecosystems. Ecol Monogr 62:447–502
Holling CS (1996) Surprise for science, resilience for ecosystems, and incentives for people. Ecol Appl 6:733–735
Hurt DR (1981) An Agricultural and Social History of the Dust Bowl. Nelson Hall, Chicago
Husar RB, Tratt DM, Schichtel BA, Falke SR, Li F, Jaffe D, Gasso S, et al. (2001) Asian dust storms of April 1998. J Geophys Res 106:18317–18330
Kabat P, Claussen M, Dirmeyer PA, Gash JHC, Bravo de Guenni L, Meybeck M, Pielke Sr. RA, Vörösmarty CJ, Hutjes RWA, Lütkemeier S, eds (2004) Vegetation, water, humans, and the climate: a new perspective on an interactive system. Springer-Verlag, Berlin
Kaplan D, Glass L (1995) Understanding nonlinear dynamics. Springer-Verlag, New York
Laurance WF, Williamson GB (2001) Positive feedbacks among forest fragmentation, drought, and climate change in the Amazon. Conserv Biol 15:1529–1535
Lockeritz W (1978) The lessons of the Dust Bowl. Amer Sci 66:560–569
Ludwig JA, Tongway DJ (2000) Viewing rangelands as landscape systems. Pages 39–52 In: Arnalds O, Archer SA (eds) Rangeland desertification. Kluwer Academic, Dordrecht, The Netherlands
Mack RN, Simberloff D, Lonsdale WM, Evans H, Clout M, Bazzaz FA (2000) Biotic invasions: causes, epidemiology, global consequences, and control. Ecol Appl 10:689–710
McKenzie D, Gedalof Z, Peterson DL, Mote P (2004) Climatic change, wildfire, and conservation. Conser Biol 18:890–902
Miller ER (1934) The dustfall of Novemner 12–13, 1933. Monthly Weather Rev 62:14–15
Miller C, Urban DL (2000) Connectivity of forest fuels and surface fire regimes. Landscape Ecol 15:145–154
National Research Council (2001) Economic and eological impacts of abrupt climate change. In: Committee on Abrupt Climate Change, Ocean Studies Board, Polar Research Board, Board on Atmospheric Sciences and Climate (eds) Abrupt climate change: inevitable surprises. National Research Council. Washington, D.C, pp 90–117
Nieminen RM (2003) From atomistic simulation towards multiscale modeling of materials. J Phys Conden Mat 14:285–286
Pascual M, Ellner SP (2000) Linking ecological patterns to environmental forcing via nonlinear time series models. Ecol 81:2767–2780
Peters DPC, Pielke RA Sr., Bestelmeyer BT, Allen CD, Munson-McGee S, Havstad KM (2004) Cross scale interactions, nonlinearities, and forecasting catastrophic events. Proc Nat Acad Sci 101:15130–15135
Peters DC, Schlesinger WH, Herrick JE, Huenneke LF, Havstad KM (2006) Future directions in Jornada research: applying an interactive landscape model to solve problems. Havstad KM, Huenneke LF, Schlesinger WH (eds) Structure and function of a Chihuahuan Desert ecosystem: the Jornada Basin long-term ecological research site. New York, NY, Oxford University Press, pp 369–386
Pielke Sr. RA (2001) Earth System modeling – an integrated assessment tool for environmental studies. Pages 311–337 InL Matsuno T, Kida H, eds. Present and future of modeling global environmental change: toward integrated modeling. Terra Scientific Pub Co, Tokyo
Reiners WA, Driese KL (2001) The propagation of ecological influences through heterogeneous environmental space. BioSci 51: 939–950
Reiners WA, Driese KL (2004) Transport Processes in Nature: Propagation of Ecological Influences through Environmental Space. Cambridge University Press, Cambridge
Reynolds JF, Stafford Smith DM (2002) Global desertification: Do humans cause deserts? Dahlem University Press, Berlin
Rial JA, Pielke Sr. RA, Beniston M, Claussen M, Canadell J, Cox P, Held H, de Noblet-Ducoudre N, Prinn R, Reynolds JF, Salas JD (2004) Nonlinearities, feedbacks and critical thresholds within the Earth's climate system. Clim Change 65:11–38
Rietkerk M, van de Koppel J (1997) Alternative stable states and threshold effects in semi-arid grazing systems. Oikos 79:69–76
Rietkerk M, Dekker SC, de Ruiter PC, van de Koppel J (2004) Self-organized patchiness and catastrophic shifts in ecosystems. Science 305:1926–1929
Rosenfeld D, Rudich Y, Lahav R (2001) Desert dust suppressing precipitation: A possible desertification feedback loop. Proc Nat Acad Sci 98:5975–5980
Sarawitz D, Pielke Jr. RA, Byerly Jr. R (eds) (2000) Prediction: science, decision making, and the future of nature. Island Press, Washington, DC
Scheffer M, Carpenter S, Foley JA, Folke C, Walker B (2001) Catastrophic shifts in ecosystems. Nature 413:591–596
Scheller RM, Mladenoff DJ (2005) A spatially interactive simulation of climate change, harvesting, wind, and tree species migration and projected changes to forest composition and biomass in northern Wisconsin, USA. Global Change Biol 11:307–321
Schlesinger WH, Reynolds JF, Cunningham GL, Huenneke LF, Jarrell WM, Virginia RA, Whitford WG (1990) Biological feedbacks in global desertification. Science 247:1043–48
Scholes MC, Matrai PA, Andreae MO, Smith KA, Manning MR (2003) Biosphere-atmosphere interactions. In: Brasseur GP, Prinn RG, Pszenny AAP (eds) The changing atmosphere: an integration and synthesis of a decade of tropospheric chemistry research. The International Global Atmospheric Chemistry Project (IGAC). IGBP Global Change Series, Springer-Verlag, Berlin Heidelberg New York
Schubert SD, Suarez MJ, Pegion PJ, Koste RD, Bacmeister JT (2004) On the cause of the 1930s Dust Bowl. Science 303:1855–1859
Speer JH, Swetnam TW, Wickman BE, Youngblood A (2001) Changes in Pandora moth outbreak dynamics during the past 622 years. Ecology 82:679–697
Squires VR (2001) Dust and sandstorms: an early warning of impending disaster. In: Youlin Y, Squires VR, Qi L (eds) Global alarm: dust and sandstorms from the world's drylands. UN Convention to Combat Desertification, Bangkok, pp 15–28
Steffen W, Sanderson A, Tyson PD, Jager J, Matson PA, Moore III B, Oldfield F, Richardson K, Schellnhuber HJ, Turner B II, Wasson RJ (2004) Global change and the Earth System: a planet under pressure. Springer-Verlag, Berlin
Swetnam TW, Betancourt JL (1998) Mesoscale disturbance and ecological response to decadal climate variability in the American Southwest. J Clim 11:3128–3147
Swetnam TW, Lynch AM (1993) Multicentury, regional-scale patterns of Western spruce budworm outbreaks. Ecol Monogr 63:399–424
Tegen I, Fung I (1995) Contribution to the atmospheric mineral aerosol load from land surface modifications. J Geophys Res 100:18707–18726
Tsonis AA (2001) The impact of nonlinear dynamics in the atmospheric sciences. I. J Bifurcation Chaos 11:881–902
Turner MG, Romme WH, Tinker DB (2003) Surprises and lessons learned from the 1988 Yellowstone fires. Front Ecol Environ 1:351–358
U.S. Department of Agriculture. Forest Service (1978) When the mountains roared. Coeur d'Alene, ID
Warrick RA (1980) Drought in the Great Plains: a case study of research on climate and society in the USA. In: Ausubel J, Biswas (eds) Climatic constraints and human activities. IIASA Proc Series, Vol 10, Pergamon Press, New York, pp 93–123
Weaver JE, Albertson FW (1936) Effects of the Great Drought on the prairies of Iowa, Nebraska, and Kansas. Ecology 4:567–639
Weaver JE, Albertson FW (1940) Deterioration of Midwestern ranges. Ecology 21:216–236
Whelan RJ (1995) The ecology of fire. Cambridge Univ. Press, Cambridge
Wilcox BP, Breshears DD, Allen CD (2003) Ecohydrology of a resource-conserving semiarid woodland: temporal and spatial scaling and disturbance. Ecol Model 73:223–239
Woodhouse CA, Overpeck JT (1998) 2000 years of drought variability in the Central United States. Bull Amer Meteor Soc 79:2693–2714
Worster D (1979) Dust Bowl: The Southern Great Plains in the 1930s. Oxford University Press, New York
Xue Y, Fennessy MJ (2002) Under what conditions does land-cover change impact regional climate? In: Reynolds JF, Stafford Smith DM (eds) Global desertification: do humans cause deserts? Dahlem University Press, Berlin, pp 59–74
Yearbook of Agriculture (1934) U.S. Government Printing office, Washington DC
Youlin Y, Squires VR, Qi L (eds) (2001) Global alarm: dust and sandstorms from the world's drylands. United Nations Convention to Combat Desertification, Bangkok
Zeng N, Pielke Sr. RA, Eykholt R (1993) Chaos theory and its applications to the atmosphere. Bull Amer Meteor Soc 74:631–644
Zeng N, Neeling JD, Lau LM, Tucker CJ (1999) Enhancement of interdecadal climate variability in the Sahel by vegetation interaction. Science 286:1537–1540

Chapter 15

Dynamic Global Vegetation Modeling: Quantifying Terrestrial Ecosystem Responses to Large-Scale Environmental Change

I. Colin Prentice · Alberte Bondeau · Wolfgang Cramer · Sandy P. Harrison · Thomas Hickler · Wolfgang Lucht Stephen Sitch · Ben Smith · Martin T. Sykes

15.1 Introduction

The annual exchange of carbon between the atmosphere and the land biota amounts to one-sixth of the atmospheric content of carbon dioxide (CO_2), and the average turnover time of terrestrial organic carbon (including the biota and soils, but excluding geological storages) is only about 20 years. The land biosphere therefore plays a dynamic role in the global carbon cycle on time scales relevant to human activities (Prentice et al. 2001; Schimel et al. 2001; Field et al. 2004). The land biosphere's variations in space and time also influence the fluxes of energy, momentum, water vapor, and many climatically important or reactive trace gases and aerosol precursors across the lower boundary of the troposphere. The land biota respond individualistically to local environmental factors such as photosynthetically active radiation (PAR), temperature, atmospheric humidity, soil moisture, CO_2 concentration and land management. These responses of organisms to their environment are fundamental for the continuing provision of ecosystem goods and services on which all human activities ultimately depend (MA 2003).

Among the many methods for observing the dynamics of terrestrial ecosystems, each method has a restricted window of applicability in space and time. Ground-based measurements (biomass inventories, community descriptions, productivity measurements, flux measurements) are made at single sites or across networks, but are not readily scaled up. Satellite-based measurements provide up to 20 years of global coverage with spatial resolution on the order of a few kilometres, and temporal sampling intervals of days to weeks. Satellite observations have special importance for understanding large-scale processes because they can provide comprehensive coverage, averaged over landscapes. New sensors and satellites are expanding the scope of such observations. But there are limitations on what can be observed from space, particularly with regard to biodiversity and below-ground processes. For the foreseeable future, it will be important to use multiple sources of information on terrestrial ecosystem structure and dynamics, and to use modeling techniques to link them.

The identification of mechanisms in the functioning of the land biosphere has meanwhile become a high scientific priority. On a fundamental level, many non-linearities and feedbacks in the Earth System, including processes determining changes in atmospheric composition on glacial-interglacial and longer time scales and rapid changes in ecosystems and the atmosphere during the recent geological past, originate in the incompletely understood coupling between global biogoechemical cycles and the physical climate (Prentice and Raynaud 2001; Overpeck et al. 2003). On a more practical level, anthropogenic alterations of the global environment have accelerated massively, through land-use change (Foley et al. 2005) as well as changes in atmospheric composition and climate (Houghton et al. 2001); this situation has created an urgent demand by society for tools to predict the risks of continued environmental changes for ecosystem services, and indeed for the future of climate and sustainable land management. The development of Dynamic Global Vegetation Models (DGVMs) by several research groups during the past 10–15 years has been largely a response to these dual "drivers" of interdisciplinary Earth System science.

15.2 Historical Antecedents and Development of DGVMs

The conceptual development of a DGVM dominated the activities of the Biosphere Dynamics Project, led by Allen M. Solomon, at the International Institute for Applied Systems Analysis (IIASA) during 1988–1990. GCTE subsequently adopted DGVM development as a near-term goal, and provided an umbrella for further work by several groups. DGVMs fuse research on four broad groups of processes: *plant geography*, *plant physiology and biogeochemistry*, *vegetation dynamics*, and *biophysics*, each historically pursued by a separate research community (Fig. 15.1). The early development of DGVMs concentrated on representing these processes and their interactions as they would have occurred without human influence. Lately DGVM development has expanded to include the representation of *human intervention* (agriculture, urbanization and forest management).

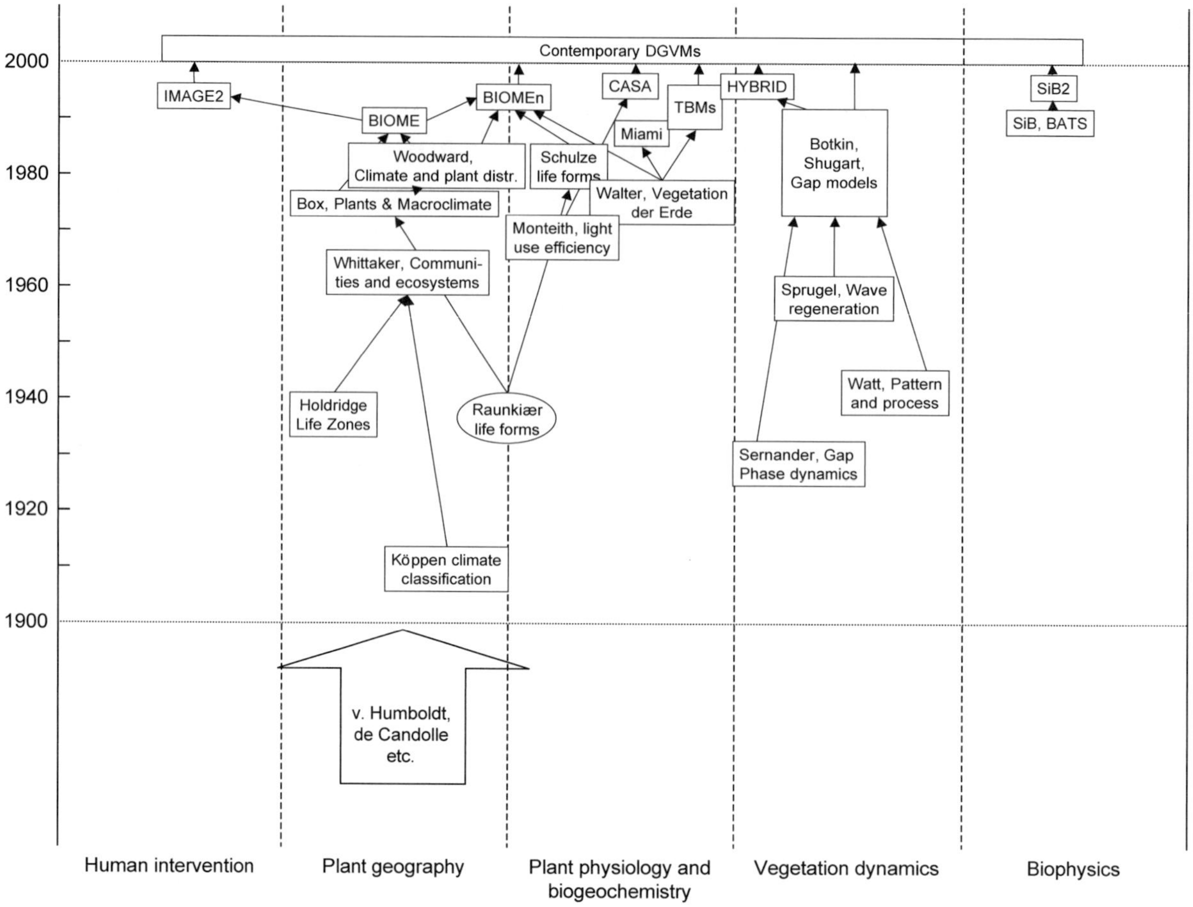

Fig. 15.1. A summary of the historical antecedents and recent development of DGVMs

15.2.1 Plant Geography

The beginnings of predictive modeling in plant geography can be traced to Köppen's (1931) world climate zones. Köppen tried to match the distribution of biomes, and included relevant seasonal aspects of climate in his classification scheme. A later (but more artificial) classification scheme based on annual climate statistics by Holdridge (1947) was used by Emanuel et al. (1985) to produce the first climate-derived map of global potential natural vegetation, and the first global projection of vegetation for a "greenhouse world" as simulated by a general circulation model (GCM). Further climate classifications designed to match biome distributions have been proposed by Whittaker (1975) and several others.

None of these schemes was based explicitly on an underlying theory of the controls on vegetation distribution, although the rudiments of a theory had been put forward by Raunkiær (1909, 1913, 1934). Raunkiær emphasized the role of mechanisms for surviving the unfavourable season in determining the distribution of different types of plants, which we would now call "plant functional types" (PFTs). Building on Raunkiær's ideas, Box (1981) created the first numerical model of global PFT distributions driven by climate. Woodward (1987) created the first explicitly process-based model of global biome distribution. The model included limits to woody PFT survival associated with cold tolerance, based on a review of experimental data. It included the dependence of leaf area index (LAI) on water availability, using an optimization principle introduced by Specht (1972). Woodward's approach was further developed in the "equilibrium biogeography models" BIOME (Prentice et al. 1992) and MAPSS (Neilson et al. 1992; Neilson and Marks 1994; Neilson 1995).

15.2.2 Plant Physiology and Biogeochemistry

General quantitative relationships between plant growth and resource availability became available during the 1960s through the International Biological Programme (IBP). Walter's *Vegetation der Erde* (Walter 1962, 1968) combined the older principles of plant geography with the new understanding of plant production. Lieth (1975) analysed IBP data statistically to create the so-called Miami model for net primary production (NPP) as a function of climate. Schulze (1982) reviewed the role of carbon, water and nutrient constraints in determining the distribution of PFTs, emphasizing the importance of competitive success as well as survival limits.

"Terrestrial biogeochemistry models" (TBMs), as they are now known, were originally developed with the main goal of simulating NPP. The first to be applied globally was the Terrestrial Ecosystem Model (TEM) (Melillo et al. 1993). Other TBMs include Century (Parton et al. 1993), Forest-BGC (Running and Gower 1991: later BIOME-BGC; Running and Hunt 1993), CASA (Field et al. 1995), G'DAY (Comins and McMurtrie 1993), CARAIB (Warnant et al. 1994), DOLY (Woodward et al. 1995) and BETHY (Knorr 2000; Knorr and Heimann 2001). The more recent TBMs use the biochemical model of Farquhar et al. (1980) for the dependence of photosythesis on external variables. This model makes explicit the dependence of photosynthesis on the leaf-internal partial pressure of CO_2, providing a key component for process-based simulation of CO_2 effects.

Several of the original TBMs are still used widely. In addition, the BIOME*n* models (Haxeltine and Prentice 1996a,b; Kaplan et al. 2003) are a hybrid of the equilibrium biogeography and TBM approaches. They predict geographic distributions of biomes by comparing the modeled NPP of different PFTs within each PFT's survival limits. They can therefore make competition-based distinctions among biomes that the earlier equilibrium biogeography models could not, and they can incorporate CO_2 effects mechanistically (Cowling 1999; Harrison and Prentice 2003).

15.2.3 Vegetation Dynamics

All of the models discussed above are restricted in their application because they cannot represent dynamic transitions between biomes (Prentice and Solomon 1991; VEMAP Members 1995; Neilson and Running 1996; Woodward and Lomas 2004). To provide this capability, DGVMs have drawn on a very different scientific tradition. Classic ecological studies of vegetation dynamics, including Sernander (1936), Watt (1947) and Sprugel (1976), laid the foundations for the modern understanding of vegetation dynamics and prepared the way for the formal description of forest dynamics in terms of individual tree establishment, growth and mortality in JABOWA (Botkin et al. 1972), FORET (Shugart and West 1977), LINKAGES (Pastor and Post 1985) and a host of descendants (Shugart 1984), including extensions to non-forest vegetation types (e.g., Prentice et al. 1987). Newer incarnations of this "gap model" concept include FORSKA (Prentice and Leemans 1990; Prentice et al. 1993) and SORTIE (Pacala et al. 1993, 1996). These models typically are applied in a small region with parameter sets based on observations for individual species. They are computationally intensive because they simulate the stochastic behavior of many individual plants on multiple replicate plots.

DGVMs struggle to represent vegetation dynamics in a computationally efficient way without losing essential features that depend on interactions between plant individuals. Friend et al. (1997) used a simplified gap model approach, representing grid cell dynamics by a sample of patches. More efficient layer- (Fulton and Prentice 1997) and cohort-based (Bugmann 1996; Bugmann and Solomon 2000) approximations for vegetation dynamics exist, but have not been widely adopted. Most DGVMs rely instead on various *ad hoc* large-area parameterizations. Smith et al. (2001) however showed that the gap model formalism continues to give more realistic estimates of PFT dynamics, at least when compared to the large-area parameterization in the Lund-Potsdam-Jena (LPJ) DGVM (Sitch et al. 2003; see also Hickler et al. 2004a). A possible route to a more rigorously "traceable" representation of individual-based processes over large areas is suggested by the work of Moorcroft et al. (2001).

15.2.4 Biophysics

GCMs include representations of the controls on the exchange of energy, water vapor and momentum between the atmosphere and the land surface. Biophysical models developed for this purpose are called "land surface schemes" or "soil-vegetation atmosphere transfer models" (SVATs). Vegetation properties needed by the GCM include rooting depth, soil porosity, surface albedo, surface roughness, fractional vegetation cover, and surface conductance. Surface albedo depends on leaf reflectance, canopy structure and vegetation structural properties (including height) that determine the "masking" of snow. Changes in vegetation that affect surface albedo can profoundly affect climate (Bonan et al. 1992). Surface conductance depends on leaf area index and stomatal conductance, and is one of the controls on evapotranspiration. Accurate simulation of exchanges between the land and the free troposphere also depends on having an adequate representation of the planetary boundary layer (PBL). PBL dynamics depend on properties of the land surface, including the latent heat flux from the canopy (Finnegan and Raupach 1987; Monteith 1995; Prentice et al. 2004).

The first GCM land-surface schemes to represent vegetation explicitly were SiB (Sellers et al. 1986) and BATS (Dickinson et al. 1993). These models represented variations in stomatal conductance by empirical functions of PAR, temperature, humidity and soil moisture (Jarvis 1976). Later models have exploited the tight coupling of CO_2 and water exchange through stomatal conductance (Collatz et al. 1991). The current trend is to replace GCM land-surface schemes with full DGVMs. For this purpose, exchanges of energy, water vapor and momentum must be modeled at a time step similar to the shortest atmospheric time step of the GCM (typically about 30 minutes). The DGVMs IBIS (Foley et al. 1996) and TRIFFID (Cox 2001) were developed for GCM coupling. Full coupling to an atmospheric GCM was first achieved by Foley et al. (1998) and Delire et al. (2002). Full physical coupling to an ocean-atmosphere GCM has been achieved by Robert J. Gallimore and others (e.g., Notaro et al. 2004) using LPJ (Sitch et al. 2003). LPJ also provided the basis for a generic vegetation dynamics component in Orchidée (Krinner et al. 2005) and several other DGVMs that are being developed for GCM coupling. All of the major climate modeling groups are now working towards full physical and carbon-cycle coupling of atmosphere, ocean and land, as first implemented by Cox et al. (2000).

15.2.5 Human Intervention

A final strand of model development addresses the changing suitability of the land for human land use and the reciprocal influence of human land use on the state of the biosphere. The most widely known example, and the most explicit in terms of representing land cover, is IMAGE2 (Alcamo 1994). IMAGE2 is widely cited and used for integrated assessment and scenario development. The land surface component of IMAGE2 was derived from BIOME (Prentice et al. 1992). However, several groups are now developing more advanced large-area representations of managed ecosystems, including explicit simulations of agricultural and forest management, as components of DGVMs (Bondeau et al. in press).

15.3 Principles and Construction of DGVMs

15.3.1 Model Architecture

The modular organization of DGVMs is rather uniform (Cramer et al. 2001; Sykes et al. 2001; Beerling and Woodward 2001; Woodward and Lomas 2004; Fig. 15.2). The design and process formulations of DGVMs are not fundamentally different from those in TBMs, which have been used to investigate some of the same questions (Heimann et al. 1998; McGuire et al. 2001). The most important unique feature of DGVMs is their ability to simulate vegetation dynamics. Within a grid cell, vegetation may be represented by fractions or strata occupied by different PFTs. Age or size classes may be distinguished, but more typically the modeled properties represent averages among the entire grid cell population of a given PFT (e.g., Sitch et al. 2003; Fig. 15.3). PFT-specific state variables, i.e., physical properties that change dynamically in the course of the model simulation, may include a description of the average geometry of individual plants, the carbon content of one or more plant biomass compartments (leaves, roots, wood), nitrogen (N) status, factors affecting resource uptake capacity (leaf area index, root density) and population density.

DGVMs implement two or three nested timing loops, calling different processes on different operational time steps (Fig. 15.2) corresponding loosely to the fastest characteristic time scale of the process. "Fast" processes varying on a diurnal cycle include energy and gas exchange

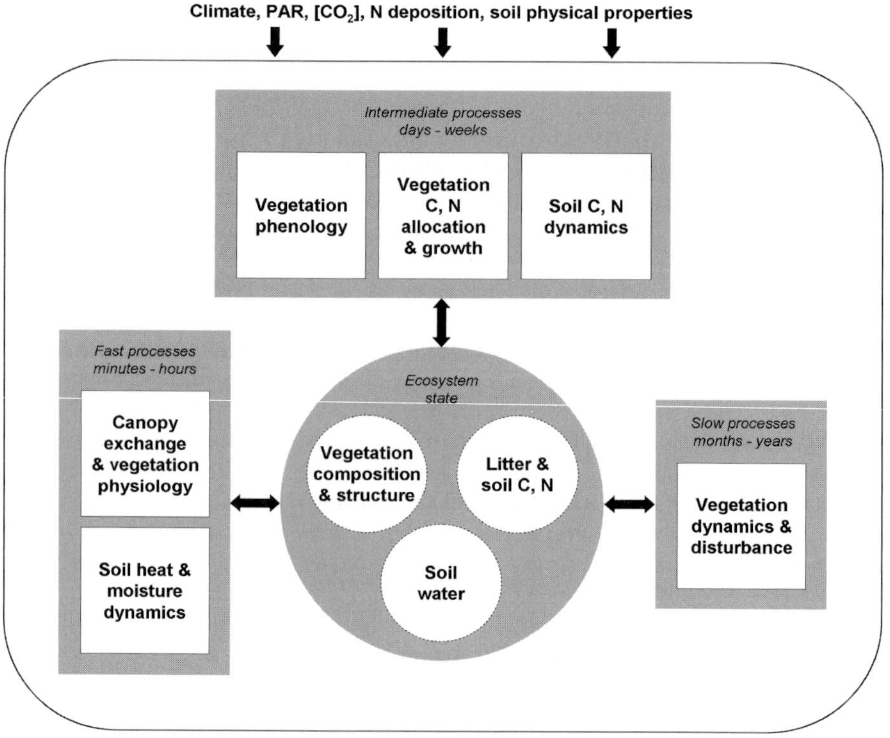

Fig. 15.2. DGVMs are structurally rather similar. This figure illustrates a typical structure, showing driving variables, main process modules (organized by operational timestep) and state variables

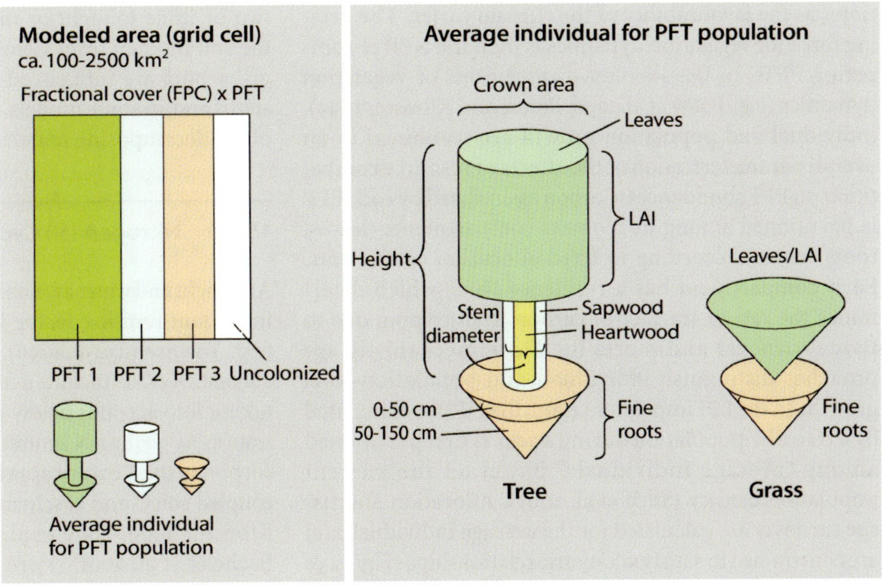

Fig. 15.3. Each DGVM has adopted a different large-area parameterization for vegetation dynamics. This figure depicts one example (Sitch et al. 2003). Here, each PFT occupies a fraction of the modeled area (grid cell). Structural properties for one *average individual* for each PFT vary dynamically depending on carbon allocation, tissue turnover and allometric relationships. Population density scales PFT properties from the average individual to the grid cell

at the canopy-atmosphere interface, photosynthesis and plant-soil water exchange. These processes are invoked either on a time step of one day, using daily integrals of driving variables such as PAR, or (more accurately, in principle) at shorter time steps of one hour or less, in models that explicitly simulate the diurnal cycle. Processes with seasonal dynamics include plant phenology, growth and soil organic matter dynamics; the typical time step used for these processes is one month. Vegetation dynamics are generally the slowest process modeled, and are typically represented with a time step of one year.

15.3.2 Net Primary Production

The currency of growth in DGVMs is NPP, the balance of carbon uptake by photosynthesis and release by autotrophic respiration. Most DGVMs use the Farquhar et al. (1980) model, or derivatives thereof (Collatz et al. 1991, 1992; Haxeltine and Prentice 1996a,b), to model photosynthesis at the leaf level. Environmental and leaf parameters are either available from the input data (e.g., air temperature and CO_2 concentration), calculated based on the current vegetation or system state (stomatal conductance, leaf nitrogen content), or prescribed (specific leaf area). DGVMs explicitly or implicitly take into account shading of leaves at lower levels in the vegetation canopy by the levels above. Nitrogen invested in leaf functional proteins is commonly assumed to distribute among canopy layers in a fashion that maximises net assimilation, i.e., photosynthesis minus leaf respiration (Haxeltine and Prentice 1996a,b; Foley et al. 1996; Friend et al. 1997; Sitch et al. 2003), at each canopy level (see Dewar 1996 and Prentice 2001 for further discussion of this hypothesis and its variants).

The rate of diffusion of CO_2 from the ambient air via the boundary layer adjacent to leaf surfaces and the stomata is controlled by aggregate stomatal conductance, and limits photosynthesis. Plants are considered to regulate stomatal conductance, within limits, to optimise CO_2 uptake in relation to water loss through transpiration (Cowan 1977; Collatz et al. 1991). Thus, DGVMs typically couple photosynthesis, canopy biophysics and soil hydrology submodels via canopy conductance, although the detailed formulations vary.

Respiration is usually separated into maintenance and growth components. Maintenance respiration is sensitive to temperature and differs among tissues (Ryan 1991). Models may adopt a tissue-specific scaling factor combined with a common temperature response function, generally a Q_{10} or Arrhenius relationship. Alternatively, a function based on tissue C/N ratio may replace the tissue-specific multiplier. Growth respiration is usually defined as a fixed fraction of NPP. Some DGVMs alternatively use more empirical approaches to estimate NPP, with a potential rate moderated by scalars standing for environmental stresses (e.g., soil water, low temperatures, shading of grasses by trees: Daly et al. 2000) and/or resource availability and uptake capacity (Pan et al. 2002; Potter and Klooster 1999).

15.3.3 Plant Growth and Vegetation Dynamics

In all DGVMs, multiple PFTs are allowed to co-occur and compete. Tolerance limits for bioclimatic variables, such as coldest-month mean temperatures and growing-season heat sums, define the climatic space each PFT can occupy (Woodward 1987; Harrison et al. submitted). PFTs may be switched "on" or "off" in a particular grid cell, through PFT-specific establishment and mortality functions, as the favourability of the climate varies. The driving force for vegetation dynamics is then the NPP of competing PFTs. In the simplest formulations of vegetation dynamics (e.g., Foley et al. 1996; Potter and Klooster 1999), individual and population growth are combined in an overall parameterization of the effects of resource competition on PFT abundances. Carbon assimilated by each PFT is partitioned among its biomass compartments (leaves, roots, stems) according to fixed allocation coefficients. Each compartment has a residence time, which determines the rate of transfer of carbon to litter pool due to tissue turnover and mortality. More mechanistic approaches distinguish individual- and population-level growth. In the LPJ implementation, the NPP accumulated by a tree PFT population during a year is first partitioned among "average individuals" based on the current population density (Sitch et al. 2003). Allocation and tissue turnover are calculated for the average individual, and are constrained to satisfy allometric relationships (Fig. 15.3). Population growth is the balance of an annual rate of establishment of new saplings, influenced by current density, and mortality, which may increase under conditions of resource limitation, crowding or disturbance.

The ability to adjust allocation patterns to maintain a balance between resource uptake and utilization is a key feature of plant competitive strategies (Field et al. 1992). Modeled allocation patterns in DGVMs can therefore be influenced by the relative supply of above- and below-ground resources. Soil water deficits in the current growing season, for example, may lead to increased investment in roots at the expense of leaves the following growing season.

15.3.4 Hydrology

DGVMs typically include some multi-layer scheme for soil water, with percolation and/or saturated flow between layers. Evaporation from the upper soil layer and the vegetation canopy (i.e., interception loss) may supplement plant transpiration. Water content in excess of field capacity is lost as runoff. Some models take account of the effects of snow and ground frost on seasonal water cycles. DGVMs have also been coupled to large-scale models for lateral water transport, in order to examine e.g., impacts of land-use change on river flow.

15.3.5 Soil Organic Matter Transformations

Carbon enters the soil as litter associated with tissue turnover and mortality. Litter and soil carbon provide the substrate for soil heterotrophs, whose respiration releases CO_2. "Pools" with different degrees of decomposability are usually distinguished. As the labile fractions are consumed, residues are transferred to pools with longer average residence times. The number of pools represented ranges from two or three to eight or more in models that implement the soil module from Century. Decomposition rates for a given pool are influenced by temperature, soil moisture status and, in some models, properties such as the C/N ratio of the decomposing material, soil texture and clay content.

15.3.6 Nitrogen (N) Cycling

After light and water availability, plant-available N is the most important limiting factor in many terrestrial ecosystems (e.g., Townsend et al. 1996). Nevertheless, only some of the current DGVMs include a full interactive terrestrial N cycle, taking into account below-ground controls on N mineralization as well as N limitations on NPP. DGVMs that incorporate the Century approach to soil processes inherit its coupled soil C and N scheme (Friend et al. 1997; Potter and Klooster 1999; Daly et al. 2000; Woodward et al. 2001; Bachelet et al. 2001). Here litter quality influences net N mineralization and decomposition rates; labile "metabolic" inputs, such as litter derived from leaves and fine roots, tend to increase net N mineralization, whereas lignin-rich "structural" material causes N immobilization and may limit N availability to plants. N limitation of production may be modeled by scaling net assimilation to plant uptake of N from the soil mineral N pool. In the Hybrid DGVM, N limitation implicitly reduces investment in Rubisco and chlorophyll, resulting in a lower maximum carboxylation rate and reduced photosynthesis (Friend et al. 1997).

15.3.7 Disturbance

The term "disturbance" is widely used to refer to processes such as fires, windstorms and floods, which rapidly destroy biomass, alter vegetation structure and alter the conditions for the growth of remaining plants and/or the establishment of new plants. This usage is illogical because "disturbances" by this definition are intrinsic to ecosystems and part of the mechanism that maintains their typical composition and character (Allen and Hoekstra 1990); however, it is entrenched in the literature. The stochastic nature of disturbance regimes makes them difficult to represent in models. Some DGVMs do not explicitly model disturbances; instead, they incorporate their effects implicitly in turnover constants for vegetation carbon (Foley et al. 1996; Friend et al. 1997).

Fire is the most important type of natural disturbance type worldwide, affecting all biomes except rainforests and deserts, at frequencies ranging from every year to once every few centuries. The most important controls on fire regimes are the frequency of ignition (whether natural or human-caused) and the amount, moisture content and flammability of biomass fuels. These controls depend on both climate and vegetation state, allowing for a variety of feedbacks in vegetation dynamics involving fire. Thonicke et al. (2001) introduced a semi-empirical fire module for use in DGVMs (Pan et al. 2002; Sitch et al. 2003). The modeled area (grid cell) is considered to be large enough that ignition sources are available, and that the *fraction* of the grid cell affected by fire in a given year is equal to the *probability* of fire affecting a randomly chosen point. This probability is estimated using empirical equations based on fuel load and moisture content (estimated from the moisture of the top soil layer). PFTs differ in their resistance to fire, so that the degree of damage caused to standing biomass depends on the vegetation composition. Fires result in vegetation mortality and volatilization of a fraction of litter and biomass over the affected area.

A more advanced approach to modeling fire dynamics has been adopted in the MC1 DGVM. This model distinguishes surface and crown fires, and fire effects are sensitive to stand structure as well as fuel load (Lenihan et al. 1998; Daly et al. 2000; Bachelet et al. 2001, 2003). Venevsky et al. (2002) and Arora and Boer (in press) have developed fire models of intermediate complexity that allow for variations in ignition rates associated with human activities.

15.4 Evaluating DGVMS

DGVMs simulate processes at a wide range of space and time scales and, accordingly, many different types of contemporary observations can be used to test their performance. The following is a non-exhaustive summary of observational "benchmarks" for DGVMs. For further examples see e.g., Kucharik et al. (2000), Beerling and Woodward (2001) and Woodward and Lomas (2004).

15.4.1 Net Primary Production

Following the Potsdam NPP Intercomparison Project (Cramer et al. 1999), which engaged mainly TBMs in a first large-scale comparison of terrestrial models driven by a common set of input variables, the Ecosystem Model-Data Intercomparison project (http://gaim.unh.edu/Structure/Intercomparison/EMDI/) ran site-specific simulations of NPP and compared them to measurements of NPP from sites in each of the major biomes. A large data synthesis effort yielded NPP values at 162 Class A sites ("well documented and intensively studied") and 2 363 Class B sites ("globally extensive but less well documented and with less site-specific information"). A tendency was found for models to over-estimate low- to mid-range production at boreal and temperate sites, and to underestimate NPP in highly productive tropical sites. Modeled NPP tended towards an asymptote ~1 000 g C m^{-2} while measurements showed some higher values. The reasons for these discrepancies remain to be established.

15.4.2 Remotely Sensed "Greenness" and Vegetation Composition

The fraction of Absorbed Photosynthetically Active Radiation (fAPAR) is the ratio of vegetation-absorbed to incident PAR. It can derived from satellite spectral reflectance data and is a measure of vegetation "greenness". The seasonal course of fAPAR provides a way to test modeled phenology (Bondeau et al. 1999). Some TBMs, known as light-use efficiency models, use remotely sensed fAPAR as input (e.g., Potter et al. 1993; Knorr and Heimann 1995; Ruimy et al. 1996). Alternatively, fAPAR observations can also be used to calibrate phenology in models (Kaduk and Heimann 1996; Botta and Foley 2002; Arora and Boer 2005). Seasonal cycles of fAPAR have also been used together with ancillary information to construct global maps of vegetation composition in terms of a few broadly defined PFTs. For example, Sitch et al. (2003) used the DeFries et al. (2000) global data set of estimated fractional PFT cover as a benchmark for vegetation composition, while Woodward and Lomas (2004) used the HYDE land-cover type data set of Klein Goldewijk (2001).

15.4.3 Atmospheric CO_2 Concentration

A different approach to large-scale evaluation of terrestrial models (Prentice et al. 2000) makes use of high-precision atmospheric measurements of CO_2 concentration (http://www.cmdl.noaa.gov/ccgg/globalview/co2/). Both the amplitude and the timing of the seasonal cycle of CO_2 vary geographically, reflecting different seasonal patterns of biospheric activity. The amplitude is greatest in northern high latitudes because of the large vegetated area in the north and the large offset in the timing of NPP and heterotrophic respiration maxima in high latitudes. Heimann et al. (1998) ran the TM2 atmospheric transport model with monthly fields of net ecosystem exchange (heterotrophic respiration and combustion minus NPP) from four terrestrial models. The output was sampled at the locations of CO_2 monitoring stations. Knorr and Heimann (1995, 2001), Dargaville et al. (2002) and Sitch et al. (2003) continued this approach. The main caveat for such comparisons is that they rely on the realism of the transport model; this is an active research area (Denning et al. 1999; Gurney et al. 2003; Law et al. 2003; Gurney et al. 2004). Inversion of tracer transport models has also been used to infer regional sources and sinks of CO_2 directly from the CO_2 concentration network (Fan et al. 1998; Bousquet et al. 2000; Kaminski and Heimann 2001; Rödenbeck et al. 2003). Peylin et al. (2005) showed good agreement between interannual carbon exchanges over broad regions as calculated by inverse models and as simulated with a DGVM and a TBM. Most of the observed interannual variability in the atmospheric CO_2 growth rate was shown to be explained by the differential responses of NPP and heterotrophic respiration to climate.

15.4.4 Runoff

As all terrestrial biosphere models simulate the interaction of the carbon and water cycles, the models can be evaluated in terms of their performance in simulating measured water fluxes (Gordon and Famiglietti 2004; Gordon et al. 2004). Over multi-annual time scales, runoff – which is measured at gauging stations on rivers – is equivalent to the difference between precipitation and evapotranspiration, averaged over the catchment upstream of the station. Gerten et al. (2004) demonstrated that LPJ showed comparable skill to existing global hydrology models in predicting global runoff patterns. They went on to model the additional effect of changing CO_2 concentration (via changes in stomatal conductance) on runoff.

15.4.5 CO_2 and Water Flux Measurements

Measurements of CO_2 and water flux from towers by the eddy covariance technique provide a temporally highly resolved record, and a powerful new tool for model evaluation (Falge et al. 2002; Baldocchi 2003). The FLUXNET global network of flux measurement stations gathers data from as many as 200 sites, although these are still very unevenly distributed across the globe (Baldocchi and Gu 2002; http://www.daac.ornl.gov/FLUXNET/fluxnet.html/). The data record diurnal, seasonal and interannual variability. CO_2, water and energy fluxes are measured simultaneously and concurrently with meteorological measurements that can be used directly to drive the models. There are two main limitations: the data are typically incomplete (for reasons discussed by Dolman et al. 2003), and the results (in common with conventional NPP measurements) are site-specific so that accurate specification of local soil conditions and disturbance history may be important. The experience obtained so far (e.g., Amthor et al. 2001; Potter et al. 2001; Gerten et al. 2004; Krinner et al. 2005; Morales et al. submitted) suggests that TBMs and DGVMs can perform well in simulating seasonal cycles and interannual variability of measured CO_2 and water exchange, but that the annually integrated carbon balance may depend on site-specific and generally unknown historical management factors.

15.5 Examples of Applications of DGVMS

DGVMs can be used alone or coupled to other types of models as tools to understand changes in the Earth System. Here we summarize a selection of DGVM studies that have helped either to explain observed phenomena, or to predict the consequences of human activities in the future.

15.5.1 Holocene Changes in Atmospheric CO_2

The causes of changes in the atmospheric concentration of CO_2 since the end of the last glacial period, about 12 ka before present (BP), are controversial. Ice-core analyses show a drop of 7 ppm from 11 to 8 ka BP, followed by a gradual rise of 20 ppm towards the pre-industrial 280 ppm (Indermühle et al. 1999; Flückiger et al. 2002). Indermühle et al. (1999) attributed both the initial drop and subsequent rise primarily to changes in terrestrial carbon storage. Broecker et al. (2001) questioned this explanation for the rise, suggesting instead that CO_2 removed from the atmosphere and surface ocean water by terrestrial carbon uptake after the glacial maximum was slowly replaced due to the precipitation of $CaCO_3$ at depth ("calcite compensation"). Ruddiman (2003) on the other hand has ascribed the CO_2 rise to deforestation. This problem has been studied with DGVMs by Brovkin et al. (2002) and by Joos et al. (2004). Joos et al. (2004) forced the Bern-CC coupled carbon cycle model (which includes the LPJ DGVM for terrestrial carbon dynamics) with palaeoclimate model simulations (Kaplan 2002). The coupled model reproduced the observed CO_2 trajectory since 11 ka BP to within a few ppm. The initial drop was explained by vegetation regrowth. The subsequent increase in CO_2 concentration was mainly due to *(a)* rising sea surface temperature, and *(b)* calcite compensation, as Broecker et al. (2001) proposed. The ice-core record of $\delta^{13}C$ (Indermühle et al. 1999) rules out any large contribution from deforestation. This model version also simulates the terrestrial $\delta^{13}C$ budget, based on Kaplan et al. (2002) and Scholze et al. (2003). The modeled $\delta^{13}C$ history was consistent with the ice-core data.

15.5.2 Boreal "Greening" and the Contemporary Carbon Balance

Spectral reflectance observations have shown a persistent greening trend in northern high latitudes through the 1980s and 1990s (Myneni et al. 1997; Zhou et al. 2001). Maximum summer LAI in the boreal zone, estimated from these observations, increased by 0.19 between 1982 and 1998. Potential explanations include vegetation response to high-latitude warming, forest regrowth due to changed management, vegetation recovery from disturbance by fire or insect attacks, CO_2 fertilization, or (just possibly) incomplete correction for drifts in the response of the sensor. Lucht et al. (2002) investigated whether the greening trend could be explained by vegetation responses to climate. Simulations driven by monthly climate data (New et al. 2000) showed that the trend, its seasonal cycles and interannual variability could all be reproduced. The simulations were entirely independent of the satellite observations. Thus, it is most likely that the observed greening trend is real and was caused by the changing climate. Further simulations showed that virtually all of the effect has been due to warming. These findings are consistent with the over-riding control of temperature on vegetation growth in high latitudes. The controls are more complex in warmer climates. Figure 15.4 shows the simulated land-atmosphere flux averaged over the whole period, compared with climate anomalies. Some regions showed a large reduction in precipitation (e.g., the Sahel), others a large increase in temperature (e.g., southern Africa), both leading to a release of carbon because of reduced NPP and/or increased heterotrophic respiration. Some regions, such as the southeastern USA, experienced increased precipitation and decreased temperatures, leading to increased carbon uptake (cf. Nemani et al. 2002; Hicke et al. 2002; Rödenbeck et al. 2003).

15.5.3 The Pinatubo Effect

Atmospheric CO_2 concentration temporarily slowed its increase after the eruption of Mount Pinatubo in 1991. Aerosols from this eruption cooled the globe by around 0.5 °C (more in northern latitudes). DGVM simulations over this period have reproduced both a temporary drop in boreal LAI (also shown by the satellite data) and an enhanced high-latitude terrestrial carbon sink (Lucht et al. 2002). Although NPP and heterotrophic respiration were both reduced, the modeled effect on respiration was stronger, producing an enhanced carbon sink. But these changes at high latitudes were not large enough to provide the full explanation. The simulated *global* patterns of simulated NPP and heterotrophic respiration anomalies are more complex, and controversial. Roderick et al. (2001) and Gu et al. (2002) have argued that an increase in the fraction of diffuse vs. direct radiation caused large-scale enhancement of NPP during the post-Pinatubo period. In other words, an enhanced sink was produced by increased NPP. But Angert et al. (2004) have shown that this hypothesis is inconsistent with observed seasonal cycles of CO_2 during these years. Jones and Cox (2001) used GCM simulations incorporating the TRIFFID DGVM to suggest that post-Pinatubo climatic anomalies overall produced enhanced NPP in the tropics, while respiration globally was reduced.

15.5.4 Future Carbon Balance Projections

Schaphoff et al. (2006) predicted the response to climate and CO_2 changes during the 21st century, as simulated by five ocean-atmosphere GCMs, all driven by a standard "business as usual" CO_2 emissions scenario. The simulated change in terrestrial carbon storage ranged from a loss of 106 Pg C to a gain of 201 Pg C (neglecting land-use changes). This finding complements Cramer et al. (2001), who found large uncertainty *within* one climate change scenario, due to differences among six DGVMs. The spatial patterns of changes in carbon content found by Schaphoff et al. were more robust than the global total. Carbon storage was enhanced due to warming in the Arctic and at high elevations, but reduced over the temperate and boreal zones. Carbon storage was also increased in many semi-arid regions due to increased vegetation water-use efficiency and woody encroachment at high CO_2, with soil carbon loss inhibited due to drought. Tropical vegetation response varied due to precipitation differences among GCMs.

15.5.5 Carbon-Cycle Feedbacks to Future Climate Change

DGVM simulations based on climate projections of the 21st century have indicated that the time course of carbon storage depends on a balance of CO_2 fertilization and the positive effect on NPP of longer and warmer growing seasons in cold climates, vs. the general increase in heterotrophic respiration rates and the negative effect of high temperatures on NPP in warm climates (Cao and Woodward 1998; Kicklighter et al. 1999; Cramer et al. 2001; Schaphoff et al. 2006). CO_2 fertilization is expected to show a "diminishing return" while the effect of warming on respiration will continue (Jones et al. 2003). As a result, future projections of terrestrial carbon storage have often shown an initial increase in terrestrial carbon storage followed by a decline. Cox et al. (2000) used a fully coupled GCM, including the Hadley Centre (HADCM3) ocean-atmosphere model and the TRIFFID DGVM, to perform a comprehensive analysis based on a "business as usual" CO_2 emissions scenario. They found that the carbon-climate feedback had generated an additional 1.5 K global warming by 2100, mostly due to increased heterotrophic respiration. A similar analysis using the IPSL ocean-atmosphere model and the SLAVE TBM (Dufresne et al. 2002) also found a positive feedback, but the change was less, and was mostly due to reduced NPP in the tropics. Reasons for the difference include stronger vertical mixing in the IPSL ocean model, and greater initial soil carbon storage in the Hadley Centre model (Friedlingstein et al. 2003). Yet despite uncertainty about the size of carbon cycle feedbacks, the largest uncertainty in the future CO_2 concentration is still the unknown future of CO_2 emissions from fossil fuels. Joos et al. (2001) used the Bern-CC model to examine the consequences of six different emissions scenarios. A positive carbon-cycle feedback to climate change was found in all cases and atmospheric CO_2 rose to between 540 and 960 ppm, depending on the scenario and on assumptions about climate sensitivity, by 2100.

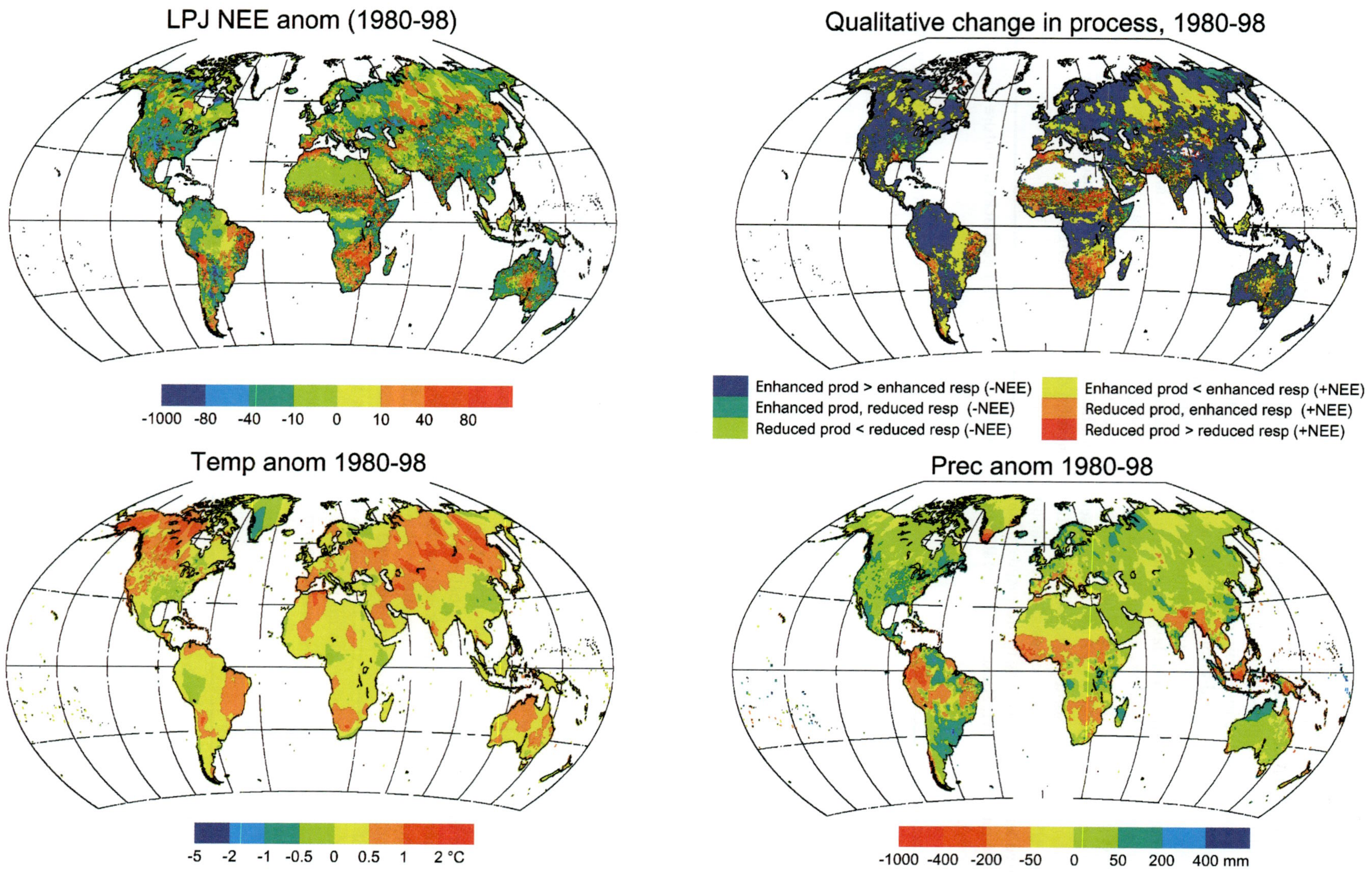

Fig. 15.4. Global patterns of the modeled terrestrial carbon balance during 1980–1998, simulated with the LPJ DGVM. The *top left panel* shows anomalies of net ecosystem exchange (*NEE* = heterotrophic respiration + combustion – *NPP*), i.e., mean *NEE* during 1980–1998 minus *NEE* for the standard climatological period (1931–1960). The *top right panel* shows the origin of the anomalies in terms of increases or decreases in "respiration" (strictly, heterotrophic respiration + combustion) and "production" (*NPP*). The corresponding annual mean anomalies of temperature and precipitation are shown in the *bottom panels*

15.5.6 Effects of Land-Use Change on the Carbon Cycle

Land-use change was the main cause of increasing atmospheric CO_2 in the early industrial period, and is still a substantial contributor. McGuire et al. (2001) used four terrestrial models to assess the relative roles of CO_2 fertilization, climate variation and land-use change through the industrial era. The simulated cumulative effect of cropland establishment and abandonment from 1920 to 1992 was a release of 56–91 Pg C. The concurrent simulated uptake, due mainly to CO_2 fertilization, was 54–105 Pg C. The modeled net terrestrial carbon budget proved broadly consistent with independent estimates from atmospheric measurements (Prentice et al. 2001; House et al. 2003). This global analysis has not yet been extended beyond 1992, nor carried into the future. Some possible consequences of future land-use change have been analysed, however. Cramer et al. (2004) used LPJ to estimate past and potential future losses of carbon from wet tropical ecosystems, which are the main site of deforestation today. During the 20th century, deforestation was estimated to have released 39–49 Pg C. Extrapolating a range of estimates for current rates of deforestation into the future yielded a projected additional loss of 158–243 Pg C by 2100. By comparison, CO_2 fertilization and climate change produced a response ranging from a gain of 80 Pg C to a loss of 50 Pg C. Direct human intervention, therefore, is likely to be the most important determinant of the fate of carbon in tropical forests.

15.6 Some Perspectives and Research Needs

The following discussion is by no means a complete overview of the aspects of DGVMs that are in need of further testing and development. However, it points to some key areas where an international collaborative effort, building on the achievements of GCTE, would very likely lead to more rapid progress than could be achieved by individual groups working alone.

15.6.1 Comparison with Field Experiments

Experimental studies of the response of terrestrial ecosystems to environmental changes has been a major focus for GCTE. For example, experimental evidence from studies with small trees in open-top chambers has shown an average stimulation of photosynthesis by ≈60% for a 300 ppmv increase in CO_2, while the annual increment in wood mass per unit leaf area increased by ≈27% (Norby et al. 1999). The Free-air CO_2 enrichment (FACE) methodology was introduced so that experiments on the effects of raised CO_2 concentrations could be conducted on intact ecosystems (Hendry et al. 1999; Nowak et al. 2004; Long et al. 2004). The first FACE study in an intact forest ecosystem was set up in a *Pinus taeda* plantation in the southeastern USA (DeLucia et al. 1999). Ambient CO_2 concentrations were increased to 560 ppmv in the replicated plots from autumn 1996. From 1997 to 2000, annual NPP was on average 23% higher in plots with elevated CO_2 than in the control plots (Delucia et al. 1999; Hamilton et al. 2002; Schäfer et al. 2003). F. I. Woodward and M. Lomas (personal communication 2004) used the SDGVM to simulate this experiment. They obtained a realistic 20% enhancement in net primary production after four years. Hickler et al. (2004b) obtained a range from 15 to 33% increase over the same period. There is considerable scope for rigorous testing of different process formulations in DGVMs using the data now available from experiments involving artificial warming, drought and N fertilization as well as an increasing range of FACE studies.

15.6.2 Plant Functional Types

PFT schemes in current DGVMs are simplistic, and the values of most PFT-specific parameters are neither agreed nor well founded. GCTE has stimulated new interest in PFT classification (Díaz and Cabido 1997; Díaz et al. 1999a; Gitay and Noble 1997; Lavorel et al. 1997; Lavorel and Cramer 1999; Lavorel and Garnier 2002; Lavorel et al. 2007, Chap. 13 of this volume), but this has not yet filtered through to influence DGVM design. Current approaches to PFT classification emphasize readily observable plant traits that confer characteristic responses to factors of the environment and disturbance or management regime (e.g., Díaz et al. 1999b; Díaz et al. 2002; Gurvich et al. 2002; Barboni et al. 2004; Díaz et al. 2004; Wright et al. 2004). This is precisely the kind of information that is needed for the more rational representation of PFTs in DGVMs. The development of an internally-consistent, global vegetation map explicitly based on PFTs and based on high-resolution multispectral reflectance data is a related but distinct goal, proposed by Nemani and Running (1996). Such a map would be extremely useful for testing DGVMs. Various satellite-based global land-cover maps are now available, but there are still considerable differences among them, and the procedures used to generate them are not entirely transparent.

15.6.3 The Nitrogen Cycle

Hungate et al. (2003) suggested that scenario analyses with current DGVMs (Cramer et al. 2001; Prentice et al. 2001) exaggerate the amount of carbon the biosphere could take up in response to a continued increase in atmospheric CO_2. In fact, the two DGVMs in Cramer et al. (2001) that explicitly allow for N cycle constraints on NPP produce lower estimates of future carbon storage than those that do not. But these estimates still fall above the range postulated by Hungate et al. (2003). Recent simulations with the LPJ model (Schaphoff et al. 2006) produce estimates within this range, even though this model does not yet include N cycle constraints on NPP. Whatever the correct view on future CO_2 uptake, improving the representation of N cycling within DGVMs is a research priority. This work is hampered by inadequate quantification of gain and loss terms in the N cycle at regional and global scales (e.g., amount of N in precipitation, controls of N_2 fixation and dissolved organic N losses, and the release and fate of N-containing trace gases). Incorporation of a realistic N cycle is also important for prediction of the influence of soil nutrient status controls on PFT distributions, and to assess the impacts of anthropogenic N deposition on the carbon cycle and ecosystems.

15.6.4 Plant Dispersal and Migration

Current DGVMs assume that the rate of plant dispersal and migration does not limit the response of vegetation to climate change. This assumption is called into question by the fact that large changes in climate could occur rapidly (i.e., over a few decades) in some regions, and by the potential barriers to plant migration caused by landscape fragmentation. The issue is hard to address observationally because of the difficulty in quantifying rare long-distance dispersal events, which are believed on theoretical grounds to be crucial to explaining how rapid, continent-wide plant migrations occurred in response to climate changes in the Quaternary (Pitelka and Plant Migration Working Group 1997; Clark et al. 1998). Model formalisms to represent plant dispersal (e.g., Higgins et al. 2003) have been devised, but not implemented in DGVMs.

15.6.5 Wetlands

Wetlands are a major carbon store and are sources of the greenhouse gases methane (CH_4) and nitrous oxide (N_2O). The lateral transport of water is generally less important as a determinant of terrestrial vegetation than the *in situ* water balance, but this is not so for wetlands. DGVMs to date treat only dryland ecosystems. Extension to wetlands will require DGVMs to be coupled to water routing models with high spatial resolution. It will also be important to account for specific wetland PFTs, and the controls on nutrient supply to different types of wetland.

15.6.6 Multiple Nutrient Limitations

Realistic simulation of nutrient constraints on vegetation productivity will require not only the incorporation of lateral transport of nutrients by water, but also lateral transport in the atmosphere. For example, on geological time scales, aeolian transport of dust is a major control on phosphorus supply to terrestrial ecosystems (Chadwick et al. 1999). Transport of sulfate-containing aerosols derived from the production of dimethylsulphide by phytoplankton is a unique natural route for the redistribution of sulfur to the land surface. Precipitation is a significant source of N even in remote, unpolluted regions. Models of plant growth have scarcely begun to address the way in which different nutrient limitations interact. Marine ecosystems are already beginning to incorporate the interactions of the cycles of nitrogen, phosphorus, iron and silicon and their consequences for competition among phytoplankton PFTs (Aumont et al. 2003; Blackford and Burkill 2002; Blackford et al. 2004; Le Quéré et al. in press) and may inspire further DGVM development in this field.

15.6.7 Agriculture and Forestry

Efforts are already underway to simulate crop productivity and yield generically using DGVMs (e.g., Kucharik and Brye 2003). One objective of this work is to predict the consequences of climate change for agriculture. Such predictions will also have to consider climatically induced changes in the suitability of different crops, and non-climatic as well as climatic drivers of changes in land use – requiring that DGVMs be embedded in an integrated assessment framework. Forest management likewise has been only partially treated in DGVMs. Global carbon cycle studies have taken into account the consequences of deforestation and reforestation (Houghton 2003; McGuire et al. 2001), but not changes in logging intensity although these are thought to have had a major role in creating a present-day carbon sink in northern temperate forests (Nabuurs et al. 2003). The representation of forest management places new demands on DGVMs to show the correct response of forest NPP to stand age and density. Progress in modeling economically important ecosystems must be matched by progress in the collection and standardization of statistical data on crop distribution, yields and farming practices, and past and present forest management regimes. Adequate representation of management is important for the assessment of practices designed to increase carbon storage in ecosystems, which the present generation of DGVMs is not well adapted to address (Bondeau et al. in press).

15.6.8 Grazers and Pests

The eventual expansion of DGVMs to cover components of the ecosystem other than autotrophic plants and heterotrophic soil organisms (bacteria and fungi) is unavoidable. The abundances of grazing animals, and of pests such as leaf- and bark-eating insects, exert an important control on vegetation productivity and disturbance in several biomes. It should be possible to simulate the impact of changes in both natural and managed grazing regimes by introducing a small number of animal functional types (AFTs). Models of marine ecosystems, where the ratio of secondary to primary production is much higher, already incorporate functional types of zooplankton grazers with different feeding preferences and population growth rates (Bopp et al. 2002; Aumont et al. 2003; Le Quéré et al. in press).

15.6.9 Biogenic Emissions of Trace Gases and Aerosol Precursors

Through various processes, the terrestrial biosphere emits the greenhouse gas N_2O, reactive gases that have a major influence on atmospheric chemistry including the greenhouse gas CH_4, carbon monoxide (CO), nitrogen oxides (NO_x), volatile organic compounds (VOCs) such as isoprene, and aerosol precursors in the form of dust, black carbon and VOCs. The extension of DGVMs to model sources and sinks of trace gases and aerosol components is a natural development. DGVMs will be called on to model CH_4 production in wetlands and oxidation in drylands (Kaplan 2002; Ridgwell et al. 1999), VOC production (e.g., Guenther et al. 1995), the N cycle including controls on the relative production of N_2, N_2O and NO in soils (Potter and Klooster 1999), ozone (O_3) uptake by vegetation, the relationships among dust emission, vegetation density and height (Tegen et al. 2002), the occurrence and intensity of fires, and the emissions of CO, CH_4, NO_x and black carbon associated with fires (Andreae and Merlet 2001; Thonicke et al. 2005). Progress has been made in most of these areas individually, but further efforts will be required to develop a comprehensive emissions model that can be coupled to an atmospheric chemistry and transport model (CTM) and ultimately to a GCM, in order to better understand the role of the biosphere in determining the atmosphere's changing chemical composition and, through this, the Earth's climate.

15.7 Summary and Conclusions

DGVMs exploit the power of modern computers and computational methods to yield a predictive description of land ecosystem processes that takes account of knowledge previously developed through long histories of separate disciplinary approaches to the study of the biosphere. The degree of interaction between the different scientific approaches still falls far short of optimal; thus, DGVM developers have a responsibility to be aware of progress in several disciplines in order to ensure that their models remain state-of-the-art. We have presented a series of case studies of the evaluation of DGVMs that demonstrate the predictive capability that current models have achieved. Nevertheless, there are plenty of unresolved issues – differences among models that are not well understood, important processes that are omitted or treated simplistically by some or all models, and sets of observations that are not satisfactorily reproduced by current models. More comprehensive "benchmarking" of DGVMs against multiple data sets is required and would be most effectively carried out through an international consortium, so as to avoid duplicating the large amount of work involved in selecting and processing data sets and model experiments. We have also presented a series of case studies that illustrate the power of DGVMs, even with their known limitations, in explaining a remarkable variety of Earth System phenomena and in addressing contemporary issues related to climate and land-use change. These case studies encourage us to believe that the continued development of DGVMs is a worthwhile enterprise. Finally, new directions in Earth System Science point to a range of aspects in which DGVMs could be improved so as to take account of recently acquired knowledge, such as experimental work on whole-ecosystem responses to environmental modification and new understanding of the functional basis of plant traits; complemented by an effort to represent semi-natural and agricultural ecosystems and the impacts of different management practices on these ecosystems; and extended to include processes such as trace-gas emissions, which are important in order to understand the functional role of the terrestrial biosphere in the Earth System. Together, these potential developments add up to an ambitious research program, requiring the economies of scale that only an international collaborative effort can provide.

References

Alcamo J (1994) IMAGE 2.0: Integrated Modeling of Global Climate Change. Kluwer Academic Press, Dordrecht, Boston, pp 314

Allen TFH, Hoekstra TW (1990) The confusion between scale-defined levels and conventional levels of organization in ecology. Journal of Vegetation Science 1:5–12

Amthor JS, Chen JM, Clein JS, Frolking SE, Goulden ML, Grant RF, Kimball A, King W, McGuire AD, Nikolov NT, Potter CS, Wang S, Wofsy SC, (2001) Boreal forest CO_2 exchange and evapotranspiration predicted by nine ecosystem process models: Inter-model comparisons and relations to field measurements. Journal of Geophysical Research, 106: D24 33623–33648

Andreae MO, Merlet P (2001) Emission of trace gases and aerosols from biomass burning. Global Biochemical Cycles 15:955–966

Angert A, Biraud S, Bonfils C, Buermann W, Fung I (2004) CO_2 seasonality indicates origins of post-Pinatubo sink. Geophysical Research Letters 31: L11103, doi:10.1029/2004GL019760

Arora VK, Boer GJ (2005) A parameterization of leaf phenology for the terrestrial ecosystem component of climate models. Global Change Biology 11:39–5

Arora VK, Boer GJ (in press) Fire as an interactive component of dynamic vegetation models. Journal of Geophysical Research – Biogeosciences

Aumont O, Maier-Reimer E, Blain S, Pondaven P (2003) An ecosystem model of the global ocean including Fe, Si, P co-limitations. Global Biochemical Cycles 17: doi:10.1029/2001GB001745

Bachelet D, Lenihan JM, Daly C, Neilson RP, Ojima DS, Parton WJ (2001) MC1: A dynamic vegetation model for estimating the distribution of vegetation and associated ecosystem fluxes of carbon, nutrients and water. USDA Forest Service General Technical Report, PNW-GTR-508:1–95

Bachelet D, Neilson RP, Hickler T, Drapek RJ, Lenihan JM, Sykes MT, Smith B, Sitch S, Thonicke K (2003) Simulating past and future dynamics of natural ecosystems in the United States. Global Biochemical Cycles 17: 1045 doi:1010.1029/2001GB001508
Baldocchi D (2003). Assessing the eddy covariance technique for evaluating carbon dioxide exchange rates of ecosystems: past, present and future. Global Change Biology 9:479–492
Baldocchi D, Gu LH (2002) Fluxnet 2000 synthesis – Foreword. Agricultural and Forest Meteorology 113:1–2
Barboni D, Harrison SP, Bartlein PJ, Jalut G, New M, Prentice IC, Sanchez-Goñi M-F, Spessa A, Davis B, Stevenson AC (2004) Relationships between plant traits and climate in the Mediterranean region: A pollen data analysis. J Vegetat Sci 15:635–646
Beerling DJ, Woodward FI (2001) Vegetation and the Terrestrial Carbon Cycle: Modelling the first 400 Million Years. Cambridge University Press
Blackford JC, Burkill PH (2002) Planktonic community structure and carbon cycling in the Arabian Sea as a result of monsoonal forcing: the application of a generic model. Journal of Marine Systems 36:239–267
Blackford JC, Allen JI, Gilbert FJ (2004) Ecosystem dynamics at six contrasting sites: a generic modeling study. Journal of Marine Systems 52:191–215
Bonan GB, Pollard D, Thompson SL (1992) Effects of boreal forest vegetation on global climate. Nature 359:716–718
Bondeau A, Kicklighter DW, Kaduk J (1999) Comparing global models of terrestrial net primary productivity (NPP): importance of vegetation structure on seasonal NPP estimates. Global Change Biology 5:35–45
Bondeau A, Smith PC, Zaehle S, Schaphoff S, Lucht W, Cramer W, Gerten D, Lotze-Campen H, Müller C, Reichstein M, Smith B (in press) Modelling the role of agriculture for the 20th century global terrestrial carbon balance. Global Change Biology
Bopp L, Le Quéré C, Heimann M, Manning AC, Monfray P (2002) Climate-induced oceanic oxygen fluxes: Implications for the contemporary carbon budget. Global Biogeochemical Cycles 16: doi:10.1029/2001GB001445
Botkin DB, Janak JF, Wallis JR (1972) Some ecological consequences of a computer model of forest growth. J Ecol 60:849–872
Botta A, Foley JA (2002) Effects of climate variability and disturbances on the Amazonian terrestrial ecosystems dynamics. Global Biogeochemical Cycles 16(4): doi:10.1029/2000GB001338
Bousquet P, Peylin P, Ciais P, Le Quéré C, Friedlingstein P, Tans P (2000) Regional changes in carbon dioxide fluxes of land and oceans since 1980. Science 290:1342–1346
Box EO (1981) Predicting physiognomic vegetation types with climate variables. Vegetatio 45:127–139
Broecker WS, Lynch-Stieglitz J, Clark E, Hadjas I, Bonani G (2001) What caused the atmosphere's CO_2 content to rise during the last 8 000 years? Geochem. Geosyst. 2: doi:2001GC00177
Brovkin V, Bendtsen J, Claussen M, Ganopolski A, Kubatzki A, Petoukhov V, Andreev A (2002) Carbon cycle, vegetation, and climate dynamics in the Holocene: Experiments with the CLIMBER-2 model Global Biogeochemical Cycles 16: 1139; doi:10.1029/2001GB001662
Bugmann H (1996) A simplified forest model to study species composition along climate gradients. Ecology 77:2055–2074
Bugmann H, Solomon AM (2000) Explaining forest composition and biomass across multiple biogeographical regions. Ecological Applications 10:95–114
Cao M, Woodward FI (1998) Dynamic responses of terrestrial ecosystem carbon cycling to global climate change. Nature 393:249–252
Chadwick OA, Derry LA, Vitousek PM, Huebert BJ, Hedin LO (1999) Changing sources of nutrient during four million years of ecosystem development. Nature 397:491–497
Clark JS, Fastie C, Hurtt G, Jackson ST, Johnson C, King GA, Lewis M, Lynch J, Pacala S, Prentice C, Schupp EW, Webb T, Wyckoff (1998) Reid's paradox of rapid plant migration – Dispersal theory and interpretation of paleoecological records. Bioscience 48:13–24
Collatz GJ, Ball JT, Grivet C, Berry JA (1991) Physiological and environmental regulation of stomatal conductance, photosynthesis and transpiration: a model that includes a laminar boundary layer. Agricultural and Forest Meteorology 54:107–136
Collatz GJ, Ribas-Carbo M, Berry JA (1992) Coupled photosynthesis stomatal conductance model for leaves of C_4 plants. Australian Journal of Plant Physiology 19:519–538
Comins HN, McMurtrie RE (1993) Long-term response of nutrient-limited forests to CO_2-enrichment; equilibrium behaviour of plant-soil models. Ecological Applications 3:666–681
Cowan IR (1977) Stomatal behaviour and environment. Advances in Botanical Research 4:117–228
Cowling SA (1999) Simulated effects of low atmospheric CO_2 on structure and composition of North American vegetation at the Last Glacial Maximum. Global Ecology and Biogeography 8:81–93
Cox PM (2001) Description of the TRIFFID dynamic global vegetation model, Tech. Note 24, Hadley Centre, Bracknell, UK, pp 16
Cox PM, Betts RA, Jones CD, Spall SA, Totterdell IJ (2000) Acceleration of global warming due to carbon-cycle feedbacks in a coupled climate model. Nature 408:184–187
Cramer W, Kicklighter DW, Bondeau A, Moore B, Churkina C, Nemry B, Ruimy A, AL S (1999) Comparing global models of terrestrial net primary productivity (NPP): overview and key results. Global Change Biology 5:1–15
Cramer W, Bondeau A, Woodward FI, Prentice C, Betts RA, Brovkin V, Cox PM, Fisher V, Foley JA, Friend AD, Kucharik C, Lomas MR, Ramankutty N, Sitch S, Smith B, White A, Young-Molling C (2001) Global response of terrestrial ecosystem structure and function to CO_2 and climate change: results from six dynamic global vegetation models. Global Change Biology 7:357–374
Cramer W, Bondeau A, Schaphoff S, Lucht W, Smith B, Sitch S (2004) Tropical forests and the global carbon cycle: impacts of atmospheric carbon dioxide, climate change and rate of deforestation. Philosophical Transactions of the Royal Society of London Series B-Biological Sciences 359:331–343
Daly C, Bachelet D, Lenihan JM, Neilson RP, Parton W, Ojima D (2000) Dynamic simulation of tree-grass interactions for global change studies. Ecological Applications 10:449–469
Dargaville RJ, Heimann M, McGuire AD, Prentice IC, Kicklighter DW, Joos F, Clein JS, Esser G, Foley J, Kaplan J, Meier RA, Melillo JM, Moore III B, Ramankutty N, Reichenau T, Schloss A, Sitch S, Tian H, Williams LJ, Wittenberg U (2002) Evaluation of terrestrial carbon cycle models with atmospheric CO_2 measurements: Results from transient simulations considering increasing CO_2, climate, and land-use effects. Global Biogeochemical Cycles 16: 1092, doi:1010.1029/2001GB001426
DeFries RS, Hansen MC, Townsend JRG, Janetos AC, Loveland TR (2000) A new global 1-km dataset of percentage tree cover derived from remote sensing. Global Change Biology 6:247–254
Delire C, Levis S, Bonan G, Foley JA, Coe M, Vavrus S (2002) Comparison of the climate simulated by the CCM3 coupled to two different land-surface models Climate Dynamics 19:657–669
DeLucia EH, Hamilton JG, Naidu SL, Thomas RB, Andrews JA, Finzi A, Lavine M, Matamala R, Mohan JE, Hendrey GR, Schlesinger WH (1999) Net primary production of a forest ecosystem with experimental CO_2 enrichment. Science 284:1177–1179
Denning AS, Holzer M, Gurney KR, Heimann M, Law RM, Rayner PJ, Fung IY, Fan S, Taguchi S, Friedlingstein P, Balkanski Y, Maiss M, Levin I (1999) Three-dimensional transport and concentration of SF6: A model intercomparison study (Transcom 2). Tellus 51B: 266–297
Dewar RC (1996) The correlation between plant growth and intercepted radiation: an interpretation in terms of optimal plant nitrogen content. Annuals of Botany 78:125–136
Díaz S, Cabido M (1997) Plant functional types and ecosystem function in relation to global change. J Vegetat Sci 8:463–474
Díaz S, Cabido M, Casanoves F (1999a) Functional implications of trait-environment linkages in plant communities. In: Weiher E, Keddy P (eds) Ecological assembly rules – Perspectives, advances, retreats. Cambridge University Press, Cambridge, pp 338–362
Díaz S, Cabido M, Zak M, Martínez Carretero E, Araníbar J (1999b) Plant functional traits, ecosystem structure and land-use history along a climatic gradient in central-western Argentina. Journal of Vegetation Science 10:651–660
Díaz S, McIntyre S, Lavorel S, Pausas JG (2002) Does hairiness matter in Harare? Resolving controversy in global comparisons of plant trait responses to ecosystem disturbance. New Phytologist 154:7–9

Díaz S, Hodgson JG, Thompson K, Cabido M, Cornelissen JHC, Jalili A, Montserrat-Marti G, Grime JP, Zarrinkamar F, Asri Y, Band SR, Basconcelo S, Castro-Diez P, Funes G, Hamzehee B, Khoshnevi M, Perez-Harguindeguy N, Perez-Rontome MC, Shirvany FA, Vendramini F, Yazdani S, Abbas-Azimi R, Bogaard A, Boustani S, Charles M, Dehghan M, de Torres-Espuny L, Falczuk V, Guerrero-Campo J, Hynd A, Jones G, Kowsary E, Kazemi-Saeed F, Maestro-Martinez M, Romo-Diez A, Shaw S, Siavash B, Villar-Salvador P, Zak MR (2004) The plant traits that drive ecosystems: Evidence from three continents. Journal of Vegetation Science 15:295–304

Dickinson RE, Henderson-Sellers A, Kennedy PJ (1993) Biosphere-Atmosphere Transfer Scheme (BATS) Version 1E as coupled to the NCAR Community Climate Model, Tech. Note NCAR/TN-383+STR, Natl. Cent. For Atmos. Res., Boulder, Colorado, pp 72

Dolman AJ, Schulze ED, Valentini R (2003) Analyzing carbon flux measurements. Science 301 (5635):916–916

Dufresne JL, Friedlingstein P, Berthelot M, Bopp L, Ciais P, Fairhead L, Le Treut H, Monfray P (2002) On the magnitude of positive feedback between future climate change and the carbon cycle. Geophysical Research Letters 29:43.41–43.44

Emanuel WR, Shugart HH, Stevenson MP(1985). Climatic change and the broad-scale distribution of terrestrial ecosystem complexes. Climatic Change 7:29–43

Falge E, Baldocchi D, Tenhunen J, Aubinet M, Bakwin P, Berbigier P, Bernhofer C, Burba G, Clement R, Davis KJ, Elbers JA, Goldstein AH, Grelle A, Granier A, Gudmundsson J, Hollinger D, Kowalski AS, Katul G, Law BE, Malhi Y, Meyers T, Monson RK, Munger JW, Oechel W, Paw UKT, Pilegaard K, Rannik U, Rebmann C, Suyker A, Valentini R, Wilson K, Wofsy S (2002) Seasonality of ecosystem respiration and gross primary production as derived from FLUXNET measurements. Agricultural and Forest Meteorology 113:53–74

Fan S, Gloor M, Mahlman J, Pacala S, Sarmiento J, Takahashi T, Tans P (1998) A large terrestrial carbon sink in North America implied by atmospheric and oceanic carbon dioxide data and models. Science 282:442–446

Farquhar GD, van Caemmerer S, Berry JA (1980) A biochemical model of photosynthetic CO_2 assimilation in leaves of C_3 species. Planta 149:78–90

Field CB, Chapin FS, Matson PA, Mooney HA (1992) Responses of terrestrial ecosystems to the changing atmosphere: a resource based approach. Annual Review of Ecology and Systematics 23:201–235

Field CB, Randerson JT, Malmström CM (1995) Global net primary production: combining ecology and remote sensing. Remote Sensing of Environment 51:74–88

Field CB, Raupach MR, Victoria R (2004) The Global Carbon Cycle: Integrating Humans, Climate, and the Natural World. In: Field CB, Raupach MR (eds) The Global Carbon Cycle: Integrating Humans, Climate, and the Natural World. Island Press, Washington, D.C., pp 1–13

Finnegan JJ, Raupach MR (1987) Transfer processes in plant canopies in relation to stomatal characteristics. In: Zeiger, E. Farquhar GD, Cowan IR (eds) Stomatal Function, Stanford University Press, Stanford, pp 385–429

Flückiger J, Monnin E, Stauffer B, Schwander J, Stocker TF, Chappellaz J, Raynaud D, Barnola J-M (2002) High-resolution Holocene N_2O ice core record and its relationship with CH_4 and CO_2. Global Biogeochemical Cycles 16: doi:10.1029/2001GB001417

Foley JA, Prentice IC, Ramankutty N, Levis S, Pollard D, Sitch S, Haxeltine A (1996) An integrated biosphere model of land surface processes, terrestrial carbon balance, and vegetation dynamics. Global Biogeochemical Cycles 10:603–628

Foley JA, Levis S, Prentice IC, Pollard D, Thompson SL (1998) Coupling dynamic models of climate and vegetation. Global Change Biology 4:561–579

Foley JA, DeFries R, Asner GP, Barford C, Bonan G, Carpenter SR, Chapin FS, Coe MT, Daily GC, Gibbs HK, Helkowski JH, Holloway T, Howard EA, Kucharik CJ, Monfreda C, Patz CJ, Prentice IC, Ramankutty N, Snyder PK (2005) Global consequences of land use. Science 309: 570–574

Friedlingstein P, Dufresne J-L, Cox PM, Rayner P (2003)How positive is the feedback between climate change and the carbon cycle? Tellus B55 692–700

Friend AD, Stevens AK, Knox RG, Cannell MGR (1997) A process-based, terrestrial biosphere model of ecosystem dynamics (Hybrid v3.0). Ecological Modelling 95:249–287

Fulton MR, Prentice IC (1997) Edaphic controls on the boreonemoral forest mosaic. Oikos 78:291–298

Gerten D, Schaphoff S, Haberlandt U, Lucht W, Sitch S (2004) Terrestrial vegetation and water balance – hydrological evaluation of a dynamic global vegetation model. J Hydrol 286:249–270

Gitay H, Noble IR (1997) What are functional types and how should we seek them? In: Smith TM, Shugart HH, Woodward FI (eds) Plant functional types: their relevance to ecosystem properties and global change. Cambridge University Press, Cambridge, pp 3–19

Gordon WS, Famiglietti JS (2004) Response of the water balance to climate change in the United States over the 20th and 21st centuries: Results from the VEMAP Phase 2 model intercomparisons, Global Biogeochemical Cycles 18: GB1030, doi:10.1029/2003GB002098

Gordon WS, Famiglietti JS, Fowler NA, Kittel TGF, Hibbard KA (2004) Validation of simulated runoff from six terrestrial ecosystem models: Results from VEMAP. Ecological Applications 14:527–545

Gu LH, Baldocchi D, Verma SB, Black TA, Vesala T, Falge EM, Dowty PR (2002) Advantages of diffuse radiation for terrestrial ecosystem productivity. Journal of Geophysical Research-Atmospheres 107: ACL 2.1–2.23

Guenther A, Hewitt CN, Erickson D, Fall R, Geron C, Graedel T, Harley P, Klinger L, Lerdau M, McKay WA, Pierce T, Scholes B, Seinbrecher R, Tallamraju R, Taylor J, Zimmerman P (1995) A global model of natural volatile organic compound emissions. Journal of Geophysical Research 100:8873–8892

Gurney KR, Law RM, Denning AS, Rayner PJ, Baker D, Bousquet P, Bruhwiler L, Chen YH, Ciais P, Fan S, Fung IY, Gloor M, Heimann M, Higuchi K, John J, Kowalczyki E, Maki T, Maksyutov S, Peylin P, Prather M, Pak BC, Sarmiento J, Taguchi S, Takahashi T, Yuen CW (2003) Transcom 3 CO_2 Inversion Intercomparison: 1. Annual mean control results and sensitivity to transport and prior flux information. Tellus 55B: 555–579

Gurney KR, Law RM, Denning AS, Rayner PJ, Pak B, TransCom3 L2 modelers (2004) TransCom3 Inversion Intercomparison: Control results for the estimation of seasonal carbon sources and sinks. Global Biogeochemical Cycles 18: GB1010, doi:10.1029/2003GB002111

Gurvich DE, Díaz S, Falczuk V, Perez-Harguindeguy N, Cabido M, Thorpe PC (2002) Foliar resistance to simulated extreme temperature events in contrasting plant functional and chorological types. Global Change Biology 8:1139–1145

Hamilton JG, George K, DeLucia EH, Naidu SL, Finzi AC, Schlesinger WH (2002) Forest carbon balance under elevated CO_2. Oecologia 131:250–260

Harrison SP, Prentice IC (2003) Climate and CO_2 controls on global vegetation distribution at the last glacial maximum: analysis based on palaeovegetation data, biome modelling and palaeoclimate simulations. Global Change Biology 9:983–1004

Harrison SP, Prentice IC, Barboni D, Kohfeld KE, Ni J, Sutra J-P (submitted) Towards a global plant functional type classification for ecosystem modelling, palaeoecology and climate impacts research. Journal of Vegetation Science

Haxeltine A, Prentice IC (1996a) BIOME3: an equilibrium terrestrial biosphere model based on ecophysiological constraints, resource availability, and competition among plant functional types. Global Biogeochemical Cycles 10:693–709

Haxeltine A, Prentice IC (1996b) A general model for the light-use efficiency of primary production. Functional Ecology 10: 551–561

Haxeltine A, Prentice IC, Creswell ID (1996) A coupled carbon and water flux model to predict vegetation structure. Journal of Vegetation Science 7:651–666

Heimann M, Esser G, Haxeltine A, Kaduk J, Kicklighter DW, Knorr W, Kohlmaier GH, McGuire AD, Melillo J, III BM, Otto RD, Prentice IC, Sauf W, Schloss A, Sitch S, Wittenberg U, Würth G (1998) Evaluation of terrestrial carbon cycle models through simulations of the seasonal cycle of atmospheric CO_2: first results of a model intercomparison study. Global Biogeochemical Cycles 12: 1–24

Hendry GR, Ellsworth DS, Lewin KF, Nagy J (1999) A free-air enrichment system for exposing tall forest vegetation to elevated atmospheric CO_2. Global Change Biology 5: doi: 10.1046/j.1365-2486.1999.00228.x

Hicke JA, Asner GP, Randerson JT, Tucker C, Los S, Birdsey R, Jenkins JC, Field C (2002) Trends in North American net primary productivity derived from satellite observations, 1982–1998. Global Biogeochemical Cycles 16: doi:10.1029/2001GB001550

Hickler T, Prentice IC, Smith B, Sykes MT (2004a) Simulating the effects of elevated CO_2 on productivity at the Duke Forest FACE experiment: a test of the dynamic global vegetation model LPJ. In: Hickler T, Towards an integrated ecology through mechanistic modelling of ecosystem structure and functioning. Meddelanden från Lunds Universitets Geografiska Institution. Avhandlingar: 153

Hickler T, Smith B, Sykes MT, Davis M, Sugita S, Walker K (2004b) Using a generalized vegetation model to simulate vegetation dynamics in northeastern USA. Ecology 85:519–530

Higgins SI, Clark JS, Nathan R, Hovestadt T, Schurr F, Fragoso JMV, Aguiar MR, Ribbens E, Lavorel S (2003) Forecasting plant migration rates: managing uncertainty for risk assessment. Journal of Ecology 91:341–347

Holdridge LR (1947) Determination of world plant formations from simple climatic data. Science 105:367–368

Houghton RA (2003) Revised estimates of the annual net flux of carbon to the atmosphere from changes in land use and land management 1850–2000. Tellus 55B: 378–390

Houghton JT, Ding Y, Griggs DJ, Noguer M, van der Linden PJ, Xiaosu D (Eds.) (2001) Climate Change 2001: The Scientific Basis. Contribution of Working Group I to the Third Assessment Report of the Intergovernmental Panel on Climate Change (IPCC). Cambridge University Press, UK

House JI, Prentice IC, Ramankutty N, Houghton RA, Heimann M (2003) Reconciling apparent inconsistencies in estimates of terrestrial CO_2 sources and sinks. Tellus 55:345–363

Hungate BA, Dukes JS, Shaw MB, Luo Y, Field CB (2003) Nitrogen and Climate Change. Science 302:1512–1513

Indermühle A, Stocker TF, Joos F, Fischer H, Smith HJ, Wahlen M, Deck B, Mastroianni D, Tschumi J, Blunier T, Meyer R, Stauffer B (1999) Holocene carbon-cycle dynamics based on CO_2 trapped in ice at Taylor Dome, Antarctica. Nature 398: 121–126

Jarvis PG (1976) The interpretation of the variances in leaf water potential and stomatal conductance found in canopies in the field. Phil. Trans. Roy. Soc. Lond. B273:593–610

Jones CD, Cox PM (2001) Modeling the volcanic signal in the atmospheric CO_2 record. Global Biogeochemical Cycles 15: 453–465

Jones CD, Cox PM, Essery RLH, Roberts DL, Woodage MJ (2003) Strong carbon cycle feedbacks in a climate model with interactive CO_2 and sulphate aerosols. Geophysical Research Letters 30: 1479, doi:1410.1029/2003GL016867

Joos F, Prentice IC, Sitch S, Meyer R, Hooss G, Plattner G-K, Gerber S, Hasselmann K (2001) Global warming feedbacks on terrestrial carbon uptake under the Intergovernmental Panel on Climate Change (IPCC) emission scenarios. Global Biogeochemical Cycles 15:891–907

Joos F, Gerber S, Prentice IC, Otto-Bliesner BL, Valdes PJ (2004) Transient simulations of Holocene atmospheric carbon dioxide and terrestrial carbon since the Last Glacial Maximum. Global Biogeochemical Cycles 18: GB2002, doi: 2010.1029/2003GB002156

Kaduk J, Heimann M (1996) A prognostic phenology model for global terrestrial carbon cycle models, Climate Research 6:1–19

Kaminski T, Heimann M (2001) Inverse modeling of atmospheric carbon dioxide fluxes. Science 294: 259a

Kaplan JO (2002) Wetlands at the Last Glacial Maximum: Distribution and methane emissions. Geophysical Research Letters 29: doi: 10.1029/2001GL013366

Kaplan JO, Prentice IC, Knorr W, Valdes PJ (2002) Modeling the dynamics of terrestrial carbon storage since the Last Glacial Maximum. Geophysical Research Letters 29: 2074, DOI: 2010.1029/2002GL015230

Kaplan JO, Bigelow NH, Prentice IC, Harrison SP, Bartlein PJ, Christensen TR, Cramer W, Matveyeva NV, McGuire AD, Murray DF, Razzhivin VY, Smith B, Walker DA, Anderson PM, Andreev AA, Brubaker LB, Edwards ME, Lozhkin AV (2003) Climate change and arctic ecosystems II: Modeling, paleodata-model comparisons, and future projections. Journal of Geophysical Research 108: 8171, doi: 8110.1029/2002JD002559

Kicklighter DW, Bruno M, Donges S, Esser G, Heimann M, Helfrich J, Ift F, Joos F, Kaduk J, Kohlmaier GH, McGuire AD, Melillo JM, Meyer R, Moore B, Nadler A, Prentice IC, Sauf W, Schloss AL, Sitch S, Wittenberg U, Wurth G (1999) A first-order analysis of the potential role of CO_2 fertilization to affect the global carbon budget: a comparison of four terrestrial biosphere models. Tellus Series B-Chemical and Physical Meteorology 51:343–366

Klein Goldewijk K (2001) Estimating global land use change over the past 300 years: the HYDE database. Global Biogeochemical Cycles 15:417–433

Knorr W (2000) Annual and interannual CO_2 exchanges of the terrestrial biosphere: process-based simulations and uncertainties. Global Ecology and Biogeography 9:225–252

Knorr W, Heimann M (1995) Impact of drought stress and other factors on seasonal land biosphere CO_2 exchange studied through an atmospheric tracer transport model. Tellus Series B-Chemical and Physical Meteorology 47:471–489

Knorr W, Heimann M (2001) Uncertainties in global terrestrial biosphere modeling. Part II: global constraints for a process-based vegetation model. Global Biogeochemical Cycles 15:227–246

Köppen W (1931) *Grundriss der Klimakunde*. Walter de Gruyter, Berlin

Krinner G, Viovy N, de Noblet-Ducoudré N, Ogée J, Polcher J, Friedlingstein P, Ciais P, Sitch S, Prentice IC (2005) A dynamic global vegetation model for studies of the coupled atmosphere-biosphere system. Global Biogeochemical Cycles 19: GB1015, doi:10.1029/2003GB002199

Kucharik CJ, Brye KR (2003) Integrated BIosphere Simulator (IBIS) yield and nitrate loss predictions for Wisconsin maize receiving varied amounts of nitrogen fertilizer. J Environ Qual 32:247–268

Kucharik CJ, Foley JA, Delire C, Fisher VA, Coe MT, Lenters JD, Young-Molling C, Ramankutty N, Norman JM, Gower ST (2000) Testing the performance of a Dynamic Global Ecosystem Model: Water balance, carbon balance, and vegetation structure. Global Biogeochemical Cycles 14:795–825

Lavorel S, Cramer W (1999) Plant functional response to land use and natural disturbance. Journal of Vegetation Science 10:604–732

Lavorel S, Garnier E (2002) Predicting changes in community composition and ecosystem functioning from plant traits: revisiting the Holy Grail. Functional Ecology 16:545–556

Lavorel S, McIntyre S, Landsberg J, Forbes TDA (1997) Plant functional classifications: from general groups to specific groups based on response to disturbance. Tree 12:474–478

Lavorel S, Díaz S, Cornelissen H, Garnier E, Harrison SP, McIntyre S, Pausas JG, Pérez-Harguindeguy N, Urcelay C (2007) Plant functional types: are we getting any closer to the Holy Grail? In: Canadell J, Pataki D, Pitelka L (eds) Terrestrial ecosystems in a changing world. IGBP Series, Springer-Verlag, Heidelberg, this volume

Law R, Chen YH, Gurney KR, Rayner P, Denning AS, TransCom3 modelers (2003) TransCom3 CO_2 inversion intercomparison: 2. Sensitivity of annual mean results to data choices. Tellus 55B: 512–521

Le Quéré C, Harrison SP, Prentice IC, Buitenhuis ET, Aumont O, Bopp L, Claustre H, Cotrim da Cunha L, Geider R, Giraud X, Klaas C, Kohfeld KE, Legendre L, Manizza M, Platt T, Rivkin R, Sathyendranath S, Uitz J, Watson AJ, Wolf-Gladrow D (in press) Ecosystem dynamics based on plankton functional types for global ocean biogeochemistry models. Global Change Biology

Lenihan JM, Daly C, Bachelet D, Neilson RP (1998) Simulating broad-scale fire severity in a dynamic global vegetation model. Northwest Science 72:91–103

Lieth H (1975) Modeling the Primary Productivity of the World. In: Lieth H, Whittaker RH (eds) Primary Productivity of the Biosphere, Springer-Verlag, Berlin, pp 237–263

Long SP, Ainsworth EA, Rogers A, Ort DR (2004) Rising atmospheric carbon dioxide. Annual Review of Plant Biology 55:591–628

Lucht W, Prentice IC, Myneni RB, Sitch S, Friedlingstein P, Cramer W, Bousquet P, Buermann W, Smith B (2002) Climatic control of the high-latitude vegetation greening trend and Pinatubo effect. Science 296:1687–1689

MA (Millenium Ecosystem Assessment) (2003) Ecosystems and Human Well-being: A Framework for Assessment. Island Press, Washington DC

McGuire AD, Sitch S, Clein JS, Dargaville R, Esser G, Foley J, Heimann M, Joos F, Kaplan J, Kicklighter DW, Meier RA, Melillo JM, Moore III B, Prentice IC, Ramankutty N, Reichenau T, Schloss A, Tian H, Williams LJ, Wittenberg U (2001) Carbon balance of the terrestrial biosphere in the twentieth century: Analyses of CO_2, climate and land use effects with four process-based ecosystem models. Global Biogeochemical Cycles 15:183–206

Melillo JM, McGuire AD, Kicklighter DW, Moore B, Vorosmarty CJ, Schloss AL (1993) Global climate change and terrestrial net primary production. Nature 363:234–240

Monteith JL (1995) Accommodation between transpiring vegetation and the convective boundary layer. J Hydrol 166:251–263

Moorcroft PR, Hurtt GC, Pacala. SW (2001) A method for scaling vegetation dynamics: the ecosystem demography model (ED). Ecological Monographs 71:557–585

Morales P, Sykes MT, Prentice IC, Smith P, Smith B, Bugmann H, Zierl B, Friedlingstein P, Viovy N, Sabate S, Sanchez A, Pla E, Gracia CA, Sitch S, Arneth A, Ogee J (submitted) Comparing and evaluating process-baserd ecosystem model predictions of carbon and water fluxes in major European forest biomes. Global Change Biology

Myneni RB, Keeling CD, Tucker CJ, Asrar G, Nemani RR (1997) Increased plant growth in the northern high latitudes from 1981 to 1991. Nature 386:698–702

Nabuurs GJ, Schelhas MJ, Mohren GMJ, Field CB (2003) Temporal evolution of the European Forest sector carbon sink 1950–1999. Global Change Biology 9:152–160

Neilson RP (1995) A model for predicting continental scale vegetation distribution and water balance. Ecological Applications 5:362–385

Neilson RP, Marks D (1994) A global perspective of regional vegetation and hydrologic sensitivities and risks from climatic change. Journal of Vegetation Science 5:715–730

Neilson RP, Running SW (1996) Global dynamic vegetation modellingg: coupling biogeochemistry and biogeography models. pp 451–465 in Walker B, Steffen W, (eds.) Global Change and Terrestrial Ecosystems. Cambridge University Press, Cambridge

Neilson RP, King GA, Koerper G (1992) Toward a rule based biome model. Landscape Ecology 7:27–43

Nemani R, Running SW (1996) Implementation of a hierarchical global vegetation classification in ecosystem function models. Journal of Vegetation Science 7:337–346

Nemani R, White M, Thornton P, Nishida K, Reddy S, Jenkins J, Running S (2002) Recent trends in hydrologic balance have enhanced the terrestrial carbon sink in the United States. Geophysical Research Letters 29: Art. No. 1468

New M, Hulme M, Jones P (2000) Representing twentieth-century space-time climate variability. Part II: Development of 1901–96 monthly grids of terrestrial surface climate. Journal of Climate 13:2217–2238

Norby RJ, Wullschleger SD, Gunderson CA, Johnson DW, Ceulemans R (1999) Tree responses to rising CO_2 in field experiments: implications for the future forest. Plant Cell and Environment 22:683–714

Notaro M, Liu Z, Gallimore R, Vavrus SJ, Kutzbach JE, Prentice IC, Jacob RL (2004) Simulated and observed pre-Industrial to modern vegetation and climate changes. Journal of Climate 18:3650–3671

Nowak RS, Ellsworth DS, Smith SD (2004) Functional responses of plants to elevated atmospheric CO_2 – do photosynthetic and productivity data from FACE experiments support early predictions? New Phytologist 162:253–280

Overpeck JT, Whitlock C, Huntley B (2003) Terrestrial biosphere dynamics in the climate system: past and future. In: Alverson K, Bradley R, Pedersen T (eds) Paleoclimate, Global Change and the Future (IGBP Synthesis Volume), Springer-Verlag, Berlin, pp 81–111

Pacala SW, Canham CD, Saponara J, Silander Jr. JA (1993) Forest models defined by field measurements: I. The design of a northeastern forest simulator. Canadian Journal of Forest Research 23:1980–1998

Pacala SW, Canham CD, Saponara J, Silander Jr. JA, Kobe RK, Ribens E (1996) Forest models defined by field measurements: estimation, error analysis and dynamics. Ecological Monographs 66:1–43

Pan Y, McGuire AD, Melillo JM, Kicklighter DW, Sitch S, Prentice IC (2002) A biogeochemistry-based dynamic vegetation model and its application along a moisture gradient in the continental United States. Journal of Vegetation Science 13:369–382

Parton WJ, Scurlock JMO, Ojima DS, Gilmanov TG, Scholes RJ, Schimel DS, Kirchner T, Menaut J-C, Seastedt T, Garcia Moya E, Apinan Kamnalrut, Kinyamario JI (1993) Observations and modeling of biomass and soil organic matter dynamics for the grassland biome worldwide. Global Biogeochemical Cycles 7:785–809

Pastor J, Post WM (1985) Development of a linked forest productivity-soil process model. ORNL/TM-9519. Oak Ridge, Tennessee

Peylin P, Bousquet P, Le Quéré C, Sitch S, Friedlingstein P, McKinley G, Gruber N, Rayner P, Ciais P, (2005) Multiple constraints on regional CO_2 flux variations over land and oceans. Global Biogeochemical Cycles 19: GB1011, doi:10.1029/2003GB002214

Pitelka LF, Plant Migration Workshop Group (1997) Plant migration and climate change. American Scientist 85:464–473

Potter CS, Klooster SA (1999) Dynamic global vegetation modelling for prediction of plant functional types and biogenic trace gas fluxes. Global Ecology and Biogeography 8:473–488

Potter CS, Randerson JT, Field CB, Matson PA, Vitousek PM, Mooney HA, Klooster S (1993) Terrestrial Ecosystem Production – A Process Model-Based On Global Satellite And Surface Data Global Biogeochemical Cycles 7:811–841

Potter CS, Wang S, Nikolov NT, McGuire AD, Liu J, King AW, Kimball JS, Grant RF, Frolking SE, Clein JS, Chen JM, Amthor JS (2001) Comparison of boreal ecosystem model sensitivity to variability in climate and forest site parameters. Journal of Geophysical Research 106: D24, 33, 671 (2000JD000224)

Prentice IC (2001) Controls on the primary productivity of terrestrial ecosystems. In: Geider RJ, DeLucia EH, Falkowski PG, Finzi A, Grime JP, Grace J, Kana TM, LaRoche J, Long SP, Osborne BA, Platt T, Prentice IC, Raven JA, Schlesinger WH, Smetacek V, Stuart V, Sathyendranath S, Thomas RB, Vogelmann TC, Williams P, Woodward FI (2001) Primary productivity of planet Earth: biological determinants and physical constraints in terrestrial and aquatic habitats. Global Change Biology 7:849–882

Prentice IC, Leemans R (1990) Pattern and process and the dynamics of forest structure: a simulation approach. Journal of Ecology 78:340–355

Prentice IC, Raynaud D (2001) Paleogeochemistry, in Global Biogeochemiscal Cycles in the Climate System (eds. Schulze ED, Heinmann M, Holland E. et al.), San Diego: Academic Press, 2001, 87–94

Prentice IC, Solomon AM (1991) Vegetation models and global change. In: Bradley RS (ed.) Global changes of the past: Papers arising from the 1989 OIES Global Change Institute; Snowmass, Colorado, 24. July–4. August 1989. UCAR, Office for Interdisciplinary Earth Studies, Boulder, Colorado. 365–384

Prentice IC, van Tongeren O, de Smidt JT (1987) Simulation of heathland vegetation dynamics. Journal of Ecology 75:203–219

Prentice IC, Cramer W, Harrison SP, Leemans R, Monserud RA, Solomon AM (1992) A global biome model based on plant physiology and dominance, soil properties and climate. Journal of Biogeography 19:117–134

Prentice IC, Sykes MT, Cramer W (1993). A simulation model for the transient effects of climate change on forest landscapes. Ecological Modelling 65:51–70

Prentice IC, Heimann M, Sitch S (2000) The carbon balance of the terrestrial biosphere: Ecosystem models and atmospheric observations Ecological Applications 10 (6):1553–1573

Prentice IC, Farquhar GD, Fasham MJR, Goulden ML, Heimann M, Jaramillo VJ, Kheshgi HS, Le Quéré C, Scholes RJ, Wallace DWR (2001) The carbon cycle and atmospheric carbon dioxide. In: Houghton JT, Ding Y, Griggs DJ, Noguer M, van der Linden PJ, Dai X, Maskell K, Johnson CA (eds.) Climate Change 2001: The Scientific Basis. Contribution of Working Group I to the Third Assessment Report of the Intergovernmental Panel on Climate Change. Cambridge University Press, Cambridge, 185–225

Prentice IC, Le Quéré C, Buitenhuis ET, House JI, Klaas C, Knorr W (2004) Biosphere dynamics: questions for Earth System modellingg. In: Hawkesworth CJ, Sparks RSJ (eds) The State of the Planet: Frontiers and Challenges. AGU Monograph

Raunkiær C (1909) Formationsundersøgelse og formationstatistik. Bot. Tidsskr. 30:20–80

Raunkiær C (1913) Formationsstatistike Undersøgelser paa Skagens Odde. Botanisk Tidsskrift. Dansk Botanisk Forening 33:197–243

Raunkiær C (1934) The life-forms of plants and statistical plant geography. Clarendon Press, Oxford

Ridgwell A, Marshall S, Gregson K (1999) Consumption of atmospheric methane by soils: a process-based model. Global Biogeochemical Cycles 13:59–70

Rödenbeck C, Houweling S, Gloor M, Heimann M (2003) CO_2 flux history 1982–2001 inferred from atmospheric data using a global inversion of atmospheric transport. Atmos. Chem. Phys. 3:1919–1964

Roderick ML, Farquhar GD, Berry SL, Noble IR (2001) On the direct effect of clouds and atmospheric particles on the productivity and structure of vegetation. Oecologia 129:213–232

Ruddiman WF (2003). The Anthropogenic Greenhouse Era Began Thousands of Years Ago. Climatic Change, 61:261–293

Ruimy A, Dedieu G, Saugier B (1996) TURC: A diagnostic model of continental gross primary productivity and net primary productivity Global Biogeochemical Cycles 10:269–285

Running SW, Gower ST (1991) FOREST-BGC, a general model of forest ecosystem processes for regional applications. 2. Dynamic carbon allocation and nitrogen budgets. Tree Physiology 9:147–160

Running SW, Hunt ER (1993) Generalization of a forest ecosystem process model for other biomes, BIOME-BGC, and an application for global-scale models. Scaling processes between leaf and landscape levels. In: Ehleringer JR, Field CB (eds) Scaling Physiological Processes: Leaf to Globe. Academic Press, San Diego, 141–158

Ryan MG (1991) Effects of climate change on plant respiration. Ecological Applications 1:157–167

Schäfer KVR, Oren R, Ellsworth DS, Lai CT, Herrick JD, Finzi AC, Richter DD, Katul GG (2003) Exposure to an enriched CO_2 atmosphere alters carbon assimilation and allocation in a pine forest ecosystem. Global Change Biology 9: 10, 1378–1400

Schaphoff S, Lucht W, Gerten D, Sitch S, Cramer W, Prentice IC (2006) Terrestrial biosphere carbon storage under alternative climate projections. Climatic Change 74:97–122, doi: 10.1007/s10584-005-9002-5

Schimel DS, House JI, Hibbard KA, Bousquet P, Ciais P, Peylin P, Braswell BH, Apps MJ, Baker D, Bondeau A, Canadell J, Churkina G, Cramer W, Denning AS, Field CB, Friedlingstein P, Goodale C, Heimann M, Houghton RA, Melillo JM, Moore III B, Murdiyarso D, Noble I, Pacala SW, Prentice IC, Raupach MR, Rayner PJ, Scholes RJ, Steffen WL, Wirth C (2001) Recent patterns and mechanisms of carbon exchange by terrestrial ecosystems. Nature 414:169–172

Scholze M, Kaplan JO, Knorr W, Heimann M (2003) Climate and interannual variability of the atmosphere-biosphere $^{13}CO_2$ flux. Geophysical Research Letters 30: 1097, doi:10.1029/2002GL015631

Schulze E-D (1982) Plant life forms and their carbon, water and nutrient relations Encyclopedia of Plant Physiology. Springer-Verlag, Berlin, Heidelberg, 615–676

Sellers PJ, Mintz Y, Sud YC, Dalcher A (1986) A simple biosphere model (SiB) for use within general circulation models. Journal of the Atmospheric Sciences 43:505–531

Sernander R (1936) The primitive forest of Granskär and Fiby. Acta Phytogeogr. Suec. 8

Shugart HH (1984) A Theory of Forest Dynamics. Springer-Verlag, New York

Shugart HH, West DC (1977) Development of an Appalachian deciduous forest succession model and its application to assessment of impact of chestnut blight. Journal of Environmental Management 5:161–179

Sitch S, Smith B, Prentice IC, Arneth A, Bondeau A, Cramer W, Kaplan JO, Levis S, Lucht W, Sykes MT, Thonicke K, Venevsky S (2003) Evaluation of ecosystem dynamics, plant geography and terrestrial carbon cycling in the LPJ dynamic global vegetation model. Global Change Biology 9:161–185

Smith B, Prentice IC, Sykes MT (2001) Representation of vegetation dynamics in the modelling of terrestrial ecosystems: comparing two contrasting approaches within European climate space. Global Ecology and Biogeography 10:621–637

Specht RL (1972) Water use by perennial evergreen plant communities in Australia and Papua New Guinea. Australian Journal of Botany 20:273–299

Sprugel DG (1976) Dynamic structure of wave-generated *Abies balsamea* forests in the northeastern United States. Journal of Ecology 64:889–911

Sykes MT, Prentice IC, Smith B, Cramer W, Venevsky S (2001) An introduction to the European Terrestrial Ecosystem Modelling Activity. Global Ecology and Biogeography 10:581–593

Tegen I, Harrison SP, Kohfeld KE, Prentice IC, Coe MT, Heimann M (2002) Impact of vegetation and preferential source areas on global dust aerosol: Results from a model study. Journal of Geophysical Research-Atmospheres 107: 4576, doi:4510.1029/2001JD000963

Thonicke K, Venevsky S, Sitch S, Cramer W (2001) The role of fire disturbance for global vegetation dynamics: coupling fire into a Dynamic Global Vegetation Model. Global Ecology and Biogeography 10:661–677

Thonicke K, Prentice IC, Hewitt C (2005) Modelling glacial-interglacial changes in global fire regimes and trace gas emissions. Global Biogeochemical Cycles 19, GB3008, doi:10.1029/2004GB002278

Townsend AR, Braswell BH, Holland EA, Penner JE (1996) Spatial and temporal patterns in terrestrial carbon storage due to deposition of fossil fuel nitrogen. Ecological Applications 6:806–814

VEMAP Members (1995) Vegetation/ecosystem modeling and analysis project: Comparing biogeography and biogeochemistry models in a continental-scale study of terresttrial ecosystem responses to climate change and CO_2 doubling. Global Biogeochemical Cycles 9:407–437

Venevsky S, Thonicke K, Sitch S, Cramer W (2002) Simulating fire regimes in human-dominated ecosystems: Iberian Peninsula case study. Global Change Biology 8:984–998

Walter H (1962) Die Vegetation der Erde in ökophysiologischer Betrachtung – Die tropischen und subtropischen Zonen, Band 1. VEB Gustav Fischer Verlag, Jena

Walter H (1968) Die Vegetation der Erde in ökophysiologischer Betrachtung – Die gemässigten und arktischen Zonen, Band 2. VEB Gustav Fischer Verlag, Jena

Warnant P, François L, Strivay D, Gérard J-C (1994) CARAIB: a global model of terrestrial biological productivity. Global Biogeochemical Cycles 8:255–270

Watt AS (1947) Pattern and process in the plant community. Journal of Ecology 1/2:1–22

Whittaker RH (1975) Communities and ecosystems, 2nd edition. Macmillan, New York

Woodward FI (1987) Climate and Plant Distribution. Cambridge University Press, Cambridge, pp 174

Woodward FI, Lomas ML (2004) Vegetation dynamics – simulating responses to climatic change, Biological Reviews 79, 1–28; DOI: 10.1017/S1464793103006419

Woodward FI, Smith TM, Emanuel WR (1995) A global land primary productivity and phytogeography model. Global Biogeochemical Cycles 9:471–490

Woodward FI, Lomas MR, Lee SE (2001) Predicting the future productivity and distribution of global terrestrial vegetation. In: Roy J, Mooney HA, Saugier B (eds) Terrestrial Global Productivity, Academic Press, San Diego, pp 521–541

Wright IJ, Reich PB, Westoby M, Ackerly DD, Baruch Z, Bongers F, Cavender-Bares J, Chapin T, Cornelissen JHC, Diemer M, Flexas J, Garnier E, Groom PK, Gulias J, Hikosaka K, Lamont BB, Lee T, Lee W, Lusk C, Midgley JJ, Navas ML, Niinemets U, Oleksyn J, Osada N, Poorter H, Poot P, Prior L, Pyankov VI, Roumet C, Thomas SC, Tjoelker MG, Veneklaas EJ, Villar R (2004) The worldwide leaf economics spectrum. Nature 428:821–827

Zhou LM, Tucker CJ, Kaufmann RK, Slayback D, Shabanov NV, Myneni RB (2001) Variations in northern vegetation activity inferred from satellite data of vegetation index during 1981 to 1999. Journal of Geophysical Research-Atmospheres 106: 20069–20083

Part D

Managing Ecosystem Services

Chapter 16

Wheat Production Systems and Global Climate Change

John R. Porter · Pete D. Jamieson · Peter R. Grace

16.1 Introduction

The production and provision of food is central to the survival and well being of humans. Globally, more people are employed in the production, processing and distribution of food than any other socio-economic sector. Agriculture contributes between 2% of GDP in developed western countries to more than 30% of GDP in many African states. From the ample sufficiency of basic commodities in the developed world, we can forget the historical importance that food and fibre have played in human social and political development. During the Napoleonic Wars in Europe in the first years of the 19th century, the price of a tonne of wheat in the UK was five times its purchasing parity price today and food riots were common. Humans have cultivated the land for about 10 000 years when the global population reached one million. Before that time and for the preponderance of human history, humans have hunted and gathered their food, fiber and energy since *Homo sapiens sapiens* diverged from its ancestors about 200 000 years ago (Evans 1998). The effect of agricultural cultivation on human political and social development was explosive. In the 10 000 years leading to AD 2000, the human population increased 100-fold from 5 to 500 million people. In the second millennium, the effect was even more profound in absolute terms as 500 million people became 5 000 million. Today more than 75% of the land area of Europe is cultivated for either crop, grassland or forestry production and humans appropriate about 40% of terrestrial net primary productivity for their own use (Vitousek 1986). In rich, demographically stable countries the effects of agriculture and forestry on biodiversity and the cycles of carbon, nutrients and water are relevant social and political agendas. Land cultivation, and particularly its management intensity, is the most influential human practice for the dynamics of the terrestrial landscape and thereby the atmospheric, biogeochemical and water cycles of the Earth System. The production of food affects the main global biogeochemical cycles and is heavily reliant on inputs of fossil fuel energy. Farmers use such products of fossil fuels as nitrogen fertilizer, herbicides and pesticides and machinery to increase the proportion of solar energy that is captured by crops to drive dry matter production and thereby harvested yield. It is only in the past 250 years out of the 10 000 years that human beings have cultivated the land for food that they have been able to swap the toil of long hours in the field to increase solar energy capture for less humanly demanding methods.

Wheat yields have been steadily rising in most parts of the world for at least the past century. New cultivars provide part of the explanation (Moot et al. 1996; Evans and Fischer 1999), where the addition of dwarfing genes has meant that newer cultivars produce shorter straw, similar total biomass but a higher harvest index. Much of the increase in crop production is attributable to the use of N fertilizer, but better knowledge and methodology leads to improvement as well. Much of the increases in crop yields in the past century are due to substantially greater use of nitrogen fertilizer. Demographic changes and urban encroachment onto peri-urban farmland will cause a change in production patterns and trade over the next 30 years or so. For instance, since the 1960s China's wheat production has nearly trebled while the land area used for its production has fallen very slightly. In the absence of any effects of atmospheric changes, Chinese wheat production is nearing stabilization. There is no sign of a decrease in demand for wheat in China, so that the shortfall is likely to be met by imports. Marathée and Gomez-MacPherson (2001) estimate that by 2030, wheat production in the developed world will have needed to increase from 308 to 440 million Mg (43%), while in the developed world the increase will need to be from 272 to 418 million Mg (54%). Most of this increase will come from increased yield per ha, rather than increased area. Accompanying the increase in production will be a change in the trade balance. Exports from the developed world are projected to increase by 91 million Mg, all into the developed world. The biggest demand increases will be from China, North Africa and the near East.

Globally, the picture is different from that seen in Europe, the USA and the affluent parts of Asia. In these areas, we have a stable overall population with an aging population structure and with some countries experiencing negative population growth. In many parts of the developing world and, particularly in south-east Asia, populations are still expanding but fortunately at a decreasing rate. There are causes for optimism that, even with a global population of 9 to 10 billion people, the world will be able to feed itself on average. However, the average conceals that the relative gap between the richest and the poorest in the world is getting larger.

The base of the human food chain is largely formed by the grasses rice, maize, wheat and tropical species such as sorghum and millet. Amthor (2001), quoting FAO data (FAO 2003) notes that of the 1.4 billion ha of land (Evans 1998) devoted to arable cultivation, wheat is harvested from about 15% of it; direct consumption of wheat contributes 20% of the calories and 22% of the protein in the human diet. As more than 30% of harvested wheat is fed to animals and thence to humans as milk or meat, the place of wheat as the most important human food source is unrivalled. The versatility of wheat, both in terms of its versatility of food end-uses and its environmental adaptability, means that its importance is likely to increase in the future. Projections based on population driven demand suggest that the total production and productivity per unit area of wheat will need to increase by about 50% during the next 30–40 years to accommodate the projected extra 3 billion humans by that time. This will partly come about as wheat continues to displace other cereals, particularly rice in Asia. This makes it important to examine the vulnerability of the world wheat crop to changes in the global environment in the foreseeable future. As the total global arable area for all crops is about 1.4 billion ha (Evans 1998), most of the production increase will need to come from increased yields per hectare, rather than increases in production area since there is little extra available land and this is often of poor quality. In terms of a global average, wheat yields needs to double to about 5 Mg per ha from the current 2.8 Mg per ha over the next 30–40 years. So far, yield increases have been maintained post Green Revolution for wheat (Evans and Fischer 1999) but this trend will need to continue in the face of a changing climate that can act as a stimulant to production via 'CO_2 fertilization'. However, rising temperatures and drier conditions in areas already at the margins of production are expected to come under enhanced agro-climatic as well as enhanced human pressure. Therefore, the interactions, defined as impacts, adaptation and mitigation, between wheat and global climate change has formed an important element in the modeling and experimental activities of GCTE Focus 3 (Agroecology and Production Systems).

For almost all the time that humans have cultivated the land the atmospheric CO_2 concentration has been between about 260 and 280 μmol mol^{-1} and the northern hemisphere temperature anomaly has had an absolute value of about 1 °C with a decreasing amplitude since about 1500. However, since 1900 both the mean and variance of the temperature anomaly have been increasing, as has the atmospheric CO_2 level. The immediate questions for arable agriculture in these circumstances can be categorised as the damage potential of climate change for the basic arable food systems and their security, possible adaptations thereof and the mitigation potential of arable farming to slow the build-up of greenhouse gases (GHG). Agriculture is situated at the interface between ecosystems and society (Olesen and Bindi 2002) and is obviously affected by changes in global environmental conditions, but it also contributes to GHG emissions, notably carbon dioxide, methane and nitrous oxide (Rosenzweig and Hillel 2000). Agriculture contributes 24% of global CO_2 emissions from energy use (4%) and anthropogenic GHG emissions (20%, as CO_2 equivalents) mainly from CH_4 and N_2O as well as carbon from land clearing (IPCC 2001).

Wheat productivity in important wheat growing regions of the world is well below its potential (Aggarwal et al. 2000). Increases in production will therefore be associated with raising the lower end of the distribution via improved management, rather than by raising the theoretical ceiling of productivity (Sinclair 1992). Simulation studies suggest that these increases, associated mostly with improved technology, will be boosted by global atmospheric changes, where CO_2 fertilization will more than compensate for the yield reducing effects of increased temperature, provided these are not too extreme (IPCC 2001). A danger will be that yield increases may be accompanied by reductions in quality via reduced protein content. However, we believe that improved understanding of the internal nitrogen economy of wheat will provide solutions to this problem, both through cultivar development and improved management.

Increases in the productivity of wheat will mean that issues such as the future of the 23 Mha of the extended Indo-Gangetic Plain that is responsible for feeding about one third of humanity will assume ever greater importance. The direction of the Green Revolution, that enabled a halving of the globally undernourished even as the global population doubled (Conway 1997), will need to be in the direction of combining the best of modern technology while paying more attention to conservation and building of the agro-ecological resource, particularly water and soil organic matter. This so-called doubly Green Revolution will both be affected by and have consequences for the global climate, an issue that is addressed later in this chapter.

This paper highlights some of the research in which members of the GCTE Focus 3 community have participated and that has contributed to raise understanding of global change and wheat and wheat cropping system productivity. We focus on four issues, as representative of the GCTE Wheat Network's activities to date. These are the experimental and modeling analysis of wheat responses to elevated CO_2 in combination with water and nitrogen treatments in Free-Air (CO_2) Concentration Enrichment (FACE) experiments in Arizona, USA; the importance of climate variability on wheat growth, development and yield; the influence of climate on the quality of wheat and the contribution of the rice-wheat cropping system of the Indo-Gangetic Plain to global climate change.

16.2 Global Atmospheric Change, Climate and Yields

The one complete certainty is that atmospheric CO_2 concentration is increasing. Just how much it will change will depend on combinations of fossil fuel use and deforestation that will occur during over the next 100 years. A near certainty is that the changes in atmospheric composition will be accompanied by changes in climate. Climate scientists refer to emissions scenarios as the drivers of climate change. Biologists, who must assess the effects that the resultant climate changes will have on patterns of production, refer to these climate changes as their scenarios for assessment. Crops respond to temperature, rainfall and its temporal and spatial spread, fertilizer levels as well as direct responses to atmospheric CO_2 concentration. A main activity of Focus 3 scientists has been to make educated assessment of whether the integrated effects will be, on balance, positive or negative. Scientific members of GCTE Focus 3 have participated in many of the experimental studies that have attempted to measure the dependence of wheat grain yield on ambient CO_2 concentration, data reviewed by Amthor (2001) for mainly non-European ones (Fig. 16.1a) and Downing et al. (2000) and Olesen and Bindi (2002) for European sites (Fig. 16.1b).

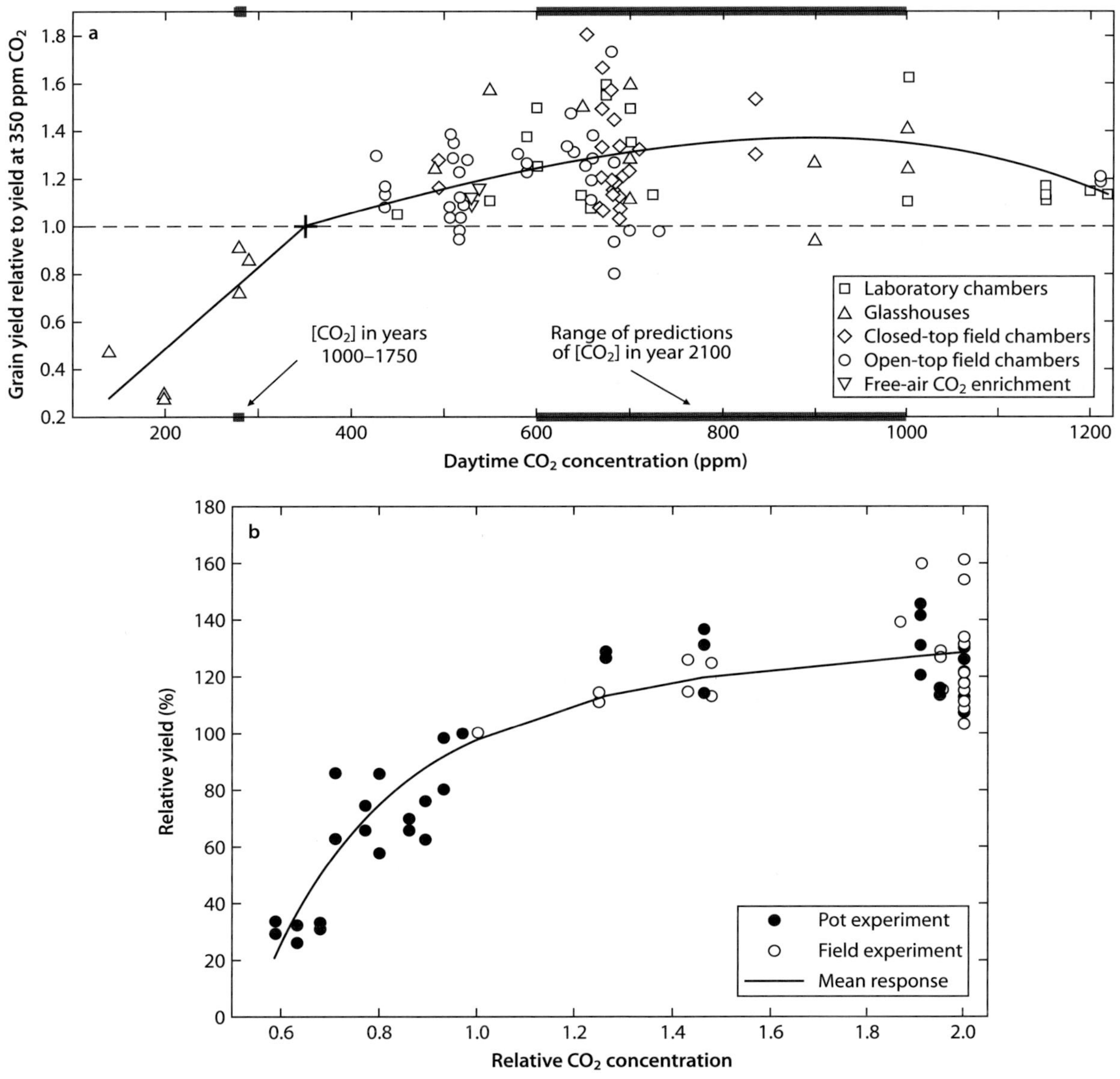

Fig. 16.1. Relative effects of CO_2 concentration on wheat grain yield in various experiments from mainly non-European (**a** Amthor 2001) and European sites (**b** Downing et al. 2001; Olesen and Bindi 2002) and where current ambient CO_2 is set to 1. In **b** *open symbols* represent data from field experiments (OTC, FACE); *filled symbols* represent data from pot or glasshouse experiments and the *solid line* shows the mean estimated response (Olesen and Bindi 2002). In both cases, values were taken from treatments in which adequate water and nitrogen were provided

Both graphs in Fig. 16.1 show that wheat grain yields, on average, rise by about 20% when the CO_2 concentration is double the pre-industrial level and it is difficult to detect an effect of experimental design in these results. There is considerable spread in the grain yield increase (or decrease) in both reviews but the curves are very congruent for ambient CO_2 levels between about 200 μmol mol^{-1} and 700 μmol mol^{-1}. The lack of dependence of the relative grain yield increase measurement on experimental design concurs with the conclusions of Kimball et al. (2002) that only two crop physiological parameters were substantially different between FACE and chamber-based approaches; namely FACE systems tended to reduce stomatal conductance by one and a half times compared with chambers and that root-to-shoot ratios were higher for FACE, as opposed to other, systems.

In both modeling and experimental studies, scientists associated with the GCTE Focus 3 Wheat group have been involved with studying the interactions between CO_2 level and temperature. Such studies have included the effects of temperature gradients on wheat responses to CO_2 (Batts et al. 1996, 1997, 1998) and have shown the extent to which rising temperatures can offset increases in growth and yield stimulated by higher CO_2, such that a whole season mean temperature increase of 2–4 °C can negate any CO_2 induced rise in yield, mainly by shortening the life-cycle of the wheat crop and speeding the rate of development through grain-filling. Episodic high temperature events, such as 30–35 °C in the few days around anthesis, have been shown to have yield reducing effects, caused by reduced seed fertility and a decline in seed number (Ferris 1998; Wollenweber et al. 2003) and the general effects of threshold temperatures (Porter and Gawith 1999) producing increased non-linear uncertainties of yield (Porter and Semenov 1999) has been an important new insight to come from work of GCTE Focus 3 and will be referred to later.

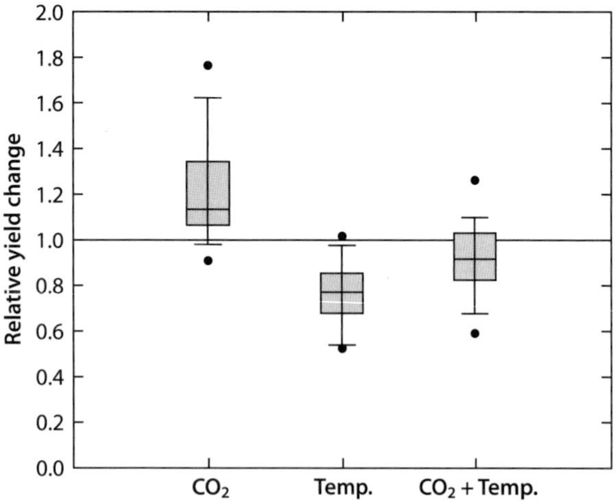

Fig. 16.2. Effects of elevated CO_2 and increased temperature, singly and in combination, on yield of wheat. The data represent the ratio of yield relative to current ambient CO_2 and temperature (relative yield change). Data are taken from the review by Amthor (2001, Table 7). *Plots* show median and standard percentiles ($n = 17$) (Fuhrer 2003)

In reviewing the interaction between temperature and CO_2 on relative yield change for wheat, Fuhrer (2003) presented a summary of data (Fig. 16.2) amassed and analysed by Amthor (2001) for 2 glasshouse, 13 closed-top chamber and two open-top chamber experiments. This showed that overall, warmer temperatures negate the CO_2 effects but that there is considerable overlap between the treatments. On average the effect of warming and higher CO_2 level on wheat yield is slightly to reduce yield in relation to current ambient conditions (Fuhrer 2003).

Studies of combinations of factors on the growth and yield of wheat have been extended by GCTE Focus 3 scientists to include the single and interactive effects of nitrogen level and irrigation. Collaborative research and modeling studies were made by using the FACE wheat experiments at Maricopa, USA and performed by Dr. Bruce Kimball and colleagues (www.uswcl.ars.ag.gov/edp/co2/co2face.htm). Although general conclusions of the effects of CO_2 enrichment on growth and yield are broadly similar, there are substantial effects of the chambers themselves on crop performance that vary among experiments. In any one experiment these are similar for both ambient and enriched treatments, but are sufficiently different from open field conditions to be difficult to simulate with models that assume uniformity over large areas and an absence of edge effects, (Ewert et al. 2002). Comparisons are facilitated when experiments are conducted in open field conditions such as those the FACE experiments at the University of Arizona Maricopa Agricultural Center (MAC), Maricopa, Arizona, USA. (Hunsaker et al. 1996; Kimball et al. 1995, 1997, 2001; Pinter et al. 1996). The experiments were conducted in replicated FACE rings of 25 m diameter with either no CO_2 enhancement or CO_2 enhancement for 24 h d^{-1} at either 550 μmol mol^{-1} (earlier experiments) or sustained at 200 μmol mol^{-1} above ambient CO_2 concentration. The FACE apparatus was placed well inside the boundaries of fields of substantial size, and therefore closely resemble the field scale that simulation models assume. The combined effects of CO_2 and water stress were studied in experiments in 1992–1993 and 1993–1994, and CO_2 and soil N restrictions in 1995–1996 and 1996–1997. A substantial scientific effort went into measurement of plant responses through time at a range of time scales and process levels within the crop, including gas exchange and micrometeorology. This allowed an analysis of the effects of elevated CO_2 on growth, biomass partitioning, yield, yield components, nutrient uptake and distribution, grain quality, evapotranspiration on time scales of less than an hour to seasonal totals plus biochemistry and other effects.

The major conclusions were that enhanced CO_2 increased production by about 10% in well watered conditions and by about 20% in drought stressed conditions – CO_2 compensated somewhat for water stress. Later analysis (Pinter et al. 2000) suggested that the CO_2 yield enhancement was probably understated by the results of the first experiments. The nitrogen experiments showed yield enhancements of about 15% at high soil N levels, but the enhancement was reduced when N was restricted, and the level differed between the two years. In general CO_2 decreased the transpiration rate although this was more obvious in the micrometeorological measurements than in long term water balance measurements. Grain protein content was decreased slightly by enhanced CO_2 unless N was not adequate; under low N conditions the protein reducing effects of enhanced CO_2 was increased.

Until the mid-1990s most studies of the impacts of climate change on agriculture (i.e., Rosenzweig and Parry 1994) focused on changes in CO_2 level and average conditions, such as a rise in mean global temperature or the amount of rainfall, on crop production. However, it was soon realized that these analyses were conceptually incomplete because *(i)* crops and plants in general, respond non-linearly to changes in their growing conditions exhibiting discontinuous threshold responses and *(ii)* because crops are often subject to combinations of stress factors that affect their growth and development (Fig. 16.3). The effect of climate variability, the frequency of extreme events and the effects of combinations of factors have since assumed greater importance (*i.e.,* Semenov and Porter 1995). IPCC reports since that time have highlighted the likely increase in extreme weather events associated with climate change. These would reveal themselves as higher maximum temperatures, hot days and heat waves over nearly all land areas and more intense precipitation events over many areas associated with El-Nino events and a likely increase in Asian summer monsoon precipitation variability. Variability at a range of spatial and temporal scales is now a key concern in impact studies on ecosystems in general and agroecosystems in particular.

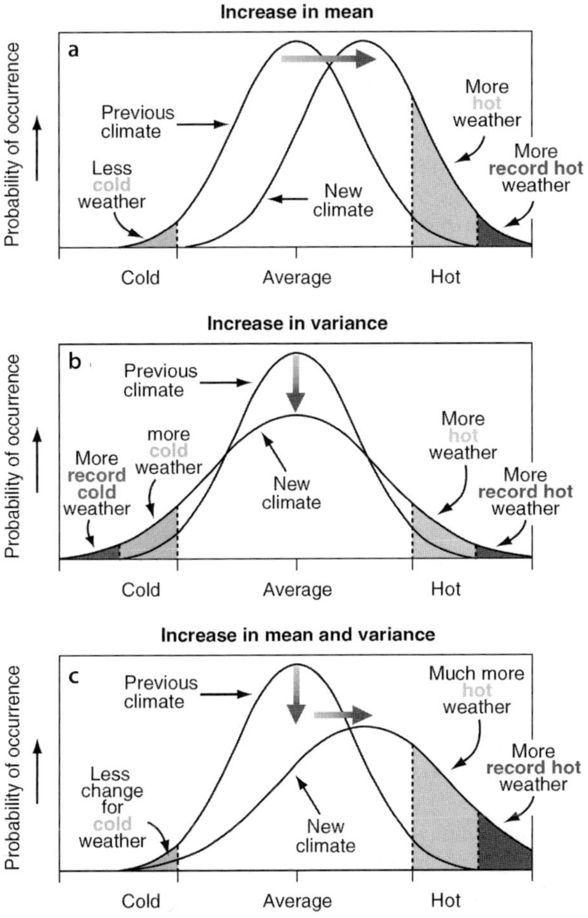

Fig. 16.3. Illustrations of the effects of increases in **a** mean temperature, **b** increase in temperature variance and **c** increase in both on the frequency of extreme hot weather

Crops models that incorporated mathematical descriptions of phenology, leaf area development, photosynthesis, respiration, water and nitrogen uptake were able to interact with climate scenarios that included changes in the variability of rather than the mean climatic driving variables. Simulation modeling of the effects of climatic variability has pointed to the general conclusion that increased annual variability in weather causes increased variation in yields. For wheat (Fig. 16.4), it was found by three simulation models that doubling the standard variation of annual temperature, while holding its mean value unchanged (*i.e.,* the scenario in Fig. 16.3b) gave the same decrease in yield as a 4 °C increase in mean temperature (Fig. 16.3a) but a more than doubled coefficient of yield variation. It remains a challenge for experimental studies to test these model predictions but there are sound reasons to expect verification of the predictions. The mechanisms that lie behind such responses are likely to be complex but will involve the non-linear relationship of respiration with temperature, and temperature threshold effects on reproductive fertility and phenology (Ferris et al. 1998; Porter and Gawith 1999).

16.3 Impacts on Wheat Productivity

With annual crops like wheat, the effects of increases in atmospheric CO_2 concentration and the accompanying temperature rises act in opposite directions. Broadly, the yield of wheat crop can be described as the product of its mean growth rate, duration or lifetime, and harvest index – that proportion of the crop that is grain. The primary effect of increased CO_2 concentration is to increase the mean growth rate by acting as an atmospheric fertilizer, provided enough N is present. A secondary effect is to reduce the transpiration rate for the same energy input, which has the effect of making more economical use of soil water. These effects are yield enhancing. However, the life cycle of a wheat crop occurs in thermal time, rather than chronological time. This means that higher temperature increases the development rate and shortens growth duration. This has the opposite effect of the

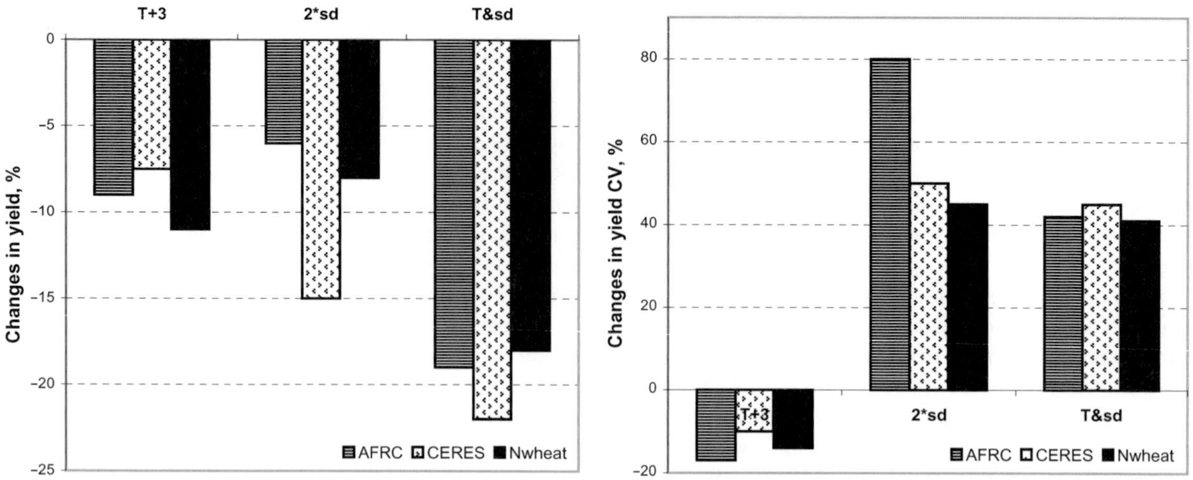

Fig. 16.4. Modeling of the effect of variation in temperature on crop yields and its variation (as *CV*) for wheat. *T*+3, mean temperature increased by 3 °C (c.f. Fig. 16.3a); 2**sd*, standard deviation of temperature doubled without change in its mean value (c.f. Fig. 16.3b); *T&sd*, combination of raised mean and standard deviation of temperature (c.f. Fig. 16.3c)

CO_2 increase – the crop has less time to grow and will hence produce less biomass. Effects on harvest index are less clear and may be positive, negative or neutral. Earlier maturity may avoid late season drought, for instance, and the yield depressing effects of shortened growth duration may be counterbalanced by a reduction in drought stress during grain growth. The main point of this discussion is that assessment of the effects of climate and atmospheric change on wheat production really needs the power of good simulation models that account for all these competing effects. Evaluation of such simulation models, particularly in comparison with data from experiments in elevated CO_2 and temperature environments, has been the core business of the GCTE International Wheat Network since its inception in 1991.

To be confident that simulation models give a reasonable assessment of likely response to changed environments, it is necessary to test them in a variety of conditions where these are well documented. If several models tend to give similar results, and at least give response in a similar sense to experiments, then some confidence can be attributed to their predictions in different circumstances. Examples of such studies, in collaborations reinforced by the GCTE network, are studies by Jamieson et al. (1998a) where the predictions of five models were compared with experiments under varying drought conditions, and Jamieson et al. (2000), where the predictions of three wheat simulation models were tested against data from wheat experiments in Arizona FACE experiments (Hendry and Kimball 1994; Grant et al. 1995; Grossman et al. 1995; Kimball et al. 1999) in which the amount of applied N and the atmospheric CO_2 concentration were both varied. This latter study complemented an earlier one (Tubiello et al. 1999) where simulations were compared with measurements from crops where CO_2 and water supply were varied.

16.4 Addressing the Yield Gap

Where are the projected yield increases to come from? Marathée and Gomez-MacPherson (2001) estimate large increases in production in both the developed and developing worlds, mostly from yield increases. Their analysis took no account of the posited increases associated with atmospheric changes discussed above. That necessarily means that increases must come via the better application of technology and knowledge. There is more opportunity to increase total average yield by raising the lowest yields, rather than by increasing the potential. Timsina and Connor (2001) noted that in the rice-wheat rotational system in the Indo-Gangetic Plain area of northern India, that "severe biological and/or technological limitations to productivity and … potential for substantial yield increase provided the environmental and management constraints can be identified and rectified, are evident". In other words, a substantial yield gap exists. The yield gap is the difference between what can be achieved in crop production systems (under climatic constraints), and what is achieved in practice. Much of the predicted increase in wheat yields is likely to be a consequence of the application of better knowledge of the causes of the yield gap and finding solutions for them.

16.5 The Protein Gap

An inevitable requirement for higher production will be increased use of nitrogen fertilizer. This is because of two reasons. The first is that maintaining the yield response to elevated CO_2 requires higher levels of the supply of Rubisco. The reasons for this are unclear but are possibly linked to the enhanced turnover of the CO_2 fixing enzyme as CO_2 level increases. A second reason is that elevated CO_2 on its own leads to an increase in crop C/N ratio (Triboï et al. 2003), meaning that the nutritional quality, in terms of grain protein will decrease in the absence of extra N. The link between climate change and food quality remains one of the main under researched areas in global change studies, yet is one of the issues most directly relevant to human needs. As new wheat varieties have been bred for improved yield, there has been a tendency for protein content to diminish. High yields call for a long-lived canopy, but most of the N in green tissue will end up as part of the storage protein in grain (Jamieson and Semenov 2000). So there is a tension between the physiologies of yield and protein formation. Recent work (Martre et al. 2003) has shown that transfer of N from vegetative to grain tissue is largely source regulated. That means that breeding attention to increase yield and quality, traditionally focused on the grain, must be directed to manipulating the N pools within the plant. Martre et al. (in press) recently showed, using simulation, that the dual aims of simultaneously increasing crop yield and protein content can be achieved by making the pool of labile N within wheat plants larger. Nevertheless, the building of protein requires a supply of N, and high yields coupled with high protein will always require N fertilizer. Such use will not make N pollution of waterways inevitable. Providing the tools so that farmers can target their N applications to periods of high demand can avoid that problem.

Over the ten years of its existence the study of managed ecosystems within GCTE has identified and been able to predict the responses of wheat to CO_2, N, H_2O and temperature alone or in combination and their effects on yield per unit area. We have identified the importance of changes in climatic variability for wheat growth and explained it as the interaction between nonlinear processes in crops and temperature thresholds. We have made progress in spatialising such predictions and also including the environmental effects of different production systems. We have started to examine and model the effects of environment on crop quality, particularly protein content and composition. The consequences of climate change for the technical, nutritional and environmental quality of food has barely been raised as a question for agricultural science to date and this situation needs to change. Finally, we have started to identify how cropping systems may respond to climate change and also may contribute to it via how they are managed. The most globally important of these is the rice-wheat system of Asia.

16.6 The Rice-Wheat System

The double or sometimes triple, cropping rice-wheat system of Asia is a regional cultivation system that is responsible for the food security of between a quarter and a third of humanity. The central Indian region of the rice-wheat system is alone responsible for the food, security and livelihood of 1.1 billion people in India, a value that is predicted to rise to about 1.5 billion by 2050. In its full extent the rice-wheat cropping system covers a region from western Pakistan to eastern China, covering 15° of latitude and more than 40° of longitude (Fig. 16.5, Timsina and Connor 2001). The major producers, with a

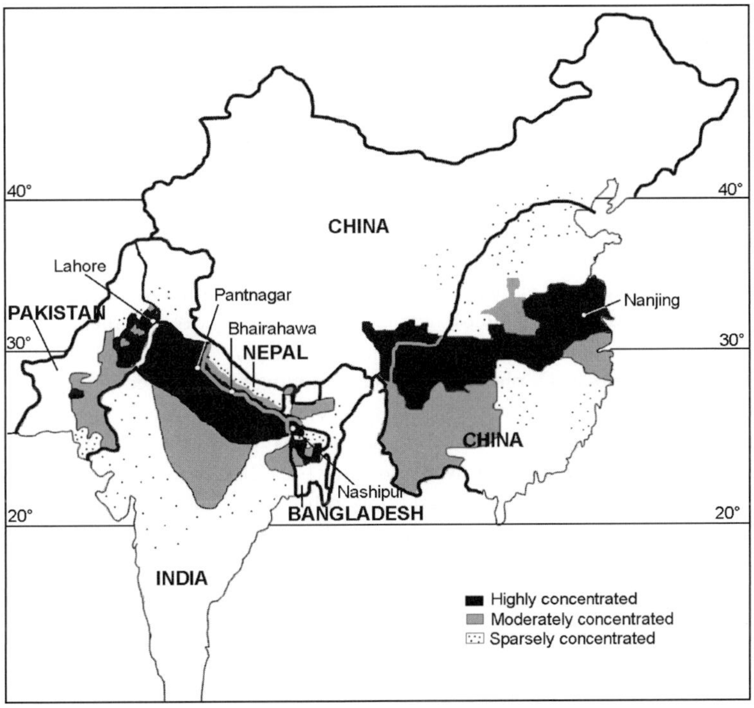

Fig. 16.5.
Distribution of the rice-wheat production areas in South Asia and China. The *curve* passing from northeast to southwest China represents the limits for growing rice-wheat sequences in China (Timsina and Connor 2001)

total area of around 23 Mha, are India and China, with *ca.* 10 and 13 Mha, respectively. Other countries in the region have rice-wheat areas that are at most 10% of the combined Indian and Chinese total. Food from rice-wheat systems contributes between 55% (China) and 95% (Bangladesh) of the national calorie intake of countries in the region (Timsina and Connor 2001). Yields of both wheat and rice have increased from the extremely low levels (ca. 1 Mg ha^{-1}) of the 1960s such that mean national yields for rice range from 2.4 Mg ha^{-1} in Nepal to 6.3 Mg ha^{-1} in China. Wheat yields less than rice and national values range from 1.6 Mg ha^{-1} in Nepal to 3.9 Mg ha^{-1} in China. National yields are in all cases superseded by yields from long-term trials in national research institutes, such that with applications of moderate amounts of nitrogen (120 kg N ha^{-1}), rice yields can increase by 50–70% and wheat yields by more than 30% (Timsina and Connor 2001). However, the trend over recent years has been for yield increases to stagnate or even decline, even in research plots (Grace et al. 2000a). Two issues considered by GCTE Focus 3 have been the impact of climate change on the rice-wheat system and the extent to which the rice-wheat system contributes to climate change via emissions of the major GHG gases, CO_2, N_2O and CH_4.

16.7 The Effect of Climate Change on the Rice-Wheat System

The medium to long-term viability of the rice-wheat system is very susceptible to global climate change for, at least, two reasons. The first is the dependence of the system on the monsoon rains that fall from June to August and provide about 600 mm (China) to 1 200 mm (India) of water that recharge the soil for rice cultivation in the autumn. Any changes in the spatial or temporal pattern of monsoon rains would be critical for production. The second reason for concern is high and rising temperatures associated with climate change. Long-term annual temperature records from sites in India and China show that temperatures in the late spring and early summer are already close to critical values for physiological processes in wheat and rice (Fig. 16.6). For example, temperature above 32 °C at anthesis induces anther sterility in wheat and rice and, as C_3 plants, both rice and wheat show large decreases in photosynthetic rates and elevated respiration and water use as temperatures rise above 35 °C (Porter and Gawith 1999). Given even the low (1–2 °C) IPCC (2001) predictions of mean global temperature increase by 2100 via the enhanced greenhouse effect, this will push temperatures, particularly in the wheat phase of the rice-wheat system, closer to known physiological thresholds for these crops. Pessimistic predictions of 5–6 °C increases in mean global temperature will mean that these are crossed. Thus, Gregory et al. (1999) concluded that wheat yields in the system are unlikely to increase by more than 10% for a doubled pre-industrial CO_2 concentration, with 5–7% being more likely under current management practices, and the negative effect of raised temperatures on crop duration and thereby yield. Major rice models concur in predicting a 5% decrease in yield per °C increase in temperature above 32 °C as a result of spikelet sterility.

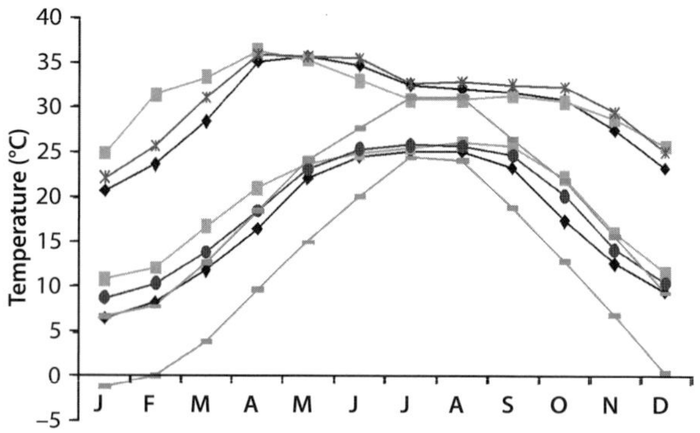

Fig. 16.6. Long-term temperature averages for a range of rice-wheat sites in south Asia and China (Timsina and Connor 2001)

In conclusion, the rice-wheat system is coming under ever increasing pressure as a result of moribund yields coinciding with population increase and income growth while, at the same time, the agro-ecological resource base is eroding, via lowering of ground water levels and decline in soil fertility. Global climate change that leads to warmer temperatures and more variable weather patterns can be expected to inflict significant limitations on the productivity of the rice-wheat system.

16.8 The Contribution of the Rice-Wheat System to Climate Change

A second important global issue is the impact that the rice-wheat system has on global biogeochemical cycles of C and N and the environment via its emissions of the GHG of CO_2, methane and nitrous oxide. Of course, these questions are linked in that crop management, such as the burning of crop residues, both reduces soil carbon levels while promoting greenhouse gas emissions. Future consideration of global land processes need to acknowledge the capacity that intensity of cultivation has to affect biogeochemical cycles (Grace et al. 2000a,b; Gregory et al. 2002). The direct cause of climate change is primarily the enhanced greenhouse effect. Different gases persist for different lengths of time in the atmosphere, and they also trap heat to different degrees. The global warming potential (GWP) is used to compare the effectiveness of each greenhouse gas to trap heat in the atmosphere relative to some standard gas, by convention CO_2. The GWP for CH_4 (based on a 100-yr time horizon) is 21, while that for N_2O is 310. Although present in concentrations lower than either CO_2 or CH_4, N_2O is a very potent greenhouse gas, accounting for about 6% of the enhanced greenhouse effect. By using GWP values for CH_4 and N_2O emissions of 21 and 310, respectively, it is possible to convert all our emission estimates to either a CO_2 or carbon equivalent (CE) basis. Carbon dioxide equivalents can be calculated by multiplying the CE values by 3.7.

Increased food-grain production in the rice-wheat system over the past 40 years has arisen mainly from the intensification of production rather its extensification, since almost all the productive arable land had been used prior to the Green Revolution. The intensification practices have included mechanical tillage of the soil, earlier sowing, increased use of inorganic fertilizer and irrigation. Crop residue burning is also used as a disease control practice. To a greater or lesser extent such management practices have an impact on the emissions of GHGs and estimates can be made of the emissions of each of the GHGs.

16.9 Carbon Dioxide

The main source of CO_2 to the atmosphere in the rice-wheat system is through tillage. Two on-site tillage sources exist: the biological decomposition of soil organic matter and the production of CO_2 as a by-product of machinery fuel usage. The burning of crop residues is not considered a CO_2 source to the atmosphere because an amount of CO_2 equivalent to the amount released will be taken up by the following crop; in other words, on an annual basis, there will be only small changes in standing C stock as a result of this activity. Carbon dioxide from this source is therefore not counted in the accounting procedures adopted by IPCC.

During tillage, soil aggregates are broken, thus increasing oxygen supply and surface area exposure of organic material. This promotes the decomposition of organic matter. An example of the effect of three to four cultivations per year on soil organic C (SOC) content can be drawn from CIMMYT's long-term wheat-maize trial at El Batán in central Mexico (Grace 2000 unpublished). Initial levels of SOC in the top 20 cm were 1.37%. After 8 yr of continuous wheat, with residue retention and conventional tillage, SOC had declined to 1.12%, a total decline of nearly 6.5 Mg C ha^{-1}, or 800 kg ha^{-1} yr^{-1}. In the adjacent no-till treatment, where yields were similar, the total decline was only 3.6 Mg C ha^{-1}, or 450 kg ha^{-1} yr^{-1}, a saving of 350 kg ha^{-1} yr^{-1}. In cultivated ecosystems, the rate of decline (or accumulation) of SOC therefore depends on several interacting factors: the amount of residue retained, the quality of residue retained, soil temperature and moisture interactions, tillage, and a reduction in fallow periods. For residue quality, high-quality (i.e., high N) residues from grain legumes or fodders can accelerate the decomposition of soil organic matter because of the additional N being available to the microbial population. Usually, the additional N being released will also be taken up by the following crop, thereby enhancing yields and biomass production; however, where high temperatures coincide with high water-holding capacity, the conditions are ideal for the promotion of SOC decomposition and the emission of CO_2. Soils therefore are generally considered to be a source of CO_2, unless the inputs outweigh the outputs, which is not usually the case in conventionally tilled low- to medium-input production systems. Fossil fuel energy sources form a large component of the total emissions from the rice-wheat system. Heavy use of diesel fuel occurs with current land preparation practices. For every liter of diesel fuel consumed, 2.6 kg of CO_2 are released to the atmosphere. Assuming that 150 l ha^{-1} yr^{-1} of fuel are used for tractor usage and irrigation pumping in conventional systems, this would amount to nearly 400 kg of CO_2 being emitted. For the entire 25 Mha under rice-wheat, this amounts to 12.0 Mg CO_2 yr^{-1} or 3.2 Mt C equivalents (MMTCE). To put this in perspective, this value is approximately three-quarters of the CH_4 emissions (expressed as MMTCE) from rice paddy fields in the entire IGP. Diesel use is a greatly underestimated source of GHGs in agriculture in general and the IGP in particular. An off-site source that is also often neglected is the production of CO_2 in the manufacture of fertilizers (Schlesinger 2000). Although not strictly relevant to on-farm environmental effects, 0.58 mol of C is produced per mol of N fixed in fertilizer production, resulting in 1.8 kg CO_2 produced per kilogram of N fertilizer manufactured. Full cost accounting that includes fertilizer transportation and application would raise this value to 1.4 mol of C per mol N fixed. In a no-till cropping system, this cost can significantly negate the mitigation gain by soil C sequestration.

16.10 Methane

Methane is produced in soil during microbial decomposition of organic matter and reduction of CO_2 under strictly anaerobic conditions. Methane emissions are regulated by two microbial processes: CH_4 production by methanogenic bacteria and CH_4 oxidation by methanotrophic bacteria. Continuously flooded rice fields are a source of CH_4 because anoxic conditions that favor methanogenesis predominate, and rice plants serve as conduits for its release to the atmosphere once it is formed in sediments (Neue et al. 1997). Under the UNDP-IRRI Inter-Regional Programmem, CH_4 emissions from rice fields were measured at eight locations in five rice-producing countries covering the main rice ecosystems – irrigated, rainfed, and deepwater systems (Wassmann et al. 2000a,b). Irrigated rice fields had the highest emission rates. Seasonal emissions varied from 1 to 50 g m^{-2}, with continuously flooded fields emitting the most (Wassmann et al. 2000a). Depending upon local precipitation, emissions from rainfed rice fields may be less than 50% of the emissions from irrigated fields. On the basis of results obtained by various rice-growing countries, IPCC has suggested that emission reduction factors under single and multiple aeration are 0.5 and 0.2 of those in continuously flooded fields. Using IPCC accounting methodology and the country default value of 10 g m^{-2}; CH_4 emissions from rice cultivation in the whole of India (45 Mha) are calculated to be 2.5 Mt yr^{-1} (14.3 MMTCE). In the 25 million ha of rice-wheat; this would equal about 1.8 Mg CH_4 yr^{-1} (or 6.5 MMTCE yr^{-1}). The main reason for relatively low CH_4 emissions from rice fields in India is that the soils of major portions of rice-growing areas have very low organic C or receive very little in the way of organic amendments (Jain et al. 2000). Incubation studies have shown that large differences in CH_4 production potential are related to organic C content (Majumdar et al. 1998). The use of organic manuring is also not very common in India, although where it occurs it can affect methane emissions (Debnath et al. 1996). The burning of crop residues also contributes to the global CH_4 budget. For each tonne of crop residue burned, 2.3 kg CH_4 are emitted, equivalent to 48 kg CO_2, using a GWP of 21. If we assume average annual residue production for rice and wheat on the IGP to be 10 Mg ha^{-1}, then, if one-half of the 25 Mha under rice-wheat is burned, a total flux of about 0.3 Mg CH_4 is emitted. This is equivalent to 50% of the total CH_4 emitted from paddy rice fields in the IGP, that has a total 1.8 Mg CH_4, as outlined above. Thus the total emission of CH_4 from the rice-wheat system is about 2.1 Mg equivalent to about 44 Mg CO_2 in terms of GWP.

16.11 Nitrous Oxide

Both fertilized and unfertilized soils emit N_2O. Although nitrogenous fertilizer is a source in the case of fertilized soils, the built-in N of the soil itself also contributes to the release of this GHG. It has long been recognized that soil water content and the availability of C enhance the production of N_2O provided a suitable NO_3 source is available. The total fluxes of N_2O have been reported to be many times higher in organic soils and mineral soils with higher contents of soil organic matter (Terry et al. 1981). The primary effect of water on N_2O production in aerobic and partially aerobic soils is the restriction of oxygen levels, which produces anaerobic conditions. Because of the relatively slow diffusion of oxygen through water (10 000 times slower than through air), denitrification in soil is highly affected by the relative amount of air- and water-filled porosity (Robertson 2000). Generally, an increase in denitrification and potential N_2O losses is observed following irrigation and precipitation. However, rice paddies are not considered an important source of atmospheric N_2O because N_2O, as the intermediary product of denitrification, would be further reduced to N_2 under the strong anaerobic conditions found in paddies (Granli and Bockman 1994). Thus, nitrous oxide is not usually detectable during the normal growing season of flooded rice (Bronson et al. 1997a,b) but is significant after mineral N fertilizer applications. During the rice-wheat dry season, Bronson et al. (1997a,b) reported N_2O fluxes that were 2.5 times higher with ammonium sulfate than with urea. The N_2O flux also increased sharply during the drainage period at mid-tillering until reflooding, when it dropped back to near zero. Experiments in rice fields conducted at the Indian Agricultural Research Institute (IARI) have shown that, as the redox potential becomes positive, CH_4 emissions decrease but N_2O emissions increase. There is much uncertainty about the amount of N_2O released; however, it is suspected that 0.1 to 6% of applied N is lost as N_2O. IPCC methodology assumes that 1.25% of fertilizer N is lost as N_2O. Crops such as wheat and maize, which receive nitrogenous fertilizers, are expected to release more N_2O than crops grown in flooded environments. The burning of crop residues also contributes to the global N_2O budget. For each tonne of crop residue that is burned, 40 g N_2O are emitted, equivalent to 12 kg CO_2, using a GWP of 310. If we assume that the average annual residue (rice plus wheat) production on the IGP is 10 Mg ha^{-1}, then, if one-half of the 23 million ha under rice-wheat is burned, a total flux of 4 000 t of N_2O is possible; that is, about 1.2 Mt CO_2 equivalent is emitted.

16.12 Comparative Greenhouse Gas Budgets for Rice-Wheat Farming Systems

The management of the rice-wheat system in the IGP is extremely diverse in terms of the amount of N applied and the rotation and tillage options that exist. As such, the GHG emission and production potential of this system is influenced by management intensity in terms of the quantity of inputs. Emissions and productivity both depend on both inputs (inorganic N, organic manure) and tillage. One useful statistic of the efficiency of resource use is overall productivity in relation to the amount of GHG emissions. To make this comparison, a carbon/productivity ratio (CPR) (Grace et al. 2000b) is used to assess how environmentally efficient various production systems are with respect to GHG emissions and how much food is placed on the table. Systems with higher CPRs are more inefficient at producing food with respect to the damage to the global environment. The CPR for Latin America, Africa, and Asia for the 30 yr of the Green Revolution was 0.52 (Grace et al. 2000b).

As an example of such a calculation for the rice-wheat system, data were analysed from a long-term rice-wheat trial in northern India at Pantnagar (Ram 2000).

Table 16.1. Agronomic details (yields, applied N fertilizer and manure and soil organic C) of a 20-yr irrigated rice–wheat–cowpea trial at Pantnagar, Uttar Pradesh, India (Ram 2000)

System	Rice ($Mg\ ha^{-1}$)	Wheat ($Mg\ ha^{-1}$)	Fodder ($Mg\ ha^{-1}$)	N fertilizer per crop ($kg\ ha^{-1}$)	Manure ($Mg\ ha^{-1}$)	Soil organic C (%)
Control	3.74	1.71	1.86	0	0	0.49
Trt. 1: 50%	5.02	3.13	1.93	60	0	0.64[a]
Trt. 2: 10%	5.67	3.97	2.36	120	0	0.84
Trt. 3: 150%	5.92	4.38	2.32	180	0	1.06[a]
Trt. 8: FYM	6.41	4.6	2.47	120	15	1.48

[a] Estimated for 0–15 cm based on published measurements in other treatments.

Table 16.2. Estimated annual (on-site) emissions of greenhouse gases (GHG) from the irrigated rice–wheat system in the Indo-Gangetic Plains based on 20-yr long-term trial data of Ram (2000). This example assumes conventional tillage with all crop residues being retained

GHG source	GHG emissions ($kg\ ha^{-1}$)				
	Control	Trt. 1	Trt. 2	Trt. 3	Trt. 8
CO_2: Soil tillage	3539	3003	2288	1502	0
CO_2: Diesel[a]	260	260	260	260	260
N_2O: Burning residues	0	0	0	0	0
N_2O: Manure application[b]	0	0	0	0	3
N_2O: N fertilizer application[b]	0	2.4	4.7	7.1	4.7
N_2O: Cereal residues retained[b]	0.1	0.1	0.1	0.1	0.1
N_2O: N-fixing crops[c]	1.1	1.1	1.4	1.4	1.5
CH_4: Burning residues	0	0	0	0	0
CH_4: Rice cultivation[d]	100	100	100	100	200
CE[e]	3496	4103	4721	5232	7137
CPR[f]	0.64	0.50	0.49	0.51	0.64

[a] 2.6 kg CO_2 released per l of diesel consumed and 100 l yr^{-1}.
[b] 1.25% of applied N.
[c] 1.25% of fixed N.
[d] 10 g $CH_4\ m^{-2}$, doubled if manure added.
[e] Carbon equivalents, global warming potentials (GWPs) of 21 and 310 used for CH_4 and N_2O, respectively.
[f] Carbon/productivity ratio, kg C emitted (equivalents) per kg of edible dry product (cereals).

This trial started in 1971 and its objective was to assess the effects of nutrient management (NPK) on productivity and soil fertility. The rice-wheat-cowpea trial consisted of 12 treatments with 50, 100, or 150% of the recommended dose of NPK fertilizer. The recommended dose of N was 120 kg ha^{-1} for each crop of rice and wheat. In the assessment, annual GHG budgets were constructed for individual gases (CO_2, CH_4, N_2O) for five of the 12 treatments (Table 16.1) on the basis of the activities outlined in the earlier part of this chapter. We included one treatment that incorporated farmyard manure.

In all budgets, we then converted this information to C equivalents to give an overall assessment of C lost to the atmosphere in relation to total edible product. Table 16.2 outlines the conventional tillage budget with all crop residues being retained, which we assume represents the experiment most closely.

The highest total emissions in terms of C equivalents were found in Treatment 8 (7 Mg C ha^{-1}), which included FYM. Treatment 2, which received the recommended dosage of fertilizer only, had emissions of 4.7 Mg C ha^{-1}, but it had the lowest CPR, which meant that it was the most efficient of all the nutrient management systems in this analysis of conventional tillage with residue retention. However, Treatments 1 and 3, which supplied 50 and 150% of the recommended dosage, respectively, had similar CPRs. In Treatment 3, the higher fertilizer N input may also result in leaching of excess fertilizer. In terms of recommendations, obviously Treatment 1, with the reduced N input (and hence reduced cost), would be the most suitable, particularly if we consider the off-site environmental costs (i.e., fertilizer production) discussed earlier. Tables 16.3 and 16.4 give GHG budgets where burning of all residues and no-tillage are alternative managements. The no-till scenario assumes complete retention of all crop residues and a hypothetical two-thirds reduction in on-site fuel use (cultivation only). Both scenarios exhibit the same trend as the earlier conventional residue-retained scenario in the final treatment analysis. The seemingly small amount of N_2O and CH_4 being emitted from residue burning in fact contributed an additional 15% to the overall C emission budgets of all treatments (ranging from 457 to 895 kg C ha^{-1} yr^{-1}). These increased emissions with residue burning exceed any benefit of moving from conventional to no-tillage,

Table 16.3. Estimated annual (on-site) emissions of greenhouse gases (GHG) from the irrigated rice–wheat system in the Indo-Gangetic Plains based on 20-yr long-term trial data of Ram (2000), assuming conventional tillage with all crop residues being burned

GHG source	GHG emissions (kg ha^{-1})				
	Control	Trt. 1	Trt. 2	Trt. 3	Trt. 8
CO_2: Soil tillage	3539	3003	2288	1502	0
CO_2: Diesel[a]	260	260	260	260	260
N_2O: Burning residues	0.4	0.6	0.6	0.7	0.7
N_2O: Manure application[b]	0	0	0	0	3
N_2O: N fertilizer application[b]	0	2.4	4.7	7.1	4.7
N_2O: Cereal residues retained[b]	0	0	0	0	0
N_2O: N-fixing crops[c]	1.1	1.1	1.4	1.4	1.5
CH_4: Burning residues	17	25.3	29.9	31.1	34.1
CH_4: Rice cultivation[d]	100	100	100	100	200
CE[e]	3953	4774	5510	6086	8032
CPR[f]	0.73	0.59	0.57	0.59	0.73

[a] 2.6 kg CO_2 released per l of diesel consumed and 100 l yr^{-1}.
[b] 1.25% of applied N.
[c] 1.25% of fixed N.
[d] 10 g CH_4 m^{-2}, doubled if manure added.
[e] Carbon equivalents, global warming potentials (GWPs) of 21 and 310 used for CH_4 and N_2O, respectively.
[f] Carbon/productivity ratio, kg C emitted (equivalents) per kg of edible dry product (cereals).

Table 16.4. Estimated annual (on-site) emissions of greenhouse gases (GHG) from the irrigated rice–wheat systems in the Indo-Gangetic Plains based on 20-yr long-term trial data of Ram (2000), assuming no-tillage and all crop residues being retained in this example

GHG source	GHG emissions (kg ha^{-1})				
	Control	Trt. 1	Trt. 2	Trt. 3	Trt. 8
CO_2: Soil tillage[a]	1769	1501	1144	751	366[h]
CO_2: Diesel[b]	86	86	86	86	86
N_2O: Burning residues	0	0	0	0	0
N_2O: Manure application[c]	0	0	0	0	3
N_2O: N fertilizer application[c]	0	2.4	4.7	7.1	4.7
N_2O: Cereal residues retained[c]	0.1	0.1	0.1	0.1	0.1
N_2O: N-fixing crops[d]	1.1	1.1	1.4	1.4	1.5
CH_4: Burning residues	0	0	0	0	0
CH_4: Rice cultivation[e]	100	100	100	100	200
CE[f]	2966	3646	4362	4981	6724
CPR[g]	0.54	0.45	0.45	0.48	0.61

[a] Assuming 66% fuel savings and soil organic C levels decrease only by 50% compared with conventional till.
[b] 2.6 kg CO_2 released per l of diesel consumed and 33 l yr^{-1}.
[c] 1.25% of applied N.
[d] 1.25% of fixed N.
[e] 10 g CH_4 m^{-2}, doubled if manure added.
[f] Carbon equivalents, global warming potentials (GWPs) of 21 and 310 used for CH_4 and N_2O, respectively.
[g] Carbon/productivity ratio, kg C emitted (equivalents) per kg of edible dry product (cereals).
[h] Assuming a C accumulation of 100 kg yr^{-1} in the topsoil.

which showed only a 9% reduction in the C emission budget. In other words, no-till must be complemented by residue retention. The use of FYM increased GHG emissions, and Treatments 1 to 3 showed similar CPRs. Again, the tendency to use a less N-intensive system would be desirable, but it is not essential provided N losses through leaching can be minimized, perhaps through better timing and placement of fertilizers (i.e., Jamieson et al. 2003). The judicial timing of application of fertilizers can produce yields similar to those from application of larger amounts. Crop simulation models developed during the lifetime of GCTE Focus 3 can mimic these systems in terms of water, N movement, and yield responses. They offer the opportunity to examine the bi-directional nature of the interaction between climate and the rice-wheat system.

16.13 Summary and Conclusions

Over the past ten years of GCTE, Focus 3 scientists have contributed to understanding of the effects of CO_2, temperature, water and nutrition on the growth and development of wheat and other crops. The effects of environmental factors have been looked at in isolation and in combination. Focus 3 scientists were among the first to realize the importance of climate variability as opposed to changes in mean climate for food production and quality and to model these effects. Challenges for the future as IGBP moves into its second phase include further modeling of cropping systems, modeling genomics and breeding strategies and genotype by environment by management relationships. Linkage to global programmes, such as GECAFS (Global Environmental Change and Food Systems), concerned with food systems and that concentrate on the effects that global environmental change on the interplay between biological and social aspects of food systems, define a clear role for the work of GCTE Focus 3 in the future. In terms of experiments, FACE systems that allow the interaction between gases such as O_3 and CO_2 need to be supported and used by the international wheat community. In agricultural systems including wheat production, non-CO_2 GHG emissions dominate (IPCC 2001) and their contributions in terms of CO_2 equivalent emissions need to be taken account of more than has been the case to date. Generally, soil organic carbon (SOC) levels in arable agricultural soils are lower than those found in undisturbed ecosystems as a result mainly of tillage and carbon removal in crops. Paradoxically, this means that a large potential C sink exists in agricultural soils that can be exploited as a mitigation option via agricultural management. The UK Royal Society (Royal Society 2001) concluded that the largest enhancement of the terrestrial carbon sink would result from changes in agricultural management (Fig. 16.7) that would include practices such as non-inversion tillage, residue return, rotational cropping and energy crops. The rate of increase of soil carbon, in all circumstances, would be very slow and finite as the SOC level moves to a new equilibrium that is potentially reversible but this depends greatly on the intensity of management (Schlesinger 2000; Smith et al. 2000). This means that management that increases SOC has to be maintained but with hopefully other benefits by way of improved soil physical and chemical quality.The vast majority of analyses of the effects of climate change on arable crops have concerned themselves with production, either per unit area or for a region. A major neglect has been studies of the influence and mechanisms by which climate change may affect crop and food quality, for either human or animal nutrition. Crop quality is a multi-faceted and complicated subject involving the growing, storage and processing pre- and post- harvest and quality has nutritional, technological and environmental definitions. For herbivores, it is possible to have a carbohydrate sufficiency but still suffer from malnutrition – as protein, mineral or vitamin deficiencies. From closed chamber experiments with wheat in which nitrogen, temperature and CO_2 level have been manipulated, we know that there is a strong weather signal in the determination of total protein content and, as importantly, its fractional composition that affects its nutritional and bread-making rheological properties (Martre et al. 2003). The effect of climate change on food quality will be an important issue for future research as we develop our perceptions of what constitutes a healthy and sustainable diet.

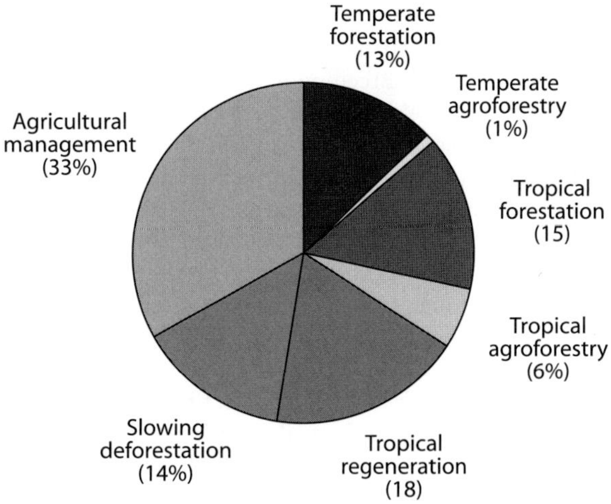

Fig. 16.7. Potential contribution of measures to increase soil carbon levels (Royal Society 2001)

Adaptation is the attempt to reduce or at least minimize the negative effects and maximise the benefits of climate change by changing management and agricultural practices, including breeding (Olesen and Bindi 2003). Management changes include earlier planting or sowing of crops that increases the length of the growing season thus increasing yield potential, provided moisture is adequate and the risk of heat stress is low. Combining the advantages of earlier planting with drought avoidance places emphasis on the importance of the crop phenological cycle in breeding. An adaptive strategy of a slowly developing vegetative period followed by rapid reproductive 'self-destruction' that transports nutrients and carbon to the grain sites as quickly as possible, thus avoiding stress, has been proposed for low latitude cereals. Profiting from increased atmospheric CO_2 for yield will mean an increase in demand for nitrogen by crops but climate change will also alter the magnitude and the balance of nitrogen losses as between leaching and gaseous losses. The links between the global carbon and nitrogen cycles will both affect the productivity of crops but will, in turn, be affected by the suite of gaseous emissions from arable and other forms of agriculture. Global warming will also likely lead to a higher incidence of pests and weeds and thus to a potentially larger use of biocides, although their use can be reduced via integrated pest management. Breeding and management have led to rises in crop yields of about 1–2% yr^{-1} since the 1960s, mainly because of changes in yield harvest index and the responsiveness of crops to nitrogen. The speed of breeding for higher yields and the release of new crop varieties is currently faster than the rate at which the climate is changing – about 0.01 °C increase in mean global temperature per year for the past several decades. It this appears that breeders may be able to stay ahead of climate change for sometime to come and incorporate climate change effects in their selections as part of the breeding process. In order to do this crop breeding needs to use both traditional and biotechnological techniques to allow the introduction of heat and drought resistant crop varieties.

Agriculture is the oldest human production industry that has withstood moderate changes in climate over the 10 000 years that humans have practiced it. It is probable that arable agricultural production can adapt to changes in mean global temperatures over the next century of the order of 1–2 °C (about 0.01 to 0.02 °C yr^{-1}) via breeding, selection and management. The agricultural genotype management environment envelope is flexible and spacious enough to accommodate small to medium changes in the weather signal for crops. However, if global warming results in mean temperature changes over the next century of 4–5 °C (i.e., 0.04 to 0.05 °C yr^{-1}) then it will be a case of global agriculture struggling to survive. Mitigation options from farming are substantial but are slow to implement and to have their effect. In the rich world, we need to keep our agricultural options open but our thinking should include a new vision for agriculture that seeks to change humankind's oldest production industry into its newest service industry. This means providing the ecosystem services of food, fiber and energy provision, biogeochemical regulation, biodiversity and soil conservation and cultural and aesthetic benefits for the landscape.

References

Aggarwal PK, Talukdar KK, Mall RK (2000) Potential yields of rice-wheat system in the Indo-Gangetic Plains of India. Rice-Wheat Consortium Paper Series 10, New Delhi, India, Rice-Wheat Consortium for the Indo-Gangetic Plains, 16 pp

Amthor JS (2001) Effects of atmospheric CO_2 concentration on wheat yield: review of results from experiments using various approaches to control CO_2 concentration. Field Crops Research 73:1–34

Batts GR, Wheeler TR, Morison JIL, Ellis RH, Hadley P (1996) Development and tillering responses of winter wheat (*Triticum aestivum*) crops to CO_2 and temperature. Journal of agricultural Science, Cambridge 127:23–35

Batts GR, Morison JIL, Ellis RH, Hadley P, Wheeler TR (1997) Effects of CO_2 and temperature on growth and yield of crops of winter wheat over four seasons. European Journal of Agronomy 7:43–52

Batts GR, Ellis RH, Morison JIL, Nkema PN, Gregory PJ, Hadley P (1998) Yield and partitioning in crops of contrasting cultivars of winter wheat in response to CO_2 and temperature in field studies using temperature gradient tunnels. Journal of agricultural Science, Cambridge 130:17–27

Bronson KF, Neue HU, Singh U, Abao Jr. EB (1997a) Automated chamber measurements of methane and nitrous oxide flux in a flooded rice. Soil. I. Residue, nitrogen, and water management. Soil Science Society of America Journal 61:981–987

Bronson KF, Neue HU, Singh U, Abao Jr. EB (1997b) Automated chamber measurement of methane and nitrous oxide flux in a flooded rice soil. II. Fallow period emission. Soil Science Society of America Journal 61:988–993

Conway G (1997) The Doubly Green Revolution. Food for All in the Twenty-First Century. Cornell University Press, 335 pp

Debnath G, Jain MC, Kumar S, Sarkar K, Sinha SK (1996) Methane emission from rice fields amended with biogas slurry and farmyard manure. Climate Change 33:97–109

Downing TE, Barrow EM, Brooks RJ, Butterfield RE, Carter TR, Hulme M, Olesen JE, Porter JR, Schellberg J, Semenov MA, Vinther FP, Wheeler TR Wolf J (2000) Quantification of uncertainty in climate change impact assessment. In: Downing TE, Harrison PA, Butterfield RE, Lonsdale KG (Eds) Climate Change, Climatic Variability and Agriculture in Europe. Environmental Change Unit. University of Oxford, UK, pp 415–434

Evans LT (1998) *Feeding the Ten Billion.* Cambridge University Press, UK

Evans LT, Fischer RA (1999) Yield potential: its definition, measurement, and significance. Crop Science 35:1544–1551

Ewert F, Rodriguez D, Jamieson PD, Semenov MA, Mitchell RAC, Goudriaan J, Porter JR, Kimball BA, Pinter PJ Jr., Manderscheid R, Weigel HJ, Fangmeier A, Fereres E, Villalobos F (2002) Modelling the effects of CO_2 and drought on wheat for different climatic conditions and cultivars. Agriculture Ecosystems and Environment 93:249–266

FAO (2003) Food and Agriculture Organization. http:// www.fao.org

Ferris R, Ellis RH, Wheeler TR, Hadley P (1998) Effect of high temperature stress at anthesis on grain yield and biomass of field-grown crops of wheat. Annals of Botany 82:631–639

Fuhrer J (2003) Agroecosystem responses to combinations of elevated CO_2, ozone, and global climate change. Agriculture, Ecosystems and Environment 97:1–20

Grace PR, Jain MC, Harrington LW (2000a) Environmental concerns in rice-wheat systems. Proceedings of the International Workshop on Developing an Action Program for Farm-level Impact in Rice-Wheat Systems of the Indo-Gangetic Plains. Rice-Wheat Consortium Paper Series 14, New Delhi, India, Rice-Wheat Consortium for the Indo-Gangetic Plains, 148 pp

Grace PR, Sanchez P, Ingram JSI, Palm C, Wassman R, Fisher M, Thomas R, Chandler F, Bowen W, Reid R, Woperies M, Waddington S (2000b) The consequences of international agricultural research on greenhouse gas emissions and global climate change. USAID, Washington, DC

Granli T, Bockman OC (1994) Nitrous oxide from agriculture. Norwegian Journal of Agricultural Science (Supplement 12). Agricultural University of Norway, Advisory Service, Ås, Norway

Grant RF, Kimball BA, Pinter Jr. PJ, Wall GW, Garcia RL, LaMorte RL, Hunsaker D (1995) Carbon dioxide effects on crop energy: testing ecosys with a free air CO_2 enrichment (FACE) experiment. Agronomy Journal 87:446–457

Gregory PJ, Ingram JSI, Campbell B, Goudriaan J, Hunt LA, Landsberg JJ, Linder S, Stafford-Smith M, Sutherst R, Valentin C (1999) Managed production systems. In: *The Terrestrial Biosphere and Global Change Implications for Natural and Managed Ecosystems. Synthesis Volume.* International Geosphere-Biosphere Program, Book Series 4, Cambridge University Press, UK, pp 229–270

Gregory PJ, Ingram JSI, Anderson R, Betts RA, Brovkin V, Chase TN, Grace PR, Gray AJ, Hamilton N, Hardy TB, Howden SM, Jenkins A, Maybeck M, Olsson M, Ortiz-Monasterio I, Palm CA, Payne T, Rummukainen M, Schulze RE, Thiem M, Valentin C, Wilkinson MJ (2002) Environmental consequences of alternative practices for intensifying crop production. Agriculture Ecosystems and Environment 88:279–290

Grossman S, Kartschall T, Kimball BA, Hunsaker D, LaMorte RL, Garcia RL, Wall GW, Pinter Jr. PJ (1995) Simulated responses of energy and water fluxes to ambient atmosphere and free-air carbon dioxide enrichment in wheat. Journal of Biogeography 22:3427–3445

Hendrey GR, Kimball BA (1994) The FACE program. Agricultural and Forest Meteorology 70:3–14

Howden SM, Reyenga PJ, Meinke H (1999) Global change impacts on Australian wheat cropping. Working paper series 99/04. Report to the Australian Greenhouse Office, 122 pp

Hunsaker DJ, Kimball BA, Pinter PJ Jr., LaMorte RL, Wall GW (1996) Carbon dioxide enrichment and irrigation effects on wheat evapotranspiration and water use efficiency. Transactions of the American Society of Agricultural Engineers 39:1345–1355

IPCC (2001) Climate Change 2001: The Scientific Basis. Contribution of Working Group I to the Third Assessment Report of the Intergovernmental Panel on Climate Change (eds) Houghton JT, Ding Y, Griggs DJ, Noguer M, van der Linden PJ, Dai X, Maskell K, Johnson KA. Cambridge: Cambridge University Press

Jain MC, Kumar S, Wassman R, Mitra S, Singh SD, Singh JP, Singh R, Yadav AK, Gupta S (2000) Methane emissions from irrigated rice fields in northern India (New Delhi). Nutrient Cycling in Agroecosystems 58:75–83

Jamieson PD, Porter JR, Goudriaan J, Ritchie JT, van Keulen H Stol W (1998a) A comparison of the models AFRCWHEAT2, CERES-Wheat, Sirius, SUCROS2 and SWHEAT with measurements from wheat grown under drought. Field Crops Research 55:23–44

Jamieson PD, Semenov MA, Brooking IR, Francis GS (1998b) Sirius: a mechanistic model of wheat response to environmental variation. European Journal of Agronomy 8:161–179

Jamieson PD, Berntsen J, Ewert F, Kimball BA, Olesen JE, Pinter PJ, Porter JR, Semenov MA (2000) Modelling CO_2 effects on wheat with varying nitrogen supplies. Agriculture Ecosystems and Environment 82:27–37

Jamieson PD, Armour T, Zyskowski RF (2003) On-farm testing of the Sirius Wheat Calculator for N fertiliser and irrigation management. Proceedings of the 11th Australian Agronomy Conference, 2–6 Feb. 2003, Geelong, Victoria. Australian Society of Agronomy

Kimball BA, Pinter PJ Jr., Garcia RL, LaMorte RL, Wall GW, Hunsaker DJ, Wechsung G, Wechsung F Kartschall T (1995) Productivity and water use of wheat under free-air CO_2 enrichment. Global Change Biology 1:429–442

Kimball BA, Pinter PJ Jr., Wall GW, Garcia RL, Hunsaker DJ, LaMorte RL (1997) Effects of elevated CO_2 on cotton and wheat as determined from FACE experiments. Journal of Agricultural Meteorology 52:787–796

Kimball BA, LaMorte RL, Pinter Jr. PJ, Wall GW, Hunsaker DJ, Adamsen FJ, Leavitt SW, Thompson TL, Matthias AD, Brooks TJ (1999) Free-air CO_2 enrichment and soil nitrogen effects on energy balance and evapotranspiration of wheat. Water Resources Research 35:1179–1190

Kimball BA, Morris CF, Pinter PJ Jr., Wall GW, Hunsaker DJ, Adamsen FJ, LaMorte RL, Leavitt SW, Thompson TL, Matthias AD, Brooks TJ (2001) Elevated CO_2, drought and soil nitrogen effects on wheat grain quality. New Phytologist 150:295–303

Kimball BA, Kobyashi K, Bindi M (2002) Responses of agricultural crops to free-air CO_2 enrichment. Advances in Agronomy 77:292–368

Majumdar D, Kumar S, Jain MC (1998) Methane entrapment in different rice soils of India, Current Science 75:951–955

Marathée JP, Gomez-MacPherson H (2001) Future world supply and demand. Chapter 43 in: Bonjean AP, Angus WJ (Eds) The World Wheat Book. A history of wheat breeding. Loivoisier Publishing, Paris. pp 1107–1116

Martre P, Porter JR, Jamieson PD, Triboi E (2003) Modeling grain nitrogen accumulation and protein composition to understand the sink/source regulations of nitrogen remobilization for wheat. Plant Physiology 133:1959–1967

Martre P, Jamieson PD, Semenov MA, Zyskowski R, Triboi E (2006). Modelling nitrogen uptake and redistribution and grain protein composition for wheat. European Journal of Agronomy, in press

Moot DJ, Jamieson PD, Henderson AL Ford MA, Porter JR (1996) Rate of change of harvest index during grain filling in wheat. Journal of agricultural Science, Cambridge. 126:387–395

Neue HU, Wassmann R, Kludze HK, Wang B, Lantin RL (1997) Factors and processes controlling methane emissions from rice fields. Nutrient Cycling in Agroecosystems 49:111–117

Olesen JE, Bindi M (2002) Review: Consequences of climate change for European agricultural productivity, land use and policy. European Journal of Agronomy 16:239–262

Pinter Jr. PJ, Kimball BA, Garcia RL, Wall GW, Hunsaker DJ, LaMorte RL (1996) Free-air CO_2 enrichment: Responses of cotton and wheat crops. In GW Koch and HA Mooney (eds.) *Carbon Dioxide and Terrestrial Ecosystems*. Academic Press, San Diego, CA 215–249

Porter JR Gawith M (1999) Temperatures and the growth and development of wheat: a review. European Journal of Agronomy 10:23–36

Porter JR, Semenov MA (1999) Climate variability and crop yields in Europe. Nature 400:724

Ram N (2000) Long-term effects of fertilizers on rice-wheat-cowpea productivity and soil properties in mollisols. In: IP Abrol et al. (ed) *Long-term soil fertility experiments in rice-wheat cropping systems*. Rice-Wheat Consortium for the Indo-Gangetic Plains, New Delhi, India, pp 50–55

Robertson GP (2000) Denitrification. In: *Handbook of Soil Science*, CRC Press, Boca Raton, Florida, USA, pp 181–190

Rosenzweig C Hillel D (2000) Soils and global climate change: challenges and opportunities. Soil Science 165:47–56

Royal Society of London (2001)

Schlesinger WH (2000) Carbon sequestration in soils: Some caution amongst optimism. Agriculture, Ecosystems and Environment 82:121–127

Semenov MA, Porter JR (1995) Climatic variability and the modelling of crop yields. Agricultural and Forest Meteorology 73: 265–283

Sinclair TR (1992) Mineral nutrition and plant growth response to climate change. Journal of Experimental Botany 43:1141–1146

Smith P, Powlson DS, Smith JU, Falloon PD, Coleman K (2000) Meeting Europe's climate change commitments: quantitative estimates of the potential for carbon mitigation by agriculture. Global Change Biology 6:525–539

Terry RE, Tate RL, Duxbury JM (1981) Nitrous oxide emissions form drained cultivated organic soils of South Florida. American Pollution Control Association Journal 31:1173–1176

The role of land carbon sinks in mitigating global climate change, The Royal Society of London, UK 27 pp

Tilman D, Cassman KG, Matson PA, Naylor R, Polasky S (2002) Agricultural sustainability and intensive production practices. Nature 418:671–676

Timsina J, Connor DJ (2001) Productivity and management of rice-wheat cropping systems: issues and challenges. Field Crops Research 69:93–132

Triboï E, Martre P, Triboï-Blondel AM (2003) Environmentally-induced changes of protein composition for developing grains of wheat are related to changes in total protein content. Journal of Experimental Botany 54:1731–1742

Tubiello FN, Rosenzweig C, Kimball BA, Pinter Jr. PJ, Wall GW, Hunsaker D, LaMorte RL, Garcia RL (1999) Testing CERES-wheat with free-air carbon dioxide enrichment (FACE) experiment data: CO_2 and water interactions. Agronomy Journal 91: 247–255

Wassmann R, Neue HU, Lantin RS, Buendia LV, Rennenberg H (2000a) Characterization of methane emissions from rice fields in Asia. I. Comparison among field sites in five countries. Nutrient Cycling in Agroecosystems 58:1–12

Wassmann R, Neue HU, Lantin RS, Makarim K, Chareonsilp N, Buendia LV, Rennenberg HR (2000b) Characterization of methane emissions from rice fields in Asia. II. Differences among irrigated, rainfed, and deepwater rice. Nutrient Cycling in Agroecosystems 58:13–22

Wollenweber B, Porter JR, Schellberg J (2003) Lack of interaction between extreme high-temperature events at vegetative and reproductive growth stages in wheat. Journal of Agronomy and Crop Science 189:142–150

Chapter 17

Pests Under Global Change – Meeting Your Future Landlords?

Robert W. Sutherst · Richard H. A. Baker · Stella M. Coakley · Richard Harrington · Darren J. Kriticos · Harald Scherm

17.1 Introduction

The term global change embraces a range of natural and anthropogenic environmental changes that are occurring around the world (Vitousek 1994), including changes in atmospheric composition, climate, land use and land cover, and the prevalence of non-indigenous species. This synthesis describes contemporary approaches to the assessment of the vulnerability of ecosystems and human societies to plant pests (arthropods, plant pathogens and weeds that are injurious to plants) in relation to global change. Animal and human health have been covered elsewhere in recent reviews (Sutherst 2001, 2004). While the detailed impacts of global change on the status of pests will only be known as events unfold, it is already clear that some pests will be able to invade new areas and become increasingly problematic for the maintenance of biodiversity, the functioning of ecosystems and the profitability of crop production. Some pests which are already present but only occur in small areas, or at low densities may be able to exploit the changing conditions by spreading more widely and reaching damaging population densities. Our objective is to discuss these potential changes in pest status and to present a framework for effective adaptation.

17.2 Methods

17.2.1 IPCC Processes

In addition to assessing the direct impacts of global change on agriculture and natural ecosystems, the global scientific community faces the task of understanding indirect impacts caused by pests of animals and plants. The numerous interacting species, the myriad of stakeholder interests, and great variation in the quality of available data make this a daunting prospect requiring generic approaches with a hierarchy of analytical tools (Sutherst et al. 1996). In order to develop a holistic approach we need to integrate the methodologies developed by the different research and policy communities into a comprehensive risk analysis framework. Pest risk analysis procedures developed by the International Plant Protection Convention (IPPC) (FAO 2004) provide an accepted standard for analysing the risks posed by pests to areas where they are not currently present or, if present, not widely distributed and under official control. At present, although assessments may focus primarily on impacts under current conditions, the risk analysis framework does allow the effects of global change to be taken into account (Baker et al. 2000; Sutherst 2000). In addition, although designed for invasive non-indigenous species, the procedures can also be used to analyze future threats posed by indigenous species.

The concept of vulnerability is used by the scientific and policy communities investigating the likely threats from climate change (Houghton et al. 1996; IPCC 2001a). *Vulnerability* is a measure of the potential *impacts* of a given change, taking into account the *adaptive capacity* available to respond to that change. The term avoids the consideration of risks in the absence of a natural or societal response, which can give an incomplete picture. *Impacts* are a combination of a change in *exposure* of ecosystems to pests and the *sensitivity* of the system to that change. *Adaptive capacity* depends on the adaptation technologies, cultural tools, and the infrastructure and resources available to implement appropriate management responses:

$$\text{Vulnerability} = \text{Impacts} \times (1 - \text{Adaptive capacity})$$

where

$$\text{Impacts} = \text{Exposure} \times \text{Sensitivity}$$

and

$$\text{Adaptive capacity} = (\text{Adaptation products and practices}) \times \text{Resources}$$

The general issues of adaptive capacity relative to climate change have been reviewed by Smith et al. (2003), who delve into both theoretical considerations and practical examples.

17.2.1.1 *Exposure*

A shift in the geographical distribution of a pest exposes new ecosystems to that pest for the first time. Climate change may enable such shifts to take place as pests track suitable climates. Immigration to new environments may occur through anthropogenic or natural pathways. Anthropogenic pest entry to new areas is already a major cause of invasion, and the numbers of species carried by traded commodities and with passengers are unlikely to be influenced directly by global change. However, entry by natural means may increase with global change if, for example, migrant species extend their range and move closer to threatened areas. While natural spread is generally undertaken by the pests themselves, movement of hosts or vectors with their associated pests may also be responsible for pest entry.

Following entry, a number of abiotic and biotic factors coupled with the intrinsic attributes of a species, e.g., its reproductive rate and adaptability, determine the success of establishment in the new area (Baker et al. 2003; FAO 2004). Climate is usually the first feature of the habitat considered in a risk assessment for pragmatic reasons, but other variables such as the spatial distribution of host plants also need to be considered. The extent to which exposure is affected by climate change will depend on the position of the particular habitat relative to the species' climatic envelope (Sutherst 2001). Populations closer to the edges of a species' range are likely to respond more quickly to climate change than populations near the center where conditions are closer to the optimum. Natural enemies of the pests may be left behind so the pest can proliferate unchecked (the enemy release hypothesis; Colautti et al. 2004), although predators and parasites may adapt to the invading species over time. Genetic observations can also contribute to understanding patterns of invasions by pests in new environments, as shown by studies of population structure of a non-pest fruit fly, *Drosophila subobscura*, in the Americas (Balanya et al. 1994; Noor et al. 2000).

Predictions of current and future distribution based on climate have been enhanced by the availability of monthly local and global gridded climatologies, e.g., the 1961–1990 global climatology interpolated to a 10' latitude/longitude grid (New et al. 2002). Where climate change scenarios are available at the same resolution as the baseline climatologies, predictions of changes in potential distribution under climate change can be made (Baker et al. 1998; Baker et al. 2003; Kriticos et al. 2003a).

17.2.1.2 *Sensitivity*

Sensitivity is the degree to which a system responds to a perturbation such as climate change (IPCC 2001a). Responses are often complex and non-linear due to a number of features of biological systems (Sutherst 1998) including thresholds (Scherm and van Bruggen 1994; Scherm 2004), discontinuities at the edges of species' ranges including hybrid zones (Barton and Hewitt 1989), multiplicative effects of population growth in multivoltine species (Sutherst 1983), the Allee effect (Stephens et al. 1999; Garrett and Bowden 2002; Taylor and Hastings 2005) and competition or predation (Silvertown 2004). A consequence of these effects is the need to augment empirical and descriptive approaches, such as statistical models, with dynamic and mechanistic simulation models (Sutherst 2001).

Once the sensitivity of the target system or species has been defined, a measure of the likely impact on the population can be derived, in the absence of any intervention. This provides a baseline measure of risk against which to estimate the benefits of alternative adaptation options.

17.2.1.3 *Impacts*

Both 'top-down' and 'bottom-up' approaches have been recommended for analyses of impacts of climate change (Parry and Carter 1998). The former is based on scenarios from global climate models (GCMs) while the latter uses sensitivity analyses (Sutherst 1998; Sutherst 2001). The two approaches have been combined into a risk management approach (Jones 2000). Integrated assessment frameworks are under development to bring some of these elements together, e.g., in a framework for analysing impacts on parasites (Sutherst 2001). The science needed to assess impacts of global change on pests is the same as for other species. However, the emphasis is different in that pests usually reproduce, disperse and colonize new habitats faster than many species of conservation value.

17.2.2 Monitoring, Benchmarks and Indicators for Measuring Impacts

We need *monitoring* data for pest-related *indicators*, covering long time-series in order to detect changes in ecology and impacts associated with global change. Such indicators need to relate to environmental, economic and social systems. We also need *benchmarks* to provide reality-checks for attribution of changes in status of populations.

The 1961–1990 climatology has become a standard from which to gauge climate change impacts, although, as climate change accelerates, its use as a baseline reflecting current climatic conditions is becoming less justifiable. Records of species presence need to include collection dates so that potential range expansion can be detected reliably. Records of absence are also useful. The difficulty of accurately determining the distribution of even well-known pest species (Yonow et al. 2004) suggests that more emphasis is needed on incorporating high-quality pest data into databases of species distributions.

A benchmark for the effects of temperature change *per se* is provided by the relationships with altitude and latitude (Linacre 1992). Each 1 °C increase in global temperature corresponds to a potential increase of ~170 m in a species' altitudinal range. As warming over the past century has been about 0.6 °C (IPCC 2001a), altitudinal range changes caused by global warming are unlikely to exceed 100 m. Similarly, in the absence of other factors, latitudinal shifts of about 1.7° (~200 km) can be expected for each 1 °C increase in temperature, corresponding to a shift of ~118 km with the observed temperature increase of 0.6 °C.

Whether changes of this order are achieved by a pest, i.e., whether the species is able to track temperature change, will depend on its rate of dispersal through fragmented habitats. In this context, it is worth drawing attention to the 'conservation-regulation' paradox. Those working on pests, i.e., species that require regulation, generally claim that, under global change, their species will become more abundant and widespread. By contrast, those working on species that require conservation generally state that their species will become less abundant and less widespread (Thomas et al. 2004). There are many exceptions on both sides, and some of the conclusions may be artefacts of the assessment tools employed (Thuiller 2003). Nonetheless, the paradox can be explained in part by the fact that the characteristics which tend to make an organism a pest, i.e., high reproductive and dispersal potentials, are the same ones that allow it to adapt to environmental change. For example, if the ability to migrate in fragmented habitats is impaired in plant species of conservation concern compared with weeds, climate change is likely to favor the weeds (Collingham et al. 2000).

Yamamura and Kiritani (1998) developed an indicator for tracking change in seasonal population dynamics of invertebrates. They devised a formula for the expected increase in the number of generations (voltinism) per year with a given temperature increase, based on values of the lower developmental threshold (T_0) and generation time (K). Using this index, those species with the largest increase in voltinism will be those with the lowest T_0 and the shortest K. Of 12 invertebrate orders examined, aphids are expected to show the greatest increase in voltinism with an extra two to three generations per 1 °C warming. Naturally, an increase in voltinism will not necessarily translate into larger population densities if, for example, required moisture conditions are not satisfied or if there is strong density-dependent parasitism, predation or lack of food.

The establishment of a global network of long-term monitoring sites will facilitate the development and implementation of indicators to monitor effects of global change. Examples of such measurements include the number of non-indigenous species in each country, the number of species extinctions, losses of agricultural production, and incidence of disease in humans (Pimentel et al. 2000b). In the context of biodiversity, suitable indicators include measures of physiological processes such as photosynthesis, metabolism and development rates, geographical distributions, phenology, and microevolutionary adaptation (Hughes 2000).

Perhaps most intuitive for policy makers is the expression of impacts in monetary terms. For agricultural crops, Oerke and Dehne (2004) gave potential yield losses for a range of commodities and regions. According to their synthesis, weeds currently cause losses of 32% with animal pests and pathogens causing 18 and 15%, respectively. Costs of pests in six countries (United States, UK, Australia, South Africa, India and Brazil) were estimated at US$336 billion, with about 9% of those representing costs of control. Extrapolated to the world the estimate was US$1.4 trillion. These figures would be multiplied several times if losses of biodiversity and ecosystem services in natural systems were included. How these costs will be impacted by global change is difficult to project, but Pimentel et al. (2000a) provide an example of how this can be estimated. Their estimate for a single global change driver (non-indigenous species) for a single country (the United States) suggests annual environmental and economic costs of nearly US$137 billion, of which almost 20% are due to non-indigenous plant pathogens and associated control costs. These figures may underestimate the problem (Perrings et al. 2002).

17.2.3 Estimating Impacts

A comprehensive assessment of impacts of pest risks under global change requires a conceptual framework that encompasses the pest and its interactions with physical and biological factors, including other species in the same or different trophic levels. It also needs a balanced program of field observations to help identify variables that affect species populations, experimental studies to elucidate mechanisms, and models to synthesize and test knowledge and understanding. Networks of researchers are also invaluable in the context of global change in order to facilitate sharing of information and comparative regional studies (Scherm et al. 1999; Hijmans et al. 2000). Some examples illustrate how these activities can contribute to ecological research to form the basis for the design of adaptation strategies for global change.

17.2.3.1 *Field Observations to Detect Population Changes*

Field observations are the grist of ecology and yet there is reluctance to fund such efforts due to the open-ended commitments necessary to obtain long time-series of observations. Many important revelations have been obtained by analysing such data. For example, long-term monitoring of the wheat stem rust pathogen (*Puccinia graminis* f. sp. *tritici*), initiated in the early 1900s in the central United States, clearly documented the effects of the deployment of specific host resistance genes and of the eradication of the alternate host, barberry (on which sexual reproduction occurs) on the pathogen's virulence dynamics (Roelfs 1982).

In community ecology, the changing correlations in rankings of abundance of each species and drifts in species compositions in communities over time provided insights into the natural behavior of species communities (Lawton 2000). Species in higher trophic levels were more sensitive to climate change (Voigt et al. 2003) suggesting that some communities may break down under global change.

There is increasing global scientific interest in developing Long-Term Ecological Research (LTER) sites and programmes, most of which are concerned with species of conservation interest. The most enduring long-term, standardised network for monitoring arthropod pest species is the suction trap network operated by the Rothamsted Insect Survey for studying aphids (Woiwod and Harrington 1994). The long run of data has enabled statistical relationships to be established between aphid phenology (e.g., annual time of first capture) and a range of climate and land-use variables. The time of first flight was strongly correlated with temperature for some species at some sites, with earlier first flights after higher winter temperatures. In the case of the important pest species *Myzus persicae* at Rothamsted, the relationship accounted for 80% of the variance.

Within the United States, several LTERs are now being used to study pest populations. In a 12-year warming experiment at a site in the Rocky Mountains of Colorado, Roy et al. (2004) documented an overall trend of increasing damage from herbivores and plant pathogens in plots with warmer temperatures and earlier snowmelt, although some pathogens and herbivores preferred the cooler plots. Results support the idea that the community composition within a particular area is likely to change with global warming. Other LTER sites in Alaska are currently being used to study how climate warming may impact pest occurrence under various temperature/moisture combinations (Mulder and Roy 2003). It would be prudent to include more climate impact studies in LTER sites in the future; assessing the effects of pest control will add to the value of biodiversity information being collected without significantly increasing the cost of maintaining these research sites.

Regardless of the species being monitored, there is a need for much more spatial distribution data coupled with geophysical measurements, including soil types, topography, vegetation type and cover and land use. Such measurements are particularly valuable if they are spatially explicit, cover a wider range of parameter values than measurements at any single location, and contain multiple combinations and permutations of environmental conditions (Sutherst 1998). In addition, spatial information is becoming valued due to the trend towards landscape-scale management of non-indigenous species (Jules et al. 2002).

17.2.3.2 *Empirical Data from Experimental Manipulations*

Reductionist experiments are traditionally employed to isolate each component of a system and examine its relative contribution. Such an approach has been used to dissect the effects of elevated CO_2 on the dynamics and intensity of anthracnose disease on the tropical legume *Stylosanthes scabra* (Chakraborty et al. 2000a; Chakraborty and Datta 2003; Pangga et al. 2004). Elevated CO_2 was found to reduce disease in growth chamber experiments by delaying pathogen spore germination, germtube growth and appressorium production and by inducing host resistance to anthracnose. In the field under elevated CO_2, however, these disease-reducing effects were counterbalanced by increased pathogen fecundity and by an enlarged canopy of *Stylosanthes* which trapped more pathogen spores and provided a more humid microclimate, leading to increased lesion numbers. Follow-up studies should include additional treatment levels in order to detect possible non-linear responses (Körner 2000).

In general, it is difficult to predict field responses from experiments involving changes in a few selected variables. For example, although the intrinsic growth rate of populations of the pest aphid *Aulacorthum solani* was enhanced in elevated CO_2 (Awmack et al. 1997a), a decrease in the insect's response to alarm pheromone could make it more vulnerable to attack by natural enemies, thus potentially offsetting the population increase (Awmack et al. 1997b). Experiments are needed to show how these and other changes interact.

Artificial ecosystems in controlled conditions, such as the Ecotron (Jones et al. 1998), can act as a useful intermediate step between laboratory experiments and the field and assist in the scaling-up process. Whole-system studies in controlled ecosystems provide insight into the complexity of simplified synthetic communities and so help to elucidate patterns and mechanisms behind specific responses of those communities to changes in environmental variables (Lawton 2000).

In one such laboratory-based study, interactions between three species of *Drosophila* fruit flies (*D. melanogaster*, *D. simulans* and *D. subobscura*) were studied in temperature clines with dispersal, with or without a hymenopteran parasitoid (*Leptopilina boulardi*) present (Landsberg 1989). As with other similar experiments, e.g., Davis et al. (1998), each of the treatments altered the balance of species numbers. The latter authors criticized the use of climatic mapping techniques (see below) on the basis of these laboratory results, and called for inclusion of population sinks and species interactions when trying to predict ranges under climate change. However, while recognising that species interactions are important in the field, Hodkinson (1999) and Baker et al. (2000) argued that experiments in confined spaces may overestimate species interactions since they do not provide species with the heterogeneous habitats that allow fauna such as these *Drosophila* spp. to coexist (Jenkins and Hoffmann 2001). A protocol has been described for ensuring that climate-based projections of species ranges do not overlook the roles of non-climatic variables (Sutherst 2003). Regardless, community-level impacts of global change will remain very uncertain due to their inherent complexity and multiple linkages.

Manipulation of whole systems in the field offers opportunities to study species interactions in more natural conditions. Free-Air CO_2 Enrichment (FACE) experiments (Hendrey 1992) have been particularly popular in global change research, but few have included pest components. A recent exception is a study on the interactive effects of CO_2 and ozone (O_3) on aspen trees infected with the rust fungus *Melampsora medusae* f.sp. *tremuloidae* (Percy et al. 2002; Karnosky et al. 2002). The results showed that O_3 predisposed the host to rust infection and that elevated CO_2 was unable to compensate for this effect. Exposure to O_3 or CO_2 increased the amount of hydrocarbons in leaf surface waxes, which aid host recognition by certain herbivorous insects; as a result, aphids were more abundant in trees treated with O_3 and/or CO_2 (Percy et al. 2002).

A complicating factor in controlled experiments is the potential for genetic adaptation to climate change. This is difficult to mimic in laboratory experiments that generally utilize populations of single strains or of a few strains with limited genetic diversity and do not investigate how populations may adapt as climate changes over long time periods with many generations. Indeed, there is now increasing evidence from field studies that some organisms can evolve rapidly in response to the strong directional selection pressures exerted by global change. For example, genetic adaptation to elevated temperature (Rodriguez-Trelles et al. 1998) and increased length of the growing season (Bradshaw and Holzapfel 2001) associated with global warming have been documented for certain insect species. The potential for genetic adaptation will be particularly important for host-parasite interactions, for which a changing climate could lead to temporal asynchrony between pest phenology and the availability of host tissue (Visser and Holleman 2001). In such cases, there will be a very strong selection pressure to maintain that close temporal synchrony (Dixon 2003).

17.2.3.3 *Modeling Tools*

Many modeling approaches have been applied to the assessment of risks from pests under global change because opportunities for large-scale experimentation are so limited. One group of models simulates the detailed population dynamics of species at a point scale. These tools have focused on software engineering to create modular and generic simulation models (Reynolds and Acock 1997; Sutherst et al. 2000). However, comprehensive, process-based models are inevitably complex and expensive to parameterise and so will be applicable only to the most intensively studied species or systems.

Simplified, inferential and dynamic simulation models (Sutherst and Maywald 1985; Sykes and Prentice 2004) can avoid the inherent problems of complex and over-parameterized models. They are also amenable to investigation of competitive interactions between species across geographical distributions. For example, in some cases where the CLIMEX model was used to project the potential range of a pest species in exotic habitats, strong evidence was found for biotic interactions truncating their ranges (Sutherst et al. 1985; Kriticos and Randall 2001; Wharton and Kriticos 2004). Inferential simulation tools that exploit the richness of spatial climatic data (Sutherst 1998) to infer the climatic responses that drive and limit species populations are highly informative, offering insights into the climatic factors that limit species distributions in different parts of their ranges (Yonow and Sutherst 1998; Kriticos et al. 2003a,b). Inference is less certain than deduction but complements mechanistic studies using traditional reductionist approaches. To date, models of combinations of species have only included species in additive form, such as the need for conditions to suit a weed and its biological control agent (Julien et al. 1995), crop tree and its arthropod pest (Wharton and Kriticos 2004), pathogen and host plants (Yonow et al. 2004), or both pathogen and vector in vector-borne disease transmission (Sutherst 2004). Recent progress has now included explicit species interactions within and between trophic levels (Sutherst et al. In press). This has allowed predation, competition and facilitation to be simulated on geographical scales. These advantages have been partially offset by the extra operational demands in estimating parameter values, a lack of probabilistic outputs, and difficulty in handling GCM climate change scenarios.

Another modeling approach consists of a heterogeneous collection of statistical, rule-based and simulation approaches with variations, such as neural networks (Berry et al. 2002) or genetic algorithms (Peterson et al. 2002), designed to predict potential changes in geographical distributions of species (Kriticos and Randall 2001; Scott et al. 2002; Cocu et al. 2005). Potential geo-

17.4 Adaptation

The IPCC has developed general guidelines for the design of adaptation measures to climate change, and these are applicable to other drivers of environmental change. Responses can be either *autonomous* (part of everyday business) or planned (requiring a new policy) (IPCC 2001b). Adaptive measures have also been classified as behavioral, engineering or administrative/legislative (Parry and Livermore 1999).

Adaptation measures rely on application of innovative approaches or products to prevent or control pests in a changing environment. The attributes of management tools that make them more likely to be available for adaptation to global change in the decades ahead are evaluated against specific objectives using criteria such as effectiveness and constraints (Parry and Carter 1998) or sustainability and robustness (Sutherst et al. 1998). Their performance needs to comply with the concept of 'triple bottom-line' accounting, i.e., social and environmental considerations need to be included in addition to economic values.

Ideally, responses to changes in pest incidence and severity under global change would be proactive rather than reactive. Unfortunately, the Earth's climate system is so complex and non-linear that we can never be sufficiently confident of predicting climate change to avoid surprises. Examples of so-called climate flip-flops (Calvin 1998), changes in regional rainfall (Hennessy et al. 1999) and temporal variation in the correlation of the Southern Oscillation Index (SOI) with tropical rainfall (Cai et al. 2001) illustrate the extent of the uncertainties. Rather than only designing adaptation responses strategically and well ahead of time, there is an additional need to respond quickly to unanticipated environmental changes. This calls for managers to be 'nimble' and respond rapidly to changes as they occur. Such a capability will have to be based on a thorough understanding of the ecology of the target pest or community of species in an ecosystem at each location. The most likely pathway to developing such a capacity is to adopt a knowledge-based, rather than a product-based, approach to pest management (Lewis et al. 1997) with sophisticated decision support systems for our major agricultural enterprises.

17.4.1 Natural Adaptations

Species rely on dispersal, phenotypic plasticity and genetic selection to respond to environmental change. In human-dominated landscapes, pests tend to disperse and be moved by humans more effectively than many less numerous species and so are less likely to be constrained from moving to track changing climates than rarer or more sessile species. Phenotypic plasticity is an important strategy for coping with a variable environment, such as the timing of onset of low temperatures in autumn (Taylor and Spalding 1989). If ecological responses are insufficiently plastic or are too slow to keep pace with the rate of environmental change, there is the potential for micro-evolutionary adaptations that are better able to exploit the new conditions. This has been documented for plants (Geber and Dawson 1993), a plant-associated mosquito (Bradshaw and Holzapfel 2001), and *Drosophila* (Rodriguez-Trelles et al. 1998). In addition, as species move with climate change they may be exposed to a new set of abiotic factors (e.g., soils and topography), as well as a modified community of competitors and predators, which will also exert new selection pressures.

17.4.2 Adaptive Management Options

Without foresight into future technologies, we can only assess the suitability of currently available practices. However, technologies, including those used in agriculture and ecosystem management, are rapidly evolving (Giampietro et al. 1999) and are not predictable. This is a major limitation, as it requires the attributes of existing and new technology to be extrapolated into the uncertain physical and biological environments of the future. While global climate models are able to generate scenarios of climate change, they can not yet be used to make predictions.

Legislative

Legislative or administrative options to adapt to environmental change include strengthening of barriers to entry of non-indigenous species, improved detection procedures for incursions, monitoring of pest abundance and geographical distribution (Baker et al. 2000), as well as enhanced knowledge-management using a hierarchy of information technologies such as expert systems, simulation models, remote sensing and GIS (Sutherst et al. 1996).

Engineering and Behavioral

Preventative measures based on environmental management are robust but depend on ongoing management inputs, which are easily downgraded in times of economic stringency. Selecting for plant species that are tolerant of drought and resistant to pests could be considered 'engineering' to the degree that biotechnology is used for genetic modification. Classic tools of breeding will also be needed to ensure plant and animal adaptability to climate change. Based on the results of an assessment of the effects of future climate change on sugar beet production in Europe, efforts are being made to incorporate appropriate disease tolerance into species that will be used increasingly in northern parts of the growing range (Jones et al. 2003).

Effective responses to ongoing or accelerating invasions by non-indigenous species require a multi-pronged approach that addresses sources, pathways and destinations of potential pests (Baker et al. 2000). The chances of such species leaving their source area can be reduced by attacking the pest populations using biological control agents at their source (Sands et al. 1993), or by strengthening inspection procedures to prevent them from entering the pathway to other countries. Prevention of international trade in horticultural plants with potentially weedy or insect-harbouring attributes (Lok 2001) necessitates a greater understanding by industry. As part of the Montreal process (http://www.mpci.org), the temperate and boreal forestry sector has adopted sustainability protocols that include features such as reductions in pesticide use. This has seen an attitudinal change to pests as well as an interest in issues such as crop species becoming weeds (e.g., wilding pines) and managed forests as sources of weed invasions into adjacent areas.

Biological control using natural enemies can be valuable in adapting to new invasions or threats. Restoration of habitats around agricultural fields can rectify the loss of natural vegetation which shelters generalist predators. Some predators have low mobility and so do not find their way into fields fast enough after each new crop to provide useful contributions to pest management. Provision of artificial shelters for generalist predators can be a useful technique for managing pests in field crops (Halaj et al. 2000).

To address the potential impacts of woody weeds on rangelands effective quarantine procedures need to be implemented to prevent weed spread, establish education programmes to assist in the early detection of new invaders, and develop more effective biological control programmes aimed at vulnerable life-stages of the weeds (Kriticos et al. 1999). Biologically-based adaptation measures have the desirable characteristic of low implementation costs per unit area.

17.4.3 Adaptation of Control Measures in Response to Global Change

Following the processes developed for increasing the sustainability of cities (AtKisson 1999) we need to address three issues in relation to adaptation:

1. Indicators of change are required in order to track pest movements and feed information back to inform the decision-making process. We have addressed this issue in relation to measuring and monitoring impacts.
2. Steps need to be taken to accelerate development and strengthen guidance of the use of new technologies and approaches to pest management. This will include knowledge-based decision support systems that measure outcomes in terms of the 'triple-bottom-line' of environmental, economic and social benefits.
3. Processes have to be implemented to accelerate the adoption of new approaches because global change is accelerating and society needs to fast-track adaptive management responses if it is to keep pace with the changes.

The ever-present risk of researchers finding solutions in isolation from those who are expected to adopt the technology can be averted by pooling the efforts of scientists, policy-makers, managers, advisers, teachers and students to develop adaptation strategies jointly. The use of modular simulation modeling networks as a vehicle for facilitating such activities has been encouraging in both developed and developing countries (Sutherst et al. 2000). It can form the basis of measures to inform the community and ensure its involvement in and ownership of adaptive measures, leading to accelerated adoption. Scenario-planning is another useful tool for generating and evaluating options to enhance sustainability (Schoute et al. 1995). This approach is easier for groups with no formal technical training to follow (Sutherst et al. 1999).

17.4.4 Threats to Sustainability of Adaptation Options

Pesticide Resistance

Synthetic pesticides are used widely to reduce damage to agriculture, and there has been a long history of development of resistant strains of arthropods, plant pathogens and weeds. Loss of susceptibility of pests to pesticides has the potential to reduce the capacity of communities to adapt to global change.

Entomologists concerned with protection of crops (Georghiou 1994) or livestock (Sutherst and Comins 1979) pioneered the development of resistance management strategies, and such strategies are now used widely in all disciplines dealing with pest management (Russell 1999; Vaughn 2003). Sutherst and Comins (1979) used the terms '*saturation*' and '*moderation*' to describe strategies aimed at either intensive control or less aggressive measures designed to reduce selection pressures. Pesticide resistance needs to be managed by avoiding a high-risk 'magic bullet' (*saturation*) approach and adopting multi-pronged measures (*moderation*).

Human Safety and Non-Target Effects

The widespread use of pesticides has had detrimental effects on both humans and animals. The early effects of

DDT on bird life are well known but the replacement chemicals were also associated with significant risks (Marco et al. 1987). Very low concentrations of some agricultural chemicals can mimic the activity of estrogen and thyroid hormones in humans and animals (Colborn et al. 1996). They are referred to as 'endocrine disrupting chemicals' and include organochlorine pesticides and some synthetic pyrethroids and herbicides (Krimsky 2000). Their disruptive effects threaten their availability for adaptation to global change. In the United States, all registered pesticides are under review under the Food Quality Protection Act (Ragsdale 2000) with an initial focus on active ingredients that have high mammalian toxicity, pose a threat to ground water resources, or have shown carcinogenic or teratogenic effects in laboratory animals.

17.5 Vulnerability

The vulnerability of affected natural and managed ecosystems to pests is assessed by combining impacts and adaptation with the adaptive capacity of the affected ecosystem. Regions or industry sectors vulnerable to pests under global change will be those that are unprepared or do not have suitable options to respond to major changes in pest status. Such changes are more likely to occur in regions adjoining endemic areas, where they rely on freedom from infestation for market access. An example is the Murrumbidgee Irrigation Area in south-eastern Australia where the citrus industry is focused on exports but has an ongoing problem with exclusion of the Queensland fruit fly. Under climate change, pest pressure is likely to increase in the temperature-limiting region, making eradication of incursions more difficult (Sutherst et al. 2000).

Tropical rangelands, where invasions with leguminous woody weeds are expected to occur with rising temperatures and concentrations of CO_2, are located mostly in less-developed countries. As such weeds spread into semi-arid rangelands, the low economic value of these lands limits human capacity to undertake adaptation measures. Economic options are restricted to biological control and in some situations, hygiene measures to stop the spread of seed by livestock. As biological control options are often limited by fear of non-target damage (Thomas and Willis 1998; Brimnen and Boland 2003) and by an apparent shortage of suitable agents, there are few options available. The problem is magnified by resource shortages and the absence of government control over animal movements in many developing countries. Indeed, all industry and public health sectors in developing countries are vulnerable to increases in impacts because they do not have the resources or infrastructure to respond effectively.

17.6 Summary and Conclusions

Pests have major effects on both natural and managed ecosystems. Their impacts are expanding as more species are spread around the world and often become an irreversible extra burden on the affected systems. Environmental change creates disturbances and the many pest species involved ensure that there will be some species available to exploit opportunities arising from disruption of native communities or vacant niches. Climatic volatility is a concern and recent predictions of the increasing frequency and intensity of extreme events demand greater attention. Environmental change results in shifts in species' geographical ranges, seasonal phenology, host preferences, population densities and population genetics. To date, results from limited experimental studies have shown a wide range of negative and positive responses by different species to the same environmental changes. This limits our ability to make generalizations about likely responses to global change. The problem is compounded by interactions between species in the same or different trophic levels in communities.

The resources necessary to enable researchers to provide managers with the information needed to adapt to global change include: baseline data on driving variables, indicators of change, strategies and tools to analyze problems and design adaptive management responses, and information delivery systems to accelerate adoption of new technologies or approaches to keep pace with the accelerating rate of change. Among the available management options, biological approaches based on either host resistance or natural enemies tend to be more robust because they have built-in flexibility and, in the most successful cases, avoid the need for extra contributions by human management, which is so often the weakest link.

There is a conundrum where strategic decisions on adaptation need to be made, but they require a sound understanding of the environmental, social and economic consequences. We support evidence-based policy and yet we acknowledge the great uncertainties associated with both the scenarios of environmental change and the responses of ecological systems to these changes. Further, we recommend *nimbleness* in adaptive management responses in order to respond to unexpected changes, and yet that relies on having sufficient understanding of the systems involved to make sound decisions. The two long-standing questions 'How much knowledge is enough?' and 'Is that knowledge attainable given the complexities of natural systems?' still have not been resolved.

References

AtKisson A (1999) Believing Cassandra. Chelsea Green Publishing Company, White River Junction, VT

Awmack CS, Harrington R, Leather SR (1997a) Host plant effects on the performance of the aphid *Aulacorthum solani* (Kalt.) (Homoptera: Aphididae) at ambient and elevated CO_2. Glob Change Biol 3:545–549

Awmack CS, Woodcock CM, Harrington R (1997b) Climate change may increase vulnerability of aphids to natural enemies. Ecol Entomol 22:366–368

Ayres MP, Lombardero MJ (2000) Assesing the consequences of global change for forest disturbance from herbivores and pathogens. Sci Total Environ 262:263–286

Baker R, MacLeod A, Cannon R, Jarvis C, Walters K, Barrow E, Hulme M (1998) Predicting the impacts of a non-indigenous pest on the UK potato crop under global climate change: reviewing the evidence for the Colorado beetle, *Leptinotarsa decemlineata*. Brighton Crop Protection Conference – Pests and Diseases, Vol. III. BCPC, Farnham, UK, pp 979–984

Baker R, Sansford C, Jarvis C, Cannon R, MacLeod A, Walters K (2000) The role of climatic mapping in predicting the potential geographical distribution of non-indigenous pests under current and future climates. Agriculture Ecosystems and Environment 82:57–71

Baker R, Cannon R, MacLeod A (2003) Predicting the potential distribution of alien pests in the UK under global climate change: *Diabrotica virgifera virgifera*. Proceedings of the British Crop Protection Conference – Crop Science and Technology, pp 1201–1208

Balanya J, Segarra C, Prevosti A, Serra L (1994) Colonization of America by *Drosophila subobscura*: the founder event and a rapid expansion. J Hered 85:427–432

Bale JS, Masters GJ, Hodkinson ID, Awmack C, Bezimer TM, Brown BK, Butterfield J, Buse A, Coulson JC, Farrar J, Good JEG, Harrington R, Hartley S, Jones TH, Lindroth RL, Press MC, Symrnioudis I, Watt AD, Wittaker JB (2002) Herbivory in global climate change research: direct effects of rising temperature on insect herbivores. Glob Change Biol 8:1–16

Barton NH, Hewitt GM (1989) Adaptation, speciation and hybrid zones. Nature 341:497–503

Berry PM, Dawson TP, Harrison PA, Pearson RG (2002) Modelling potential impacts of climate change on the bioclimatic envelope of species in Britain and Ireland. Glob Ecol Biogeogr 11:453–462

Bradshaw WE, Holzapfel CM (2001) Genetic shift in photoperiodic response correlated with global warming. Proceedings National Adacemy of Sciences U.S.A. 98: 14509–14511

Brasier CM (2001) Rapid evolution of introduced plant pathogens via interspecific hybridization. Bioscience 51:123–133

Brimnen TA, Boland GJ (2003) A review of the non-target effects of fungi used to biologically control plant diseases. Agriculture Ecosystems and Environment 100:3–16

Brown JH, Stevens GC, Kaufman DM (1996) The geographic range: size, shape, boundaries, and internal structure. Annu Rev Ecol Syst 27:597–623

Cai W, Whetton PH, Pittock B (2001) Fluctuations of the relationship between ENSO and northeast Austalian rainfall. Climate Dyamics 17:421–432

Calvin WH (1998) The great climate flip-flop. The Atlantic Monthly 281:47–64

Chakraborty S, Datta S (2003) How will plant pathogens adapt to host plant resistance at elevated CO_2 under a changing climate. New Phytol 159:733–742

Chakraborty S, Pangga IB, Lupton J, Hart L, Room PM, Yates D (2000a) Production and dispersal of *Colletotrichum gloeosporioides* spores on *Stylosanthes scabra* under elevated CO_2. Environ Pollut 108:381–387

Chakraborty S, von Tiedemann A, Teng PS (2000b) Climate change: Potential impact on plant diseases. Environ Pollut 108:317–326

Coakley SM, Scherm H, Chakraborty S (1999) Climate change and plant disease management. Annu Rev Phytopathol 37:399–426

Cocu N, Harrington R, Hulle M, Rounsevell MDA (2005) Spatial autocorrelation as a tool for identifying the geographical patterns of aphid annual abundance. Agricultural and Forest Entomology 7:31–43

Colautti R, Ricciardi A, Grigorovich I, MacIsaac H (2004) Is invasion success explained by the enemy release hypothesis? Ecol Lett 7:721–733

Colborn T, Dumanoski D, Myers JP (1996) Our Stolen Future: Are we Threatening our Fertility, Intelligence and Survival? A Scientific Detective Story. Dutton, New York

Collingham YC, Wadsworth RA, Huntley BH, Hulme PE (2000) Predicting the spatial distribution of non-indigenous riparian weeds: issues of spatial scale and extent. J Appl Ecol 37:13–27

D'Antonio CM, Vitousek PM (1992) Biological invasions by exotic grasses, the grass/fire cycle, and global change. Annu Rev Ecol Syst 23:63–87

Davis AJ, Lawton JH, Shorrocks B, Jenkinson L (1998) Individualistic species responses invalidate simple physiological models of community dynamics under global environmental change. J Anim Ecol 67:600–612

Dixon AFG (2003) Climate change and phenological asynchrony. Ecol Entomol 28:380–381

FAO (2004) Pest risk analysis for quarantine pests including analysis of environmental risks and living modified organisms. International Standards for Phytosanitary Measures. No. 11. FAO, Rome

Farquhar, G (1997) Where could Australia's forests move with change in atmospheric composition: some ideas from plant physiology and the paleo-record CSIRO and Bureau of Resource Science Canberra pp 1–7

Fleming RA, Tatchell GM (1995) Shifts in the flight periods of British aphids: a response to climate warming? In: Harrington R, Stork NE (eds) Insects in a changing environment. Proceedings of the 17th Royal Entomological Society Symposia on I, 7–10 September 1993, Harpenden, Hertfordshire. Academic Press, London, pp 505–508

Fuhrer J (2003) Agroecosystem responses to combinations of elevated CO_2, ozone, and global climate change. Agriculture Ecosystems and Environment 97:1–20

Garrett K, Bowden R, (2002) An Allee effect reduces the invasive potential of *Tilletia indica*. *Phytopathology* **92**, 1152–1159

Gavazzi M, Seiler J, Aust W, Zedaker S (2000) The influence of elevated carbon dioxide and water availability on herbaceous weed development and growth of transplanted loblolly pine (*Pinus taeda*). Environ Exp Bot 44:185–194

Geber MA, Dawson TE (1993) Evolutionary responses of plants to global change. In: Kareiva PM, Kingsolver JG, Huey RB (eds) Biotic Interactions and Global Change. Sinauer Associates, Sunderland, MA, pp 179–197

Georghiou GP (1994) Principles of insecticide resistance management. Phytoprotection 75: Suppl. 51–59

Giampietro M, Pimentel D, Bukkens SGF (1999) General trends of technological changes in agriculture. Crit Rev Plant Sci 18:261–282

Gregory PJ, Ingram JSI, Campbell B, Goudriaan J, Hunt LA, Landsberg JJ, Linder S, Stafford-Smith M, Sutherst RW, Valentin C (1999) Managed productions systems. In: B. Walker, W. Steffen, J. Canadell, J. Ingram (eds) The terrestrial biosphere and global change. Implications for natural and managed ecosystems. Cambridge University Press, London, pp 229–270

Halaj J, Cady A, Uetz. GW (2000) Modular habitat refugia enhance generalist predators and lower plant damage in soybeans. Environ Entomol 29:83–393

Harrington, R (2003) Turning up the heat on pests and diseases: a case study for Barley yellow dwarf virus. Proceedings of the British Crop Protection Conference – Crop Science and Technology, pp 1195–1200

Harrington R, Tatchell GM, Bale JS (1990) Weather, life cycle strategy and spring populations of aphids. Acta Phytopathol Entomol Hung 25:423–432

Harrington R, Bale JS, Tatchell GM (1995) Aphids in a changing climate. In: Harrington R, Stork NE (eds) Insects in a Changing Environment. Proceedings of the 17th Royal Entomological Society Symposia on I, 7–10 September 1993, Harpenden, Hertfordshire. Academic Press, London, pp 125–155

Harrington R, Clark SJ, Welham SJ, Verrier SJ, Denholm CH, Hullé M, Maurice D, Rounsevell MDA, Cocu N (in press) EU EXAMINE Consortium. Environmental change and the phenology of European aphids. Glob Change Biol

Harrington R, Verrier P, Denholm C, Hullé M, Maurice D, Bell N, Knight J, Rounsevell M, Cocu N, Barbagallo S, Basky Z, Coceano P-G, Derron J, Katis N, Lukášová H, Marrkula I, Mohar J, Pickup J, Rolot J-L, Ruszkowska M, Schliephake E, Seco-Fernandez M-V, Sigvald R, Tsitsipis J, Ulber, B (2004) 'EXAMINE' (EXploitation of Aphid Monitoring in Europe): An EU Thematic Network for the study of global change impacts on aphids. In: Simon JC, Dedryver CA, Rispe C, Hullé M (eds) Aphids in a New Millennium Proceedings 6th International Aphid Symposium. INRA, Versailles, pp 45–49
Hartley S, Jones T (2003) Plant diversity and insect herbivores: effects of environmental change in contrasting model systems. Oikos 101:6–17
Hendrey GR (1992) Global greenhouse studies: need for a new approach to ecosystem manipulation. Critical Reviews in Plant Science 11:61–74
Hennessy KJ, Suppiah R, Page CM (1999) Australian rainfall changes. Aust Met Mag 48:1–13
Hijmans RJ, Grünwald NJ, van Haren RJF, MacKerron DKL, Scherm H (2000) Potato late blight simulation for global change research. GILB Newsletter 12:1–3
Hódar JA, Zamora R (2004) Hebivory and climatic warming: a Mediterranean outbreaking caterpillar attacks a relict, boreal pine species. Biodivers Conserv 13:493–500
Hódar JA, Castro G, Zamora R (2002) Pine processionary caterpillar *Thaumetopoea pityocampa* as a new threat for relict Mediterranean Scots pine forests under climatic warming. Biol Conserv 110:123–129
Hodkinson I (1999) Species response to global environmental change or why ecophysiological models are important: a reply to Davis et al. Journal of Animal Ecology 68, 1259–1262
Hogg EH, Brandt JP, Kochtubajda B (2002) Growth and dieback of aspen forests in northwestern Alberta, Canada, in relation to climate and insects. Can J For Res 32:823–832
Houghton JT, Meira Filho LG, Callander BA, Harris N, Kattenberg A, Maskell K (1996) Climate change 1995: The science of climate change. Contribution of Working Group I to the Second Assessment Report of the Intergovernmental Panel on Climate Change. Cambridge University Press, Cambridge
Hughes L (2000) Biological consequences of global warming: is the signal already apparent? TREE 15:56–61
IPCC (2001a) Climate change 2001 Third Assessment Report of the Intergovernmental Panel on Climate Change. Cambridge University Press, Cambridge
IPCC (2001b) Technical summary. Climate change (2001) Impacts, Adaptation and Vulnerability. A report of working group II of the Intergovernmental Panel on Climate Change. IPCC, Geneva
Jenkins N, Hoffmann A (2001) Distribution of *Drosophila serrata* Malloch (Diptera: Drosophilidae) in Australia with particular reference to the southern border. Australian Journal of Entomology 40, 41–48
Jones R (2000) Analysing the risk of climate change using an irrigation demand model. Clim Res 14:89–100
Jones T, Thompson L, Lawton J, Bezemer T, Bardgett R, Blackburn T, Bruce K, Cannon P, Hall G, Hartley S, Howson G, Jones C, Kampichler C, Kandeler E, Richie D (1998) Impacts of rising atmospheric carbon dioxide on model terrestrial ecosystems. Science 280:441–443
Jones PD, Lister DH, Jaggard KW, Pidgeon JD (2003) Future climate impact on the productivity of sugar beets (*Beta vulgaris*) in Europe. Clim Change 58:93–108
Jules ES, Kauffman MJ, Ritts WD, Carroll AL (2002) Spread of an invasive pathogen over a variable landscape: A nonnative root rot on Port Orford cedar. Ecology 83:3167–3181
Julien MH, Skarratt B, Maywald GF (1995) Potential geographical distribution of alligator weed and its biological control by *Agasicles hygrophila*. J Aquat Plant Manag 33:55–60
Kamata N, Esaki K, Kato K, Igeta Y, Wada K (2002) Potential impact of global warming on deciduous oak dieback caused by ambrosia fungus *Raffaelea* sp. carried by ambrosia beetle *Platypus quercivorus* (Coleoptera: Platypodidae) in Japan. Bull Entomol Res 92:119–126
Karl TR, Jones PD, Knight RW (1993) A new perspective on global warming: asymmetric trends of daily maximum and minimum temperatures. Bull. Am. Meteorol. Soc. 74:1007–1023
Karnosky DF, Percy KE, Xiang B, Callan B, Noormets A, Mankovska B, Hopkin A, Sober J, Jones W, Dickson RE, Isebrands JG (2002) Interacting elevated CO_2 and tropospheric O_3 predisposes aspen (*Populus tremuloides* Michx.) to infection by rust (*Melampsora medusae* f.sp. *tremuloidae*). Glob Change Biol 8:329–338
Körner C (2000) Biosphere responses to CO_2 enrichment. Ecol Appl 10:1590–1619
Körner C, Morgan J, Richard Norby R (2007) CO_2 fertilization: when, where, how much? In: Canadell J, Pataki D, Pitelka L (eds) Terrestrial ecosystems in a changing world. The IGBP Series, Springer-Verlag, Berlin
Krimsky, S (2000) Hormonal Chaos. The Scientific and Social Origins of the Environmental Endocrine Hypothesis John Hopkins University Press, Baltimore, MD
Kriticos D, Randall P (2001) A comparison of systems to analyse potential weed distributions. In: Groves RH, Panetta FD, Virtue J (eds) Weed Risk Assessment. CSIRO, Melbourne, pp 61–79
Kriticos DJ, Brown JR, Radford ID, Nicholas M (1999) Plant population ecology and biological control: *Acacia nilotica* as a case study. Biological Control. 16:230–239
Kriticos DJ, Sutherst RW, Brown JR, Adkins SW, Maywald GF (2003a) Climate change and biotic invasions: a case history of a tropical woody vine. Biol Invasions 5:147–165
Kriticos DJ, Sutherst RW, Brown JR, Adkins SW, Maywald GF (2003b) Climate change and the potential distribution of an invasive alien plant: *Acacia nilotica* ssp. *indica* in Australia. J Appl Ecol 40:111–124
Landsberg, JJ (1989) The greenhouse effect: Issues and directions for Australia. Occasional Paper No 4 CSIRO, Melbourne
Lawton JH (2000) Community Ecology in a Changing World. Inter Research, Oldendorf/Luhe, Germany
Levine JM, D'Antonio C M (2003) Forecasting biological invasions with increasing international trade. Conserv Biol 322–326
Lewis WJ, van Lenteren JC, Phatak SC, Tumlinson JHI (1997) A total system approach to sustainable pest management. Proc Natl Acad Sci USA 94: 12243–12248
Linacre E (1992) Climate data and resources: a reference and guide. Routledge, London
Lok C (2001) Unlucky bamboo. Asian mosquitoes stow away on plant shipment. Nature
Marco GJ, Hollingworth RM, Durham W (1987) Silent Spring Revisited. American Chemical Society, Washington, DC
Mulder C, Roy B (2003) Climate change and invertebrate herbivory on boreal understory plants: A survey. Ecological Society of America Annual Meeting. Abstracts. pp 88–245
Nakicenovic N, Alcamo J, Davis G, de VB, Fenhann J, Gaffin S, Gregory K, Grübler A, Jung T, Kram T, Lebre La Rovere E, Michaelis L, Mori S, Morita T, Pepper W, Pitcher H, Price L, Riahi K, Roehrl A, Rogner H, Sankovski A, Schlesinger M, Shukla P, Smith S, Swart R, van Rooijen S, Victor N, Dadi Z (2000) Special Report on Emissions Scenarios. Cambridge University Press, Cambridge, 599 pp
New M, Lister D, Hulme M, Makin I (2002) A high-resolution data set of surface climate over global land areas. Clim Res 21:1–25
Noor M, Pascual M, Smith K (2000) Genetic variation in the spread of *Drosophila subobscura* from a nonequilibrium population. Evolution 54:696–703
Oerke EC, Dehne HW (2004) Safeguarding production losses in major crops and the role of crop protection. Crop Prot 23:275–285
Pangga IB, Chakraborty S, Yates D (2004) Canopy size and reduced resistance in *Stylosanthes scabra* determine anthracnose severity at high CO_2. Phytopathology 93:221–227
Parmesan C, Yohe G (2003) A globally coherent fingerprint of climate change impacts across natural systems. Nature 421:37–42
Parry M, Carter T (1998) Climate impact and adaptation assessment. Earthscan, London
Parry ML, Livermore M (1999) A new assessment of the global effect of climate change. Global Environmental Change 9: S1–S107
Peperzak L (2003) Climate change and harmful algal blooms in the North Sea. Acta Oecol 24: S139–S144
Percy KE, Awmack CS, Lindroth RL, Kubiske ME, Kopper BJ, Isebrands JG, Pregitzer KS, Hendrey GR, Dickson RE, Zak DR, Oksanen E, Sober J, Harrington R, Karnosky DF (2002) Altered performance of forest pests under atmospheres enriched by CO_2 and O_3. Nature 420:403–407

Perrings P, Williamson M, Barbier E, Delfino D, Dalmazzone S, Shogren J, Simmons P, Watkinson A (2002) Biological invasion risks and the public good: an economic perspective. Conserv Ecol 6: 1 [online] URL: http://www.consecol.org/vol6/iss1/art1/

Peterson AT, Ortega-Huerta MA, Bartley J, Sanchez-Cordero V, Soberon J, Buddemeier RH, Stockwell DRB (2002) Future projections for Mexican faunas under global climate change scenarios. Nature 416:626–629)

Pimentel D, Lach L, Zuniga R, Morrison D (2000a) Environmental and economic costs of nonindigenous species in the United States. Bioscience 50:53–65

Pimentel D, McNair S, Janecka J, Wightman J, Simmonds C, O'Connell C, Wong E, Russel L, Zern J, Aquino T, Tsomondo T (2000b) Economic and environmental threats of alien plant, animal, and microbe invasions. Agriculture, Ecosystems and Environment 84:1–20

Ragsdale NN (2000) The impact of the Food Quality Protection Act on the future of plant disease management. Annu Rev Phytopathol 38:577–596

Ramakrishnan PS, Vitousek PM (1989) Ecosystem-level processes and the consequences of biological invasions. In: Drake JA, Mooney HA, di Castri F, Groves RH, Kruger FJ, Rejmánek M, Williamson M (eds) Biological Invasions: a Global Perspective. John Wiley & Sons, Chichester, pp 281–300

Reynolds JF, Acock B (1997) Modularity and genericness in plant and ecosystem models. Ecol Model 94:7–16

Rodriguez-Trelles F, Rodriguez MA, Scheiner SM (1998) Tracking the genetic effects of global warming: *Drosophila* and other model systems. Conserv Ecol 2 (2): [online] URL: http://www.consecol.org/vol2/iss2/art2

Roelfs AP (1982) Effects of barberry eradication on stem rust in the United States. Plant Dis 66:177–181

Root TL, Price JT, Hall KR, Schneider SH, Rosenzweig C, Pounds JA (2003) Fingerprints of global warming on wild animals and plants. Nature 421:57–60

Roy BA, Güsewell S, Harte J (2004) Response of plant pathogens and herbivores to warming experiment. Ecology 85:2570–2581

Russell P (1999) Fungicide resistance management into the next millennium. Pesticide Outlook 10:213–215

Sands DPA, Bakker P, Dori FM (1993) *Cotesia erionotae* (Wilkinson) (Hymenoptera:Braconidae), for biological control of banana skipper, *Erionota thrax* (L.) (Lepidoptera:Hesperiidae) in Papua New Guinea. Micronesica Suppl 4:99–105

Scherm H (2004) Climate Change: Can we predict the impacts on plant pathology and pest management. Can J Plant Pathol 26:267–273

Scherm H, Coakley SM (2003) Plant pathogens in a changing world. Australas Plant Pathol 32:157–165

Scherm H, van Bruggen A (1994) Global warming and non-linear growth: How important are changes in average temperature? Phytopathology 84:1380–1384

Scherm H, Sutherst RW, Harrington R, Ingram JSI (1999) Global networking for assessing impacts of global change on plant pests. Environ Pollut 107:1–9

Schoute JF, Finke PA, Veeneklaas FR, Wolfert HPE (1995) Scenario studies for the rural environment. Kluwer Academic Publishers, Dordrecht, The Netherlands

Scott JM, Heglund PJ, Morrison MLE (2002) Predicting Species Occurences: Issues of Accuracy and Scale. In: Townsend Peterson A, DRB Stockwell, DA Kluza (eds) Distributional Prediction Based on Ecological Niche Modeling of Primary Occurence Data. Island Press, Washington, DC

Silvertown J (2004) The ghost of competition past in the phylogeny of island endemic plants. J Ecol 92:168–173

Smith JB, Klein RJT, Huq S (2003) Climate Change, Adaptive Capacity and Development. Imperial College Press, London

Stephens P, Sutherland W, Freckleton R (1999) What is the Allee effect? *Oikos* **87**, 185–190

Sutherst RW (1983) Variation in the numbers of the cattle tick, *Boophilus microplus* (Canestrini), in a moist habitat made marginal by low temperatures. J Aust Entomol Soc 22:1–5

Sutherst RW (1998) Implications of global change and climate variability for vector-borne diseases: generic approaches to impact assessments. Int J Parasitol 28:935–945

Sutherst RW (2000) Climate change and invasive species – a conceptual framework. In: HA Mooney, RJ Hobbs (eds) Invasive species in a changing world. Island Press, Washington DC, pp 211–240

Sutherst RW (2001) The vulnerability of animal and human health to parasites under global change. Int J Parasitol 31:933–948

Sutherst RW (2003) Prediction of species geographical ranges. Guest Editorial. J Biogeogr 30:1–12

Sutherst RW (2004) Global change and human vulnerability to vector-born diseases. Clin Microbiol Rev 17:136–173

Sutherst RW, Comins HN (1979) The management of acaricide resistance in the cattle tick, *Boophilus microplus* (Canestrini) (Acari: Ixodidae), in Australia. Bull Entomol Res 69:519–537

Sutherst RW, Maywald GF (1985) A computerised system for matching climates in ecology. Agriculture Ecosystems and Environment 13:281–300

Sutherst R, Yonow T, Chakraborty S, O'Donnell C, White N (1996) A generic approach to defining impacts of climate change on pests, weeds and diseases in Australasia. In: Bouma W, Pearman G, Manning M (eds) Greenhouse: coping with climate change. CSIRO, Melbourne, pp 281–307

Sutherst RW, Ingram J, Scherm H (1998) Global change and vector-borne diseases. Parasitol Today 14:297–299

Sutherst RW, Murdiyarso D, Widayati A (1999) Modelling global change impacts on pests Report No. 7. Biotrop-GCTE IC-SEA Bogor, Indonesia pp 108

Sutherst RW, Collyer BS, Yonow T (2000a) The vulnerability of Australian horticulture to the Queensland fruit fly, *Bactrocera (Dacus) tryoni*, under climate change. Aust J Agric Res 51: 467–480

Sutherst RW, Maywald GF, Russell BL (2000b) Estimating vulnerability under global change: modular modelling of pests. Agric Ecosyst Environ 82:303–319

Sutherst RW, Maywald GF, Bourne AS (in press) Including species-interactions in risk assessments for global change. Glob Change Biol

Sykes MT, Prentice IC (2004) Boreal forest futures. Modelling the controls on tree species range limits and transient responses to climate change. Water, Air, and Soil Polution 82: 415–428

Taylor CM, Hastings A (2005) Allee Effects in Biological Invasions. Ecology Letters 8[8], 895–908

Taylor F, Spalding JB (1989) Timing of diapause in relation to temporally variable catastrophes. Journal of Evolutionary Bioilogy 2:285–297

Teng PS, Heong KL, Kropff MJ, Nutter FW, Sutherst RW (1996) Linked pest-crop models under global change. In: Walker B, Steffen W (eds) Global change and terrestrial ecosystems. Cambridge University Press, Cambridge

Thomas MB, Willis AJ (1998) Biocontrol – risky but necessary? Trends in Ecology and Evolution 13:325–329.\

Thomas D, Cameron A, Green R, Bakkenes M, Beaumont L, Collingham Y, Erasmus B, De SM, Grainger A, Hannah L, Hughes L, Huntley B, Van JA, Midgley GML, Ortega-Huerta M, Peterson A, Phillips O, Williams S (2004) Extinction risk from climate change. Nature 427:145–148

Thuiller W (2003) Biomod – optimizing predictions of species distributions and projecting potential future shifts under global change. Glob Change Biol 9:1353–1362

USDA-ERS (2002) Floriculture and nursery crop situation and outlook yearbook Economic Research Service, US Department of Agriculture, Washington, DC

Vaughn KC (2003) Herbicide resistance work in the United States Department of Agriculture-Agricultural Research Service. Pest Manag Sci 59:764–769

Visser ME, Holleman LJM (2001) Warmer springs disrupt the synchrony of oak and winter moth phenology. Proc. Royal Soc. Lond. B 268:289–294

Vitousek PM (1994) Beyond global warming: Ecology and global change. Ecology 75:1861–1876

Voigt W, Perner J, Davis AJ, Eggers T, Schumacher J, Bahrman R, Fabian B, Heinrich W, Kohler G, Lichter D, Marstaller R, Sander FW (2003) Trophic levels are differentially sensitive to climate. Ecology 84:2444–2453

Wayne P, Foster S, Connolly J, Bazzaz F, Epstein P (2002) Production of allergenic pollen by ragweed (*Ambrosia artemisiifolia* L.) is increased in CO_2-enriched atmospheres. Annals of Allergy, Asthma, and Immunology 88:279–282

Wharton T, Kriticos D (2004) The fundamental and realised niche of the Monterey pine aphid, *Essigella californica* (Essig) (Hemiptera: Aphididae: implications for managing softwood plantations in Australia. Divers Distrib 10:253–262

Woiwod I, Harrington R (1994) Flying in the face of change: the Rothamstead insect survey. In: Leigh RA, Johnston AE (eds) Long-term Experiments in Agricultural and Ecological Sciences. CAB International, Wallingford, UK, pp 321–342

Woodward FI (1987) Climate and Plant Distribution Cambridge University Press, Cambridge

Yamamura K, Kiritani K (1998) A simple method to estimate the potential increase in the number of generations under global warming in temperate zones. Appl Entomol Zool 33: 289–298

Yonow T, Sutherst RW (1998) The geographical distribution of the Queensland fruit fly, *Bactrocera* (*Dacus*) *tryoni,* in relation to climate. Aust J Agric Res 49:935–953

Yonow T, Kriticos DJ, Medd RW (2004) The potential geographic range of *Pyrenophora semeniperda*. Phytopathology 94:805–812

Ziska LH (2003a) Climate change, plant biology and public health. World Resource Review 15:271–288

Ziska LH (2003b) Evaluation of the growth response of six invasive species to past, present and future carbon dioxide concentrations. J Exp Bot 54:395–404

Ziska LH (2003c) The impact of nitrogen supply on the potential response of a noxious, invasive weed, Canada thistle (*Cirsium arvense*) to recent increases in atmospheric carbon dioxide. Physiol Plant 119:105–112

Ziska LH (2003d) Rising carbon dioxide and weed ecology. In: Inderjit (ed) Weed Biology and Management. Kluwer Academic Publishers, Dordrecht, pp 159–176

Ziska LH, Caufield FA (2000) Rising CO_2 and pollen production of common ragweed (*Ambrosia artemisiifolia*) a known allergy-inducing species: implications for public health. Aust J Plant Physiol 27:893–898

Ziska LH, Gebhard DE, Frenz DA, Faulkner S, Singer BD (2003) Cities as harbingers of climate change: Common ragweed, urbanization, and public health. J Allergy Clin Immunol 111:290–295

Chapter 18

Greenhouse Gas Mitigation Potential in Agricultural Soils

Pete Smith · Pete Falloon · Uwe Franko · Martin Körschens · Rattan Lal · Keith Paustian · David Powlson Vladimir Romanenkov · Ludmila Shevtsova · Jo Smith

18.1 Introduction

Agricultural soils can act as sources and sinks for carbon dioxide and other greenhouse gases. Whether soils act as a sink or source, and the sink/source strength depends critically upon the management of the soil. Agricultural soils emit carbon dioxide (CO_2; through soil and root respiration / decomposition of soil organic matter), nitrous oxide (N_2O; formed during nitrification and denitrification processes) and may either emit or remove (oxidize) methane (CH_4) from the atmosphere. The greenhouse gas mitigation potential for agricultural soils results from reducing these emissions. For CO_2 this entails reducing the CO_2 efflux from the soil or sequestering carbon in the soil. For N_2O, this entails reducing N_2O emissions. For CH_4, this entails reducing CH_4 emissions from soils emitting CH_4 (e.g., rice paddy soils) and maximising the methane oxidation potential of other soils.

18.1.1 Soil Carbon and Carbon Dioxide

Globally, soils contain about 1 500 Pg (1 Pg = 1 Gt = 10^{15} g) of organic carbon to a depth of 1 m (Batjes 1996), which is about three times the amount of carbon in vegetation and twice the amount in the atmosphere (IPCC 2001). The annual fluxes of CO_2 from atmosphere to land (global Net Primary Productivity [NPP]) and land to atmosphere (respiration and fire) are of the order of 60 Pg C yr^{-1} (IPCC 2000b). During the 1990s, fossil fuel combustion and cement production emitted 6.3 ±1.3 Pg C yr^{-1} to the atmosphere, while land-use change emitted 1.6 ±0.8 Pg C yr^{-1} (Schimel et al. 2001; IPCC 2001). Atmospheric C increased at a rate of 3.2 ±0.1 Pg C yr^{-1}, the oceans absorbed 2.3 ±0.8 Pg C yr^{-1} and there was an estimated terrestrial sink of 2.3 ±1.3 Pg C yr^{-1} (Schimel et al. 2001; IPCC 2001).

The amount of carbon in the soil is therefore large compared to gross and net annual fluxes of carbon to and from the terrestrial biosphere. Manipulating the size of the soil carbon pools is at the heart of soil carbon sequestration.

Soil carbon pools are smaller now than they were before human intervention. Historically, soils have lost between 40 and 90 Pg C globally through cultivation and disturbance (Paustian et al. 1998; Houghton et al. 1999; Lal 1999). Soil carbon sequestration can be achieved by increasing the net flux of carbon from the atmosphere to the terrestrial biosphere, by increasing global NPP (thus increasing carbon inputs to the soil), by storing a larger proportion of the carbon from NPP in the longer-term carbon pools in the soil, or by slowing decomposition. For soil carbon sinks, the best options are to increase C stocks in soils that have been depleted in carbon, i.e., agricultural soils and degraded soils (Lal 2004a; Smith 2004a).

Estimates of the potential for additional soil carbon sequestration vary widely. The most recent global estimate is that of Lal (2004a) of 0.9 ±0.3 Pg C yr^{-1}. Over 50 years, this level of C sequestration would restore a large part of the carbon lost from soils historically. Cropland soils can be a large source of carbon dioxide (e.g., Janssens et al. 2003). There is significant potential to reduce the efflux of carbon from agricultural soils, and to sequester carbon in them. However, soil carbon sequestration rates have a limited duration and cannot be maintained indefinitely. Mitigation options for agricultural soil carbon are given in Table 18.1.

18.1.2 Trade-Offs between GHGs in Agriculture

In addition to carbon dioxide, nitrous oxide and methane are important greenhouse gases emitted from agriculture. Nitrous oxide is formed primarily from nitrification (the conversion of ammonium to nitrate, via nitrite) and denitrification (the conversion of nitrate, ultimately to N_2 gas). N_2O is a by-product of nitrification and an intermediate product during denitrification. N_2O fluxes from agricultural soils (0.53 Pg C equivalents yr^{-1}) account for more than 50% of the global anthropogenic N_2O flux (Robertson 2004). Options for N_2O mitigation are given in Table 18.2.

The majority (52%) of the methane flux from agriculture arises from enteric fermentation (microbial fermentation in the rumen of ruminants), with biomass burning (19%) and animal waste treatment (8%) accounting for other significant proportions (Robertson 2004). Most aerobic soils actually oxidize CH_4 but there is good evidence that the capacity of soils as a sink for methane is reduced upon cultivation. The only significant soil source of methane in agriculture arises from rice cultivation

Table 18.1. Soil carbon sequestration measures in agricultural land (after Smith 2004c)

Cropland	Grazing land
Zero/reduced tillage	Improve efficiency of animal manure use
Set-aside/conservation reserve program	Improve efficiency of crop residue use
Convert to permanent crops	Improve livestock management to reduce soil disturbance
Convert to deep-rooting crops	Improve livestock management to maximize manure C returns
Improve efficiency of animal manure use	Agricultural use of sewage sludge
Improve efficiency of crop residue use	Convert to deeper-rooting species
Agricultural use of sewage sludge	Application of compost to land
Application of compost to land	Fertilizer use
Rotational changes	Irrigation
Fertilizer use	Extensification/de-intensification of farming
Irrigation	Improved management to reduce wind and water erosion
Bioenergy crops	
Extensification/de-intensification of farming	
Organic farming (a combination of many different individual practices)	
Convert cropland to grassland	
Management to reduce wind and water erosion	

Table 18.2. Nitrous oxide mitigation options for agricultural soils (after Smith et al. 2004e)

Practices	
Inorganic fertilizer	▪ Suitable type and characteristics e.g. – ammonium providing fertilizer vs. nitrate fertilizer – slow-release – inhibitors ▪ Application techniques – synchronisation/timing (e.g., split-application to coincide with crop demand) – placement, burial may reduce flux ▪ Amount/rate of application, appropriate fertilizer recommendations required
Organic fertilizer	▪ Type – farm yard manure, (flux dependent on degree of maturity/degradation, moisture content) – liquid fertilizer/slurry (give higher fluxes than dried material; avoid spreading with mineral fertilizer) – industrial waste – household waste – biogas residue – fermented manure ▪ Timing (less critical than inorganic) ▪ Application techniques ▪ Storage, processing, and handling (temperature, duration, capacity, cover, etc.) ▪ Amount/rate of application (apply at recommended rate to coincide with crop demand)
Biological N fixation	▪ Avoid mono cropping, flux dependent on legumes/grass mixture ratio and is mostly after ploughing (net effect is unknown)
Crop residue	▪ (Important for C sequestration) ▪ Quality and size (C:N ratio, total N) ▪ Quantity (weight) ▪ Application/incorporation techniques (note the interaction with tillage; priming effect on soil N_2O flux mainly with incorporation) avoid wet conditions
Farming system and management	▪ Extensive cropping may be better than intensive – further research needed for to assess this difference and the effect of the arable/livestock ratio. Also compare to intensive areas with other land use on spare land
Tillage	▪ Important for C sequestration but interacts with soil physical conditions ▪ Ploughing or deep ploughing may reduce trace gas emissions but decrease carbon sequestration ▪ Conservation (reduced) tillage may be intermediate but more information required
Crop rotation	▪ Catch crops reduce bare soils (possible link to BNF) ▪ Amelioration crops (crop type i.e. deep rooting or shallow rooting)
Water management	▪ Drainage useful but irrigation should be avoided, flooding water buffers may have an effect
Compaction status	▪ Appropriate timing and size of machinery required ▪ Manage grazing to avoid poaching

(0.25 Pg C equivalents yr^{-1}), which accounts for 22% of agricultural emissions or 12% of total anthropogenic fluxes (Robertson 2004). Mitigation of CH_4 fluxes from agricultural soils needs to focus almost entirely on this rice land flux. Rice crop management to reduce CH_4 emissions includes yield improvement with well-managed, high yield rice crops have significantly lower CH_4 emissions due to more C being allocated to the grain than to the rhizosphere where it can undergo methanogenesis. Other mitigation options in rice include residue management and irrigation scheduling (Robertson 2004). Methane mitigation is not discussed further in this chapter.

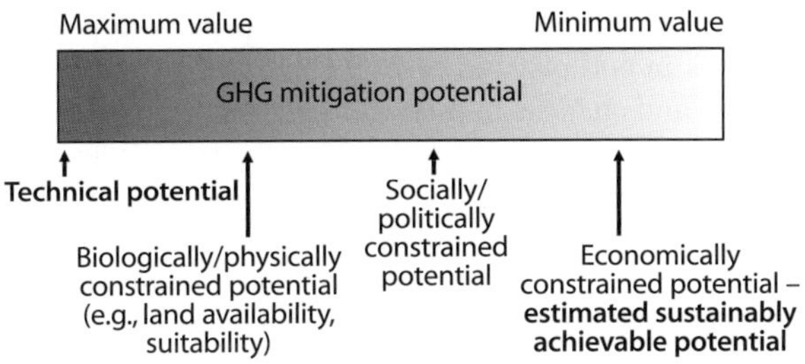

Fig. 18.1. How different constraints reduce the GHG mitigation potential from its theoretical biological maximum to realistically achievable potentials that are much lower (adapted from Smith 2004a)

Robertson (2004) emphasises the needs for a systems approach for assessing GHG mitigation potential in agriculture. For example, increasing soil carbon stocks in the soil through reduced tillage can lead to anaerobic zones in some soils and thereby increase N_2O emissions (MacKenzie et al. 1998; Smith et al. 2001; Six et al. 2004). Similarly, management to reduce CH_4 emissions in paddy rice fields might increase N_2O emissions. Trade-offs between the GHGs are complex (Robertson et al. 2000), but should always be considered.

18.2 What Is Meant by GHG Mitigation Potential?

Estimation of mitigation potential is often confounded by the choice of constraints. Some authors quote biological potentials (Metting et al. 1999), others quote potentials as limited by available land or resources (Smith et al. 2000), and others also consider economic and social constraints (Cannell 2003; Freibauer et al. 2004). Smith (2004a) provided a figure showing how these mitigation potential estimates differ and how the potential is reduced by a number of constraints (Fig. 18.1).

Table 18.3. Carbon sequestration potentials limited only by availability of land, biological resources and land-suitability, and those estimated to be achievable by 2012 (Smith 2004c; after Freibauer et al. 2004)

Practice	Soil carbon sequestration potential ($t\,C\,ha^{-1}\,yr^{-1}$)	Estimated uncertainty	Total soil carbon sequestration potential for EU15 ($Mt\,C\,yr^{-1}$)[a]	Realistic soil carbon sequestration potential for EU15 ($Mt\,C\,yr^{-1}$) by 2012
Zero-tillage	0.38 (0.29)*	>50%	24.4	2.4
Reduced tillage	<0.38	≫50%	<24.4	<2.4
Set-aside	<0.38	≫50%	2.4 (maximum)	0
Permanent crops	0.62	≫50%	0?	0?
Deep-rooting crops	0.62	≫50%	0?	0?
Animal manure	0.38 (1.47)*	≫50%	23.7	?
Cereal straw	0.69 (0.21)*	≫50%	5.5	?
Sewage sludge	0.26	≫50%	2.1	?
Composting	0.38	≫50%	3	3?
Improved rotations	>0	Very high	0?	0?
Fertilization	0	Very high	0	0
Irrigation	0	Very high	0	0
Bioenergy crops	0.62	≫50%	4.5	0.9
Extensification	0.54	≫50%	11	?
Organic farming	0–0.54	≫50%	3.9	3.9
Convert cropland to grassland	1.2–1.69 (1.92)*	≫50%	8.7–12.3	0
Convert cropland to woodland	0.62	≫50%	4.5	4.5 (maximum)

[a] Carbon sequestration potentials limited only by availability of land, biological resources and land-suitability. All estimates based on extrapolation from Smith et al. (2000) except those marked *, where the figure in brackets is derived from Vleeshouwers and Verhagen (2002). For full list of assumptions, limitations and sources, see Freibauer et al. (2003).

An analysis of the estimates presented in Freibauer et al. (2004) and the assumptions used by Cannell (2003) suggest that the realistic sustainable (or conservative) achievable potential of GHG mitigation (taking into account limitations in land use, resources, economics, and social and political factors) may be about 10–20% of the biological potential (Table 18.3).

Although this value is derived predominantly from expert judgment, it may be useful in assessing how different estimates of GHG mitigation potential can be compared and how they might realistically contribute to GHG stabilization. More robust methods to develop estimates of socioeconomic potential are being developed, e.g., McCarl and Schneider (2001).

In a study of European agricultural GHG mitigation potential, a range of estimates of carbon sequestration potential were compared (Smith et al. 2004e). The results (Fig. 18.2) show that by including more constraints in the calculation, the estimated mitigation potential is drastically reduced.

Estimates on the left of the graph are for biological potential while those on the right extrapolate from current trends. These findings show that, despite a high biological potential for carbon sequestration in European agricultural soils, lack of policy incentives to implement carbon sequestration measures means that virtually none of this potential will be realized by the end of the first Kyoto commitment period in 2012.

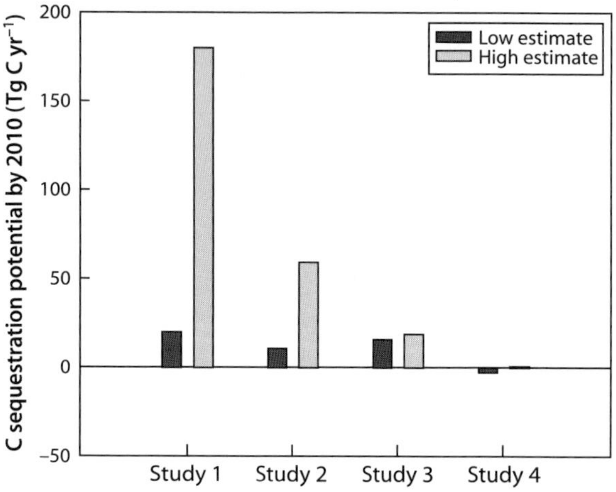

Fig. 18.2. Estimates of cropland soil carbon sequestration potential in EU-15 from previous studies and from this study. *Study 1* is from Vleeshouwers and Verhagen (2002) with the low estimate for straw incorporation and the high figure for conversion of all cropland to grassland. *Study 2* is from Smith et al. (2000a) with figures scaled from geographical Europe (including Baltic States but excluding Russia) to EU-15 as per Smith et al. (1997). The low estimate is from the combined land management scenario with extensification of surplus arable land and straw incorporation; the high estimate is for the combined "optimal" scenario (see Smith et al. 2000 for further details). *Study 3* is from Freibauer et al. (2004) with values assessed for realistically achievable potential by 2010 (about 1/5 of the estimated biological potential). *Study 4* uses figures based on measured trends 1990 to 2000 and extrapolations to 2010 (Smith et al. 2004e)

18.3 Regional Case Studies

Greenhouse gas mitigation potential can be estimated at a range of scales, from the region to the continent to the globe. In this section, we provide some examples of regional estimates of carbon sequestration.

18.3.1 Sustainable Soil Management in the Moscow Region to Enhance Soil Carbon

In the Moscow region, the soil-weather C static model, which is based on an equilibrium approach to predict changes in soil organic carbon (SOC) as a function of climate and changes in management practices, was used for the zone of podzoluvisols of the Russian Federation (Shevtsova et al. 2003).

Analysis of different initial SOC stocks, crop rotations and fertilization rates has revealed that the most important resource for preventing SOM decrease in the long-term for intensive agricultural systems is an increase in the percentage of grasses used in crop rotations at the expense of row crops. This is especially true on coarse-textured soils. Farm-yard manure (FYM) application has a smaller potential of increasing soil sustainability while changes in the rates of mineral N fertilization are the least important (Romanenkov et al. 2001).

Simulation of the influence of climate change using the IPCC-SRES A1FI scenario (IPCC 2000a) to 2050 (as implemented by the HadCM3 climate model) will increase soil C gain as a result of a net increase in the amount of residue returned to the soil (higher plant productivity with higher temperature), and a reduced rate of C loss with declining soil moisture, resulting in slower decomposition of residues and SOC. These regional studies in the Moscow region are a good example of how GHG mitigation studies can advise policy requirements as they show, for most of the regions, that FYM application rates can be decreased and percentage of perennials in crop rotation in 2050 can be reduced by 10–20% to sustain current SOC level in arable soils (Sirotenko et al. in press). Increases of SOC of up to 30% are possible with increased addition of FYM, but the necessary application rates are both economically unprofitable and ecologically unsafe. Change in the ratio of perennial grasses to row crops in rotations is a more realistic measure.

Similar results were found using the CANDY model (Franko 1997; Franko and Schenk 2001) for the whole territory of Russia. The average increase of SOC storage in arable soils from 1990 to 2070 for the business as usual scenario was about 0.09 kg m^{-2} ranging from a decrease of 0.4 kg m^{-2} to an increase of 1.8 kg m^{-2} (Fig. 18.3). The majority of the arable land (62%) was classified as invariant (±0.1 kg m^{-2}).

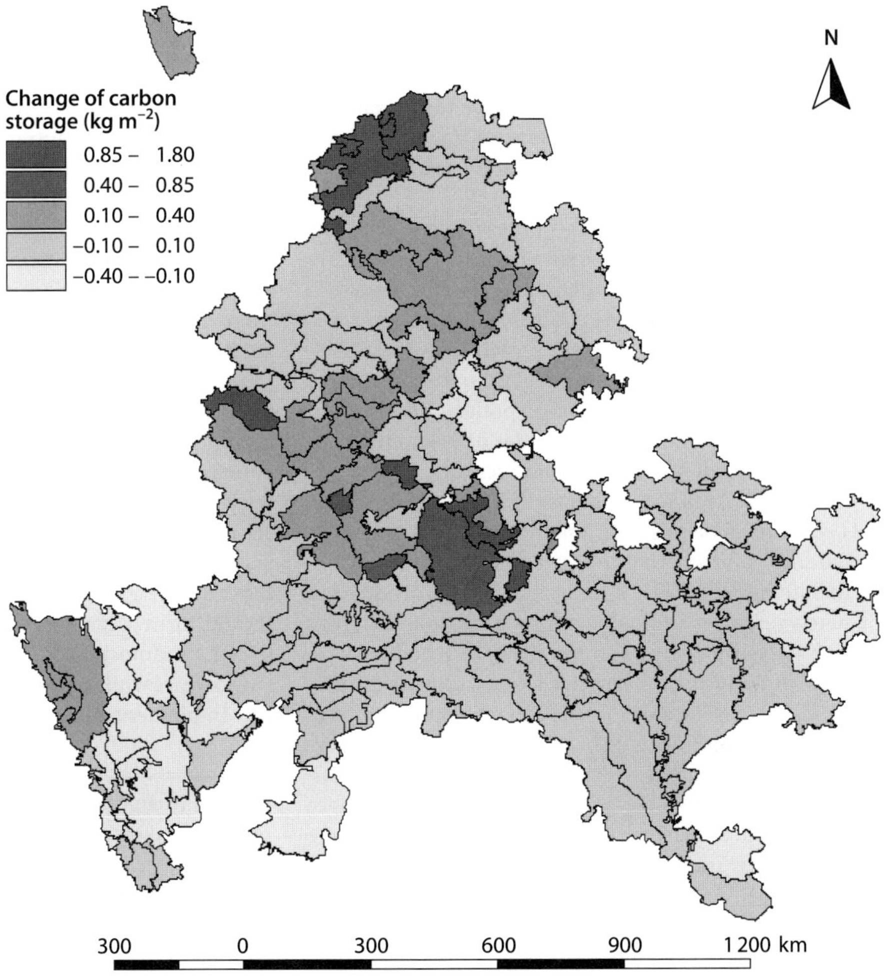

Fig. 18.3. Change of carbon storage in Russian cropland soils (1990 to 2070) assuming no land-use change and the SRES B1 climate change scenario

18.3.2 Soil Carbon Sequestration Potential in the US

Lal and colleagues estimated that total C sequestration potential in the soils of the U.S. is 144 to 432 Tg C yr^{-1} or an average of 288 Tg C yr^{-1} (Table 18.4). Of the total potential, 25% is from croplands, 22% from forest lands, 14% from grazing lands, 15% from land restoration, 17% from land conversion and 7% from improving other lands. In comparison, CO_2 emissions from agricultural activities in the U.S. are 43 Tg C yr^{-1}, and the hidden, additional carbon costs of adopting recommended management practices (e.g., increased fossil fuel use) are 27 to 32 Tg C yr^{-1} (Lal et al. 2003).

A model-based assessment of soil C sequestration potential for US agriculture was conducted by Sperow et al. (2003). To represent regional differences in sequestration potential, soil, climate and land-use/management attributes were broken out for approximately 240 major ecoregions in the country. To assess potentials, scenarios entailing adoption of 'best management practices' for C sequestration were designed. These included conversion of all highly erodible annual cropland to conservation set aside, eliminating summer fallow in semi-arid regions, adding winter cover crops to annual crop rotations in humid climates and adopting conservation tillage practices on annual cropland. Potential sequestration rates totalled about 100 Tg C yr^{-1}, compared to current (1990s) estimated rates of C accumulation of around 10–15 Tg C yr^{-1}. Potential increases in C storage above baseline rates come from adoption of no-till (47 Tg C yr^{-1}), use of cover crops (22 Tg C yr^{-1}), set-asides (11 Tg C yr^{-1}) and elimination of summer fallow (3 Tg C yr^{-1}).

Table 18.4. Soil carbon sequestration potential in the U.S. (adapted from Lal et al. 2003)

Strategy	Total potential (Tg C yr^{-1})
1. Land use conversion	21 - 77
2. Land restoration	25 - 60
3. Cropland	45 - 98
4. Grazing land	13 - 70
5. Forest land	25 - 102
6. Other land uses in forestry	15 - 25
Total	144 - 432

18.4 Carbon Sequestration in the Future

Smith (2004b) recently calculated how future carbon emissions and CO_2 stabilization targets might influence the relevance of soil carbon sequestration as a GHG mitigation measure.

The future trajectory of carbon emissions over the next century depends upon many factors. The IPCC recently developed a range of standard reference emission scenarios (SRES) to provide estimates of possible emissions under a range of different possible futures (IPCC 2000a). These possible futures depend upon the degree to which greenhouse gas mitigation policies become global and upon whether environmental or economic concerns take precedence over the next century.

Among the A1 family of scenarios (global – free market), a number of possible emissions trajectories exist depending upon whether the energy sector remains fossil fuel intensive (A1FI), whether the rapid introduction of new energy technologies allows a move away from carbon intensive energy sources (A1T) or whether there is a mix of fossil fuel/alternative energy sources (A1B).

In all of these scenarios, the global population will grow, the population will become wealthier and per-capita energy demand will increase over the next century (IPCC 2000a). The extent to which these changes will occur differs between different scenarios, with some showing larger increases than others, but in all of them, these trends are observed. For each of the scenarios carbon emission trajectories have been determined (IPCC 2000a). Annual carbon emissions (Pg C yr^{-1}) by 2100 would be A1FI ~ 30, A1B ~ 17, A1T ~ 7, A2 ~ 28, B1 ~ 6, B2 ~ 18.

Emissions trajectories can also be calculated for a range of atmospheric CO_2 stabilization targets (e.g., 450, 550, 650, 750 ppm). For each stabilization target, the allowed carbon emission trajectories, which cannot be exceeded if the target is to be reached, can be calculated. The difference between the allowed emission trajectory for stabilization at a given target concentration, and the emissions associated with the estimated global energy demand is the carbon emission gap. For a stabilization target of 550 ppm, carbon emission gaps for each scenario by 2100 (Pg C yr^{-1}) are A1FI = 25, A1B = 12, A1T = 2, A2 = 22, B1 = 1, B2 = 13 (IPCC 2001).

The current annual emission of CO_2-carbon to the atmosphere is 6.3 ±1.3 Pg C yr^{-1}. Carbon emission gaps by 2100 could be as high as 25 Pg C yr^{-1} meaning that the carbon emission problem could be up to 4 times greater than at present. The maximum annual global C sequestration potential for all soils (mostly agricultural) is 0.9 ±0.3 Pg C yr^{-1} meaning that even if these rates could be maintained until 2100, soil carbon sequestration would contribute a maximum of 2–5% towards reducing the carbon emission gap under the highest emission scenarios. When we also consider the limited duration of carbon sequestration options in removing carbon from the atmosphere, we see that carbon sequestration could play only a minor role in closing the emission gap by 2100. It is clear from these figures that if we wish to stabilize atmospheric CO_2 concentrations by 2100, the increased global population and its increased energy demand can only be supported if there is a large-scale switch to non-carbon emitting technologies for producing energy.

Given that soil carbon sequestration can play only a minor role in closing the carbon emission gap by 2100, is there any role for carbon sequestration in climate mitigation in the future? The answer is yes. If atmospheric CO_2 levels are to be stabilized at reasonable concentrations by 2100 (e.g., 450–650 ppm), drastic reductions in emissions are required over the next 20–30 years (IPCC 2000a). During this critical period, all measures to reduce net carbon emissions to the atmosphere would play an important role – there will be no single solution (IPCC 2000a). Given that carbon sequestration is likely to be most effective in its first 20 years of implementation, it should form a central role in any portfolio of measures to reduce atmospheric CO_2 concentrations over the next 20–30 years while new energy technologies are developed and implemented (Smith 2004b).

18.5 Win-Win Strategies for Greenhouse Gas Mitigation by Agricultural Soils

Soil carbon sequestration is a process under the control of human management and, as such, the social dimension needs to be considered when implementing soil carbon sequestration practices. Since there will be increasing competition for limited land resources in the coming century, soil carbon sequestration cannot be viewed in isolation from other environmental and social needs. The IPCC (2001) have noted that global, regional and local environmental issues such as climate change, loss of biodiversity, desertification, stratospheric ozone depletion, regional acid deposition and local air quality are inextricably linked. Soil carbon sequestration measures clearly belong in this list. The importance of integrated approaches to sustainable environmental management is becoming ever clearer.

In any scenario, there will be winners and losers. The key to increasing soil carbon sequestration, as part of wider programmes to enhance sustainability, is to maximise the number of winners and minimise the number of losers. One possibility for improving the social/cultural acceptability of soil carbon sequestration measures would be to include compensation costs for losers when costing implementation strategies. By far the best option however, is to identify win-win measures (Lal 2004a), i.e., those which increase carbon stocks while at the same time improving other aspects of the environment, e.g., improved soil fertility, decreased erosion, or greater profitability by improved yield of agricultural or forestry products. There are a number of management practices available that could be implemented to protect and enhance existing C sinks now, and in the future, i.e., a no regrets policy. Smith and Powlson (2003) developed these arguments for soil sustainability, but the no regrets policy option is equally applicable to soil carbon sequestration. Since such practices are consistent with, and may even be encouraged by, many current international agreements and conventions, their rapid adoption should be encouraged as widely as possible.

18.6 Future Challenges

18.6.1 Improving Carbon Sequestration

The soil carbon sequestration strategy is the common link between three U.N. Framework Conventions: Climate Change, Biodiversity and Desertification Control. Soil C sequestration could potentially reduce the rate of enrichment of atmospheric CO_2 by 0.6 to 1.2 Pg C yr^{-1} over about 50 years, while enhancing biodiversity and controlling desertification (Lal 2004b). Restoration of degraded/desertified soils and ecosystems is an important ancillary benefit of soil C sequestration.

Of all the ancillary benefits, the impact of soil C sequestration in advancing the global food security can neither be ignored nor over-emphasized. Severe soil degradation is a common problem among all regions threatened by food insecurity including sub-Saharan Africa, Central and South Asia, Andean region, the Caribbean etc. The soil C pool in these regions has been strongly depleted because of the extractive agricultural practices such as complete residue removal for fodder and fuel, uncontrolled and excessive grazing, nutrient mining caused by low external input of chemical fertilizers and organic amendments etc. Increasing the soil C pool is necessary to improving its quality and enhancing productivity of crops, pastures and tree plantations grown on them. The soil C pool of many soils of South Asia and sub-Saharan Africa is as low as 0.1 to 0.2%, in contrast to the critical limit of the soil C concentration of 1.1% for maintenance of an adequate level of soil quality (Aune and Lal 1998). The entry point for breaking the vicious cycle of: degraded soils-depleted soil C pool-low crop yields-poverty-hunger-further degraded soils, is through adopting land-use and management systems which restore soil C pools (Lal 2004b). Lal (2004b) argues that because of the beneficial role of soil organic matter content and soil fertility, there is a very close link between soil C sequestration and the global food security. Indeed, soil C sequestration is an important ancillary benefit of the inevitable necessity of enhancing soil quality for feeding the global population of 6 billion in 2000, which is expected to reach 7.5 billion by 2020 and 9.4 billion by 2050.

18.6.2 Monitoring Soil Carbon Sequestration

Since there is a significant role for carbon sequestration in climate mitigation in the next 20–30 years, it will be essential to accurately monitor the amounts of carbon sequestered in order to estimate its role in closing the carbon emission gap. Signatory nations to the United Nations Framework Convention on Climate Change (UNFCCC) are required to report detailed inventories of greenhouse gas emissions due to Forestry and Land-Use Change, while Article 3.4 of the Kyoto Protocol lays the foundations for inclusion of human-induced changes in fluxes of C from agricultural soils and Land-Use Change and forestry in Annex I countries' commitments. Under Article 3.4, parties electing to include cropland management, grazing land management and re-vegetation need to account for changes in these soil carbon sinks and sources on a net-net basis, comparing the net flux of carbon from a given activity during the commitment period with the equivalent net flux of carbon in the baseline year. Sinks and sources of carbon must be accounted for 'taking into account uncertainties, transparency in reporting, verifiability'. Table 18.5 summarises measurement methods for assessing losses or accumulation of carbon on land (compiled from IPCC 2000b).

There are methodological, and other issues, such as cost, associated with each of the assessment methods described in Table 18.5 (Smith 2004d). For example, levels of soil sampling adequate to represent spatial variation, overcome resolution problems (detectable levels of change) and satisfy verification requirements could be extremely costly, depending on the scale and design of C sequestration projects. To achieve the highest levels of certainty, any chosen method (such as stock changes) would need to be backed up by a second independent assessment method such as flux methods, or models. However, the IPCC currently consider that flux methods are not sufficiently reliable to be used as a primary estimation method, and in addition, flux measurement is expensive and it is difficult to factor out the individual contributions of soil, roots and above-ground vegetation. Further, with any measurement system, it is very difficult to demonstrate that changes in carbon stocks are directly due to human induced activity rather than other causes such as nitrogen deposition or carbon dioxide fertilization – although models could have a role to play in this regard.

At its most stringent, verifiability under Kyoto activities would require the sampling of each geo-referenced piece of land subject to an Article 3.4 activity at the beginning and end of a commitment period, using a sampling regime that gives adequate statistical power. Soil and vegetation samples would be archived and the data from each piece of land aggregated to produce a national figure. Separate methods would be required to deliver a second set of independent verification data (Smith 2004d). Such an undertaking at the national level would likely be prohibitively expensive. At its least stringent, verifiability would entail

Table 18.5. Measurement methods for assessing losses or accumu-lations of carbon on land. Compiled from information in IPCC (2000b) after Smith (2004d)

Stock change measurements methods	Vegetation inventory Stemwood volume – forest inventory Total tree biomass – allometry Wood products – models of wood products Soil and litter Woody debris – volume and mass measured Litter – sampling and carbon analysis – highly spatially variable Mineral soil – sampling and carbon analysis – highly spatially variable *(Sampling strategy, methods and sampling depth all need to be considered)*
Flux measurement methods	Chambers, eddy covariance – for scales less 1 km^2 Tall towers, balloons for convective boundary layer budgeting – landscape, regional scale Flask measurements and flux measurements from aircraft; coupled with inversion analysis – continental scale
Remote sensing to determine geographic extent and change	Current resolution (NOAA-AVHRR) is 1 km^2 but 30 m possible soon Geographic extent possible, vegetation type possible, residue over, tillage, and perhaps soil organic carbon and moisture content of bare soil will become possible in near future
Models	To be used in combination with the above methods

the reporting of areas under a given practice (without geo-referencing) and the use of default values for a carbon stock change for each practice, to infer a change for all areas under that practice. A less stringent definition of verifiability would allow simple methods, such as those derived from IPCC default values for CO_2 fluxes from soil, to be used for estimating changes in soil carbon. These may enable low-level verifiability to be achieved by most parties by the beginning of the first commitment period (2008–2012).

The need to provide information on uncertainty in carbon sequestration estimates presents further challenges, not least given that IPCC (2000b) recommended the use of models in combination with direct sampling methods. For example, consider the use of a dynamic modeling approach for carbon sequestration estimates. Such models are usually tested against data sets of long-term changes in SOC, but many data sets have only mean SOC values available at each sample date, with no estimates of error about the mean. Falloon and Smith (2003) showed that when using data sets that do not include estimates of error about the mean, it is not possible to reduce the error (root mean squared error) between modeled and measured values below 6.8–8.5%, even with site-specific model calibration. Using error as an indicator of the certainty that can be attached to model projections, a significant reduction in uncertainty would be needed for Kyoto Protocol accounting. This could be achieved by better replication of soil measurements at benchmark sites, which would allow model error to be separated from measurement error. Thus more comprehensive model testing could be completed, and more certainty could be attached to model projections. This would be a practical option for benchmark testing of models at a small number of global sites, but would be prohibitively expensive to implement were it needed for more routine monitoring assessments at a large number of sites. At the national scale, information on uncertainty is even scarcer (Smith 2004d).

The challenges for the future lie in developing robust, transparent carbon accounting and verification systems which combine models with a variety of direct and indirect measurement techniques which are repeatable and can be performed quickly and cheaply, allowing for projections to be verified by two independent data sets where possible. Collection of information to contribute to uncertainty estimates in these assessments is crucial, and replication of soil measurements at global benchmark sites is an example of how this might be achieved. Modeling techniques which can separate the influence of human-induced changes in activity on changes in carbon stocks from changes driven by other causes will help contribute to these aims. These issues should be met with urgency if carbon sequestration is to play the significant role it deserves in climate mitigation in the next 20–30 years.

18.7 Summary and Conclusions

There is considerable biological potential to reduce greenhouse gas emissions from agricultural soils but many factors prevent the full biological potential being realized. When considering greenhouse gas mitigation, it is important to consider all of the greenhouse gases together as a management practice suitable for reducing one gas may increase emissions of another. Successful greenhouse gas mitigation options for agricultural soils will likely be those that provide other economic and environmental benefits and win-win strategies should be targeted. In the long term, soil-based greenhouse gas mitigation options (including carbon sequestration) can play only a small role in reducing the gap between projected emissions and the reduction in emissions necessary to achieve atmospheric CO_2 stabilization. Nevertheless, since it is critical to reduce greenhouse gas emissions over the next 20–30 years to achieve CO_2 stabilization within a century, and since there is no single solution, soil-based greenhouse gas mitigation options should form part of a broad portfolio of measures aimed at reducing greenhouse gas emissions.

References

Aune J, Lal R (1998) Agricultural productivity in the tropics and critical limits of Oxisols, Ultisols and Alfisols. Tropical Agriculture 74:96–103

Batjes NH (1996) Total carbon and nitrogen in the soils of the world. European Journal of Soil Science 47:151–163

Cannell MGR (2003) Carbon sequestration and biomass energy offset: Theoretical, potential and achievable capacities globally, in Europe and the UK. Biomass and Bioenergy 24:97–116

Falloon P, Smith P (2003) Accounting for changes in soil carbon under Kyoto: Long-term data-sets need to be improved to reduce uncertainty associated with model projections. Soil Use and Management 19:265–269

Franko U (1997) Modelling approaches of soil organic matter turnover within the CANDY system. In: Powlson DS, Smith P, Smith JU (eds) Evaluation of soil organic matter models. Springer-Verlag, Berlin, Heidelberg, New York, pp 247–254

Franko U, Schenk S (2001) Modelling of Carbon Dynamics in a Rural Area of Central Germany. In Sustainable Management of Soil Organic Matter. Eds. Rees RM, Ball BC, Campbell CD, Watson CA. pp 60–66. CABI, Wallingford, UK

Freibauer A, Rounsevell M, Smith P, Verhagen A (2004) Carbon sequestration in European agricultural soils. Geoderma 122:1–23

Houghton RA, Hackler JL, Lawrence KT (1999) The US carbon budget: contributions form land-use change. Science 285:574–578

IPCC (2000a) Special Report on Emissions Scenarios. Cambridge University Press, Cambridge, UK

IPCC (2000b) Special Report on Land Use, Land-Use Change and Forestry (SR-LULUCF), Cambridge University Press, Cambridge, UK

IPCC (2001) Climate change 2001: The scientific basis (Contribution of Working Group I to the third assessment report of the Intergovernmental Panel on Climate Change), Houghton JT, Ding Y, Griggs DJ, Noguer M, van der Linden PJ, Dai X, Maskell K, Johnson CA (eds) Cambridge University Press, Cambridge, UK

Janssens IA, Freibauer A, Ciais P, Smith P, Nabuurs G-J, Folberth G, Schlamadinger B, Hutjes RWA, Ceulemans R, Schulze E-D, Valentini R, Dolman H (2003) Europe's biosphere absorbs 7–12% of anthrogogenic carbon emissions. Science 300:1538–1542

Lal R (1999) Soil management and restoration for C sequestration to mitigate the accelerated greenhouse effect. Progress in Environmental Science 1:307–326

Lal R (2004a) Soil carbon sequestration to mitigate climate change. Geoderma 123:1–22

Lal R (2004b) Soil carbon sequestration impacts on global climate change and food security. Science 304: 11 June 2004

Lal R, Follett RF, Kimble JM (2003) Achieving soil carbon sequestration in the U.S.: A challenge to the policy makers. Soil Science 168:1–19

MacKenzie AF, Fan MX, Cadrin F (1998) Nitrous oxide emission in three years as affected by tillage, corn-soybean-alfalfa rotations, and nitrogen fertilization. Journal of Environmental Quality 27:698–703

McCarl BA, Schneider UA (2001) Climate change – Greenhouse gas mitigation in US agriculture and forestry. Science 294:2481–2482

Metting FB, Smith JL, Amthor JS (1999) Science needs and new technology for soil carbon sequestration. In: Rosenberg NJ, Izaurralde RC, Malone EL (eds) Carbon sequestration in soils: Science, monitoring and beyond. Battelle Press, Columbus, OH. pp 1–34

Paustian K, Cole CV, Sauerbeck D, Sampson N (1998) CO_2 mitigation by agriculture: An overview. Climatic Change 40:135–162

Robertson GP (2004) Abatement of nitrous oxide, methane and other non-CO_2 greenhouse gases: the need for a systems approach. In: Field CB, Raupach MR (eds) The Global Carbon Cycle. Integrating humans, climate, and the natural world, Scope 62, Island Press, Washington DC, pp 493–506

Robertson GP, Paul EA, Harwood RR (2000) Greenhouse gases in intensive agriculture: Contributions of individual gases to the radiative forcing of the atmosphere. Science 289:1922–1925

Romanenkov VA, Sirotenko OD, Shevtsova LK (2001) Modelling carbon balance of the agroecosystems based on the database of long-term experiments. In: Proc. of International Symposium Functions of Soils in the Geosphere-Biosphere Systems, 2001. Moscow State University Press. pp 170–171

Schimel DS, House JI, Hibbard KA, Bousquet P, Ciais P, Peylin P, Braswell BH, Apps MJ, Baker D, Bondeau A, Canadell J, Churkina G, Cramer W, Denning AS, Field CB, Friedlingstein P, Goodale C, Heimann M, Houghton RA, Melillo JM, Moore B, Murdiyarso D, Noble I, Pacala SW, Prentice IC, Raupach MR, Rayner PJ, Scholes RJ, Steffen WL, Wirth C (2001) Recent patterns and mechanisms of carbon exchange by terrestrial ecosystems. Nature 414:169–172

Shevtsova L, Romanenkov V, Sirotenko O, Smith P, Smith JU, Leech P, Kanzyvaa S, Rodionova V (2003) Effect of natural and agricultural factors on long-term soil organic matter dynamics in arable soddy-podzolic soils – modelling and observation. Geoderma 116:165–189

Sirotenko OD, Abashina EV, Romanenkov VA (in press) Simulation the effect of climate change on soil carbon dynamics in arable soils, CO_2 emission and agroecosystems productivity. Meteorology and Hydrogeology 4

Six J, Ogle SM, Breidt FJ, Conant RT, Mosier AR, Paustian K (2004) The potential to mitigate global warming with no-tillage management is only realized when practiced in the long term. Global Change Biology 10:155–160

Smith P (2004a) Engineered biological sinks on land. In: Field CB, Raupach MR (eds) The Global Carbon Cycle. Integrating humans, climate, and the natural world, Scope 62, Island Press, Washington DC, pp 479–491

Smith P (2004b) Soils as carbon sinks – the global context. Soil Use and Management 20:212–218

Smith P (2004c) Carbon sequestration in croplands: the potential in Europe and the global context. European Journal of Agronomy 20:229–236

Smith P (2004d) Monitoring and verification of soil carbon changes under Article 3.4 of the Kyoto Protocol. Soil Use and Management 20:264–270

Smith P, Goulding KW, Smith KA, Powlson DS, Smith JU, Falloon PD, Coleman K (2001) Enhancing the carbon sink in European agricultural soils: Including trace gas fluxes in estimates of carbon mitigation potential. Nutrient Cycling in Agroecosystems 60:237–252

Smith P, Powlson DS (2003) Sustainability of soil management practices – a global perspective. In: Abbott LK, Murphy DV (eds) Soil Biological Fertility – A Key To Sustainable Land Use In Agriculture. Kluwer Academic Publishers, Dordrecht, Netherlands, pp 241–254

Smith P, Powlson DS, Glendining MJ, Smith JU (1997) Potential for carbon sequestration in European soils: preliminary estimates for five scenarios using results from long-term experiments. Global Change Biology 3:67–79

Smith P, Powlson DS, Smith JU, Falloon PD, Coleman K (2000) Meeting Europe's climate change commitments: Quantitative estimates of the potential for carbon mitigation by agriculture. Global Change Biology 6:525–539

Smith P, Ambus P, Amézquita MA, Andrén O, Arrouays D, Ball B, Boeckx P, Brüning C, Buchmann N, Buendia L, Cellier P, Cernusca A, Clifton-Brown J, Dämmgen U, Ewert F, Favoino E, Fiorelli JL, Flechard C, Freibauer A, Hacala S, Harrison R, Hiederer R, Janssens I, Jayet PA, Jouany JP, Jungkunst H, Karlsson T, Kuikman PJ, Lagreid M, Leffelaar PA, Leip A, Loiseau P, Milford C, Neftel A, Oenema O, Ogle S, Olesen JE, Perälä P, Pesmajoglou C, Petersen SO, Pilegaard K, Raschi A, Regina K, Rounsevell M, Saletes S, Schils R, Seguin B, Sezzi E, Soussana JF, Stefani P, Stengel P, van Amstel A, Van Cleemput O, van Putten B, van Wesemael B, Verhagen AJ, Viovy N, Vuichard N, Weigel HJ, Weiske A, Willers HC (2004) CarboEurope GHG: Greenhouse gas emissions from European croplands. CarboEurope GHG, University of Tuscia, Viterbo, Italy pp 68

Sperow M, Eve MD, Paustian K (2003) Potential soil C sequestration on U.S. agricultural soils. Climatic Change 57:319–339

Vleeshouwers LM, Verhagen A (2002) Carbon emission and sequestration by agricultural land use: a model study for Europe. Global Change Biology 8:519–530

Chapter 19

Carbon and Water Tradeoffs in Conversions to Forests and Shrublands

Robert B. Jackson · Kathleen A. Farley · William A. Hoffmann · Esteban G. Jobbágy · Rebecca L. McCulley

19.1 Introduction

Carbon sequestration programs on land and in the oceans are gaining attention globally as a means to offset increasing fossil fuel emissions and atmospheric carbon dioxide concentrations (e.g., DeLucia et al. 1999; Caldeira and Duffy 2000; Schimel et al. 2000; Pacala et al. 2001; Hoffert et al. 2002; Jackson et al. 2002; Hungate et al. 2003; McNeil et al. 2003). Many industrialized nations now have national plans to foster land-based sequestration; Australia's Plantations 2020 program is one example of such a national effort (Polglase et al. 2000). Ocean based sequestration, particularly deep ocean pumping of CO_2 and iron fertilization, is also receiving considerable attention, although it remains even more controversial than land-based programs (e.g., Chisholm et al. 2001; Lawrence 2002; Buesseler and Boyd 2003; Tsuda et al. 2003). Despite uncertainties about the size and sustainability of sinks and markets, programs for emissions trading and carbon credits are underway, including the Chicago Climate Exchange and the European Union Greenhouse Gas Emission Trading Scheme.

On land, many biological sequestration programs emphasize storing carbon in soil organic matter in agricultural fields, in woody encroached sites, and in the soils and wood of plantations. Land-based sequestration in agricultural soils restores all or part of the soil organic carbon (SOC) lost with plowing and intensive agriculture (Gebhart et al. 1994; Lal et al. 1999). No-till and low-till management are additional approaches proposed for increasing soil organic carbon in croplands.

For plantations, the most controversial factors for carbon sequestration and management include the feasibility and permanence of the carbon sequestered, the scale of management needed to offset anthropogenic emissions, and the accompanying biogeochemical changes that would occur. As an example, a carbon sequestration rate of 3 000 kg C ha^{-1} annually in plantations would require an area the size of Texas or Pakistan to offset 0.2 Pg C yr^{-1} of emissions; such a rough calculation ignores economic and biophysical limitations to storage and downstream losses of carbon as the wood is processed. A more complete evaluation of the feasibility of carbon storage by vegetation management, both scientifically and economically, is needed, including a more complete biogeochemical accounting of the consequences. The biogeochemical interaction that we examine in this chapter is water availability; other key interactions, such as with nitrogen, are beyond the scope of this chapter (e.g., Vejre et al. 2001; Dalal et al. 2003).

Here we will examine some of the potential benefits of biological sequestration programs on land, some of the uncertainties surrounding them, and some unintentional consequences if they are initiated broadly. We will also address a related land-cover change, woody plant encroachment, which has important consequences for carbon and water cycling. Woody plant encroachment differs from afforestation and abrupt land-cover changes because it occurs over many decades. However, its global extent, potential for carbon sequestration, and similarities to afforestation make it important to address. For these land cover and land-use changes we will estimate potential carbon sequestration rates, explore key biophysical interactions, and discuss examples of other biogeochemical and hydrological changes that may occur. For example, plantations may be the most beneficial environmentally when they are used to ameliorate groundwater upwelling, but they may also decrease water yield (defined as the amount of water from a unit area of watershed) (Herron et al. 2002; Farley et al. 2005). Our long-term goal is to identify these biogeochemical and hydrological costs and benefits that accompany sequestration scenarios.

19.2 Afforestation

19.2.1 Afforestation: Carbon Storage Potential

For purposes of this chapter, afforestation is defined as the conversion of rangelands and agricultural lands to tree plantations. Forest plantations grown on rangelands or agricultural lands have large sequestration potentials because carbon can be stored rapidly in biomass pools that are large, have relatively slow turnover times compared to the previous vegetation, and have wider carbon to nutrient ratios (e.g., Paul et al. 2002; Halliday et al.

2003). Depending on land-use history, plantations can also increase or decrease carbon storage in the soil (e.g., Post and Kwon 2000; Guo and Gifford 2002; Farley et al. 2004). Summary data across plantations for total carbon storage in soil and wood reveal an average rate of C storage of ~3 600 kg C $ha^{-1} yr^{-1}$ (e.g., Hamilton et al. 2002), an order of magnitude larger than that in agricultural soils (Post and Kwon 2000). Based on this coarse estimated rate of carbon storage, ~280 million ha of plantations would be needed globally to reach a target of 1 Pg C yr^{-1}, approximately one-sixth of global carbon emissions in fossil fuels.

Sequestration rates a little higher than 3 600 kg C $ha^{-1} yr^{-1}$ are possible in some locations and in the short term, but significantly higher rates are unlikely over large areas (Jackson and Schlesinger 2004). Moreover, back-of-the-envelope estimates such as the one described above do not address issues such as permanence of carbon storage and leakage, defined as activities shifted to locations outside of a sequestration program that counteract some of its carbon benefits (e.g., Murray et al. 2004). One alternative that acknowledges these uncertainties is carbon "rental" payments, where landowners contract to store carbon for specific periods of time (Lewandrowski et al. 2004).

Additional costs also need to be included to estimate the technical and economic potentials of carbon storage, including site preparation and planting, potential carbon losses from disturbance (e.g., storms, pests, and fires), and post-harvest losses in timber use and processing (Jackson and Schlesinger 2004; Murray et al. 2004). Carbon stored in the soil or wood must be protected from plowing, fire, storm damage, and decomposition to keep the carbon from returning to the atmosphere. If long-term uses for the wood are not found, some of the plantation carbon will almost certainly return to the atmosphere after harvesting through uses such as burning for fuel or pulp and paper supply. Economic competition for land use must also be considered. Recent economic models for U.S. agriculture and forestry suggest that C prices would need to be ~U.S.$125 to $400 per metric ton C equivalents for potential sequestration in plantations to approach 0.2 Pg C yr^{-1} (McCarl and Schneider 2001; Lewandrowski et al. 2004).

19.2.2 Afforestation: Evapotranspiration and Water Yield

National policies promoting afforestation can lead to considerable carbon storage for decades, but the amount stored, the economic subsidies needed, and the environmental changes that would result require careful evaluation (e.g., McCarl and Schneider 2001). If economic incentives are provided to convert agricultural or range lands to plantations, what other social, economic, or biogeochemical changes might occur?

To provide a concrete example of such "ancillary" changes with afforestation, we examine changes in water yield. We explore the evidence for this interaction at a range of scales, including long-term catchment measurements and potential atmospheric feedbacks operating over large areas. The outcomes of those feedbacks depend on the initial land use, climate, and location of afforestation in complex but predictable ways.

Large changes in evapotranspiration (ET) and water yield are common with shifts between herbaceous and woody vegetation (Bosch and Hewlett 1982; Zhang et al. 2001; Brown et al. 2005; Farley et al. 2005). The effects of deforestation on water balance have been examined extensively at a range of scales that include eddy covariance measurements of canopy fluxes, multi-year catchment manipulations, and general circulation models examining regional mass and energy fluxes (e.g., Bosch and Hewlett 1982; Rannik et al. 2002; Werth and Avissar 2002). As just one example, Grace et al. (1998) found that daily evapotranspiration from a Brachiaria pasture in Brazil was 39% lower than in a nearby undisturbed rain forest, 2.74 mm d^{-1} compared with 4.48 mm d^{-1}, respectively. However, research has clearly shown that deforestation and afforestation are not opposite and reversible processes in terms of water yield (Robinson et al. 1991; Scott and Lesch 1997; Vertessy 1999).

In grasslands and agricultural lands, afforestation can reduce water yield from rivers and streams because canopy interception and evapotranspiration generally both increase (e.g., Herron et al. 2002; Farley et al. 2005). Interception tends to be relatively small in grasslands but can account for 10–20% of rainfall in hardwood systems and 20–40% in conifer plantations (e.g., Rao et al. 1979; Le Maitre et al. 1999; Levia and Frost 2003). For evapotranspiration, afforestation can alter leaf area, stomatal characteristics, surface roughness (a measure of the aerodynamic properties of the land surface), albedo, and the depth and density of root systems (Vertessy 2001; Brooks et al. 1997; Jackson et al. 2000). Holmes and Sinclair (1986) and Zhang et al. (1999) suggested that annual evapotranspiration from catchments planted to eucalypts or other plantation species could increase anywhere from ~50–250 mm compared to a grassland catchment. This change in ET could then affect other terms in the water budget, potentially decreasing annual stream flow, deep drainage, and rates of baseflow, as well as altering salinity.

In humid grasslands like those in the Pampas of Argentina, evapotranspiration is less than precipitation (P), and hydrological recharge takes place through deep drainage. Where groundwater is available, trees reverse the relationship between ET and P, initiating a net discharge regime (ET > P) in which groundwater use exceeds deep drainage (Fig. 19.1). The discharge regime is sustained by groundwater recharge and lateral transport from the surrounding grassland. Ground-

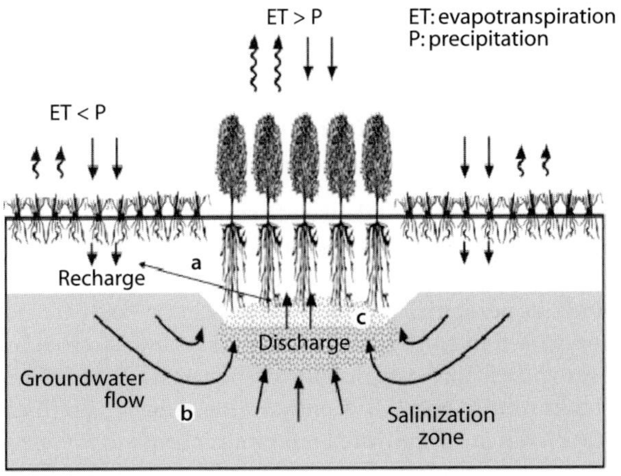

Fig. 19.1. Mechanism of groundwater use and salinization after afforestation of grasslands. Tree establishment reverses the net vertical flux of groundwater when absorption exceeds recharge. Groundwater exits the system through transpiration, leaving salts in the afforested stand (from Jobbagy and Jackson 2004)

water use through transpiration leaves dissolved salts behind that accumulate in the system if this net discharge regime is sustained. In the framework in Fig. 19.1, salinization is predicted to occur when three ecohydrological conditions are met: *(a)* grasses allow a moderate net hydrological recharge that is interrupted by trees, *(b)* the terrain and soil conductivity allow a sustained horizontal flow of groundwater and dissolved salts towards the tree stands, and *(c)* trees access groundwater. In the Pampas, afforested plots (10–100 ha in size) showed 2–19-fold increases in groundwater salinity (Jobbágy and Jackson 2004) compared to adjacent grasslands; in addition, surface soil pH typically decreased by a full pH unit, a change comparable in magnitude to the effects of acid rain (Jobbágy and Jackson 2003).

Globally, one of the best tools for examining the effects of afforestation on streamflow is long-term catchment datasets (e.g., Holmes and Sinclair 1986; Zhang et al. 2001). More recently, Farley et al. (2005) compiled catchment data in afforested grasslands and shrublands from 26 catchment studies with 504 annual observations, in part to stratify changes by original vegetation type and climate, and to quantify the effects of plantation age. Across the dataset, annual runoff decreased consistently in afforested catchments (Fig. 19.2, grassland data only; $P < 0.001$ for both grasslands and shrublands). Reductions in annual runoff in afforested grasslands and shrublands were similar in the first 5 years after tree establishment (–16% and –15%, respectively) but diverged as the plantations aged. Afforested grasslands reached a 50% reduction in runoff within a decade compared with a one-third decrease in afforested shrublands. Moreover, the data from grasslands show faster losses of runoff with eucalypt afforestation, sometimes within a decade (Fig. 19.2 and Farley et al. 2005).

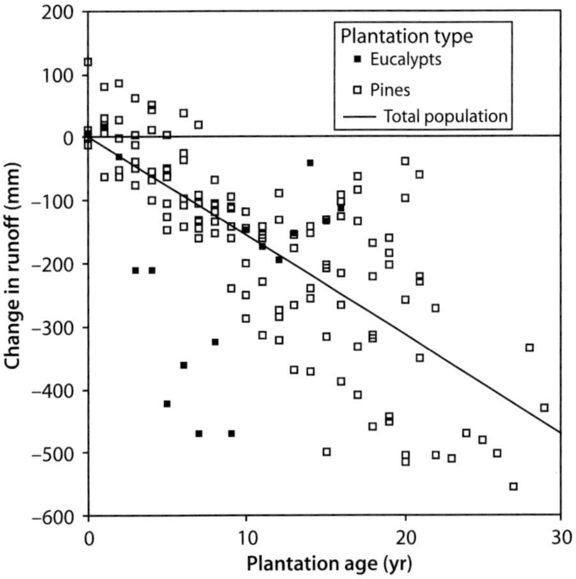

Fig. 19.2. Change in runoff with plantation age ($r^2 = 0.77$; $P < 0.001$). Data come from 26 catchment studies of afforestation globally and a database of 504 yearly observations (from Farley et al. 2005)

This synthesis clearly demonstrates that a loss of 100 to 500 mm (or one-third to three-quarters) of annual streamflow can be expected with afforestation at sites with mean annual precipitation >750 mm. Eucalypts tend to reduce streamflow more than pines do (Fig. 19.2 and Farley et al. 2005). Dye (1996) compared eucalypts and pines in South Africa and found that the rate of increase in evapotranspiration following afforestation was more rapid under eucalypts because of their faster growth and canopy closure; better data for poplar plantations would be helpful to compare with the more extensive pine and eucalypt datasets. While there is likely to be some variation in species effects by region, generalizations regarding the effects of different plantation species on runoff should be useful for planning afforestation projects and the tree species that will be used in them.

19.2.3 Afforestation: Potential Atmospheric Feedbacks

While afforestation typically increases evaporative water losses and reduces runoff locally, climate feedbacks at regional scales could also affect precipitation and water use. These regional effects on water availability depend on the location, extent, and patchiness of afforestation and operate through changes in albedo, roughness length, and water transport properties from soil to the atmosphere, including leaf area index, stomatal conductance, and rooting depth. These variables influence climate because they help determine the total amount of energy transferred between the vegetation and the atmosphere as well as the fraction of that energy driving evapotranspiration (latent heat) and warming of the air (sensible heat). Since little attention has been given to the direct climatic effects of afforestation, our understanding here must derive primarily from studies of deforestation.

Most simulations of tropical deforestation have concluded that the effect of the forest, relative to pasture, on precipitation is greater than its effect on evapotranspiration (Hahmann and Dickinson 1997; Hoffmann et al. 2003), such that runoff should be higher under forest than pasture. Similar effects of tree cover were observed in simulations of tropical savanna regions (Hoffmann and Jackson 2000, 2002; Clark et al. 2001). Analogously, at coarse scales the conversion of agricultural lands and grasslands to homogeneous forest cover might be predicted to increase precipitation slightly, in part because of increased water recycling through convective storms. Whether this increase is enough to offset completely the increased evapotranspiration of the forests is unclear.

In temperate regions, however, published simulations suggest otherwise. Although simulation studies have sometimes found that tree cover increases precipitation in temperate regions, this effect has been consistently smaller than the effect on ET, indicating that complete afforestation should result in reduced runoff (e.g., Xue et al. 1996; Bounoua et al. 2000). Furthermore, under some conditions, afforestation is likely to *reduce* precipitation, particularly near bodies of water, where actively transpiring vegetation might reduce the strength of the sea breeze and associated convection and cloud formation (Pielke and Avissar 1990). Simulations of land-use change in the US indicate that summer precipitation should decline with afforestation of the southern coastal plain, Florida, and the Great Lakes region (Copeland et al. 1996; but see Pielke et al. 1999). In such cases, the climate feedbacks resulting from afforestation should magnify the local effects of plantations on water yield, resulting in even lower water yield than predicted by catchment studies alone.

The pattern and scale of land-cover change in some cases may be as important as the switch between trees and herbaceous vegetation. Circulation patterns similar to a sea breeze can arise at the interface between forest and non-forest vegetation (Pielke and Avissar 1990; Pielke et al. 1997), which is particularly relevant over heterogeneous landscapes typical of afforestation. Where heterogeneity promotes these mesoscale circulations, convection cells develop where ET is lowest (Pielke and Avissar 1990; Pielke et al. 1997), leading to the rather counterintuitive observations that cloud formation can be greater over nonforest patches (Roy and Avissar 2002; Weaver et al. 2002). A patchwork of plantation and nonforest vegetation may therefore generate more precipitation than homogeneous forest vegetation, particularly under conditions of weak synoptic forcing, such as when prevailing winds are weak (Pielke et al. 1997).

19.3 Woody Encroachment and Agriculture

19.3.1 Grassland Conversions with Woody Plant Encroachment and Agriculture

Shifts among grasslands, woody communities, and croplands represent another form of vegetation change important for carbon cycling today. The conversion of croplands to perennial vegetation consistently leads to carbon stored in soil organic matter (Post and Kwon 2002; Guo and Gifford 2002). This carbon storage benefit is recognized in the U.S. Conservation Reserve Program (CRP), which has provided economic incentives for farmers to convert crops to perennials in ~15 million ha to date (Skold 1989; Jackson and Schlesinger 2004).

At the regional scale, potential carbon storage in croplands can be estimated by combining observed sequestration rates through programs like the CRP or no-till agriculture with the extent of agricultural lands in a region. Recent reviews of more than 100 observations show that SOC increased ~450 kg C ha^{-1} yr^{-1} after croplands converted to pastures or no-till management (Post and Kwon 2000; West and Post 2002), peaking five to ten years after conversion and slowing considerably within two decades (West and Post 2002). Taking the U.S. as an example, if all ~132 000 000 ha of croplands in production in 2001 were converted to no-till agriculture or, less likely, were retired through the CRP, potential sequestration rates of 0.059 Pg C yr^{-1} might be possible for several decades. This amount would be ~4% of total U.S. fossil fuel emissions of ~1.6 Pg yr^{-1}. Currently, the ~15 million ha enrolled in the CRP store an order of magnitude less than this technical potential, ~0.005 Pg C yr^{-1} (Jackson and Schlesinger 2004).

Conversion of croplands back to perennial grassland is not the only change in land use likely to affect C pools in grasslands. Many regions of the world are currently undergoing woody plant encroachment, a process enhanced by management practices such as intensive grazing and fire suppression and possibly by interactions with climate change (Kucera 1960; Bragg and Hulbert 1976; Briggs et al. 2002). The soil C consequences of woody encroachment into native grasslands vary considerably, ranging from gains of soil C in some ecosystems (Archer et al. 2001) to no net change or losses in others (Kieft et al. 1998; Schlesinger and Pilmanis 1998; Gill and Burke 1999; Tilman et al. 2000; Smith and Johnson 2003). Recent work suggests that the changes in soil carbon storage are related to climate, with drier grasslands more likely to gain soil C and productive sub-humid/humid grasslands to lose it (Jackson et al. 2002).

From a soil C perspective, the primary difference between woody plant encroachment into native grasslands and into old agricultural fields is the quantity, quality, and distribution of soil C at the onset of woody plant

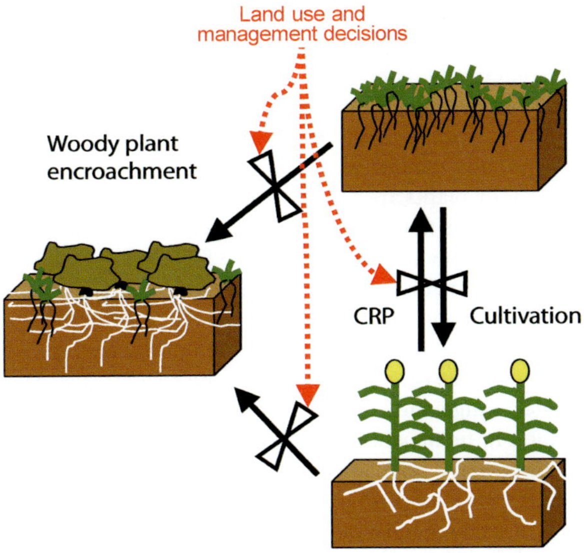

Fig. 19.3. Conceptual diagram of how land-use and management decisions alter the structure and function of grassland ecosystems (*CRP* = Conservation Reserve Program)

establishment (McCulley et al. 2005). In tallgrass prairie, for example, agricultural cultivation significantly reduces soil C pools (Burke et al. 1989; Huggins et al. 1998; Knops and Tilman 2000). Thus, unlike woody plant encroachment into native grasslands where the soil C response varies depending on the climate, land management, and species involved, woody plant establishment and dominance in many old agricultural fields should replenish soil C pools. Indeed, this hypothesis is supported by Guo and Gifford (2002) who found that converting cropland to secondary forest or plantations increased soil C pools by 53% and 18%, respectively.

19.3.2 Processes Controlling Soil C Storage: Grassland Vs. Woodland

The amount of C stored within the soil and the rate at which C accumulates and turns over is determined largely by the quantity and quality of C inputs, soil properties such as clay or sand content, and possibly the microbial community present. Edaphic properties such as texture are unlikely to change dramatically within the time-scale of plant establishment on abandoned cropland or conversion of grassland to woodland. However, major changes in other controlling variables, such as the quantity and quality of C inputs through net primary productivity as well as microbial community composition are likely.

In terms of carbon inputs to the soil, many differences between herbaceous and woody species exist. For example, grassland species tend to allocate a higher proportion of fixed C to belowground root systems than woody species do (Jackson et al. 1997). In addition, the two vegetation types differ in root morphologies and distributions, with grasses having a shallower, more densely fibrous root system and woody plants typically having deeper, coarser roots (Canadell et al. 1996; Schenk and Jackson 2002a,b). Thus, woody dominated systems may deliver more C than grasslands to the soil surface via litter while increasing C inputs deep within the soil profile. The quality of litter inputs, both above- and belowground, often differs significantly between the two types as well, with woody material containing more lignin and secondary compounds. All of these factors interact to alter the rate of decomposition, the soil microbial community, and ultimately the quantity, quality, and depth distribution of soil C storage.

Previous studies showing that differences in carbon inputs between grasslands and woodlands alter soil C storage come primarily from global analyses and woody encroachment studies in rangelands. Global analyses of root distributions and soil organic C contents suggest that converting croplands to grasslands or woodlands alters soil C profiles through changes in the amount and depth of root C inputs (Jackson et al. 1996; Jobbágy and Jackson 2000). More specifically, a comparison of grasslands and forests in sub-humid climates, such as tallgrass prairies experience, found that the proportion of total soil organic C stored in the top 20 cm of the soil profile was higher in forests; in contrast grasslands stored proportionally more soil C deeper within the profile (Jackson et al. 2000). Using the fact that woody and herbaceous plants often differ in photosynthetic pathways and produce organic matter with different ^{13}C signatures (Smith and Epstein 1971), woody plant encroachment studies have shown that C contributions to the soil are highest at the surface soil layers (Gill and Burke 1999; Smith and Johnson 2003), perhaps resulting from greater inputs of litter at the soil surface (Connin et al. 1997; Boutton et al. 1999; Hibbard et al. 2001). Moreover, in woodlands or brush, the presence of a thicker litter layer may result in a relatively high proportion of aboveground litter decomposing before entering the mineral soil. Alternatively, if such material had high C/N ratios, it might be slow to decompose and enhance longer-term soil C sequestration.

Besides altering the depth distribution of soil C storage, grasslands and woodlands differ in soil C quality, as measured by soil organic matter C/N ratios, organic C chemical constituents, and the relative percentage residing in active, slow, and passive soil C pools. In general, woody dominated areas have higher total soil C/N ratios, greater quantities of complex organic molecules within soil organic matter, higher percentages of slowly turning over soil C pools, and, consequently, longer mean residence times for C within the soil than grasslands have (Connin et al. 1997; Boutton et al. 1998; Gill and Burke 1999; Martens et al. 2003; Smith and Johnson 2003; McCulley et al. 2005).

19.3.3 Uncertainties in Water and Carbon Balances with Woody Plant Encroachment

The ecohydrological consequences of woody plant encroachment are much less clear than for afforestation (Wilcox 2002; Petheram et al. 2002; Huxman et al. 2005). Because the ratio of evaporation to transpiration increases with aridity, plants have a proportionally smaller role in regulating total water fluxes in arid systems. In addition, streamflow and evapotranspiration are differentially affected by woody encroachment, depending on the degree and seasonality of aridity and the availability of subsurface water (Huxman et al. 2005).

Despite uncertainties, a few generalizations are possible. Streamflow may decline with woody encroachment in semi-arid landscapes dominated by subsurface flow, including systems with available groundwater or karst systems where water flows quickly into the subsoil (Jackson et al. 1999; Seyfried et al. 2005). There, the deeper root systems of woody plants can capture water that is otherwise unavailable to herbs. Similarly, woody encroachment may increase bare soil evaporation in semiarid ecosystems, because of the increasingly patchy vegetation cover (Huxman et al. 2005). Explicitly considering the ecohydrology associated with vegetation change provides important information on the consequences of woody plant encroachment for carbon storage.

In addition to woody plant encroachment, woody plant or brush "thickening" is likely to decrease stream flow. Brush thickening is defined as an increase in woody plant density at sites where woody species were already present. Like woody encroachment into grasslands, thickening has occurred in many regions of the world, including the western U.S. and Australia (Van Auken 2000; Fensham and Fairfax 2003; Hicke et al. 2004). In practice, brush thickening will affect stream flow through similar mechanisms to those occurring with afforestation and woody encroachment. All three land-cover changes can increase leaf area, evapotranspiration, and rainfall interception, potentially decreasing stream flow. However, the more arid a system is, the more uncertain the outcome for water yield will be, as evapotranspiration from the soil plays an increasingly important role relative to plant transpiration at drier sites.

How much more carbon is stored in shrublands and woodlands compared to grasslands remains uncertain. Decades of research have described the formation of "islands of fertility" under single plants and plant clusters as woody species invade grasslands. This process increases heterogeneity in cover and net primary productivity, concentrating soil nutrients and carbon under the woody plants and depleting them by erosion and other processes in the interspaces. Using such shrub islands and herbaceous interspaces as endpoints to compare encroached and intact grassland systems may underestimate the carbon stored in the original grasslands and overestimate it in the expanding woodlands. This is especially true when the microsites first invaded by the woody plants are the deepest, most productive soils. When examined for entire ecosystems, data from 242 sites show increasingly more SOC for grasslands than for shrublands/woodlands as precipitation increases to 1 000 mm (Fig. 19.4).

Land-use history is a crucial lens for viewing the consequences of vegetation and land-use change. Degraded rangelands can be actively restored, and store carbon in turn, using any type of vegetation. Transformations in the opposite direction, however, from shrubland or woodlands to grasslands, are also enlightening. Recent meta-analysis of >100 studies emphasizes the potential for SOC gains at sites originally dominated by woodlands when converted to grasslands (Conant et al. 2001) and soil carbon losses when pastures are converted to plantations (Guo and Gifford 2002). Grasslands are deceptively productive ecosystems at both short and long time scales (e.g., Retallack 2001).

U.S. carbon budgets estimate the sequestration due to woody plant encroachment to be ~0.12 to 0.13 Pg C yr^{-1} but acknowledge large uncertainties in this estimate (e.g., Houghton et al. 1999). We believe the correct estimate will eventually be positive but smaller, both because of the relationship between soil carbon stor-

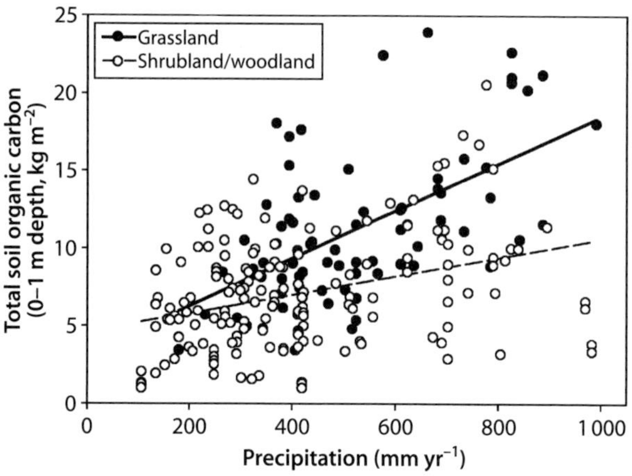

Fig. 19.4. An analysis of soil organic carbon (kg m^{-2}) vs. mean annual precipitation (mm) for the top meter of soil. The data are derived from the National Soil Characterization Database of the U.S. Department of Agriculture and the World Inventory of Soil Emission Potential Database of the International Soil Reference and Information Centre. Across the global dataset, the slope of the relationship of SOC vs. precipitation was 2.6 times higher for grassland vegetation than for shrublands/woodlands ($p = 0.001$). Whereas grassland SOC was statistically indistinguishable from values for woody plants at 200 mm MAP (Mean Annual Precipitation), woodlands had 43% less total SOC than grasslands at 1 000 mm ($p < 0.01$) (from Jackson et al. 2002)

age and precipitation described earlier (Jackson et al. 2002) and because the land area attributed to woody encroachment is over-estimated – 220 million ha, or two-thirds of the continental U.S. excluding forests and croplands (Houghton et al. 1999; Pacala et al. 2001). Better constraining the carbon sink for woody encroachment will entail a combination of field research, remote sensing, and spatially explicit modeling that takes into account precipitation (and climate in general), land-use history, erosion, and other disturbances. Patch-based comparisons of vegetation yield important ecological insight but must be scaled up appropriately for conclusions at ecosystem, landscape, and regional scales.

19.4 Summary and Conclusions

The environmental consequences of afforestation and other biologically based sequestration programs have received much less attention than their carbon sequestration potential (e.g., Herron et al. 2002; Jackson and Schlesinger 2004; Farley et al. 2005). Potential sequestration rates are high but will require managing enormous land areas. One of the most important issues in how to implement land-use change projects within the Clean Development Mechanism of the Kyoto Protocol is to understand their total effects on local livelihoods and environments (Pedroni 2003). For plantations, the synthesis described above clearly shows that a reduction in runoff can be expected after afforestation of grasslands and shrublands. In a few locations such as part of Australia where lower runoff can ameliorate salinity and groundwater upwelling, this may be a positive change. In other regions, the reduction in runoff could cause water shortages, a tradeoff that should be acknowledged before policies promoting afforestation are implemented. One challenge and opportunity to the research community is to help policy makers choose locations for afforestation that are as economical and environmentally beneficial as possible (Pimentel et al. 2004).

The ability to predict the likely effects of afforestation in specific places will be the biggest challenge to zoning and planning for these projects (Farley et al. 2005). Catchment data are collected over decades and are unavailable in most places. However, we can use indicators, such as the change in runoff as a percent of mean annual precipitation at a site, to gain insight into the likely severity of the loss of runoff. Synthesis data suggest that trees are able to use 15 to 20% more precipitation than grasses, so that in a region where natural runoff is less than 10% of mean annual precipitation, afforestation may result in a complete loss of runoff, whereas in places where natural runoff is 30% of precipitation, it may be cut in half when trees are planted (Farley et al. 2005). This information can be useful to land managers and policy makers in guiding the location of plantations with respect to other water demands. There are also many other changes that need to be examined through sequestration scenarios. For example, the potential effects on biodiversity would depend in part on the previous land use. If landowners choose to afforest agricultural lands, the effects on biodiversity would probably be minimal; if instead they choose to afforest native grasslands and rangelands, the consequences would be larger.

For woody plant encroachment in arid and semi-arid systems, the biggest uncertainties are the consequences for the water balance and groundwater recharge and its regional and global role in the carbon balance. Understanding the consequences for water yield is especially important because essentially all of the places where woody encroachment occurs are water limited for both plants and people.

Finally, this book summarizes some of the scientific progress in more than a decade of GCTE's existence (GCTE-Global Change and Terrestrial Ecosystems). As GCTE ends and the scientific community looks forward, the biggest need for global change research is to do a better job of combining research in the social and natural sciences. The natural science community cannot hope to predict the consequences of global change without input from economic and policy models. The plantation scenarios discussed above are a good example. In turn, economists and other social scientists need input on the environmental changes that are likely to affect people and economies in the future. This need is reflected in the next generation of GCTE, the Global Land Project (see Ojima et al. 2007, Chap. 25 of this volume). The Global Land Project is a joint effort of the International Geosphere-Biosphere Programme (IGBP) and the International Human Dimensions Programme (IHDP).

Fortunately, many good examples of such coordination are underway. One of these is the global desertification network, as presented in such projects as ARIDnet, a research network for studies of global desertification (Reynolds and Smith 2003; Reynolds et al. 2007, Chap. 20 of this volume; http://www.biology.duke.edu/aridnet/). ARIDnet and other related projects recognize the important dual effects of humans and the environment in understanding global change. Another example is SAHRA (Sustainability of semi-Arid Hydrology and Riparian Areas) based in Arizona (http://www.sahra.arizona.edu/). This project seeks to develop an integrated, multidisciplinary understanding of the hydrology of semi-arid regions, and to build partnerships with a broad spectrum of stakeholders (both public agencies and private organizations). With the multi-disciplinary philosophy that these and other projects display, we are well on our way to making our research stronger and more relevant to society.

References

Archer S, Boutton TW, Hibbard K (2001) Trees in grasslands: biogeochemical consequences of woody plant expansion. In: Schulze ED, Heimann M, Harrison S, Holland EA, Lloyd J, Prentice I, Schimel DS (eds) Global biogeochemical cycles in the climate system. Academic Press, San Diego, Calif., pp 115–138

Bosch JM, Hewlett JD (1982) A review of catchment experiments to determine the effect of vegetation changes on water yield and evapotranspiration. Journal of Hydrology 55:3–23

Bounoua L, Collatz GJ, Los SO, Sellers PJ, Dazlich DA, Tucker CJ, Randall DA (2000) Sensitivity of climate to changes in NDVI. Journal of Climate 13:2277–2292

Boutton TW, Archer S, Midwood AJ, Zitzer SF, Bol R (1998) Delta ^{13}C values of soil organic carbon and their use in documenting vegetation change in a subtropical savanna ecosystem. Geoderma 82:5–41

Boutton TW, Archer S, Midwood AJ (1999) Stable isotopes in ecosystem science: Structure, function and dynamics of a subtropical savanna. Rapid Communications in Mass Spectrometry 13:1263–1277

Bragg TB, Hulbert LC (1976) Woody plant invasion of unburned Kansas bluestem prairie. Journal of Range Management 29: 19–23

Briggs JM, Hoch GA, Johnson LC (2002) Assessing the rate, mechanisms, and consequences of the conversion of tallgrass prairie to *Juniperus virginiana* forest. Ecosystems 5:578–586

Brooks KN, Ffolliott PF, Gregersen HM, DeBano LF (1997) Hydrology and the management of watersheds, 2nd edition. Iowa State University Press, Ames, Iowa

Brown AE, Zhang L, McMahon TA, Western AW, Vertessy RA (2005) A review of paired catchment studies for determining changes in water yield resulting from alterations in vegetation. Journal of Hydrology 310:28–61

Buesseler KO, Boyd PW (2003) Will ocean fertilization work? Science 300:67–68

Burke IC, Yonker CM, Parton WJ, Cole CV, Flach K, Schimel DS (1989) Texture, climate, and cultivation effects on soil organic matter content in U.S. grassland soils. Soil Science Society of America Journal 53:800–805

Caldeira K, Duffy PB (2000) The role of the Southern Ocean in uptake and storage of anthropogenic carbon. Science 287:620–622

Canadell J, Jackson RB, Ehleringer JR, Mooney HA, Sala OE, Schulze ED (1996) Maximum rooting depth of vegetation types at the global scale. Oecologia 108:583–595

Chisholm SW, Falkowski PG, Cullen JJ (2001) Dis-crediting ocean fertilization. Science 294:309–310

Clark DB, Xue Y, Harding RJ, Valdes PJ (2001) Modeling the impact of land surface degradation on the climate of tropical North Africa. Journal of Climate 14:1809–1822

Conant RT, Paustian K, Elliott ET (2001) Grassland management and conversion into grassland: Effects on soil carbon. Ecological Applications 11:343–355

Connin SL, Virginia RA, Chamberlain CP (1997) Carbon isotopes reveal soil organic matter dynamics following arid land shrub expansion. Oecologia 110:374–386

Copeland JH, Pielke RA, Kittel TGF (1996) Potential climatic impacts of vegetation change: A regional modeling study. Journal of Geophysical Research 101:7409–7418

Dalal RC, Wang WJ, Robertson GP, Parton WJ (2003) Nitrous oxide emission from Australian agricultural lands and mitigation options: a review. Australian Journal of Soil Research 41:165–195

DeLucia EH, Hamilton JG, Naidu SL, Thomas RB, Andrews JA, Finzi A, Lavine M, Matamala R, Mohan JE, Hendrey GR, Schlesinger WH (1999) Net Primary Production of a Forest Ecosystem with Experimental CO_2 Enrichment. Science 284:1177–1179

Dye PJ (1996) Climate, forest, and streamflow relationships in South African afforested catchments. Commonwealth Forestry Review 75(1):31–38

Farley KA, Kelly EF, Hofstede RGM (2004) Soil organic carbon and water retention following conversion of grasslands to pine plantations in the Ecuadorian Andes. Ecosystems 7:729–739

Farley KA, Jobbagy EG, Jackson RB (2005) Effects of afforestation on water yield: a global synthesis with implications for policy. Global Change Biology 11:1565–1576

Fensham RJ, Fairfax RJ (2003) Assessing woody vegetation cover change in north-west Australian savanna using aerial photography. International Journal of Wildland Fire 12:359–367

Gebhart DL, Johnson HB, Mayeux HS, Polley HW (1994) The CRP increases soil organic-carbon. Journal of Soil and Water Conservation 49:488–492

Gill RA, Burke IC (1999) Ecosystem consequences of plant life form changes at three sites in the semiarid United States. Oecologia 121:551–563

Grace J, Lloyd J, Miranda AC, Miranda H, Gash JHC (1998) Fluxes of carbon dioxide and water vapour over a C_4 pasture in southwestern Amazonia (Brazil). Australian Journal of Plant Physiology 25:519–530

Guo LB, Gifford RM (2002) Soil carbon stocks and land use change: a meta analysis. Global Change Biology 8:345–360

Hahmann AN, Dickinson RE (1997) RCCM2-BATS Model over tropical South America: Applications to tropical deforestation. Journal of Climate 10:1944–1964

Halliday JC, Tate KR, McMurtrie RE, Scott NA (2003) Mechanisms for changes in soil carbon storage with pasture to *Pinus radiata* land-use change. Global Change Biology 9:1294–1308

Hamilton JG, DeLucia EH, George K, Naidu SL, Finzi AC, Schlesinger WH (2002) Forest carbon balance under elevated CO_2. Oecologia 131:250–260

Herron N, Davis R, Jones R (2002) The effects of large-scale afforestation and climate change on water allocation in the Macquarie River catchment, NSW, Australia. Journal of Environmental Management 65:369–381

Hibbard K, Archer S, Schimel DS, Valentine DW (2001) Biogeochemical changes accompanying woody plant encroachment in a subtropical savanna. Ecology 82:1999–2011

Hicke JA, Sherriff RL, Veblen TT, Asner GP (2004) Carbon accumulation in Colorado ponderosa pine stands. Canadian Journal of Forest Research 34:1283–1295

Hoffert MI, Caldeira K, Benford G, Criswell DR, Green C, Herzog H, Jain AK, Kheshgi HS, Lackner KS, Lewis JS, Lightfoot HD, Manheimer W, Mankins JC, Mauel ME, Perkins LJ, Schlesinger ME, Volk T, Wigley TML (2002) Advanced Technology Paths to Global Climate Stability: Energy for a Greenhouse Planet. Science 298:981–987

Hoffmann WA, Jackson RB (2000) Vegetation-climate feedbacks in the conversion of tropical savanna to grassland. Journal of Climate 13:1593–1602

Hoffmann WA, Schroeder W, Jackson RB (2002) Positive feedbacks of fire, climate, and vegetation and the conversion of tropical savanna. Geophysical Research Letters 29:2052-doi:10.1029/2002GL015424

Hoffmann WA, W Schroeder, Jackson RB (2003) Regional feedbacks among fire, climate, and tropical deforestation. Journal of Geophysical Research 108, No. D23, 4721, doi: 10.1029/2003JD003494

Holmes JW, Sinclair JA (1986) Water yield from some afforested catchments in Victoria. Hydrology and Water Resources Symposium: River Basin Management, Griffith University, Brisbane, 25–27 November 1986. Barton, A.C.T., Australia: Institution of Engineers, Australia

Houghton RA, Hackler JL, Lawrence KT (1999) The U.S. carbon budget: Contributions from land-use change. Science 285: 574–578

Huff FA, Changnon SA (1973) Precipitation modification by major urban areas. Bulletin of the American Meteorological Society 54:1220–1232

Huggins DR, Buyanovsky GA, Wagner GH, Brown JR, Darmody RG, Peck TR, Lesoing GW, Vanotti MB, Bundy LG (1998) Soil organic C in the tallgrass prairie-derived region of the corn belt: Effects of long-term crop management. Soil & Tillage Research 47:219–234

Hungate BA, Dukes JS, Shaw MR, Luo YQ, Field CB (2003) Nitrogen and climate change. Science 302:1512–1513

Huxman TE, Wilcox BP, Breshears DD, Scott RL, Snyder KA, Small EE, Hultine K, Pockman WT, Jackson RB (2005) Ecohydrological implications of woody plant encroachment. Ecology 86:308–319

Jackson RB, Schlesinger WH (2004) Curbing the U.S. carbon deficit. Proceedings of the National Academy of Sciences USA 101: 15827–15829

Jackson RB, Canadell J, Ehleringer JR, Mooney HA, Sala OE, Schulze ED (1996) A global analysis of root distributions for terrestrial biomes. Oecologia 108:389–411

Jackson RB, Mooney HA, Schulze E-D (1997) A global budget for fine root biomass, surface area, and nutrient contents. Proceedings National Academy of Sciences, USA 94:7362–7366

Jackson RB, Moore LA, Hoffmann WH, Pockman WT, Linder CR (1999) Ecosystem rooting depth determined with caves and DNA. Proceedings of the National Academy of Sciences, USA 96: 11387–11392

Jackson RB, Schenk HJ, Jobbágy EG, Canadell J, Colello JD, Dickinson RE, Field CB, Friedlingstein P, Heimann M, Hibbard K, Kicklighter DW, Kleidon A, Neilson RP, Parton WJ, Sala OE, Sykesm MT(2000) Belowground consequences of vegetation change and their treatment in models. Ecological Applications 10:470–483

Jackson RB, Banner JL, Jobbágy EG, Pockman WT, Wall DH (2002) Ecosystem carbon loss with woody plant invasion of grasslands. Nature 418:623–626

Jobbágy EG, Jackson RB (2000) The vertical distribution of soil organic carbon and its relation to climate and vegetation. Ecological Applications 10:423–436

Jobbágy EG, Jackson RB (2003) Patterns and mechanisms of soil acidification in the conversion of grasslands to forests. Biogeochemistry 64:205–229

Jobbágy EG, Jackson RB (2004) Groundwater use and salinization with grassland afforestation. Global Change Biology 10: 1299–1312

Kieft TL, White CS, Loftin SR, Aguilar R, Craig JA, Skaar DA (1998) Temporal dynamics in soil carbon and nitrogen resources at a grassland-shrubland ecotone. Ecology 79:671–683

Knops JMH, Tilman D (2000) Dynamics of soil nitrogen and carbon accumulation for 61 years after agricultural abandonment. Ecology 81:88–89

Kucera CL (1960) Forest encroachment in native prairie. Iowa State College Journal of Science 34:635–640

Lal R, Follett RF, Kimble J, Cole CV (1999) Managing US cropland to sequester carbon in soil. Journal of Soil and Water Conservation 54:374–381

Lawrence MG (2002) Side Effects of Oceanic Iron Fertilization. Science 297: 1993

Le Maitre DC, Scott DF, Colvin C (1999) A review of information on interactions between vegetation and groundwater. Water SA 25:137–152

Levia DF, Frost EE (2003) A review and evaluation of stemflow literature in the hydrologic and biogeochemical cycles of forested and agricultural ecosystems. Journal of Hydrology 274:1–29

Lewandrowski J, Peters M, Jones C, House R, Sperow M, Eve M, Paustian K (2004) USDA Economic Resource Service Report TB-1909

Martens DA, Reedy TE, Lewis DT (2003) Soil organic carbon content and composition of 130-year crop, pasture and forest land-use managements. Global Change Biology 10: 65–78

McCarl BA, Schneider UA (2001) Greenhouse gas mitigation in U.S. agriculture and forestry. Science 294:2481–2482

McCulley RL, Archer S, Boutton TW, Hons FM, Zuberer DA (2005) Soil respiration and nutrient cycling of wooded communities developing in grassland. Ecology 85:2804–2817

McNeil BI, Matear RJ, Key RM, Bullister JL, Sarmiento JL (2003) Anthropogenic CO_2 uptake by the ocean based on the global chlorofluorocarbon data set. Science 299:235–239

Murray BC, McCarl BA, Lee H (2004) Estimating leakage from forest carbon sequestration programs. Land Economics 80: 109–124

Ojima D, William O, McConnell J, Moran E, Turner III B, Canadell JG, Lavorel S (2007) The Future Research Challenge: The Global Land Project. In: Canadell J, Pataki D, Pitelka L (eds) Terrestrial ecosystems in a changing world. The IGBP Series, Springer-Verlag, Berlin

Pacala SW, Hurtt GC, Baker D, Peylin P, Houghton RA, Birdsey RA, Heath L, Sundquist ET, Stallard RF, Ciais P, Moorcroft P, Caspersen JP, Shevliakova E, Moore B, Kohlmaier G, Holland E, Gloor M, Harmon ME, Fan S-M, Sarmiento JL, Goodale CL, Schimel D, Field CB (2001) Consistent Land- and Atmosphere-Based U.S. Carbon Sink Estimates. Science 292:2316–2320

Paul KI, Polglase PJ, Nyakuengama JG, Khanna PK (2002) Change in soil carbon following afforestation. Forest Ecology and Management 168:241–257

Pedroni, L. 2003. Ruling on the "crunch issues" of LULUCF: impacts on project viability. International Journal of Global Energy Issues 20:75–94

Petheram C, Walker G, Grayson R, Thierfelder T, Zhang L (2002) Towards a framework for predicting impacts of land use on recharge: 1. A review of recharge studies in Australia. Australian Journal of Soil Resources 40:397–417

Pielke RA, Avissar R (1990) Influence of landscape structure on local and regional climate. Landscape Ecology 4:133–155

Pielke RA, Lee TJ, Copeland JH, Eastman JL, Ziegler CL, Finley CA (1997) Use of USGS-provided data to improve weather and climate simulations. Ecological Applications 7:3–21

Pielke RA, Walko RL, Steyaert LT, Vidale PL, Liston GE, Lyons WA, Chase TN (1999) The influence of anthropogenic landscape changes on weather in south Florida. Monthly Weather Review 127:1663–1673

Pimentel D, Berger B, Filiberto D, Newton M, Wolfe B, Karabinakis E, Clark S, Poon E, Abbett E, Nandagopal S (2004) Water resources: agricultural and environmental issues. BioScience 54:909–918

Polglase PJ, Paul KI, Khanna PK, Nyakuengama JG, O'Connell AM, Grove TS, Battaglia M (2000) Changes in soil carbon following afforestation or reforestation. National Carbon Accounting System Technical Report No. 20, CSIRO Forestry and Forest Products, Australia

Post WM, Kwon KC (2000) Soil carbon sequestration and land-use change: processes and potential. Global Change Biology 6:317–327

Rabin RM, Stadler S, Wetzel PJ, Stensrud DJ, Gregory M (1990) Observed effects of landscape variability on convective clouds. Bulletin of the American Meteorological Society 71:272–280

Rannik U, Altimir N, Raittila J, Suni T, Gamana A, Hussein T, Teemu Hölttä T, Lassila H, Latokartano M, Lauri A, Natsheh A, Petäjä T, Orjamaa R, Ylä-Mella H, Keronen P, Berninger F, Vesala T, Hari P, Kulmala M (2002) Fluxes of carbon dioxide and water vapour over Scots pine forest and clearing. Agricultural and Forest Meteorology 111:187–202

Rao AR, McKerchar AI, Pearce AJ (1979) Modeling and analysis of data from catchment studies of land-use change. Progress in Water Technology 11:579–597

Retallack GJ (2001) Cenozoic expansion of grasslands and climatic cooling. Journal of Geology 109:407–426

Reynolds JH, Stafford Smith M (eds) (2003) Global Desertification: Do Humans Cause Deserts? Dahlem Workshop Report 88, Dahlem University Press, Berlin, 437 pp

Reynolds JF, Maestre FT, Kemp PR, Stafford-Smith DM, Lambin E (2007) Natural and human dimensions of land degradation in drylands: causes and consequences In: Canadell J, Pataki D, Pitelka L (eds) Terrestrial ecosystems in a changing world. The IGBP Series, Springer-Verlag, Berlin

Robinson M, Gannon B, Schuch M (1991) A comparison of the hydrology of moorland under natural conditions, agricultural use and forestry. Hydrological Sciences Journal 36:565–577

Roy S, Avissar R (2002) Impact of land use/land cover change on regional hydrometeorology in Amazonia. Journal of Geophysical Research – Atmospheres 107: DOI 10.1029/2000JD000266

Schenk HJ, Jackson RB (2002a) The global biogeography of roots. Ecological Monographs 72:311–328

Schenk HJ, Jackson RB (2002b) Rooting depths, lateral root spreads and below-ground/above-ground allometries of plants in water-limited ecosystems. Journal of Ecology 90:480–494

Schimel D, Melillo J, Tian H, McGuire AD, Kicklighter D, Kittel T, Rosenbloom N, Running S, Thornton P, Ojima D, Parton W, Kelly R, Sykes M, Neilson R, Rizzo B (2000) Contribution of Increasing CO_2 and Climate to Carbon Storage by Ecosystems in the United States. Science 287:2004–2006

Schlesinger WH, Pilmanis AM (1998) Plant-soil interactions in deserts. Biogeochemistry 42:169–187

Scott DF, Lesch W (1997) Streamflow responses to afforestation with *Eucalyptus grandis* and *Pinus patula* and to felling in the Mokobulaan experimental catchments, South Africa. Journal of Hydrology 199:360–377

Seyfried MS, Schwinning S, Walvoord MA, Pockman WT, Newman BD, Jackson RB, Phillips FM (2005) Ecohydrological control of deep drainage in semiarid regions. Ecology 86:277–287

Skold MD (1989) Cropland retirement policies and their effects on land use in the Great Plains. Journal of Production Agriculture 2:197–201

Smith BN, Epstein S (1971) Two categories of $^{13}C/^{12}C$ ratios for higher plants. Plant Physiology 47:380–384

Smith DL, Johnson LC (2003) Expansion of *Juniperus virginiana* L. in the Great Plains: Changes in soil organic carbon dynamics. Global Biogeochemical Cycles 17: 1062

Tilman D, Reich PB, Phillips H, Menton M, Patel A, Vos E, Peterson D, Knops J (2000) Fire suppression and ecosystem carbon storage. Ecology 81:2680–2685

Tsuda A, Takeda S, Saito H, Nishioka J, Nojiri Y, Kudo I, Kiyosawa H, Shiomoto A, Imai K, Ono T, Shimamoto A, Tsumune D, Yoshimura T, Aono T, Hinuma A, Kinugasa M, Suzuki K, Sohrin Y, Noiri Y, Tani H, Deguchi Y, Tsurushima N, Ogawa H, Fukami K, Kuma K, Saino T (2003) A Mesoscale Iron Enrichment in the Western Subarctic Pacific Induces a Large Centric Diatom Bloom. Science 300:958–961

Van Auken OW (2000) Shrub invasions of North American semiarid grasslands. Annual Review of Ecology and Systematics 31:197–215

Vejre H, Ingerslev M, Raulund-Rasmussen K (2001) Fertilization of Danish forests: a review of experiments. Scandinavian Journal of Forest Research 16:502–513

Vertessy RA (1999) The impacts of forestry on streamflows: A review. In: Croke J, Lane P, (eds) Forest management for water quality and quantity. Proceedings of the Second Forest Erosion Workshop, May 1999, Warburton, Australia. Report 99/6. Cooperative Research Centre for Catchment Hydrology, CSIRO Land and Water, Canberra, Australia, pp 69–76

Vertessy RA (2001) Impacts of plantation forestry on catchment runoff. In: Nambiar EKS, Brown AG (eds) Plantations, farm forestry, and water. Water and salinity issues in agroforestry no. 7, RIRDC publication no. 01/20. RIRDC, Kingston, Australia, pp 9–19

Weaver C, Roy S, Avissar R (2002) Sensitivity of simulated mesoscale atmospheric circulations resulting from landscape heterogeneity to aspects of model configuration. Journal of Geophysical Research – Atmospheres 107: DOI 10.1029/2001JD000376

Werth D, Avissar R (2002) The local and global effects of Amazon deforestation. Journal of Geophysical Research – Atmospheres 107 (D20): Art. No. 8087

West TO, Post WM (2002) Soil organic carbon sequestration rates by tillage and crop rotation: A global analysis. Soil Science Society of America Journal 66:1930–1946

Wilcox BP (2002) Shrub control and streamflow on rangelands: A process based viewpoint. Journal of Range Management 55: 318–326

Xue Y, Fennessy MJ, Sellers PJ (1996) Impact of vegetation properties on U.S. summer weather prediction. Journal of Geophysical Research – Atmospheres 101:7419–7430

Zhang L, Dawes WR, Walker GR (1999) Predicting the effect of vegetation changes on catchment average water balance. Cooperative Research Centre for Catchment Hydrology, CSIRO Land and Water

Zhang L, Dawes WR, Walker GR (2001) Response of mean annual evapotranspiration to vegetation changes at catchment scale. Water Resources Research 37:701–708

Chapter 20

Natural and Human Dimensions of Land Degradation in Drylands: Causes and Consequences

James F. Reynolds · Fernando T. Maestre · Paul R. Kemp · D. Mark Stafford-Smith · Eric Lambin

20.1 Introduction

Land degradation in drylands, which is referred to as desertification, is viewed by many as one of the most critically important issues facing many countries (e.g., Darkoh 1998; Dregne 1996; Kassas 1995). The United Nations portrays it as one of the "most important global change issues facing mankind" (UNCCD 1994). Land degradation is a vital societal concern because of its impacts on human populations (food security, economics, sustainability, etc.) and environment quality (dust storms, trace gas emissions to the atmosphere, soil erosion, etc.) (Vitousek et al. 1997). Like global climate change and biodiversity, desertification is the subject of an international framework convention, the Convention to Combat Desertification (CCD), the aim of which is to "target poverty, drought and food insecurity in dryland countries experiencing desertification, particularly those in Africa" (United Nations 1994). The CCD was established by the United Nations in order to facilitate the role of national governments in enacting policies to combat land degradation. The Convention provides for a large infrastructure (e.g., a Secretariat and the Global Mechanism), which is designed to mobilize and channel financial resources, including the transfer of technology to developing countries (details in Chasek and Corell 2002).

However, in spite of its high profile and acknowledged importance, desertification has remained stubbornly intractable. Part of the problem is that desertification tends to stir up more disagreement and controversy than consensus (e.g., Leach and Mearns 1996; Reynolds and Stafford Smith 2002a; Thomas and Middleton 1994). While the reasons underlying the uncertainty and confusion associated with this topic are numerous (see reviews by Grainger et al. 2000; Reynolds 2001; Reynolds and Stafford Smith 2002a), the bottom line is that desertification has proven to be a complex problem that is not amenable to simple solutions. A second aspect of the problem appears to reside with the CCD itself. The CCD model has been roundly criticized for its deficiency of directed research efforts, for lacking connections to "real-world" problems, and for serving solely as a mechanism for some countries to elicit funds from donor nations (Toulmin 2001). In spite of these criticisms, the CCD is the primary vehicle for addressing desertification and has inspired some successful spin-off activities that engage scientists and various stake-holders (see Corell 1999).

In this chapter we provide a brief overview of the most significant issues of land degradation in global drylands, with an emphasis on the interaction between human and natural dimensions of the problem. Despite considerable work on case studies of land degradation in individual regions of the globe, there is little integration between social and biophysical sciences, and there are great opportunities for comparative studies across the many different social and biophysical systems. We discuss some of the underlying causes of land degradation and their consequences. Further, we discuss a joint initiative on desertification of the Global Change and Terrestrial Ecosystems (GCTE) and Land-Use and Land-Cover Change (LUCC) programs, which provided a framework to facilitate directed research effort and progress on this important global environmental issue. Land degradation is an excellent topic for such an initiative given that LUCC is geared to improve the understanding of land-use and land-cover change dynamics, GCTE was focused on synthesis activities on critical topics in the terrestrial biosphere, and both programs are keen to engage both the physical and social science communities for the development of science relevant to global change.

20.2 Drylands, Desertification, Drivers, and Scales

20.2.1 Distribution of People and Land-Cover Types

In general, drylands are characterized by low and variable rainfall, extreme air temperatures, and seasonally high potential evaporation. Technically, drylands are defined as regions that have an index of aridity (ratio of mean annual precipitation to mean annual potential evapotranspiration) of 0.05 to 0.65 (see Middleton and Thomas 1997). Drylands can be further subdivided as arid (0.05–0.20 index of aridity; ca. 30% of drylands), semiarid (0.20–0.50 index of aridity, ca. 45% of drylands), and sub-humid (0.50–0.65 index of aridity, ca. 25% of drylands) and together cover approximately 5.2 billion hectares or about 40% of the land surface of the globe (Table 20.1a). Drylands have two primary types of human uses: the overwhelming majority serve as rangelands (>75% of drylands), while nearly 20% are rainfed or irrigated cropland. In terms of a land-cover classification, shrubland is the dominant cover type (24% of drylands), followed by cropland (20%), savanna (15%), grassland (13%), forest (8%) and urban (3%) (Table 20.1b).

Drylands are home to over two billion people: 42% of the Asian population, 41% of Africans, and 25 to 30% of the rest of the world (Table 20.1c). Combined, Asia and Africa contain 84% of all global drylands, dwarfing the amount of dryland area on other continents. In terms of importance, however, these numbers can be somewhat misleading. While Europe contains only ca. 6% of the world's drylands, this represents about 32% of its land mass and is home to 25% of its population. Similarly, Australia contains about 13% of the world's drylands but they cover over 75% of the continent and are home to 25% of its population. Hence in both Europe and Australia, drylands are crucial determinants of the economy, culture and climate. Some of the highest densities of human populations are in the dry sub-humid and semiarid areas of cropland, e.g., those in India, eastern China, and Europe, and in the savannas of Africa (White et al. 2003).

20.2.2 Defining Land Degradation and Desertification

Land degradation and desertification are composite phenomena that have no single, readily identifiable attribute. Perhaps this is why there are so many conflicting and confusing definitions (see reviews by Reynolds 2001;

Table 20.1. Distribution of global drylands, their major land cover types, and human populations

a Distribution of drylands (million ha)[a]	Arid	Semiarid	Dry sub-humid	% of global drylands	% of continental land area
Asia	625.7	693.4	352.7	32.3	39.3
Africa	503.5	513.8	268.7	24.9	43.4
Europe	11.0	105.2	183.5	5.8	31.5
South America	44.5	264.5	207.0	10.0	29.2
North America	81.5	419.4	231.5	14.2	33.4
Oceania	303.0	309.0	51.3	12.8	75.2
World totals	1 569 (30%)	2 305 (45%)	1 295 (25%)	100.0	39.7% of globe
b Distribution of land cover types (million ha)[b]	**Arid**	**Semiarid**	**Dry sub-humid**	**Total**	**% of continental land area**
Shrubland	683	534	50	1 268	24
Cropland	47	530	475	1 052	20
Savanna	83	402	303	788	15
Grassland	181	473	65	719	13
Forest	11	140	284	436	8
Urban/developed	26	82	66	173	3
Other[c]	559	213	149	922	17
World totals	1 591	2 374	1 391	5 356	100
c Distribution of human population (%)[b]	**Arid**	**Semiarid**	**Dry sub-humid**	**Percent of population living in drylands**	
Asia (Incl. Russia)	5	18	19	42	
Africa	6	18	17	41	
Europe	0	5	20	25	
South America	2	16	12	30	
North America	2	16	5	23	
Central America and Caribbean	6	11	8	25	
Oceania	1	5	19	25	
World totals	4	4	17	37	

[a] See Reynolds (2001) for citations.
[b] Based on GLCCD 1998; ESRI 1993; UNEP/GRID 1991; see White and Nackoney (2003) for citations.
[c] Includes barren or sparsely vegetated land; open water; permanent wetlands; snow and ice; and islands.

Thomas 1997). For example, a common misunderstanding among land managers and stakeholders is to equate land degradation solely with soil degradation. The definition of Stocking and Murnaghan (2001) emphasizes changes in biophysical variables and implies an impact on human populations: land degradation is how one or more land resources (soils, vegetation, water, landforms, etc.) has changed "for the worse," signifying a temporary or permanent decline in the productive capacity of the land. While heuristic, we favor the definitions used by the CCD (UN 1994), which go a step further by making it clear that while biophysical components of ecosystems and their properties are involved, the interpretation of change as 'loss' is dependent upon the integration of these components within the context of the *socio-economic* activities of human beings. The CCD's definition states that land degradation is the reduction or loss of the biological and economic productivity and complexity of terrestrial ecosystems, including soils, vegetation, other biota, and the ecological, biogeochemical, and hydrological processes that operate therein. In drylands, this involves soil erosion and sedimentation, shifts in natural fire cycles, the disruption of biogeochemical cycling, and a reduction of native perennial plants and associated microbial and animal populations. The CCD's definition of desertification explicitly focuses on the linkages between humans and their environments that affect human welfare in arid and semi-arid regions.

20.2.3 What Drives Land Degradation and Desertification?

Desertification is caused by a relatively large number of factors that vary from region to region, and that often act in concert with one another in varying degrees. Geist and Lambin (2004) carried out a worldwide review of the causes of desertification, and from 132 case studies identified four major categories of proximal causal agents: (1) increased aridity; (2) agricultural impacts, including livestock production and crop production; (3) wood extraction, and other economic plant removal; and (4) infrastructure extension, which could be separated into irrigation, roads, settlements, and extractive industry (e.g., mining, oil, gas). They concluded that only about 10% of the case studies were driven by a single cause (with about 5% due to increased aridity and 5% to agricultural impacts). About 30% of the case studies were attributable to a combination of two causes (primarily increased aridity and agricultural impacts), while the remaining cases were combinations of three or all four proximal causal factors.

A primary objective of Geist and Lambin's (2004) review was to identify general global patterns in causation of desertification. As such, the study identified specific agents as more or less important in particular regions, and indicated that these agents derive from underlying forces associated with particular combinations of socio-economic (including technology) and biophysical factors characteristic of particular regions. For example, two underlying forces, climate and technological factors (either new technologies or deficiencies in technology) were the key drivers of desertification in the majority (54%) of the case studies in southern Europe. In Africa, climate, alone or acting in concert with population demography, was a key driver in the 38% of the case studies. In the United States, 50% of the case studies were attributable to a combination of climate and technology drivers or these two factors interacting with economic forces. Desertification processes in Asia, Latin America, and Australia could only be attributed to more complex interactions among four or more underlying forces.

20.2.4 Estimating the Extent of Desertification

Not surprisingly, obtaining accurate estimates of the amount of drylands affected by desertification is a difficult task, fraught with numerous obstacles and complications. Nevertheless, the extent of global desertification is routinely reported as high as 70% of all drylands (UNCCD 2000)! Thomas and Middleton (1994) and others view such estimates as suspect because they are largely derived from the subjective judgments of scientists and laypersons, surveys completed by local governments, qualitative assessments, and data of varying authenticity and consistency. Unfortunately, the CCD definition of desertification (Sect. 20.2.2), which we consider most authoritative at present, is not amenable to easy quantification, especially as a single number or synthetic index.

A variety of problems confound estimates of the extent of desertification. For instance, observations made on short-term ecosystem dynamics are often cited as evidence of desertification ignoring the fact that drylands are highly variable over time (both intra- and inter-annually) and that a temporary loss of vegetation cover due to a short-term drought or to local land use (e.g., grazing) is distinct from – and not necessarily related to – a permanent loss of vegetation associated with desertification (Reynolds 2001). Inaccurate estimations are also fueled by technological barriers that preclude the monitoring of relevant variables, such as the cover of dry vegetation, that cannot be easily estimated using traditional approaches (Asner et al. 2003). Another crucial, but often overlooked, concern is that desertification is usually promoted by two or more causal agents (see Sect. 20.2.3). In spite of the CCD definition, most estimates of desertification are derived solely from either biophysical factors (e.g., soil erosion, loss of plant cover, change in albedo) or socio-economic factors (decreased production, economic loss, population movements, etc.), but rarely both types simultaneously (Stafford Smith and Reynolds 2002). When assessments are made without good knowledge of the underlying causes, it brings into question the validity of the variables or sets of variables being used in the assessment.

Over the years, in different arid and semiarid regions of the world, there has been a concerted effort to categorize and map various forms of land degradation at various scales, but these efforts have failed to include a careful, systematic identification of the critical variables that cause the observed dynamics (Stafford Smith and Reynolds 2002). This problem lies at the base of the confusion about how much 'desertification' there really is (see Batterbury et al. 2002). Stafford Smith and Reynolds (2002) argued that much of this confusion could be eliminated by focusing on a small number of critical variables that contribute to an understanding of the cause, rather than effect, of desertification. Of course, this is all the more problematic when we try to account for the causal factors driving desertification in different regions of the world and at different times: approaches developed to estimate desertification in one region may not be effective in others. The failure to recognize these issues has led to the disparities of estimates of desertification in the literature and is responsible for many of the disagreements alluded to above (Stafford Smith and Pickup 1993; Stafford Smith and Reynolds 2002).

20.2.5 Consequences of Desertification

There are few disagreements that desertification has a large number of biophysical and socio-economic consequences, which range across a wide spectrum of spatial and temporal scales. An in-depth treatment of the different consequences is beyond the scope of this chapter, but some relevant ones are presented in Table 20.2 (and see group reports in Reynolds and Stafford Smith 2002b). From the socio-economic point of view, most consequences (especially in pastoral systems) are a direct consequence of the decline in 'productivity' or the capacity of the land to support plant growth and animal production. During early stages of desertification such losses are compensated by the social resilience of the local human populations, especially in developing countries, or by economic inputs from government (Vogel and Smith 2002). However, when certain thresholds are crossed, social resilience or government subsidies may not be enough to compensate for the loss of productivity, and this fuels a battery of socio-economic changes that range from modifications in trade promoted by lower agricultural production to large population migrations (Fernández et al. 2002).

Virtually all of the biophysical consequences start with the loss of vegetation and soil (Table 20.2). These losses have a 'cascading' effect on other components and processes, leading to a progressive deterioration of the ecological structure and functioning of the system (Fernández et al.

Table 20.2. Some biophysical and socio-economic consequences of desertification in drylands and approximate spatial scales at which they are most relevant. The list is not exhaustive. See Sect. 20.2.5

Dimension	Consequence	Scale[a]	Examples
Socio-economic	Reduction in crop yield	F, C, D	Zaman (1997)
	Reduction in animal production	F, C, D	Fredrickson et al. (1998)
	Loss of economical important species	C, N	Latchininsky and Gapparov (1996)
	Migratory movements	C, D, N	Pamo (1998)
	Loss of local environmental knowledge	C, D, N	Bollig and Schulte (1999)
	Loss of traditional agricultural structures	D	Gallart et al. (1994)
	Changes in land use patterns	C, D, N	Zhao et al. (2005)
Biophysical	Loss of soil and nutrients	P, L	Schlesinger et al. (1999)
	Decreased infiltration	P	Sharma (1998)
	Modification of geomorphology	P, R	Lavee et al. (1998)
	Addition of sediments to water bodies	P, L, R	Kelley and Nater (2000)
	Reduction of plant cover	P, L, R	Asner et al. (2003)
	Shifts in species composition and richness	P, L, R	Gonzalez (2001)
	Changes in primary net productivity	P, L, R	Huenneke et al. (2002)
	Changes in the spatial pattern of resources	P, L, R	Schlesinger et al. (1990)
	Loss of biodiversity	R, G	Whitford (1993)
	Loss of biological soil crusts	P, L	Belnap and Eldridge (2001)
	Depletion of soil carbon stocks	P, L, R, G	Jackson et al. (2002)
	Depletion of soil nitrogen stocks	P, R	Asner et al. (2003)
	Reduction in ecosystem resilience	P, L, R	Von Handenberger et al. (2001)
	Modification of climate	R, G	Rosenfeld et al. (2001)

[a] Socio-economic: *F:* farm/household, *C:* community, *D:* district/provincial, *N:* national/international; biophysical: *P:* patch, *L:* landscape, *R:* regional, *G:* global.

2002). The specific biophysical consequences of desertification differ substantially between geographical areas of the globe as a function of the intensity and number of driving forces at work, the extent of the impacted area, the duration of the deterioration, and the resilience of the system components (especially vegetation). Even within a particular area we may find that there are different consequences depending upon the unique characteristics of the system. For instance, while Krogh et al. (2002) found that some keystone species associated with grasslands are negatively affected by shrub encroachment into former grasslands (a form of desertification, Schlesinger et al. 1990), recent studies in the southwestern United States have shown that shrub encroachment is associated with an increase in the species richness of birds (Pidgeon et al. 2001), mammals (Whitford 1997) and ants (Bestelmeyer 2005). Thus, some generalizations regarding biophysical consequences of desertification may be misleading or incorrect when applied to specific situations.

20.2.6 Scale and Hierarchy

The importance of scale is manifest in a number of ways when evaluating the extent and effects of desertification (Sect. 20.2.4). Obviously, humans are most concerned with the local subset of degradation that impacts them personally. Reynolds and Stafford Smith (2002a) use a hypothetical case-study of gully formation (overgrazing by cattle leads to loss of vegetative cover and, ultimately, soil erosion) to illustrate how different segments of society (the stake-holders) see such problems with differing degrees of concern. Whereas an ecologist might view erosion gullies as a breakdown in ecosystem function, this will resonate with a farmer only if the gullies have a demonstrable impact on his values, i.e., meat production by his cattle. If not, the farmer will not consider this as 'degradation'. Alternatively, other stake-holders, such as urban dwellers in a nearby town, may consider this localized soil erosion and gully formation a more serious problem because of the potential for silt runoff into the town's reservoir and its adverse effects on water quality.

Because local activities often have regional consequences (e.g., localized erosion gullies impacting regional water quality) and regional issues can, in turn, have local impacts, it behooves us to assume a multifactor, multiscale, hierarchical view of desertification. In Table 20.3, we illustrate representative scales of interest in desertification viewed from both socio-economic and biophysical perspectives. Coupled socio-economic and biophysical systems must be hierarchically nested in order to avoid errors that will undoubtedly occur if we attempt to extrapolate understanding over a range of scales that is too great, e.g., trying to predict what will happen at the

Table 20.3. Examples of various socio-economic and biophysical scales of concern in desertification and associated key variables. The socio-economic scales are from Stafford Smith and Reynolds (2002) and the biophysical scales approximate those used by Prince (2002). The list of key variables associated to each scale is not exhaustive and some are relevant at more than one scale

Socio-economic		Biophysical	
Scale	**Some key variables**	**Scale**	**Some key variables**
Farm/ household	Household size Labor attributes Characteristics of food supply (e.g. security, flexibility, etc.) Technological development Land management patterns	Patch	Plant cover Soil nutrients Soil infiltration and water holding capacity Abiotic attributes (e.g., slope, aspect)
Community or village	Land tenure and ownership Local government attributes Population size Flexibility of job market	Landscape	Sediment transport Species composition and richness Net primary productivity
District/ provincial	District land planning Decentralization of planning for communal land Land reform Conflicts between groups (ethnic, social and economical) Financial constraints (subsidies) Economic opportunities 'Natural' disasters (e.g., floods, drought, fire, pests, diseases)	Regional	Geomorphology Soil nutrient stocks (e.g. carbon and nitrogen) Ecosystem resilience
National/ international	Human population growth Macro-economic indicators National policies (e.g., resettlement programs, economic incentives) National and international conflicts (e.g., war)	Global	Climate (e.g., rainfall, radiation, albedo) Biodiversity Cover of vegetation types

household level based on observations made at the national scale. Sørbø (2003) describes examples from east Africa to show the importance of scale in pastoral herding communities. These pastoral communities are networked into various localized units, which function in a complex interplay of local and regional social, economic, and political factors, all of which have evolved over many years against a backdrop of severe environmental instability and unpredictable contingencies. It is not surprising that 'top-down' attempts to 'manage' these systems fall short of expectations. This is consistent with Batterbury et al.'s (2002) observations that, in arid and semiarid lands, the highest levels of the 'management hierarchy' are invariably quite remote from marginal lands and, consequently, have weak political and economic feedbacks. An important objective of institutions such as the CCD should be to provide a context within which levels in the hierarchy become more integrated and aware of the issues (Batterbury et al. 2002; Lambin et al. 2002).

20.3 Joint GCTE-LUCC Desertification Initiative

The simultaneous assessment of biophysical and socioeconomic drivers (and consequences) of desertification has been recognized as one of the most challenging – but potentially rewarding – topics for further research (see review by Reynolds 2001). In an attempt to address this challenge, the GCTE and LUCC programs of the International Geosphere-Biosphere Programme joined forces to establish an initiative on desertification. The intent of this initiative was to bring together researchers from the various global change programs, representing both natural and human-influenced systems, with the objective of stimulating, developing, and refining a new paradigm to bear on this important global change concern.

One of the key, initial products of this GCTE-LUCC initiative was a book on global desertification (*Global Desertification. Do Humans Cause Deserts?*), which explicitly addresses many of the significant interactions and feedbacks between natural and human-influenced dryland systems. Focusing on the multitude of interrelationships within coupled human-environment systems that cause desertification, and drawing heavily from the chapters of this book, Stafford Smith and Reynolds (2002) proposed the *Dahlem Desertification Paradigm* (DDP). The DDP is a new synthetic framework that is unique in two ways:

- First, the DDP attempts to capture the multitude of interrelationships within human-environment systems that cause desertification, within a *single*, synthetic framework; and
- Second, the DDP is testable, which ensures that it can be revised and improved upon as a *dynamic* framework.

In this section we briefly present an outline of the key elements or assertions of the DDP and describe one method we are using for testing it. As is the case for many paradigms, the constituent ideas contained within the DDP themselves are generally not new, but rather, they bring together much of the previous work on this difficult topic in a way that reveals new insights.

20.3.1 Dahlem Desertification Paradigm

The DDP consists of nine assertions (Table 20.4), which embrace a hierarchical view of land degradation and highlight key linkages between socio-economic and biophysical systems at different scales. The first three assertions relate to the working framework of the DDP while the remaining ones focus on its implementation, limitations, and potentials. The main points of the DDP are:

1. that an integrated approach, which simultaneously considers both biophysical and socio-economic attributes in these systems, is absolutely essential to understand land degradation (assertions #1, #7);
2. that the biophysical and socio-economic attributes that govern or cause land degradation in any particular region are invariably 'slow' (e.g., soil nutrients) relative to those that are of immediate concern to human welfare – the 'fast' variables (e.g., crop yields). It is necessary to distinguish these in order to identify the causes of land degradation from its effects (assertion #2);
3. that socio-ecological systems in drylands of the world are not static (assertions #3, #6);
4. that while change is inevitable, there does exist a constrained set of ways in which these socio-ecological systems function, thereby allowing us to understand and manage them (assertion #9);
5. that restoring degraded socio-ecological systems to more productive, sustainable states requires outside intervention (assertion #4)
6. that socio-ecological systems in drylands of the world are hierarchical (assertion #8). Hence, scale-related concerns abound and desertification itself is a regionally-emergent property of localized degradation (assertion #5)

The strength of the Dahlem Desertification Paradigm is in its cross-scale conceptual holism. While the term desertification is only useful at the higher levels of aggregation, and degradation (appropriately refined) at the lower levels, the DDP framework embraces all these levels of concern. For example, at the international level, implementation of the CCD must be framed in terms of changes in coupled human-environment systems that matter to humans, which dramatically changes the meaning of the "extent of desertification" (Sect. 20.2.4) and both the timing and distribution of funding for inter-

Table 20.4. The nine assertions of the Dahlem Desertification Paradigm, and some of their implications. From Stafford Smith and Reynolds (2002)

	Assertion	Implications
1.	Desertification always involves human and environmental drivers	Always expect to include both socio-economic and biophysical variables in any monitoring or intervention scheme
2.	'Slow' variables are critical determinants of system dynamics	Identify and manage for the small set of 'slow' variables that drive the 'fast' ecological goods and services that matter at any given scale
3.	Thresholds are crucial, and may change over time	Identify thresholds in the change variables at which there are significant increases in the costs of recovery, and quantify these costs, seeking ways to manage the thresholds to increase resilience
4.	The costs of intervention rises non-linearly with increasing degradation	Intervene early where possible, and invest to reduce the transaction costs of increasing scales of intervention
5.	Desertification is a regionally emergent property of local degradation	Take care to define precisely the spatial and temporal extent of and processes resulting in any given measure of local degradation. But don't try to probe desertification beyond a measure of generalized impact at higher scales
6.	Coupled human-environment systems change over time	Understand and manage the circumstances in which the human and environmental sub-systems become 'de-coupled'
7.	The development of appropriate local environmental knowledge (LEK) must be accelerated	Create better partnerships between LEK development and conventional scientific research, employing good experimental design, effective adaptive feedback and monitoring
8.	Systems are hierarchically nested (manage the hierarchy!)	Recognize and manage the fact that changes at one level affect others; create flexible but linked institutions across the hierarchical levels, and ensure processes are managed through scale-matched institutions
9.	A limited suite of processes and variables at any scale makes the problem tractable	Analyze the types of syndromes at different scales, and seek the investment levers which will best control their effects – awareness and regulation where the drivers are natural, changed policy and institutions where the drivers are social

vention. Similarly, at the household or community level, where concern is on the specific type of land degradation that is occurring and its local socio-economic consequences, the DDP channels resources towards identifying those essential biophysical and socio-economic slow variables that really matter in terms of quantifying current and future risk.

20.3.2 Initiatives to Test the Dahlem Desertification Paradigm

The joint GCTE-LUCC initiative on desertification is embodied within the ARID*net*[1] research network (Reynolds et al. 2003). The general objectives of this network are to foster international cooperation, discussion, and exchange of ideas about global desertification (as summarized in the DDP), to conduct case studies, representing a range of biophysical/socio-economic land degradation types around the world, and to facilitate communication between researchers to foster more practical, field-level interactions with stakeholders in sustainable land management.

To accomplish these objectives, ARID*net* is organized into three nodes (North/South America; Asia/Australia; Europe/Africa) and is pursuing four specific tasks (Fig. 20.1):

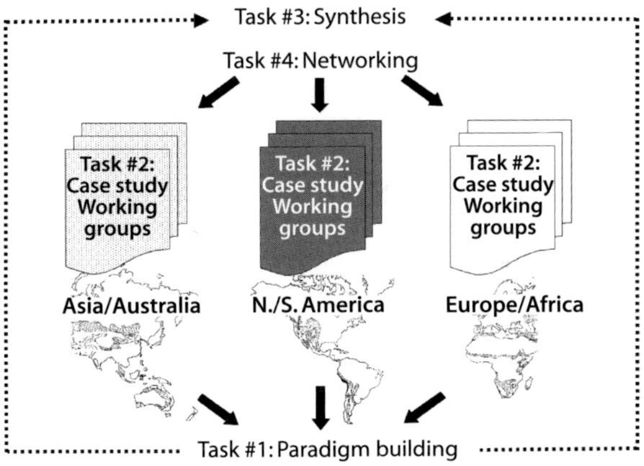

Fig. 20.1. ARID*net* is an international research network organized into three nodes (North/South America; Asia/Australia; Europe/Africa) and is pursuing four tasks (see text): paradigm-building, conducting case studies, developing a synthesis, and to facilitate participation via network-building

- *Paradigm-building.* Using workshops and symposia conducted in different parts of the world, ARID*net* will develop and refine the contents of the DDP through the joint participation of the international community of desertification researchers, stakeholders, and policy-makers;
- *Case studies.* Working Groups are specifically formed to develop case studies throughout the world in order to test the DDP in a well-stratified, comparative manner. These case studies are based on existing data and specific stakeholders, and represent a wide range of

[1] Assessment, Research, and Integration of Desertification.

the biophysical and socio-economic conditions existing in drylands;

- *Synthesis.* The numerous case studies will feed into a quantitative assessment of what really matters in desertification. This synthesis will especially focus on those interactions between key biophysical and socio-economic variables; and
- *Network-building.* ARID*net* strives to recruit and foster the participation of a diversity of researchers from different fields and countries in the activities of the network.

20.4 Management of Desertified Drylands

In preceding sections we provided examples of the magnitude and importance of desertification as a major local and global environmental problem. In order to prevent further degradation and to restore degraded lands, a number of countries have enacted environmental policies to establish management actions to combat desertification. These actions are diverse but can be grouped under three headings: avoidance, monitoring, and restoration.

20.4.1 Avoidance

Management actions to avoid desertification are rarely proactive. If they exist, they normally are focused on the human drivers of desertification. Such actions vary widely according to the socio-economic conditions of a particular country, but often include the use of economic subsidies to promote changes in land use or crops (Harou 2002), the diversification of human activities in the areas affected (Pamo 1998), and the establishment of educational programs to improve education and social welfare (Vogel and Smith 2002). The latter activity is of crucial importance since one of the core causes of desertification in developing nations is the extreme pressure on the land resulting from high population growth (Geist and Lambin 2004; Le Houérou 1996). However, there are examples of success in reducing demographic growth and in implementing sustainable production systems in desertified areas worldwide (Arkutu 1995; Vogel and Smith 2002), indicating that with appropriate resources and political willingness, some of the most important desertification drivers can be controlled.

20.4.2 Monitoring

The monitoring of desertification is an increasingly important development in the management of dryland areas. The establishment of long-term and rigorous monitoring programs is an effective way to assess the status of natural resources and the evolution of desertification processes. Such programs could provide an "early warning" of pending concerns, e.g., the detection of changes in ecosystem attributes and processes at stages where management actions would be most cost-effective (see Fernández et al. 2002).

It is encouraging to see increased research dedicated to the development of easily accessible monitoring methods based on simple soil and vegetation indicators (e.g., Pyke et al. 2002; Tongway and Hindley 1995). These methods are based on the collection of basic information of those vegetation attributes and soil properties (e.g., cover, spatial pattern, resistance to penetration and texture) that largely determine the ecosystem's resilience to erosive forces and its ability to use water and nutrients. An important goal of these methods is to minimize the training and equipment required so as to increase their availability to nations with low economic resources. These ground-based methods are complemented with the use of remote sensing data, which have been successfully employed to monitor desertification processes in the U.S. (Asner and Heidebrecht 2005), South America (Asner and Heidebrecht 2003), Africa (Prince et al. 1998), Australia (Bastin et al. 2002) and Europe (Imeson and Prinsen 2004). Remote-sensing approaches are often based on measuring the same vegetation and soil attributes as ground-based methods, but they allow the establishment of monitoring programs at larger spatial scales. However, they often require the use of expensive equipment and appropriate training, two factors not available in many countries. Furthermore, recent studies suggest that traditional remote sensing measurements (e.g., NDVI measured during the growing season) do not provide an adequate indication of biophysical degradation (Asner et al. 2003). Further advancement in desertification monitoring depends upon a coordination between ground- and remote sensing-based research and the establishment of sound and cost-effective methodologies appropriate for particular regions.

20.4.3 Restoration

While we might hope that future actions of land owners, communities, regions, and nations will begin to adopt management practices and policies that minimize or avoid desertification, the sad fact is that vast areas of drylands are already in a wretched state, with varying degrees of reduced plant cover, impoverished species diversity, and depleted or eroded soils (Whisenant 1999). Restoration actions in these lands have traditionally focused on the biophysical variables, especially those aimed at increasing plant cover and halting soil erosion. Often, and irrespective of the underlying drivers of desertification, restoration efforts involve the establishment of woody plants (Whisenant 1999), which is deemed crucial to stop further degradation, and to foster recovery of ecosystem structure, composition, and functioning (Reynolds 2001). Such afforestation programs, which have been carried out since the 1800s, have resulted in millions of hectares of conifer trees (mainly) planted in drylands of Spain, Turkey, Morrocco, Algeria, Argentina, China, and many other countries (Richardson 1998; Pausas et al. 2004; Sánchez Martínez and Gallego Simón 1993; Gao et al. 2001). While some of these efforts have been effective in controlling desertification, many have failed (Odera 1996), and in some cases have exacerbated the degradation process (see review by Maestre and Cortina 2004).

In other instances, restoring desertified lands has involved instituting changes in land management practices (e.g., Pyke et al. 2002; Sørbø, 2003). Again, these may be undertaken without full understanding of the causal mechanisms involved, or without an appreciation of the socio-economic conditions (Sørbø, 2003), and again are prone to failure. This history of lack of success suggests a clear need to address fundamentals. First, as we have emphasized throughout this chapter, it is important to have a sound understanding of both the biophysical and human dimensions of causality. Second, we must incorporate as much of our knowledge of dryland structure and functioning as possible into our management actions. In this respect, important advances have been made in the application of conceptual and practical ecosystem models. Third, we must carefully evaluate and incorporate existing socio-economic attributes. Fourth, we must assess the potential for successful rehabilitation. Lastly, we must strive for improved restoration methodologies, including the use of new understanding of key ecosystem processes, e.g., plant-plant interactions (Maestre et al. 2001) and the role of soil heterogeneity in plant establishment (Maestre et al. 2003).

20.5 Summary and Conclusions

While conceptual, methodological and technological advances have been made during the past several decades to help land managers establish appropriate strategies for combating desertification, there have been relatively few successes in desertification abatement (Le Houérou 1996). As discussed in this chapter, desertification may have numerous underlying causes, which involve a complex interplay among biophysical and human dimensions. We propose that a key aspect of the prevention and remediation of desertification is the development of a integrative, theoretical framework in which biophysical and socioeconomic components are treated as coupled processes fostering desertification.

We further suggest that the Dahlem Desertification Paradigm provides the necessary theoretical framework for bringing researchers and policy-makers together for the purpose of developing testable hypotheses and methodologies regarding the monitoring, prevention and remediation of desertification. The DDP emphasizes the need to address all socio-economic levels (local, regional, national, international) in the development of effective desertification policy decisions. The DDP emphasizes moving beyond isolated studies of various parts of the desertification problem and toward establishing an integrated program of causal links of dryland degradation, from climate dynamics to ecological impacts to policy response strategies, which can be applied to a wide range of temporal and spatial scales. The challenges of building the necessary political and scientific bridges are enormous, but so is the need for urgent action to understand and manage desertified drylands worldwide.

Acknowledgments

The authors acknowledge the helpful comments and critiques provided by Greg Asner and an anonymous reviewer. This research was supported by the National Science Foundation (grant NSF-DEB-02-34186 to ARID*net*) and the Center for Integrated Study of the Human Dimensions of Global Change, through a cooperative agreement between the NSF (SBR-9521914) and Carnegie Mellon University. FTM acknowledges support via a Fulbright fellowship from the Spanish Ministry of Education and Science, funded by Secretaría de Estado de Universidades and Fondo Social Europeo. JFR acknowledges partial support by the Alexander von Humboldt-Stiftung during the preparation of this manuscript and the excellent hospitality of the Lehrstuhl für Pflanzenökologie at the University of Bayreuth.

References

Arkutu A (1995) Family planning in Sub-Saharan Africa: Present status and future strategies. Int J Gyn Obst 50: S27–S34

Asner GP, Heidebrecht KB (2003) Imaging spectroscopy for desertification studies: Comparing AVIRIS and EO-1 Hyperion in Argentina drylands. IEEE Trans Geosci Rem Sens 41:1283–1296

Asner GP, Heidebrecht KB (2005) Desertification alters regional ecosystem-climate interactions. Global Change Biol 11:182–194

Asner GP, Borghi CE, Ojeda RA (2003) Desertification in central Argentina: Changes in ecosystem carbon and nitrogen from imaging spectroscopy. Ecol Appl 13:629–648

Asner GP, Elmore AJ, Olander LP, Martin RE, Harrias T (2004) Grazing systems, ecosystem responses, and global change. Annu Rev Environ Resour 29:261–299

Bastin G, Ludwig JA, Eager RW, Chewings VH, Liedloff AC (2002) Indicators of landscape function: comparing patchiness metrics using remotely-sensed data from rangelands. Ecol Indic 1:243

Batterbury SPJ, Behnke RH, Döll PM, Ellis JE, Harou PA, Lynam TJP, Mtimet A, Nicholson SE, Obando JA, Thornes JB (2002) Responding to desertification at the national scale: detection, explanation, and responses. In: Reynolds JF, Stafford Smith M (eds) Global Desertification: Do Humans Cause Deserts? Dahlem University Press, Berlin, pp 357–386

Belnap J, Eldridge DJ (2001) Disturbance and recovery of biological soil crusts. In: Belnap J, Lange OL (eds) Biological Soil Crusts: Structure, Function, and Management Springer-Verlag, Berlin, pp 363–384

Bestelmeyer BT (2005) Does desertification diminish biodiversity? Enhancement of ant diversity by shrub invasion in south-western USA. Divers Distrib 11:45–55
Bollig M, Schulte A (1999) Environmental change and pastoral perceptions: Degradation and indigenous knowledge in two African pastoral communities. Human Ecol 27:493–514
Chasek P, Corell E (2002) Addressing desertification at the international level: the institutional system. In: Reynolds JF, Stafford Smith M (eds) Do Humans Create Deserts? The Ecological, Meteorological, and Human Dimensions of Global Desertification. Dahlem University Press, Berlin, pp 275–296
Corell E (1999) The Negotiable Desert: Expert Knowledge in the Negotiations of the Convention to Combat Desertification. Linköping Studies in Arts and Science No. 191. Linköping University, Linköping, Sweden
Darkoh MBK (1998) The nature, causes and consequences of desertification in the drylands of Africa. Land Degrad Develop 9:1–20
Dregne HE (1996) Desertification: challenges ahead. Ann Arid Zone 35:305–311
Fernández RJ, Archer ERM, Ash AJ, Dowlatabadi H, Hiernaux PHY, Reynolds JF, Vogel CH, Walker BH, Wiegand T (2002) Degradation and recovery in socio-ecological systems: a view from the household/farm level. In: Reynolds JF, Stafford Smith M (eds) Do Humans Create Deserts? The Ecological, Meteorological, and Human Dimensions of Global Desertification. Dahlem University Press, Berlin, pp 297–324
Fredrickson E, Havstad KM, Estell R, Hyder P (1998) Perspectives on desertification: South-western United States. J Arid Environ 39:191–207
Gallart F, Llorens P, Latron J (1994) Studying the role of old agricultural terraces on runoff generation in a small Mediterranean mountainous basin. J Hydrol 159:291–303
Gao J, Zha Y, Ni S (2001) Assessment of the effectiveness of desertification rehabilitation measures in Yulin, north-western China using remote sensing. Int J Rem Sens 22:3783–3795
Geist HJ, Lambin EF (2004) Dynamic causal patterns of desertification. Bioscience 54:817–829
Gonzalez P (2001) Desertification and a shift of forest species in the West African Sahel. Clim Res 17:217–228
Grainger A, Stafford Smith M, Squires V, Glenn E (2000) Desertification and climate change: the case for greater convergence. Mitig Adapt Strat Global Change 5:361–377
Harou PA (2002) What is the role of markets in altering the sensitivity of arid land systems to perturbation? In: Reynolds JF, Stafford Smith DM (eds) Global Desertification: Do Humans Cause Deserts? Dahlem University Press, Berlin, pp 253–274
Huenneke LF, Anderson JP, Remmenga M, Schlesinger WH (2002) Desertification alters patterns of aboveground net primary production in Chihuahuan ecosystems. Global Change Biol 8:247–264
Imeson AC, Prinsen HAM (2004) Vegetation patterns as biological indicators for identifying runoff and sediment source and sink areas for semi-arid landscapes in Spain. Agric Ecosyst Environ 104:333–342
Jackson RB, Banner JL, Jobbagy EG, Pockman WT, Wall DH (2002) Ecosystem carbon loss with woody plant invasion of grasslands. Nature 418:623–626
Kassas M (1995) Desertification: a general review. J Arid Environ 30:115–128
Kelley DW, Nater EA (2000) Historical sediment flux from three watersheds into Lake Pepin, Minnesota, USA. J Environ Qual 29:561–568
Krogh SN, Zeisset MS, Jackson E, Whitford WG (2002) Presence/absence of a keystone species as an indicator of rangeland health. J Arid Environ 50:513–519
Lambin EF, Chasek PS, Dowing TE, Kerven C, Kleidon A, Leemans R, Lüdeke M, Prince SD, Xue Y (2002) The interplay between international and local processes affecting desertification. In: Reynolds JF, Stafford Smith M (eds) Do Humans Create Deserts? The Ecological, Meteorological, and Human Dimensions of Global Desertification. Dahlem University Press, Berlin, pp 387–402
Latchininsky A, Gapparov FA (1996) Consequences of the Aral sea's drying up on the regional locust situation. Secheresse 7:109–113
Lavee H, Imeson AC, Sarah P (1998) The impact of climate change on geomorphology and desertification along a mediterranean-arid transect. Land Degrad Develop 9:407–422
Le Houérou HN (1996) Climate change, drought and desertification. J Arid Environ 34:133–185
Leach M, Mearns R (eds) (1996) The Lie of the Land: Challenging Received Wisdom on the African Environment. Heinemann Press, Portsmouth, N.H
Maestre FT, Cortina J (2004) Are *Pinus halepensis* plantations useful as a restoration tool in semiarid Mediterranean areas? For Ecol Manage 198:303–317
Maestre FT, Bautista S, Cortina J, Bellot J (2001) Potential for using facilitation by grasses to establish shrubs on a semiarid degraded steppe. Ecol Appl 11:1641–1655
Maestre FT, Cortina J, Bautista S, Bellot J, Vallejo R (2003) Small-scale environmental heterogeneity and spatiotemporal dynamics of seedling establishment in a semiarid degraded ecosystem. Ecosystems 6:630–643
Middleton NJ, Thomas DSG (eds) (1997) World Atlas of Desertification. U.N. Environment Programmem, Edward Arnold, New York
Odera J (1996) L'état de dégradation actuel des écosystèmes fragiles dans les zones sèches et le rôle de la foresterie dans la restauration, Note de secrétariat nº 1. Réunion Internationale d'experts sur la restauration des ecosystèmes forestiers dégradés. Portuguese Ministry of Agriculture, UNP/UNSO, IPF, CCD & RIOD, Lisbon
Pamo ET (1998) Herders and wildgame behaviour as a strategy against desertification in northern Cameroon. J Arid Environ 39:179–190
Pausas JG, Bladé C, Valdecantos A, Seva J, Fuentes D, Alloza J, Vilagrosa A, Bautista S, Cortina J, Vallejo R (2004) Pines and oaks in the restoration of Mediterranean landscapes of Spain: New perspectives for an old practice. Plant Ecol 171:209–220
Pidgeon AM, Mathews NE, Benoit R, Nordheim EV (2001) Response of avian communities to historic habitat change in the northern Chihuahuan Desert. Cons Biol 15:1772–1788
Prince SD (2002) Spatial and temporal scales for detection of desertification. In: Reynolds JF, Stafford Smith M (eds) Global Desertification: Do Humans Cause Deserts? Dahlem University Press, Berlin, pp 23–40
Prince SD, de Colstoun EB, Kravitz LL (1998) Evidence from rain-use efficiencies does not indicate extensive Sahelian desertification. Global Change Biol 4:359–374
Pyke DA, Herrick JE, Shaver PL, Pellant M (2002) Rangeland health attributes and indicators for qualitative assessment. J Range Manage 55:584–597
Reynolds JF (2001) Desertification. In: Levin S (ed) Encyclopedia of Biodiversity, Volume 2. Academic Press, San Diego, CA, pp 61–78
Reynolds JF, Stafford Smith M (2002a) Do humans cause deserts? In: Reynolds JF, Stafford Smith M (eds) Global Desertification: Do Humans Cause Deserts? Dahlem University Press, Berlin, pp 1–22
Reynolds JF, Stafford Smith M (eds) (2002b) Global Desertification: Do Humans Cause Deserts? Dahlem University Press, Berlin
Reynolds JF, Stafford Smith DM, Lambin EF (2003) ARIDnet: Seeking novel approaches to desertification and land degradation. IGBP Global Change Newsletter 54, 5–9 (PDF available at: http://www.igbp.kva.se)
Richardson DM (1998) Ecology and biogeography of *Pinus*. Cambridge University Press, Cambridge
Rosenfeld D, Rudich Y, Lahav R (2001) Desert dust suppressing precipitation: A possible desertification feedback loop. Proc Natl Acad Sci USA 98:5975–5980
Sánchez Martínez JD, Gallego Simón VJ (1993) La Política de repoblación forestal en España, siglos XIX y XX: Planteamientos, Actuaciones y resultados, estado de la cuestión y recopilación bibliográfica. MAPA. Secretaria General Técnica. Serie Recopilaciones Bibliográficas nº 10, Madrid
Schlesinger WH, Reynolds JF, Cunningham GL, Huenneke LF, Jarrell WM, Virginia R, Whitford WG (1990) Biological feedbacks in global desertification. Science 247:1043–1048
Schlesinger WH, Abrahams AD, Parsons AJ, Wainwright J (1999) Nutrient losses in runoff from grassland and shrubland habitats in Southern New Mexico: I. rainfall simulation experiments. Biogeochemistry 45:21–34

Sharma KD (1998) The hydrological indicators of desertification. J Arid Environ 39:121–132
Sørbø GM (2003) Pastoral ecosystems and the issue of scale. Ambio 32:113–117
Stafford Smith DM, Pickup G (1993) Out of Africa, looking in: understanding vegetation change. In: Behnke RH Jr., Scoones I, Kerven C (eds) Range Ecology at Disequilibrium: New Models of Natural Variability and Pastoral Adaptation in African Savannas. Overseas Development Institute and International Institute for Environment and Development, London, pp 196–244
Stafford Smith DM, Reynolds JF (2002) The Dahlem Desertification Paradigm: A new approach to an old problem. In: Reynolds JF, Stafford Smith M (eds) Global Desertification: Do Humans Cause Deserts? Dahlem University Press, Berlin, pp 403–424
Stocking M, Murnaghan N (2001) Handbook for the Field Assessment of Land Degradation. Earthscan, London; Sterling, VA
Thomas DSG (1997) Science and the desertification debate. J Arid Environ 37:599–608
Thomas DSG, Middleton N (1994) Desertification: Exploding the Myth. John Wiley & Sons, Chichester
Tongway DJ, Hindley N (1995) Assessment of soil condition of tropical grasslands. CSIRO Ecology and Wildlife, Canberra
Tongway DJ, Ludwig JA (1996) Rehabilitation of semiarid landscapes in Australia. I. Restoring productive soil patches. Restor Ecol 4:388–397
Toulmin C (2001) Lessons from the theatre: Should this be the final curtain call for the Convention to Combat Desertification? WSSD Opinion Series. International Institute for Environment and Development. Available from: www.iied.org
UNCCD (1994) United Nations Convention to Combat Desertification, Intergovernmental Negotiating Committee For a Convention to Combat Desertification, Elaboration of an International Convention to Combat Desertification in Countries Experiencing Serious Drought and/or Desertification, Particularly in Africa. U.N. Doc. A/AC.241/27, 33 I.L.M. 1328. United Nations, New York
UNCCD (2000) Fact Sheet 2: The Causes of Desertification. United Nations Secretariat of the Convention to Combat Desertification. Available from: http://www.unccd.int/publicinfo/factsheets/showFS. php?number=2 [Geo-2-171]
United Nations (1994) Earth Summit. Convention on Desertification. United Nations Conference on Environment and Development, Rio de Janeiro, Brazil, 3–14 June 1992. In. United Nations, New York, p 44
Vitousek PM, Mooney HA, Lubchenco J, Melillo JM (1997) Human domination of Earth's ecosystems. Science 277:494–499
Vogel CH, Smith J (2002) Building social resilience in arid ecosystems. In: Reynolds JF, Stafford Smith DM (eds) Global Desertification: Do Humans Cause Deserts? Dahlem University Press, Berlin, pp 149–166
von Hardenberg J, Meron E, Shachak M, Zarmi Y (2001) Diversity of vegetation patterns and desertification. Phys Rev Let 87: 198101-1–198101-4
Whisenant SG (1999) Repairing Damaged Wildlands. Cambridge University Press, Cambridge
White RP, Nackoney J (2003) Drylands, people, and ecosystem goods and services: A web-based geospatial analysis. World Resources Institute, Washington, DC
White R, Tunstall D, Henninger N (2003) An Ecosystem Approach to Drylands: Building Support for New Development Policies. World Resources Institute, EarthTrends Publication Series. Available online: http://earthtrends.wri.org/features/view_feature.cfm?fid=37&theme=39
Whitford WG (1993) Animal feedbacks in desertification: An overview. Rev Chil Hist Nat 66:243–251
Whitford WG (1997) Desertification and animal biodiversity in the desert grasslands of north America. J Arid Environ 37:709–720
Zaman S (1997) Effects of rainfall and grazing on vegetation yield and cover of two arid rangelands in Kuwait. Environ Cons 24: 344–350
Zhao WZ, Xiao HL, Liu ZM, Li J (2005) Soil degradation and restoration as affected by land use change in the semiarid Bashang area, northern China. Catena 59:173–186

Part E

Regions under Stress

Chapter 21

Southeast Asian Fire Regimes and Land Development Policy

Daniel Murdiyarso · Louis Lebel

21.1 Introduction

Fires have long been an important tool for land development and management in tropical Southeast Asia. Fire disturbance regimes vary with forest structure, climate, topography and soils. These regimes have influenced, and been influenced by, the land-use systems of diverse cultures over centuries and millennia, producing diverse histories of fire and landscape dynamics. Low and moderate intensity fires are a regular event in the seasonally dry deciduous or savannah forests of Southeast Asia. These forests contain fire-adapted species and are frequently settled and used by swidden – cultivators. In contrast, fires in the moister evergreen forests of the humid tropics are much more irregular and typically associated with unusual events, for example droughts caused by dry phases of the El Niño – Southern Oscillation (ENSO). In the last 50–100 years the rate and extent of change in landscapes and disturbance regimes has increased substantially as societies throughout Southeast Asia have successfully grown in size, expanded and intensified the agricultural and forestry foundations of their economies, and adopted new management strategies to fire.

One of the side-effects of this success has been a more frequent incidence of synchronous fires, especially in very dry years, when deliberately lit and managed fires get out of control to become wildfires. Depending on wind and atmospheric conditions these events can lead to episodes of haze affecting major cities in other regions (Murdiyarso et al. 2004a; Tomich et al. 1998b). The most extreme episodes in Indonesia and Malaysia over the last 20 years have been associated with dry phases of ENSO and resulted in vast parts of Southeast Asia being covered in smoke for weeks at a time. In the northern parts of Thailand, Vietnam, Myanmar and Lao PDR, fires are an annual event associated with a predictable long dry season. There, the intensification of land-use practices and the creation of protected area networks in upland areas may now be reducing the frequency of fires.

At the simplest level fire regimes are affected by ecosystem conditions, land management practices and climate (Fig. 21.1). Agricultural and forest practices are, in turn, influenced by systems of incentives and constraints on land development, the most important of which are systems of local and state property rights to land and land-derived resources, and diverse sets of policies with strong impacts on land development (Fig. 21.1).

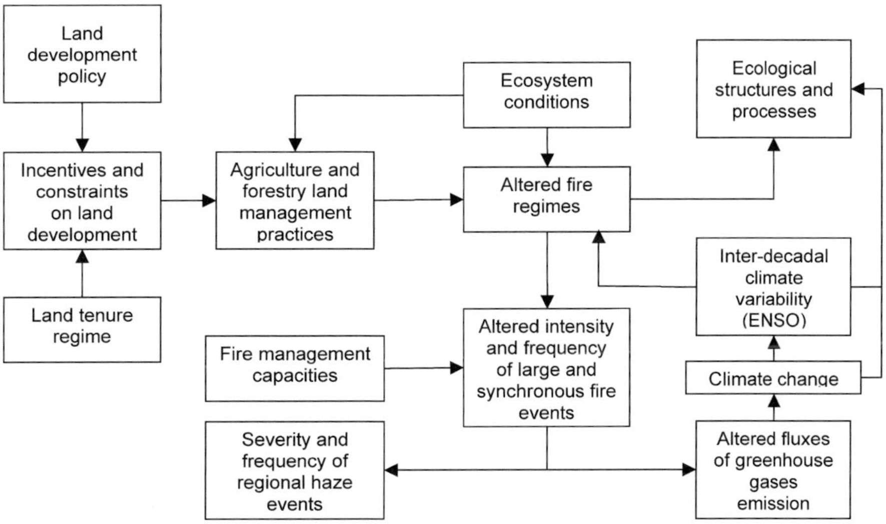

Fig. 21.1. Conceptual framework for how fire regimes are being altered in Southeast Asia

In this chapter we address two overarching questions about land development policies and vegetation fires in Southeast Asia, namely:

1. How have land development policies and changes in land management practices influenced fire disturbance regimes in Southeast Asia? What is the significance of recent changes in fire regimes for local ecosystems and livelihoods, regional air quality and global emissions?
2. How could insights from research about fire ecology, emissions and livelihood impacts be incorporated into development strategies and institutional responses that would reduce the risks of large-scale fire and haze episodes? Are there ways to mitigate undesirable feedbacks on the regional and global atmosphere which would not unfairly disrupt the aspirations of societies in the region to improve their well-being through land development?

The chapter is organized around the broad framework of how human activities are altering fire regimes in the region and the implications this has for ecosystems, livelihoods and feedbacks on the regional and global atmosphere (Fig. 21.1). First, we start with a consideration of what is known about the underlying causes of fires in the region. Second, we consider the immediate ecological and biogeochemical consequences at various scales. And third, we consider how fire has many positive and few negative implications for livelihoods. The chapter ends with a summary of our responses to the two questions.

21.2 Underlying Causes of Land Fires

21.2.1 Explaining Fire Occurrence

The likelihood of fires has been usefully conceptualized as an interplay between *pre-disposing conditions* and *human causes* (Stolle et al. 2003). A multi-factor analysis of factors associated with the distribution of fires from NOAA-AVHRR satellite imagery for 1992–1993 in Jambi province documented interactions between condition- or context-creating factors like climate, elevation, vegetation cover and suitability of land for rubber with more direct human factors like the presence of transmigration projects and timber concessions (Stolle et al. 2003). What is noteworthy from this study is that few fires were accidental. Land development policies and land-use practices were largely driving vegetation fires.

Repeated burning alters the nutrient balance of soils (Crutzen and Andreae 1990) and fires are frequently used as tools by farmers in nutrient-poor soils to take advantage of short-term nutrient releases from burned vegetation. Major fires as well as even more modest understorey fires in closed canopy forests generally increase the likelihood of subsequent fires both in Southeast Asia and the Amazon (Cochrane et al. 1999).

To this we add that past human practices, for example those which make ground conditions drier, alter fuel loads and fragment vegetation, over-time may themselves feedback to alter the "pre-disposing" conditions.

21.2.2 Land Development Policies

A wide range of land development policies have been important in choices of land use, in use of fire for land management, and in fire management practices. Some of the land policies with the most direct effects include forest timber concessions, re-zoning of land for human settlement and cultivation, subsidies for land improvements for specific crops, and irrigation infrastructure projects to stimulate expansion or intensification of production. In Indonesia, Myanmar and Lao PDR, in particular, but in all countries of the region at some stages, resettlement policies backed by varying degrees of coercion and force have been used against swidden farmers, in part, because of perceptions about the negative impacts of their land management practices using fire.

Several other policy areas are not so easy to label as "land development" per se, but their effects may be even larger than some of those listed above because, by altering economic incentives, these policies have much broader reach than rules and regulations that need to be backed by agency actions. In this set, those policies aimed at stimulating (or not) export-oriented agricultural development appear to be particularly important. These include rural credit schemes, trade tariffs, investments in and special deals for agribusiness, for forestry and for forest product enterprises, and monetary policies that influence foreign exchange rates and domestic interest rates.

Fire is often used in conversion of land from forest to agriculture. In Indonesia, much of this expansion has been driven by export-oriented agricultural development policies in recent decades, especially for pulpwood, palm oil and rice, but also by a need to open new land for growing rural populations. Fires may lead or follow other land-use activities like logging (Eva and Lambin 2000).

The history of major fires in Sumatra and Kalimantan, Indonesia, illustrate the interactions with land development policies well. In 1981 the Government of Indonesia released Decree No. 682/Kpts/Um/8/1981 that designated 20–30 Mha of forestland as Conversion Forests (World Bank 1999). Among others, 5 Mha of forest lands were allocated for timber plantations in Sumatra and Kalimantan. This policy sparked extensive fires mainly in those two islands as fire was used as a tool to clear land before plantation activities were established. In 1982 3.6 Mha were burned. The area of oil palm plantation more than tripled between 1981 and the fires of 1997. More than 2 Mha of new plantation was established in less than a decade.

At least as important as the policies on paper has been the rich history of *de facto* policies of the Suharto era that ensured land conversion permits, lucrative timber concessions, and deals for paper mills and associated tree plantations went to friends and family in the military, business and government (Dauvergne 2001; Pasong and Lebel 2000). Throughout this history, fire was often a tool for intentionally degrading land in insurance scams and a weapon of legitimization for land claims or of resistance by displaced people.

Another controversial policy-driven land development was the conversion of around 1.4 Mha of peat-swamp forests in Central Kalimantan (Murdiyarso and Adiningsih 2006; Tacconi 2003; World Bank 1999). The project, legalized by Presidential Decree No. 82/1995 and known as the Mega Rice Project, planned to create some 500 000 ha of rice fields and other tree crops. The construction of 700 km of primary canals (25 m wide and 6 m deep) and of numerous secondary canals has caused substantial drainage and peatland subsidence (Fig. 21.2). The drained area which was prepared to accommodate some 1.5 million transmigrants became susceptible to fire. Around 200 000 transmigrants who had settled there had to leave the area since it was no longer suitable to support their livelihood. Transmigration schemes which aimed to move people away from the densely populated island of Java often failed to live up to their original goals, but nevertheless, had important impacts on land use in many parts of Indonesia. In the strong 1997–1998 ENSO event, 11.6 Mha were burned.

Furthermore, in line with an attempt to decentralize authority, the government of Indonesia launched a new law in 1999 (Law No. 22/1999). As a result many regional regulations were enacted to boost revenues. Unfortunately, the human resources at regional levels could not meet the challenges when extraction of natural resources created new social and environmental problems.

21.2.3 Land Management Practices

In Fig. 21.1 we highlighted fire management capacities. This should be understood to mean not just technical capacities in terms of number of trucks or planes, but also understanding of farmers and local government agencies about how to handle fire regimes which have changed or with which, as new settlers, they may be unfamiliar (Colfer 2001).

Furthermore, at least six distinct interest groups have a stake in the trajectory of land-use change in Sumatra, but there are crucial differences among them in the weights they place on the various economic and environmental outcomes (Tomich et al. 1998a). These groups include the international community, hunter-gatherers, small-scale farmers, large-scale public and private estates, absentee farmers, and public policymakers.

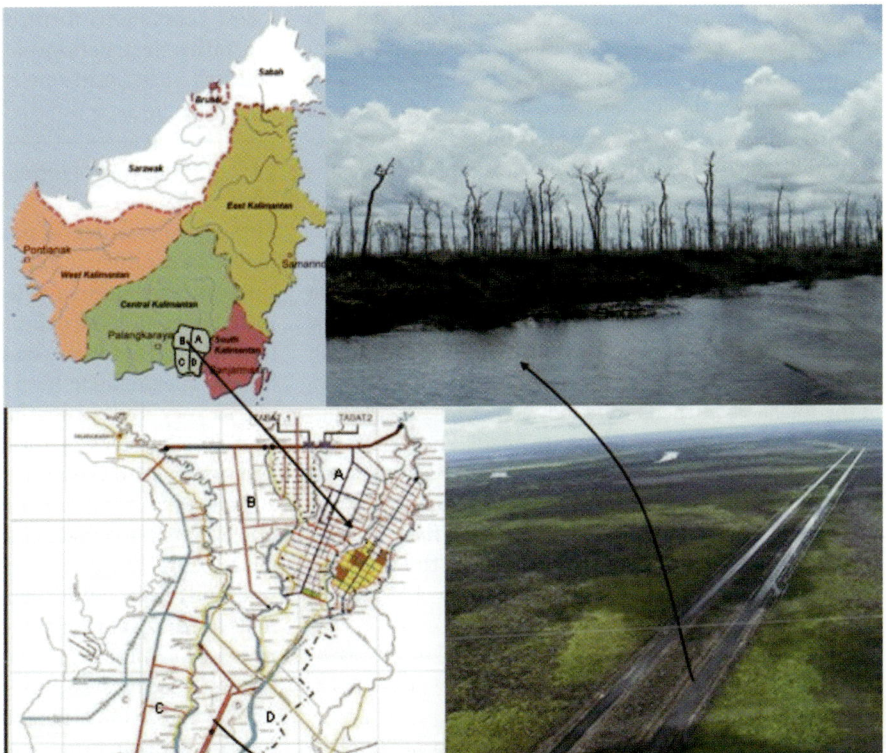

Fig. 21.2. Mega Rice Project in Central Kalimantan, Indonesia (photos: Wetland International-Indonesia Program 2002)

Small-Holder Farmers

Fire is often used as an effective tool for land clearing (Tomich et al. 1998a). Land management practices and the use of fires, however, fall into different categories depending on the intension of land-users. Wibowo et al. (1997) categorized four different types of Indonesian farmers practising shifting cultivation: (1) traditional communities with strong local institutions restricting the actions of individual farmers; (2) communities that are more integrated in the market economy, often combined with a loosening of traditional restrictions; (3) spontaneous settlers/migrants who may have skills but who are usually landless; and (4) government-sponsored transmigrants, who are granted land with titles. Except for the first category, these groups of farmers practice non-sustainable agriculture and bring no significant conservation skills to the management of their agricultural lands. Most of them viewed that burning is the cheapest and quickest way for land clearing.

Fires are used to provide space since they offer cheap and effective means to remove the debris before new plantations are established. Furthermore, fires also provide ash which contains readily available nutrients and controls pests and diseases. This may not be the case for large-scale plantations, especially as to the long-term economic advantages (Simorangkir 2006). In eastern Malaysia exclusion of fire and minimum soil disturbance during plantation forest establishment could double production of wood and strongly reduce environmental impacts (Nykvist et al. 1994). However, this has not automatically led to a reduced use of fire in the region. Slash-and-burn fires are still a common tool for halting or re-directing succession.

The practice is slightly different compared with that used in the peatlands. There, the fires burn much more below-ground biomass and, as the combustion is usually incomplete, they tend to produce a lot more haze over longer periods (Murdiyarso et al. 2004b). Fuller and Fulk estimated that during the 1997/1998 fires, 1.45 Mha of peatlands were burned involving 300 000 ha in Sumatra and 750 000 ha in Kalimantan. The most recent assessment shows that the total area of peatland burned was 2.12 Mha (Tacconi 2003). This is about 10% of Indonesia's total peatland area or 20% of the total area burned in the 1997/1998 fires.

Clearly, policy interventions are needed but, given the variety of fires, they cannot be generalized since each type requires a specific intervention. This confirms the earlier finding that quick fixes by banning fires will not be successful unless they properly address the underlying causes and provide alternatives (Tomich et al. 1998c).

21.2.3.1 *Logging*

Logging and tree plantation forestry have had a major impact on fire disturbance regimes. On the one hand logging has made much of the landscape susceptible to fire during periods of drought, and on the other, management for timber has often involved fire suppression policies. A history of unsustainable logging practises has been reinforced by state corruption that together have exacerbated fire management challenges (Dauvergne 2001).

Logging in closed canopy tropical forest creates canopy gaps and a pulse of leaf litter. The resulting drier conditions are likely to increases the risk of fires (Nepstad et al. 1999). Moreover, after fires occur in a concession the normal restriction on the 20-year cutting cycle is lifted for salvage harvesting, thus creating a perverse incentive for poor fire management (van Nieuwstadt et al. 2001). Logging in the logged-over forests appears to increase the risk of disastrous fires (Siegert et al. 2001).

21.2.3.2 *Forest Management*

Forestry departments in Southeast Asia have adopted models of "land management" based on moist, temperate forest ecologies. The key attitude is that fires are natural disasters and management is about fire suppression to protect property, especially valuable timber, from being burned. Although this may make sense in some forest ecosystems near urban areas, ecologically, and eventually economically, it doesn't make sense in many others. In seasonally dry tropical forests, fire suppression leads to higher fuel loads, more intense fires and the possibility of devastating fires, which take a long time for forest ecosystems to recover from because of massive mortality, destruction of seed banks, and alteration of soil structure (Stott 1988, 1996).

21.2.3.3 *Shifting Cultivation*

Fire is also an integral part of shifting cultivation practices that are diverse and in many ways qualitatively distinct from using fire to clear forests (Dove and Kammen 1997; Fox et al. 2000; Schmidt-Vogt 1998). Some practices, for instance, appear to be particularly effective in maintaining intermediate-level forest cover at landscape scales, with a very diverse spatial mosaic of landscape patches in different stages of succession. Whether or not swidden systems are viewed as destructive to forest biodiversity depends very much on what components are emphasized and to which land-uses such practices are compared (Forsyth 1998, 2003; Schmidt-Vogt 1998). The mosaic landscapes do not support many of the larger mammals needing intact tracts of forest, but on the other hand much other vegetation and small animals may persist in these landscapes and many other watershed functions may be maintained (Forsyth 1996; Thomas 2002). In comparison to plantations or conventional tillage agriculture such systems of course preserve a lot more of the original biodiversity (Schmidt-Vogt 1998; Thomas 2002).

This is brought home by comparing such practices to for example the impacts of conventional lowland agriculture which leads to largely irreversible conversion of land cover. In longer rotational systems there can be very intentional pre-burn preparation of land including lopping and stacking of branches and so on.

Intentional fires to prepare land for upland cultivation of rice and a diversity of other products is common in Lao PDR, in northern Vietnam, and in some scattered locations of northern Thailand, Borneo Indonesia, northern Myanmar and Yunnan province of China.

Lao PDR has pursued a policy of resettling ethnic minorities that utilize shifting cultivation practices from the uplands to the lowlands with the rationale that such practices cause deforestation and erosion. By 1997 the government had established 62 "sedentarisation" sites targeting some 320 000 people. In practice local officials have varied widely in their implementation efforts, and villagers may move to the lowlands for economic reasons rather than as a result of government intervention. Either way, the Akha in Muang Sing who moved to lower slopes were often not able to secure better livelihoods, received little or no government assistance, and instead became a cheap and exploited labor force for lowland Tai (Cohen 2000). In some locations the Akha remain under pressure to continue opium cultivation.

Government policies to limit or end shifting cultivation are found throughout montane mainland Southeast Asia. In combination with macro-economic changes accompanying improved road infrastructure and emerging markets for cash crops apart from opium, this has undoubtedly lead to major shifts in human use of fire in the uplands. After several decades of expansion of areas under shifting cultivation as a result of migration and population growth, there has been a large decline in areas with such practices over the past two decades. As far as we know, however, there has been no detailed study of fire regime changes and their ecological consequences. For example, is the occurrence of fires less frequent now, but when they occur are they more intense? Has the proportion of all fires that are accidental or lightning-lit increased?

21.2.4 Property Rights and Conflicts

The likelihood of destructive fires, both intentional and accidental, appears to be often related to investments in fire protection and management by communities, which in turn is affected by land tenure security and conflicts over land access rights. Insecure tenure and poor relations reduce the incentives for careful fire management. In serious conflicts fire may be used as a weapon, both by smallholders and larger commercial enterprises (Colfer 2001). Thus, in the 1997–1998 fires, estimates suggest that as many as 14% of the fires were related to social conflicts between plantation developers and local communities (Ganz 2003).

In most remote areas it is not the existence of formal state-recognized property rights that matters so much as whether the rights, customary or formal, are recognized by the various local actors. Upland swidden farmers throughout Southeast Asia typically have lacked formal title to the land they cultivate, but nevertheless have continued with well-defined land-use systems in which controlled fires are a crucial tool for food production.

In conflicts between large firms and small-holders, on the other hand, as has been the case frequently in Indonesia, clear systems of property rights appear to help reduce conflicts (Tomich et al. 1998b). Procedures also matter. Communities complain about unheard claims, unfair judicial systems, and non-transparent decision-making processes (Suyanto 2005). When push comes to shove, customary or "*adat*" rights are often not fully recognized by legal systems in the region. In the Philippines, fires in reforestation projects were also caused by conflicts between management (typically Department of Environment and Natural Resources, DENR) and field labourers (typically from local communities). It was not uncommon for newly planted trees to be burned because of some grievance, real or imagined (R. Lasco pers. comm. 2005).

21.3 Landscape, Regional and Global Interactions

21.3.1 Ecosystem Dynamics

Land fire regimes across Southeast Asia vary among locations because of differences in climate, types of vegetation and intensity and form of human activities. Some woody savannas burn very frequently, whereas other moist tropical communities are very rarely burned (van der Werf et al. 2003).

Knowledge about the ecological consequences of fires on tropical forests in Southeast Asia is modest, with most research coming from work on moist but ENSO-impacted forests in Indonesia. For many key processes influenced by humans activities the most detailed understanding so far is based on studies done in Central and South America (Cochrane et al. 1999; Nepstad et al. 2001), and more recently in northern Australia (Edwards et al. 2001; Horstman and Wightman 2001; Yibarbuk et al. 2001). This matters greatly for generalizations because the ecological impacts vary widely across different ecosystems (Eva and Lambin 2000) which is probably reflected by the cultural variation in land-use practices (see Sect. 21.2.1).

Nutrient losses due to volatilization during the burning of residual biomass are generally higher than the losses by leaching (Bruijnzeel 1998). This is not only for nitrogen, which comprises more than 90% of the lost biomass (Note: the sentence reads like if 90% of the biomass burned (presumably by weight) is nitrogen; I assume this is incorrect as N is a much smaller fraction of the biomass; please rewrite accordingly), but often also for mineral nutrients. These high atmospheric losses have been used to suggest the reduction of burning in land-clearing but very little reduction has been implemented. Burning also increases losses by leaching when compared to non-burning practices (Malmer et al. 1994).

As the carbon content of biomass fuels, apart from charcoal, varies over only a narrow range (37–54%) emission factors are largely determined by the combustion process. However, many other chemical species are emitted as trace gases and aerosols and these can vary substantially across both types of vegetation and fires (Andreae and Merlet 2001).

Southeast Asia is home to about 60% of the world's tropical peatlands. The impacts of fire and ecological change in this landscape are probably the area of greatest concern from the perspective of both regional haze episodes and global greenhouse gas emissions. The impacts of fires in Kalimantan were described in detail earlier in this chapter. Changing fire regimes in peat wetlands are an issue not only for Indonesia. Similar problems have arisen in the Melaleuca peat forests of U Minh in the Mekong delta of Vietnam, an area targeted for conservation but where new and enlarged canals have altered the water regime exacerbating the seasonal drying and facilitating the human access that increase the risks of ignition at critical times (Sanders 2002). Likewise, in peninsular southern Thailand road construction, as well as infrastructure for shrimp farms and agriculture, has contributed to some major peat forest fires by altering water regimes. Such fires produce relatively large amounts of smoke and greenhouse gas emissions per unit area.

Ecological interactions with fire are likely, but not well understood in Southeast Asia. Pest outbreaks by causing crown loss and tree dieback may increase susceptibility to fires. There are likely to be important interactions between fire, pest regimes, harvesting and other land management practices (Gower 2003). Post-fire undergrowth and seedbanks are sensitive to additional disturbances. Thus, the damage caused by conventional machinery used in "*salvage*" logging after fires reduces the forest's potential for recovery (van Nieuwstadt et al. 2001).

21.3.2 Regional Haze Episodes

Regional haze episodes are primarily a consequence of synchrony of fire events, strong winds and atmospheric inversions that expose large, sometimes distant, populations to hazardous levels of smoke and haze (Murdiyarso et al. 2004a). Trans-boundary haze episodes pose significant economic and public health burdens. For this reason they attract the attention of the international media and policy makers alike, at least until the next monsoonal rains clear the skies.

In the 1997–1998 fire events in Indonesia there were some indications that permits for clearing land had been held onto in the hope of drier, and thus easier to use, fire conditions eventuating and that these permits were then "used" simultaneously by many farmers when the dry ENSO phase unfolded.

Long-term fire histories in regions of Argentina and the US in which the climate is affected by the El Niño-Southern Oscillation have indicated that major fire years are those after a switch from wetter conditions that built up fuel loads to drier conditions that then rapidly dried out the forests (Kitzberger et al. 2001). In Southeast Asia human activities appear to be a more critical part of the causal fire cluster and to modify the pre-disposing conditions.

21.3.3 Greenhouse Gas Emissions

The 1997–1998 fire events in Indonesia were well studied and provide some of the better estimates of carbon fluxes (Table 21.1).

Default estimates of carbon released from deforestation could be based upon the decomposition of biomass over the lifetime of the wood products, but fire is the most effective and direct means to convert carbon stored in the biomass into the gaseous phase. Fire could also release the enormous below ground biomass stored in the peatlands. Page et al. (2002) made a detailed study of a 2.5 million hectare area in Central Kalimantan from which they estimated 0.81 to 2.57 Gt of carbon were released to the atmosphere in 1997 as a result of peat and vegetation fires in Indonesia. This represents 13–40% of total emissions in that year from fossil fuel burning. The contribution from peat fires was very significant.

The stability of wetlands overlying peat deposits is very important for the global carbon balance as they contain huge deposits. In 1990 Sumatran peatlands covered an area of 7.2 Mha, but this had reduced to 6.5 Mha in 2002. The estimated carbon released during that period was 3.47 Gt (Murdiyarso et al. 2004b). This is one of the more problematic aspects of land development and how it affects fires. Changes in these wetland peat landscapes which result in carbon emissions may be much harder to reverse. Due to the uniqueness of the ecosystems, car-

Table 21.1. Estimates of the carbon fluxes attributable to land fires in various locations during the 1997–98 dry phase of ENSO in Indonesia. Numbers in brackets indicate emissions just from fires in peatlands (Murdiyarso and Adiningsih 2006)

El Nino year	Area burned (Mha)	Estimated C-released (Gt)
1982	3.6	0.45
1987	0.1	0.01
1991	0.5	0.06
1997	11.6 (2.1)	1.45 (0.47)

bon sequestration elsewhere may not compensate the loss in their biological diversity.

Van der Werf et al. (2003) analysed a 4-year infrared satellite dataset from the TRMM (Tropical Rainfall Measuring Mission) and combined this with a large-scale biogeochemical model to estimate global carbon fluxes from fires in the tropics and sub-tropics. The model made estimates of emissions from combustion losses and decomposition due to fire-induced mortality. They estimated that overall the annual direct and indirect carbon losses from fires amounted to about 9% of net primary carbon production in the tropics and sub-tropics.

21.3.4 Interactions with Climate Variability and Change

How precipitation regimes across Southeast Asia will be affected by climate change is still highly uncertain at the spatial resolutions relevant for land management and vegetation community dynamics. Nevertheless, historical observations, experimental studies manipulating soil moisture and modeling, all suggest that climate changes could have major impacts on terrestrial ecosystems (Weltzin et al. 2003). Short-term changes in soil moisture can impact risks of ignition, and longer-term changes in rainfall may influence vegetation structure and litter densities that in turn influence fire regimes. Interactions between precipitation and temperature changes are likely.

Regional land-surface interactions that dominate weather patterns in the Amazon seem less likely to be important in most parts of Southeast Asia because of the predominance of ocean influences on climate in SEA.

21.4 Human Well-Being

21.4.1 Economic and Health Impacts

The regional economic impacts of fires have rarely been assessed, but the 1997–1998 event was so large and clear that several estimates were soon made (Bapenas 1999; Glover and Jessup 1998). These highlighted the importance of losses to the transport and tourism sectors within and outside Indonesia. The most worrisome impact in the public mind, however, was on health which can help explain the declines in tourist visits. The health impacts on inhabitants near fires were likely to have been much higher than on distant populations but longer-term health effects of recurrent episodes of haze and smoke are not well understood and most of the emphasis has been on the discrete transboundary episodes.

For example, in Singapore effective computerized patient information systems and air quality monitoring networks allowed researchers to analyze the association between air quality measurements and diagnoses for haze-related outpatient visits to hospital (Emmanuel 2000). They found that an increase in levels of fine particulate matter, PM_{10} from 50 μg m^{-3} to 150 μg m^{-3} was associated with increases of 12–26% in upper respiratory tract illness, asthma and rhinitis. No significant increases of admissions or mortality were noted. Emmanuel concluded that the health effects on Singapore of the 1997 smoke haze were generally mild. A thorough assessment in Malaysia, reporting at the same time as the Singapore study, acknowledged the significant economic costs of the transboundary haze, but concluded that the most serious sources of air pollution for human health were domestic, especially motor vehicles and oil and gas works (Awang et al. 2003).

21.4.2 Livelihoods

Fires have both positive and negative impacts on the livelihoods of people in affected and nearby areas. Most fires in Southeast Asia, as we have seen, are not accidental, but deliberately lit to achieve land conversion and management goals.

The drivers for land-use change, however, are often in part embedded in national development policies that are reinforced by preferential allocation of land for different activities. It is through these mechanisms that large negative impacts on the livelihoods of the poor are to be found, and fire is mostly a side-effect of these rather than a cause. In many remote areas, it cannot be ignored that fire is a tool for land improvements that help stake a land claim. In addition, it is not insignificant the involuntary risks that neighbouring and sometimes distant populations are placed under by smoke from vegetation fires, but neither are they unprecedented disasters – many of the citizens with the largest complaints are in cities filled with exhaust fumes of cars and the investments and consumption patterns of these citizens are driving some of the land-conversions in those distant places.

Road-building in remote areas is a controversial topic throughout the tropics. Obviously, improved infrastructure is important to market-integrated development and provision of basic services. Improved access can also accelerate land-conversion in un-intended places as well as the risks of accidental fires (Nepstad et al. 2001).

The importance of macro-economic conditions is underlined by the interactions between the 1997–1998 ENSO dry event and the Asian financial crisis which simultaneously impacted on livelihoods in the region. Sunderlin et al. (2001) found that dependence on forest resources increased markedly during the 1997–1999 economic and political crisis in Indonesia as unemployed city workers returned to their home villages.

One of the difficult ideas for policy and rural development advisors from developed countries is to understand that the conventional separation of agriculture and forestry activities so firmly embedded in their natural resource management bureaucracies is not the way many smallholder livelihoods have been organized in the tropical landscapes of Southeast Asia. Forestry, including collection, active management and cultivation of useful plants within forests, and farming activities that often include tree-crops, is much more integrated than the rural extension and new state-centric models of property rights can handle. On the other hand, people in the region are remarkably adaptable and willing to try new technologies given appropriate incentives. This often undermines the guiding, somewhat nostalgic, assumptions of non-government actors about aspirations for cash-crops of many smallholders. Much of the debate about alternative fire and related land management practices begins with assumptions about appropriateness of different livelihood and lifestyle activities, a debate which unfortunately is not always well-informed about or with representation of the interests of the farmers concerned. The appropriate use of fire in the tropics is ultimately a political issue to which research-based knowledge, about impacts and likely sustainability, can make a contribution, but not predetermine deliberation around trade-offs.

21.5 Informed Decision-Making and Better Governance

21.5.1 Role of Expertise

Research-based knowledge has been summarized several times as inputs into policy. Nevertheless, there appears to still be a large gap between the information which is needed to strengthen and re-design institutions at various scales to address changes in fire regimes and, in particular, large fire and haze episodes and the contributions which various expert communities have been able to deliver. In part this reflects the tendency for both science and policy to de-politicize issues by emphasizing the technical aspects of fire fighting and prevention or the alternative land preparation techniques, rather than by addressing directly the underlying causes which lie primarily in land development policies and investments. It also is an acknowledgment of some of the inherent complexities in an issue where the drivers, consequences and institutional responses are each significantly multi-scale (Fig. 21.3).

Multilateral and bilateral technical assistance by organizations and governments, such as the Asian Development Bank, the European Union, FAO, UNEP, The World Bank, Germany, Canada, Japan, USA, UK, and Australia has prepared global, regional, and national fire assessments. Capacities of national and local governments have been built up to suppress and mitigate fires. International and national NGOs, such as IUCN and WWF, have also been devoting increased attention to fires and highlighting the underlaying causes and impacts. Some 40 fire projects and missions costing well over US$30 million have worked in Indonesia over the last 20 years (Dennis 1999). Despite the money and effort spent on them, fires continue to burn every year. Thus it may appear to some that efforts to address the "fire problem" have not been effective (Tacconi et al. 2003).

Several other policy initiatives are possible and have in part been taken up. These include development of drought and fire information systems, medium-term forecasting of ENSO drought conditions, and reviews of

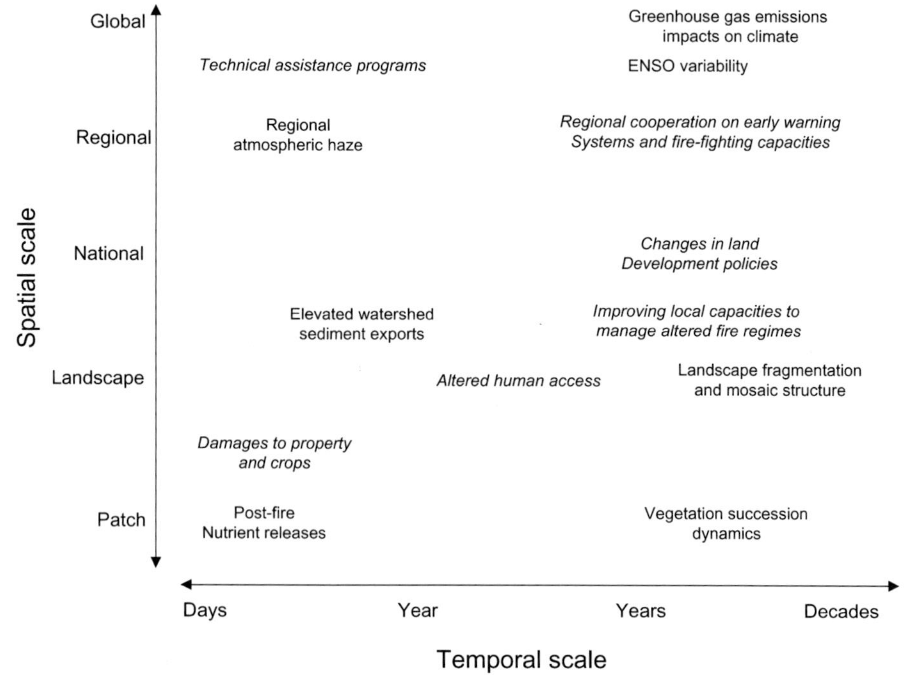

Fig. 21.3. Multi-scale view of the large-scale fire and haze events in Southeast Asia as an environmental and social "problem"

land-use zoning and clearing regulations (Applegate et al. 2001; Murdiyarso et al. 2004a). Proposed zero-burning options do not appear likely to be adopted widely (Murdiyarso et al. 2004a). There are also opportunities to improve the utilization of waste wood after logging by the removal of policies distorting prices (Murdiyarso et al. 2004a).

State capacities are in general limited, given the size of the areas of forest and cultivated lands under their jurisdictions. However, for communities to play an important role in fire management they must have control over forest resources (Ganz and Moore 2002). Where ownership is weak there is no incentive for communities to help. Incentives, however, can also act perversely: in parts of Indonesia communities are paid to fight fires, creating an incentive to light them (Karki 2002). On the other hand, without control of and rights to forest resources, pushes for "community-based" fire management may be seen by governments as a way to obtain "cheap labor" to protect their "forestry" resources. The private sector, in general, has not had responsibilities for fire management of forests commensurate with the benefits it has extracted.

In wetlands, greater attention to the impacts of infrastructure projects on hydrology could remove some of the underlying fire problems. In the provinces of Jambi, South Sumatra and Central Kalimantan, Wetlands International promotes canal blocking in the peatlands to prevent further peat dome subsidence, to improve hydrology and to avoid fires (Murdiyarso et al. 2004b). Controlling the water table is a key issue of peatland management in which local communities are heavily involved. Alternative livelihoods are also promoted by re-introducing local timber species with a high economic value and traditional fishery practices (*tabat*) wherever the water table is elevated.

21.5.2 Regional Cooperation

The Association of Southeast Asian Nations (ASEAN) has been at the forefront of mostly rhetorical attempts to promote regional cooperation on trans-boundary pollution issues, including those arising from vegetation fires. In 1999, after several years of discussions a Regional Haze Action Plan was developed by a Haze Technical Task Force with support from the Asian Development Bank. The widely accepted ASEAN response was expected to be revisited annually (Qadri 2001). The emphasis in implementation, as for example laid out in the Operationalized Regional Action Plan, remains firmly on fire suppression and mitigation. The underlying causes of fires are not addressed (Murdiyarso et al. 2004a).

The legally binding Agreement on Transboundary Haze Pollution (ATHP) adopted in 2002 entered into force in November 2003 following ratification by six member countries. However tension and pessimism with regards to the ATHP remain since Indonesia, the most likely source of such pollution, has not ratified the agreement. Indonesia faces tough institutional challenges in trying to bring together central and increasingly autonomous provincial governments for the implementation of international agreements of this type.

The ASEAN Peatland Management Initiative (APMI) with technical support from the Malaysia-based Global Environmental Centre (GEC) is a promising venture as it has been based on the concept of broad consultation with diverse stakeholders.

Improved fire-fighting capacities and preventative measures, especially with respect to "scaling-down-activities" in seasons with high risks of unintentional fire-spread, are both needed. Research has provided many new insights into the proximate causes of fires, their spread and the climactic and local conditions under which people prefer to light them. All this is useful background understanding to assist institutional and educational interventions. So far, however, this cumulative understanding has not had anywhere near enough impact. In the last few years, the transfers of responsibilities by central governments to local communities and local governments have risen but they have usually not being accompanied by the resources required to build the capacity to understand and manage new fire regime. While the barriers are not insurmountable the challenges are large and will require investments. These should be the focus of regional cooperation and broader technical assistance.

21.6 Summary and Conclusions

Forest and land fires in Southeast Asia are not the apocalyptic ecological and social disasters that are sometimes portrayed by the international media and some experts. However, fire regimes for most landscapes in the region have undoubtedly been altered by human activities, sometimes in ways detrimental to the ecosystems upon which livelihoods partly depend. In the past two decades the changes in Indonesia, in particular, have been rapid and associated with conversion of forest lands to agricultural crops and pulpwood plantations. In this latter context fire remains an important land-conversion and management tool for both larger corporations and small holders.

The challenge is largely one of governance. Alternative visions, objectives and paradigms for the development of the rural hinterlands of Southeast Asia persist (Lebel et al. 2004). Each adopts a narrow set of assumptions about the appropriateness of different ways to manage the complex agricultural and forested landscapes of the region, but without adequate regard for how the knowledge and interests of various stakeholders are to be considered nor how those vested with authorities are to be held accountable (Pasong and Lebel 2000).

At the same time other forces are making the institutional challenges even more difficult. Macro-economic and political changes are placing more of the critical drivers beyond the control of governments and societies in the developing countries within the region. The fire regimes in the tropics of Southeast Asia have been altered by human actions for millennia and undoubtedly will continue to be, but in the future these will increasingly be confounded by the short-term and cumulative impacts of the consumption behavior of distant actors on crop choices, land-use, atmospheric composition and climate change.

Acknowledgments

The authors would like to express their thanks and appreciation to those involved in the commissioned studies organized by the Global Change Impacts Centre for Southeast Asia (ICSEA) which was financially supported by AusAID through GCTE. Grateful thanks are also due to those involved in the Alternative to Slash and Burn program, administered by World Agroforestry Centre (ICRAF) with financial support from GEF/UNDP. Continuing support for regionally-focused global change research in Southeast Asia given by the START program is gratefully acknowledged.

References

Andreae MO, Merlet P (2001) Emission of trace gases and aerosols from biomass burning. Global Biogeochemical Cycles 15:955–966

Applegate G, Smith R, Fox JJ, Mitchell A, Packham D, Tapper N, Baines G (2001) Forest fires in Indonesia: impacts and solutions. In: Colfer CJP,Resosudarmo IAP (eds) Which way forward?: people, forests and policymaking in Indonesia. Center for International Forestry Research, Bogor, pp 293–308

Awang M, Jaafar AB, Abdullah AM, Ismail M, Hassan MN, Abdullah R, Johan S, Noor H (2003) Air quality in Malaysia: Impacts, management issues and future challenges. Respirology 5:183

BAPPENAS (1999) Planning for Fire Prevention and Drought Management Project. Planning for Fire Prevention and Drought Management in Indonesia. BAPPENAS, Jakarta

Bruijnzeel LA (1998) Soil chemical changes after tropical forest disturbance and conversion: The hydrological perspective. In: Schulte A, Ruhiyat D (eds) Soils of tropical forest ecosystems: characteristics, ecology and management. Springer-Verlag,, Berlin, pp 45–61

Cochrane MA, Alencar A, Schulze MD, Souza CMJ, Nepstad DC, Lefebvre P, Davidson EA (1999) Positive feedbacks in the fire dynamics of closed canopy tropical forests. Science 284: 1832–1835

Cohen PT (2000) Resettlement, opium and labour dependence: Akha-Tai relations in northern Laos. Development and Change 31:179–200

Colfer CJP (2001) Ten propositions to explain Kalimantan's fires. In: Colfer CJP, Resosudarmo IAP (ed) Which way forward?: people, forests and policymaking in Indonesia. Center for International Forestry Research, Bogor, pp 309–324

Crutzen PJ, Andreae MO (1990) Biomass burning in the tropics: impacts on atmospheric chemical and biogeochemical cycles. Science 250:1669–1678

Dauvergne P (2001) Loggers and degradation in the Asia-Pacific: corporations and environmental management. Cambridge University Press, Cambridge

Dennis R (1999) A review of Fire Projects in Indonesia (1982–1998). Center for International Forestry Research, Bogor

Dove MR, Kammen DM (1997) The epistemology of sustainable resource use: managing forest products, swiddens, and high-yielding variety crops. Human Organization 56:91–101

Edwards A, Hauser P, Anderson M, McCartney J, Armstrong M, Thackway R, Allan G, Hempel C, Russell-Smith J (2001) A tale of two parks: contemporary fire regimes of Litchfield and Nitmiluk National Parks, monsoonal northern Australia. International Journal of Wildland Fire 10:79–89

Emmanuel SC (2000) Impact to lung health of haze from forest fires: the Singapore experience. Respirology 5:175–182

Eva H, Lambin EF (2000) Fires and land-cover change in the tropics: a remote sensing analysis at the landscape scale. Journal of Biogeography 27:765–776

Forsyth T (1996) Science, myth and knowledge: testing Himalayan environmental degradation in Thailand. Geoforum 27:275–292

Forsyth T (1998) Mountain myths revisited: integrating natural and social environmental science. Mountain Research and Development 18:126–139

Forsyth T (2003) Critical Political Ecology: The Politics of Environmental Science. Routledge, London

Fox J, Dao Minh Truong, Rambo AT, Nghiem Phuong Tuyen, Le Trong Cuc, Leisz, S (2000) Shifting cultivation: a new old paradigm for managing tropical forests. Bioscience 50:521–528

Ganz D (2003) Framing Fires: A country-by-country analysis of forest and land fires in the ASEAN nations. PFFSEA-WWF-IUCN-EU

Ganz D, Moore, PF (2002) Precis of Communities Conference on the Involvement of Communities in Fire Management. Project Fire Fighting South East Asia, Jakarta

Glover D, Jessup T (1998) The Indonesian fires and haze of 1997: the economic toll. EEPSEA, Singapore

Gower ST (2003) Patterns and mechanisms of the forest carbon cycle. Annu Rev Environ Resour 28:169–204

Horstman M, Wightman G (2001) Karparti ecology: Recognition of Aboriginal ecological knowledge and its application to management in north-western Australia, Ecological Management and Restoration 2:99–109

Karki S (2002) Community involvement in fire management in East Asia. Project Forest Fire South East Asia, Jakarta

Kitzberger T, Swetnam TW, Veblen, TT (2001) Inter-hemispheric synchrony of forest fires and the El Nino – Southern Oscillation. Global Ecology and Biogeography 10:315–326

Lebel L, Contreras A, Pasong S, Garden P (2004) Nobody knows best: alternative perspectives on forest management and governance in Southeast Asia: Politics, Law and Economics. International Environment Agreements 4:111–127

Malmer A, Wong FK (1994) Converting tropical rainforest to forest plantation in Sabah, Malaysia. II. Changes of nutrient dynamics and net losses in streams due to treatments. Hydrological Processes 8:195–209

Murdiyarso D, Adiningsih E (2006) Climatic variability, Indonesian vegetation fires and terrestrial carbon emissions. Journal of Mitigation and Adaptation Strategy for Global Change, DOI 10.1007/s11027-006-9047-4

Murdiyarso D, Lebel L, Gintings AN, Tampubolon SMH, Heil A, Wasson M (2004a) Policy responses to complex environmental problems: Insights from a science-policy activity on transboundary haze from vegetation fires in Southeast Asia. Journal of Agriculture, Ecosystems, and Environment 104:47–56

Murdiyarso D, Suryadiputra IN, Wahyunto (2004b) Tropical peatlands management and climate change: a case study in Sumatra, Indonesia. Proceedings of the 12th International Peat Congress, Tampere: Finland, pp 696–706

Nepstad DC, Verissimo A, Alencar A, Nobre C, Lima E, Lefebvre P, Schelsinger P, Potter C, Moutinho P, Mendoza E, Cochrane M, Brooks V (1999). Nature 398:505–508

Nepstad DC, Carvalho G, Barros AC, Alencar A, Copobianco JP, Bishop P, Moutinho P, Lefebvre P, Silva ULJ, Prins E (2001) Road paving, fire regime feedbacks, and the future of Amazon forests. Forest Ecology and Management 154:395–407

Nykvist N, Grip H, Sim BL, Malmer A, Wong FK (1994) Nutrient losses in forest plantations in Sabah, Malaysia. Ambio 23:210–215
Page SE, Siegert F, Rieley JO, Boehm HDV, Limin S (2002) The amount of carbon released from peat and forest fires in Indonesia during 1997. Nature 420:61–65
Pasong S, Lebel L (2000) Political transformation and the environment in Southeast Asia. Environment 42:8–19
Qadri S (2001) Fire, Smoke, Haze: The ASEAN response strategy. ASEAN-ADB
Sanders BM (2002) Fire Incident Assessment U Minh Ha Forest and U Minh Thuong National Park Ca Mau and Kien Giang Provinces, Vietnam 20 April–26 April 2002. GTZ Deautsche Gesslshaft fur Technische Zusammenarbeit GmBH
Schmidt-Vogt D (1998) Defining degradation: the impacts of swidden on forests in northern Thailand. Mountain Research and Development 18:135–149
Siegert F, Ruecker G, Hinrichs A, Hoffman AA (2001) Increased damage from fires in logged forests during droughts caused by El Nino. Nature 414:437–440
Simorangkir D (2006) Fire use: Is it really the cheaper land preparation method for large-scale plantations? Journal of Mitigation and Adaptation Strategy for Global Change, DOI 10.1007/s11027-006-9049-2
Stolle F, Chomitz KM, Lambin EF, Tomich TP (2003) Land use and vegetation fires in Jambi Province, Sumatra, Indonesia. Forest Ecology and Management 179:277–292
Stott P (1996) The spatial pattern of dry season fires in the savanna forests of Thailand. Biogeog 13:345–358
Stott P (1988) The forest as Phoenix: towards a biogeography of fire in mainland South East Asia. The Geographical Journal 154:337–350
Sunderlin WD, Angelsen A, Resosudarmo DP, Dermawan A, Rianto E (2001) Economic crisis, small farmer well-being, and forest cover change in Indonesia. World Development 29:767–782
Suyanto S (2005) Land tenure conflicts: An underlying cause of fire in Sumatra. Journal of Mitigation and Adaptation Strategy for Global Change. In press
Tacconi L (2003) Fires in Indonesia: causes, costs and policy implications. Center for International Forestry Research, Bogor
Tacconi L, Moore PF, Kaimowitz D (2003) Fires in Tropical Forests – throwing good money after bad? Lessons from Indonesia. Proceedings of International Congress of Wild Fires, Sydney
Thomas D (2002) Landscape agroforestry in northern Thailand. ICRAF, Chiang Mai
Tomich T, van Noordwijk M, Budidarsono S, Gillison A, Kusumanto T, Murdiyarso D, Stolle F, Fagi AM (1998a) Alternatives to Slash-and-Burn in Indonesia. Summary Report and Synthesis Phase II ICRAF SE Asia, Bogor
Tomich TP, Fagi AM, de Foresta H, Michon G, Murdiyarso D, Stolle F, van Noordwijk M (1998b) Indonesia's fires: smoke as problem, smoke as a symptom. Agroforestry Today:4–7
Tomich TP, van Noordwijk M, Vosti SA, Witcover J (1998c) Agricultural development with rainforest conservation: methods for seeking best bet alternatives to slash-and-burn, with applications to Brazil and Indonesia. Agricultural Economics 19: 159–174
van der Werf GR, Randerson JT, Collatz GJ, Giglio L (2003) Carbon emissions from fires in tropical and subtropical ecosystems. Global Change Biology 9:547–562
van Nieuwstadt MGL, Sheil D, Kartawinata K (2001) The ecological consequences of logging in the burned forests of East Kalimantan, Indonesia. Conservation Biology 15:1183–1186
Weltzin JF, Loik ME, Schwinning S, Williams DG, Fay PA, Haddad BM, Harte J, Huxman TE, Knapp AK, Lin G, Pockman WT, Shaw MR, Small EE, Smith MD, Smith SD, Tissue DT, Zak JC (2003) Assessing the response of terrestrial ecosystems to potential changes in precipitation. Bioscience 53:941–952
Wibowo AM, Suharti, Sagala APS, Hibani H, van Noordwijk M (1997) Fire management on Imperata grasslands as part of agroforestry development in Indonesia. Agroforestry Systems 36:203–217
World Bank. 1999. Deforestation in Indonesia: A preliminary view of the situation in Sumatra and Kalimantan. World Bank, Washington, DC
Yibarbuk D, Whitehead PJ, Russell-Smith J, Jackson D, Fisher A, Cooke P, Choquenot D, Bowman DMJS 2001. Fire ecology and Aboriginal land management in central Arnhem Land, northern Australia: a tradition of ecosystem management. Journal of Biogeography 28:325–344

Chapter 22

Global Change Impacts on Agroecosystems of Eastern China

Guangzhou Zhou · Shaoling Peng · Li Zhou · Yuhui Wang · Zhenzhu Xu

22.1 Introduction

Global environmental change and its direct impacts on the living environment and human societies have attracted the attention of international scientific bodies and governments of many countries (Zhang 1994; Ye and Chen 1992; Peng 1997; Peng and Ren 1998; Koch et al. 1995). Research shows that global change is already exerting a detectable influence on the structure and function of natural and agricultural ecosystems, including forest, grassland and farmland ecosystems in China (Liu 1993; Feng et al. 1999; Wang 1965; Xu et al. 1997). Some of the most relevant drivers of global change are the increase in atmospheric CO_2 which has a "fertilization effect" on net primary productivity. In addition, climate change resulting from increased greenhouse gases (GHGs) will also affect the energy balance of the regions, causing changes in water and heat exchanges which largely control the biotic zonation and productivity of ecosystems (Melillo et al. 1993; Zhang 1993; Martin 1990; Li and Wang 1997; Lorius et al. 1990). The resulting changes in the energy balance and biogeochemical cycles will also exert feedbacks on the climate system either enhancing or damping the changes due to increased GHGs (Parton et al. 1993; Neilson 1995; Peng and Ren 1996).

During the last fifty years, the rapid increase in mean annual temperature in China took place mainly in North China. The increasing temperature in the northern part of Northeast China, Inner Mongolia and west basin regions reached a maximum of 0.8 °C per decade. An observable increase in precipitation in China took place in the western basin regions, with a maximum of about 10–15% per decade. However, the precipitation has decreased in North China and the southern part of Northeast China. The projections of climate change in China for the next 100 years based on the China Regional Climate model (RCM) indicates that climate warming will continue, and the temperature will increase about 3.9–6.0 °C by the year of 2100. The maximal increase in temperature will happen in north China during winter season. The mean annual precipitation may increase 11–17% by the year of 2100, and 10–25% in the northwest, northeast and south regions of China (Qin et al. 2005).

Under climate and atmospheric change, agricultural systems will undergo major impacts which will influence the quantity and quality of regional food production. This will require the development of major adaptation strategies to diminish negative impacts and take advantage of changing climate conditions (Luo and Peng 1996). However, in some regions over-cultivation, over-tending, and deforestation have caused soil degradation to become the major obstacle to adaptation and sustainable development of agriculture in China. In addition, urbanization and industrialization in other regions have taken large areas of farmland, rangeland, forest, and water supplies, and increasing environmental pollution will further aggravate the future productivity and viability of agricultural ecosystems (Luo and Peng 1996). Therefore, the study of possible impacts of global change on the agricultural ecosystem of China can provide insights into the development of adaptation strategies under future environmental conditions. This is a prerequisite to ensure the continuous provision of food with sustainable production systems (Parton et al. 1993; Prentice et al. 1992; Running et al. 1993).

Research on the impacts of global change on terrestrial ecosystems has taken many forms and used a variety of experimental and modeling approaches. One such approach has been the use of a large transect expanding different vegetation types and management practices under gradients of climate and other environmental drivers. Transects bridge the gap between discrete observation in plots and comprehensive spatial analysis by fully coupling observations and modeling and forcing them through environmental gradients as test beds of their robustness and appropriateness for detecting non-linear and threshold-like behavior.

The International Geosphere-Biosphere Programme (IGBP) has a core project, Global Change and Terrestrial Ecosystem (GCTE) which emphasized the importance of the ecological transect as a method in the study of terrestrial ecosystem and global change. The principles of transect selection are that the gradient of a particular ecological factor significantly changes along the transect, and the other ecological factors remain almost constant or their changes are quite small, so that the possible process of global change could be revealed by way of temporal-spatial substitution (Zhang 1994; GCTE 1995).

GCTE supported the establishment of fifteen global change terrestrial transects distributed in different key regions of the Earth (Canadell et al. 2002). Two of these are located in China; the water-driven Northeast China Transect (NECT) and the heat-driven North-South Transect of Eastern China (NSTEC). This chapter will report on the highlights of more than 10 years' researches on the impacts of global change (climate, elevated CO_2, and land use) on productivity of agroecosystems, including crops and grasslands from these two Chinese transects.

This paper reports on (1) physiological and plant responses to multiple global change forcing, (2) productivity and its responses to global change, and (3) carbon budget and its responses to global change.

22.2 Chinese Terrestrial Transects

Both transects, NECT and NSTEC, cover one of the regions in China where there are the most people and most food is being produced to sustain an ever growing human population. The transects include diverse ecosystems ranging from remaining native forests to a large diversity of man-made agricultural systems under different management intensities. The region is also under the influence of the Monsoon climate and relies heavily on the rain it brings for food production. Thus, the region covered by the transects is among the most sensitive regions in China for food security reasons.

NECT was established in 1994 and is mainly a rainfall-driven transect centering at 43.5° N and extending 1 400 km from 132° E in the east to 108° E in the west (Fig. 22.1) (IGBP 1995). Annual precipitation decreased sharply from around 800 mm in the east to less than 100 mm in the west. Vegetation types along the transect are determined largely by rainfall gradient and shift from dark conifer forest, conifer-broadleaf mixed forest, deciduous broadleaf forest, woodlands and shrublands in the east, to meadow steppe and cropland in the middle, and to typical steppe, desert steppe in the west (Gao and Zhang 1997). Functional types of plant species, jointly determined by life forms and carbon pathways, are quite different due to the rainfall gradient (Jiang et al. 1999). As a result of increasing human population and fast economic development, the region covered by NECT has been subject to increasing human pressure, including cultivation and overgrazing. Progressive degradation has been reported in the semiarid and arid grassland along

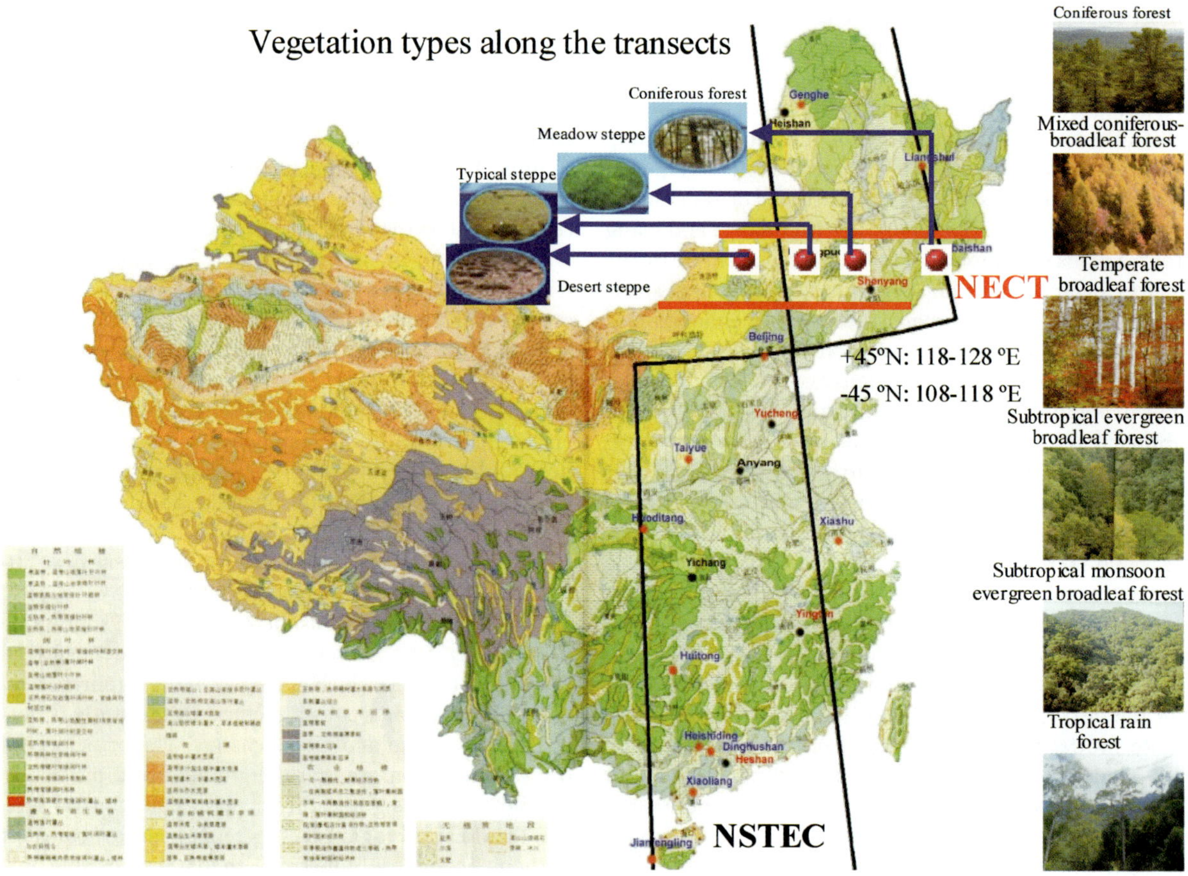

Fig. 22.1. Vegetation types along NECT and NSTEC

NECT during the past 30 years. A general question guiding the NECT is how does water availability influence the composition of plant functional types, soil organic matter, net primary production, trace gas flux, and land-use distribution?(Koch et al. 1995)

NSTEC is composed of two parts: 110–120° E, 15–40° N and 118–128° E, 40–57° N. It covers about 2 889 100 km^2, about 30.1% of the total land area in China. The geographical range covers 1 323 counties from 25 provinces, municipalities, and autonomous regions (Fig. 22.1). The main driving forces along NSTEC are temperature and land use (Teng et al. 2000). Eight vegetation types from south to north have been identified: mountain tropical rain forest, tropical seasonal rain forest, low subtropical monsoon evergreen broadleaf forest, mid-latitude subtropical evergreen broadleaf forest, high latitude subtropical evergreen-deciduous broadleaf forest, warm temperate broadleaf deciduous forest, temperate coniferous and broadleaf mixed forest, and cold temperate coniferous forest.

22.3 Physiological and Plant Responses to Multiple Global Change Forcing

Grasslands are one of the most widespread ecosystem types worldwide; they cover ca. 25% of the natural land surface and account for ca. 16% (18.9 Gt yr^{-1}) of the terrestrial global net annual primary production (Tieszen and Detling 1983). Thus they have a major influence on the functioning of the terrestrial biosphere. The *Leymus chinensis* grassland is a renewable resource to support livestock farming in the northern China. It is widely distributed in the east of the Euro-Asia continent grassland region, and more than half is located in China, especially in the Northeastern China plain and Inner Mongolia Plateau. The western and central portions of NECT are dominated by *Leymus chinensis* grassland. The physiological and whole-plant level responses of *Leymus chinensis* to multiple global change drivers were studied in order to understand how the ecosystem will respond to global change.

In laboratory cabinet experiments, the responses of *Leymus chinensis* to two different treatments: drought (30–45%, 45–60% and 60–80% of field water-holding capacity) and elevated CO_2 (350 ppm and 650–700 ppm) and their combinations were done from May 11 to October 15, 2001 (Zhou et al. 2002). The results indicated that doubling atmospheric CO_2 concentration resulted in increases of 40.9% in total biomass, 35.1% in root biomass, 27.5% in stem biomass and 57.7% in leaf biomass of *Leymus chinensis* during the whole growing season. Drought stress increased leaf water potential and stomatal resistance, and decreased transpiration and photosynthetic rates. However, when elevated CO_2 interacted simultaneously with drought stress, the "CO_2 fertilization" effect was suppressed and biomass production declined by 47.3% when compared with non drought treatments (Zhou et al. 2002). It is different from the more classic thinking that CO_2 fertilization effect will be proportionally bigger in water-stressed regions because of the increased water use efficiency of plants under elevated CO_2. This result implies that the effects of concurrent elevated CO_2 and water stress on water use efficiency of plants are interactive and that the beneficial CO_2 fertilization effects on water use efficiency will not be enough to compensate the negative effects of droughts in the region.

The responses of wheat, corn and soybean to two different treatments: temperature (Table 22.1) and elevated CO_2 (350 ppm and 650–700 ppm) and their combinations were done from November 4, 2000 to May 10, 2001 in laboratory cabinet experiments (Zhou et al. 2002). The results indicated that global warming and elevated atmospheric CO_2 concentration would accelerate crop growth and development and shorten the growth period, provided water is not limiting. The photosynthetic rate of crop and stomatal resistance would increase, and the transpiration rate decrease. During the same phase, the leaf area and biomass of leaf, root and stalk of the crop decreased because of insufficient growing time. Moreover, elevated atmospheric CO_2 and high temperature resulted in a decrease in trace elements (Zn, Fe, Mg, Ca, S) in wheat, soybean and corn (leaf, stalk and grain) with large differences among crops and crop parts, therefore lowering their nutritive values. Basically, the high temperature is a positive effect to the trace elements in wheat, but there is negative effect on soybean grain and positive effect on leaf and stalk (Zhou et al. 2002).

The combination of elevated CO_2, drought and warming will result in little gains despite the benefits of the CO_2 fertilization effect. In fact, results show decrease not only in the productivity of *Leymus chinensis* grassland and crop (wheat, corn and soybean) but also in trace elements (Zn, Fe, Mg, Ca, S) in wheat, soybean and corn.

The photosynthetic physiological characteristics of tree species will affect their ecological distribution and succession status. The photosynthesis light response curves of the potted seedlings of *Pinus koraiensis*, *Picea koraiensis*, and *Larix olgensis*, the dominant coniferous tree species in Changbai Mountains grown under doubled CO_2 (700 ppm) and ambient CO_2 (400 ppm) for two years were determined under a series of light intensities from 0 to 2 500 µmol $m^{-2}\,s^{-1}$. The comparison of dark respiration rate (DRR), maximum net photosyn-

Table 22.1. Temperature difference (°C) between treatment and CK

Crops	Nov	Dec	Jan	Feb	1st decade of Mar
Wheat and corn	2.5	4.8	5.8	5.2	0.9
Soybean	1.2	1.1	0.8	0.3	0.0

thetic rate (A_{max}), quantum use efficiency (QUE), light compensation point (LCP), light saturation point (LSP) and photo inhibition point (PP) for those coniferous seedlings indicated that *Larix olgensis* is a sun-loving species with the most obvious response to elevated CO_2 concentration, followed by *Pinus koraiensis*, and the shade – tolerant species *Picea koraiensis* with the least response (Zhou et al. 2002). It implied that *Larix olgensis* might be apt to grow and develop from seeds in the course of ecosystem succession due to the sun-loving characteristics and high photosynthetic rate under elevated CO_2 concentration, followed by *Pinus koraiensis. Picea koraiensis* acclimatizes itself to shade environment. However, *Picea koraiensis* shows the most obvious sensitivity to the photosynthetic physiological characteristics of doubled CO_2 compared with those under the current situation, followed by *Pinus koraiensis* and *Larix olgensis.* Those changes will affect the capacity in the competition among tree species, and result in the change in ecosystem composition and tree diversity in temperate forest ecosystems.

We also examined the ecological responses of different tree species saplings (potted) to three soil moistures: 85%~100% (high water, CK), 65%~85% (Medium water, MW) and 45%~65% (low water, LW) of field water-holding capacity. The species selected were *Pinus koraienes, Fraxinus mandshurica, Juglans mandshurica, Tilia amurensis,* and *Quercus mongolica* which are five dominant species in the korean pine-broadleaf forest at Changbai mountain located in the east end of NECT. The results showed that drought treatments significantly decreased the single-leaf area, number of fine root, root length, and individual biomass of seedlings. The root weight ratio was higher than the control. There were different responses to soil water stress for different species. The root weight ratios of *Tilia amurensis* and *Fraxinus mandshurica* increased with increasing soil water content, but the leaf weight ratio of *Juglans mandshurica* and *Quercus mongolica* were higher under MW compared with CK and LW. Biomass was significantly different among the 5 tree species. The biomass of *Tilia amurensis, Fraxinus mandshurica, Juglans mandshurica, Pinus koraienes* decreased by 32.8%, 43.6%, 26.2%, 23.8%, 4.8% respectively, while *Quercus mongolica* showed an increase in 4.8%. It is suggested that *Quercus mongolica* had the highest drought-tolerance among the tested species. It implied that *Quercus mongolica* might replace *Pinus koraienes* to become the dominant species of the temperate forest ecosystems if this region becomes drier in the future because of climate change.

22.4 Productivity and Its Responses to Global Change

The productivity of terrestrial ecosystems is an important index for its health and its response to multiple global change forcing, and it also directly connects with carbon sequestration and its feedback on climatic change. Plant species number, soil C, soil N and aboveground biomass of grassland ecosystems along a precipitation gradient in the west part of NECT showed positive linear relationships with precipitation. In addition, plant species number, soil C and total soil N had strong positive linear relationships with above-ground biomass (Zhou et al. 2002). This suggests that global warming might result in the decrease in grassland productivity, affecting the development of stockbreeding in this area.

The responses of Northeast China Transect to global change were shown by a remote sensing driven regional vegetation model (Gao et al. 1997). It indicated that the potential vegetation of the transect was very sensitive to variation of temperature and CO_2 concentration. With a temperature increase of 4 °C, the induced increase in evapotranspiration could reduce the average biomass and net primary productivity (NPP) over the whole transect by 32.1% and 41.9% respectively. In contrast, a 20% increase in precipitation alone could lead to an increase of the average biomass and NPP by 8.1% and 13.4% respectively. Without changing the climatic conditions, CO_2 doubling could increase the average biomass and NPP by 12.2% and 17.1% respectively. Because of compensation between the positive effects of CO_2 and precipitation increase, and the negative effect of temperature increase, a doubling of atmospheric CO_2, a 20% increase in precipitation and a 4 °C increase of temperature would lead to a 2% reduction in the biomass and NPP of the natural vegetation over the transect.

Yu et al. (2002) modified the remote sensing driven regional vegetation model developed by Gao et al. (1997) to include land use as a constraint to spatio-temporal vegetation dynamics, and applied it to NSTEC in order to examine the combined effects of climate and land use on vegetation distribution, primary production, and nitrogen cycling. The responses of vegetation in NSTEC to global change, under different climatic scenarios, were obtained by subjecting the modified model to different climatic scenarios derived from 7 general circulation models (GCMs): GFDL, GISS, OSU, LLNL, MPI, UKML and UKMH. The simulated vegetation distributions and relative increases in areas of the 8 natural vegetation classes under the altered climate scenarios from 7 GCMs are given in Fig. 22.2 and Fig. 22.3, respectively. The simulations indicated that there were significant increases in the area of deciduous broadleaf forests but unchanged extent of evergreen broadleaf forests under all 7 GCM scenarios. The distributions of all other natural vegetation types decreased. CO_2 fertilization and increases in annual precipitation both enhanced assimilation and exerted positive effects on the competitive ability of broadleaf forests, exceeding the negative effects of increased temperature that increased the evapotranspiration and reduced water use efficiency. Under contemporary climate and present atmospheric CO_2 concentration, de-

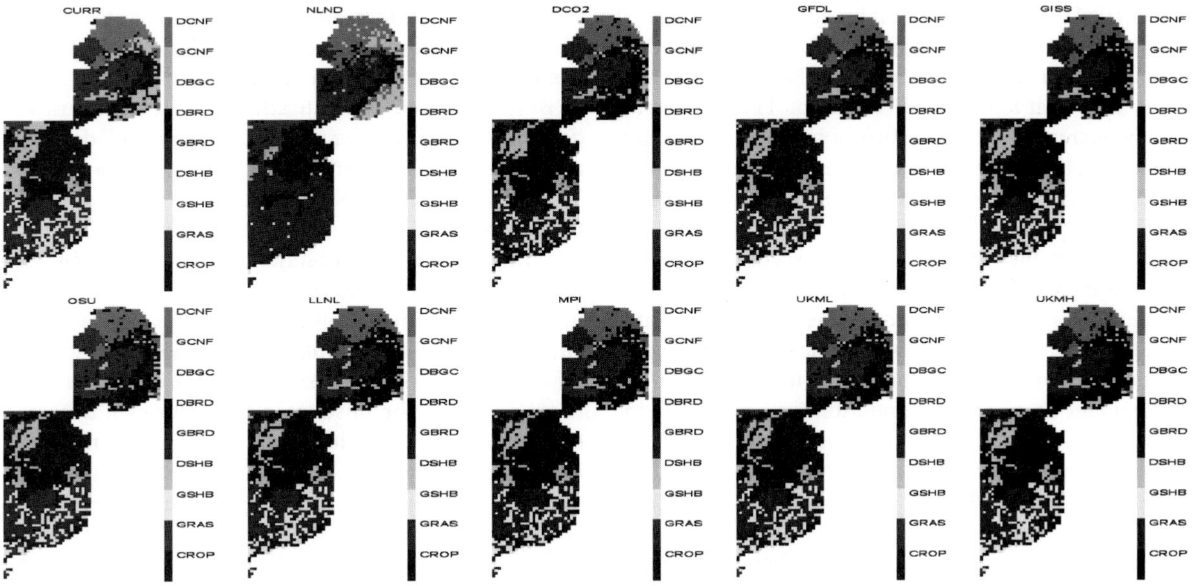

Fig. 22.2. Simulated vegetation distributions of NSTEC under contemporary climate with 350 ppm CO_2 with (CURR) and without (NLND) land-use constraint, with doubled CO_2 concentration but unchanged precipitation and temperature (DCO_2), and with doubled CO_2 concentration and altered precipitation and temperature as predicted by 7 GCMs (Yu et al. 2002)

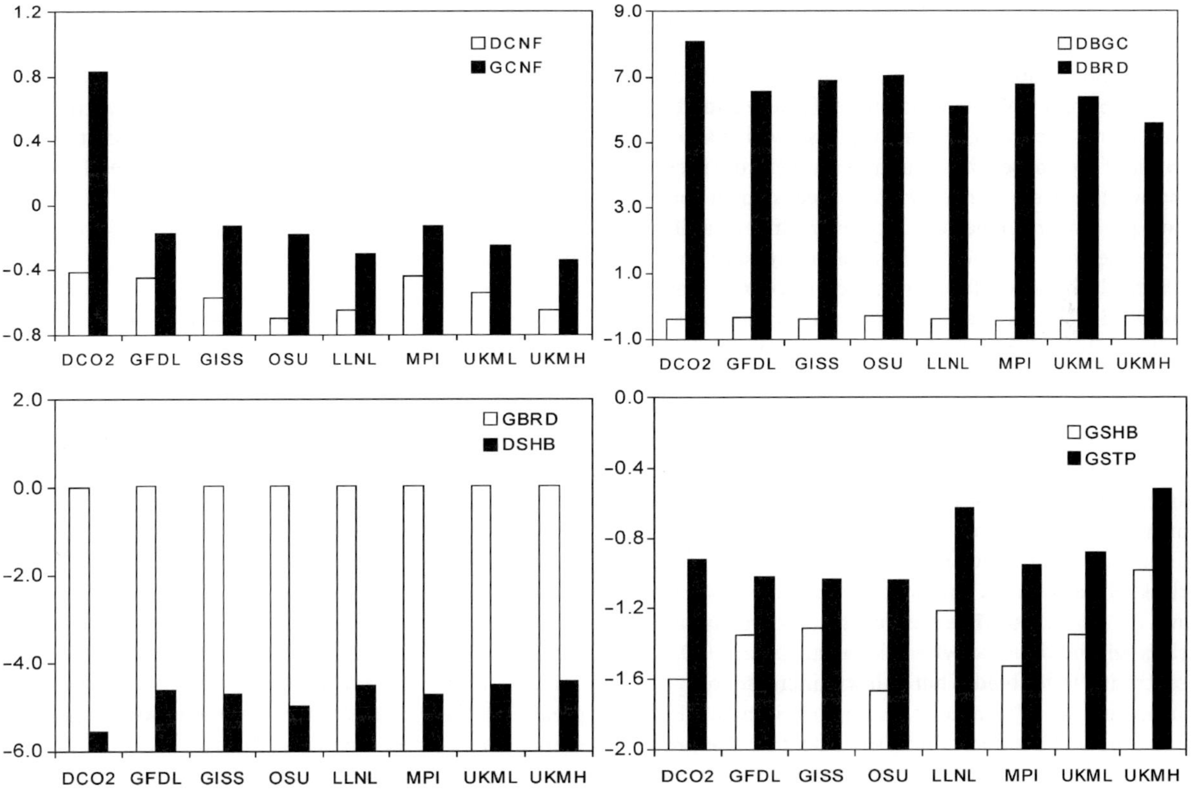

Fig. 22.3. Simulated distribution of 8 natural vegetation classes under doubled CO_2 concentration with unchanged precipitation and temperature (DCO_2) and altered precipitation and temperature as predicted by 7 GCMs (Yu et al. 2002)

ciduous broadleaf forests (DBRD) in the northern NSTEC are more likely to experience drought than evergreen broadleaf forests (GBRD) in southern China. Among the 7 GCMs, OSU and GFDL simulate the highest and lowest increases in precipitation respectively, but with approximately the same low temperature increases. UKMH and LLNL had the highest temperature increases among 7 GCMs, but with similar increases in precipitation. Figure 22.3 illustrates that deciduous broadleaf forests (DBRD) had the greatest increase in occupied area with the OSU scenario, whereas the increase for GFDL scenario was lower than that for the OSU scenario. Therefore, with similar increases in temperature, a higher precipitation increase would lead to a greater increase in area for DBRD because more water was available for assimilation at higher precipitation. The increases in area of DBRD were the lowest for the UKMH and LLNL scenarios because the large increases in temperature for these two scenarios increased evapo-transpiration and hence reduced water use efficiency. This effect of increasing precipitation and temperature on evergreen broadleaf forests (GBRD) is less evident because this vegetation type is mainly located in southern NSTEC where the baseline precipitation is much higher than in the north.

The agricultural climatic divisions of the NSTEC fall into the eastern monsoon agricultural climatic area. The suitability for growing winter wheat and maize based on sown-maturation temperature index were simulated in NSTEC. This was based on monthly and annual temperature and precipitation during 1961–1990 from 1 015 meteorological stations in and neighboring NSTEC (climatic baseline) and the future climatic scenarios given by the climatic baseline plus the change of climatic parameters (temperature and precipitation herein) between the 2030, and 2056 from the HADCM2 output (future climate projections) in Arc/Info GIS. The results show that at present 48.2% of the transect has a suitable temperature for growing winter wheat during winter wheat's sown-maturation phonological period. However, the suitability of wheat declines to 35.6% and 35.3% in 2030 and 2056, respectively, based on future climate projections from the HADCM2. Meanwhile, the proportion of unsuitable land will also be reduced. The land area for the moderately suitable class will increase greatly, from 23.7% at present to 44.28% in 2030 and 44.36% in 2056. If other production conditions could be met, there will be more land area suitable for growing winter wheat. The suitability assessment results for winter wheat production, if based on heading-maturation mean temperature index, are different from the above-mentioned results. The suitable class will drop from 84.5% at present to 67.9% in 2030 and to 61% in 2056. The area for the moderately suitable class will be doubled when the atmospheric CO_2 concentration is doubled. That is, it will increase from 15.2% at present to 32% in 2030 and 38.9% in 2056. Future warming will produce a negative effect on the production suitability of winter wheat in the NSTEC transect. Water is another important factor affecting winter wheat production. Based on the heading-maturation precipitation index, most of the land in the transect will not be very suitable for winter wheat production if there is no irrigation. Both of the classes of suitable and moderately suitable areas will decrease in the future, whereas unsuitable land will increase. This implies that the winter wheat production in the transect will increasingly become more depended on irrigation.

Based on the sown-maturation mean temperature index, the suitable land for maize production will decrease about 2.6% in 2030, and will increase to 72.5% in 2056. There will be little unsuitable land for maize production in the NSTEC transect in 2056. Negative effects of warming on maize production suitability will be less than that on winter wheat production. But, the heading-maturation mean temperature index tells a different story. Based on the index, the suitable land will increase from 12.5% at present to 13.8% in 2030 and to 16.2% in 2056. The unsuitable class land will decrease from 12.2% at present down to 10.5% in 2030, and it will increase to 23.5% in 2056. The moderately suitable class land will remain nearly unchanged to 2030, but it will decrease about 15% in 2056.

The above results show that global change impacts on winter wheat and maize production suitability are different. The direct reasons for this phenomenon are twofold: (1) maize has a wider distribution range than winter wheat, and hence is more adaptive to environmental change; (2) the growing seasons for winter wheat and maize are different. Winter wheat is usually sown in October and harvested in the next June, whereas maize grows in May–October. Studies show that the global warming effect in winter is stronger than in summer, which is why global change will greatly affect the growing suitability of winter wheat and will have less impact on the production suitability of maize in the future.

22.5 Carbon Budget and Its Responses to Global Change

Carbon fluxes from and into agroecosystems mainly include CO_2 and CH_4 emitted from cropland and grassland practices, ruminant animals and their waste treatment, biomass burning and carbon sequestration caused by corresponding land management activities. Of particular interest and importance is the role of soils in the global carbon cycle which affects the composition of the atmosphere and degree of climate change. Whether the soil acts as a source or sink of carbon gases depends greatly on the type and intensity of activities of human management on the land. Generally, intensive use of ecosystem leads to a net depletion of carbon storage compared with lightly exploited ecosystems. For example, conversion of grassland to cropland typically results in a decline in carbon stocks. Conversion of land to agricultural purposes often involves such a decline due to land clearing, draining, sod breaking, cultivating, and replacing perennial vegetation for annual vegetation. For agricultural systems, reduced tillage, reduced bare fallow, irrigation and fertilization are practices that are most likely to achieve substantial rates of carbon gain. Land-use changes, often conversion for agricultural development, accounts for an additional 14% decrease (Lin et al. 2002).

Based on CO_2 fluxes measured along the NSTEC and NECT during 1998–2000, the fluctuation of CO_2 emissions from soils showed obvious seasonal variation and was positively correlated with soil temperature at 5 cm soil depth (Lin et al. 2002; Zhou et al. 2002). The seasonal variation is mainly because the soil temperature is an important factor controlling the microbial activities in grassland soils. The highest emission rate happened in summer. Lin et al. (2002) pointed out that there was an exponential relationship between soil temperature and CO_2 fluxes of natural grasslands ($n = 95$, $y = 1.97e^{0.0681x}$, $P < 0.01$) and also maize fields at the depth of 5 cm during 1997 and 1998 in the Dalate Experimental Station. At Huangpuchuan Experimental Station, correlation coefficients between soil temperature and emission rates were 0.73 and 0.72 ($P < 0.01$, $n = 49$) for natural grassland and restoration grassland in 5 cm soil depth, respectively. For forest soils, CO_2 emission rate is also closely related to soil temperature. At the Dinghu Mountain Station, the CO_2 emission rate was higher in spring, autumn and winter than at Maoer Mountain Station because the soil temperature was lower in northern forests during these seasons than in southern forest area. In the summer season, however, CO_2 emission rates were lower at Dinghu Mountain than at Maoer Mountain because of lower soil temperatures. The variation range of CO_2 flux in north temperate forest soils were larger than in subtropical forest soils due to the wider temperature differences among seasons in northern areas. However, soil types, in addition to disturbance history, vegetation, and topography, modulate a great deal of the relative quantitative relationship between temperature and CO_2 emissions. Conversion of natural ecosystems to managed systems usually results in soil carbon emissions. Continuous observations of carbon fluxes were conducted in conversion lands from natural grassland to other land uses such as grazing land and cropland. In 1998 the data showed that conversion of natural grassland into farmland increased CO_2 emission. The average emission rate of CO_2 from natural grassland was 10.42 kg C ha^{-2} d^{-1} during the growing seasons for grasses, while it was 20.02 kg C ha^{-2} d^{-1} during the growing seasons for maize six years after conversion. The emission rate increased about 100% after the native grassland was converted to farmland. In 1999 and 2000, the CO_2 emission rates from the natural grassland were 10.0 and 9.12 kg C ha^{-2} d^{-1} respectively. In a maize field, the CO_2 emissions were 14.27 and 14.20 kg C ha^{-2} d^{-1}, respectively. Conversion of natural grassland to cropland increased CO_2 emission by 43–100% in sixth to eighth years since conversion. The conversion of natural grassland to cropland not only increased the CO_2 emissions, but also decreased the CH_4 oxidation rate. Measurements at the Yingtan Red Soil Experiment Station during the growing seasons showed that large CH_4 emissions occurred from rice producing flooded soils. Within the same experimental station these emissions show a striking contrast with those from uplands and forests (*Pinus massoniana*) which absorbed CH_4 from the atmosphere. Even upland and grassland acted as a CH_4 sink during the period of measurements, although uptake strength decreased by 41–56% when grassland was converted into farmland. The observations along NSTEC showed that sensitivity of soil carbon emission to temperature, water regime and direct human induced activities under current climate condition would continue in a warmer climate.

Terrestrial ecosystems can absorb large quantities of carbon dioxide from the atmosphere, but some observations suggest that land-use changes are greatly impairing this ability. When land is first converted into agricultural uses, such as conversion of grassland into cropland, conversion of forest into grassland or conversion of forest into cropland, the organic matter in the soil is oxidized quickly at the beginning, slowing over time. In addition, much of the originally existing vegetative biomass is released to the atmosphere as CO_2. Deforestation, biomass burning, drainage, plowing, cultivation, and overgrazing all promote the decomposition of organic matter and the release of CO_2 into atmosphere. Soil degrading processes, such as erosion, crusting and compaction, acidification, and salinification, further exacerbate the loss of soil carbon.

However, measurements from the national soil surveys during 1958–1961 and 1979–1982 as well as the continuous observational record from the observation network established after the mid-1980s by the former National Soil Survey Office of China, show that soil organic carbon (SOC) increased at a level of 11.6 g kg^{-1} in the plowing layer by appropriate management practices through 1985–1995 in cropland soils, an increase in 22% of the area in China (Lai et al. 1999). This suggests that appropriate management in cropland could indeed increase soil carbon stocks, particularly after SOC levels have been decided due to intense land use.

Lin et al. (2002) calculated the potential for agricultural and forest management activities to sequester carbon in the next fifty years in China (Table 22.2). They estimated that carbon sequestration potential is about 77.2 Mt C yr^{-1} (ranging from 26.1–128.3 Mt C yr^{-1}) for these activities during the next fifty years in China. The sustainability of carbon accumulated by management practices will strongly depend on the situation in which they are implemented and the duration of these activities. Due to the wide range of carbon aging rate and ini-

Table 22.2. Carbon sequestration potential of land use practices in China in the next fifty years (Lin et al. 2002)

Activities	Area (Mhm^2)	Feasibility (% of area)	Rate ($t C hm^{-2} yr^{-1}$)	Duration (yr)	C gain potential ($Mt C yr^{-1}$)	Other GHGs	Confidence
A	*B*	*C*	*D*	*E*	$F = B \cdot C \cdot D$	G	H
Cropland managenemt							
Boreal	18.2	30	0.3–0.6	40	1.6 – 3.3	$+N_2O$	M
Temp. dry	23.7	30	0.1–0.3	30	0.7 – 2.1	$+N_2O$	H
Temp wet	18.7	30	0.2–0.6	25	1.1 – 3.4	$+N_2O$	H
Tropical wet	9.4	30	0.2–0.8	15	0.6 – 2.3	$+N_2O$	M
Sub-total	70.1				4.0 – 11.1 (7.5)		
Rice paddy management	9.4	30	0.2–0.8	25	1.5 – 6.0 (3.7)	$+N_2O$, $++CH_4$	L
All agroforest shelterbelt forest	10.0	30	0.1–2.5	50	0.3 – 7.5	$+N_2O$	M
Economic forest	14.9	30	0.1–1.3	8	0.4 – 5.8	$+N_2O$	M
Sub-total grassland management	24.9				0.7 – 13.3 (7.0)		
Temp. wet	35.5	15	0.4–2.0	50	2.1 – 10.7	$-N_2O-$	M
Temp. dry	257.7	15	0–0.3	50	0.0 – 11.6	CH_4	M
Tropical wet	66.6	15	0.3–3.0	40	2.0 – 30.0	$-N_2O-$, CH_4	M
Sub-total	360.0				4.1 – 52.3 (28.2)	$-N_2O-$, CH_4	M
Forest management							
Cold temperate	38.8	20	0.1–0.8	80	0.8 – 6.2		L
Warm temperate(wet)	8.2	20	0.1–3.0	50	0.2 – 4.9		L
Warm temperate(dry)	16.6	20	0.1–0.8	80	0.3 – 2.7	$-N_2O$, $-CH_4$	L
Tropic and sub-tropic	49.7	20	1.6–3.8	60	15.9 – 37.8		L
Sub-total	113.3				17.2 – 51.6 (34.4)		L
Total					26.1 – 128.3 (77.2)		

tial assumptions used to make these calculations, the uncertainty may be great (listed in the last column of Table 22.2). Most often, activities that decrease the loss or increase the stocks of carbon affect the emissions or removals of other greenhouse gases such as CH_4; relative influence of these activities on other greenhouse gases are also addressed in Table 22.2.

In the absence of disturbance, the simulated temporal dynamics of NPP and total soil carbon by the CENTURY model (Parton et al. 1993) were different at the three sites along NECT (Fig. 22.4). For each of the sites, inter-annual precipitation had a strong effect on NPP, which declined continuously due to decreased precipitation. The relationship between NPP and precipitation was linear (Fig. 22.5), suggesting that precipitation is a main limiting factor for plant growth along NECT. NPP was highest at the mixed coniferous-broadleaf forest ecosystem and lowest at the typical steppe ecosystem, following the mean annual precipitation gradient along NECT. Moreover, the mean annual carbon budgets also showed the same pattern. The value of the carbon budget was largest for the mixed coniferous broadleaf forest ecosystem (503.2 g C m^{-2} yr^{-1}), followed by the meadow steppe ecosystem (227.1 g C m^{-2} yr^{-1}), with the lowest being the typical steppe ecosystem (175.8 g C m^{-2} yr^{-1}). This also indicates that the precipitation gradient is a main driving force along NECT.

The simulated temporal dynamics of total soil carbon along NECT was independent of inter-annual precipitation (Fig. 22.4–22.5). This might be due to the smaller fluctuation of inter-annual precipitation during the simulation. However, the values of total soil carbon decreased from the mixed coniferous- broadleaf forest ecosystem, through the meadow steppe ecosystem, to the typical ecosystem, following the mean annual precipitation gradient along NECT. This suggests some control of mean annual precipitation on soil carbon along NECT implying that to some extent future carbon sequestration will be driven by future changes in precipitation for which large uncertainties still remain.

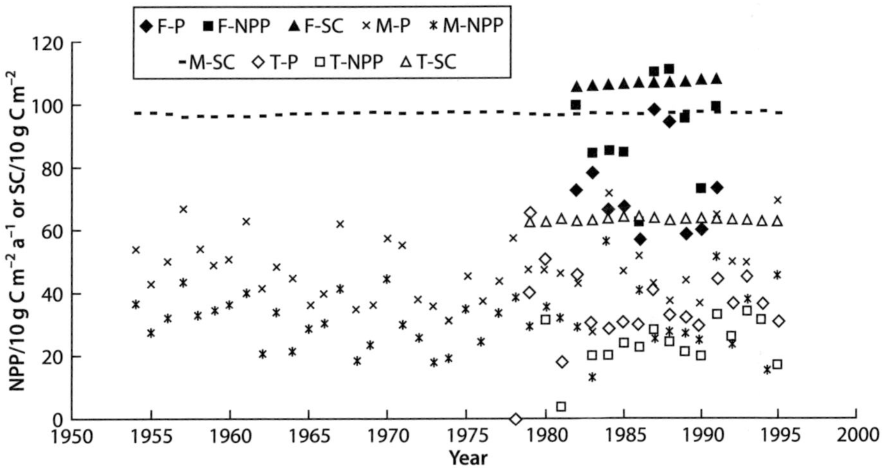

Fig. 22.4. Simulation of temporal dynamics of NPP and total soil carbon using actual weather records from 1982 to 1991 at Changbai Mountains Forest Ecosystem Research Station, from 1954 to 1995 at Changling Grassland Station, and from 1979 to 1998 at Inner Mongolia Grassland Ecosystem Research Station. *F:* Changbai Mountains Forest Ecosystem Research Station; *M:* Changling Grassland Station; *T:* Inner Mongolia Grassland Ecosystem Research Station; *NPP:* 10 g C m^{-2} yr^{-1}; *SC:* total soil carbon (10 g C m^{-2}) (Zhou et al. 2002)

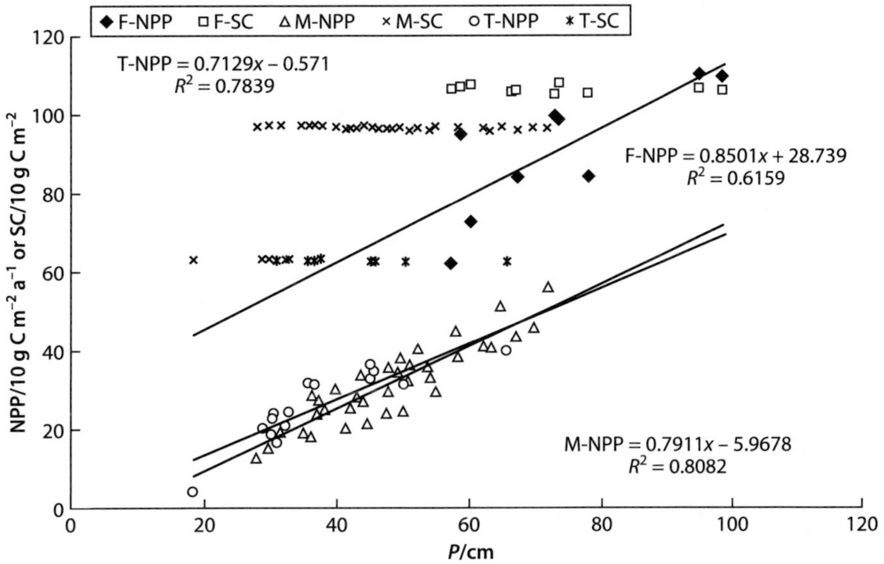

Fig. 22.5. Effects of precipitation (*P*) on NPP and total soil carbon at three simulated sites. *F:* Changbai Mountains Forest Ecosystem Research Station; *M:* Changling Grassland Station; *T:* Inner Mongolia Grassland Ecosystem Research Station; *P:* precipitation (cm); *NPP:* 10 g C m^{-2} yr^{-1}; *SC:* total soil carbon (10 g C m^{-2}) (Zhou et al. 2002)

22.6 Summary and Conclusions

Our limited knowledge on the dynamics of managed and unmanaged ecosystems has curtailed our ability to predict the effects of global change on the ecosystem goods and services that societies rely upon for their wealth and development. Global change terrestrial transects have proved to be an important and useful scientific approach to study the spatial and temporal dynamics of multiple drivers and complex responses. Although a number of studies on global change and terrestrial ecosystems have been done during the recent decade, much remains to be learned of the interactive effects of multiple drivers and their spatial and temporal dynamics which this chapter has presented. In addition, it is recognized that the study of global change not only needs fully effective cooperation of scientists in China, but also needs connection with relevant studies world-wide.

In order to improve integrative global change studies in China and develop the capability to predict the responses of terrestrial ecosystems to global change in China, the following research fields will require further development over the next decade: (1) mechanisms driving responses of terrestrial ecosystems to global change; (2) shifts and adaptation of vegetation and ecosystems to global change; (3) index system of plant functional types of China combined with remote sensing techniques; (4) soil responses to global change; (5) parameterization and scaling techniques in models to study ecosystem dynamics from landscape to regional scales; (6) development of coupled models of the carbon balance, and vegetation dynamics, and atmospheric circulation and climate in order to simulate responses and feedbacks of terrestrial ecosystems to atmosphere and climate changes; and (7) establishment of an information system of terrestrial ecosystems including continuous observations, databases, ecosystem models, and expert systems to observe and predicts changes in terrestrial ecosystems, and to inform policy makers.

Acknowledgments

We would like to thank Prof. Hai Ren (South China Institute of Botany, Chinese Academy of Sciences), Prof. Xiaofeng Zhou (College of Forest Resources and Environments, North East Forestry University), Prof. Erda Lin (Institute of Climate, Chinese Academy of Agricultural Sciences), Prof. Peijun Shi and Prof. Qiong Gao (MOE Key Lab of Environmental Change and Natural Disaster, Institute of Resources Science), Prof. Huajun Tang and Prof. Bin Xu (Institute of Natural Resources and regional Planning, Chinese Academy of Agricultural Sciences) for their contributions to this manuscript. We also want to thank Pep Canadell for encouraging and helping to prepare this paper.

References

Canadell JG, Steffen W, White PS (2002) IGBP/GCTE terrestrial transects: dynamics of terrestrial ecosystems under environmental change. Journal Vegetation Science 13:297–448

Feng ZW, Wang XK, Wu G (1999) Biomass and productivity of forest ecosystems in China, Science Press, Bejing. PRC

Gao Q, Zhang XS (1997) A simulation study of responses of the northeast China transect to elevated CO_2 and climate change. Ecological Applications 7(2):470–483

Gao Q, Yu M, Zhang X, Guan F (1997) Dynamic modeling on responses of Northeast China Transect to global change – a remote sensing driven regional vegetation model. Acta Botanica Sinica 39:800–810

GCET (1995) The terrestrial transacts: science plan. IGBP of ICSU, report 36, Stockholm

IGBP (1995) Spatial extrapolation and modeling on IGBP transects. Global Change Report 36:15–20

Jiang GM, Tang HP, Yu M, Dong M, Zhang X (1999) Response of photosynthesis of different plant functional types to environmental changes along Northeast China Transect. Trees – Structure and Function 14:72–82

Koch GW, Scholes RJ, Steffen WL, Vitousek PM, Walker BH (eds) (1995) The IGBP Terrestrial Transects: Science Plan. IGBP Report No. 36, The International Geosphere-Biosphere Programme, Stockholm, 61 pp

Li QS, Wang ZQ (1997) Influence of climate change on Chinese agroecological system and our countermeasures. Research on and Probe into Ecology, edited by Yan Lijiao. China Environmental Science Press, Beijing, PRC

Lai R, Kimble JM, Follett RF (1999) Agricultural practices and policies for carbon sequestration in soil. In: Recommendation and Conclusions of the International Symposium, 19–23 July 1999, Columbus, OH, USA, 12

Lin E, Li Y, Guo L (2002) Carbon emissions and sinks in agroecosystems of China. Science in China(Series C) 45(Supp.):30–39

Liu C (1993) The impacts of climate change on water resources and its agricultural countermeasures. In: Deng G (ed) The Impact of Climate Change on Agriculture in China. Beijing SciTech Publishing House, Beijing, PRC, pp 167–183

Lorius C, Jouzel J, Raynaud D, Hansen J, Treut HL (1990) The ice-core record: climate sensitivity and future greenhouse warming. Nature 347:139–145

Luo SM, Peng SL (1996) Agroecosystem Analysis. Guangzhou: Guangdong Science and Technology Press, Guangzhou, PRC

Martin Parry (1990) Climate Change And World Agriculture. London: Earthscan Publications Limited

Melillo JM, McGuire AD, Kicklighter DW, Moore IIIB, Vorosmarty CJ, Schloss AL (1993) Global climate change and terrestrial net primary production. Nature 363:234–240

Neilson R P (1995) A model for predicting continental-scale vegetation distribution and water balance. Ecol. Appl. 5:362–285

Parton WJ, Scurlock JM, Ojima DS, Gilmanov TG, Scholes RJ, Schimel DS, Kirchner T, Menaut JC, Seastedt T, Garcia Moya E, Apinan K, Kinyamario JI (1993) Observations and modeling of biomass and soil organic matter dynamics for the grassland biome worldwide, Global Biogeochemical Cycles 7:785–809

Peng SL (1997) Global change and its effects. Ecologic Science 15: 1–9

Peng SL, Ren H (1996) South Subtropical Forest Community Dynamic. Academical Press, Beijing, PRC

Peng SL, Ren H (1998) The ecological energetics of lower subtropical forest ecosystem. Meteorological Press, Beijing, PRC

Prentice IC, Cramer W, Harrison SP, Leemans R, Monserud RA, Solomon AM (1992) A global biome model based on plant physiology and dominance, soil properties and climate. Journal of Biogeography 19:117–134

Qin D, Ding Y, Su J, Ren J, Wang S, Wu R, Yang X, Wang S, Liu S, Dong G, Lu Q, Huang Z, Du B, Luo Y (2005) Assessment of climate and environment changes in China (I): climate and environment changes in China and their projection. Adv. Clim. Change Res 1:4–9

Running SW, Hunt Jr. ER (1993) Generalization of a forest ecosystem process model for other biomes, BIOME-BGC, and an application for global-scale models, pp 141–158: *In*: Ehleringer and C. Field [eds], Scaling Physiological Processes: Leaf to Globe, Academic Press, Orlando, USA

Teng L, Ren H, Peng SL (2000) The Natural Situation of North South Transect of Eastern China. Ecologic Science 19:1–10

Tieszen LL, Detling J (1983) Productivity of grassland and tundra. In: Lange OL, Nobel PS, Osmond CB, Ziegler H (eds) Encyclopedia of plant physiology. Vol. 13D. Physiological plant Ecology IV. Ecosystem processes: mineral cycling, productivity and man's influence, pp 173–202. Springer-Verlag, Berlin, DE

Wang WJ (1965) A preliminary study on structure and ecological characteristic of the main forest communities in Nanwutaishan, north of Qinling Mountain. Acta Phytoecologica et Geobatanica *Sinica* 3:298–306

Xu DY, Guo QS, Yan H (1997) Study on the impacts of climate change on forests in China. Chinese Science and Technology Press, Beijing, PRC

Ye DZ, Chen PQ (eds) (1992) The pilot study of global change in China. Earthquake Press, Beijing, PRC

Yu M, Gao Q, Zhang X (2002) Responses of vegetation structure and primary production of a forest transect in eastern China to global change. Global Ecology and Biogeography 11:223–236

Zhang XS (1993) Study on the system of climate-vegetation classification in global change. Fourth Epoch Research 2:157–169

Zhang XS (1994) The terrestrial ecosystem research under global change. Annual report of Laboratory of Quantitative Vegetation Ecology, 41–52

Zhou G, Wang Y, Wang S (2002) Responses of grassland ecosystems to precipitation and land use along the Northeast China Transect. Journal of Vegetation Science 13:361–368

Chapter 23

Terrestrial Ecosystems in Monsoon Asia: Scaling up from Shoot Module to Watershed

Takashi Kohyama · Jotaro Urabe · Kouki Hikosaka · Hideaki Shibata · Takahito Yoshioka · Eiichi Konohira Jun Murase · Eitaro Wada

23.1 Introduction

We carried out research on "Global Change Impacts on Terrestrial Ecosystems in Monsoon Asia" (TEMA) as a GCTE core research project from 1995 to 2003. The purpose of TEMA was to predict the effects of elevated CO_2 and climate change on the distribution and structure of forest ecosystems in eastern Asia, and to determine the associated feedback effects to the global carbon cycle (Hirose et al. 1998). The target region is characterized by continuous forested biomes from tropic to sub-arctic zones under prevailing humid climates, and by the ongoing pressures of land use and pollution. The comparison between a latitudinal gradient where seasonality is amplified with latitude, and an altitudinal gradient in tropical mountains without seasonal thermal change enables us to analyze and describe the climate-vegetation linkage. We made detailed investigations mainly in such core sites as Mount Kinabalu (Malaysian Borneo), the Lake Biwa basin (central Japan), and the Tomakomai Forest (northern Japan), and in many satellite sites scatteed across the target region. The interim achievements of TEMA have been published in separate papers as well as special issues (Hirose and Walker 1996; Nakashizuka et al. 1999). The overall synthesis will be published as a journal special issue (Kohyama et al. 2005). We do not summarize these overall results here.

Our purpose in this chapter is to emphasize certain unique aspects of TEMA, which were not emphasized in other GCTE activities. We particularly focused on (1) the linkage of physiological processes of foliage canopies to landscape-scale processes of population demography and community dynamics, (2) integration of forest ecosystem processes into watershed-scale budgets, and (3) attention to lake ecosystems characterizing watershed-scale land systems. Other GCTE research has focused heavily on responses of physiological and ecosystem processes to global change. However, the prediction of the long-term response of forest systems to global change requires the coupling of ecosystem physiology and tree population demography. To interface the gap between them, TEMA focused on the branch-module-based reconstruction of tree foliage development at smaller scale, and the shift of vegetation zones determined by the dispersion of propagules and the suppression by resident vegetation at larger spatial scales.

Another aspect of TEMA was watershed-scale integration. Recently growing evidence shows that the majority of lotic and lentic waters are supersaturated in terms of CO_2 (Cole et al. 1994), suggesting that the inland aquatic systems function to return C that was fixed in the terrestrial ecosystems to the atmosphere through the biological activities. The inland aquatic systems also function to bury terrigenous organic carbon. According to Dean and Gorham (1998), lakes, reservoirs and wetlands bury organic carbon more than the oceans do, although these cover less than 2% of Earth's surface. Thus, the aquatic systems are likely "hot spots" for carbon processing in terrestrial ecosystems (Hanson et al. 2004).

We therefore focused on carbon flux along the atmosphere-forest-stream-lake continuum at a watershed scale. We found that the stream chemistry reflects biogeochemical metabolism at upstream forest ecosystems. In addition, we found that the carbon metabolism and biogeochemical cycling of inland aquatic systems are tightly linked with allochthonous organic matter input from the watershed and function to determine the fate of terrigenous carbon from the forests. These achievements from TEMA suggest that the effects of global and regional human impacts on terrestrial ecosystems should be examined at the scale of a watershed, as a unit of spatial integration.

23.2 Responses of Plant Communities to the Global Change: Scaling from Leaf to Landscape Through Individual Plant

In many natural plant communities, the local environment one individual experiences is different from that for others. Plants in a stand compete for resources with each other, and competition alters morphological and physiological traits of plants. These alterations in turn change the microclimate and resource acquisition of individuals in the stand, and consequently influence the growth of individuals (Hikosaka et al. 1999). However, studies on plant individuals growing in a stand are very few, which may make it difficult to predict responses of plants to CO_2 at a stand level.

One of the important foci of TEMA was a process-based analysis of environmental responses in population demography in plant communities. We adopted several approaches. We carried out an experimental study of monospecific stands as the simplest plant community. We also constructed a shoot-module-based simulation model, and an individual-plant-based model, to predict the response of plant communities to global change at different spatial scales.

23.2.1 Competition among Individual Plants in Even-Aged Monospecific Stands at Elevated CO_2

Competition among individuals in plant populations is categorized into two modes, i.e., symmetric and asymmetric competition (Weiner 1990). Symmetric competition indicates that individuals in a stand acquire resources in proportion to their sizes while in asymmetric competition large individuals acquire resources more than proportionately. For instance competition for light is asymmetric in dense stands (Ford and Diggle 1981; Weiner 1986; Jurik 1991; Nagashima 1999; Hikosaka et al. 1999), while that for nutrients is more symmetric (Wilson 1988; Weiner et al.1997; Hikosaka and Hirose 2001). The mode of competition is critical in development of size inequality in the stand. Size inequality is generally assessed with the coefficient of variation (Weiner 1990). Symmetric competition, where plant growth is proportional to the size, does not alter size inequality, while asymmetric competition increases size inequality in the stand.

Since diffusion of CO_2 within plant stands is very fast, competition for CO_2 is unlikely to occur among individuals (Jones 1992). Even though elevated CO_2 may benefit all individuals in the stand, the enhancement of growth by elevated CO_2 may indirectly alter the mode of competition (Wayne and Bazzaz 1997). There are two alternative hypotheses in this respect. One is that elevated CO_2 makes the competition more asymmetric and increases size inequality in the stand, because enhanced growth of larger individuals suppresses light acquisition of smaller individuals. The other is that elevated CO_2 reduces the degree of asymmetry in competition and consequently size inequality. This is because the end-product inhibition of photosynthesis due to elevated CO_2 (Stitt and Krapp 1999) may be stronger in larger plants exposed to high light, and because the reduction of the light compensation point of photosynthesis at elevated CO_2 may benefit smaller individuals growing in the shade (Osborne et al. 1997).

To test these hypotheses, we established even-aged monospecific stands of an annual, *Chenopodium album*, at ambient and doubled CO_2 with high and low nutrient availabilities in open top chambers (Nagashima et al. 2003). The growth of individual plants was monitored non-destructively every week until flowering. Elevated CO_2 significantly enhanced plant growth at high nutrients, but did not at low nutrients. The size inequality expressed as the coefficient of variation tended to increase at elevated CO_2. Size structure of the stands was analyzed by the cumulative frequency distribution of plant size (Fig. 23.1). At early stages of plant growth, CO_2 elevation benefited all individuals and shifted the whole size distribution of the stand to larger size classes. At later stages, dominant individuals were still larger at elevated than at ambient CO_2, but the difference in small subordinate individuals between two CO_2 levels became smaller. Although these tendencies were found at both nutrient availabilities, the difference in size distribution between CO_2 levels was larger at high nutrients. The CO_2 elevation did not significantly enhance the growth rate as a function of plant size except for the high nutrient stand at the earliest stage. This indicates that the higher biomass at elevated CO_2 at later stages in the high nutrient stand was caused by the larger size of individuals at the earliest stage. Therefore, the effects of elevated CO_2 on stand structure and size inequality are strongly depended on the growth stage and nutrient availability.

The emerging question is the ecophysiological basis of competition at elevated CO_2. As mentioned above, the difference in size structure results from different size-dependent growth rates of individuals within the stand. In a dense stand, large dominant individuals are better at capturing light. Small subordinate individuals, on the other hand, need less investment for biomass in support tissues because they maintain leaves at lower positions (Givnish 1982), and they can increase biomass allocation for leaf expansion, and partly ameliorate the limited light availability (Anten and Hirose 1998). To indicate the efficiency of biomass use to capture light, Hirose and Werger (1995) introduced the index Φ_{mass}, that is photon flux captured per unit aboveground mass. They suggested that Φ_{mass} might not differ between dominant and subordinate species in multispecies systems. However, plant growth is determined not only by the amount of acquired resources, but also by the efficiency of resource use (growth per unit amount of resource acquired). Hikosaka et al. (1999) defined light-use efficiency of photosynthesis (LUE) as photosynthesis per unit photon interception, and described the photosynthesis of individuals as the product of Φ_{mass} and LUE: $RPR = \Phi_{mass} \times LUE$, where RPR is the relative photosynthetic rate (photosynthetic rate per unit aboveground mass). Provided that plant growth is proportional to leaf photosynthesis, RPR is closely related to the relative growth rate. With a modification of the canopy photosynthesis model of Hirose and Werger (1987), Hikosaka et al. (1999) estimated the photosynthetic rate of individuals in a natural monospecific stand of an annual, *Xanthium canadense*. They found that Φ_{mass} was higher in larger individuals, while LUE was the highest in intermediate individuals. As a conse-

Fig. 23.1.
Comparisons of cumulative frequency distributions of biomass in the *Chenopodium album* stands grown at ambient (360 μmol mol^{-1}, *open symbols*) and elevated (700 μmol mol^{-1}, *closed symbols*) CO_2 levels, at high (**a**) and low (**b**) nutrients (modified from Nagashima et al. 2003)

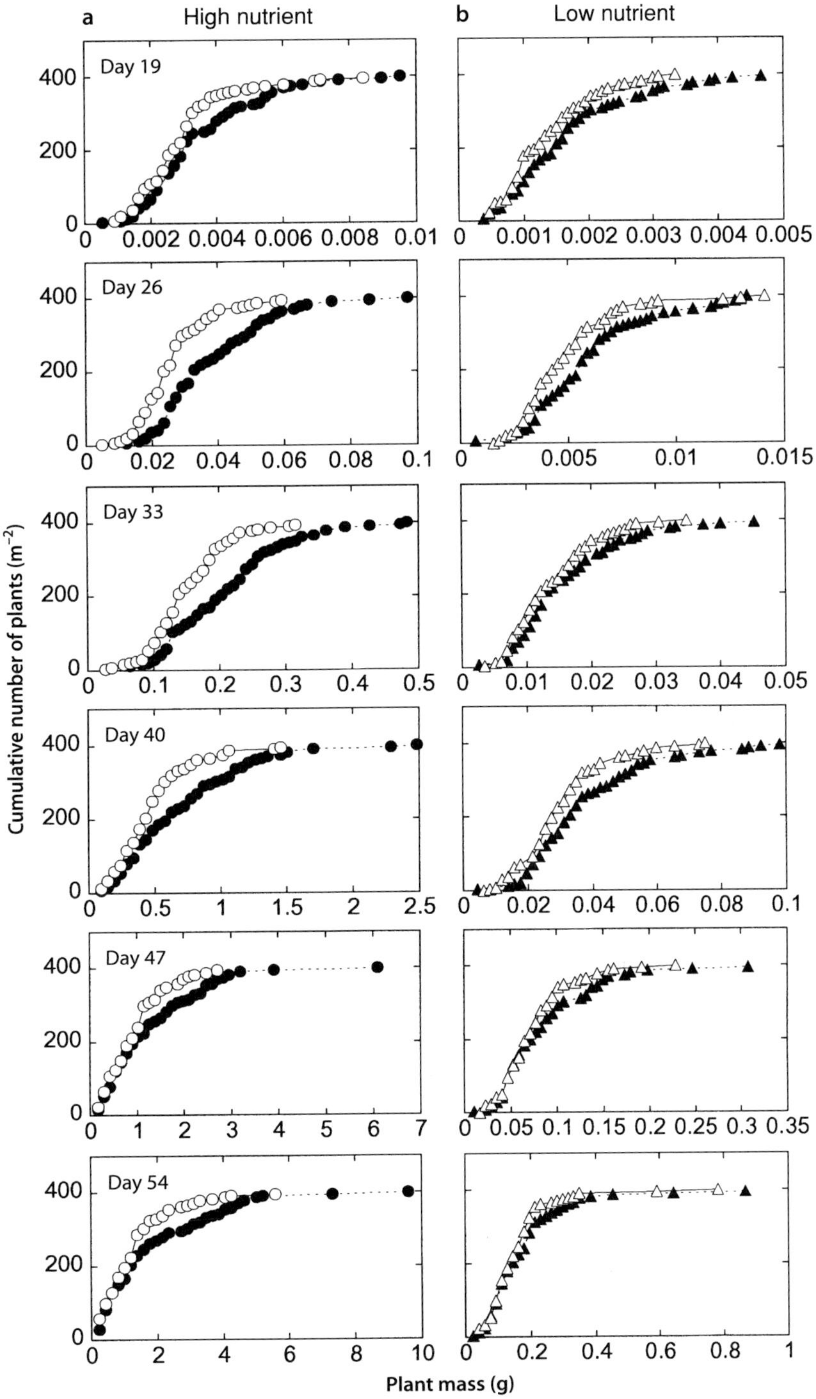

quence, RPR was high in intermediate and larger individuals, and lowest in smaller individuals.

We then applied the method of Hikosaka et al. (1999) to analyze acquisition and use of light in individuals in monospecific stands at elevated CO_2 (Hikosaka et al. 2003). As in the previous study, we established even-aged monospecific stands of an annual, *Chenopodium album*, at ambient and doubled CO_2 concentrations in open-top chambers (only high nutrient availability). The whole-plant photosynthesis of every individual in the stand was calculated from *(i)* the distribution of light and leaf nitrogen and *(ii)* the relationships between photosynthetic parameters and leaf nitrogen content per area. Elevated CO_2 increased light-saturated rates of photosynthesis by 10–15% and the initial slope of the light-response curve by 11%, but had no effect on the convexity of the light-response curve and dark respiration. The relative rate of photosynthesis (RPR) was analyzed as the product of

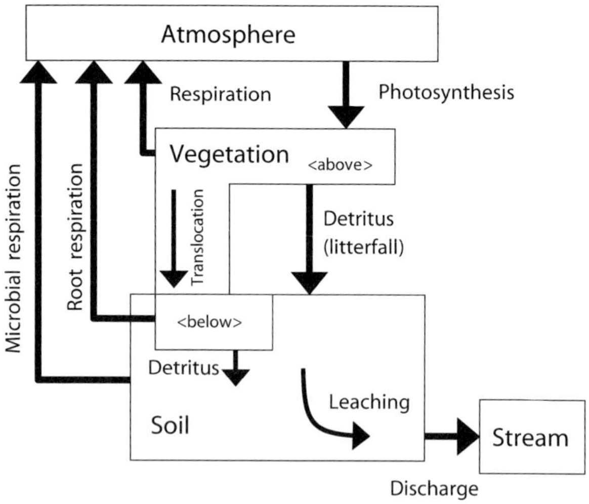

Fig. 23.4. Compartment model of carbon flow in a forested watershed

and streams (Fig. 23.4). Here, we focus on current results and a synthesis of the biogeochemical processes of carbon in forest watersheds based mainly on work conducted during TEMA.

23.3.1 Carbon Exchange between Atmosphere-Forest-Stream Boundaries

Eddy-correlation techniques for assessing CO_2 flux in the forest canopy provide quantitative information on net photosynthesis and respiration (for both vegetation and microorganisms), also referred to as net ecosystem exchange (NEE). NEE is a critical index for the quantitative assessment of the extent to which atmospheric carbon dioxide has been sequestrated by the ecosystem. In the Horonai River basin located in northern Japan, the NEE flux observed using the eddy correlation method (1999–2001) was 258 g C m^{-2} yr^{-1} (SD = 36), with atmospheric carbon dioxide sequestrated mainly during the growing season from June to September (Tanaka and Tanaka 2002; Shibata et al. 2005). Yamamoto et al. (1999) reported that NEE flux ranged from 65 to 136 g C m^{-2} yr^{-1} (mean = 106) in a temperate forest located in central Japan, which suggests that the inter-annual variation in regional climate is reflected in the inter-annual variation in NEE flux. Carbon exchange between forests and the atmosphere is driven by biotic factors such as photosynthesis, respiration, and nutrient conditions, and abiotic factors like solar radiation, temperature, air stability and others. While the expected increase in anthropogenic nitrogen pollution in the Asian region over the next few decades (Galloway and Cowling 2002) is likely to affect the future dynamics of NEE in this region, the effect of nitrogen deposition on carbon sequestration is still under discussion (Lloyd 1999; Nadelhoffer et al. 1999).

Table 23.1. The annual carbon budget of a cool-temperate forest basin (1999–2001; Horonai River in Tomakomai Experimental Forest, Hokkaido University, northern Japan)

	g C m^{-2} yr^{-1}
Net ecosystem exchange (NEE) = a	258
Increment of biomass = b	108
Leaching to stream = c	4
Net storage of carbon in the ground (= $a - b - c$)	146

In the NEE budget analyses, increases in above-ground biomass and leaching in streams are useful measures that can be employed to better understand the role of vegetation and soil in the carbon sequestration from a quantitative perspective. Since the budget for NEE (input) and leaching in streams (output) is indicative of the net storage or release of carbon (so called net ecosystem production (NEP)) in the watershed, respectively, the difference between NEP and the carbon accumulated in the vegetation corresponds to the net carbon storage (or release) for the soil system. In the Horonai River basin, the annual rate of increase of carbon in vegetation, the extent of leaching to streams, and the net carbon storage in the soil were 108, 4 and 146 g C m^{-2} yr^{-1}, respectively (Table 23.1, Hiura 2005; Shibata et al. 2001; Shibata et al. 2005). Furthermore, it was found that 41% and 57% of the NEE were retained in vegetation and soil, respectively. Since the mean residence time of carbon is markedly different for vegetation and soil, the relative partitioning of sequestrated carbon in the ecosystem is a very important consideration when predicting the effect of carbon dynamics on environmental change. However, the cumulative effect of possible sampling errors means that budget analyses are inherently difficult. These include: *(i)* difficulties in measuring eddies during periods of high atmospheric stability and the irregularity of the canopy surface, *(ii)* the large spatial heterogeneity in the vegetation of the watershed, *(iii)* the large temporal and spatial variation in residence times associated with the leaching of carbon into streams. Consequently, indirect or direct estimates of the carbon contributions from above and below-ground biomass, combined with the quantification of fine root production and mortality, will be one of the most effective means of ameliorating the effect of the aforementioned uncertainties associated with assaying carbon storage in the ground (Shutou and Nakane 2004; Satomura et al. 2003).

23.3.2 Transport of Dissolved Organic Carbon Associated with Dissolved Nitrogen from Terrestrial to Aquatic Ecosystems

Although the annual flux of dissolved carbon into streams was generally very small when compared with that associated with gas exchange between the terrestrial ecosys-

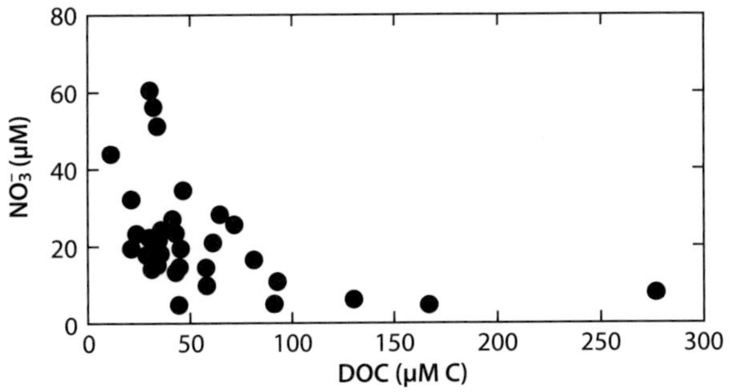

Fig. 23.5. The relationship between dissolved organic carbon (DOC) and nitrate (NO_3^-) concentrations in 35 natural streams in the Lake Biwa watershed, central Japan (modified from Konohira and Yoshioka 2005)

tem and the atmosphere, the supply of dissolved organic carbon to lotic and lentic ecosystems is an important source of energy for aquatic biota and as a source of secondary emissions of carbon dioxide from water surfaces. The quantitative and qualitative fate of dissolved organic carbon was different for each watershed. This is due to the heterogeneity expressed in terms of microbiological activities, reaction to the solid phase, hydrological flowpass in the ground, and the interrelationship between carbon and nitrogen cycling in forest ecosystems (Shibata et al. 2001; Neff and Asner 2001; Katsuyama and Ohte 2002; Konohira and Yoshioka 2005). Consequently, the flux and concentration of dissolved organic carbon in stream water can be a useful biogeochemical signature of carbon cycling in catchment ecosystems. Nitrate (NO_3^-) concentrations in streams are also important biogeochemical signatures of nitrogen cycling. In the Lake Biwa watershed in central Japan, an inverse relationship was observed between dissolved organic carbon (DOC) and NO_3^- concentrations in 35 natural streams (Fig. 23.5; Konohira and Yoshioka 2005). In a regional analysis of the variability in DOC and NO_3^- concentrations in stream water collected from two regions outside the Lake Biwa watershed, it was found that forest streams have high and variable NO_3^- concentrations with low DOC concentrations near Tokyo (an urban area in central Japan), while in northern Hokkaido (a cold, rural area in northern Japan) streams had high and variable DOC, and low NO_3^- concentrations, respectively (Konohira and Yoshioka 2005). These results suggested that the forest biogeochemical processes that affect DOC and NO_3^- concentrations are closely related to each other.

The relationship between DOC and nitrates in forest streams has also been reported in England (Harriman et al. 1998). Since the inorganic nitrogen in the soil is produced by the decomposition of organic nitrogen and then consumed by nitrogen immobilization, the decrease in nitrogen immobilization that arose as a consequence of a decrease in available carbon explains the increase of NO_3^- in soil (Hart et al. 1994). A carbon (energy) deficit with an excess of nitrogen available in the soil will result in low DOC and high NO_3^- in streams. Conversely, a nitrogen deficit in the presence of excess carbon will result in leachate that has high DOC and low NO_3^- concentrations being transported to streams. Stream DOC and NO_3^- concentrations will therefore be useful indices of carbon and nitrogen availability in forest catchments.

The transport of carbon and nitrogen from terrestrial ecosystems to streams also tends to fluctuate seasonally depending on each watershed environment. Shibata et al. (2001) indicated that direct input of litterfall as POC from riparian forest to stream was strongly affected by the phenology of the riparian canopy, while the DOC and DIC in the stream was influenced by the water infiltration rate the ground, with a longer retention time in Horonai River watershed, northern Japan. Seasonality of the climate and biological productivity are important drivers for the leaching of the dissolved nitrogen and carbon to the stream. Ohte et al. (2001) revealed that high temperature and precipitation during summer increase stream NO_3^- concentrations in central Japan, while higher discharge during snowmelt was important for NO_3^- leaching in snow-dominated regions of northern Japan. They emphasized the importance of the seasonal interrelation between hydrological processes in the ground and biogeochemical production of dissolved nitrogen (mainly nitrate) in soil for the temporal fluctuation of dissolved nitrogen in stream water in monsoon regions. Typhoons, which are episodic hydrological events, also enhance export of DOC to stream, especially in subtropical to tropical regions in Monsoon Asia (Kao and Liu et al. 1997).

23.3.3 Dynamics of Dissolved Organic Carbon at the Interface of Stream and Lake Ecosystems

Riverine export and watershed processes are important for modifying the terrestrial sink of anthropogenic CO_2 emission (Kling et al. 1991; Ludwig et al. 1996). The DOC concentration in rivers is usually higher than that of particulate organic carbon (POC) (Ludwig et al. 1996; Webster and Meyer 1997). Furthermore, DOC is an important component in stream and lake waters not only as an energy source, but also as a mediator of optical properties of waters. Since humic substances, which are major parts of DOC being exported from forest ecosystems, absorb sunlight, DOC controls the penetration of sunlight in rivers and lakes, and filters UV-B radiation for aquatic organisms (Schindler et al. 1996). Dynamics of DOC are, thus, important for understanding the biogeochemistry in watershed-scale land ecosystems.

Stream DOC concentrations exhibited a linear correlation with humus-like fluorescence intensity in the lake watersheds of Lakes Hovsgol, Baikal, Biwa and Shumarinai, although the slopes of these regressions differed slightly between the different watersheds (Hayakawa et al. 2003; Yoshioka 2000). The results suggest that humic substances were predominant in stream DOC, and the chemical composition of the stream dissolved organic matter was relatively homogeneous among watersheds. However, such a correlation was not found in the pelagic areas of these lakes. The surface DOC concentrations of Lake Biwa mainly fluctuated during the period at which stratification occurred (from March to November), with a concomitant gradual decrease being observed in the humus-like fluorescence intensity. These findings indicate the production of autochthonous DOC and that the humic substances were derived from the watershed (allochthonous) and subsequently decomposed in the lake. The increase in the protein-like fluorescence observed in the surface layers of Lake Biwa also suggested the autochthonous production of dissolved organic matter (Mostofa et al. in review). Since humic substances are highly resistant to biological degradation, the change in the humus-like fluorescence intensity observed in the lake may instead be attributed to photodegradation. The change in the position of the peak humus-like fluorescence, determined using a three-dimensional excitation emission matrix, suggested that the alteration of the humic substances could be attributed to solar irradiation (Hayakawa et al. 2003; Mostofa et al. in review).

23.4 Carbon Budget and Functions of the Lake Biwa Ecosystem

To understand key processes determining the function of aquatic habitats in terrestrial carbon balance, we measured and analyzed the budget and the fate of C in a lake ecosystem in the TEMA program. We chose Lake Biwa for this model study. Lake Biwa is the largest lake in Japan and one of the ancient lakes, with a surface area of 674 km^2, a mean depth of 41 m, a watershed of 3 848 km^2 and a water retention time of ca. 5.5 years. The lake was originally oligotrophic. However, due to development in the watershed for the past 40 years, the lake has become moderately eutrophic. At present, the lake is mesotrophic and supplies water resources for 14 million peoples living in the Kinki District of Japan. The following estimates were made in the north basin, which accounts for 99% of the water volume of this lake.

23.4.1 Carbon Budget in the Lake

The concentration of particulate organic C (POC) in Lake Biwa ranged from 0.2 to 1.1 g C m^{-3} in the surface mixing layer (0–15 m deep), but was less than 0.3 g C m^{-3} in the hypolimnion (>15–20 m deep). The pattern coincided with seasonal and vertical changes in phytoplankton abundance (Yoshida et al. 2001). Similarly, the concentration of dissolved organic C (DOC) was high in summer in the surface layer (1.4 g C m^{-3}) but became vertically homogeneous in winter (~1.0 g C m^{-3}). Based on monitoring data, the annual input of allochthonous organic C (POC + DOC) per unit area of Lake Biwa was 9.6 g C m^{-2} yr^{-1}. The net primary production rate in the surface layer in the Lake Biwa ranged from 0.3 to 2.7 g C m^{-2} d^{-1} and algae fixed 323 g C m^{-2} annually (Urabe et al.1999; Yoshimizu et al. 2001). Thus, autochthonous input of organic C is 30 times higher than the allochthonous input (Fig. 23.6). Nonetheless, the share of allochthonous C was <40% of DOC in the surface mixing layer and 50–60% in the hypolimnion (Yoshioka 2000), implying that allochthonous organic C is less respired compared with autochthonous organic C. Interestingly, organic C corresponding to 90% of the allochthonous input flowed out annually from the north basin of Lake Biwa (Yoshimizu et al. 2002). Since most organic C from the inflow rivers was humic substance and since its concentration was seasonally stable in the lake water (Yoshioka 2000), a substantial fraction of allochthonous organic matter seemed to flow out from the lake without being consumed efficiently within the lake.

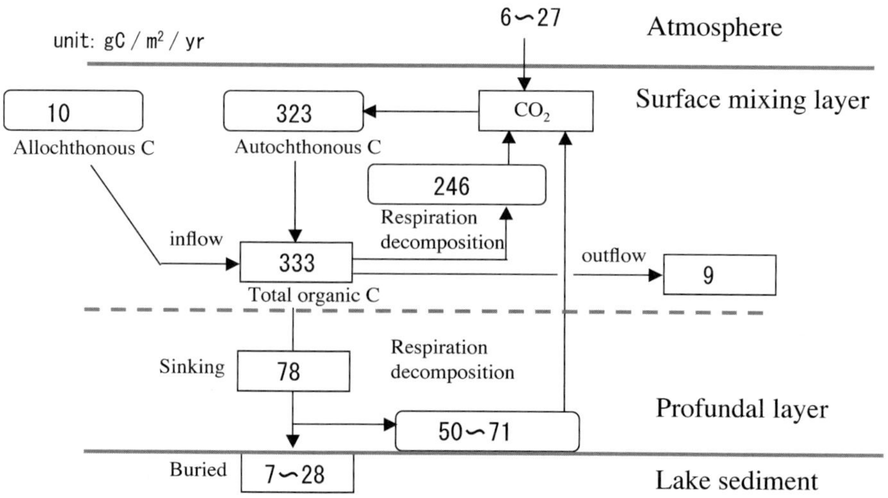

Fig. 23.6. Carbon budget at an offshore site of Lake Biwa

Of the organic C in the surface water, 78 g C m^{-2} yr^{-1} was transported annually to the profundal layer (Yoshimizu et al. 2001), implying that 75% of the organic C was respired within the surface layer. From the age of lake sediments determined by ^{210}Pb and ^{137}Cs and the abundance of organic C at the lake bottom (Murase and Sakamoto 2000; Urabe et al. 2005), the annual rate of organic C burial at the lake bottom was estimated to be 7–28 g C m^{-2} yr^{-1}. This implies that 64–90% of the organic C exported from the surface layer is respired or decomposed at the profundal layer and/or the lake sediments. This number is smaller than burial rates in lakes with surface areas < 500 km^2 but larger than those in larger lakes such as Lake Micigana and Lake Bikal (e.g., Dean and Gorham 1998). In addition, it is in accordance with the oxygen consumption rate (~180 g O_2 m^{-2}) during the stagnant period in the hypolimnion of the Lake Biwa (Kumagai and Fushimi 1995). According to its $\delta^{15}N$ signature, 40% of organic matter in the lake sediments was estimated to be terrestrial in origin (Murase and Sakamoto 2000). This suggests that allochthonous organic matter is selectively stored in the lake sediments due to less efficient utilization by living organisms.

These estimates show that the *P*/*R* ratio of Lake Biwa is 1.02~1.09, implying that C fixation slightly exceeds CO_2 production. This is because the allochthonous organic matter is much less discharged compared with primary production and less efficiently utilized by organisms. Note that the present C budgets did not consider horizontal transports of organic matters within the lake. In this lake, pCO_2 at the surface water was less than saturation level from spring to fall, but its annual mean was 397 ppmv because CO_2 in the water was supersaturated in winter (Urabe et al. 2005). Thus, as a whole, the Lake Biwa ecosystem functions as a C source. Indeed, there are estimates that the community respiration rate of heterotrophs exceeds the net primary production at an offshore site. The slight discrepancy between the C budgets of the biological community and the pCO_2 observations may be due to decomposition of organic matter produced at the littoral area.

23.4.2 Metabolism in the Lake Sediments

The lake sediments are characterized by the concentrated accumulation of organic carbon. Standing stocks of organic carbon in the upper most layer (2 cm) of the sediment were estimated to be 9.2×10^{10} g C for the whole lake (Murase and Sakamoto 2000). This value is 2.5 times higher than the estimated amount of carbon in the whole lake water (3.6×10^{10} g C). The littoral sediments and the deeper layer of the profundal sediments showed higher contribution of terrestrial organic matter than the surface layer of the profundal sediments (Murase and Sakamoto 2000).

Except for the top few millimeters, the bottom sediments of the lake are anoxic and methanogenesis is the major terminal carbon metabolism. Potential methane production rates of the profundal sediment (0–25 cm in depth) ranged from 2.4–24 mg C m^{-2} d^{-1} (Murase and Sugimoto 2002). Carbon isotopic data for methane and CO_2 in the sediment suggested that hydrogenotrophic methanogenesis was the major pathway (Murase and Sugimoto 2001). A part of the produced methane was stored in adsorbed form in the lake sediment (Sugimoto et al. 2003), but some was diffused up to the surface, where 80–90% of diffused methane was oxidized (Murase et al. 2005). The methane oxidation rates at the sediment surface are comparable to the rate of oxic carbon mineralization (17–20 mg C m^{-2} d^{-1}) and these rates accounted for 50–160% of the estimated carbon sedimentation rate (Murase, unpublished). Our estimates suggested that the increase in oxygen consumption rate by 50% potentially makes the surface sediment anoxic causing the catastrophic change in gas metabolism of the sediment. In addition to eutrophication, an increase in temperature of the bottom water due to the global warming (Hayami and Fujiwara 1999), which caused enhancement of microbial activities, prolonged stagnant periods, and lower solubility of oxygen, would accelerate the oxygen depletion in the sediment.

The potential methane production rate in the littoral sediments was approximately 5 to 10 times higher than that in the profundal sediments (Murase and Sugimoto 2002). At the littoral area, substantial methane diffused from the sediment across the oxic-anoxic boundary into the lake water during the stagnant period (Murase et al. 2003). Disturbance of the surface sediment by internal waves of the lake seemed to accentuate transportation of methane from the sediment to the lake (Sakai et al. 2002), which can be major sources of dissolved methane in the whole lake water (Murase et al. 2003).

23.4.3 Terrestrial Environment and Function of Lake Ecosystems

Considering the watershed and lake surface areas, the discharge rate of allochthonous organic carbon to, and carbon burial rate in, lake Biwa are two orders of magnitude lower than the NEE in the temperate forests in Japan, but almost comparative to the outflow rate of organic carbon to streams from the forests (Sect. 23.3.1 and Table 23.1). Thus, standing water is a potentially important arena determining the fate of organic carbon leaching to streams from the forests. This suggests that even slight changes in the carbon budget in the forests may have serious impacts on lake ecosystems.

According to del Giorgio et al. (1994), the *P*/*R* ratio of lake ecosystems becomes > 1 with increasing the trophic condition. This implies that the function of lake ecosys-

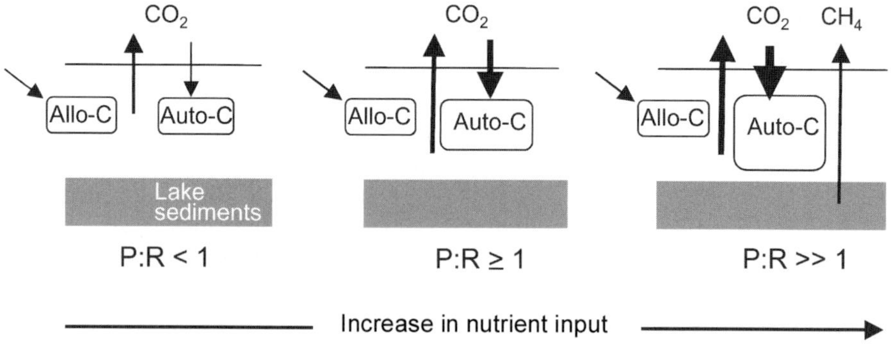

Fig. 23.7. Changes in lake ecosystem functions in terms of C sinks/sources as a function of increases in nutrient loadings. *Auto-C:* autochthonous organic carbon, and *Allo-C:* allochthonous organic carbon

tems is changed from C sink to C source with increasing nutrient loading from the watershed. In Lake Biwa, the primary production rate is nutrient limited (Urabe et al. 1999). Therefore, increase in nutrient loadings would increase C sequestration from the atmosphere through increased C fixation rate and push this lake toward a C sink. Progress of further eutrophication due to anthropogenic activities in the watershed, however, may turn the lake ecosystem to vent gasses that are more efficient for warming (Fig. 23.7). An increase in the sinking flux of organic C due to increased primary production promotes oxygen consumption in the profundal layer and lake sediments. Global warming would accelerate the oxygen consumption at the lake bottom by increasing water temperature at the profundal zone and reducing vertical mixing in winter. This may, in turn, promote the emission of methane, a more efficient gas for global warming than CO_2, from the lake sediments to the atmosphere.

As such, it is most likely that the function of lake ecosystems in terms of being a C sink or a C source depends on the land cover and land use, and thus the anthropogenic activities, in the watershed. Unfortunately, only a few studies have included aquatic habitats when carbon balance of terrestrial ecosystems is examined in Asian area (Kao and Liu 1997; Ludwig et al. 1996; Sarin et al. 2002; Shibata et al. 2001). Considering future anthropogenic development in Monsoon Asia and the potential importance of inland aquatic systems as a hot spot where terrigenous carbon is processed, studies at a watershed scale are essential to assess impacts of changes in carbon budgets in terrestrial ecosystems under elevated CO_2 and climate change.

23.5 Summary and Conclusions

A GCTE core project "Global change impacts on terrestrial ecosystems in Monsoon Asia" (TEMA) has been carried out from 1995 to 2003. This chapter overviews the TEMA-employed unique approach of integrating across different scales, i.e., from a plant leaf to watershed budgets, targeting on the eastern Asian region. We particularly focused on the linkage between physiological processes of foliage canopy and landscape-scale processes of plant demography and plant community dynamics, where individual plant processes were integrated from physiology, and we projected the change in geographic pattern from individual plant processes. We evaluated the watershed unit where freshwater chemistry provides a signature of biogeochemical characteristics of terrestrial ecosystems. Stream chemistry controls the trophic condition of lake ecosystems, which can contribute to global change particularly through methane emission. Integration at the scale of watersheds will contribute, within the scope of the new GLP in relation to LOICZ, to the validation of the impact of environmental change on human society, and the impact of human activities on watershed-scale environments.

References

Anten NPR, Hirose T (1998) Biomass allocation and light partitioning among dominant and subordinate individuals in *Xanthium canadense* stands. Ann Bot 82:665–673
Bugmann HKM (2001) A review of forest gap models. Climatic Change 51:259–305
Carignan RD, Planas, Vis C (2000) Planktonic production and respiration in oligotrophic shield lakes. Limnol Oceanogr 45:189–199
Cole JJ, Caraco NF, Kling GW, Kratz TK (1994) Carbon dioxide supersaturation in the surface waters of lakes. Science 265:1568–1570
Dean WE, Gorham E (1998) Magnitude and significance of carbon burial in lakes, reservoir, and peatlands. Geology 26:535–538
del Giorgio, Paul A, Peters RH (1994) Patterns in planktonic P:R ratio in lakes: influence of lake trophy and dissolved organic carbon. Limnol Oceanogr 39:772–787
Ford ED, Diggle PJ (1981) Competition for light in a plant monoculture modelled as a spatial stochastic process. Ann Bot 48: 481–500
Galloway JN, Cowling EB (2002) Reactive nitrogen and the world: 200 years of Change. Ambio 31:64–71
Givnish TJ (1982) On the adaptive significance of leaf height in forest herbs. Am Nat 120:353–381
Gurung TB, Urabe J (1999) Temporal and Vertical difference in factors limiting growth rate of heterotrophic bacteria in Lake Biwa. Microb Ecol 38:136–145
Hanson PC, Pollard A, Bade DL, Predick K, Carpenter SR, Foley J (2004) A model of carbon evasion and sedimentation in temperate lakes. Global Change Biology 10:1285–1298
Harriman R, Curtis C, Edwards AC (1998) An empirical approach for assessing the relationship between nitrogen deposition and nitrate leaching from upland catchments in The United Kingdom using runoff chemistry. Water Air Soil Pollut 105:193–203
Hart SC, Nason GE, Myrold DD, Perry DA (1994) Dynamics of gross nitrogen transformation in an old-growth forest: The carbon connection. Ecology 75:880–891

Hayakawa K, Sekino T, Yoshioka T, Maruo M, Kumagai M (2003) Dissolved organic carbon and fluorescence in Lake Hovsgol: factors reducing humic content of the lake water. Limnology 4:25–33

Hayami Y, Fujiwara T (1999) Warming of hypolimnetic water in Lake Biwa. Oceanography in Japan 8:197–202 (in Japanese)

Hikosaka K, Hirose T (2001) Nitrogen uptake and use of competing individuals in a *Xanthium canadense* stand. Oecologia 126:174–181

Hikosaka K, Sudoh S, Hirose T (1999) Light acquisition and use by individuals competing in a dense stand of an annual herb, *Xanthium canadense*. Oecologia 118:388–396

Hikosaka K, Yamano T, Nagashima H, Hirose T (2003) Light-acquisition and use of individuals as influenced by elevated CO_2 in even-aged monospecific stands of *Chenopodium album*. Funct Ecol 17:786–795

Hirose T, Walker B (ed) (1996) Global change and terrestrial ecosystems in monsoon Asia. Vegetatio 121:1–191

Hirose T, Werger MJA (1987) Maximizing daily canopy photosynthesis with respect to the leaf nitrogen allocation pattern in the canopy. Oecologia 72:520–526

Hirose T, Werger MJA (1995) Canopy structure and photon flux partitioning among species in a herbaceous plant community. Ecology 76:466–474

Hirose T, Kohyama T, Oshima Y (1998) GCTE activities in Japan. Global Environmental Research 1:19–24

Hiura T (2005) Estimation of aboveground biomass and net biomass increment in a cool temperate forest on a landscape scale. Ecol Res 20:271–277

Jones HG (1992) Plants and microclimate, 2nd edn. Cambridge University Press, Cambridge

Jurik TW (1991) Population distribution of plant size and light environment of giant ragweed (*Ambrosia trifida* L.). Oecologia 87:539–550

Kao SJ, Liu KK (1997) Fluxes of dissolved and nonfossil particulate organic carbon from an Oceania small river (Lanyang Hsi) in Taiwan. Biogeochemistry 39:255–269

Katsuyama M, Ohte N (2002) Determining the sources of stormflow from the fluorescence properties of dissolved organic carbon in a forested headwater catchment. J Hydrol 268:192–202

Kling GW, Kipphut GW, Miller MC (1991) Arctic lakes and streams as gas conduits to the atmosphere: Implications for tundra carbon budgets. Science 251:298–301

Kohyama T (2005) Scaling up from shifting gap mosaic to geographic distribution in the modeling of forest dynamics. Ecol Res 20:305–312

Kohyama T, Shigesada N (1995) A size-distribution-based model of forest dynamics along a latitudinal gradient. Vegetatio 121:117–126

Kohyama T, Canadell J, Ojima DS, Pitelka L (ed) (2005) Forest ecosystems and environments: scaling up from shoot module to watershed. Ecol Res 20:241–385

Konohira E, Yoshioka T (2005) Dissolved organic carbon and nitrate concentrations in stream. A useful index indicating carbon and nitrogen availability in catchments – Ecological Research 20:359–365

Kubo T, Kohyama T (2005) *Abies* population dynamics simulated by a functional-structural tree model. Ecol Res 20:255–269

Kumagai M, Fushimi H (1995) Inflows due to snowmelt: Physical processes in a lakrge lake, Lake Biwa, Japan. Coastal and Estuarine Studies 48:129–139

Lloyd J (1999) The CO_2 dependence of photosynthesis, plant growth responses to elevated CO_2 concentrations and their interaction with soil nutrient status, II. Temperate and boreal forest productivity and the combined effects of increasing CO_2 concentrations and increased nitrogen deposition at a global scale. Funct Ecol 13:439–759

Ludwig W, Probst JL, Kempe S (1996) Predicting the oceanic input of organic carbon by continental erosion. Global Biogeochem Cycles 1:23–41

Mostofa KMG, Yoshioka T, Konohira E, Tanoue E, Hayakawa K, Takahashi M (2005) Three-dimensional fluorescence as a tool for investigating the dynamics of dissolved organic matter in the Lake Biwa watershed. Limnology 6:101–115

Murase J, Sakamoto M (2000) Horizontal distribution of carbon and nitrogen and their isotopic compositions in the surface sediment of Lake Biwa. Limnology 1:177–184

Murase J, Sugimoto A (2001) Spatial distribution of methane in the Lake Biwa sediments and its carbon isotopic compositions. Geochemical J 35:257–263

Murase J, Sugimoto A (2002) Seasonal and spatial variations of methane production in mesotrophic lake sediments (Lake Biwa, Japan) Verh Int Ver Limnol 28:971–974

Murase J, Sakai Y, Sugimoto A, Okubo K, Sakamoto M (2003) Sources of dissolved methane in Lake Biwa. Limnology 4:91–99

Murase J, Sakai Y, Kametani A, Sugimoto A (2005) Dynamics of methane in a mesotrophic Lake Biwa, Japan. Ecol Res 20: 377–385

Nadelhoffer KJ, Emmett BA, Gundersen P, Kjonaas OJ, Koopmans CJ, Schleppi P, Tietema A, Wright RF (1999) Nitrogen deposition makes a minor contribution to carbon sequestration in temperate forests. Nature 398:145–148

Nagashima H (1999) The processes of height-rank determination among individuals and neighborhood effects in *Chenopodium album* L. stands. Ann Bot 83:501–507

Nagashima H, Yamano T, Hikosaka K, Hirose T (2003) Effects of elevated CO_2 on the size structure of even-aged monospecific stands of annual, *Chenopodium album*. Global Change Biol 9: 619–629

Nakashizuka T, Kohyama T, Whitmore TC, Ashton PS (ed) (1999) Tree diversity and dynamics of western Pacific and eastern Asian forests. J Veg Sci 10:763–860

Neff JC, Asner GP (2001) Dissolved organic carbon in terrestrial ecosystems: synthesis and a model. Ecosystems 4:29–48

Norby RJ, Ogle K, Curtis PS, Badeck F-W, Huth A, Hurtt GC, Kohyama T, Penuelas J (2001) Aboveground growth and competition in forest gap models: an analysis for studies of climatic change. Climatic Change 51:415–447

Ohte N, Tokuchi N, Shibata H, Tsujimura M, Tanaka T, Mitchell MJ (2001) Hydrobiogeochemistry of Forest ecosystems in Japan. Hydrol Process 15:1771–1789

Osborne CP, Drake BG, LaRoche J, Long SP (1997) Does long-term elevation of CO_2 concentration increase photosynthesis in forest floor vegetation? Indian strawberry in a Maryland forest. Plant Physiol 114:337–344

Sakai Y, Murase J, Sugimoto A, Okubo K, Nakayama E (2002) Resuspension of bottom sediment by an internal wave in Lake Biwa. Lakes and Reservoirs 7:339–344

Sarin MM, Sudheer AK, Balakrishna K (2002) Significance of riverine carbon transport: A case study of a large tropical river, Godavari (India). Science in China (Series C) 45 supplement: 97–108

Satomura T, Nakatsubo T, Horikoshi T (2003) Estimation of the biomass of fine roots and mycorrhizal fungi: a case study in a Japanese red pine (*Pinus densiflora*) stand. J Forest Res 8: 221–225

Schindler DW, Curtis PJ, Parker BR, Stainton MP (1996) Consequences of climate warming and lake acidification for UV-B penetration in North American boreal lakes. Nature 379: 705–708

Shibata H, Mitsuhashi H, Miyake Y, Nakano S (2001) Dissolved and particulate carbon dynamics in a cool-temperate forested basin in northern Japan. Hydrol Process 15:1817–1828

Shibata H, Hiura T, Tanaka Y, Takagi K, Koike T (2005) Carbon cycling and budget in a forested basin of southwestern Hokkaido, northern Japan. Ecol Res 20:325–331

Shinozaki K, Yoda K, Hozumi K, Kira T (1964) A quantitative analysis of plant form – the pipe model theory I. basic analyses. Jpn J Ecol 14:97–105

Shutou K, Nakane K (2004) Change in soil carbon cycling for stand development of Japanese Cedar (*Cryptomeria japonica*) plantations following clear-cutting. Ecol Res 19:233–244

Stitt M, Krapp A (1999) The interaction between elevated carbon dioxide and nitrogen nutrition: the physiological and molecular background. Plant, Cell and Environment 22:583–621

Sugimoto A, Dan J, Kumai T, Murase T (2003) Adsorption as a methane storage process in natural lake sediment. Geophys Res Lett Doi: 10.1029/2003GL018162

Takenaka A (2005) Local coexistence of tree species and the dynamics of global distribution pattern along environmental gradient: a simulation study. Ecol Res 20:297–304

Tanaka Y, Tanaka N (2002) Carbon dioxide exchange and ecosystem respiration in a cool-temperate deciduous forest, In: Wada E (eds.): Response of terrestrial ecosystem to global climate change, Annual synthesis report of IGBP-MEXT, pp 111–118 (in Japanese)
Urabe J, Sekino T, Nozaki K, Tsuji A, Yoshimizu C, Kagami M, Koitabashi T, Miyazaki T, Nakanishi M (1999). Light, nutrients and primary productivity in Lake Biwa: an Evaluation of the current ecosystem situation. Ecol Res 14:233–242
Urabe J, Yoshida T, Gurung TB, Sekino T, Tsugeki N, Nozaki K, Maruo M, Nakayama E, Nakanishi M (2005) The production to respiration ratio and itsi implication in Lake Biwa. Ecol Res 20:367–375
Wayne PM, Bazzaz FA (1997) Light acquisition and growth by competing individuals in CO_2-enriched atmospheres: consequences for size structure in regenerating birch stands. J Ecol 85:29–42
Webster JR, Meyer JL eds (1997) Stream organic matter budgets. J North Am Benth Soci 16:3–161
Weiner J (1986) How competition for light and nutrients affects size variability in *Ipomoea tricolor* populations. Ecology 67:1425–1427
Weiner J (1990) Asymmetric competition in plant populations. Trend Ecol Evol 5:360–364
Weiner J, Wright DB, Castro S (1997) Symmetry of below ground competition between *Kochia scoparia* individuals. Oikos 79: 85–91
Wilson JB (1988) The effect of initial advantage on the course of plant competition. Oikos 51:457–472
Yamamoto S, Murayama S, Saigusa N, Kondo H (1999) Seasonal and inter-annual variation of CO_2 flux between a temperate forest and atmosphere in Japan. Tellus 51B:402–413
Yoshida T, Kagami M, Gurung TB, Urabe J (2001) Seasonal succession of zooplankton in Lake Biwa. Aquat Ecol 35:19–29
Yoshimizu C, Yoshida T, Nakanishi M, Urabe J (2001) Effects of zooplankton on the sinking flux of organic carbon in Lake Biwa. Limnology 2:37–43
Yoshimizu C, Urabe J, Sugiyama S, Maruo M, Nakayama E, Nakanishi M (2002) Carbon and phosphorus budgets in the pelagic area of Lake Biwa, the largest lake in Japan. Verh Int Verein Limnol 24:1409–1414
Yoshioka T (2000) Characterization of dissolved organic matter using chemical, isotopic and optical properties. In: Report of the Grant-in-Aid for Scientific Research (C), No. 10640474,from the Ministry of Education, Culture, Sports, Science, and Technology, Japan, pp 53 (in Japanese)
Yoshioka T, Tanoue E, Ohta K, Masuzawa T, Konohira E, Wu F, Mostofa KG, Ohte N, Kawasaki M, Okazaki R, Katsuyama M, Takahashi M, Hayakawa K, Ueda T, Boontanon N, Yoh M, Ueda S (2001) Response of the terrestrial ecosystem from the viewpoint of material cyclings in freshwater system. In Wada E (ed) Report of Response of Terrestrial Watershed Ecosystem to Global environmental Change, Scientific Research of Priority Area B (No. 11213101), the Ministry of Education, Culture, Sports, Science and Technology, Japan.175–194 (in Japanese)

Chapter 24

Responses of High Latitude Ecosystems to Global Change: Potential Consequences for the Climate System

A. David McGuire · F. S. Chapin III · Christian Wirth · Mike Apps · Jagtar Bhatti · Terry Callaghan · Torben R. Christensen Joy S. Clein · Masami Fukuda · Trofim Maximov · Alexander Onuchin · Anatoly Shvidenko · Eugene Vaganov

24.1 Introduction

Terrestrial ecosystems of high latitudes occupy approximately one-fourth of the Earth's vegetated surface. Substantial climatic warming has occurred in many high latitude areas during the latter half of the 20th Century (Serreze et al. 2000), and evidence continues to mount that this warming has been affecting the structure and function of terrestrial ecosystems in this region (Stow et al. 2004; Hinzman et al. 2005). It is important to understand these changes because they may have consequences for the functioning of the climate system, particularly in the way that *(a)* radiatively active gases are exchanged with the atmosphere, *(b)* water and energy are exchanged with the atmosphere, and *(c)* fresh water is delivered to the Arctic Ocean (Chapin et al. 2000a; McGuire et al. 2003). The exchange of water and energy has implications for regional climate that may influence global climate, while the exchange of radiatively active gases and the delivery of fresh water to the Arctic Ocean are processes that could directly influence climate at the global scale.

Over the past decade the IGBP-GCTE high latitude transects have become important foci for research on responses of high latitude terrestrial regions to global change (Steffen and Shvidenko 1996; McGuire et al. 2002). This network of transects (Fig. 24.1) includes two in Siberia, the Far East Siberian Transect (FEST) and the East Siberian Transect (EST); one in Scandinavia (SCANTRAN/SCANNET), which has been augmented by carbon storage studies along a transect in Finland; one in Canada, the Boreal Forest Transect Case Study (BFTCS); and one in Alaska.

The high latitude transects generally span substantial temperature gradients (mean annual temperature of 5° to –15°C) both within and among transects (McGuire et al. 2002). Temperature along each transect co-varies with precipitation and photosynthetically active radiation. Disturbance regimes including fire and insects are also variable among the high latitude transects. For example, fire is essentially non-existent in much of Scandinavia, but burns annually an average of approximately 1% of the boreal forest along the EST (McGuire et al. 2002; Fig. 24.2). Similarly, land-use and land-cover change also varies among the high latitude transects (Kurz and Apps 1999; McGuire et al. 2002, 2004).

Each of the transects provides a different perspective into the responses of high latitude ecosystems to global change. In this chapter we first summarize how climate, disturbance regimes, and land cover in high latitudes have changed during the last several decades. We then summarize the results of ecological research along these transects that have contributed towards a richer understanding of high latitude terrestrial responses to these changes. We conclude with a discussion of challenges and opportunities for integration.

24.2 Recent Changes in Climate, Disturbance Regimes, and Land Cover

Temperature. While temperature has changed substantially in high latitudes during recent decades (Serreze et al. 2000), changes have not been uniform. Warming has been most pronounced in continental Siberia and in Alaska, with most of the warming occurring in winter (December–February) and spring (March–May) (Serreze et al. 2000; McBean et al. 2005). During the last century, northern Scandinavia experienced warming during the 1920s and 1930s, cooling from the 1940s until the 1960s, and warming since the 1970s; this pattern mirrors the global mean temperature trend over the last century (IPCC 2001). In recent decades, air temperatures have increased substantially in western Canada (Serreze et al. 2000). While eastern Canada generally cooled since the 1970s (Serreze et al. 2000), the cooling appears to have ceased since the late 1990s (McBean et al. 2005).

Precipitation. The increase in the hydrological discharge of northern Eurasian rivers into the Arctic Ocean over the last century, as documented by Peterson et al. (2002), is largely explained by increased moisture transport into high latitudes (McClelland et al. 2004). Consistent with this observation, Vaganov et al. (1999) documented increased winter precipitation along the FEST during the last century. While precipitation has remained stable in Alaska, several lines of evidence indicate that Alaska is

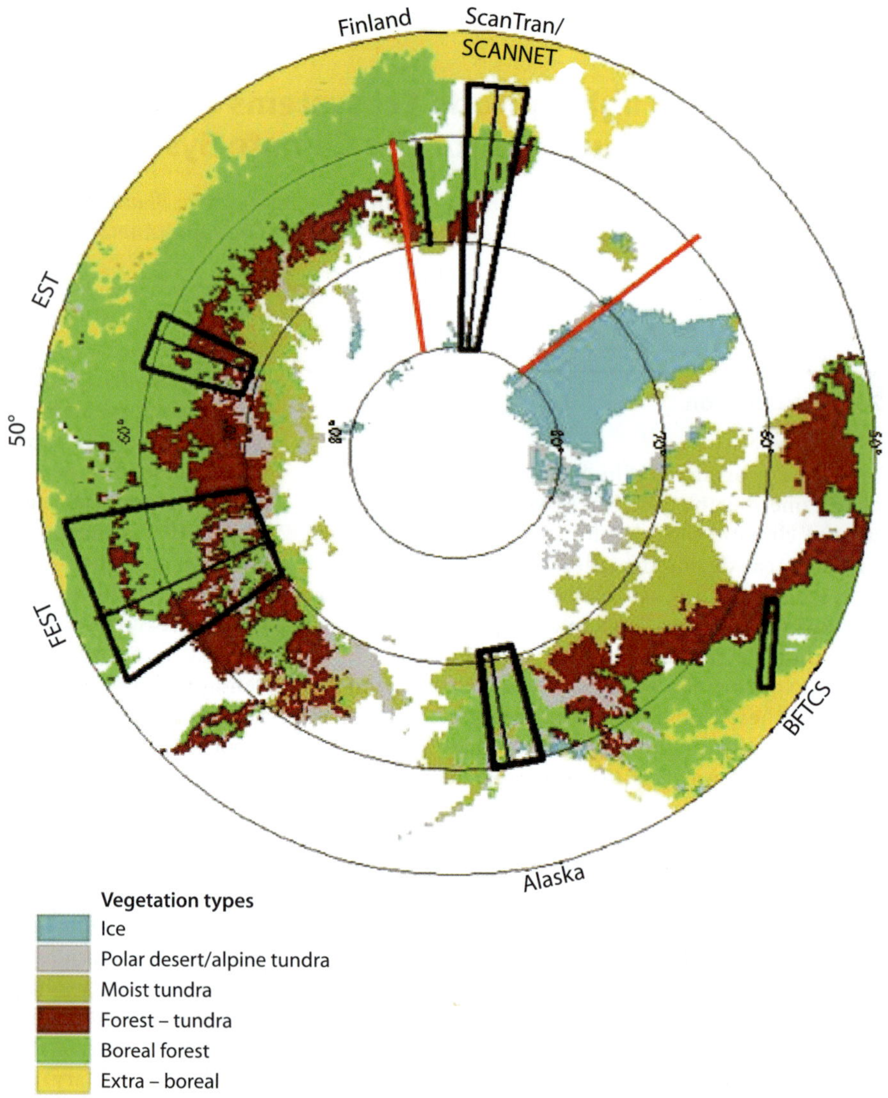

Fig. 24.1. Polar projection vegetation map indicating the location of high latitude transects. This network of transects includes two in Siberia, the Far East Siberian Transect (*FEST*) and the East Siberian Transect (*EST*); one in Scandanvia (*SCANNET*), which has been augmented by carbon storage studies along a transect in Finland; one in Canada, the Boreal Forest Transect Case Study (*BFTCS*), and one in Alaska

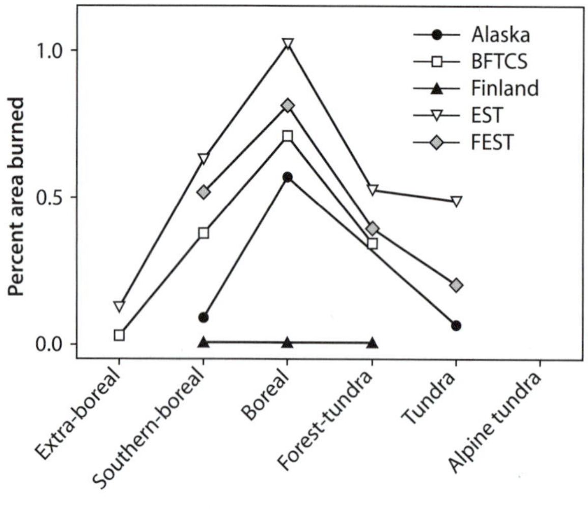

Fig. 24.2. Patterns of historical area burned along each of the high latitude transects as a function of vegetation distribution (reprinted with permission from McGuire et al. 2002)

experiencing increased drought stress because of increased summer water deficits (Oechel et al. 2000; Barber et al. 2000). There is also increased variability in climate in Siberia, and long periods of warm dry weather have become more frequent.

Permafrost. A wealth of evidence indicates that permafrost is warming in the FEST (Hinzman et al. 2005), Canada (Vitt et al. 2000), Alaska (Fig. 24.3; Romanovsky and Osterkamp 1997; Osterkamp and Romanovsky 1999; Hinzman et al. 2005), and Fennoscandia (Christensen et al. 2004; Luoto et al. 2004). While permafrost warming is consistent with regional increases in air temperature, deeper snow cover also plays a roll (Osterkamp and Romanovsky 1999; Stieglitz et al. 2003). In Alaska, warming of permafrost may be causing a significant loss of open water across the landscape, as thawing of permafrost connects closed watersheds with groundwater (Yoshikawa and Hinzman 2003). However, the reduction

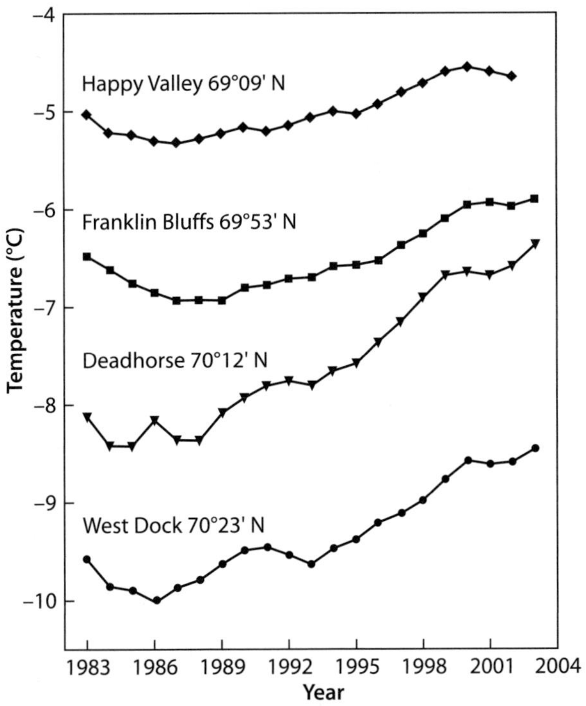

Fig. 24.3. Changes in soil temperature at 20 m depth for several northern sites of the Alaska transect (data courtesy of Tom Osterkamp; see also Romanovsky and Osterkamp 1997)

of open water bodies may also reflect increased evaporation under a warmer and effectively drier climate in Alaska as the loss of open water has also been observed in permafrost-free areas (Klein 2005).

Growing Season. Several studies based on remote sensing indicate that growing seasons are changing in high latitudes regions (Dye 2002; McDonald et al. 2004; McGuire et al. 2004; Smith et al. 2004). These studies identify earlier onset of thaw in both northern North America and northern Eurasia, but the magnitude depends on the study. Putting together the trends in the onset of both thaw and freeze, Smith et al. (2004) indicate that the trend for longer growing seasons in northern North America (3 days per decade) is primarily because of later freezing, while the trend in northern Eurasia (1 day per decade) is because the trend for earlier thaw is slightly greater than the trend for earlier freezing. However, in the EST Vaganov et al. (1999) found delays in the onset of the growing season associated with increases in winter precipitation. Data from the Scandinavian transect also indicate that that growing season changes may be complex at the regional scale (Callaghan et al. 2004).

Fire Disturbance. Important changes in land cover that have occurred in high latitudes include changes associated with disturbance (e.g., fire, insect outbreaks, timber harvest, cropland establishment/abandonment, and industrial activities like oil and gas extraction) and large-scale changes in the distribution of vegetation (e.g., the advance of tree line in regions now occupied by tundra). During the 1970s and 1980s, the area burned annually in northwest Canada increased substantially (Kurz and Apps 1999; Stocks et al. 2000, 2003). There is also evidence that fire frequency in Northern Eurasia has increased in recent decades. Analyses of fire frequency data from Russia suggest a long-term average of annual area burned of about 10 million ha yr^{-1} including low-severity surface fires (Conard and Ivanova 1997; Wirth 2004). Satellite-based analyses also suggest increased area burned with an average exceeding 10 million ha yr^{-1} from 1998 through 2003, with a peak of 22 million ha in 2003 (Soja et al. 2004). After fire, soil temperatures typically warm and the active layer becomes deeper, but soils cool again as mosses grow back during succession. However, severe fires or more frequent fires can lead to the degradation of permafrost, which may result in substantial mortality of forests.

Insect Disturbance. On average insect infestations annually affect an area almost as large as does fire in the forests in Canada (Kurz and Apps 1999) and Alaska (Werner et al. 2006). Since approximately 1920, between 1–2 million ha yr^{-1} of forests in Canada have annually experienced insect-induced stand mortality (Kurz and Apps 1999), while the long-term annual fire area in Canada averages 2 million ha yr^{-1} (Stocks et al. 2003). Large outbreaks seem to occur at intervals of approximately thirty years. In Russia, the area affected by insect disturbance is about 2 million ha yr^{-1} (Shvidenko and Nilsson 2003), which is less than the area affected by fire. Similarly, northern Fennoscandia experiences outbreaks approximately every decade during which thousands of square kilometers of mountain birch forests are defoliated (Tenow 1996). There is concern that some of the recent large outbreaks that have been observed in Siberia (Siberian gypsy moth), Canada (Mountain pine beetle), and Alaska (Spruce bark beetle, Larch sawfly, Eastern spruce budworm; Werner et al. 2006) are associated with warm dry weather in the summer. For example, the Siberian gypsy moth is estimated to have affected 10 to 11 million ha of forests in the taiga of Siberia in 2001 and 2002, which is much higher than previously observed rates of infestation. Thus, insect outbreaks that are linked to climate change appear to be increasing in a number of areas spanned by the high latitude transects.

Forest Harvest. Annual forest harvest in Canada approximately doubled from ~0.5 million ha in 1970 to ~1 million ha in 1990 (Kurz and Apps 1999). More dramatically, timber harvest in Alaska increased over sixfold from 1952 to 1992 (McGuire et al. 2004). Recent trends of forest harvest rates in Canada and Alaska are substantially influenced by economics of the global forest sector, as much of the harvested wood is exported out of the region to markets in Asia and the conterminous United States. Concern over conservation issues and the collapse of Asian economies in the 1990s have had substantial impacts in decreasing forest harvest in Alaska and the U.S. Pacific northwest during the 1990s. In Russia, forest harvest between 1950 and 1990 was relatively steady at about 2 million ha yr^{-1}. The harvested areas were mostly concentrated in the European North (about two-thirds of the total) and in the most populated regions of Siberia. With the breakup of the Soviet Union, forest harvest during the last 15 years decreased substantially to around 1 million ha yr^{-1} for the years 2000–2002. This estimate of recent harvest rates should probably be increased by 15 to 20% because of illegal harvest in the Russian Far East.

Agricultural Land Use. In the prairie provinces of Canada, there was an estimated net deforestation of 12.5 million ha between 1860 and 1992 (based on Ramankutty and Foley 1999). Since 1950, Canadian forests have had a net gain of 3.0 million ha at the expense of agriculture. While this is a small proportion of the total forest base (<1%), it is important to recognize that most of the afforestation has occurred in eastern Canada and deforestation continues to occur in western Canada. In Russia, abandonment of arable lands between 1988 and 2001 was approximately 30 million ha. Abandonment is most pronounced in the zone of the boreal forest because the low productivity lands were unprofitable in the transition to a market economy. Recent land-use changes in Fennoscandia primarily include expansion of some of the larger towns in the region, but at the cost of the rural areas where depopulation is at an advanced stage as 32 000 Norwegian northern coastal farms have been abandoned since 1955 (Eilertsen 2002). Habitat fragmentation as a result of increased infrastructure and transport route development has proceeded rapidly since the 1940s and is expected to accelerate.

Treeline responses. The transects differ in the presence or absence of barriers to coniferous forest movement northward into areas occupied by tundra. The Brooks Range separates boreal forest in interior Alaska from arctic tundra on the north slope of Alaska, and there are ocean barriers to vegetation shifts in the SCANNET region. In contrast, geography is not a significant barrier to treeline migration along the Siberian and Canadian transects. The replacement of tundra with boreal forest occurred in earlier warm periods of the Holocene in northern Eurasia (MacDonald et al. 2000) and western Canada (Spear 1993). Over the last half century, treeline advances into tundra have been documented in Russia (Esper and Schweingruber 2004), Canada (Scott et al. 1987; Lavoie and Payette 1994), and Alaska (Suarez et al. 1999; Lloyd et al. 2003; Lloyd and Fastie 2003). However, fire and human activities in Russia have moved treeline to the south (Vlassova 2002). In Fennoscandia, treeline advanced at some sites during the first half of the 20th Century (Hustich 1958; Kullman 1986), but changed tree growth form at treeline at other sites (e.g., shifts from stunted krummholz trees to upright trees) without substantial changes in position of the tree species' limit (Kullman 1995). In mountainous areas of Scandinavia, several studies documented that elevational treeline has moved upslope during the last half of the 20th century in association with increases in temperature (Juntunen et al. 2002). However, in Scandinavia there are other issues besides climate affecting the position of treeline such as land use and browsing by reindeer and moose (Callaghan et al. 2004).

24.3 Responses of Radiatively Active Gases

24.3.1 General Issues

High latitude ecosystems contain approximately 30% of the world's vegetation carbon (McGuire et al. 1995) and about 40% of the world's soil carbon (Melillo et al. 1995). Much of the soil carbon in high latitude ecosystems is highly labile and has accumulated simply because of cold and/or anaerobic soils conditions. Thus, high latitude ecosystems could substantially affect atmospheric concentrations of CO_2 and CH_4. Likely changes in the fluxes of CO_2 and CH_4 could both enhance warming (positive feedbacks) and mitigate warming (negative feedbacks) (Smith and Shugart 1993; Chapin et al. 2000a; Clein et al. 2002; Zhuang et al. 2004). As summarized below, studies of carbon and methane dynamics along the high latitude transects have provided new insights on how CO_2 and CH_4 exchange respond to changes in climate and disturbance.

24.3.2 Responses of CO_2 Exchange to Climatic Change

Aerobic vs. Anaerobic Decomposition. Warming could cause release of carbon as CO_2 from aerobic boreal soils, i.e., soils that are not saturated with water, through enhanced decomposition (McGuire et al. 1995, Arneth et al. 2002). In anaerobic boreal soils, warming could affect carbon storage by altering soil drainage patterns. Although soil drainage may be especially vulnerable to the response of permafrost to climatic warming, the net effect on CO_2 exchange is not clear because drainage can either be enhanced or retarded by permafrost degradation. For example, the release of CO_2 from aerobic decomposition is likely to be enhanced if permafrost warming results in a drop of the water table (Oechel et al. 1995; Christensen et al. 1998) or thaws soil in areas of discontinuous permafrost (Goulden et al. 1998). In contrast, CO_2 emissions from soils are likely to be reduced if permafrost thaws in situations where drainage is impeded and decomposition is diminished because of aneorobic conditions (Christensen et al. 1998, 2004) and moss production is increased (Turetsky et al. 2000).

Experimental Warming of Aerobic Soils. In general, the warming of aerobic soils is expected to increase decomposition in high latitude ecosystems. This warming effect may be constrained by increases in leaf area index, which shade the soil surface. For this reason summer soil temperatures in arctic sites can be higher than those in the shade of subarctic birch forests (Callaghan et al. 2005). Although the warming of aerobic soils will tend to increase the release of CO_2 from high latitude ecosystems, the net effect of warming depends on the balance between production and decomposition. A recent meta-analysis of experimental warming studies indicates that an increase in productivity is approximately compensated by an increase in decomposition rates (Rustad et al. 2001).

Responses to Extension of the Growing Season. Climate warming is lengthening the growing season throughout much of the region occupied by high latitude ecosystems (Dye 2002; McDonald et al. 2004; Smith et al. 2004; Euskirchen et al. 2006). In temperate forests, annual carbon storage is enhanced by approximately 6 g C m^{-2} for every day that the growing season is lengthened (Baldocchi et al. 2001), and modeling analyses suggest that carbon storage in high latitude ecosystems may have a similar level of sensitivity to changes in growing season length (Euskirchen et al. 2006). The start of the growing season, as defined by photosynthetic activity of the canopy, is tightly coupled to the thawing of the soil in conifer stands of the BFTCS in Canada because frozen soil prevents transpiration (Frolking et al. 1999). Eddy covariance studies of Scots pine stands in the EST indicate that photosynthetic capacity at the beginning of the growing season is less than expected from relationships of assimilation vs. light and temperature during the peak of the growing season (Lloyd et al. 2002). One possible cause is a high respiration cost in needles for restructuring the photosynthetic apparatus (Shibistova et al. 2002) and for repairing damage of early season photoinhibition (Ensminger et al. 2004). Deciduous stands begin net carbon uptake following leafout, which is also sensitive to the timing of snowmelt. Studies in Siberia and Alaska indicate that deciduous forests become net CO_2 sinks much later in the season than coniferous stands, but that the delay is compensated for by higher assimilation rates and results in annual carbon balances that are similar between deciduous and coniferous stands (Röser et al. 2002; Lui et al. 2005).

Evidence from Inventory and Remote Sensing Studies. While site-specific studies do not clearly show whether experimental warming or lengthening of the growing season augments carbon storage, analyses of forest inventory data for Russia suggest that Russian forests are generally sinks for carbon (Schulze et al. 1999, 2002; Shvidenko and Nilsson 2002, 2003). Analyses based on satellite data suggest that both production and vegetation carbon storage have generally been enhanced in high latitude ecosystems during recent decades (Myneni et al. 1997, 2001; Randerson et al. 1999; Zhou et al. 2001; Nemani et al. 2003; Jia et al. 2003), although there are extensive areas in high latitudes that exhibit decreases in production (Goetz et al. 2005).

The Role of Soil Nitrogen Cycling. One hypothesis for the mechanism of increased production is that warming increases decomposition of soil organic matter to release nitrogen in forms that can be taken up by plants. Since production is often limited by plant nitrogen supply in boreal forests (Van Cleve and Zasada 1976; Van Cleve et al. 1981; Chapin et al. 1986; Vitousek and Howarth 1991; Schulze et al. 1995; Wirth et al. 2002a), an increase in nitrogen availability to plants should increase production. Several warming experiments and modeling studies have provided support for this mechanism (Van Cleve et al. 1990; Bonan and Van Cleve 1992; Bergh et al. 1999; Stromgrem and Linder 2002; Clein et al. 2002). One hypothesis is that the transfer of soil nitrogen released by decomposition to plants should result in greater carbon storage in plants because plants have a higher carbon to nitrogen ratio than soils (Shaver et al. 1992). Whether the capacity for increased plant growth can offset decomposition losses largely depends on the degree to which nitrogen released through enhanced decomposition is transferred to plants vs. immobilized in soil organic matter or lost from the terrestrial ecosystems in aquatic or gaseous pathways (McGuire et al. 1992; Stieglitz et al. 2000). If warming enhances production of high latitude ecosystems, soil carbon storage could increase if the transfer of carbon from vegetation to the soil is greater than the enhancement in decomposition from warming. If this condition occurs, then the long-term rate of soil carbon storage depends on whether the carbon that is transferred to the soil decomposes quickly or slowly (Hobbie et al. 2000; Clein et al. 2000). Our understanding of soil carbon and nitrogen transformations in response to warming is incomplete and is a key gap that limits our ability to make projections of the long-term response of soil carbon to warming in high latitude ecosystems (Clein et al. 2000).

Drought Stress. Warming-induced increases in production may not occur if other factors limit production. For example, forests on coarse-textured soils are among the most severely nitrogen-limited boreal ecosystems, but are also frequently subject to drought stress because of low water retention in the soil and superficial root systems (Kelliher et al. 1999). Warming has reduced growth in old white spruce trees growing on south-facing aspects in interior Alaska because of drought stress (Barber et al.

2000), and remote sensing analyses suggest that drought stress may be affecting a substantial portion of the North America boreal forest (Goetz et al. 2005). At treeline in Alaska, the growth of trees located in warm dry sites below the forest margin declined in response to recent warming, whereas the growth of trees located at treeline, particularly in wet regions, increased (Lloyd and Fastie 2002). Thus, there appears to be substantial spatial variability in the response of white spruce growth to recent warming in Alaska, and studies conducted elsewhere on other species throughout the boreal forest suggest that growth responses of warming depend on interactions between temperature and the timing and amount of precipitation (Briffa et al. 1998; D'Arrigo and Jacoby 1993; Jacoby and D'Arrigo 1995; Linderholm et al. 2003).

Winter Decomposition. Warming can also promote the loss of carbon as CO_2 from high latitude terrestrial ecosystems through higher rates of winter decomposition and through the increased decomposition of terrestrially derived carbon in aquatic ecosystems. A number of studies, which have primarily been conducted in the vicinity of the FEST and Alaska transects, have concluded that winter decomposition represents an important component of the annual budget of CO_2 exchange between high latitude ecosystems and the atmosphere (Coyne and Kelley 1974; Waelbroeck and Louis 1995; Hobbie and Chapin 1996; Zimov et al. 1993, 1996; Oechel et al. 1997; Fahnestock et al. 1999; Grogan and Chapin 1999; Shibistova et al. 2002; Michaelson and Ping, 2003). In general, winter decomposition is expected to increase with increases in soil temperature. One particularly interesting hypothesis involves the degree to which the heat of microbial activity might further enhance decomposition from high latitude soils (Zimov et al. 1993, 1996). The nitrogen released by winter decomposition may be less accessible to plants than nitrogen released in summer because of plant dormancy in winter, immobilization of nitrogen by soil microbes during winter, and loss of nitrogen from terrestrial ecosystems in spring runoff.

Terrestrial-Aquatic Linkages. High latitude streams and lakes can act as conduits for CO_2 via the decomposition of dissolved and particulate carbon derived from terrestrial ecosystems (Kling et al. 1991). After spring runoff, concentrations of dissolved and particulate organic carbon in high latitude aquatic ecosystems are highly correlated with precipitation as water is flushed through the organic layer (Prokushkin et al. 2005). There is also a significant increase in the carbon concentrations of streams after fire. Therefore, increases in precipitation or increases in the frequency of fire disturbance in high latitudes might enhance delivery of soil organic carbon to and subsequent decomposition in aquatic ecosystems. Arctic rivers also deliver a substantial amount of organic carbon to the Arctic Ocean (Romankevich and Vetrov 2001). A key uncertainty about increases in this flux is whether this will increase the release of CO_2 from immediate decomposition in coastal ecosystems or whether the carbon will be sequestered in marine sediments. About half of the carbon entering the Arctic Ocean from terrestrial ecosystems is from river inputs and about half from the erosion of coastal soils along the Arctic Ocean. While some of this carbon may become buried in ocean sediments, some of this material will likely be immediately decomposed in coastal Arctic ecosystems. Coastal erosion has increased in recent decades (Are 1999) associated with reduced summer cover of sea ice on the Arctic Ocean. It is expected that erosion of organic matter from soils along the coast of the Arctic Ocean will increase over the next century if sea ice continues to retreat and that this will enhance the CO_2 flux to the atmosphere from the Arctic.

24.3.3 Responses of CH_4 Exchange to Climatic Change

In general, CH_4 emissions of wetlands are expected to increase dramatically in response to warming (Zhuang et al. 2004). A study of CH_4 emissions from wetlands in Greenland, Iceland, Scandinavia and Siberia showed that annual mean emissions were strongly dependent on temperature (Fig. 24.4; Christensen et al. 2003), which indicates that high latitude CH_4 emissions could increase in response to climate warming. Emissions of CH_4 will also depend on changes in the water table. For example, while the release of CO_2 from aerobic decomposition is likely to be enhanced if permafrost warming results in a drop of the water table (Oechel et al. 1995; Christensen et al. 1998), emissions of CH_4 are likely to decrease because methanogenesis is an anaerobic process (Roulet 2000). In contrast, if the thawing of permafrost results in the expansion of lakes and wetlands, then releases of CH_4 are likely to be enhanced (Zimov et al. 1997; Christensen et al. 2004).

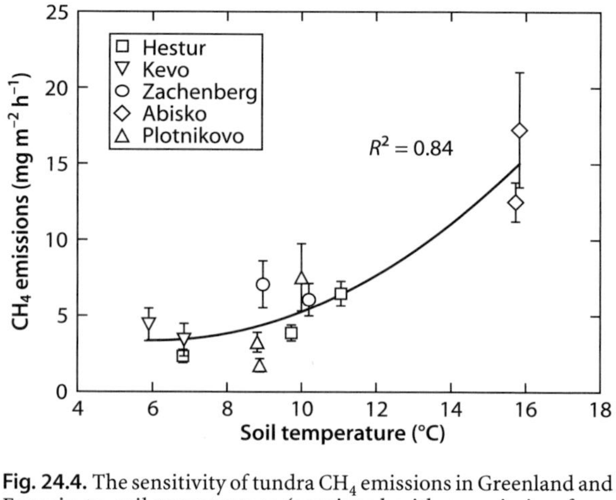

Fig. 24.4. The sensitivity of tundra CH_4 emissions in Greenland and Eurasia to soil temperature (reprinted with permission from Christensen et al. 2003)

The balance of changes in CO_2 exchanges vs. CH_4 emissions in terms of their radiative forcing is, however, complicated and difficult to assess. The current emissions of CH_4 from high latitude ecosystems, which are in the range 20–60 Tg CH_4 yr^{-1} (Zhuang et al. 2004), play a significant role in the global methane budget. In eastern Canadian peatlands, the enhanced CH_4 emissions associated with the creation of wetlands will likely result in a positive feedback to warming for up to 500 years until the enhanced storage of carbon in the wetlands (i.e., uptake of CO_2 from the atmosphere) offsets the enhanced radiative forcing associated with CH_4 emissions (Roulet 2000). In the southern end of the EST, recent studies of wetland trace gas exchange have shown that the radiative forcing of the CH_4 emissions are stronger than the substantial uptake of CO_2 (Friborg et al. 2003). For tundra regions of the high latitude transects, the balance of evidence suggests that tundra is currently a source of greenhouse warming due to substantial CH_4 emissions (Callaghan et al. 2005; Zhuang et al. 2004) that represent a radiative forcing effect that is much greater than small source/sink activity associated with the exchange CO_2 (Sitch et al. 2006). It is also likely that this source strength will increase in the future due to enhanced CH_4 emissions regardless of the strength of the carbon sink in tundra regions (Callaghan et al. 2005).

24.3.4 Responses to Changes in Disturbance and Land Cover

General Issues. A number of regional analyses suggest that carbon storage in high latitude forests is largely determined by how stand-age distribution changes at the regional scale. For stand-replacing disturbance, the temporal course of carbon storage in forests is largely controlled by wood increment and decomposition of coarse woody debris. Disturbance is generally characterized by a period of ecosystem carbon loss, during which production is less than decomposition, followed by a period of ecosystem carbon gain once production exceeds decomposition. If disturbances in high latitude regions become more frequent or more severe, carbon could be released from some ecosystem carbon pools (Kasischke et al. 1995; Wirth et al. 2002b), although the carbon stored in dead wood initially increases after disturbance (Shvidenko and Nilsson 2002).

Changes in Fire and Insect Disturbance Regimes. Together, both fire and insect disturbance have likely released substantial amounts of carbon into the atmosphere from forests in Canada (Kurz and Apps 1999; Chen et al. 2000; Amiro et al. 2001) and Russia (Shvidenko and Nilsson 2000a,b; Kajii et al. 2003). The degree to which increased fire frequency could release carbon from high latitude ecosystems depends, in part, on fire severity. Fires in most of the southern Eurasian boreal forest, which is dominated by Scots pine and Sibierian larch, tend to be surface fires in which most trees survive because of thick bark. In contrast, fires in the permafrost zone of far eastern Russia (Gmelin's larch), in Siberian dark taiga forests (spruce, fir, stone pine) and in boreal North America (mainly spruce), tend to be stand-replacing fires (Wirth et al. 1999; Shvindenko and Nilsson 2000a; McGuire et al. 2002; Wirth 2004; Csiszar et al. 2004). Analyses of the effects of climate change projections on fire weather suggest that climate change could increase fire frequency in Canada (Flannigan et al. 2001; Csiszar et al. 2004). In contrast, palaeoecological work in Canada revealed that colder and wetter periods were associated with higher fire frequencies (Lesieur et al. 2002). Nevertheless, of disturbance regime responses to climate change, fire is the disturbance agent that has the greatest potential to quickly release large amounts of carbon from high latitude regions.

The Influence of Disturbance Severity. Disturbance severity may substantially affect the temporal dynamics of carbon release and storage of high latitude ecosystems. For example, severe fires can lead to a complete destruction of organic soils. In these cases there is a large loss of up to 60 t C ha^{-1} CO_2 in fire emissions, but the ecosystem could come into positive carbon balance sooner if post-fire decomposition rates are low and vegetation recovery is high. The severity of fire disturbance also could affect the trajectory of vegetation succession after fire. Insect disturbance that causes partial or complete stand mortality leads to immediate post-fire carbon loss because of lowered production. In stands that have suffered substantial mortality, subsequent additional loss of carbon may occur if stands are then salvage logged. Also, stands affected by insect disturbance may be more vulnerable to fire because tree mortality generally increases the flammability of forest stands.

Changes in Forest Harvest Regimes. Forests of high latitude regions represent a wood resource of global significance. In general, forest harvest and management results in lower vegetation and soil carbon stocks than equivalent unmanaged forests. In the Russian Far East, carbon loss from illegal logging in the transboundary areas (Rosencranz and Scott 1992) results in the export of wood to China and other Asian countries, but this activity has not offset the drop in legal commercial logging associated with the breakup of the Soviet Union. Thus, it is expected that the change in the degree of forest harvest in the Russian boreal forest will result in net carbon storage over the next few decades unless harvest rates return to previous levels.

Changes in Treeline. Although rapid tree migration rates of up to 25 km yr^{-1} (Ritchie and MacDonald 1986) have been suggested for warming periods in the early Holocene, there are major uncertainties concerning the future rate of forest movement and the extent of range expansion that can take place. It is estimated that the potential increase in ecosystem carbon storage by replacing tundra with boreal forest is likely to proceed at a very slow pace because of inertia associated with the ability of boreal forests to migrate into regions of arctic tundra (Starfield and Chapin 1996; Chapin and Starfield 1997; Lloyd et al. 2003). The replacement of arctic tundra with boreal forest could increase ecosystem carbon storage substantially (Smith and Shugart 1993; Betts 2000).

24.4 Responses of Water and Energy Exchange

24.4.1 General Issues

Most of the energy that heats Earth's atmosphere is first absorbed by the land surface and then transferred to the atmosphere. The energy exchange properties of the land surface therefore have a strong direct influence on climate. High latitude ecosystems differ from more southerly biomes in having a long period of snow cover, when white surfaces might be expected to reflect incoming radiation (high albedo) and therefore absorb less energy for transfer to the atmosphere. Observed winter albedo in the boreal forest varies between 0.11 (conifer stands) and 0.21 (deciduous stands) (Betts and Ball 1997). This is much closer to summer albedo (0.08–0.15) than to the winter albedo of tundra (0.6–0.8), which weather models had previously assumed to be appropriate for boreal forests. The incorporation of true boreal albedo into climate models led to substantial improvements in medium-range weather forecasting and in climate re-analyses (Viterbo and Betts 1999). There is substantial spatial variability in winter albedo within high latitude ecosystems due to the spatial mosaic of conifer forests, deciduous forests, non-forested wetlands, burn scars, and tundra (Chapin et al. 2000a). The latter three have an albedo of approximately 0.6 when the short-statured vegetation is snow covered.

During summer, albedo of boreal vegetation is lower than in winter, with deciduous stands and boreal non-forested wetlands having approximately twice the albedo of conifer forests (Chapin et al. 2000a; Chambers and Chapin 2003). This difference in albedo leads to fluxes of sensible heat in conifer stands that are 2 to 3 times those of deciduous stands, whereas the latent energy fluxes (i.e., evapotranspiration) of deciduous forest stands in the boreal forest are 1.5 to 1.8 times greater than those of conifer forest stands (Schulze et al. 1999; Baldocchi et al. 2000; Chapin et al. 2000a). Because transpiration is tightly linked to photosynthesis, latent heat exchange tends to be dominated by transpiration in boreal forest stands with high productivity (e.g., deciduous forests). In contrast, evaporation plays a more important role than transpiration in the latent energy exchange of forest stands with low productivity (e.g., black spruce and pine forests), where surface evaporation from mosses or lichens can account for up to half of total evaporation (Baldocchi et al. 2000; Kelliher et al. 1999). The substantial sensible heating over conifer stands leads to thermal convection and may contribute to the frequency of thunderstorms and lightning (Dissing and Verbyla 2003), which plays an important role in the fire regime of the boreal forest as a source of ignition.

24.4.2 Responses to Changes in Climate, Disturbance, and Land Cover

Responses to Changing Growing Seasons. Responses of high-latitude ecosystems to global change could influence water and energy exchange with the atmosphere in several ways. Because there are substantial seasonal and spatial differences in sensible and latent energy exchange in high latitude ecosystems, climate warming could affect regional climate by altering both positive and negative feedbacks. One positive feedback associated with climate warming may result from lengthening of the growing season, which leads to earlier snowmelt and later snow cover. This effectively reduces annual albedo and should lead to substantial heating of the atmosphere (Chapin et al. 2005). Besides the extension of the snow-free period, the extension of ice-free periods on lake surfaces and the reduction in the area occupied by glaciers and continental ice sheets may also enhance atmospheric warming.

Responses to Changes in Vegetation. Positive feedbacks that involve changes in vegetation include more shrubs in tundra (Sturm et al. 2001; Silapswan et al. 2001), expansion of boreal forest into regions now occupied by tundra (Chapin et al. 2005), and replacement of summergreen conifers (larch) with evergreen conifers (pine; Kharuk et al. 2005). These changes would lead to substantial heating of the atmosphere, a response that could possibly accelerate the replacement of tundra by boreal forest (Table 24.1; Bonan et al. 1995; McFadden et al. 1998; Chapin et al. 2000a,b). Studies conducted with general circulation models indicate that the position of northern treeline has a substantial influence on global climate, with effects extending to the tropics (Bonan et al. 1992; Thomas and Rowntree 1992; Foley et al. 1994).

Table 24.1. Energy budget feedbacks to regional summer climate

Energy feedbacks	W m^{-2}
Feedback from vegetation change: tussock to shrub transition	3.9
Feedback from vegetation change: tundra to forest transition	5.0
2% change in solar constant	4.6
Doubling atmospheric CO_2	4.4

Responses to Changes in Disturbance Regimes. Disturbance and logging may also affect energy exchange with the atmosphere. For example, while fire disturbance often reduces albedo shortly after the fire, it also provides the opportunity for herbs, shrubs, and eventually deciduous broadleaf trees to develop, which will generally raise albedo. Thus, disturbance regimes that increase the proportion of non-forested lands and deciduous forests could reduce energy absorption and represent a negative feedback to atmospheric warming (Chapin et al. 2000a). In Siberia, self-replacing pine and larch in light taiga forests occur on dry upland soils and on permafrost, respectively (Furyaev et al. 2001). After these vegetation types are disturbed by fire or logging, a sometimes sparse post-fire vegetation and the lack of a deciduous pioneer phase result in a high and sustained production of sensible heat (Schulze et al. 1999). Severe fires in the Russian Far East can cause the collapse of permafrost and prevent the recovery of trees, effectively increasing albedo by converting conifer forests into ecosystems dominated by deciduous herbs and shrubs for hundreds of years. In general, post-fire deciduous broadleaf stands have a higher summer albedo (0.14) than do conifer stands (0.09) which they replace and therefore transfer less energy to the atmosphere (Chambers et al. 2003), as described above. Thus, the degree to which the response of vegetation dynamics to climate warming influences regional climate depends on the interaction of factors that may both enhance and mitigate warming.

24.5 Delivery of Freshwater to the Arctic Ocean

24.5.1 General Issues

The delivery of freshwater from high latitude ecosystems is of special importance because the Arctic Ocean, which contains only about 1% of the world's ocean water and receives about 11% of world river runoff (Forman et al. 2000; Shiklomanov et al. 2000), is the most river-influenced and land-locked of all oceans. Changes in freshwater inflow, which currently contribute as much as 10% to the upper 100 meters of the water column for the entire Arctic Ocean (Barry and Serreze 2000), could alter salinity and sea ice formation to affect the strength of the North Atlantic Deep Water Formation (Aagaard and Carmack 1989; Broecker 1997). Modeling studies suggest that maintenance of thermohaline circulation is sensitive to fresh-water inputs to the North Atlantic (Manabe and Stouffer 1995).

High latitude ecosystems play a significant role in the hydrology of the circumpolar north because they dominate the land-mass that contributes to the delivery of freshwater to the Arctic Ocean. Over the past 70 years there has been a 7% increase in the delivery of freshwater from the major Russian rivers to the Arctic Ocean (Peterson et al. 2002; Serreze et al. 2003). The analysis of McClelland et al. (2004) evaluated three mechanisms for these changes: (1) the construction of dams on the major Siberian rivers, (2) the thawing of permafrost, and (3) an acceleration of the fire regime. Of these possible explanations, an increase in fire frequency during the 20th century has the greatest potential to influence trends in runoff. However, the changes in the fire regime cannot fully explain the magnitude of increase in the delivery of freshwater to the Arctic Ocean, and therefore it appears that a poorly detected increase in precipitation may be the primary cause of the increased discharge. Nevertheless, a major challenge is to better understand and quantify the role of disturbance regime dynamics in the discharge dynamics of freshwater into the Arctic Ocean.

24.6 Summary and Conclusions

While it is clear that changes in high latitude regions have consequences for the climate system via a number of possible pathways (Fig. 24.5), we do not completely understand whether the net effect of changes will enhance or mitigate warming. Responses of water, energy, and trace gas exchange may result in either positive or nega-

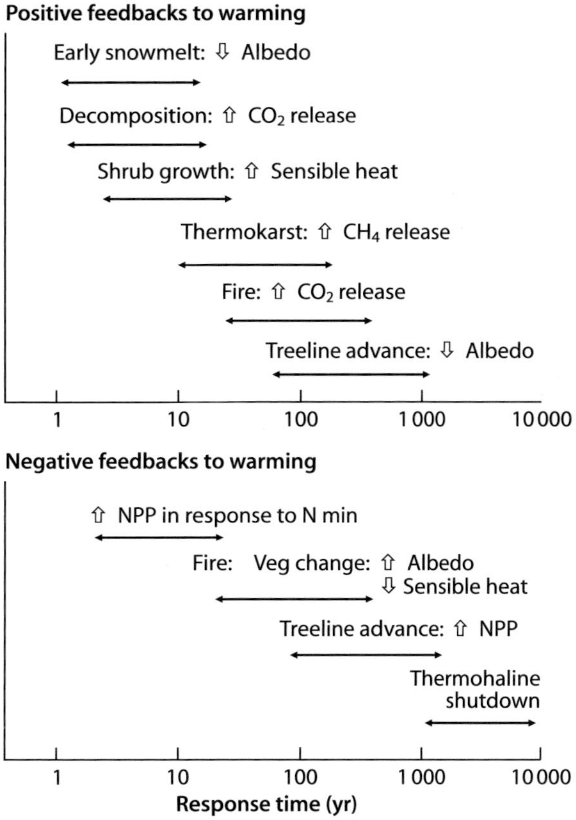

Fig. 24.5. The response times over which different positive and negative feedbacks to climate are most pronounced (reprinted with permission from Chapin et al. 2000a)

tive feedbacks to both regional and global warming. Of particular concern is whether the net response of high latitude ecosystems could lead to positive feedbacks that greatly enhance the rate of regional and global warming. While the responses of carbon storage in high latitude ecosystems have important implications for the rate of CO_2 accumulation in the atmosphere and international efforts to stabilize the atmospheric concentration of CO_2 (Smith and Shugart 1993; Betts 2000), it is important to understand how simultaneous changes in other trace gas exchanges and albedo of high latitude ecosystems also influence regional and global energy balance. For example, the reduction in radiative forcing associated with enhanced carbon storage from an expansion of the boreal forest may be exceeded by the warming effects of lower albedo (Betts 2000). Also, current responses of fire regimes to climate change suggest that fire is likely to increase in frequency and severity in the future, which has implications for both carbon storage and albedo of high latitude ecosystems. Increased delivery of freshwater from the high latitudes to the Arctic Ocean also has substantial implications for climate if it disrupts thermohaline circulation by weakening the formation of North Atlantic Deep Water, a response to warming that could ironically launch the Earth into another ice age (Manabe and Stouffer 1995). The exchange of water, energy, and trace gases among high latitude ecosystems, the atmosphere and the ocean are linked. Therefore, analyses of the response of high latitude ecosystems to global change will require an integrated understanding of how the response of these linkages will manifest themselves at a spectrum of spatial and temporal scales. The studies that have been conducted within and among the high latitude transects have laid the foundation for integration of ecological research with climate system research. Further development of this integrated understanding is relevant to identifying the implications for how the responses of high latitude ecosystems will influence the climate system. These insights are important for the development of mitigation and adaptation strategies in high latitude regions and in regions outside of the high latitudes.

References

Aagaard K, Carmack EC (1989) The role of sea ice and other fresh water in the arctic circulation. J Geophys Res 94: 14485–14498

Amiro BD, Todd JB, Wotton BM, Logan KA, Flannigan MD, Stocks BJ, Mason JA, Martell DL, Hirsch KB (2001) Direct carbon emissions from Canadian forest fires, 1959–1999. Can J For Res 31:512–525

Are FE (1999) The role of coastal retreat for sedimentation in the Laptev Sea. In: Kassens H, Bauch H, Dmitrenko I, Eicken H, Hubberten H-W, Melles M, Thiede J, Timokhov LA (eds) Land-ocean systems in the Siberian Arctic: Dynamics and history. Springer-Verlag, Berlin, pp 287–295

Arneth A, Kurbatova J, Kolle O, Shibistova OB, Lloyd J, Vygodskaya NN, Schulze ED (2002) Comparative ecosystem-atmosphere exchange of energy and mass in a European Russian and central Siberian bog. II. Interseasonal and interannual variability of CO_2 fluxes. Tellus 54B: 514–530

Baldocchi D, Kelliher FM, Black TA, Jarvis PG (2000) Climate and vegetation controls on boreal zone energy exchange. Global Change Biol 6(Suppl 1):69–83

Baldocchi D, Falge E, Gu L, Olson R, Hollinger D, Running S, Anthoni P, Bernhofer C, Davis K, Evans R, Fuentes J, Goldstein A, Katul G, Law B, Lee X, Malhi Y, Meyers T, Munger W, Oechel W, Pilegaard KUK, Schmid H, Valentini R, Verma S, Vesala T, Wilson K, Wofsy S (2001) FLUXNET: A new tool to study the temporal and spatial variability of ecosystem-scale carbon dioxide, water vapor, and energy flux densities. Bull Amer Meteor Soc 82:2415–2434

Barber VA, Juday GP, Finney BP (2000) Reduced growth of Alaska white spruce in the twentieth century from temperature-induced drought stress. Nature 405:668–673

Barry RG, Serreze MC (2000) Atmospheric components of the Arctic Ocean freshwater balance and their interannual variability. In: Lewis EL (ed) The freshwater budget of the Arctic Ocean. Kluwer Academic Publishers, Netherlands, pp 45–56

Bergh J, Linder S, Lundmark T (1999) The effect of water and nutrient availability on the productivity of Norway spruce in northern and southern Sweden. For Ecol Manage 119:51–62

Betts RA (2000) Offset of the potential carbon sink from boreal forestation by decreases in surface albedo. Nature 408:187–190

Betts AK, Ball JH (1997) Albedo over the boreal forest. J Geophys Res 102: 28901–28909

Bonan GB, Van Cleve K (1992) Soil temperature, nitrogen mineralization, and carbon source-sink relationships in boreal forest. Can J For Res 22:629–639

Bonan GB, Pollard D, Thompson SL (1992) Effects of boreal forest vegetation on global climate. Nature 359:716–718

Bonan GB, Chapin FS III, Thompson SL (1995) Boreal forest and tundra ecosystems as components of the climate system. Climatic Change 29:145–167

Briffa KR, Schweingruber FH, Jones PD, Osborn TJ, Shiyatov SG Vaganov EA (1998) Reduced sensitivity of recent northern tree-growth to temperature at northern high latitudes. Nature 391:678–682

Broecker W (1997) Thermohaline circulation, the achilles heel of our climate system: Will man-made CO_2 upset the current balance? Science 278:1582–1588

Callaghan TV, Johansson M, Heal OW, Sælthun NR, Barkved LJ, Bayfield N, Brandt O, Brooker R, Christiansen HH, Forchhammer M, Høye TT, Humlum O, Järvinen A, Jonasson C, Kohler J, Magnusson B, Meltofte H, Mortensen L, Neuvonen S, Pearce I, Rasch M, Turner L, Hasholt B, Huhta E, Leskinen E, Nielsen N Siikamäki P (2004) Environmental changes in the North Atlantic region: SCANNET as a collaborative approach for documenting, understanding and predicting changes. Ambio Special Report 13:39–50

Callaghan TV, Bjorn LO, Chapin FS III, Chernov Y, Christensen TR, Huntley B, Ims R, Johnasson M, Jolly D, Jonasson S, Matveyeva N, Oechel WC, Panikov N, Shaver GR (2005) Tundra and polar desert ecosystems. In: Corell R (ed) Arctic Climate Impact Assessment. Cambridge University Press, Cambridge, UK, pp 243–352

Chambers SD, Chapin FS III (2003) Fire effects on surface-atmosphere energy exchange in Alaskan black spruce ecosystems. J Geophys Res 108(D1):8145, 10.1029/2001JD000530

Chapin FS III, Starfield AM (1997) Time lags and novel ecosystems in response to transient climatic change in arctic Alaska. Climatic Change 35:449–461

Chapin FS III, Vitousek PM, Van Cleve K (1986) The nature of nutrient limitation in plant communities. American Naturalist 127:48–58

Chapin FS III, McGuire AD, Randerson J., Pielke Sr. R, Baldocchi D, Hobbie SE, Roulet N, Eugster W, Kasischke E, Rastetter EB, Zimov SA, Oechel WC, Running SW (2000a) Feedbacks from arctic and boreal ecosystems to climate. Global Change Biol 6(Suppl 1):211–223

Chapin FS III, Eugster W, McFadden JP, Lynch AH, Walker DA (2000b) Summer differences among arctic ecosystems in regional climate forcing. J Climate 13:2002–2010

Chapin FS III, Sturm M, Serreze MC, McFadden JP, Key JR, Lloyd AH, McGuire AD, Rupp TS, Lynch AH, Schimel JP, Beringer J, Epstein HE, Hinzman LD, Jia G, Ping C-L, Tape K, Chapman WL, Euskirchen SE, Thompson CD, Walker DA, Welker JM (2005) Role of Land-Surface Changes in Arctic Summer Warming. Science 310:657–660

Chen J, Chen W, Liu J, Cihlar J (2000) Annual carbon balance of Canada's forests during 1895–1996. Global Biogeochem Cycles 14:839–849

Christensen TR, Jonasson S, Michelsen A, Callaghan TV, Hastrom M (1998) Environmental controls on soil respiration in the Eurasian and Greenlandic Arctic. J Geophys Res 103: 29,015–29,021

Christensen TR, Joabsson A, Ström L, Panikov N, Mastepanov M, Öquist M, Svensson BH, Nykänen H, Martikainen P, Oskarsson H (2003) Factors controlling large scale variations in methane emissions from wetlands. Geophys Res Lett 30: 1414

Christensen TR, Johansson T, Åkerman HJ, Mastepanov M, Malmer N, Friborg T, Crill P, Svensson BH (2004) Thawing sub-arctic permafrost: Effects on vegetation and methane emissions Geophys Res Lett 31: L04501, doi:10.1029/2003GL018680

Clein J, Kwiatkowski B, McGuire AD, Hobbie JE, Rastetter EB, Melillo JM, Kicklighter DW (2000) Modeling carbon responses of tundra ecosystems to historical and projected climate: A comparison of a plot- and a global-scale ecosystem model to identify process-based uncertainties. Global Change Biol 6 (Suppl 1):127–140

Clein JS, McGuire AD, Zhuang X, Kicklighter DW, Melillo JM, Wofsy SC, Jarvis PG, Massheder JM (2002) Historical and projected carbon balance of mature black spruce ecosystems across North America: The role of carbon-nitrogen interactions. Plant and Soil 242:15–32

Conard SG, Ivanova GA (1997) Wildfire in Russian boreal forests – Potential impacts of fire regime characteristics on emissions and global carbon balance estimates. Environ Pollut 98:305–313

Coyne PI, Kelley JJ (1974) Variations in carbon dioxide across an arctic snowpack during spring. J Geophys Res 79:799–802

Csiszar I, Justice CO, McGuire AD, Cochrane MA, Roy DP, Brown F, Conard SG, Frost PGH, Giglio L, Elvidge C, Flannigan MD, Kasischke E, McRae DJ, Rupp TS, Stocks BJ, Verbyla DL (2004) Land use and fires. Chapter 19 In: Gutman G, Janetos AC, Justice CO, Moran EF, Mustard JF, Rindfuss RR, Skole D, Turner II BL, Cochrane MA Dordrecht (eds) Land Change Science: Observing, Monitoring, and Understanding Trajectories of Change on the Earth's Surface. Kluwer Academic Publishers, Netherlands, pp 329–350

D'Arrigo RD, Jacoby GC (1993) Secular trends in high northern latitude temperature reconstructions based on tree rings. Climatic Change 25:163–177

Dissing D, Verbyla DL (2003) Spatial patterns of lightning strikes in Interior Alaska and their relations to elevation and vegetation. Can J For Res 33:770–782

Dye DG (2002) Variability and trends in the annual snow-cover cycle in Northern Hemisphere land areas, 1972–2000. Hydrol Proc 16:3065–3077

Eilertsen SM (2002) Utilization of abandoned coastal meadows in northern Norway by reindeer. Ph.D. thesis, University of Tromsø. 25 pp

Ensminger I, Sveshnikov D, Campbell DA, Funk C, Jansson S, Lloyd J, Shibistova O, Oquist G (2004) Intermittent low temperatures constrain spring recovery of photosynthesis in boreal Scots pine forests. Global Change Biol 10:995–1008

Esper J, Schweingruber FH (2004) Large-scale treeline changes recorded in Siberia. Geophys Res Lett 3: L06202

Euskirchen SE, McGuire AD, Kicklighter DW, Zhuang Q, Clein JS, Dargaville RJ, Dye DG, Kimball JS, McDonald KC, Melillo JM, Romanovsky VE, Smith NV (2006) Importance of recent shifts in soil thermal dynamics on growing season length, productivity, and carbon sequestration in terrestrial high-latitude ecosystems. Global Change Biol 12:731–750

Fahnestock JT, Jones MH, Welker JM (1999) Wintertime CO_2 efflux from arctic soils: Implications for annual carbon budgets. Global Biogeochem Cycles 13:775–779

Flannigan M, Campbell I, Wotton M, Carcaillet C, Richard P, Bergeron Y (2001) Future fire in Canada's boreal forest: Paleoecology results and general circulation model – regional climate model simulations. Can J For Res 31:854–864

Foley JA, Kutzbach JE, Coe MT, Levis S (1994) Feedbacks between climate and boreal forests during the Holocene epoch. Nature 371:52–54

Forman SL, Maslowski W, Andrews JT, Lubinski D, Steele M, Zhang J, Lammers R, Peterson B (2000) Researchers explore arctic freshwater's role in ocean circulation. Eos Trans AGU 81(16):169–174

Friborg T, Soegaard H, Christensen TR, Lloyd CR, Panikov NS (2003) Siberian Wetlands: Where a sink is a source. Geophys Res Lett 30:2129

Frolking S, McDonald KC, Kimball J, Way JB, Zimmermann R, Running SW (1999) Using the space-borne NASA scatterometer (NSCAT) to determine the frozen and thawed seasons of a boreal landscape. J Geophys Res 104: 27,895–27,907

Furyaev VV, Vaganov EA, Tchebakova NM, Valendik EN (2001) Effects of fire and climate on succession and structural changes in the Siberian boreal forest. Eur J For Res 2:1–15

Goetz SJ, Bunn AG, Fiske GJ, Houghton RA (2005) Satellite-observed photosynthetic trends across boreal North America associated with climate and fire disturbance. Proc Nat Acad Sci 102:13521–13525

Goulden ML, Wofsy SC, Harden JW, Trumbore SE, Crill PM, Gower ST, Fries T, Daube BC, Fan S-M, Sutton DJ, Bazazz A, Munger JW (1998) Sensitivity of boreal forest carbon balance to soil thaw. Science 279:214–217

Grogan P, Chapin FS III (1999) Arctic soil respiration: effects of climate and vegetation depend on season. Ecosystems 2:451–459

Hinzman LD, Bettez ND, Bolton WR, Chapin FS III, Dyurgerov MB, Fastie CL, Griffith DB, Hollister RD, Hope A, Huntington HP, Jensen AM, Jia GJ, Jorgenson T, Kane DL, Klein DR, Kofinas G, Lynch AH, Lloyd AH, McGuire AD, Nelson FE, Nolan M, Oechel WC, Osterkamp TE, Racine CH, Romanovsky VE, Stone RS, Stow DA, Sturm M, Tweedie CE, Vourlitis GL, Walker MD, Walker DA, Webber PJ, Welker J, Winker KS, Yoshikawa K (2005) Evidence and implications of recent climate change in terrestrial regions of the Arctic. Climatic Change 72:251–298

Hobbie SE, Chapin FS III (1996) Winter regulation of tundra litter carbon and nitrogen dynamics. Biogeochemistry 35:327–338

Hobbie SE, Schimel JP, Trumbore SE, Randerson JR (2000) A mechanistic understanding of carbon storage and turnover in high-latitude soils. Global Change Biol 6 (Suppl 1):196–210

Hustich I (1958) On the recent expansion of the Scotch Pine in northern Europe. Fennia 82:3–23

IPCC (2001) Climate Change 2001: The Scientific Basis. Contribution of Working Group I to the Third Assessment Report of the Intergovernmental Panel on Climate Change. In: Houghton JT, Ding Y, Griggs DJ, Noguer M, van der Linden PJ, Dai X, Maskell K, Johnson CA (eds) Cambridge University Press, Cambridge, United Kingdom and New York, NY, USA, 881 pp

Jacoby GC, D'Arrigo RD (1995) Tree ring width and density evidence of climatic and potential forest change in Alaska. Global Biogeochem Cycles 9:227–234

Jia GJ, Epstein HE, Walker DA (2003) Greening of Arctic Alaska, 1981–2001. Geophys Res Lett 30: 2067, doi:10.1029/2003GL018268

Juntunen V, Neuvonen S, Norokorpi Y, Tasanen T (2002) Potential for timberline advance in northern Finland, as revealed by monitoring during 1983–1999. Arctic 55:348–361

Kajii Y, Kato S, Streets D, Tsai N, Shvidenko A, Nilsson S, McCallun J, Minko N, Abushenko N, Altynsev D, Khozder T (2003) Vegetation fire in Russia in 1998: Estimation of area and emissions of pollutants by AVHRR satellite data. J Geophys Res 108, doi:10.1029/2001JD001078

Kasischke E, Christensen NL Jr., Stocks BJ (1995) Fire, global warming, and the carbon balance of boreal forests. Ecol Appl 5:437–451

Kelliher FM, Lloyd J, Baldocchi DD, Rebmann C, Wirth C, Schulze E-D (1999) Evaporation in the boreal zone during summer – physics and vegetation. In: Schulze E-D, Heimann M, Harrison S, Holland E, Lloyd J, Prentice C, Schimel D (eds) Global biogeochemical cycles in the climate system. Academic Press, San Diego, pp 151–165

Kharuk VI, Dvinskaya ML, Ranson KJ, Im ST (2005) Expansion of evergreen conifers to the larch-dominated zone and climatic trends. Russian Journal of Ecology 36:186–193

Klein E, Berg EE, Dial R (2005) Wetland drying and succession across the Kenai Peninsula Lowlands, south-central Alaska. Can J For Res 35:1931–1942

Kling GW, Kipphut GW, Miller MC (1991) Arctic lakes and streams as gas conduits to the atmosphere: Implications for tundra carbon budgets. Science 251:298–301

Kullman L (1986) Late Holocene reproductional patterns of *Pinus sylvestris* and *Picea abies* at the forest limit in central Sweden. Can J Botany 64:1682–1690

Kullman L (1995) New and firm evidence for mid-Holocene appearance of *Picea abies* in the Scandes Mountains, Sweden. J Ecology 83:439–447

Kurz WA, Apps MJ (1999) A 70-year retrospective analysis of carbon fluxes in the Canadian forest sector. Ecol Appl 9:526–547

Lavoie C, Payette S (1994) Recent fluctuations of the lichen-spruce forest limit in subarctic Quebec. J Ecology 82:725–734

Lesieur D, Gauthier S, Bergeron Y (2002) Fire frequency and vegetation dynamics for the south-central boreal forest of Quebec, Canada. Can J For Res 32:1996–2009

Linderholm HW, Solberg BØ, Lindholm M (2003) Tree-ring records from central Fennoscandia: the relationship between growth and climate across an east-west transect. The Holocene 13:887–895

Liu H, Randerson JT, Lindfors J, Chapin FS III (2005) Changes in the surface energy budget after fire in boreal ecosystems of interior Alaska: An annual perspective. J Geophys Res 110: D13101, doi 10.1029/2004JD005158

Lloyd AH, Fastie CL (2002) Spatial and temporal variability in the growth and climate response of treeline trees in Alaska. Climatic Change 52:481–509

Lloyd AH, Fastie CL (2003) Recent changes in treeline forest distribution and structure in interior Alaska. Ecoscience 10:176–185

Lloyd J, Shibistova O, Zolotoukhine D, Kolle O, Arneth A, Wirth C, Styles JM, Tchebakova NM, Schulze ED (2002) Seasonal and annual variations in the photosynthetic productivity and carbon balance of a central Siberian pine forest. Tellus 54B: 590–610

Lloyd AH, Rupp TS, Fastie CL, Starfield AM (2003) Patterns and dynamics of treeline advance in the Seward Peninsula, Alaska. J Geophys Res 108 (D2):8161, 10.1029/2001JD000852

Luoto M, Heikkinen RK, Carter TR (2004) Loss of palsa mires in Europe and biological consequences. Environ Cons 31:1–8

MacDonald GM, Velichko AA, Kremenetsi CV, Borisova OK, Goleva AA, Andreev AA, Cwynar LC, Riding RT, Forman SL, Edwards TWD, Aravena R, Hammarlund D, Szeicz JM, Gattaulin VN (2000) Holocene treeline history and climate change across northern Eurasia. Quat Res 53:302–311

Manabe S, Stouffer RJ (1995) Simulation of abrupt climate change induced by freshwater input to the North Atlantic Ocean. Nature 378:165–167

McBean G, Alekseev G, Chen D, Forland E, Fyfe J, Groisman PY, King R, Melling H, Vose R, Whitefield PH (2005) Arctic climate – past and present. In: Corell R (ed) Arctic Climate Impact Assessment. Cambridge University Press, Cambridge, UK, pp 21–60

McClelland JW, Holmes RM, Peterson BJ, Stieglitz M (2004) Increasing river discharge in the Eurasian Arctic: Consideration of dams, permafrost thaw, and fires as potential agents of change. J Geophys Res. 109, D18102, doi:10.1029/2004JD004583

McDonald K, Kimball J, Njoku E, Zimmermann R, Zhao M (2004) Variability in springtime thaw in the terrestrial high latitudes: Monitoring a major control on the biospheric assimilation of atmospheric CO_2 with spaceborne microwave remote sensing. Earth Interactions 8: 20.1–20.23

McFadden JP, Chapin FS III, Hollinger DY (1998) Subgrid-scale variability in the surface energy balance of arctic tundra, J Geophys Res 103: 28,947–28,961

McGuire AD, Melillo JM, Joyce LA, Kicklighter DW, Grace AL, Moore III B, Vorosmarty CJ (1992) Interactions between carbon and nitrogen dynamics in estimating net primary productivity for potential vegetation in North America. Global Biogeochem Cycles 6:101–124

McGuire AD, Melillo JM, Kicklighter DW, Joyce LA (1995) Equilibrium responses of soil carbon to climate change: Empirical and process-based estimates. J Biogeo 22:785–796

McGuire AD, Wirth C, Apps M, Beringer J, Clein J, Epstein H, Kicklighter DW, Bhatti J, Chapin FS III, de Groot B, Efremov D, Eugster W, Fukuda M, Gower T, Hinzman L, Huntley B, Jia GJ, Kasischke E, Melillo J, Romanovsky V, Shvidenko A, Vaganov E, Walker D (2002) Environmental variation, vegetation distribution, carbon dynamics, and water/energy exchange in high latitudes. J Veg Sci 13:301–314

McGuire AD, Sturm M, Chapin FS III (2003) Arctic Transitions in the Land-Atmosphere System (ATLAS): Background, objectives, results, and future directions. J Geophys Res 108: 8166, doi:10.1029/2002JD002367

McGuire AD, Apps M, Chapin FS III, Dargaville R, Flannigan MD, Kasischke ES, Kicklighter D, Kimball J, Kurz W, McRae DJ, McDonald K, Melillo J, Myneni R, Stocks BJ, Verbyla DL, Zhuang Q (2004) Land cover disturbances and feedbacks to the climate system in Canada and Alaska. Chapter 9 In: In: Gutman G, Janetos AC, Justice CO, Moran EF, Mustard JF, Rindfuss RR, Skole D, Turner II BL, Cochrane MA Dordrecht (eds) Land Change Science: Observing, Monitoring, and Understanding Trajectories of Change on the Earth's Surface. Kluwer Academic Publishers, Netherlands, pp 139–161

Melillo JM, Kicklighter DW, McGuire AD, Peterjohn WT, Newkirk KM (1995) Global change and its effects on soil organic carbon stocks. In: Zepp RG, Sonntag C (eds) Role of nonliving organic matter in the Earth's carbon cycle. John Wiley and Sons, New York, pp 175–189

Michaelson GJ, Ping CL (2003) Soil organic carbon and CO_2 respiration at subzero temperature in soils of Arctic Alaska. J Geophys Res 108(D2):8164

Myneni RB, Keeling CD, Tucker CJ, Asrar G, Nemani RR (1997) Increased plant growth in the northern high latitudes from 1981 to 1991. Nature 386:698–702

Myneni RB, Dong J, Tucker CJ, Kaufmann RK, Kauppi PE, Liski J, Zhou L, Alexeyev V, Hughes MK (2001) A large carbon sink in the woody biomass of northern forests. Proc Natl Acad Sci USA 98: 14784–14789

Nemani RR, Keeling CD, Hashimoto H, Jolly WM, Piper SC, Tucker CJ, Myneni RB, Running SW (2003) Climate-driven increases in global terrestrial net primary production from 1982 to 1999. Science 300:1560–1563

Oechel WC, Vourlitis GL, Hastings SJ, Bochkarev SA (1995) Change in arctic CO_2 flux over two decades: Effects of climate change at Barrow, Alaska. Ecol Appl 5:846–855

Oechel WC, Vourlitis GL, Hastings SJ (1997) Cold season CO_2 emission from arctic soil. Global Biogeochem Cycles 11:163–172

Oechel WC, Vourlitis GL, Hastings SJ, Zuleta RC, Hinzman L, Kane D (2000) Acclimation of ecosystem CO_2 exchange in Alaskan Arctic response to decadal climate warming. Nature 406:978–981

Osterkamp TE, Romanovsky VE (1999) Evidence for warming and thawing of discontinuous permafrost in Alaska. Perm Periglal Proc 10:17–37

Peterson BJ, Holmes RM, McClelland JW, Vorosmarty CJ, Lammers RB, Shiklomanov AI, Shiklomanov IA, Rahmstorf S (2002) Increasing river discharge to the Arctic Ocean. Science 298:2171–2173

Prokuskin AS, Kajimoto T, Prokushkin SG, McDowell WH, Abaimov AP, Matsuura Y (2005) Climatic factors influencing fluxes of dissolved organic carbon from the forest floor in a continuous permafrost Siberian watershed. Can J For Res 35:2130–2140

Ramankutty N, Foley JA (1999) Estimating historical changes in global land cover: Croplands from 1700 to 1992. Global Biogeochem Cycles 13:997–1027

Randerson JT, Field CB, Fung IY, Tans PP (1999) Increases in early season net ecosystem uptake explain changes in the seasonal cycle of atmospheric CO_2 at high northern latitudes. Geophys Res Lett 26:2765–2768

Ritchie JC, MacDonald GM (1986) The patterns of post-glacial spread of white spruce. Journal of Biogeography 13:527–540

Romankevich EA, Vetrov AA (2001) Cycle of Carbon in the Russian Arctic Seas, Nauka, Moscow, 302 pp, in Russian

Romanovsky VE, Osterkamp TE (1997) Thawing of the active layer on the coastal plain of the Alaskan Arctic. Perm Periglac Proc 8:1–22

Rosencranz A, Scott A (1992) Siberia's threatened forests. Nature 355:93–294

Röser C, Nontagnani L, Kolle O, Meroni M, Mollicone D, Papale D, Marchesini LB, Federici S, Schulze E-D, Valentini R (2002) CO_2-exchange rates of three differently structured stands in central Siberia during one vegetation period. Tellus 54B: 642–654

Roulet NT (2000) Peatlands, carbon storage, greenhouse gases, and the Kyoto Protocol: Prospects and significance for Canada. Wetlands 20:605–615

Rustad LE, Campbell JL, Marion GM, Norby RJ, Mitchell MJ, Hartley AE, Cornelissen JHC, Gurevitch J (2001) A meta-analysis of the response of soil respiration, net nitrogen mineralization, and aboveground plant growth to experimental ecosystem warming. Oecologia 126:543–562

Schulze E-D, Schulze W, Kelliher FM, Vygodskaya NN, Ziegler W, Kobak KI, Koch H, Arneth A, Kusnetsova WA, Sogatchev A, Issajev A, Bauer G, Hollinger DY (1995) Aboveground biomass and nitrogen nutrition in a chronosequence of pristine Dahurian larix stands in Eastern Siberia. Can J For Res 25:943–960

Schulze E-D, Lloyd J, Kelliher FM, Wirth C, Rebmann C, Lühker B, Mund M, Knohl A, Milyukova I, Schulze W, Ziegler W, Varlagin A, Valentini R, Dore S, Grigoriev S, Kolle O, Vygodskaya NN (1999) Productivity of forests in the Eurosiberian boreal region and their potential to act as a carbon sink – A synthesis. Global Change Biol 5:703–722

Schulze E-D, Wirth C, Heimann M (2002): Carbon fluxes in the Eurosiberian Region. Environmental Control in Biology 40(3):249–258

Scott PA, Hansell RIC, Fayle DCF (1987) Establishment of white spruce populations and responses to climate change at the treeline, Churchill, Manitoba, Canada. Arct Alp Res 19:45–51

Serreze MC Walsh JE, Chapin III FS, Osterkamp T, Dyurgerov M, Romanovsky V, Oechel WC, Morison J, Zhang T, Barry RG (2000) Observational evidence of recent change in the northern high-latitude environment. Climatic Change 46:159–207

Serreze MC, Bromwich DH, Clark MP, Etringer AJ, Zhang T, Lammers R (2003) The large-scale hydro-climatology of the terrestrial arctic drainage. J Geophys Res 107: 8160, doi:10.1029/2001JD000919

Shaver GR, Billings WD, Chapin FS, Giblin AE, Nadlehoffer KJ, Oechel WC, Rastetter EB (1992) Global change and the carbon balance of arctic ecosystems. Bioscience 42:433–441

Sheingauz AS (2001) Forest Complex of Khabarovsk Kray. RIOTIP, Kabarovsk, 201 pp 103, in Russian

Shibistova O, Lloyd J, Evgravova S, Savushkina N, Zrazhevskaya G, Arneth A, Knohl A, Kolle O, Schulze E-D (2002) Seasonal and spatial variability in soil CO_2 efflux rates for a central Siberian *Pinus sylvestris* forest. Tellus 54B: 552–567

Shiklomanov IA, Shiklomanov AI, Lammers RB, Peterson BJ, Vorosmarty CJ (2000) The dynamics of river water inflow to the Arctic Ocean. In: Lewis EL (ed) The freshwater budget of the Arctic Ocean. Kluwer Academic Publishers, Netherlands, pp 281–296

Shvidenko A, Nilsson S (2000a) Extent, distribution, and ecological role of fire in Russian forests. In: Kasischke ES, Stocks BJ (eds) Fire, climate change, and carbon cycling in the boreal forest, Ecological Studies Series, Springer-Verlag, New York, pp 132–150

Shvidenko A, Nilsson S (2000b) Fire and carbon budget of Russian forests. In: Kasischke ES, Stocks BJ (eds) Fire, climate change, and carbon cycling in boreal forest, Ecological Studies Series, Springer-Verlag, New York, pp 289–311

Shvidenko A, Nilsson S (2002) Dynamics of Russian forests and the carbon budget in 1961–1998: An assessment based on long-term forest inventory data. Climatic Change 55:5–37

Shvidenko A, Nilsson S (2003) A synthesis of the impact of Russian forests on the global carbon budget for 1961–1998. Tellus 55B: 391–415

Silapaswan CS, Verbyla DL, McGuire AD (2001) Land cover change on the Seward Peninsula: The use of remote sensing to evaluate the potential influences of climate warming on historical vegetation dynamics. Can J Rem Sens 27:542–554

Sitch S, McGuire AD, Kimball J, Gedney N, Gamon J, Engstrom R, Wolf A, Zhuang Q, Clein JS (2006) Assessing the circumpolar carbon balance of arctic tundra with remote sensing and process-based modeling approaches. Ecological Applications, In Press

Smith TM, Shugart HH (1993) The transient response of terrestrial carbon storage to a perturbed climate. Nature 361:523–526

Smith NV, Saatchi SS, Randerson JT (2004) Trends in high northern latitude soil freeze and thaw cycles from 1988 to 2002. J Geophys Res 109: D12101, doi:10.1029/2003JD004472

Soja AJ, Sukhinin AI, Cahoon DR Jr., Shugart HH, Stackhouse PW Jr. (2004) AVHRR-derived fire frequency, distribution and area burned in Siberia. Internl J Rem Sens 25:1939–1960

Spear RW (1993) The palynological record of Late-Quaternary arctic tree-line in northwest Canada. Rev Paleobot Palynol 79:99–111

Starfield AM, Chapin FS III (1996) Model of transient changes in arctic and boreal vegetation in response to climate and land use change. Ecol Appl 6:842–864

Steffen W, Shvidenko A (1996) IGBP Northern Eurasia Study: Prospectus for an Integrated Global Change Research Project. IGBP Global Change Reports, No 37, 95 pp

Stieglitz M, Giblin A, Hobbie J, Kling G, Williams M (2000) Effects of climate change and climate variability on the carbon dynamics in arctic tundra. Global Biogeochem Cycles 14:1123–1136

Stieglitz M, Dery SJ, Romanovsky VE, Osterkamp TE (2003) The role of snow cover in the warming of arctic permafrost. Geophys Res Lett 30: 1721, doi:10.1029/2003GL017337

Stocks BJ, Fosberg MA, Wotten MB, Lynham TJ, Ryan KC (2000) Climate change and forest fire activity in North American Boreal Forests. In: Kasischke ES, Stocks BJ (eds) Fire, climate change, and carbon cycling in the boreal forest. Ecological Studies Series, Springer-Verlag, New York, pp 368–376

Stocks BJ, Mason JA, Todd JB, Bosch EM, Wotton BM, Amiro BD, Flannigan MD, Hirsch HG, Logan KA, Martell DL, Skinner WR (2003) Large forest fires in Canada 1959–1997. J Geophys Res 108: D1, 8149, doi:10.1029/2001JD000484

Stow D, Hope A, McGuire AD, Verbyla D, Gamon J, Huemmrich K, Houston S, Racine C, Sturm M, Tape K, Yoshikawa K, Hinzman L, Tweedie C, Noyle B, Silapaswan C, Douglas D, Griffith B, Jia G, Epstein H, Walker D, Daeschner S, Petersen A, Zhou L, Myneni R (2004) Remote sensing of vegetation and land-cover changes in Arctic tundra ecosystems. Rem Sens Environ 89:281–308

Stromgren M, Linder S (2002) Effects of nutrition and soil warming on stemwood production in a boreal Norway spruce stand. Global Change Biol 8:1195–1204

Sturm M, Racine C, Tape K (2001) Increasing shrub abundance in the Arctic. Nature 411:546–547

Suarez F, Binkley D, Kaye M, Stottlemyer R (1999) Expansion of forest stands into tundra in the Noatak National Preserve, Northwest Alaska. Ecoscience 6:465–470

Tenow O (1996) Hazards to a mountain birch forest – Abisko in perspective. Ecol Bull 45:104–114

Thomas G, Rowntree PR (1992) The boreal forests and climate. Quart J Royal Meteor Soc 118:469–497

Turetsky MR, Wieder RK, Williams CJ, Vitt DH (2000) Organic matter accumulation, peat chemistry, and permafrost melting in peatlands of boreal Alberta. Ecoscience 7:379–392

Vaganov EA, Hughes MK, Kirdyanov AV, Schweingruber FH, Silkin PP (1999) Influence of snowfall and melt timing on tree growth in subarctic Eurasia. Nature 400:149–151

Van Cleve K, Zasada J (1976) Response of 70-year-old white spruce to thinning and fertilization in interior Alaska. Can J For Res 6:145–152

Van Cleve K, Barney R, Schlentner R (1981) Evidence of temperature control of production and nutrient cycling in two interior Alaska black spruce ecosystems. Can J For Res 11:258–273

Van Cleve K, Oechel WC, Hom JL (1990) Response of black spruce (*Picea mariana*) ecosystems to soil temperature modification in interior Alaska. Can J For Res 20:1530–1535

Viterbo P, Betts AK (1999) Impact on ECMWF forecasts of changes to the albedo of the boreal forests in the presence of snow. J Geophys Res 104: 27,803–27,810

Vitousek PM, Howarth RW (1991) Nitrogen limitation on land and in the seas: How can it occur? Biogeochemistry 13:87–115

Vitt DH, Halsey LA, Zoltai SC (2000) The changing landscape of Canada's western boreal forest: the current dynamics of permafrost. Can J For Res 30:283–287

Vlassova TK (2002) Human impacts on the tundra-taiga zone dynamics: the case of the Russian lesotundra. In: Callaghan TV (ed) Dynamics of the tundra-taiga interface. Ambio Special Report 12, pp 30–36

Waelbroeck C, Louis JF (1995) Sensitivity analysis of a model of CO_2 exchange in tundra ecosystems by the adjoint method. J Geophys Res 100:2801–2816

Werner RA, Raffa KF, Illman BL (2006) Insect and pathogen dynamics in the Alaskan boreal forest. In: FS Chapin III, Oswood MW, K Van Cleve, LA Viereck, and DL Verbyla (eds) Alaska's changing boreal forest. Oxford University Press, New York

Wirth C (2004) Fire regime and tree diversity in boreal and high elevation forests: Implications for biogeochemical cycles. In: Scherer-Lorenzen M, Körner Ch, Schulze E-D (eds) The ecological significance of forest diversity. Ecological Studies, Springer-Verlag New York, Berlin, Heidelberg account the socio-cultural factors leading to disturbance, or feedback between ecosystem function and decision-making by individuals and institutions. However, decision-making, ecosystem processes, disturbance, and ecosystem services are inextricably intertwined, and a consideration of ecosystems as coupled human-environment systems is a critical next step in understanding the changing Earth System.

Global environmental changes affect the coupled human-environment system differently in different regions of the world. Biophysical alterations, such as increased atmospheric carbon dioxide concentrations or enhanced erosion of soils, and social forces, such as globalization of markets, generate different responses in Northern vs. Southern hemispheres, urban vs. rural environments, and developed vs. developing countries. These changes, in turn, affect local land-use decisions and the delivery and maintenance of ecosystem services. Links between decision-making, ecosystem services, and global environmental change define important pathways of coupled human-environment activities at the local and regional scale to and from global scales (Fig. 25.2).

The focus of the Global Land Project is largely land-centric which includes the people, biota, and natural resources (air, water, plants, animals, and soil). In addition, critical feedbacks and interactions between the land and the atmosphere, and between the land and oceans will be researched (GLP 2005). This research strategy is designed to bring together elements resulting from over 10 years of research of the individual IGBP and IHDP core projects, Global Change and Terrestrial Ecosystems (GCTE) and Land-Use and Land-Cover Change (LUCC), and assessment efforts such as the Millennium Ecosystem Assessment (Mooney 2003). From a programmatic sense, the GLP is the evolution of the GCTE and LUCC core projects into a new phase of more integrated research and coupled developments between natural and social sciences. This legacy of research and recent efforts in integrated environmental research have opened the opportunity to undertake a more integrated human-environment system set of studies in ways not possible in the past.

Given this focus on the human-environment system, the research activities need to be formulated to account for place-based differences and to consider the time and spatial scale of changes in human-environmental properties. The human-environment system approach provides a context to better understand the cultural landscape in which changes in the land environment are taking place and the manner in which regional differences in social structures at multiple scales affect biogeochemical cycles, biodiversity, and biophysical processes. The development of the research activities will require the joint efforts of scientists from various communities, including the social, economic, and environmental sciences.

25.2 Research Objectives

The Global Land Project has three objectives that provide the framework for the research:

- to identify the agents, structures and nature of change in coupled socio-environmental systems on land and quantify their effects on the coupled system;

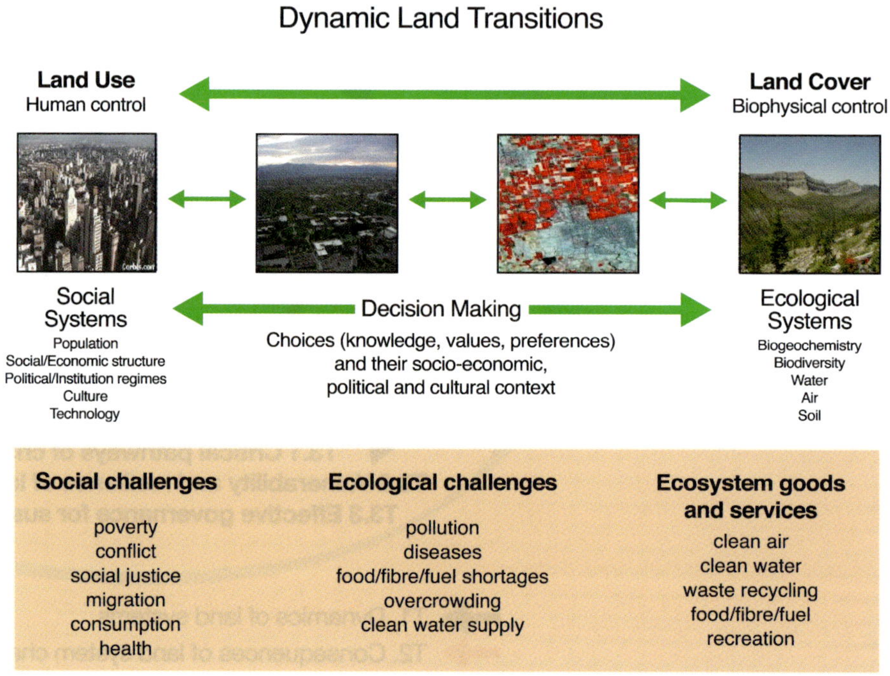

Fig. 25.2. The continuum of states resulting from the interactions between societal and natural (GLP 2005)

- to assess how the provision of ecosystem services is affected by the changes in the coupled socio-environmental system;
- to identify the character and dynamics of vulnerable and sustainable coupled socio-environmental LAND-SYSTEMS to interacting perturbations, including climate change.

Three thematic research areas define the Global Land Project research framework: (1) Dynamics of Land-Systems; (2) Consequences of Land-System Change; and (3) Integrating Analysis and Modeling for Land Sustainability. These research themes are depicted in Fig. 25.1, and additional research areas are indicated related to *(a)* factors affecting decision making, *(b)* the implementation of land-use management, *(c)* the effects on ecosystem and environmental dynamics, *(d)* the provisioning of ecosystem services, and *(e)* the evaluation of the coupled human-environmental land system's vulnerability to global environmental changes.

25.3 Emergent Concepts

The scope of the Global Land Project includes research on the effects of atmospheric, climatic, and land-use change on ecosystems, but in addition, several emergent concepts provide new directions for land research. These emergent concepts are associated with a greater focus on the social and environmental aspects of land system decision making, the valuation and provisioning of ecosystem services, and factors affecting vulnerability and sustainability of the coupled human-environment system. The following section provides a brief overview of these emergent concepts.

25.3.1 Land-Use Decision Making and Adaptive Management

Our increased understanding of the decision-making process related to land management provides a basic foundation for evaluating the interactions between human activities and their impacts on ecosystem structure and function. In this context, the decision making processes affecting land management are of particular interest in understanding how the range of settlement patterns, extraction of natural resources and production practices, land conversion methods, and conservation practices affect ecosystem dynamics.

The turn of the century has been marked by political and economic processes labeled globalization. The term refers to the worldwide reach and dominance of market institutions, engendered by the virtual instantaneous delivery of information, whose spread carries with it political and cultural implications (Stiglits 2002). These changes, associated with demographic and labor-flow trends, are occurring concurrently with climate change, CO_2 effects, and N deposition across different regions around the world. One outcome of globalization is the increasing separation of places of consumption from places of production, such that land systems often cannot be adequately understood without considering linkages to decisions made elsewhere (Blaikie and Brookfield 1987; Kasperson et al. 2001). For example, commodity crops throughout Africa are directly contracted from Europe (Bassett 2001) and large-scale deforestation in Borneo is driven by Japanese timber and pulp wood industries, although the translation of the demand to land systems is typically mediated by local policies and institutions (Angelsen and Kaimowitz 1999). In either case, the dynamics in question are embedded within and often directly tied to decisions made halfway around the world or processes operating globally.

Research is needed on the ways in which the socio-economic forces of globalization are tied to specific land-use practices and the role of institutions in mediating their outcome, including their impacts on ecosystem services. Analysis of economics, institutions, and the role of multiple stakeholders in land-use decision-making in regions undergoing socio-economic and environmental change is critical for various kinds of land change models. This research will provide insight into what types of coping strategies might be utilized at the local and regional scale. An area of emphasis will be socio-economic and political changes resulting in the relocation of significant numbers of people, both between regions, and within them. Understanding these interactions and how they affect ecosystem services (which are in many cases altered by changing land-use practices) will facilitate the development of coping and mitigation strategies to offset further perturbations.

Demographic factors, including growth, density, fertility, mortality, age, and sex composition of households, are known to be important factors influencing land-use and cover change. Research in the past decade has shown that while population growth is strongly related to land-cover dynamics, such as deforestation (Allen and Barnes 1985), this relationship is mediated by many other factors, such as land settlement policies and market forces (Geist and Lambin 2002). A key aspect of demographic dynamics is human migration, including shifts to and from rural and urban areas, and migratory flows linked to globalization processes. While urban areas continue to draw labor from rural agricultural areas, global labor markets induce the movement of large numbers of people across countries and continents (Massey 1998). Meanwhile, large-scale planned re-settlement continues in Amazonia and Indonesia, where land-use patterns have been found to depend on household composition, income, soil fertility, distance to markets, and the duration of tenancy (Pichon 1997; Perez and Walker 2002). Land tenure itself can affect fertility decision-making at the household level, with more secure tenure, *ceteris paribus*, resulting in lower fertility rates (Moran 1993). Establishing the generality of such findings is an important research challenge.

25.3.2 Ecosystem Services

Interactions between ecosystem structure and function, and with society, have consequences for the provisioning of ecosystem goods and services. There is much progress to be made in explicitly linking ecosystem processes as they have been traditionally evaluated and the services that ecosystems provide to society. Because of this focus on ecosystem services, it is essential that natural and social scientists develop reasonable sets of metrics of these services. Recent research has heightened awareness among scientists, policy makers and the public of the vital role of ecosystem services not only in the provisioning of key economic goods but also in the services that sustain, regulate and support life on Earth (Constanza et al. 1997; Daily 1997; Daily et al. 2000; Mooney 2003). Human well-being depends on a wide array of goods and services that terrestrial ecosystems provide. Whereas consumptive goods provided by land systems, such as grains, animal protein, and fiber and wood products, are typically valued through well developed markets, the contributions to human well-being and ecosystem functioning of the 'underpinning' services provided by ecosystems often remain 'invisible' and unvalued (or undervalued).

The array of such services is broad, from those services that regulate critical human-environment processes (e.g., climate, disease, flooding, detoxification) to services that support economic activity (e.g., soil formation, primary productivity, nutrient cycling, pest control, pollination). Indeed, our reliance on ecosystem services extends well beyond economic welfare, encompassing income, assets and capabilities, to health, security, food and nutrition, as well as, cultural identity, aesthetics and spirituality. Changes in land systems (in land-use practices and land cover) can disrupt the provisioning of services, degrade the quality of ecosystem services, and reduce the ability of ecosystems to maintain delivery of essential services upon which human well-being depends. One major feature of coupled human-environment systems is the notion that ecosystem services are not independent from each other, and that any change results in positive and negative outcomes depending on services considered, i.e., on the trade-offs.

Changes in the delivery and the maintenance of ecosystem services are also being affected by modifications of ecological systems through alterations of community composition or structure brought about by human activities. Inadvertent or sometimes purposeful actions by humans have led to creation of new ecosystems with unique combinations of organisms under modified environmental conditions. Examples of these emergent ecosystems can be found along a range of conditions ranging from urban environments to more natural settings. Examples include invasive species introduced in South Florida which have altered fire and hydrological regimes (Ewel et al. 1986) landscape alteration of hedge row communities in agricultural landscapes of Europe affecting predator-prey and pollinator communities, and salinization of Australian woodlands due to hydrological shifts resulting in changing forest dynamics (Hobbs et al. 2006). These "emergent ecosystems" may lead to modification of ecosystem structure and function in novel ways. However, management regimes can change abruptly as a result of environmental and/or human perturbations. For example, expansion of agricultural lands due to market or local demands for food production, sudden changes in prices of agricultural inputs (e.g., oil prices), policy shifts which alter environmental valuation of certain ecosystem properties, or political crises can result in rapid transitions in land management regimes. This set of examples point out the need to develop an analysis of the ways in which ecosystem processes are affected in terms of changes in biogeochemical cycles, biodiversity characteristics, and biophysical properties along urban-wildland continuums.

25.3.3 Vulnerability and Sustainability Science

The interaction of the coupled human-environment system with global change effects determines the vulnerability or the ability to take advantage of opportunities emerging from global change and access to markets and technologies. Vulnerability is the degree to which a system is likely to experience harm due to exposure to a hazard (Kasperson et al. 1995, 2001; Turner et al. 2003a). Resilience refers to the ability of a system to recover to a reference state following a disturbance and/or its capacity to maintain certain structures and functions despite disturbance (Carpenter et al. 2001; Gunderson 2000; Gunderson and Holling 2002; Harrison 1979). Emerging from risk-hazard studies and ecology, respectively, vulnerability and resilience have been incorporated in frameworks applicable to land systems (Downing et al. 2001; Kasperson and Kasperson 2001; McCarthy et al. 2001; Turner et al. 2003a,b; Watson et al. 1997). These frameworks evaluate how hazards-disturbances and exposure to them affect the sensitivity and resilience of the land system, including the consequences of adjustments and adaptations resulting from the hazards. Their applications have shown, for example, how the changing political economic structures in Mexico due to economic liberalization interact with drought to change the vulnerability of land systems on *ejidos* (communal lands) in different parts of the country (Turner et al. 2003b).

Vulnerability-resilience assessment is central to the Global Land Project for at least two reasons: (1) land systems are exposed and respond to hazards and disturbances, with the resulting mechanisms sustaining the systems or placing them at risk of change; and (2) the identification of those components of the land system most at risk and the mechanisms that enhance risk mitigation are central societal concerns. Information associated with these two reasons will be useful to decision makers. Most vulnerability-resilience work to date, however, has not addressed the land system per se. Risk-hazards approaches have focused on human vulnerability hazards (e.g., hurricanes) or on natural resources (i.e., the provisioning component of ecosystem services). Resilience research, in turn, has focused on ecosystem functioning. Recent work indicates that a more integrated approach is needed and the coupled human-environmental system perspective provides a framework to evaluate these linkages between ecosystem services and society (Adger et al. 2000; Luers et al. 2003).

Multiple perturbations and stresses operating at different spatio-temporal scales affect goods and services, challenging the science community to develop methods and metrics to achieve effective vulnerability analysis. Vulnerability involves more than the exposure of the system to a hazards, such as drought. It also includes the coupled human-environment system's sensitivity to the exposure and its resilience to the consequences. Resilience involves the copying capacity, adjustments, and adaptation of the system. Vulnerability analysis, therefore, must be holistic and integrative in design, and owing to the complexity and variation of coupled human-environment systems and their impacts on environmental goods and services, should be placed-based but explicitly linked to other places, drivers, and consequences to which it is connected.

25.4 Research Framework

The research approach of the Global Land Project is designed around the challenges of integrating the social and biophysical sciences. Three major themes have been developed to enable the research community to better engage in different facets of the research framework. The following section provides an introduction to this research which is being implemented through a network of research activities.

25.4.1 Theme 1: Dynamics of Land System

Research Questions

- How does globalization and population change affect regional and local land-use decisions and practices?
- How do changes in land management decisions and practices affect biogeochemistry, biodiversity, biophysical properties, and disturbance regimes of terrestrial and aquatic ecosystems?
- How do the atmospheric, biogeochemical and biophysical dimensions of global change affect ecosystem structure and function?

Understanding of global change is dependent on better understanding of the role of human activities in altering the structure and functioning of terrestrial and aquatic ecosystems, and the effect of changes in the Earth System on coupled human-environment systems operating at smaller scales. Increased understanding of the decision-making processes related to land-use management provides a foundation for evaluating the interactions between factors influencing human activities and feedbacks within the coupled human-environment system. Particularly dramatic are the impacts of land-use decision-making on the human use of land and the consequent changes in land cover and ecosystem dynamics.

This theme contributes to understanding the mechanisms by which human activities and global environmental changes affect terrestrial and aquatic ecosystems. The research undertaken for this theme will develop and synthesize knowledge on the proximate and underlying causes of land-use change, and hence ecosystems, with particular attention to the role of broader social, demographic and economic forces in shaping land-use decisions. The effects of land-use practices (e.g., agricultural, silvicultural, and pastoral systems) and global change effects on ecosystem services and the resulting feedbacks to the Earth System have not been well studied or quantified. Particular attention is given to the effects of human domination of landscapes, as these effects relate to urbanized areas, habitat management, increasing impact of invasive species, and other environmental characteristics. Likewise, this theme seeks a deeper understanding of the functioning of ecosystems within the context of global environmental change, assessing the effects of changes in atmospheric composition and physics on hydrological and biogeochemical cycles, biodiversity, and ecological disturbance regimes (Box 25.1). The greatest challenge, then, is the integration of this knowledge to forge an understanding of the combined and interactive effects of land-use and broader global environmental change on ecosystem structure and function.

Box 25.1. Fire and Pests

Fire and pest outbreaks have been part of the natural cycle of many forested ecosystems world wide. Fire has often a necessary component of a forested ecosystem to reset the successional cycle of these ecosystems. However, recently studies have indicated the frequency of pest outbreaks have increased due to increased growing season temperatures resulting in large areas of dead or damaged trees (Logan et al. 2003). These expanded areas of insect damaged forests have the potential to enhance the fire extent and intensity under these climate altered conditions.

Fig. 25.3.
a Pine forest in Eastern Oregon, two years after the catastrophic fires of 2002. Note the lack of regeneration in detrimentally burned soils. Provided by Kathy Hibbard. **b** Pine beetle damage (provided by Kathy Hibbard); with inset showing mountain pine beetle (provided by USDA Forest Service)

25.4.2 Theme 2: Consequences of Land-System Change

Research Questions

- What are the critical feedbacks from changes in ecosystems to the coupled Earth System?
- How do changes in ecosystem structure and functioning affect the delivery of ecosystem services?
- How can ecosystem services be linked to human well-being?
- How do people respond to changes in ecosystem service provision, considering the various scales and contexts of decision making?

Land ecosystems, both terrestrial and aquatic, provide a multiplicity of ecosystem services that are vital to human well-being. Agricultural systems, for example, both rely on and contribute to services, such as, soil formation and fertility renewal, freshwater, genetic crop resources, and pollination that result from agro-ecosystem processes. Changes in the availability of ecological services affect the viability, productivity and stability of the coupled human-environmental systems upon which humans rely for sustenance and economic livelihood.

Decisions about land management often result in trade-offs in the delivery of different ecosystem services. A number of factors may change through time which can affect the decision-making process and the evaluation of particular ecosystem services. A first objective of the research for this theme will be to identify these trade-offs, their causes, and the consequences they impose on land management in the context of global change. An effort will be undertaken to quantify how reductions in ecosystem services caused by changes in disturbance or management regimes may provoke shifts in land productivity by modifying physical inputs and economic returns.

Theme 2 addresses the consequences of changes in ecosystems brought about by land-use and global environmental changes studied under Theme 1. These consequences include feedbacks to the people within an ecosystem, and to the broader Earth System. Feedbacks to people are understood as changes in the delivery of a broad range of ecosystem services, such as agricultural productivity, clean air, potable water, and many others. Feedbacks to the broader Earth System occur through biogeochemical cycles, biodiversity, and natural disturbance regimes (Box 25.2). Finally, Theme 2 examines the combined effects of such ecosystem feedbacks, along with the effects of broader social, demographic and economic forces, in shaping local land-use decisions.

25.4.3 Theme 3: Integrating Analysis and Modeling for Land Sustainability

Research Questions:

- What are the critical pathways of change in LAND-SYSTEMS?
- How do the vulnerability and resilience of LAND-SYSTEMS to hazards and disturbances vary due to changes in human and environment interactions?
- Which institutions and policies enhance land sustainability?

Box 25.2. Urban Carbon Dioxide

Just as most ecosystems on Earth are now influenced by human activities, human dominated land systems such as cities remain influenced by ecological processes, including plant and soil biogeochemistry. The Biosphere-Atmosphere Stable Isotope Network (BASIN) organized within IGBP uses isotopic tracers to detect the influence of ecosystem processes on the atmosphere and how these processes are modified by global change. In this figure from Pataki et al. (2003), measurements of CO_2 isotopes in Salt Lake City, Utah, USA were used to separate nighttime CO_2 concentrations into a biogenic component originating from respiration of the urban forest and anthropogenic components originating from fossil fuel combustion. The results show that despite the large influence of fossil fuel emissions on the urban atmosphere, biological processes are easily detectable. These processes contribute to the urban carbon cycle and provide a variety of services for urban residents, including carbon sequestration, removal of atmospheric pollutants, and the cooling effects of transpiration and altered albedo.

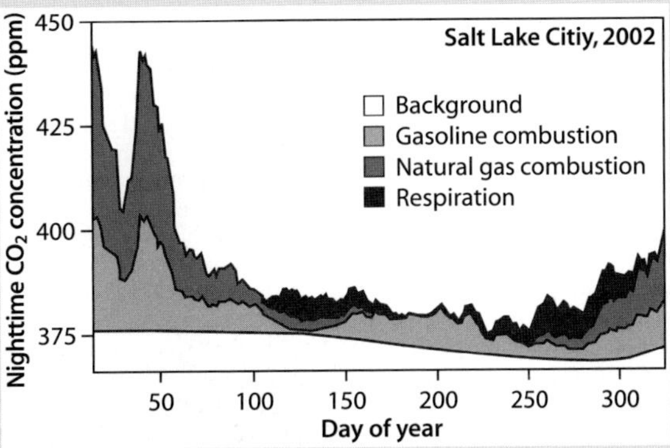

Fig. 25.4. Partitioning of 2002 night time carbon dioxide concentrations for Salt Lake City, Utah (Pataki et al. 2003)

Theme 3 seeks to integrate the dynamic interactions of human and environment subsystem characteristics (i.e., integrated land-change science) for assessment of vulnerability, resilience, and adaptation towards sustainable land systems and to provide this understanding in ways that are meaningful to decision making and policy (Box 25.3). To accomplish this goal, Theme 3 integrates findings from Themes 1 and 2 of the Global Land Project.

The sciences addressing complex systems have revealed the roles of emergent and path-dependent properties of coupled human-environment systems, and the thresholds in these systems that change their structure and function (e.g., Berkes and Folke 1998; Holling 1978; Levin 1998; Schellnhuber and Wenzel 1998). The sciences addressing vulnerability, resilience, and ecosystem services have demonstrated the nature of threats to land systems, especially in regard to the provisioning of food, fiber, and water (Daily et al. 2000; Dow and Downing 1995; Folke et al. 2002; Kasperson et al. 1995; Raskin et al. 1996; Rosenzweig 2003; Turner et al. 2003a,b). Studies of social learning and decision making have improved understanding of how coupled human-environmental systems, including land systems, are sustained or cope with forces of change (Cash et al. 2003; Kates et al. 2001; Lubchenco 1998; Mooney 2003; NRC 1999; Raven 2002). Recent advances in agent-based and other integrated modeling permit these complex factors to be treated systematically and holistically, providing land-based outcomes and near-term projections (e.g., Parker et al. 2001, 2003). The complexity of land systems, the variability in the forcing functions acting on them (e.g., Berkes and Folke 1998; Lambin et al. 2003; Lambin et al. 2001; Levin 1998) and the synergy of the human and environmental subsystems (Schellnhuber et al. 1997) enhance the need for place-based analysis (e.g., production unit, ecosystem, landscape) to address vulnerability, resilience, and sustainability (Cutter et al. 2000; Cutter 2001; NRC 1999; Wilbanks and Kates 1999). Yet, profound scalar dynamics (Parker et al. 2003; Schellnhuber and Wenzel 1998; Steffen et al. 2004) in land systems and the multiple needs of science and society regarding these systems also require that integrative analysis and assessment address multiple spatio-temporal resolutions to the problem.

25.5 Implementation Strategy

The GLP will engage in a variety of approaches to achieve its research goals. Working groups and networks of researchers will serve a fundamental role in the implementation of the science plan. The guiding principle associated with the GLP is that studies will *(a)* be place-based research studies, *(b)* require the establishment of interdisciplinary teams, *(c)* be cognizant of the need to be able to scale up and down and across disciplines, and *(d)* define the relationship of the research to the broader coupled human-environmental framework.

The research activities will include:

- case studies, manipulative studies and comparative studies
- networks of experimental and case studies across gradients of land systems
- long-term observations/experiments (remote sensing, sites, cross-site analysis)
- process models (e.g., vegetation/ecosystem, agroecosystem, agent-based models)
- land-use meta-analyses
- integrated analytical tools, not only models but also advances in field techniques
- decision-making models
- integrated regional studies
- interdisciplinary database development and archival systems

Box 25.3. Vulnerability

Determining and examining linkages between vulnerability and resilience with reference to LAND SYSTEMS and among the varying perspectives of the participating sciences remains an exciting challenge.

The vulnerability theme emerged from the social and application sciences dealing with risk-hazards. It invariably addresses the characteristics of individuals or groups in terms of their capacity to anticipate, cope with, resist, and recover from the impacts of a hazard (Blaikie et al. 1994), natural hazards for LAND SYSTEMS. Resilience, implying the opposite of vulnerability, is used in the wider ecological community (e.g., Folke et al. 2002) to understand how components of an ecosystem are configured to enable it to rebound after a perturbation of stress (for vulnerability, both are hazards). To date, natural systems are transparent in most vulnerability assessments, while ecosystem resilience focuses primarily on the biophysical processes in question.

The vulnerability and resilience of LAND SYSTEMS is determined by complex interactions among ecosystem and a suite of political, economic and social conditions and processes. Not only do shocks or perturbations (e.g., war, conflict, and climate change) and 'every day' stressors (e.g., economy, land use, and nutrient cycling) affect the environmental and human components of the LAND SYSTEM, but the consequences on either component interact in ways that change vulnerability-resilience. LAND SYSTEMS are coupled human-environment systems that require their vulnerability-resilience to be treated in an integrated manner. The vulnerability and resilience of land systems must be studied in an integrated manner: a research challenge for GLP.

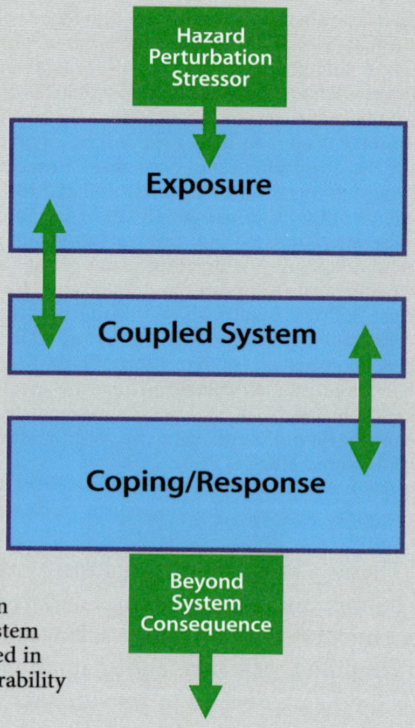

Fig. 25.5. The linkages between perturbation and system response encapsulated in the concept of vulnerability (GLP 2005)

Initial implementation Steps

Taking advantage of existing case study examples and initiatives, e.g.:

- On urban-rural gradient, linking GLP efforts to studies of the effects of conversion of rural to urban land on ecosystem functioning, and the subsequent effects and feedbacks related to ecosystem services.
- On arid lands to address dryland problems in a comprehensive an integrative way.
- On mountain ecosystems providing opportunities for comparative regional studies and for analyses of regional differentiation of environmental change processes in fragile ecosystems threatened by both systemic and cumulative human impacts.
- On managing the carbon cycle by contribution to the assessment of carbon sequestration potential; this is of the potential gain in carbon stocks in biomass and in soils within a given land area resulting from a change in land use, land cover or land management.
- On agricultural land, responsible for growing food, fiber and energy as primary provisional services for humankind, and the impact of global environmental change on agricultural production.
- On aquatic ecosystems, which closely interact with terrestrial ecosystems in providing habitat for diverse flora and fauna species, transport of nutrients and sediments and numerous other elements, and support diverse biogeochemical activity.

25.6 Summary and Conclusions

The Global Land Project (GLP) represents the joint, land-based research agenda of two major global change science programmes: *(i)* the International Geosphere-Biosphere Programme (IGBP), which originally focused mainly on biophysical processes in the Earth System through its Global Change and Terrestrial Ecosystems (GCTE) core project, and *(ii)* the International Human Dimensions Programme through its core project on Land-Use and Land-Cover Change (LUCC). The focus of the new project includes people, biota, and other natural resources (air, water, and soil). The strategy presented here critically emphasizes changes in the coupled human and environmental system, which is an extension of the ecosystem concept to explicitly include human actions and decision-making. The research planning builds upon the extensive heritage of global change research including the research discussed in the other chapters in this volume. The Global Land Project is designed to promote greater integration of social and biophysical sciences to meet the current challenges to coping and adapting to global change impacts the world is facing today and the near future. The sustainability of the coupled human-environment system and of ecosystem services is highly vulnerable to global change impacts as we move toward Earth System dynamics not yet faced by our societies.

References

Adger WN, Brooks N, Kelly M, Bentham S, Eriksen S (2004) New Indicators of Vulnerability and Adaptive Capacity. Tyndall Centre Technical Report 7. 126 pp

Allen MJ, Barnes MR (1985) The Causes of Deforestation in Developing Countries. Annals Assoc of Amer Geograp 75:163–184

Angelsen A, Kaimowitz D (1999) Rethinking the Causes of Deforestation: Lessons from Economic Models. The World Bank Research Observer. 14:73–98

Bassett, T (2001) The Peasant Cotton Revolution in West Africa. Côte d'Ivoire, 1880–1995. Cambridge University Press, Cambridge

Berkes F, Folke C (1998) Linking Social and Ecological Systems. Cambridge University Press, Cambridge

Blaikie P, Brookfield HC (1987) Land degradation and society. Methuen: London

Blaikie P, Cannon T, Davis I, Wisner B (1994) At Risk: Natural hazards, People's Vulnerability and Disasters. Routledge, London

Bohle H-G (2001) Conceptual Approaches to Vulnerability: Searching for Common Ground. In: Kasperson JX, Kasperson RE (eds), International Workshop on Vulnerability and Global Environmental Change, Stockholm Environment Institute, October, 4–17

Brookfield H, Potter L, Byron Y (1995) In Place of the Forest: Environmental and Socio-economic Transformations in Borneo and the Eastern Malay Peninsula. UNU Press, Tokyo. 310 pp

Carpenter S, Walker B, Anderies JM, Abel N (2001) From metaphor to measurement: resilience of what to what? Ecosystems 4:765–781

Cash DW, Clark WC, Alcock F, Dickson NM, Eckley N, Guston DH, Jager J, Mitchell RB (2003) Knowledge Systems for Sustainable Development. PNAS 100:8086–8091

Costanza R, d'Arge R, De Groot RS, Farber S, Grasso M, Hannon B (1997) The value of the world's ecosystem services and natural capital. Nature 387:253–260

Cutter S (2001) American Hazardscapes: The Regionalization of Hazards and Disasters. Washington, D.C., Joseph Henry Press

Cutter S, Mitchell JT, Scott MS (2000). Revealing the vulnerability of people and places: a case study of Georgetown County, South Carolina. Annals of the Association of American Geographers 90:713–737

Daily GC (ed) (1997) Nature's Services: Societal Dependence on Natural Ecosystems: Island Press, Washington, D.C

Daily GC, Soderquist T, Aniyar S, Arrow K, Dasgupta P, Ehrlich PR, Folke C, Jansson AM, Jansson BO, Kautsky N, Levin S, Lubchenco J, Maler KG, David S, Starrett D, Tilman D, Walker W (2000) The value of nature and the nature of value. Science 289:395–396

Dow K, Downing TE (1995) Vulnerability research: Where things stand. Human Dimen Quart 1:3–5

Downing TE, Butterfield R, Cohen S, Huq S, Moss R, Rahman A. Sokona and Stephen L (2001) Climate Change Vulnerability: Linking Impacts and Adaptation. Environmental Change Institute, University of Oxford, Oxford

Ewel JJ (1986) Invasibility: Lessons from south Florida. In: Mooney, Harold A.; Drake, James A (eds). Ecology of biological invasions of North America and Hawaii. Ecol Studies 58:214–230, Springer-Verlag, New York

Folke C, Carpenter S, Elmqvist T, Gunderson L, Holling CS, Walker B, Bengtsson J, Berkes F, Colding J, Danell K, Falkenmark M, Gordon L, Kasperson R, Kautsky N, Kinzig A, Levin S, Maler K, Moberg F, Ohlsson L, Olsson P, Ostrom E, Reid W, Rockstrom J, Svenije H, Svendin U (2002) Resilience and sustainable development: building adaptive capacity in a world of transformations. ICSU Series for Sustainable Development, No. 3 www.resalliance.org/reports/resilience_and_sustainable_development.pdf

Geist HJ, Lambin EF (2002) Proximate Causes and Underlying Driving Forces of Tropical Deforestation. BioScience. 52:143–150

GLP (2005) Science Plan and Implementation Strategy. IGBP Report No. 53/IHDP Report No. 19. IGBP Secretariat, Stockholm. 64 pp

Gunderson LH (2000) Ecological resilience – In theory and application. Annual Review of Ecol and System 31(2000):425–439

Gunderson L, Holling CS (2002) Panarchy: Understanding Transformations in Human and Natural Systems. Island Press, Washington, D.C

Harrison G (1979) Stability under environmental stress; Resistance, resilience, persistence and variability. American Naturalist 13:659–669

Hobbs RJ, Arico S, Aronson J, Baron JS, Bridgewater P, Cramer VA, Epstein R, Ewel JJ, Klink CA, Lugo A, Norton D, Ojima DS, Richardson DM, Sanderson EW, Valladares F, Vilà M, Zamora R, Zobel M (2006) Novel ecosystems: theoretical and management aspects of the new ecological world order. Global Ecol Biogeogr 15:1–7

Holling CS (1978) Adaptive Environmental Assessment and Management. Wiley, London

Kasperson JX, Kasperson RE, Turner BL II (1995) Regions at Risk: Comparisons of Threatened Environments. United Nations University Press, Tokyo. 582 pp

Kasperson JX, Kasperson RE, Turner BL, III (2001) International Workshop on Vulnerability and Global Environmental Change 17–19 May 2001. Stockholm Environment Institute (SEI). Stockholm, Sweden: A workshop summary. SEI Risk and Vulnerability Programme Report 2001-01, SEI, Stockholm

Kates RW, Clark WC, Corell R, Hall JM, Jaeger CC, ILowe I, McCarthy JJ, Schellenhuber HJ, Bolin B, Dickson NM, Faucheaux S, Gallopin GC, Grübler A, Huntley B, Jäger J, Jodha NS, Kasperson RE, Mabogunje A, Matson P, Mooney H, Moore B, III, O'Riordan T, Svedin U (2001) Sustainability science. Science 292:641–642

Lambin EF, Turner BL II, Geist HJ, Agbola SB, Angelsen A, Bruce JW, Coomes OT, Dirzo R, Fischer G, Folke C, George PS, Homewood K, Imbernon J, Leemans R, Li X, Moran EF, Mortimore M, Ramakrishnan PS, Richards JF, Skanes H, Steffen W, Stone GD, Svedin U, Veldkamp TA, Vogel C, Xu J (2001) The causes of land-use and land-cover change: moving beyond the myths. Global Environ Change 11:261–269

Lambin EF, Geist HJ, Lepers E (2003) Dynamics of Land-use and Land-Cover Change in Tropical Regions. Annual Review Environ Resour 28:205–241

Levin SA (1998) Ecosystems and the biosphere as complex adaptive systems. Ecosystems 1:431–436

Logan JA, Regniere J, Powell J (2003) Assessing the impacts of global warming on forest pest dynamics. Frontiers Ecol Environ 1:130–137

Lubchenco L (1998) Entering the Century of the Environment: A New Social Contract for Science. Science 279:491–497

Luers AL, Lobell DB, Sklar LS, Addams CL, Matson PA (2003) A method for quantifying vulnerability applied to the agricultural system of the Yaqui Valley. Global Environmental Change (Part A) 13:255–267

Massey D (1998) Worlds in Motion: Understanding International Migrations at the end of the millennium. University Press, Oxford

McCarthy JJ, Canziani OF, Leary NA, Dokken DJ, White KS (Eds) (2001) Climate Change 2001: Impacts, Adaptation and Vulnerability. Intergovernmental Panel on Climate Change: Working Group II. Cambridge University Press, Cambridge, UK, 1042 pp

Melillo JM, Field CB, Moldan B (2003) Interactions of the major biogeochemical cycles. Island Press, Washington, D.C

Mooney, H (ed) (2003) Ecosystems and Human Well-Being: A Framework for Assessment: Island Press, Washington, D.C

Moran, EF (1993) Deforestation and Land Use in the Brazilian Amazon. Human Ecology 21:1–22

NRC (1999) Our Common Journey: A Transition Toward Sustainability. National Academy Press, Washington, D.C

Parker D, Manson S, Janssen MA, Hoffmann MF, Deadman P (2001) Multi-agent systems for the simulation of land-use and land-cover change: a review. Center for the Study of Institutions, Population and Environmental Change, Indiana University, Bloomington, IN

Parker DC, Manson SM, Janssen MA, Hoffmann MF, Deadman P (2003) Multi-agent systems for the simulation of land use and land cover change: a review. Annals Assoc Amer Geograp 93: 316–340

Pataki DE, Bowling DR, Ehleringer JR (2003) Seasonal cycle of carbon dioxide and its isotopic composition in an urban atmosphere: anthropogenic and biogenic effects. J Geophys Res –Atmospheres 108: 4735
Perez S, Walker RT (2002) Household life cycles and secondary forest cover among small farm colonists in the Amazon. World Development 30:1009–1027
Pichon F (1997) Colonist land-allocation decisions, land use, and deforestation in the Ecuadorian Amazon frontier. Economic Develop Cultural Change 45:707–744
Raskin P, Chadwick M, Jackson T, Leach G (1996) The Sustainability Transition: Beyond Conventional Development. Stockholm Environment Institute, Stockholm
Raven PH (2002) Science, Sustainability, and the Human Prospect. Science 297:954–958
Rosenzweig ML (2003) Win-Win Ecology: How the Earth's Species Can Survive in the Midst of Human Enterprise. Oxford University Press, New York, NY. 211 pp
Schellnhuber H-J, Wenzel V (eds) (1998) Earth System Analysis: Integrating Science for Sustainability. Springer-Verlag, Berlin. 530 pp
Schellnhuber HJ, Black A, Cassel-Gintz M, Kropp J, Lammel G, Lass W, Lienenkamp R, Loose C, Ludeke MKB, Moldenhauer O, Petschel-Held G, Plochl M, Reusswig F (1997) Syndromes of global change. GAIA 6:19–34
Steffen W, Sanderson A, Tyson PD, Jäger J, Matson PA, Moore B III, Oldfield F, Richardson K, Schnellhuber H-J, Turner BL II, Wasson RJ (2004) Global Change and the Earth System: A Planet Under Pressure. Springer-Verlag, Berlin, pp 336
Stiglits J (2002) Globalization and its Discontent. New York, NY WW Norton Publisher
Turner BL II, Matson PA, McCarthy JJ, Corell RW, Christensen L, Eckley N, Hovelsrud-Broda GK, Kasperson JX, Kasperson RE, Luers A, Martello ML, Mathiesen S, Naylor RL, Polsky C, Pulsipher A, Schiller A, Selin H, Tyler N (2003a) Illustrating the coupled human-environment system for vulnerability analysis: three case studies. PNAS 100:8080–8085
Turner BL II, Kasperson RE, Matson PA, McCarthy JJ, Corell RW, Christensen L, Eckley N, Kasperson JX, Luers A, Martello ML, Polsky C, Pulsipher A, Schiller A (2003b) A framework for vulnerability analysis in sustainability science. PNAS 100:8074–8079
Vitousek PM, Mooney HA, Lubchenco J, Melillo JM (1997) Human domination of Earth's ecosystems. Science 277:494–499
Watson RT, Zinyowera MC, Moss RH (eds) (1997) Climate Change 1995: Impacts, Adaptations and Mitigation of Climate Change: Scientific-Technical Analysis. International Panel on Climate Change. Cambridge University Press, Cambridge, pp 878
Wilbanks TJ, Kates RW (1999) Global Change in Local Places. Climatic Change 43:601–628

Index

A

Abies veitchii 289
aboveground
–, respiration 38
–, net primary productivity 31–32
Abutilon theophrasti 62
Acacia nilotica 96
Acer
–, *rubrum* 26–28
–, *saccharum* 26–28
adaptation 130, 150, 152, 220
adaptive
–, capacity 211
–, management 315
–, option 220
Adelges tsugae 97
Aegilops triuncialis 88
aerosol 187, 266
afforestation 66, 237, 239
–, and evapotranspiration 238
–, carbon storage potential of 237
–, potential atmospheric feedbacks of 239
–, program 255
afforested catchment 239
Africa 233, 248
African Sahara 168
AFT *see animal functional type*
Agreement on Transboundary Haze Pollution 269
agricultural
–, development 262
–, disturbance 154
–, intensification 67
–, land 228, 238
–, conversion 46
–, use 300
–, shift 67
–, soil 71, 227–228
–, system 273
agriculture 2, 186, 240
agroecosystem 199, 273, 278
air
–, isotopic composition 37
–, pollution 16, 64
–, temperature 48
airborne pollutant 16
Alaska 297–298, 302
–, transect 299
albedo 167, 172, 304
–, of boreal vegetation 304
Alberta 144
alien
–, invasive species 134
–, plant species, density of 98
–, species 93, 96, 97, 99
allochthonous
–, carbon 292
–, organic matter 293
alpine
–, grassland 11
–, vegetation 217
Amazon River Basin 46
Amazonia 17
Ambrosia artemisiifolia 217
ambrosia beetle 218
ammonia-oxidizing bacterium 30
Ampelodesmos mauritanica 85– 87
anaerobic condition 203
ANCOVA model 29
Andean region 233
Andropogon gerardii 104
animal functional type 187
annual net primary productivity 94, 156
anoxic
–, microsite 53
–, sediment 293
ANPP *see aboveground net primary productivity* or *annual net primary productivity*
anthropogenic change 93
aphid 214–215, 218
–, monitoring network 218
APMI *see Association of Southeast Asian Nations Peatland Management Initiative*
aquatic system 285
Arabidopsis thaliana 105
Arctic Ocean 297, 305
arctic tundra 85
ARIDnet research network 253
Arizona 196
Arrhenius relationship 179
arthropod 211, 218
ASEAN *see Association of Southeast Asian Nations*
Asia 195, 201, 248
Asian long-horned beetle 49
Association of Southeast Asian Nations 269
–, Peatland Management Initiative 269
asymmetric competition 286
ATHP *see Agreement on Transboundary Haze Pollution*
atmosphere-land flux 71
atmospheric
–, carbon dioxide 216
–, concentration 85, 93, 181, 196
–, holocene change in 182
–, chemical 49
–, composition 47, 81–82
–, change 60
–, dust loading 167
–, General Circulation Model 178
–, humidity 14
–, nitrogen 96
–, pollutant 49
Aulacorthum solani 214
Australia 83, 85
Australian Capital Territory 144
autochthonous carbon 292
Avena 97

B

Bactrocera tryoni 219
Baltimore 46, 55
–, County 50–51
–, Ecosystem Study 46, 48–49, 51, 54–55
–, Long Term Ecosystem Research Study 51
Bangladesh 202
Barcelona 86–87
barley yellow dwarf virus 218
Basel 12
BASIN *see Biosphere-Atmosphere Stable Isotope Network*
BATS *see Biosphere-Atmosphere Transfer Scheme*
BEF *see biodiversity and ecosystem functioning*
belowground
–, carbon
–, flux 26
–, allocation 31
–, net primary productivity 31–32
–, respiration 38
benchmark 212
Berberis thunbergii 50
BETHY terrestrial ecosystem model 177
BFTCS *see Boreal Forest Transect Case Study*
BioCON project 107–108
biodiversity 2–3, 16–17, 50, 107, 113–115, 219
–, and biomass 114
–, and ecosystem function 114–115
–, change in 114
–, crisis 97
–, effect 17, 87
–, loss 82, 107, 113–114
–, ecosystem consequence of 113
–, luxury effect of 50
Biodiversity and Ecosystem Functioning 116
–, relationship 115–116
–, research framework 115–116
biogeochemical
–, cycle 45
–, cycling, disruption of 249
–, effect 54
–, interaction 31
–, signature of carbon cycling 291
biogeochemistry 99, 156, 175–176
biogeography 116
biological
–, activity 54
–, diversity 113
–, invasion 98
biomass 10, 28
–, accumulation 10
–, burning 68
–, fuel 266
–, gain 10
–, loss 114
–, production 84, 88
–, response 10–11, 108
biome 176–177
BIOME model 63, 159, 176–177
BioMERGE *see Biotic Mechanisms of Ecosystem Regulation in the Global Environment Project*
biophysical
–, and socio-economic attribute 252
–, model 177
biophysics 175, 177
biosphere 10
Biosphere Dynamics Project 175
Biosphere-Atmosphere Stable Isotope Network 2, 42, 319
Biosphere-Atmosphere Transfer Scheme 177
–, model 177
biotic
–, exchange 81
–, interaction 28, 82
–, invasion 93
Biotic Mechanisms of Ecosystem Regulation in the Global Environment project 116–117, 119, 121
–, conceptual framework 117
–, framework 113, 116–117, 122
black
–, blizzard 166
–, carbon 187
blowup 170
BNPP *see belowground net primary productivity*
boreal 182
–, forest 33, 68
–, leaf area index 183
Boreal Forest Transect Case Study 297–298, 301
Borneo 285
bottom-up 212
Boulder 54
Brassica 105
–, *juncea* 105
Brazilian Amazon 70
Bromus 17
–, *erectus* 104, 107
–, *madritensis* 94–96, 104, 107
–, *madritensis* ssp. *rubens* 17, 94
–, *mollis* 97
–, *tectorum* 93–94
Brooks Range 300
brown plant hopper 219
bryophyte 96
building 55
built
–, environment, effect of 52
–, structure 55
bulk density 52
burning 262
–, of crop residue 203
BYDV *see barley yellow dwarf virus*

C

C *see carbon*
C/N *see carbon to nitrogen ratio*
C_3 104
–, and C_4 photosynthesis 40
–, grass 108
–, plant 39, 41, 103
–, versus C_4 plant 40
–, vine 94
C_4 104
–, -dominant grassland 32
–, grass 108
–, plant 39, 103
–, species 95
C4MIP global circulation model 71
–, family 71–72
calcareous grassland 11, 109
–, experiment 105
calcite compensation 182
California 28, 97, 154
Californian grassland 11
Canada 68, 94, 297–298
Canadian Forest Fire Behavior Prediction System 143
Canadian Forestry Fire Weather Index System 143
canopy 12, 37
–, interception 32
–, photosynthesis 32
–, ventilation 15
CARAIB terrestrial biogeochemistry model 177
carbohydrate supply 16
carbon 9
–, and nitrogen
–, allocation 32
–, cycling 32
–, and water tradeoff 237
–, assimilation efficiency 94

–, burial 69
–, budget 242, 278, 289–290
–, -climate feedback 59
–, commercial export of 72
–, contemporary balance 182
–, cycle 2
–, effects of land use change 185
–, feedback 183
–, cycling 110
–, emission 59
–, equivalent 203
–, exchange 81, 290
–, fixation 293
–, flow 290
–, flux 109, 278
–, in frozen ground 73
–, isotope fractionation 37
–, pool 10, 13, 16
–, saturation 60
–, -productivity ratio 204
–, sequestration 5, 9, 51, 70, 113, 232–234
–, capacity 85
–, potential 229, 280
–, uncertainty in estimates 234
–, sink 72, 243
–, process 60, 63
–, source 67
–, stock 13
–, storage 13, 32, 51, 231
–, to nutrient ratio 13
–, to nitrogen ratio 13–14, 26, 31, 63, 155, 241
–, terrestrial sink
–, location 59
–, sink saturation 59
–, turnover 10
–, uptake 179
carbon dioxide 66, 103, 203
–, atmospheric 216
–, concentration 85, 93, 181, 196
–, concentration change 2
–, enrichment 9–10
–, exchange 300
–, between atmosphere and land 175
–, fertilization 9, 60, 64, 196, 275
–, effect 60, 63
–, flux 26, 290
–, measurement 182
–, holocene change in 182
–, production 293
–, spring 12
–, stabilization target 232
Carex flacca 104–105, 107
Caribbean 233
Carpinus 16
CASA terrestrial biogeochemistry model 177
cascading effect 165
cash crop 265
Catalonia 85–86, 97–98
catastrophic
–, event 166
–, fire 68
–, shift 165
catchment 239
CBD *see Convention on Biological Diversity*
CCD *see Convention to Combat Desertification*
CE *see carbon equivalent*
Centaurea solstitialis 94
Central and South Asia 233
Central Kalimantan 263, 266
CENTURY terrestrial biogeochemistry model 30, 51, 177, 281
cereal 196
CH_4 *see methane*
Changbai Mountains 275–276
–, Forest Ecosystem Research Station 281
change
–, in disturbance regime 305
–, in land-use/cover 97–98
–, in vegetation 304
changing global environment 166
Changling Grassland Station 281
cheatgrass 94
chemistry and transport model 187
Chenopodium album 286–288
Chesapeake Bay 16
Chicago Climate Exchange program 237
Chihuahuan Desert 171
China 201–202, 273–274
Chinese Terrestrial Transect 274
circumpolar north 305
Clean Development Mechanism 110
climate 82, 154, 216, 297
–, change 64, 81, 86, 110
–, driven process 64
–, experiment 30
–, impact 129
–, scenario 30
–, controlled chamber 24
–, envelope 110
–, feedbacks 305
–, model 14
–, regulation 85
–, system 85
–, recent change in 297
–, -vegetation linkage 285
climatic
–, effect of afforestation 240
–, event 166
–, filter 85
–, treeline 13
–, warming 23
CLIMEX model 215
CO_2 *see carbon dioxide*
coarse scale 54
coexistence mechanism 87
Colorado 17, 32
Columbus 52
combustion loss 267
common ragweed 217
community
–, assembly 155
–, biomass 106
–, composition 29, 105
–, disassembly 110
–, diversity 107
–, dynamics 105, 285
–, ecology 214
competition 286
–, for CO_2 286
–, ecophysiological basis of 286
complex gradient 154
complexity 30, 103
composite phenomena 248
compost pile 54
concentrated flow 53
conifer tree 255
connectivity 52, 168
–, of erosion processes 171
conservation tillage 69
consumer 14
–, response 13
Convention on Biological Diversity 113
Convention to Combat Desertification 247, 249, 252
copper 49
Corsica 144
cottonwood 47
coupled socio-environmental system 314
CPR *see carbon-productivity ratio*
critical threshold 165

crop
–, management 69
–, model 199
–, production 249
–, yields 200
–, drought sensitive 166
cropland 71
–, abandoned 67
–, soil carbon sequestration potential 230
–, system 51
cross-ecosystem correlational studies 114
cross-scale
–, conceptual holism 252
–, interaction 166, 172
–, linkage 171
cross-system comparison 56
Cryptococcus fagisuga 97
Cryptostegia grandiflora 96
CTM *see chemistry and transport model*
cultivation 51, 69
–, system 201

D

$\delta^{13}C$ 39
–, and environmental variables 38
–, and soil moisture 38
–, of phloem sugar 38
–, of soil-respired carbon dioxide 38
Dahlem Desertification Paradigm 252–253
dark respiration rate 275
DAYCENT ecosystem model 30–31
–, description 30
–, simulation and result 31
DDP *see Dahlem Desertification Paradigm*
deciduous broadleaf forest 278
decision-making 268, 314
–, process 315
decomposition 14
–, aerobic vs. anaerobic 300
deforestation 59, 70, 240, 265–266, 300
delineation 54
–, of ecosystem boundary 54
demographic
–, factor 315
–, rate 134
denitrification 24, 227
desert FACE experiment 13, 16
desertification 171, 247–252
–, avoidance 254
–, biophysical scale of 251
–, consequence of 250
–, drivers of 247
–, dynamics 171–172
–, extent estimating of 249
–, monitoring 254
–, rate of 172
–, restoration 254
–, scale of 247
–, socio-economic scale of 251
determinate growth 10
DGVM *see Dynamic Global Vegetation Model*
DIC *see dissolved inorganic carbon*
diesel-plant 13
diffusive uptake 16
Dinghu Mountain 279
dinitrogen oxide 187
–, flux 204
Dirty Thirties 166
discontinuity 54
dispersal
–, capacity 289
–, coefficient 289
–, process 289
dissolved
–, inorganic carbon 291
–, nitrogen 290–291
–, organic carbon 66, 71, 290–292
–, dynamics 291
–, transport 290
disturbance 3, 47, 50, 66, 68, 86, 139, 152, 165, 180, 299, 303
–, event 68
–, regime 3, 81, 82, 93, 154, 261, 297, 303
–, response 157
–, scenario 313
–, severity 303
–, shift 68
DIVERSITAS program 116
diversity
–, -function transition matrix 121
–, -productivity relationship 109
DOC *see dissolved organic carbon*
DOLY terrestrial biogeochemistry model 177
dominant plant species 84
Douglas fir 16
Drosophila 212–215
–, *melanogaster* 214
–, *simulans* 214
–, *subobscura* 212, 214
drought 16, 167
–, mitigating effect 16
–, stress 301
DRR *see dark respiration rate*
dryland 247
–, land degradation 247
–, global distribution 248
–, management of desertified 254
Duke Forest 12
Dust Bowl 166–169
dust
–, generation 171
–, particle 49
–, rising 167
–, storm 166–168, 247
Dutch elm disease 49
dynamic global vegetation model 30, 72, 94, 131, 139, 157–158, 175–180, 186–187
–, and hydrology 180
–, and nitrogen cycle 185
–, and plant functional type 185
–, application 182
–, construction 175, 178
–, design 185
–, evaluation 181
–, forecast 132
–, parameterization 179
–, perspective 185
–, principle 178
–, simulation 132
–, structure 178

E

Earth System 1, 165, 175
–, Science Partnership 5, 74
earthworm 50
East Siberian Transect 297–298, 301, 303
Eastern China 273
Eastern Oregon 318
eastern spruce budworm 299
ecological
–, diversity 114
–, footprint 45
–, niche modelling 98
economic plant removal 249
ecophysiological trade-off 150
ecosystem
–, carbon

–, flux 108
–, cycle 37–38
–, dioxide response 10
–, characteristics 53
–, climate change impact on 129
–, dynamics 166, 265
–, engineer 89, 114
–, flammability 157
–, function 114–115
–, functioning 3, 81, 113–114, 318
–, change in 114
–, good 175
–, model 65
–, physiology 2, 38
–, process 24
–, property 84
–, response 23, 55
–, service 4, 84, 88, 113, 175, 316, 318
–, provision 318
–, value 85
–, structure 2, 318
Ecosystem Model-Data Intercomparison project 181
Ecotron 214
eddy
–, correlation 290
–, covariance 182, 301
–, flux tower 48
–, measurement 55
Edson 144
effect 88
–, of built environment 52
effective
–, impervious area 52
–, imperviousness 52
EIA *see effective impervious area*
El Niño 68–69, 71, 94
–, and La Niña cycle 69
–, Southern Oscillation 261, 265–268
–, dry event 267
–, event 41, 263
electrical heat-resistance ground cable 24
elevated carbon dioxide 2, 11, 14–15, 28–29, 103, 105, 198, 286
–, and plant invasion 93
Elevated Carbon Dioxide Network 2
EMBYR model 141, 145
emission
–, of greenhouse gas 205–206
–, trajectory 232
empirical data 214
encroachment 60, 66–67
endocrine disrupting chemical 222
energy
–, budget feedback 304
–, exchange 304
–, flow 52
engineered
–, flow path 53
–, pathway 52
England 291
ENSO *see El Niño Southern Oscillation*
environmental
–, change 151, 313
–, dataset 98
–, filter 81–82, 151
–, gradient 152
epiphenomenon 114
equilibrium biogeography model 176
Erodium cicutarium 96
erosion 265
EST *see East Siberian Transect*
establishment 82
ET *see evapotranspiration*
eucalypt 16
Eucalyptus 142
Euphorbia lathyris 13–14
Europe 195
European Union Greenhouse Gas Emission Trading Scheme 237
eutrophication 293
evapotranspiration 15, 238–239, 242
evenness 106–107
evolution 105
evolutionary
–, response 105
–, trade-off 150
experimental
–, approach 23
–, carbon dioxide research 17
–, manipulation 214
–, warming 23, 301
exposure 211
extinction 82, 110
–, risk 82
extreme
–, climate 166
–, weather event 199

F

FACE *see Free Air Carbon Dioxide Enrichment*
facilitation 88
Fagus 16
Fallopia japonica 96
fAPAR *see fraction of Absorbed Photosynthetic Active Radiation*
Far East Siberian Transect 297–298, 302
farmer 264
farm-yard manure 230
FATE model 157
FATELAND model 157
FD *see functional diversity*
feedback mechanism 165–166
fertilization 97
–, by nitrogen deposition 63
–, effect 273
fertilizer 11, 50, 203
FEST *see Far East Siberian Transect*
field experiment 106
filter 155
fine root
–, biomass 27–28, 108
–, dynamics 24
Finland 297–298
fire 68, 71, 73, 303, 261, 266, 318
–, activity 68
–, and succession 140
–, behavior 170
–, carbon flux from 266
–, disturbance 299, 305
–, ecological interaction 266
–, ecology 262
–, effect 140
–, emission 68
–, frequency 68, 86, 299, 303
–, global carbon flux from 267
–, human cause of 262
–, ignition 140–141
–, -induced mortality 267
–, model 139, 145–146
–, ignition module 146
–, occurrence 262
–, pre-disposing conditions 262
–, regime 68, 139, 142, 261, 266
–, shift 68
–, spread 140–141, 169
–, succession model 141
–, suppression 240
–, policy 68
FIRESCAPE model 141–142, 145
Flakaliden 32–33

flow path 47, 52
flowering 25
FLUXNET network 39, 182
foliage canopy 285
food 273
 –, insecurity 233
 –, production 195, 274
 –, security 201, 247, 274
forb species 29
forecasting
 –, catastrophic event 172
 –, spatial nonlinearity 172
forest
 –, canopy 290
 –, carbon stocking 16
 –, fire 68
 –, harvest 299
  –, regime 303
 –, management 67, 264
 –, regrowth 67
 –, thickening 68
 –, tree growth 71
 –, zonation 289
Forest-BGC terrestrial biogeochemistry model 177
forestry 2, 186
FORET terrestrial biogeochemistry model 177
FORSKA terrestrial biogeochemistry model 177
fossil fuel burning 59
fraction of Absorbed Photosynthetically Active Radiation 181
Fraxinus mandshurica 276
Free Air Carbon Dioxide Enrichment 10, 12, 16–17, 28, 62, 95, 185, 196–198, 215
 –, experiment 12, 23
freshwater 305
frost resistance 16
fruit fly 212
fuel
 –, connectivity 169
 –, pattern 145
functional
 –, composition 84
 –, diversity 81–82, 87–89, 114
  –, affected by environmental filter 82
  –, and climate 85
  –, and land use/disturbance regime 85
  –, effect 82
  –, quantification of 83, 89
 –, group 29
  –, richness 108–109
  –, of plant species 103
 –, marker 150
 –, richness 84, 86
 –, trait 81–82, 149–150
future carbon balance projection 183
FYM *see farm-yard manure*

G

G'DAY terrestrial biogeochemistry model 30, 32, 177
gap model 177
GCM *see General Circulation Model*
GCTE *see Global Change and Terrestrial Ecosystems project*
GEC *see Global Environmental Centre*
General Circulation Model 59, 65, 71–72, 176–177, 212, 276–277
 –, coupling 178
 –, fully coupled 66
 –, HadCM3 132–133
 –, land-surface schemes 177
 –, time step 178
genetic
 –, selection 130
 –, variation 105
geographical distribution of pest 211
GHG *see greenhouse gas*
GILB *see Global Initiative for Late Blight*
Ginninderra 144
glacial period 182
Glacier National Park 98, 144
global 249
 –, carbon cycle 37, 40, 175
 –, change 9, 93, 317
  –, driver 23, 81–82
  –, ecology 113
  –, effect 316
  –, factor 23
  –, simulation 30
 –, climate
  –, change 195
  –, model 71, 212
 –, environmental
  –, change 93, 98, 273
  –, driver 82
 –, fire dynamics 139
 –, food security 233
 –, plant functional type distribution 176
 –, warming 59
  –, potential 202
Global Carbon Project 5, 74
Global Change and Terrestrial Ecosystems project 1–2, 18, 23, 37, 42, 103, 107, 110, 116, 139, 149, 153, 175, 185, 196, 207, 216, 247, 273, 285, 320
 –, decade 9
 –, International Wheat Network 200
 –, Network of Ecosystem Warming Studies 24
  –, synthesis 24
Global Change Impacts on Terrestrial Ecosystems in Monsoon Asia Research Project 285, 288–289, 292, 294
Global Environmental Centre 269
Global Initiative for Late Blight 216
Global Land Project 1, 313–315, 317, 319
 –, concept 315
 –, implementation strategy 319
 –, research framework 315–317
globalization 315, 317
GLP *see Global Land Project*
Glycine max. 62
goatgrass 88
gopher 88
governance 268
GPP *see gross primary productivity*
gradient 154
grass degradation 171
grassland 11, 16, 51, 67, 238, 275
 –, conversion 240
 –, plot 28
grazer 186
grazing 152, 240
Great Plains 166–167
Green Revolution 196, 203
greenhouse 24
 –, effect 51, 85
 –, gas 23, 64, 71, 196, 202, 204–206
  –, concentration 23
  –, emission 266
  –, mitigation measure 232
  –, mitigation potential 227, 229–230
  –, mitigation, win-win strategy 232
  –, sink 227
  –, source 227
  –, trade-off in agriculture 227
 –, world 176
GHG *see greenhouse gas*
Greenland 302
greenness 181
Grime's competitive syndrome 155
gross primary productivity 60, 64
groundwater use 239
growing season 25, 64, 299, 304
 –, extension of 301

growth 13, 25, 179
 –, chamber 106
 –, form 152
 –, response 27, 61
 –, stimulation 17
GWP *see global warming potential*
gypsy moth 28, 97

H

habitat destruction 110
hard trait 151
Harvard Forest Soil Warming Experiment 26
harvest index 200
Hayman fire 169
heat
 –, island 47
 –, loading 48
 –, stress 28
heated plot 26
heavy
 –, isotope 37
 –, metal 49
  –, contamination 49
hemlock wooly adelgid 97
herbivore 14
herbivory 14, 155
heterotrophic respiration 65
 –, temperature effect on 65
 –, water effect on 65
hierarchy 251
high carbon-density region 74
high latitude 66, 85, 297
 –, ecosystem 297, 300–301, 305
 –, region 303
 –, peatland 71
 –, transect 297–298
higher-order response 28
Hill ratio 106
historical
 –, area burned 298
 –, global carbon sink 71
homogenization of biological system 93
Homogyne alpina 62
hormone 222
Horonai River 290
 –, basin 290
 –, watershed 291
horticultural management 50
 –, practice 55
human
 –, -caused disturbance 50
 –, -environment system 316
 –, intervention 175, 178
 –, -modified landscape 139
 –, population density 97
 –, well-being 267
humid
 –, grassland 238
 –, tropics 17
Hutchinson's realised niche 132
HYDE land-cover type data set 181
hydrogenotrophic methanogenesis 293
hydrological discharge 297
hydrology 14, 16, 180
hypolimnion 292

I

IBIS dynamic global vegetation model 178
IBP *see International Biological Programme*
ice-core analysis 182
Iceland 302
IGBP *see International Geosphere-Biosphere Programme*
IHDP *see International Human Dimensions Project*
IIASA *see International Institute for Applied Systems Analysis*
impact 211–213
 –, estimation
in situ persistence 130
India 202, 204–205
indirect interaction 88–89
individual
 –, plant
  –, function 149
  –, structure 149
 –, tree based model 288
Indo-Gangetic Plain 196, 205–206
Indonesia 262, 266
industrialization 273
infiltration rate 48
infrared
 –, heat lamp 28
 –, heater 24
infrastructure extension 249
Inner Mongolia Grassland Ecosystem Research Station 281
insect
 –, disturbance 299, 303
 –, outbreak 299
 –, pest 49
integrated
 –, approach 252
 –, impact assessment 219
intercropping 114
interface of stream and lake ecosystem 291
Intergovernmental Panel on Climate Change 232
 –, process 211
 –, Standard Reference Emission Scenario 70
internal
 –, carbon housekeeping 9
 –, wave 293
International Biological Programme 176
International Geosphere-Biosphere Programme 1, 252, 273, 320
International Human Dimensions Project 1, 320
International Institute for Applied Systems Analysis 175
International Plant Protection Convention 211
international trade 93
International Tundra Experiment 25
 –, synthesis 25
interspecific
 –, competition 87, 155
 –, hybridization 218
introduction of exotic species 49
invader 86, 88
invasion 99
 –, resistance 88
invasive species 50, 86, 93, 171
 –, response to elevated carbon dioxide 94
inverse modeling 59
invertebrate population dynamics 213
IPCC *see Intergovernmental Panel on Climate Change*
IPCD approach 158
Ipomea sp. 105
isotopic
 –, composition 37
 –, discrimination in photosynthesis 39
 –, signature 38
 –, tracer 37, 319
Israel 10
Italy 11, 144
ITEX *see International Tundra Experiment*

J

JABOWA forest model 177
Japan 289, 292
Japanese honeysuckle 94
Jasper Ridge Biological Preserve 105
Jasper Ridge Global Change Experiment 23, 28–30, 106

–, summary 30
Joint GCTE-LUCC Desertification Initiative 252
JRGCE *see Jasper Ridge Global Change Experiment*
Juglans mandshurica 276
Juniperus 68
Juniperus virginiana 68

K

Kalimantan 262–263
key
–, ecosystem process 255
–, species 107
keystone species 89, 114, 251
King Mountain fire 169
knowledge integration 71
Kyoto Protocol 233

L

LAI *see leaf area index*
Lake Biwa 291–294
–, basin 285
–, carbon budget 292
–, ecosystem 292
–, primary production rate of 294
–, watershed 291
lake ecosystem 285
–, function 293–294
–, sediment metabolism 293
LAMOS landscape modelling platform 86, 141, 145, 157
land
–, abandonment 67
–, atmosphere feedback 171
–, biosphere 175
–, clearing 264
–, cover 219, 297, 303
–, change 69
–, recent change in 297
–, shift 70
–, type 172
–, type distribution 247
–, degradation 247–248
–, development policy 261–262
–, eutrophication 96
–, fire regimes 265
–, management 66, 315, 317
–, practice 262–263
–, sustainability 318
–, system change 318
–, transformation 97
–, use 166, 219, 313
–, and cover 97–99
–, and cover change 247
–, change 59, 66, 69–70, 81, 97, 185, 240
–, change legacy 72
–, decision making 315
–, history 238, 242
Land Use Change and Cover project 1, 247, 320
landscape
–, carbon gain 13
–, complexity 165
–, context 54
–, dynamics 157
–, scale vegetation 15
Landscape Fire Succession Model 139–141, 157
Landscape Fire Working Group 139
LANDSUM Landscape Succession Model 141–145
Lao 262, 265
larch sawfly 299
large herbivore grazing 171
large scale
–, biodiversity and ecosystem functioning 122
–, environmental change 175
large-plant-scenario 103, 108
Larix olgensis 275–276
Larrea 68
Larrea tridentata 96
Las Vegas 96
Last Glacial Maximum 133
latent heat 239
latex 14
lawn 50
–, care management practice 51
LCP *see light compensation point*
LDMC *see leaf dry matter content*
lead 49
leaf
–, area index 12, 176
–, biomass 31
–, bud burst 25
–, conductance 11
–, dry matter content 152
–, lifespan 152
–, quality 14
–, stomata 14
legume 114, 214
Leptopilina boulardi 214
Leymus chinensis 275
LFSM *see Landscape Fire Succession Model*
LGM *see Last Glacial Maximum*
liana 16, 17
lichen 96
life cycle 105, 152
light compensation point 276
light saturation point 276
light-use efficiency 288
LINKAGES growth model 177
Liquidambar 16
–, *styraciflua* 12, 62
Liriodendron tulipifera 62
litter 29
–, decomposition 24, 29, 32
–, lignin 29
–, quality 29, 241
livelihood 267–269
livestock production 249
loblolly 55
–, pine 12
local community 82
lodgepole pine 130
logging 264
Lolium
–, *multiflorum* 97
–, *perenne* 11
Long Term Ecological Research 46, 214
–, program 172
long-term catchment datasets 239
Lonicera japonica 93–94
Los Angeles 96
Lotus corniculatus 107
lowland tropical peatland 71
LPJ *see Lund-Potsdam-Jena dynamic global vegetation model*
LSP *see light saturation point*
LTER *see Long Term Ecological Research* 46, 214
LUC *see land use and cover*
LUCC *see Land Use Change and Cover project*
LUE *see light-use efficiency*
Lund-Potsdam-Jena dynamic global vegetation model 177, 184
Lymantria dispar 97

M

major axe of plant specialisation 150
Malaysia 267
management
–, effect 50
–, effort 51

–, hierarchy 252
–, practice 70
–, regime 47
Maoer Mountain Station 279
MAPSS equilibrium biogeography model 176
mass migration 166
mass-ratio hypothesis 89
material flow 52
maximum net photosynthetic rate 275
MDE *see model descriptive element*
mean global temperature 23
Mediterranean 12, 86, 154, 218
–, Basin 98
Mega Rice Project 263
Mekong Delta 266
Melaleuca peat forest 266
Melampsora medusae f. sp. *tremuloidae* 215
mesic grassland 11
meta-analysis 24, 25, 65, 242
metabolic theory of ecology 120
methane 66, 186–187, 203, 294
–, emission 24, 302
–, exchange 302
–, oxidation 203
–, rate 293
–, production 203
–, rate 293
methanogenesis 203, 302
methanogenic bacterium 203
methanotrophic bacterium 203
Mexico 203
microbial process 24
Microstegium vimineum 50
migration 134, 186
–, rate 133
–, measurement 133
–, model 133
mineral resource availability 152
mitigation potential 229
mixed-model ANOVA 29
model 30
–, architecture 178
–, classification 144
–, comparison 145
–, descriptive element 140–141
–, ecosystem 13
–, prediction 71
–, simulation 23
Mojave Desert 17, 94, 96
Mongolia 273
monitoring data 212
monospecific stand 286
monsoon 202, 274, 278, 285, 291
Montana 98, 144
montane vegetation 217
morpho-functional trait 84
mosaic of landscape patches 264
Moscow 230
Mount Kinabalu 285
Mount Pinatubo 183
mountain pine beetle 299
mowing 154
MTE *see metabolic theory of ecology*
multiple
–, factor imperative 23
–, global change forcing 275–276
–, nutrient limitation 186
–, perturbation 317
multi-species community 103
Murrumbidgee Irrigation Area 219
Myanmar 262
mycorrhiza 155
–, association strategy 156
Myzus persicae 214

N

N *see nitrogen*
N_2O *see dinitrogen oxide*
NAR *see net assimilation rate*
National Ecological Observatory Network 172
native plant species 93
natural
–, adaptation 220
–, enemy 212
–, resource 314
–, vegetation class 277
NBP *see net biome productivity*
NDVI see Normalized Differential Vegetation Index
NECT see Northeast China Transect
Nectria 97
NEE see net ecosystem exchange
Negev Desert 10, 17
NEP see net ecosystem productivity
Nepal 202
net
–, assimilation rate 94
–, biome
–, production 64
–, productivity 60
–, carbon
–, fixation 103
–, sequestration 62
–, sink 59, 67
–, ecosystem
–, exchange 39, 184, 290
–, production 30, 290
–, productivity 30, 32–33, 60, 62, 64, 67, 72, 109–110, 290
–, photosynthesis 290
–, primary production 11, 60, 64, 109, 179, 181
–, precipitation effect 64
–, temperature effect 64
–, primary productivity 11, 29–33, 50, 60, 62, 64, 70, 106, 109–110, 177, 179, 273, 281
–, terrestrial sink 64
Network of Ecosystem Warming Studies 2
neutral landscape 140, 143
Nevada Desert 13, 16
–, free air carbon dioxide enrichment facility 94–95, 107
New Mexico 171
New York City 48–49
–, Central Park 49
niche 82, 133
–, based model 132–134
–, partitioning 87
nickel 49
night time carbon dioxide concentration 319
Nilaparvata lugens 219
nitrate 291
nitrification 24, 227
–, rate 51
nitrogen
–, addition 51
–, concentration 29
–, cycle 96, 185
–, cycling 180
–, deposition 28, 30, 63–64
–, fertilization 51, 60, 63
–, fixation 96, 114
–, fixing
–, legume 108
–, plant 103
–, input 96
–, limitation 12, 51, 72
–, mineralization 24, 26, 65
–, oxide 49
nitrogenous fertilizer 204
nitrous oxide 186–187, 227
–, emission 204

–, mitigation 228
NO_3^- *see nitrate*
non structural carbohydrate 13
–, accumulation 13
non-equilibrium dynamics 139
non-legume 114
nonlinear
–, behavior 165
–, change 169
–, ecosystem response 110
nonliving organic matter 71
Normalized Differential Vegetation Index 254
North Carolina 12
Northeast China Transect 274, 276, 281
northern high latitude 182
North-South Transect of Eastern China 274–276, 279
Norway spruce 32
NPK fertilizer 205
NPP *see net primary productivity*
NSC *see non structural carbohydrate*
NSTEC *see North-South Transect of Eastern China*
nutrient 11
–, cycling dynamics 93
–, limited ecosystem 104
–, loading 294
–, loss 265
–, management 205
–, poor environment 155
–, uptake capacity 103
–, use efficiency 103, 151

O

$O_3 \times CO_2$ FACE experiment 16
Oak 11
Oak Ridge 12
–, free air carbon dioxide enrichment site 94
–, National Environmental Research Park 26
–, National Laboratory 12
oasis effect 47
ocean-atmosphere general circulation model 178
open-top chamber 24–27, 95
operational biota 119
Opuntia 97
organic
–, carbon 292–293
–, chemical 49
–, manuring 204
oxygen
–, consumption 294
–, rate 293
–, isotopic composition of carbon dioxide 41
oxygen-18 37
ozone 16
–, exposure 47

P

P-R ratio *see photosynthesis-respiration ratio*
P *see phosphorus*
PAH *see polycyclic aromatic hydrocarbon*
Pakistan 201
palaeoecology 158
Palaeovegetation Mapping Project 158
paleo-data 158
Pampas 238
Panama 17
Pantnagar 204–205
PAR *see photosynthetically active radiation*
parcelization 47
particulate organic carbon 291–292
passive night-time warming 24
patch
–, dynamics 157
–, interrelationship 54
pathogen 49
peat 266
–, fire 266
peatland 71, 266, 303
–, drainage 71
–, methane emission 71
–, carbon dioxide emission 71
people distribution 247
percolation theory 165
perennial grassland 153
permafrost 66, 298
–, boundary 66
–, thawing 65
persistence 82
persistent seed bank 153
perturbation 212, 315
pest 186, 211, 318
–, control measure 221
–, entry 212
–, outbreak 266
–, related indicator 212
–, risk modelling 215
pesticide 50
–, resistance 221
PFT *see plant functional type*
Phleum 106–107
–, *pratense* 106
Phoenix 96
–, Metropolitan Area 45
phosphorus 238
photo inhibition point 276
photosynthesis 9, 60
–, -respiration ratio 293
photosynthetic
–, response curve 17
–, stimulation 9
photosynthetically active radiation 175, 181
physical environment 103
–, alterations of 103
phytoplankton abundance 292
Picea koraiensis 275, 276
Pinatubo effect 183
pine 12
–, beetle 318
–, forest 318
–, processionary caterpillar 218
Pinus 16
–, *koraiensis* 275–276
–, *massoniana* 279
–, *sylvestris nevadensis* 218
–, *taeda* 12, 62, 185
PipeTree functional tree simulator 289
plant
–, architecture 94
–, biodiversity 103
–, community
–, response to global change 285
–, structure 104
–, competitive strength 151
–, dispersal 186
–, diversity 103, 109
–, ecophysiology 95
–, ecosystem 32
–, functional
–, classification 149
–, group 17
–, response 152
–, trait 82, 84, 149, 154
–, type 149, 157, 176–177, 179, 185
–, type classification scheme 158
–, fungus interaction 156
–, geography 175–176
–, growth 179

–, hopper 219
–, invasion 93, 94, 96–98
–, migration 133
–, mortality 166–167
–, nutrient uptake 24
–, nutrition 14
–, pathogen 211, 218
–, performance 95
–, pest 211
–, phenology 25
–, physiology 175–176
–, -plant
–, facilitation 88
–, interaction 255
–, productivity 24, 50
–, species
–, migration 129
–, richness 86
–, turnover 133
–, water
–, consumption 14
–, relation 14
Plantago major 106
plantation age 239
Platypus quercivorus 218
Poa pratensis 106
POC *see particulate organic carbon*
podzoluvisol 230
pollen 217
pollution 47
–, atmospheric 49
–, soil 49
polycyclic aromatic hydrocarbon 49
population
–, change 213, 317
–, demography 285
–, structure 195
post-fire vegetation development 139
post-glacial migration rate 131
potential
–, evaporation 247
–, migration rate 134
Potsdam NPP Intercomparison Project 181
prairie 11
preagricultural ecosystem 46
precipitation 28, 64, 238, 297
–, regime 267
presence/absence 107
prickly-pear cactus 97
primary production rate 294
productivity 250, 276
–, of terrestrial ecosystem 276
propagation, nonlinear 166
propagule persistence capacity 153
property
–, conflict 265
–, right 265
Prosopis 68
–, *glandulosa* 62
protein gap 200
Puccinia graminis f. sp. *tritici* 214
pupal weight 28

Q

quantum use efficiency 276
Quebec 98
–, experiment 106
Queensland fruit fly 219
Quercus
–, *alba* 62
–, *crispula* 218
–, *ilex* 11
–, *mongolica* 276

R

radiation 85, 304
radiatively active gas 300
Raffaelea sp. 218
randomized complete block design 27
range of traits 87
Raphanus sp. 105
rapid genetic change 130
rat 219
Rattus argentiventer 219
red brome 94
reductionist experiment 214
reforestation 66
regeneration 318
regional
–, case study 230
–, climate 83
–, cooperation 269
–, haze episode 266
–, species pool 82
–, summer climate 304
relative
–, abundance 119
–, growth rate 152, 155
–, photosynthetic rate 288
remote sensing 301
reproduction 25, 105
reproductive allocation 94
research coordinating network 116
re-settlement 315
resource
–, availability 105, 154
–, use complementarity 87–88
respiration 32, 179, 290
–, flux 65
response trait 155
RGR *see relative growth rate*
Rhamnus cathartica 50
Rhododendron ponticum 93
Rhopalosiphum maidis 218
ribulose-1,5-biphosphate 61
rice 196
–, ecosystem 203
–, field 203
–, rat 219
–, -wheat
–, production 201
–, system 201–202
–, system, contribution to climate change 202
–, system, effect of climate change 202
riparian soil 53
riverine export 291
road 55
–, building 267
Rocky Mountains 214
root morphology 150
Rothamsted Insect Survey 214
RPR *see relative photosynthetic rate*
Rubisco 61, 200
RuBP carboxlyase 61
runoff 182, 239
Russia 68
rust fungus 215

S

salinity 238
salinization 239
Salt Lake City 319
San Bernardino Mountains 49
–, Gradient Study 47
San Francisco Bay Area 28
SANPP *see specific annual net primary productivity*

saturation
–, dynamics 62
–, of the terrestrial carbon sink 59
savanna 67, 265
–, fire 68
scale 54, 251
–, importance in assessing biogeochemical effect 54
scaling 13, 155–157, 285
Scandanvia 297–298, 302
SCANNET transect 297–298, 300
SCANTRAN transect 297
Scots pine 130, 218
secondary biochemical compound 155
sediment 72
seed dispersal 25
self organized criticality 165
SEMLAND landscape fire succession model 141, 143, 145
sensible heat 240
sensitivity 211–212
septic
–, load 55
–, system 54
sequestration 26
–, rate 238
Sheffield Differential Global Vegetation Model 131–132
shifting cultivation 264–265
short-grass steppe 32–33
shrub
–, encroachment 251
–, invasion 171
–, proliferation 171
SiB model 177
Siberia 65, 85, 297–298, 302
Siberian gypsy moth 299
Singapore 267
SLA *see specific leaf area*
SLAVE terrestrial biogeochemistry model 183
SOC *see soil organic carbon*
socio-ecological system 252
soft trait 150
SOI *see Southern Oscillation Index*
soil
–, and root respiration 227
–, bulk density 51
–, carbon 227, 230, 281
–, level 207
–, pool 10, 65, 227, 241
–, storage 241, 301
–, carbon sequestration 227, 241
–, measure 228
–, potential 231
–, practice 232
–, strategy 233
–, degradation 171, 233
–, disturbance 50, 153
–, erosion 69, 171, 247
–, moisture 14, 48
–, nitrogen
–, cycling 301
–, feedback 32
–, nutrient availability 103
–, organic carbon 50, 65, 70, 203, 230, 237, 242, 279
–, organic matter 32, 227
–, transformation 180
–, plant continuum 52
–, pollution 49
–, property 254
–, respiration 24, 26, 38, 64–65, 99
–, response to warming 24
–, restoration 233
–, surface evaporative water loss 32
–, temperature 299, 302
–, vegetation atmosphere transfer model 177
–, water 16
–, availability 55
–, balance 104
–, content 104
Sonora Desert 96
SORTIE model 177
South America 88
South Asia 201, 233
South Dakota 167
Southeast Asia 169, 261, 265–268
Southern Oscillation Index 220
Southern Québec 105
spatial
–, interaction 165
–, nonlinearity 165
–, cross scale interaction 166
–, threshold behavior 166
–, feedback mechanism 166
spatially
–, explicit model for landscape dynamics 143
–, nonlinear response 165
species
–, abundance 16
–, based
–, approach 132
–, model 132
–, diversity 107
–, extinction 132
–, risk 132
–, poor communities 108
–, richness 29, 50, 82, 84, 107–109, 114
specific
–, annual net primary productivity 156
–, leaf area 150, 152, 155
–, root length 150, 152
split-plot ANCOVA model 29
spread 82
spruce bark beetle 299
SRES *see Standard Reference Emission Scenario*
SRL *see specific root length* 150
stable isotope 37
Standard Reference Emission Scenario 231–232
standing stock 11
stand-scale simulation 30
steppe 17
stinging nettle 217
stomata 15, 217
stomatal conductance 14, 38, 177
stormwater retention basin 54
stream
–, ecology 53
–, flow 242
–, temperature 48
–, monitoring 48
stress resistance 16
structural diversity 85
Stylosanthes scabra 214
suburban fringe 55
sulfur oxide 49
Sumatra 262
summer carbon gain 85
susceptibility 97
sustainability 316
sustainable soil management 230
Sweden 32
sweet gum 12
Swiss
–, Calcareous Grassland Study 107
–, Grassland Experiment 107
symmetric competition 286

T

TACIT experiment 26–28
taiga 299

Tamarix 95
Taraxacum 107
  –, *officinale* 106
taxonomic
  –, diversity 82
  –, richness 114
technology 313
TEM *see Terrestrial Ecosystem Model*
TEMA *see Global Change Impacts on Terrestrial Ecosystems in Monsoon Asia Research Project*
temperate herbfield community 16
temperature 64–65, 297
  –, and carbon dioxide interactions in trees 26
template heterogeneity 166
Tennessee 12
terrestrial
  –, -aquatic linkage 302
  –, biogeochemistry model 177
  –, biosphere 59
  –, biospheric model 59
  –, carbon
    –, balance 184
    –, sink 30, 59
    –, sink saturation 60
  –, ecosystem
    –, dynamics 175
    –, response 175
  –, net carbon dioxide sink 60
  –, net primary production 30
  –, organic carbon 175
  –, vascular plant 149
Terrestrial Ecosystem Model 177
Thaumetopoea pityocampa 218
thawing of permafrost 302
threshold 166
  –, behavior 166
TIA *see total impervious area*
Tilia amurensis 276
tillage 203
Tomakomai Forest 285, 290
top-down 212
total impervious area 52
toxic algal bloom 217
trace gas 266
  –, emission 247
trade 219
trade-off 150
trait 81, 87, 149
  –, based extinction scenario 122
  –, biased subset of 82
  –, of dominants 82
  –, range 87
  –, relative abundance 87
  –, response 154
transect 4, 273–282
transient change 139
transpiration 11, 15
transport processes 165
transportation corridor 98
travel 93, 219
tree
  –, establishment 239
  –, growth 28
    –, response 13
  –, migration rate 303
treeline 303
    –, response 300
    –, change in 303
TRIFFID model 178, 183
Trifolium
  –, *pratense* 105
  –, *repens* 11
TRMM *see Tropical Rainfall Measuring Mission*
trophic condition 293
tropical
  –, peatland 266
  –, sea surface temperature 166
Tropical Rainfall Measuring Mission 267
tropospheric ozone 47, 63
Tucson 96
tunable diode laser 42
tundra 300, 303
turfgrass productivity 51
tussock grass 86
typhoon 291

U

UNDP-IRRI Interregional Programme 203
ungulate grazing 153
United Kingdom (UK) 218
United Nations Framework Convention on Climate Change (UNFCCC) 233
urban
  –, area 55
    –, disturbed soil 48
  –, carbon dioxide 319
  –, core 53
  –, ecosystem 45–46, 52
  –, environmental factor 47
  –, hydrology 52–53
  –, land
    –, transformation 54
    –, use change 45–46, 50
    –, use conversion 47
  –, settlement 45
urbanization 273
USA 185
Utah 98, 319
UV-B radiation 291

V

VA *see Vital Attributes model*
vapor pressure deficit 38
vegetation
  –, attribute 254
  –, community dynamics 267
  –, composition 181
  –, distribution 298
  –, dynamics 175, 177, 179
  –, map 298
  –, model 139
  –, shift 66, 289
  –, succession 140
  –, thickening 67
  –, type model 131
Vietnam 266
Vital Attributes model 157
Vitousek-Hooper framework 115, 117–118
volatile organic compound 187
volatilisation 265
VPD *see vapor pressure deficit*
Vulnerability 211, 222, 316, 320
  –, of ecosystem 211
  –, resilience assessment 317
vulnerable carbon pool 59
Vulpia octoflora 94, 95

W

warming 23, 28, 59–68
  –, climate 217
  –, experiment 24, 26, 33, 64, 65, 214, 301
  –, induced nitrogen mineralization 26
water 11, 65
  –, balance 14, 32
    –, model 32

–, ballast 93
–, cycle 2
–, exchange 304
–, feedback 17
–, flux measurement 182
–, shortage 11, 16
–, yield 238
–, use efficiency 32
watershed 48, 52, 54, 285, 290, 293
–, area 55
–, process 291
–, scale
–, budget 285
–, integration 285
weed 211, 216
–, approach 46
–, ecosystem invasion 217
wetland 266, 269, 302
–, ecosystem 16
wheat 196
–, grain yield 197–198
–, growing region 196
–, production system 195
–, productivity 196, 199
–, impacts on 199
–, stem rust pathogen 214
wildfire 68, 168–170
–, and patch connectivity 169
–, behavior 169
–, cost of 169
wildland fire 139
wind
–, erosion 166–169
–, shift 169
winter
–, albedo 304
–, carbon loss 85
–, decomposition 302
wood
–, extraction 249
–, product 71
woodland 67
woody
–, plant encroachment 67–68, 71, 85, 171, 240–241
–, ecohydrological consequence 242
–, species 10
world climate zone 176

X

Xanthium canadense 286

Y

yellow starthistle 94
yield 197
–, gap 200
Yingtan Red Soil Experiment Station 279